Handbook of Cyanobacterial Monitoring and Cyanotoxin Analysis

Handbook of Cyanobacterial Monitoring and Cyanotoxin Analysis

Edited by

JUSSI MERILUOTO

Biochemistry, Faculty of Science and Engineering, Åbo Akademi University, Turku, Finland, and Laboratory for Paleoenvironmental Reconstruction, Faculty of Sciences, University of Novi Sad, Serbia

LISA SPOOF

Biochemistry, Faculty of Science and Engineering, Åbo Akademi University, Turku, Finland

GEOFFREY A. CODD

Biological and Environmental Sciences, University of Stirling, Scotland, United Kingdom, and School of the Environment, Flinders University, Adelaide, Australia

WILEY

COST EUROPEAN COOPERATION IN SCIENCE AND TECHNOLOGY

CYANOCOST

This edition first published 2017

Registered Office
John Wiley & Sons, Ltd, The Atrium, Southern Gate, Chichester, West Sussex, PO19 8SQ, United Kingdom

For details of our global editorial offices, for customer services and for information about how to apply for permission to reuse the copyright material in this book please see our website at www.wiley.com.

Library of Congress Cataloging-in-Publication Data

Names: Meriluoto, Jussi, editor. | Spoof, Lisa, editor. | Codd, G. A. (Geoffrey A.), editor.
Title: Handbook of cyanobacterial monitoring and cyanotoxin analysis /
editors, Jussi Meriluoto, Lisa Spoof and Geoffrey A. Codd.
Other titles: Cyanobacterial monitoring and cyanotoxin analysis
Description: Chichester, West Sussex : John Wiley & Sons, Inc., 2016. |
Includes bibliographical references and index.
Identifiers: LCCN 2016024735 | ISBN 9781119068686 (cloth) | ISBN 9781119068747 (epub)
Subjects: LCSH: Cyanobacteria. | Cyanobacterial blooms–Monitoring. | Cyanobacterial toxins.
Classification: LCC QR99.63 .H3594 2016 | DDC 579.3/9–dc23
LC record available at https://lccn.loc.gov/2016024735

A catalogue record for this book is available from the British Library.

Cover Images: Cyanobacterial image (main), sampling image (top left) and bloom image (second photo from top left) Dr. Tore Lindholm/ Åbo Akademi University; Further two images courtesy of the editors.

Set in 10/12pt Times by SPi Global, Pondicherry, India

10 9 8 7 6 5 4 3 2 1

MIX
Paper from
responsible sources
FSC® C013604
FSC
www.fsc.org

Contents

List of Contributors

Ondřej Adamovský
Faculty of Science, RECETOX, Masaryk University, Brno, Czech Republic

Reyhan Akçaalan
Faculty of Fisheries, Istanbul University, Laleli-Istanbul, Turkey

Meriç Albay
Faculty of Fisheries, Istanbul University, Laleli-Istanbul, Turkey

Andreas Ballot
Norwegian Institute for Water Research (NIVA), Oslo, Norway

Sandra A. Banack
Institute for Ethnomedicine, Jackson, USA

Cécile Bernard
UMR 7245 MCAM CNRS-MNHN, Muséum National d'Histoire Naturelle, Paris, France

Luděk Bláha
Faculty of Science, RECETOX, Masaryk University, Brno, Czech Republic

Lucie Bláhová
Faculty of Science, RECETOX, Masaryk University, Brno, Czech Republic

Agata Błaszczyk
Institute of Oceanography, University of Gdansk, Gdynia, Poland

Milena Bruno
Department of Environment and Primary Prevention, Istituto Superiore di Sanità, Rome, Italy

Josep Caixach
Mass Spectrometry Laboratory/Organic Pollutants, IDAEA-CSIC, Barcelona, Spain

Cristiana Callieri
CNR – Institute of Ecosystem Study, Verbania, Italy

Ana Maria Cameán
Area of Toxicology, Faculty of Pharmacy, University of Sevilla, Spain

Camilla Capelli
IASMA Research and Innovation Centre, Fondazione Edmund Mach-Istituto Agrario di S. Michele all'Adige, Trento, Italy

Shmuel Carmeli
Raymond and Beverly Sackler School of Chemistry, Tel Aviv University, Tel Aviv, Israel

Arnaud Catherine
UMR 7245 MCAM CNRS-MNHN, Muséum National d'Histoire Naturelle, Paris, France

Leonardo Cerasino
IASMA Research and Innovation Centre, Fondazione Edmund Mach-Istituto Agrario di S. Michele all'Adige, Trento, Italy

Geoffrey A. Codd
Biological and Environmental Sciences, University of Stirling, Scotland, United Kingdom, and School of the Environment, Flinders University, Adelaide, Australia

Rossella Colagrossi
Department of Prevention, Ministry of Health, Rome, Italy

Audrey Combes
LSABM, UMR CBI 8231 CNRS/ESPCI-ParisTech, ESPCI, Paris, France

Roberta Congestri
Department of Biology, University of Rome 'Tor Vergata,' Italy

Paul A. Cox
Institute for Ethnomedicine, Jackson, USA

Dunja Đenić
Council for Environmental Protection and Energy Efficiency, Užice, Serbia

Jean-Pierre Descy
Research Unit in Environmental and Evolutionary Biology (URBE), University of Namur, Belgium

Damjana Drobac
Laboratory for Paleoenvironmental Reconstruction, Faculty of Sciences, University of Novi Sad, Serbia

Tamara Dulić
Laboratory for Paleoenvironmental Reconstruction, Faculty of Sciences, University of Novi Sad, Serbia

Isidora Echenique-Subiabre
UMR 7245 MCAM CNRS-MNHN, Muséum National d'Histoire Naturelle, Paris, France

Christine Edwards
School of Pharmacy and Life Sciences, Robert Gordon University, Aberdeen, United Kingdom

Tina Elersek
Department of Genetic Toxicology and Cancer Biology, National Institute of Biology, Ljubljana, Slovenia

Johan Eriksson
Department of Botany, Stockholm University, Sweden

Elisabeth J. Faassen
Aquatic Ecology and Water Quality Management Group, Wageningen University, The Netherlands

Jutta Fastner
Section Drinking Water Resources and Water Treatment, German Environment Agency, Berlin, Germany

Valérie Fessard
Unit Toxicology of Contaminants, French Agency for Food, Environmental and Occupational Health and Safety, Fougères laboratory, France

Cintia Flores
Mass Spectrometry Laboratory/Organic Pollutants, IDAEA-CSIC, Barcelona, Spain

Ambrose Furey
Mass Spectrometry Research Centre, Department of Physical Sciences, Cork Institute of Technology (CIT), Ireland

William B. Glover
Institute for Ethnomedicine, Jackson, USA

Daniel Gutiérrez-Praena
Area of Toxicology, Faculty of Pharmacy, University of Sevilla, Spain

Remedios Guzmán-Guillén
Area of Toxicology, Faculty of Pharmacy, University of Sevilla, Spain

Kerstin Häggqvist
Biochemistry, Faculty of Science and Engineering, Åbo Akademi University, Turku, Finland

Anastasia Hiskia
Institute of Nanoscience & Nanotechnology, National Center for Scientific Research "DEMOKRITOS," Athens, Greece

Mária Horecká
Slovak Limnological Society, Slovak Academy of Sciences, Bratislava, Slovakia

Jean-François Humbert
Institute of Ecology and Environmental Sciences, UPMC, Paris, France

Peter D. Hunter
Biological and Environmental Sciences, School of Natural Sciences, University of Stirling, United Kingdom

Lutz Imhof
DVGW Technologiezentrum Wasser TZW, Branch Office Dresden, Germany

Katarzyna Izydorczyk
European Regional Centre for Ecohydrology of the Polish Academy of Sciences, Łódź, Poland

Iwona Jasser
Faculty of Biology, University of Warsaw, Poland

Ángeles Jos
Area of Toxicology, Faculty of Pharmacy, University of Sevilla, Spain

Tomasz Jurczak
Department of Applied Ecology, Faculty of Biology and Environmental Protection, University of Łódź, Poland

Triantafyllos Kaloudis
Water Quality Control, Athens Water Supply and Sewerage Company – EYDAP SA, Athens, Greece

Kunimitsu Kaya
Center for Environmental Measurement and Analyses, National Institute for Environmental Studies, Tsukuba, Japan

Jiří Kohoutek
Faculty of Science, RECETOX, Masaryk University, Brno, Czech Republic

Latife Köker
Faculty of Fisheries, Istanbul University, Laleli-Istanbul, Turkey

Mikołaj Kokociński
Department of Hydrobiology, Adam Mickiewicz University, Poznań, Poland

Jiří Komárek
Department of Plant Ecology, Institute of Botany, Academy of Sciences of the Czech Republic, Třeboň, Czech Republic

Svetislav Krstić
Faculty of Natural Sciences and Mathematics, "St. Cyril and Methodius" University, Skopje, Macedonia

Tiit Kutser
Estonian Marine Institute (LOMI), University of Tartu, Tallinn, Estonia

Liliana La Sala
Department of Prevention, Ministry of Health, Rome, Italy

Linda A. Lawton
School of Pharmacy and Life Sciences, Robert Gordon University, Aberdeen, United Kingdom

Tore Lindholm
Environmental and Marine Biology, Faculty of Science and Engineering, Åbo Akademi University, Turku, Finland

Luca Lucentini
Department of Environment and Primary Prevention, Italian National Institute of Health, Rome, Italy

Selma Maloufi
UMR 7245 MCAM CNRS-MNHN, Muséum National d'Histoire Naturelle, Paris, France

Joanna Mankiewicz-Boczek
European Regional Centre for Ecohydrology of the Polish Academy of Sciences, and Department of Applied Ecology, Faculty of Biology and Environmental Protection, University of Łódź, Poland

Benjamin Marie
UMR 7245 MCAM CNRS-MNHN, Muséum National d'Histoire Naturelle, Paris, France

Mark W. Matthews
CyanoLakes (Pty) Ltd., Cape Town, South Africa

Hanna Mazur-Marzec
Institute of Oceanography, University of Gdansk, Gdynia, Poland

Craig McKenzie
School of Pharmacy and Life Sciences, Robert Gordon University, Aberdeen, United Kingdom

Annick Mejean
LIED UMR 8236 CNRS, University Paris Diderot, and Chimie ParisTech, Paris, France

Jussi Meriluoto
Biochemistry, Faculty of Science and Engineering, Åbo Akademi University, Turku, Finland, and Laboratory for Paleoenvironmental Reconstruction, Faculty of Sciences, University of Novi Sad, Serbia

James S. Metcalf
Institute for Ethnomedicine, Jackson, USA

Biljana Mijović
Institute of Public Health, Užice, Serbia, and Medical Faculty Foca, University of East Sarajevo, Foca, Republic of Srpska

Sílvia Monteiro
Instituto Superior Técnico, Laboratorio de Análises, Lisboa, Portugal

Isabel Maria Moreno
Area of Toxicology, Faculty of Pharmacy, University of Sevilla, Spain

Federica Nigro Di Gregorio
Department of Environment and Primary Prevention, Italian National Institute of Health, Rome, Italy

Sonja Nybom
Biochemistry, Faculty of Science and Engineering, Åbo Akademi University, Turku, Finland

Antonije Onjia
Anahem Laboratory, Belgrade, Serbia

Tamara Palanački Malešević
Laboratory for Paleoenvironmental Reconstruction, Faculty of Sciences, University of Novi Sad, Serbia

Dijana Pantelić
Laboratory for Paleoenvironmental Reconstruction, Faculty of Sciences, University of Novi Sad, Serbia

Barbara Pawlik-Skowronska
Department of Hydrobiology, University of Life Sciences in Lublin, Poland

Paulo Baptista Pereira
Instituto Nacional de Saúde Dr Ricardo Jorge (INSA), Lisboa, Portugal

Carlos Joao Pestana
School of Pharmacy and Life Sciences, Robert Gordon University, Aberdeen, United Kingdom

Silvia Pichardo
Area of Toxicology, Faculty of Pharmacy, University of Sevilla, Spain

Renata Pilkaityte
Open Access Centre for Marine Research, Klaipeda University, Klaipeda, Lithuania

Olivier Ploux
LIED UMR 8236 CNRS, University Paris Diderot, and Chimie ParisTech, Paris, France

Ana Isabel Prieto
Area of Toxicology, Faculty of Pharmacy, University of Sevilla, Spain

Mlađan Protić
Medical Faculty, University of Novi Sad, Novi Sad, Serbia, and Oncology Institute of Vojvodina, Sremska Kamenica, Serbia

María Puerto
Area of Toxicology, Faculty of Pharmacy, University of Sevilla, Spain

Antonio Quesada
Departamento de Biología, Universidad Autónoma de Madrid, Spain

Catherine Quiblier
UMR 7245 MCAM CNRS-MNHN, Muséum National d'Histoire Naturelle and Département Sciences du Vivant, Université Paris Diderot, Paris, France

Nico Salmaso
IASMA Research and Innovation Centre, Fondazione Edmund Mach-Istituto Agrario di S. Michele all'Adige, Trento, Italy

Tomoharu Sano
Center for Environmental Measurement and Analyses, National Institute for Environmental Studies, Tsukuba, Japan

Ricardo Santos
Instituto Superior Técnico, Laboratorio de Análises, Lisboa, Portugal

Wido Schmidt
DVGW Technologiezentrum Wasser TZW, Branch Office Dresden, Germany

Jelica Simeunović
Laboratory for Paleoenvironmental Reconstruction, Faculty of Sciences, University of Novi Sad, Serbia

Lisa Spoof
Biochemistry, Faculty of Science and Engineering, Åbo Akademi University, Turku, Finland

Maya Petrova Stoyneva-Gärtner
Department of Botany, Sofia University "St. Kliment Ohridski," Sofia, Bulgaria

Assaf Sukenik
Kinneret Limnological Laboratory, Israel Oceanographic & Limnological Research, Migdal, Israel

Stamenko Šušak
Medical Faculty, University of Novi Sad, Serbia

Zorica Svirčev
Laboratory for Paleoenvironmental Reconstruction, Faculty of Sciences, University of Novi Sad, Serbia, and Biochemistry, Faculty of Science and Engineering, Åbo Akademi University, Turku, Finland

Solène Thomazeau
UMR 7245 MCAM CNRS-MNHN, Muséum National d'Histoire Naturelle, Paris, France

Nada Tokodi
Laboratory for Paleoenvironmental Reconstruction, Faculty of Sciences, University of Novi Sad, Serbia

Andrea Törökné
Knot Ltd., Budapest, Hungary

Theodoros M. Triantis
Institute of Nanoscience & Nanotechnology, National Center for Scientific Research "DEMOKRITOS", Athens, Greece

Andrew N. Tyler
Biological and Environmental Sciences, School of Natural Sciences, University of Stirling, United Kingdom

Gábor Vasas
Department of Botany, University of Debrecen, Hungary

Tamara Važić
Laboratory for Paleoenvironmental Reconstruction, Faculty of Sciences, University of Novi Sad, Serbia

Emanuela Viaggiu
Department of Biology, University of Rome 'Tor Vergata', and AlgaRes srl, Spin-off of University of Rome 'Tor Vergata', Italy

Milka Vidović
Institute of Chemistry, Technology and Metallurgy, Department of Ecology and Technoeconomics, University of Belgrade, Serbia

Marina Vukašinović
Waterworks Užice, Water treatment plant "Petar Antonijević," Public Utility Company Užice, Serbia

Kyari Yates
School of Pharmacy and Life Sciences, Robert Gordon University, Aberdeen, United Kingdom

Claude Yéprémian
UMR 7245 MCAM CNRS-MNHN, Muséum National d'Histoire Naturelle, Paris, France

Bojana Žegura
Department of Genetic Toxicology and Cancer Biology, National Institute of Biology, Ljubljana, Slovenia

Sevasti-Kiriaki Zervou
Institute of Nanoscience and Nanotechnology, National Center for Scientific Research "DEMOKRITOS," Athens, Greece

Preface

In addition to presenting researchers with fascinating material with which to study microbial cell processes, cell structure and communication, molecular evolution and ecology, cyanobacteria are impacting widely on the daily activities, practical needs and lives of the general public. The ability of many terrestrial and especially of aquatic cyanobacteria to produce mass populations (blooms, scums and mats), including by strikingly pigmented and odorous species in natural and controlled environments, readily brings these microbes to the direct attention of the general public. Perceptions of cyanobacterial mass populations range from 'nuisance' to 'harmful', since public awareness is also driven by the occurrence of poisonings to domestic pets, farm livestock and wildlife and by occasional human illnesses, from mild to serious, after contact with waterborne cyanobacteria. At official level, in public and private institutions, cyanobacteria are being increasingly recognised as agents which should be monitored and managed to avoid or reduce the likelihood of their adverse impacts on water resources and human and animal health although gaps remain.

Institutional recognition and obligations range from international (e.g. World Health Organization, UNESCO, European Union), though national authorities, to regional and local organisations, with responsibilities to uphold international and national guidelines, and in some cases national legislation, for the risk management of water resources containing or prone to supporting cyanobacteria. The range of institutions involved, or which need to be involved, in cyanobacterial risk management reflects the diverse impacts of cyanobacteria and their toxins (cyanotoxins) on the direct and indirect uses of water resources which can contain cyanobacteria. Relevant sectors include public health, drinking water supply, environmental protection and conservation, aquaculture and fisheries, agriculture and, not least, recreation and tourism. However, although the risk management of cyanobacteria and cyanotoxins is recognised and implemented at multiple levels, the extent of awareness and experience at national, regional and local level varies widely. This variation exists amongst managers and their field operatives and laboratory personnel.

The need for increased awareness raising across Europe and neighbouring countries for the dissemination of knowledge and experience and sharing of best practices, for the risk management of cyanobacteria and cyanotoxins, has been recognised and supported by the COST Action ES1105 – 'CYANOCOST: Cyanobacterial blooms and toxins in water resources: Occurrence, impacts and management', running from 2012 to 2016. The Action involves 32 countries, reflecting the extent of cyanobacterial mass population developments across Europe and beyond. Recent findings of increases in the geographical distribution, population sizes and seasonal duration of annual mass populations and of invasive species of cyanobacteria in European waters (aspects of climate change) attest to the timeliness of CYANOCOST and of this handbook.

In addition to dissemination, technology and skills sharing and validation, the effective sampling and identification and quantification of cyanobacteria and cyanotoxins closely follow behind primary research. Virtually all areas of cyanobacterial monitoring and cyanotoxin analysis continue to be researched, for example, as new instrumentation becomes available (both high-technology laboratory equipment and ease-of-use methods), molecular biology methods are applied in the field, cyanobacterial identification and nomenclature undergo change, and further cyanotoxins are recognised. This handbook provides reviews on growth areas in this research in addition to surveying the more established aspects of cyanobacterial ecology and monitoring and of cyanotoxins and their analysis. For example, after an earlier period of decline in the use of bioassays, when physico-chemical methods for cyanotoxin analysis became more widely available, new bioassays are again being seen as a valuable asset for cyanobacterial toxicity assessment where toxicity cannot be adequately accounted for by the analysis for known cyanotoxins. Twenty-nine standard operating procedures are provided which include detailed instructions to enable the methods to be used for the environmental sampling, examination and quantification of cyanobacteria, the known cyanotoxins, and for toxicity assessment by bioassay. Whilst useable, reliable, repeatable and robust methods are required, the value of such field and laboratory monitoring and analysis in guiding risk management also depends, as discussed here, upon the suitable choice of methods and on the effective reporting and interpretation of the results. We hope that this handbook will contribute to the effective widening of cyanobacterial and cyanotoxin risk management for the welfare of environmental resources and protection of human health.

Jussi Meriluoto, Lisa Spoof and Geoffrey A. Codd
January 2016

Acknowledgements

This publication is based on the work from the European Cooperation in Science and Technology (COST) Action ES1105 (CYANOCOST: *Cyanobacterial blooms and toxins in water resources: Occurrence, impacts and management*), supported by COST. CYANOCOST is acknowledged for adding value to this work through networking and knowledge sharing with European experts. We thank all contributors for their hard work and texts of high quality. The authors created a professional, inspirational and inclusive working atmosphere, and we know that many new collaborations were established during handbook preparation. Such fruitful contacts are likely to result in new European research projects and other exchanges between the authors in coming years. We give our special thanks to those responsible for the management of CYANOCOST: Chair Dr. Triantafyllos Kaloudis, Vice-Chair Prof. Ludek Blaha and the working group leaders and deputy leaders are gratefully acknowledged for their positive attitude and encouragement. Science Officer of the Action Dr. Deniz Karaca, Administrative Officer Ms. Tania Gonzalez Ovin and Rapporteur Mr. Dick Blaauboer are also thanked for their support.

Research grants from the European Union Seventh Framework Programme (to J.M./L.S.; MAREX – Exploring Marine Resources for Bioactive Compounds: From Discovery to Sustainable Production and Industrial Applications 2010–2014, grant no. FP7-KBBE-2009-3-245137) and Svenska Litteratursällskapet i Finland (to L.S. in 2015) are gratefully acknowledged. These grants have enabled the active participation of J.M. and L.S. in the ongoing research which underpins this handbook.

COST (European Cooperation in Science and Technology) is a pan-European intergovernmental framework. Its mission is to enable break-through scientific and technological developments leading to new concepts and products and thereby contribute to strengthening Europe's research and innovation capacities. It allows researchers, engineers and scholars to jointly develop their own ideas and take new initiatives across all fields of science and technology, while promoting multi- and interdisciplinary approaches. COST aims at fostering a better integration of less research intensive countries to the knowledge hubs of the European Research Area. The COST Association, an international not-for-profit association under Belgian law, integrates all management, governing and administrative functions necessary for the operation of the framework. The COST Association has currently 36 member countries (www.cost.eu).

Jussi Meriluoto, Lisa Spoof and Geoffrey A. Codd
January 2016

COST is supported by the EU Framework Programme
Horizon 2020

Section I

Introduction

1

Introduction: Cyanobacteria, Cyanotoxins, Their Human Impact, and Risk Management

Geoffrey A. Codd[1,2], *Jussi Meriluoto*[3,4], *and James S. Metcalf*[5]

[1] *Biological and Environmental Sciences, University of Stirling, Scotland, United Kingdom*
[2] *School of the Environment, Flinders University, Adelaide, Australia*
[3] *Biochemistry, Faculty of Science and Engineering, Åbo Akademi University, Turku, Finland*
[4] *Laboratory for Paleoenvironmental Reconstruction, Faculty of Sciences, University of Novi Sad, Serbia*
[5] *Institute for Ethnomedicine, Jackson, USA*

1.1 Introduction

Cyanobacteria (blue-green algae) are ancient microorganisms with a global distribution. Via their oxygen-producing photosynthesis, they were responsible for the creation of the Earth's aerobic atmosphere 2200–2400 million years ago, and today they remain major agents in the biological cycling of carbon, nitrogen, and minerals. They are natural inhabitants of diverse environments including fresh, brackish, and marine waters, and the illuminated surfaces of rocks and soils. Due to their photosynthetic mode of growth, to the ability of many species to fix gaseous nitrogen into ammonia and amino acids, and to an ability to withstand adverse and extreme environmental conditions, cyanobacteria are primary colonisers. In times of abundant resources, cyanobacteria can store essential nutrients (phosphorus, nitrogen, carbon, iron) to permit growth under nutrient-limiting conditions. Some apply further strategies to withstand environmental extremes (high salinity, high and low temperatures, intermittent desiccation, high solar irradiation) to survive and grow in extreme environments. Cyanobacteria form the basis of numerous

Handbook of Cyanobacterial Monitoring and Cyanotoxin Analysis, First Edition.
Edited by Jussi Meriluoto, Lisa Spoof and Geoffrey A. Codd.

aquatic and terrestrial food chains and are finding increasing applications in biotechnology, from the production of biofertilizers to potential pharmaceuticals [1].

In aquatic environments, cyanobacteria may occur in diverse habitats: suspended in dispersed form or as aggregates in the water, on the water surface, on the bottom sediment, or attached to shoreline rocks and sediments and to plants. Mass populations of cyanobacteria in these locations can constitute blooms, scums, and biofilms or mats. Such mass populations can occur in pristine waterbodies, uninfluenced by human activity, where water enrichment with the nutrients necessary for cyanobacterial growth occurs via natural geological and hydrological processes. However, it is recognised today that this increase in the trophic condition of a waterbody (eutrophication) is particularly susceptible to anthropogenic pressures. These include increased nutrient loading from human, domestic, agricultural, and industrial sources, increased erosion in the waterbody catchment, and increased water abstraction. With favourable conditions of temperature, light penetration of the water column, water pH, in-lake water residence time, or river flow, massive increases in cyanobacterial bloom, scum, and mat formation can occur [2, 3].

In addition to deleterious effects on ecosystem biodiversity, cyanobacterial mass populations can adversely impact the availability, aesthetic quality, health/safety, and cost of water resources for human use. Human requirements and water-based activities that can be adversely affected include drinking water supplies, livestock watering, crop irrigation, aquaculture, industrial processing, recreation, and tourism [4]. Specific effects on water requirements and activities include the increased costs of drinking water treatment (e.g. filter blockage and increased treatment trains) and the presentation of human and animal health risks due to the common production of potent toxins, cyanotoxins, by the cyanobacteria [4, 5].

1.2 Cyanotoxins

Evidence from environmental poisoning events and bioassays with cyanobacterial biomass and extracts indicated the existence of specific cyanotoxins long before the first purification and toxicological and structural characterisations of the toxin molecules. Among the most widely implicated in intoxications and most commonly reported in ongoing analyses, the structures of anatoxin-a and of several microcystins were elucidated about 40 and 30 years ago, respectively. The cyanotoxins can be grouped into families according to chemical structure (Table 1). Most are low-molecular-weight molecules, ranging from 118 Da (β-*N*-methylamino-L-alanine [BMAA] and 2,4-diaminobutyric acid [DAB]) to ~1000 Da (microcystins), although the lipopolysaccharides (LPS) range from 10 to 20 kDa. Knowledge of the toxicity of the cyanotoxins, plus environmental, pathological, and histological data, with quantitative analysis, has identified most of the cyanotoxins in Table 1.1 as causes, whether principal or contributory, of human and animal intoxications. This list, however, cannot be regarded as exhaustive because toxicity testing of cyanobacterial cells and crude extracts in bioassays, both *in vitro* and *in vivo,* continues to indicate the presence of toxic compounds that cannot be accounted for by the known cyanotoxins [e.g. 6, 7]. Indeed, in terms of the vast array of secondary metabolites and other bioactive products that is emerging from cyanobacterial screening programmes [e.g. 8], additional toxic products of cyanobacteria of environmental and health significance should be anticipated. This handbook, therefore, includes a chapter on the role, performance, and interpretation of bioassays, with the potential to indicate the presence of novel toxins, to complement the practical guidelines for the analysis of the named cyanotoxins.

Table 1.1 *Cyanotoxins: summary of chemical structures, principal examples of sources, and modes of action*

Cyanotoxin	Chemical structure	Cyanobacteria producers	Modes of action
Microcystins	Cyclic heptapeptides	*Microcystis, Anabaena, Nostoc, Planktothrix, Phormidium, Oscillatoria, Radiocystis, Gloeotrichia, Anabaenopsis, Rivularia, Tolypothrix, Hapalosiphon, Plectonema*	Hepatotoxic, tumour promoting, inhibition of eukaryotic protein phosphatase PP1, PP2A, and phosphoprotein phosphatases PPP4, PPP5
Nodularins	Cyclic pentapeptides	*Nodularia spumigena, Nostoc* (symbiotic)	As for microcystins, plus weak carcinogenicity
Cylindrospermopsins	Tricyclic guanidine alkaloids	*Cylindrospermopsis, Umezakia, Anabaena, Oscillatoria, Raphidiopsis, Aphanizomenon*	Multiple organ toxicity, neurotoxic, genotoxic, protein synthesis inhibitor
Anatoxin-a	Bicyclic alkaloids	*Anabaena, Phormidium, Aphanizomenon*	Neurotoxic, binds competitively at acetylcholine receptors
Anatoxin-a(*S*)	Phosphorylated cyclic *N*-hydroxyguanine	*Anabaena*	Neurotoxic, inhibits acetylcholine esterase
Saxitoxins	Alkaloids	*Aphanizomenon, Anabaena, Lyngbya, Cylindrospermopsis, Planktothrix*	Neurotoxic, blocks voltage-gated sodium channels
Lyngbyatoxins, Aplysiatoxins	Indole alkaloids	*Lyngbya, Oscillatoria, Schizothrix*	Tumour-promoting, binds to eukaryotic protein kinase C
BMAA, DAB	Diamino acids	Many genera	Neurotoxic, developmental toxin, erroneous insertion into proteins
LPS	Lipopolysaccharides	All genera	Inflammatory, promotion of cytokine secretion

List is not exhaustive: see [5] and Appendix 2.

1.3 Exposure Routes, Exposure Media, and At-Risk Human Activities

The successful risk management of the hazards presented by cyanobacteria and cyanotoxins requires adequate recognition of the exposure routes and exposure media via which humans and animals can be placed at risk. These are summarised in Table 1.2. The principal exposure media currently recognised in Europe at present are water and foods harvested or caught from waterbodies with cyanobacterial mass populations. Because most people in the developed world drink treated drinking water, exposure via drinking raw water is most likely to occur incidentally during recreational activities. Whether the ingestion of cyanobacteria and cyanotoxins via treated water occurs (in the event of their presence in the water abstracted for treatment) depends on the removal efficiency of the drinking water treatment process. A range of both traditional and advanced drinking water treatment methods has the potential to remove cyanobacterial cells and cyanotoxins. Thus, effective monitoring is essential to inform water engineers of the operating efficiency of their treatment plants during periods of cyanobacterial development and persistence and for periods after the decline of mass populations, so that informed decisions can be made on the continuation or disruption of supply or upgrading of treatments to protect health. Terrestrial plant crops can constitute an exposure medium if they are spray-irrigated because mucilaginous cyanobacterial cells can adhere tenaciously to plant leaves and cyanotoxins can be taken up by plant tissues. The uptake of cyanobacterial cells and cyanotoxins by finfish, shellfish, and prawns can lead to these foods being a significant exposure medium, although the bioaccumulation of cyanotoxins by

Table 1.2 Exposure routes, exposure media, and activities at risk of contact with or ingestion of cyanobacteria and cyanotoxins

Exposure route	Exposure medium	At-risk activities
Oral	Water	Daily drinking of raw water or inadequately treated water. Incidental drinking of water during water-based recreational activities. Animal livestock watering.
	Food	Consumption of finfish, shellfish, crustaceans; plant foods (e.g. salad) if spray-irrigated with water containing cyanobacteria/cyanotoxins.
	Dietary supplements, drinks ("smoothies")	If consisting of or containing cyanobacteria or cyanotoxins.
Dermal	Water	Skin and mucosal contact with blooms, scums during recreation and work practices. Showering with raw water or inadequately treated water.
Inhalation	Water, aerosols	Spray containing cyanobacteria/cyanobacterial extracts/cyanotoxins during recreation, showering, work.
	Dust	Inhalation of dust particles from air-dried and lyophilised cyanobacterial biomass, biogenic desert crusts during work practices, military deployment.
Haemodialysis	Water	Contact with inadequately treated haemodialysis water.

Updated from [5].

these animals can be counteracted to some degree by cyanotoxin elimination and biotransformation (enzymic detoxification). The latter process can also occur in plant tissue. Cyanotoxin analyses of exposed food materials should, therefore, distinguish between detection of the native cyanotoxins and cyanotoxin detoxification products. Understanding the significance of dermal and inhalation exposure to spray, aerosols, or dust is less advanced than that via ingestion, although these routes may be predominant in some recreational and occupational situations. The documented mass mortalities and severe illness among haemodialysis patients due to exposure to cyanotoxins in ineffectively treated haemodialysis water serve as tragic reminders of the potential of this exposure medium versus vulnerable subjects [5].

1.4 Cyanobacterial Blooms and Cyanotoxins in Relation to Human Pressures on Water Resources and Climate Change

With a rapidly growing human population, the influence of human activities (agricultural, urban, and industrial) in increasing the size, geographical distribution, and seasonal persistence of cyanobacterial mass populations seems set to continue at global level. In developed countries with increasing affluence and declining household occupancy and where human population growth is negligible or, as in parts of Europe, decreasing, domestic per capita water consumption remains high [e.g. 9]. This is despite the use of water metering, tariffs, and public education programmes.

Climate change already appears to have negatively affected some ecosystems and ecosystem services, including provisioning (water, crops, food, energy) and supportive services (water and nutrient cycling, habitat provision). Further projected changes including human migration and European desertification would exacerbate the pressures on the availability, quality, and safety of water [10]. Many of the hydrological changes in aquatic ecosystems caused by climate change can specifically favour cyanobacterial mass development: a rise in water temperature, decrease in vertical mixing and flushing, and extreme weather events (droughts, floods). It will, thus, be increasingly necessary for waterbody managers (potable supply, recreation, aquaculture, irrigation) and health authorities to address the problems presented by cyanobacterial mass populations, using reactive and proactive strategies. For these to be applied, reliable and relevant environmental data, supplied without delay, are essential.

1.5 Aims of the Handbook

Methods for the monitoring of cyanobacterial populations and the analysis of cyanotoxins have been developed, applied, refined, and reapplied for many years. They have been driven by operational and risk management needs, including human and animal intoxications, and enabled by the combination of the latest knowledge of cyanobacterial form and function, together with both available and emerging technology. Adjustments to existing methods and the introduction of new methods (microscopic; physicochemical; immuno-, bioassay-, and biosensor-based; and molecular biological) are a constant feature of the primary literature and these continue. The aims of this handbook are to:

1. Provide practical assistance in the planning and performance of monitoring and analytical procedures in the field and laboratory.
2. Provide guidance in the calculation and interpretation of the results.

3. Facilitate the transfer of accumulated experience and best practices from long-established practitioners and researchers in the cyanobacterial and cyanotoxin fields to laboratories that are newly entering these fields and to end-users who may not be specialists.
4. Encourage the development of further interlaboratory method validation studies.
5. Contribute to the protection of water resources and health from hazards presented by cyanobacterial mass populations and cyanotoxins.

References

[1] Whitton, B.A. (ed.) (2012) *Ecology of Cyanobacteria II: Their Diversity in Space and Time*, Springer, Dordrecht/Heidelberg/New York/London.

[2] Schindler, D.W (2006) Recent advances in the understanding and management of eutrophication. *Limnology and Oceanography* **51** (2), 1356–1363.

[3] Lampert, W. and Sommer, U. (2007) Limnoecology, 2nd ed., Oxford University Press, Oxford, UK.

[4] Codd, G.A., Azevedo, S.M.F.O., Bagchi, S.N. *et al.* (2005) *A Global Network for Cyanobacterial Bloom and Toxin Risk Management, Technical Documents in Hydrology No. 76*, UNESCO, Paris.

[5] Metcalf, J.S. and Codd, G.A. (2012) Cyanotoxins. In *Ecology of Cyanobacteria II: Their Diversity in Space and Time* (Whitton, B.A., ed.), Springer, Dordrecht/Heidelberg/New York/London, pp. 651–675.

[6] Bernard, C., Froscio, S., Campbell, R. *et al.* (2011) Novel toxic effects associated with a tropical *Limnothrix/Geitlerinema*-like cyanobacterium. *Environmental Toxicology* **26** (3), 267–270.

[7] Nováková, K., Kohoutek, J., Adamovský, O. *et al.* (2013) Novel metabolites in cyanobacterium *Cylindrospermopsis raciborskii* with potencies to inhibit gap junctional intercellular communication. *Journal of Hazardous Materials* **262**, 571–579.

[8] Gademann, K. (2011) Out in the green: biologically active metabolites produced by cyanobacteria. *Chimia* **65** (6), 416–419.

[9] Environment Agency (2008) *Climate Change: Approach in Water Resources Planning – Overview of New Methods*. Report SC090017/R3. Environment Agency, Bristol, UK.

[10] Gosling, S.N. (2013) The likelihood and potential impact of future change in the large-scale climate-earth system on ecosystem services. *Environmental Science Policy* **27** (Suppl. 1), S15–S31.

Section II
Cyanobacteria

2

Ecology of Cyanobacteria

Jean-François Humbert[1] *and Jutta Fastner*[2]

[1] *Institute of Ecology and Environmental Sciences, UPMC, Paris, France*
[2] *Section Drinking Water Resources and Water Treatment, German Environment Agency, Berlin, Germany*

2.1 Introduction

Cyanobacteria are ubiquitous, colonizing many different habitats worldwide and under all climatic conditions [1]. They occur suspended in freshwater, brackish water, or saltwater (planktonic) and can be attached to the bottom of lakes and rivers (benthic) and to plants and stones. Some species live in symbiosis with higher plants or fungi (e.g. lichens).

The morphology of cyanobacteria comprises unicellular, colonial, and multicellular filamentous forms (for details, see Chapter 6). Some groups have specialized cells, including heterocysts for fixing atmospheric nitrogen or akinetes, spore-like cells, to survive adverse conditions. Single cells are usually microscopic, but larger forms like gelatinous colonies (e.g. *Microcystis*) or benthic mats (e.g. *Phormidium*) are easily visible to the naked eye.

Cyanobacteria are able to growth under low-light conditions due to their accessory pigments (phycobilins, in particular phycocyanin and phycoerythrin). In addition, their floatation capacities (due to gas vesicles) allow scum-forming cyanobacteria (e.g. *Microcystis* spp.) to occupy the first centimetres of the water column. At the same time, these scum-forming cyanobacteria and benthic forms that are exposed to very high amounts of light and ultraviolet (UV) radiation possess protection mechanisms to avoid photoinhibition or UV damage, such as carotenoid pigments and scytonemin. Finally, the capacity of some of species to fix free nitrogen (N_2) is considered important by allowing some species to outcompete others that are not able to do so under N-limiting conditions and high phosphorus (P) concentrations, although some N_2-fixing species have been found blooming under equally high levels of fixed N and P [2].

Handbook of Cyanobacterial Monitoring and Cyanotoxin Analysis, First Edition.
Edited by Jussi Meriluoto, Lisa Spoof and Geoffrey A. Codd.

The nutrient enrichment of many waterbodies due to human activities (eutrophication) has led to proliferations of cyanobacteria throughout the world. Since many species of potentially toxic cyanobacteria tend to bloom formation, knowledge on the specific traits suspected to favour their dominance in eutrophic waterbodies and their ecological behaviour is necessary for adequate monitoring and management in the context of risk assessment. Overall, many physicochemical and biological factors and processes are known to have an impact on the population dynamics of cyanobacteria and on their vertical and horizontal distribution in waterbodies (see later). The multiple interactions between all these factors and processes make the prediction of the medium- and long-term (several weeks or months) development of cyanobacterial blooms almost impossible, highlighting the need for an accurate monitoring.

2.2 Environmental Conditions Leading to Cyanobacterial Blooms

Cyanobacteria are primary producers that use light energy to synthesize organic matter from mineral nutrients and CO_2 (i.e. oxygenic photosynthesis like true algae and higher plants). Consequently, the main factors affecting the growth of these microorganisms are nutrient (in particular, phosphorus) and light availabilities. Other physical and biological processes such as the thermal stratification of the water column, competition with other organisms, predation, and parasitism have also an impact on the population dynamics of cyanobacteria in freshwater ecosystems.

It is now well established that blooms of planktonic cyanobacteria are most prevalent in eutrophic ecosystems [3], that is, in lakes and ponds with total phosphorus concentrations >50 $\mu g\,L^{-1}$. Substantial biomasses of cyanobacteria can sometimes also be found in mesotrophic ecosystems (total phosphorus concentrations between 20 and 50 $\mu g\,L^{-1}$), such as *Planktothrix rubescens* (up to 30–50 $\mu g\,L^{-1}$ eq. chlorophyll-*a* for *P. rubescens*) growing in deep sub-alpine lakes. On the other hand, in flowing waters, proliferations of benthic cyanobacteria can be found in oligotrophic conditions in rivers and mountain lakes with low phosphorus concentrations.

In temperate regions cyanobacterial blooms occur generally during the summer time, when the water temperature is above 20°C and the thermal stratification of the water column is established (for lakes >3-m depth). However, some species are able to proliferate in colder water, such as *P. rubescens*, which is able to bloom even under the ice of lakes in winter [4], and *Planktothrix agardhii*, which is able to form perennial blooms in ponds (sometimes for several years).

Many studies have been performed on the influence of thermal stratification of the water column on the dynamics of cyanobacterial blooms. Stratification enables buoyant cyanobacteria (e.g. *Microcystis*) to adjust their position in the water column for light accessibility, for example, and thereby outcompete other organisms such as microalgae. Further, thermal stratification in eutrophic waterbodies can lead to the release of phosphorus from the sediment under anoxic conditions in the hypolimnion and thus indirectly influence the dynamics of cyanobacterial populations.

In addition to these physicochemical factors and processes, several biological processes have been described as acting on the population dynamics of cyanobacteria. In non-limiting nutrient conditions (eutrophic conditions), competition occurs for access to light between photosynthetic microorganisms (i.e. between cyanobacteria and microalgae) and between cyanobacteria and floating plants. Concerning the latter, it is known that depending on the turbidity and on some other processes such as water level, alternation between floating plants and phytoplankton (including cyanobacteria) dominance can occur in lakes and ponds [5].

The abundance of large zooplankton feeding on cyanobacteria can also play a role in the appearance of cyanobacterial blooms. Their abundance, however, is determined to a large extent by the composition and the biomass of the fish community, in particular by the abundance of planktivorous species.

2.2.1 What Species for Which Types of Environments?

From the first report of Reynolds et al. [6] to the recent one from Dolman et al. [2], numerous studies have tried to identify the environmental conditions that determine the distribution of the different bloom-forming cyanobacteria in eutrophic freshwater ecosystems. Indeed, it has been found that for some waterbodies, the dominant species are always the same, whereas in others, different species can bloom at the same time or successively. While one of the main bloom-forming species, *P. agardhii*, is known to proliferate preferentially in turbid, mixed lakes and ponds, *Microcystis aeruginosa* instead blooms in the epilimnic layer of well-stratified lakes in summer. Some genera, including *Microcystis* and *Aphanizomenon*, are frequently found in the same lakes, sometimes successively and sometimes associated with each other. Similarly, *Planktothrix*, *Limnothrix*, and *Pseudanabaena* spp. are frequently associated in turbid lakes (e.g. [7]).

Recently, newly-observed, invasive species have been described in European lakes. For example, *Cylindrospermopsis raciborskii*, a typical cyanobacterium from tropical and subtropical areas (e.g. Australia, South America) has become increasingly abundant in numerous European lakes since the end of the last century, probably in relation to climate warming [8].

2.3 Population Dynamics of Cyanobacteria

2.3.1 How Is a Bloom Defined?

A bloom is an increase of cyanobacterial biomass in a lake (measured, for example, by chlorophyll-*a* concentration) over a relatively short time (between a few days and 1 or 2 weeks) and is characterized by the dominance (>80%) of only one or a few species within the phytoplankton community. In mesotrophic or less-eutrophic lakes and ponds, biomasses from 30 to 50 $\mu g\,L^{-1}$ chlorophyll-*a* correspond to large blooms, whereas in eutrophic and hypereutrophic lakes, biomasses exceeding 300 to 400 $\mu g\,L^{-1}$ chlorophyll-*a* can be found (Fig. 2.1).

2.3.2 Seasonality in the Dynamics of Cyanobacterial Populations

The population dynamics of cyanobacterial populations can display very different features between species but also for the same species from year to year. As shown in Fig. 2.2, cyanobacterial species can display contrasting population dynamics, one genus (*Aphanizomenon*) being characterized by a very chaotic dynamics with two peaks of abundance, whereas another (*Microcystis*) was characterized by a more regular population dynamics [9]. These differences in population dynamics make the choice of appropriate sampling strategies for monitoring very difficult. For example, it was shown in the same study that if a monthly sampling was sufficient to provide a good overall estimation of the population dynamics of *Microcystis* in the pond, at least a weekly sampling was necessary for *Aphanizomenon*.

As described previously, the population dynamics of planktonic cyanobacteria are mainly influenced by meteorological events, with the influence on water temperature and thermal stratification of the water column being considered most important. In the same way, river flow and water

Figure 2.1 *Bloom of* Microcystis aeruginosa *in a French reservoir. The chlorophyll-a concentrations exceeded 200 μg L*$^{-1}$ *in the surface layer*

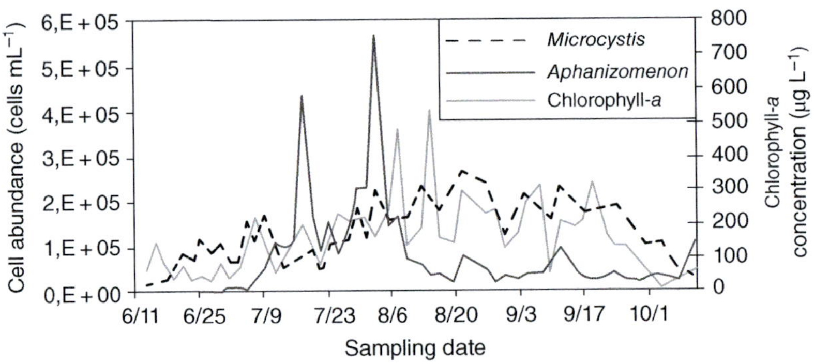

Figure 2.2 *Variations in chlorophyll-a concentration and in the abundance of two cyanobacterial genera (*Microcystis and Aphanizomenon*) blooming in a small French pond*

temperature are known to be to main factors driving the population dynamics of benthic cyanobacteria (10). Further, a lot of biological factors and processes, which are not fully understood, influence population dynamics of cyanobacteria. Consequently, only short-term predictions (4–5 days) can be made on the development and persistence evolution of blooms based on weather forecasts.

2.4 Spatial Distribution of Cyanobacteria in Freshwater Ecosystems

In addition to temporal distribution patterns, heterogeneity in the spatial distribution of cyanobacteria can occur in both lakes and rivers. Differences in both the horizontal and vertical distribution of cyanobacteria has been observed in deep, stratified lakes. Under conditions of thermal stratification, several of the bloom-forming, potentially toxic cyanobacteria, including *Microcystis*, *Anabaena*, and *Aphanizomenon flos-aquae*, can float to the water surface by means of buoyancy regulation, forming "scums" (i.e. surface layers of cells). These scums can drifted by slight wind action to downwind shorelines, where they (and their toxins) can accumulate to concentrations up to several orders of magnitude higher than that in the open water (Fig. 2.3). Such phenomena can appear and disappear within hours, and they depend primarily on wind speed and direction, as well as on the morphology of the waterbody itself. At wind-exposed sites, there may be a high temporal variability, whereas in sheltered bays, for example, scums may persist for weeks (Fig. 2.3). Also, in less deep, only temporarily stratified lakes, the formation of scums may be very dynamic.

Vertical distribution patterns in stratified lakes in temperate climatic zones are mainly associated with *Planktothrix rubescens* (Fig. 2.4). This potentially toxic cyanobacterium grows in the metalimnion of mesotrophic, stratified lakes in summer using nutrient-rich deep water layers. During autumnal turnover, this species is entrained in the whole water column and can eventually form dense surface blooms under suitable conditions. It can persist during the whole year, even under ice.

Other cyanobacteria, which have been occasionally reported for epilimnetic abundance maxima at 1- to 3-m depth, are *Anabaena* spp., *Aphanizomenon* spp., and *C. raciborskii* (e.g. [11]). In contrast, shallow (<3-m depth) lakes are usually frequently mixed by wind, leading to a homogeneous distribution of cyanobacteria throughout the entire water column (e.g. [12]). Several filamentous, potentially toxic cyanobacteria, including *Planktothix agardhii*, often together with varying proportions of *Aphanizomenon gracile*, *Cuspidothrix issatschenkoi*, and *Limnothrix*, are favoured under these conditions.

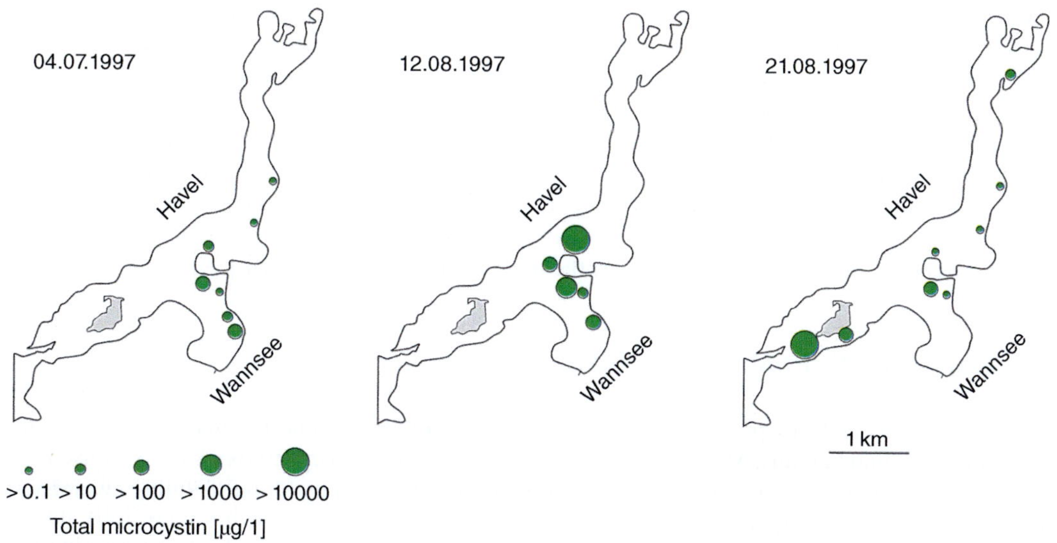

Figure 2.3 Concentrations of total microcystin (circles) along the shoreline of the River Havel, Berlin, on 3 different days; maximal microcystin concentrations amounted to around 25,000 µg L^{-1}

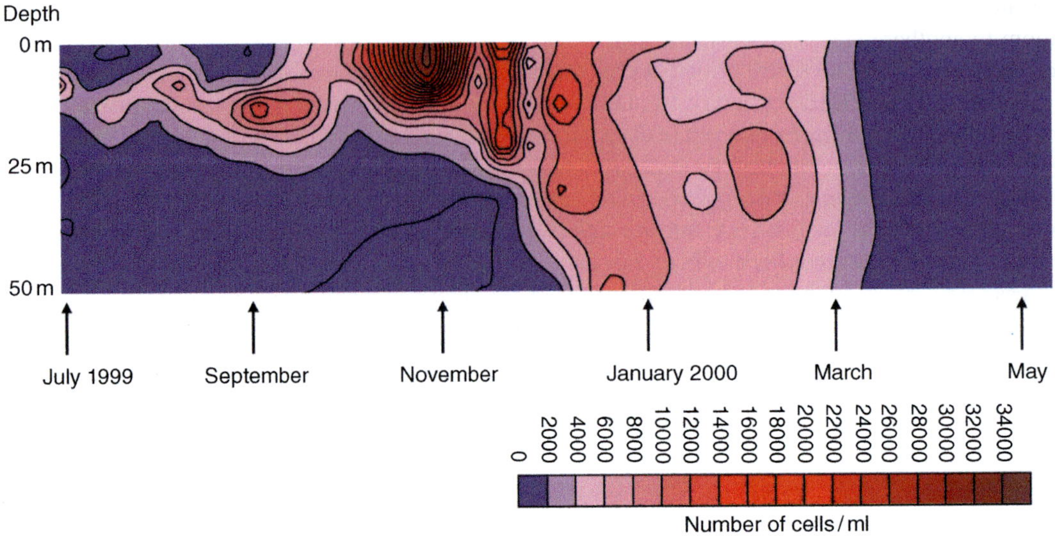

Figure 2.4 *Vertical distribution of* Planktothrix rubescens *during an annual cycle in Lake Bourget, France*

Also, benthic cyanobacteria (i.e. attached to sediments along shorelines and the bottom of shallow waters) in lakes and rivers can show pronounced heterogeneity in their horizontal distribution. Though cyanobacterial mats often consist primarily of one or a few cyanobacterial species, other organisms such as bacteria, phototrophic algae, and fungi are also typical biofilm inhabitants. The main factors in effecting growth of mats are physical disturbance (wet/dry cycles, wave action, or shear stress), light, water temperature, nutrients, and grazing pressures. It has been shown recently that the occurrence of massive growths of potentially toxic *Phormidium* in New Zealand rivers mainly depends on river flow rates and water temperature. Moreover, concerning the nutrients, in contrast to planktonic cyanobacteria, blooms of benthic cyanobacteria can be found in oligotrophic lakes and rivers (see review by Quiblier et al. [10]).

In addition to spatial heterogeneity in the distribution of benthic cyanobacteria in rivers, the detachment of biofilms from the substrate and their transportation and occasional accumulation in some areas of the river must also be considered as sources of heterogeneity in biomass distribution. The factors favouring these phenomena are strong wind or increasing flow velocity, in combination with trapping of oxygen bubbles within the mats, giving them buoyancy [10].

2.5 Ecology of the Production of Toxins by Cyanobacteria

Cyanobacteria are able to produce many secondary metabolites, some of them being toxic to plants, animals, and humans. Most of the investigations dealing with cyanotoxins have focused on microcystins, which appear to be the most common toxins produced by bloom-forming cyanobacteria in freshwater ecosystems.

Patterns of microcystin distribution and concentration in a waterbody are determined by (i) the biomass of the toxin-producing cyanobacteria; (ii) the amount and type of toxins in the cells; (iii) the share of toxigenic genotypes in a population; and (iv) the release of extracellular microcystins in water. Microcystin concentrations per unit of biomass can vary considerably both from one bloom to another and during the course of a single bloom, but can they also can be rather stable both between as well as within years (e.g. [13]).

Numerous research has been conducted in the past 10 years to better understand the impact of environmental conditions on the toxin content of cyanobacterial cells. Though environmental factors (including trace metals, nutrients, and light availability) can have an influence on the proportion of microcystin-producing and non–microcystin-producing cells and on the production of microcystins within cells, it has been problematic to predict with high confidence the potential toxicity of a bloom and the health risks associated [14].

Less data are available for other cyanotoxins, but some recent reports on homo/anatoxin-a produced by benthic cyanobacteria (in particular, by *Phormidium*) have shown that there was a great heterogeneity when comparing the toxin concentrations in different mats collected in the same area, probably due to the existence of toxic and nontoxic genotypes in *Phormidium* populations but also to the concentrations of anatoxin-a produced by the toxic genotypes (see review by Quiblier et al. [10]).

2.6 General Conclusions

Though knowledge on the ecology and toxicity of cyanobacteria has advanced significantly in the past 20 years, it remains very difficult to make predictions for more than several days in advance on the population dynamics and on the potential toxicity of cyanobacterial blooms. Moreover, it appears that the population dynamics and potential toxicity of blooms are mainly influenced by environmental pressures acting on various scales of the ecosystem. Consequently, blooms of cyanobacteria occurring in several ecosystems in the same area and their potential toxicity can display very different trajectories. Moreover, for a given ecosystem, large variations can be found from year to year. Thus, it appears that sampling strategy (including the number of sampling points, their locations, sampling frequency, and sampling depth) is the most critical part of the monitoring of cyanobacteria and that this strategy must be adapted to local conditions and with reference to the aim of the monitoring (whether for scientific purposes, health protection, etc.).

References

[1] Whitton, B.A. (ed.) (2012) *Ecology of Cyanobacteria. II Their Diversity in Space and Time*, Springer Dordrecht.

[2] Dolman, A.M., Rücker, J., Pick, F.R. *et al.* (2012) Cyanobacteria and cyanotoxins: The influence of nitrogen versus phosphorus. *PLoS ONE* **7** (6), e38757.

[3] Conley, D.J., Paerl, H.W., Howarth, R.W. *et al.* (2009) Controlling eutrophication: Nitrogen and phosphorus. *Science* **323** (5917), 1014–1015.

[4] Blikstad Halstvedt C., T. Rohrlack, T., Andersen, O. *et al.* (2007) Seasonal dynamics and depth distribution of *Planktothrix* spp. in Lake Steinsfjorden (Norway) related to environmental factors. *Journal of Plankton Research* **29** (5), 471–482.

[5] O'Farrell, I., Izaguirre, I., Chaparro, G. *et al.* (2011) Water level as the main driver of the alternation between a free-floating plant and a phytoplankton dominated state: a long-term study in a floodplain lake. *Aquatic Sciences* **73** (2), 275–287.

[6] Reynolds, C.S., Huszar, V., Kruk, C. *et al.* (2002) Towards a functional classification of freshwater phytoplankton. *Journal of Plankton Research* **24** (5), 417–428.

[7] Nixdorf, B., Mischke, U. and Rücker, J. (2003) Phytoplankton assemblages and steady state in deep and shallow eutrophic lakes – an approach to differentiate the habitat properties of Oscillatoriales. *Hydrobiologia* **502** (1), 111–121.
[8] Briand, J.F., Leboulanger, C., Humbert, J.F. *et al.* (2004) *Cylindropsermopsis raciborskii* (Cyanobacteria) invasion at mid-latitudes: Selection, wide physiological tolerance or global warming? *Journal of Phycology* **40** (2), 231–238.
[9] Pobel, D., Robin, J. and Humbert, J.F. (2011) Influence of sampling strategies on the monitoring of cyanobacteria in shallow lakes: Lessons from a case study in France. *Water Research* **45** (3), 1005–1014.
[10] Quiblier, C., Wood, S., Echenique, I. *et al.* (2013) A review of current knowledge on toxic benthic freshwater cyanobacteria – Ecology, toxin production and risk management. *Water Research* **47** (15), 5464–5479.
[11] Everson S., Fabbro, L., Kinnear, S. *et al.* (2009) Distribution of the cyanobacterial toxins cylindrospermopsin and deoxycylindrospermopsin in a stratified lake in north-eastern New SouthWales, Australia. *Marine and Freshwater Research* **60** (1), 25–33.
[12] Mischke U. (2003) Cyanobacteria associations in shallow polytrophic lakes: influence of environmental factors. *Acta Oecologica* **24** (Supplement 1), 11–23.
[13] Briand, J.F., Jacquet, S. Flinois, C. *et al.* (2005) Variations in the microcystin production of *Planktothrix rubescens* assessed from a four-year survey of Lac du Bourget (France) and from laboratory experiments. *Microbial Ecology* **50**, 418–428.
[14] Neilan, B.A., Pearson, L.A., Muenchhoff, J. *et al.* (2013) Environmental conditions that influence toxin biosynthesis in cyanobacteria. *Environmental Microbiology* **15** (5), 1239–1253.

3

Picocyanobacteria: The Smallest Cell-Size Cyanobacteria

Iwona Jasser[1] and Cristiana Callieri[2]

[1] Faculty of Biology, University of Warsaw, Poland
[2] CNR – Institute of Ecosystem Study, Verbania, Italy

3.1 Introduction

3.1.1 General Characteristics of Picocyanobacteria

Picocyanobacteria are cyanoprokaryotes, cells smaller than 2 to 3 μm in diameter, that live as solitary cells or, under particular conditions, form microcolonies and colonies in marine and freshwater habitats [1]. Freshwater single-celled picocyanobacteria have been divided into five genera: *Synechococcus, Cyanobium, Synechocystis, Cyanothece*, and *Cyanobacterium*. Even if the cell dimension is by definition in the pico-size range (0.2–2.0 μm), some genera can exceed the upper limit. Further, under specific conditions, some picocyanobacteria can develop mucilage and remain near the mother cell, forming microcolonies or larger colonies [1]. The most common non–bloom-forming, colonial picocyanobacteria in freshwater are species of *Aphanocapsa, Aphanothece, Chroococcus, Coelosphaerium, Cyanobium, Cyanodictyon, Merismopedia, Romeria, Snowella*, and *Tetracercus*.

Picocyanobacteria have a special cell wall type that combines structural elements of both the bacterial Gram-negative cell wall type (presence of an outer membrane composed of lipopolysaccharides (LPS), lipids, proteins, and carotenoids) and the Gram-positive cell wall type (thick and highly cross-linked, eight-layered peptidoglycan network with covalently linked polysaccharide) [2]. Frequently, picocyanobacteria have a paracrystalline surface layer (S layer formed by a glycoprotein), which has a protective function and is involved in cell adhesion and motion.

Handbook of Cyanobacterial Monitoring and Cyanotoxin Analysis, First Edition.
Edited by Jussi Meriluoto, Lisa Spoof and Geoffrey A. Codd.

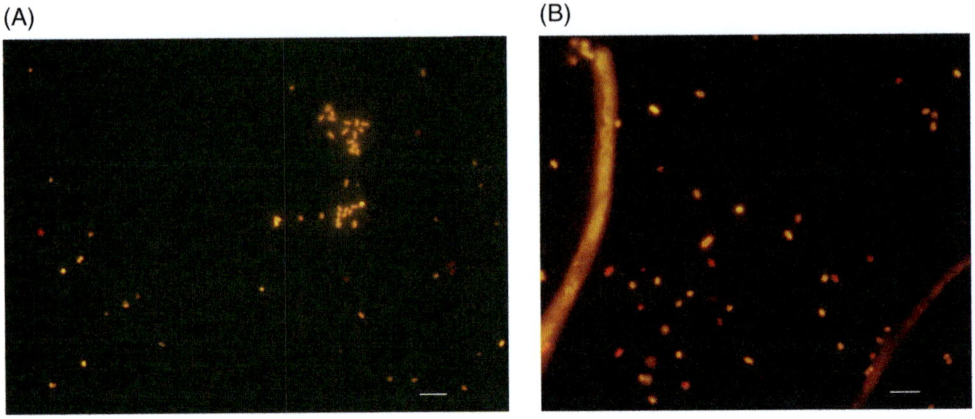

Figure 3.1 *Picocyanobacteria from environmental samples under an epifluorescence microscope. (A) PE cells – solitary and micro-colonies fluorescing yellow and eukaryotic cells, fluorescing red (blue filter). (B) PE cells fluorescing yellow and PC cells fluorescing red (green-CY3 filter). Bar denotes 5 μm. Photographs by (A) Callieri and (B) Jasser*

3.1.2 Detection and Identification

Samples of picocyanobacteria for microscopic observation or for counting by flow cytometry should be preserved or processed immediately. Even if preservatives are added, the samples should be counted within 1 week, to ensure that there is no loss of pigment fluorescence. Alternatively, they can be immediately filtered onto a 0.2-μm polycarbonate membrane, mounted on a slide, and preserved at –20°C.

Picocyanobacterial ecotypes exhibit differences in their accessory pigments that affect their adaptation to spectral light quality. It is helpful to classify picocyanobacteria into two cell types: one with yellow autofluorescing phycoerythrin (PE) and the other with red autofluorescing phycocyanin (PC) as the major light-harvesting pigments. They can be easily observed with the use of epifluorescence microscopy with different filters: blue, green, and narrow banded green (Fig. 3.1). Old samples can be visualized by using DAPI staining and observed in UV light; however, one has to keep in mind that by using this method, heterotrophic bacteria are also visualised. Details concerning sampling, visualisation, and enumeration of picocyanobacteria can be found in SOP 4.

3.1.3 Phylogenetic Position

There are three picocyanobacteria genera more commonly found in phylogenetic trees: *Synechococcus* and *Cyanobium* in freshwaters and *Prochlorococcus* (and *Synechococcus*) in marine waters. The genus *Synechococcus* is polyphyletic, but the marine clade 5.1 (A and B) and the *Prochlorococcus* clade are monophyletic [3]. Outside these well-defined marine clades, there are a number of freshwater clades formed by cosmopolitan picocyanobacteria widely dispersed in nonmarine environments. Endemic strains have been found in Lake Superior (USA) [4], Mazurian lakes (Poland) [5], Austrian lakes [6], and Patagonian deep lakes [7]. Recently, novel clades of halotolerant and freshwater picocyanobacteria have been found close to the marine *Synechococcus* subclusters 5.2 and 5.3, opening new perspectives of phylogenetic affiliation with different habitats [7].

3.1.4 Occurrence in Freshwater and Marine Environments

Picocyanobacteria are numerous and ubiquitous primary producers in freshwater and marine environments. In lakes of temperate regions, single-cell picocyanobacteria maxima generally conform to a typical bimodal pattern, with a peak in spring or early summer and a second peak during late summer/autumn. The second peak coincides with a maximum of microcolonies. Such a variety of morphotypes reflects a genotypic diversity among picocyanobacterial communities that accounts for different composition according to trophic status, season, and the vertical profile observed in different lakes and oceans [1].

In the oceans, *Prochlorococcus* abundance generally exceeds that of *Synechococcus* and extends to greater depths. Conversely, *Synechococcus* has a wider distribution as a consequence of a higher growth rate in a wider temperature range and at elevated nitrate concentrations. Peak concentrations of *Synechococcus* occur more consistently in the southern Atlantic. High concentrations have also been found in the Baltic Sea, while persistent blooms were noted in coastal lagoons of the Adriatic Sea [8–10].

The abundance of picocyanobacteria varies between 10^3 and 10^6 cells mL^{-1}, generally increasing along with increasing trophic status. It has been also widely accepted that the percentage of picocyanobacteria in the total phytoplankton biomass increases with decreasing ecosystem trophic state, based on chlorophyll *a* content. This means that picocyanobacteria may dominate in ultraoligotrophic regions of the marine environment and in oligotrophic lakes [1, 11]. This model works particularly well for large deep lakes, whereas in shallow hypertrophic lakes and in coastal waters, picocyanobacterial abundance is more difficult to predict, without considering the physical and chemical characteristics of the environment.

Underwater light quality and quantity influence the presence of picocyanobacteria and their prevalent pigment composition. It was found that in highly coloured (humic) lakes, non–phycoerythrin-rich cells (PC) dominated numerically, while in clearer, hardwater lakes, phycoerythrin-rich cells (PE) were the most abundant [12], which is similar to what happens in marine systems.

3.1.5 Ecological Role of Picocyanobacteria

Together with bacteria, picocyanobacteria form the base of the microbial foodweb, of which picocyanobacteria are the autotrophic components. They serve as prey for protozooplankton, with heterotrophic and mixotrophic nanoflagellates and small ciliates being the main grazers. The nanoflagellates (both heterotrophic and mixotrophic) can remove up to 90% of picocyanobacterial and other bacterial biomass, while ciliates are supposed to be responsible only for about 10% of this removal [13]. The grazing of protozoans influences not only the abundance but also the diversity of *Synechococcus* at the strain-specific level [14]. Picocyanobacteria may also, to some extent, be grazed on by filter-feeding metazooplankton, nauplii, and early copepodite stages of copepods; by planktonic rotifers feeding on suspended particles; and by mixotrophic algae [1]. Thus, they may enter higher trophic levels directly, when grazed on by metazooplankton, or indirectly, when nanoflagellates, ciliates, and mixotrophic algae serve as a trophic link between them and metazooplankton. In either case, picocyanobacteria should not be ignored in considerations of material and energy flux in planktonic foodwebs.

3.2 Records of Toxic Picocyanobacteria

The reports of the toxicity of picocyanobacteria are very rare, because most previous works have assumed a lack of toxin production by these cyanobacteria and, additionally, some sampling procedures (using plankton nets) do not retain picocyanobacteria. There are, however, several records providing evidence from laboratory and field studies concerning various aspects of the potential and actual toxicity of picocyanobacteria. There are also results demonstrating that picocyanobacteria are capable of the production of various bioactive compounds.

3.2.1 Occurrence of Microcystins in Picocyanobacteria

The most frequently found toxins produced by cyanobacteria are the cyclic heptapeptide hepatotoxins of the microcystin group. The first records on *Synechocystis* sp. and *Synechococcus* sp. toxicity consistent with microcystins come from Lincoln and Carmichael [15] and Mitsui et al. [16]. However, the first widely cited account of a possibility that picocyanobacteria produce these hepatotoxins was provided by Blaha and Marsálek [17]. The authors presented results of analyses of seven picocyanobacterial strains belonging to the genera *Cyanobium*, *Synechococcus* (picoplankton), and *Cyanobacterium* (metaphyton), which were tested to determine if they were capable of microcystin and other cyanotoxin production. High-performance liquid chromatography (HPLC) analysis demonstrated that they produced several low-molecular metabolites and that in the extracts of two strains—*Synechococcus nidulans* and *Cyanobium rubescens*—one and two types of microcystins, respectively, were detected. The results indicated that the amount of products detected in studied strains was smaller than usually recorded in cyanobacterial blooms. Still, the toxicity tests in vitro in cell cultures showed that these products were hepatotoxic and tumour-promoting microcystins as well as cytotoxic. Lymphocyte-targeted immunotoxic effects for some of the strains were also demonstrated.

At the same time, Domingos et al. [18] reported a possible association between picoplanktonic cyanobacteria present in reservoirs used as water supplies for Cararau in Brazil and human poisoning involving microcystins, which occurred in inadequately treated water used for human haemodialysis in February 1996. The authors tested seven strains of colonial and single-celled picocyanobacteria isolated from Caruaru's Tabocas and Sr. Jose Maria reservoirs for toxicity. Enzyme-linked immunosorbent assay (ELISA) demonstrated that each of the seven strains was able to produce very low to low amounts of microcystins, with concentrations varying between 0.08 and 3.7 μg g dry wt^{-1}. HPLC analysis revealed a peak with a similar UV spectrum to microcystin-LR in extracts from one of the strains, a single-celled *Synechococcus* NPCA-15. Two of the strains, single-celled NPCA-15 and colonial *Aphanocapsa cumulus* (NCPA-23), exhibited toxicity, when tested by mouse bioassay with an estimated LD_{100} of 600 mg kg body wt^{-1}, which is considerably low. Despite such low production of microcystins, the two strains were shown to be lethal to a mouse assay, with signs of poisoning characteristic of microcystins. The authors suggested that other toxins, in addition to microcystins, might be responsible for the lethal effect.

Further results from environmental and laboratory investigations for microcystin production by *Synechocystis* concerned two strains (Syn-WTP93 and Syn-WTP97) isolated from wastewater treatment plants of Marrakech, Morocco [19]. The strains showed overall microcystin contents detected by ELISA of 15 and 56 μg g^{-1}, respectively. HPLC analysis demonstrated five variants of microcystins of the Syn-WTP97 strain, with a total of 842 μg g^{-1}. The toxicity (LD_{50}) of these two *Synechocystis* strains was 350 and 150 mg kg^{-1}, respectively. The results demonstrated that some *Synechococcus* strains were capable of microcystin production and indicated clearly that these cyanobacteria may also be a source of other bioactive compounds. These results point to potential health risks associated with the smallest phytoplanktonic fraction as a new source of cyanotoxins in water supplies.

On the other hand, results from the Salton Sea, a saline lake in California and from picocyanobacteria isolates from this lake, indicated that *Synechococcus* might produce microcystins dominated by microcystin-LR and microcystin-YR [20]. The analysis of 16S rRNA demonstrated that one of the microcystin-producing strains was closely related to marine *Synechococcus*. Because *Synechococcus* may be a dominant component of marine phytoplankton, especially in well-mixed, nutrient-enriched waters of lower salinity (also brackish waters), it seems that microcystin presence in the oceans may be more common than assumed before and that the source of this toxicity may not necessarily be via the inflow from freshwaters. Subsequent results showed that *Synechococcus* sp. strain 63a-1 isolated from the water column in the Florida Keys is a potent microcystin-LR producer under laboratory conditions [21]. Recent studies [22] suggested on the basis of indirect evidence that the cosmopolitan *Synechococcus* and *Synechocystis* genera could be responsible for the presence of microcystins in Mediterranean marine ecosystems.

Although studies described so far have concentrated on the analysis of microcystins picocyanobacteria, genes from the operon encoding biosynthetic pathways responsible for microcystin production in *Synechococcus* strains have been recently found: Bukowska et al. [23] detected *mcy* A, B, and E genes in one of the strains isolated from freshwater lakes in Mazurian lakes. These genes, separated by DGGE analysis and sequenced, showed a high homology with the *Planktothrix* microcystin operon.

3.2.2 Other Bioactive Compounds in Picocyanobacteria

Among other bioactive compounds that may influence various organisms including humans, are neurotoxins, LPS, taste and odour compounds, and other compounds, which were not identified but were shown to exhibit strong antimicrobial and cytotoxic effects and were considered to take part in allelopathic relationships inhibiting the growth of prokaryotic and eukaryotic organisms (Table 3.1).

3.2.2.1 Neurotoxins

According to Cox et al. [24], a single neurotoxin, β-*N*-methylamino-L-alanine (BMAA), a nonprotein amino acid, may be produced by members of all known cyanobacterial groups, including picocyanobacteria. This is in contrast to the case of hepatotoxins (microcystins, described earlier, or nodularins), which seem to be synthesised by a limited number of taxa. BMAA is hypothesised to be an environmental cause of human neurodegenerative diseases including amyotrophic lateral sclerosis/parkinsonism–dementia complex apart from undisputable genetic causes of these diseases. Cianca et al. [25] confirmed the finding that most cyanobacterial groups may produce BMAA by screening 18 strains of belonging to the Chroococcales, Oscillatoriales, and Nostocales from the Portuguese coast. Among them were unicellular *Synechococcus* and *Synechocystis* from planktonic and benthic habitats. The authors established further that the content of BMAA, which varied between 0.1 and 69 $\mu g\ g^{-1}$ depending on the cyanobacterial taxon and extraction technique (trichloroacetic acid, methanol/acetone, and HCl), seemed to not be related to the taxonomic affiliation or to the habitat, from which the strains were isolated. The potential ability of all cyanobacterial groups to produce BMAA suggests that human exposure to this neurotoxin, which may accumulate in trophic chains, may be more common than expected. This is especially relevant if we take into consideration that picocyanobacteria are ubiquitous and abundant in freshwater and marine environments. However, no other neurotoxins have been shown to be produced by picocyanobacteria until now.

3.2.2.2 LPS

LPS are a constituent of the outer cell wall of Gram-negative bacteria, including cyanobacteria. They are considered endotoxins, because one of the main components of LPS, lipid A, is responsible for developing symptoms such as fever, diarrhoea, vomiting, and hypotension in people after

Table 3.1 *Toxins and bioactive compounds, which were proved to be produced by picocyanobacteria*

Taxon	Hepatotoxins	Neurotoxins	Dermatotoxins	Taste and odour compounds		Allelopathic compounds	
	Microcystins	BMAA	LPS	Geosmin	MIB	Microcin-like	Other*
Synechococcus	17,18, 19, 20, 21, 22, 23	24, 25	26, 28	29, 30	29, 30	33	31, 32
Synechocystis	19	25	27				32

BMAA, neurotoxin β-*N*-methylamino-L-alanine
LPS, lipopolysaccharides
MIB, 2-methylisoborneol
Other*, unidentified compounds, which exerted negative effect on Gram-positive bacteria or eukaryotic cells
Numbers indicate publications in the References section.

exposure. Results of Schmidt et al. [26, 27], who analysed eight strains of *Synechococcus* and four strains of *Synechocystis*, confirmed that unicellular picocyanobacteria also contain LPS with lipid A in their cell wall LPS. More recent research has demonstrated that picocyanobacterial LPS are very simplified structures, differing from those of other Gram-negative bacteria by lacking typical components including 3-deoxy-D-manno-2-octulosonic acid, heptose, and phosphate [28]. The effect and toxicity of picocyanobacterial LPS are yet to be studied.

3.2.2.3 Taste and Odour Compounds

Another group of nuisance compounds produced by cyanobacteria that were also shown to be produced by picocyanobacteria are taste and odour compounds. These are not toxic to other organisms, but they influence the quality of water and organisms living there, by causing malodours or unpalatable drinking water and fish flesh. The presence of these substances in the water results in rising costs for drinking water treatment and causes losses in recreation and fishing areas affected by these compounds. Geosmin and MIB, the most widely known taste and odour compounds, are detectable by humans at very low levels (of 5–10 ng L^{-1}) and can be detected earlier than their potential producers. *Synechococcus* is among the genera known to produce geosmin and MIB [29] and was one of three genera dominating the phytoplankton throughout the clear water phase during maxima of geosmin and MIB concentrations in lake waters [30]. Because picocyanobacteria can be to some extent consumed by *Daphnia*, in contrast to filamentous cyanobacteria like *Oscillatoria* or *Aphanizomenon*, which are difficult to be consumed by zooplankton, it has been suggested that *Synechococcus* might not only produce these compounds but also take part in their transfer to higher trophic levels, including fish [30].

3.2.2.4 Antimicrobial and Cytotoxic Compounds: Allelopathic Relationships

Cyanobacteria including picocyanobacteria exhibit antimicrobial and cytotoxic activities, which influence the growth of prokaryotic and eukaryotic organisms. This form of ecological interaction is known as allelopathy and is considered to act against potential competitors or predators. Noaman et al. [31] found that *Synechococcus leopoliensis* produced antimicrobial compounds, which were active against the Gram-positive bacterium *Staphylococcus aureus,* causing growth inhibition. Later, Martins et al. [32], working on benthic and planktonic *Synechococcus* and *Synechocystis*, demonstrated that cell extracts caused apoptosis in eukaryotic cells and inhibited the growth of Gram-positive bacteria, whilst exerting no inhibitory effect on Gram-negative bacteria and fungi. The differing activities of the picocyanobacterial extracts suggested that organic compounds involved were of dissimilar polarities. Another study has revealed that marine *Synechococcus* CC9605 negatively influenced the growth of two *Synechococcus* strains CC9311 and WH8102, when they were cultivated together [33]. A gene cluster in *Synechococcus* CC9605, which showed homology to genes encoding Microcin-C (McC) in *Escherichia coli,* appears to be responsible for this allelopathic inhibition [33].

3.3 Summary

The potential toxicity of picocyanobacteria is an important issue, because most previous investigations concerning the toxicity of cyanobacteria have assumed that genera belonging to the picocyanobacteria size group were nontoxic. However, data are accumulating on the ability of picocyanobacteria to produce hepatotoxins, BMAA, LPS, and other bioactive metabolites, including bacteriocins compounds inhibiting growth of other cyanobacteria and eukaryotic cells. However, there are presently too few data concerning the actual contribution of picocyanobacteria to overall cyanotoxicity as well as allelopathy in natural environments. Nevertheless, because picocyanobacteria are among the most common cyanobacteria in freshwater and marine environments and are known to form persistent blooms in some coastal waters, the first results of their capacity to produce toxic substances highlight the potential risks involved with their massive presence in waterbodies. The role of picocyanobacteria as a source of toxins in freshwaters (including water supplies) and in marine environments has been until now overlooked, as well as microcystin occurrence in the marine environment. Thus, we recommend that picocyanobacteria are monitored carefully, similarly to larger cyanobacteria, whenever there is a suspicion of cyanotoxins or other bioactive cyanobacterial compound occurrence in waters. It should be borne in mind that because of the very small size of picocyanobacterial cells, their identification requires special equipment and procedures and, in the case of water treatment, their removal by traditional methods may be ineffective.

References

[1] Callieri, C., Cronberg, G. and Stockner, J. (2012) Freshwater Picocyanobacteria: Single Cells, Microcolonies and Colonial Forms. In: *Ecology of Cyanobacteria: Their Diversity in Time and Space*, 2nd ed. (ed. B. Whitton), Springer Publishers, pp. 229–269.

[2] Hoiczyk, E. and Hansel, A. (2000) Cyanobacterial cell walls: news from an unusual prokaryotic envelope. *Journal of Bacteriology* **182**, 1191–1199.

[3] Honda, D., Yokota, A. and Sugiyama, J. (1999) Detection of seven major evolutionary lineages in cyanobacteria based on the 16S rRNA gene sequence analysis with new sequences of five marine *Synechococcus* strains. *Journal of Molecular Evolution* **48**, 723–739.

[4] Ivanikova, N.V., Popels, L.C., McKay, R.M.L. and Bullerjahn, G.S. (2007) Lake Superior supports novel clusters of cyanobacterial picoplankton. *Applied and Environmental Microbiology* **73**, 4055–4065.

[5] Jasser, I., Królicka, A. and Karnkowska-Ishikawa, A. (2011) A novel phylogenetic clade of picocyanobacteria from the Mazurian lakes (Poland) reflects the early ontogeny of glacial lakes. *FEMS Microbiology Ecology* **75**, 89–98.

[6] Crosbie, N.D., Pöckl, M. and Weisse, T. (2003) Dispersal and phylogenetic diversity of nonmarine picocyanobacteria, inferred from 16S rRNA gene and *cpc BA*-intergenic spacer sequence analyses. *Applied and Environmental Microbiology* **69**, 5716–5721.

[7] Callieri, C., Coci, M., Corno, G. *et al.* (2013) Phylogenetic diversity of nonmarine picocyanobacteria. *FEMS Microbiology Ecology* **85**, 293–301.

[8] Scanlan, D. (2012) Marine picocyanobacteria. In: *Ecology of Cyanobacteria: Their Diversity in Time and Space*. 2nd edn (ed B. Whitton), Springer Publishers, pp. 503–533.

[9] Mazur-Marzec, H., Sutryk, K., Kobos, J. *et al.* (2013) Cyanobacterial blooms, cyanotoxin production and accumulation in biota from the southern Baltic Sea. *Hydrobiologia* **701**, 235–252.

[10] Sorokin, Y.I. and Zakuskina, O.Y. (2010) Features of the Comacchio ecosystem transformed during persistent bloom of picocyanobacteria. *Journal of Oceanography* **66**, 373–387.

[11] Bell, T. and Kalff, L. (2001) The contribution of picophytoplankton in marine and freshwater systems of different trophic status and depth. *Limnology and Oceanography* **46**, 1243–1248.

[12] Stomp, M., Huisman, J., Vörös, L. *et al.* (2007) Colourful coexistence of red and green picocyanobacteria in lakes and seas. *Ecology Letters* **10**, 290–298.

[13] Pernthaler, J., Šimek, K., Sattler, B. *et al.* (1996) Short-term changes of protozoan control on autotrophic picoplankton in an oligo-mesotrophic lake. *Journal of Plankton Research* **18**, 443–462.

[14] Zwirglmaier, K., Spence, E., Zybkov, M.V. *et al.* (2009) Differential grazing of two heterotrophic nanoflagellates on marine *Synechococcus* strains. *Environmental Microbiology* **11**, 1767–1776.

[15] Lincoln, E.P and Carmichael, W.W. (1981) Preliminary tests of toxicity of *Synechocystis* sp. growth on wastewater medium. In *The Water Environment: Algal Toxins and Health*, (ed. W.W. Carmichael) Plenum Press, New York. pp. 223–230.

[16] Mitsui, A., Rosner, D., Goodman, A., Reyes-Vasquez, G., Kusumi, T. Kodama, T. and Nomoto, K. (1987). Hemolytic toxins in a marine cyanobacterium *Synechococcus* sp. Proceedings of the International Red Tide Symposium. Takamatsu, Japan.
[17] Bláha, L. and Marsálek, B. (1999) Microcystin production and toxicity of picocyanobacteria as a risk factor for drinking water treatment plants. *Algological Studies* **92** (60), 95–108.
[18] Domingos, P., Rubim, T.K., Molica, R.J.R. *et al.* (1999) First report of microcystin production by picoplanktonic cyanobacteria isolated from a northeast Brazilian drinking water supply. *Environmental Toxicology* **14**, 31–35.
[19] Oudra, B., Loudiki, M. Vasconcelos, V.M. *et al.* (2002) Detection and quantification of microcystins from cyanobacteria strains isolated from reservoirs and ponds in Morocco. *Environmental Toxicology* **17**, 32–39.
[20] Carmichael, W.W. and Li, R.H. (2006) Cyanobacteria toxins in the Salton Sea. Saline systems **2**:5 doi:10.1186/1746-1448-2-5.
[21] Gantar, M., Sekar, R. and Richardson, L.L. (2009) Cyanotoxins from black band disease of corals and from other coral reef environments. *Microbial Ecology* **58**, 856–864.
[22] Vareli, K., Jaeger, W., Touka, A. *et al.* (2013) Hepatotoxic seafood poisoning (HSP) due to microcystins: a threat from the ocean? *Marine Drugs* **11**, 2751–2768.
[23] Bukowska, A., Karnkowska-Ishikawa, A. and Jasser, I. (2014) Microcystin-encoding gene cluster in *Synechococcus* strain isolated from Great Mazurian Lakes. In *Harmful Algae 2012, Proceedings of the 15th International Conference on Harmful Algae*. International Society for the Study of Harmful Algae (eds. H.G. Kim, B. Reguera, G.M. Hallegraeff *et al.*) pp. 181–183.
[24] Cox, P. A., Banack, S. A., Murch, S. J. *et al.* (2005) Diverse taxa of cyanobacteria produce β-*N*-methylamino-l-alanine, a neurotoxic amino acid. *PNAS, Proceedings of the National Academy of Sciences U S A* **102** (14), 5074–5078.
[25] Cianca, R.C.C., Baptista, M. S., Lopes, V.R. and Vasconcelos, V.M. (2011) The non-protein amino acid β-methylamino-L-alanine in Portuguese cyanobacterial isolates. *Amino Acids*, doi:10.1007/s00726-011-1057-1.
[26] Schmidt, W., Drews, G., Weckesser, J. and Fromme, I. (1980) Characterization of the lipopolysaccharides from eight strains of the cyanobacterium *Synechococcus*. *Archives of Microbiology* **127**, 209–215.
[27] Schmidt, W., Drews, G., Weckesser, J. and Mayer, H. (1980) Lipopolysaccharides in four strains of the unicellular cyanobacterium Synechocystis. *Archives of Microbiology* **127**, 217–222.
[28] Snyder, D.S., Brahamsha, B., Azadi, P. and Palenik, B. (2009) Structure of compositionally simple lipopolysaccharide from marine *Synechococcus*. *Journal of Bacteriology* **191**, 5499–5509.
[29] Graham J.L., Loftin, Keith, A. *et al.* (2008) Guidelines for design and sampling for cyanobacterial toxin and taste-and-odor studies in lakes and reservoirs Reston, Va. U.S. Dept. of the Interior, U.S. Geological Survey, http://purl.access.gpo.gov/GPO/LPS97055.
[30] Journey, C.A., Beaulieu, K.M. and Bradley, P.M. (2013) Environmental factors that influence cyanobacteria and geosmin occurrence in reservoirs. In *Current Perspectives in Contaminant Hydrology and Water Resources Sustainability* (ed. P.M. Bradley) pp. 27–55, doi:10.5772/54807.
[31] Noaman, N.H., Fattah, A., Khaleafa, M. and Zaky, S.H. (2004) Factors affecting antimicrobial activity of *Synechococcus leopoliensis*. *Microbial Research* **159**, 395–402.
[32] Martins, R.F., Ramos, M.F., Herfindal, L. *et al.* (2008) Antimicrobial and cytotoxic assessment of marine cyanobacteria: *Synechocystis* and *Synechococcus*. *Marine Drugs* **6**, 1–11 (2008).
[33] Paz-Yepes, J., Brahamsha, B. and Palenik, B. (2013) Role of a Microcin-C–like biosynthetic gene cluster in allelopathic interactions in marine *Synechococcus*. *Proceedings of the National Academy of Sciences* **110**, 12030–12035.

4

Expansion of Alien and Invasive Cyanobacteria

Mikołaj Kokociński[1], *Reyhan Akçaalan*[2], *Nico Salmaso*[3], *Maya Petrova Stoyneva-Gärtner*[4], *and Assaf Sukenik*[5]

[1]*Department of Hydrobiology, Adam Mickiewicz University, Poznań, Poland*
[2]*Faculty of Fisheries, Istanbul University, Laleli-Istanbul, Turkey*
[3]*IASMA Research and Innovation Centre, Fondazione Edmund Mach-Istituto Agrario di S. Michele all'Adige, Trento, Italy*
[4]*Department of Botany, Sofia University "St. Kliment Ohridski," Sofia, Bulgaria*
[5]*Kinneret Limnological Laboratory, Israel Oceanographic & Limnological Research, Migdal, Israel*

4.1 Introduction

The expansion of some cyanobacteria species is a worldwide phenomenon that attracts the attention of scientists seeking their environmental and biological origins. Moreover, occurrence of new cyanobacteria outside of their native range is of increasing interest due to their potential ability to produce toxins and the associated threats they may pose. However, many terms are used to describe cyanobacteria that expand their distribution, including "exotic," "nonindigenous," "non-native," "alien," and "invasive" species, which makes the study associated with biological invasions difficult and confusing. Nevertheless, proper identification of the expansion process of potentially toxic cyanobacteria and determining their status in this context are required to better understand their ecology, biogeography, and rational management of water resources. In this chapter, the invasive species concept in cyanobacteria is proposed and discussed. Moreover, within the proposed terminology framework, examples of expansion of invasive and alien cyanobacteria are provided that include information about environmental factors that enhance their expansion and have an impact on the ecosystem.

Handbook of Cyanobacterial Monitoring and Cyanotoxin Analysis, First Edition.
Edited by Jussi Meriluoto, Lisa Spoof and Geoffrey A. Codd.

4.2 Definition of Invasive/Alien Species: Nomenclature Problems

During the past 20 years, there has been a significant growth in research on invasion biology but a common, unified model or concept for the invasion processes of microorganisms and macroorganisms has not been defined [1, 2]. The terminology associated with biological invasions has been debated frequently, especially in the context of subjectivity and objectivity, and it seems to be still confusing or even misleading [3, 4]. For example, terms other than "invasive species" are frequently used in the literature, including "exotic species," "nonindigenous species," and "alien" or "non-native species," to signify that a species is outside of its native range. Moreover, these terms are used sometimes synonymously to describe the same invasion concept; in other cases, the same term is used to represent different invasion processes.

Among the many definitions of "invasive species," the most common one outlines it as "a widespread non-indigenous species (NIS) that have adverse effects on the invaded habitat" [5]. This description has been accepted by the International Union for Conservation of Nature (IUCN) that defines invasive species as "animals, plants or other organisms introduced by man into places out of their natural range of distribution, where they become established and disperse, generating a negative impact on the local ecosystem and species." In addition, documents used in the Convention on Biological Diversity (CBD) Guiding Principles (CBD Decision VI/23) and the European Strategy on Invasive Alien Species (IAS), classify invasive alien species as "species whose introduction and/or spread threaten biological diversity." All of these definitions integrate the concept of environmental harm caused by invasive species. The economic implications have been recognized also in the definitions of laws and regulations, as in Executive Order 13112 (1999) in the United States [6], aimed to minimize the economic, ecological, and human health impacts of invasive species.

The CBD has also proposed a separate term for alien species—"a species, subspecies or lower taxon, introduced outside its natural past or present distribution; includes any part, gametes, seeds, eggs, or propagules of such species that might survive and subsequently reproduce."

These definitions are based on a biogeographical concept of well-defined geographical ranges of the species and on knowledge of the history of introduction to the new environment. A problem arises, however, when the species expands its range or increased its abundance very slowly and it is colonizing neighbouring areas [7]. The definitions of "invasive" or "alien" species thus very often require a subjective judgment of two factors, its spatial and temporal scale of expansion, which are still open to debate [7].

Another broadly discussed topic is related to fact that the terms "invasive" or "alien" species tend to be applied to entire taxonomic groups instead of the particular groups (populations, strains) that are responsible for the ecological phenomena of spreading and causing harm [5].

Finally, it is important to evaluate invasions within a time frame according to the paradigm that potential invaders may pass through a series of stages including transport, establishments, and further expansion of geographical range [4, 5, 8] (please see conceptual model presented in Fig. 4.1).

4.2.1 Invasive Species Concept in Cyanobacteria

In the light of the discussed considerations, the concepts of invasive or alien species in cyanobacteria seem to be even more difficult to define. The major problem is related to species concept in modern cyanobacterial taxonomy complicated by infrageneric diversification including ecospecies, morphospecies, and the occurrence of different chemotypes [9]. Thus, we recommend

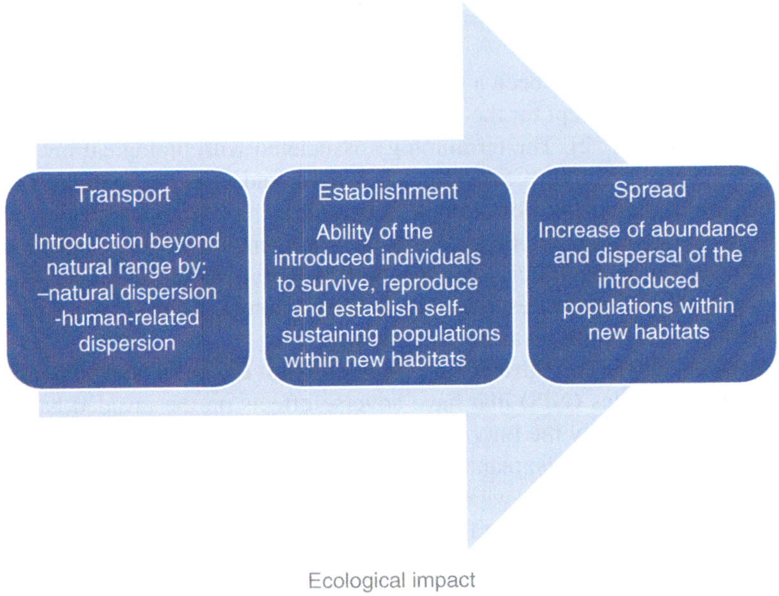

Figure 4.1 *Conceptual model of stages during invasion process*

restriction of the use of the term "invasive" for all new occurring non-native species or populations of cyanobacteria that enter, proliferate, and spread in habitats where they had not resided before. The term "alien" or "non-native" refers to the occurrence of species or populations of cyanobacteria in a new environment, irrespective of their abundance. Implicitly, the use of these terms assumes the absence of resident inocula before the first identification. This cannot be assumed unambiguously, due to the possible failure of correct identification in past studies, or simply because of correlations between the sampling effort and the report of new species. In such cases, the term "cryptogenic" species should be used, referring to species that are neither native nor exotic and whose origins are unknown or not proven unambiguously [10–12].

Therefore, major concerns when considering the invasion of cyanobacteria are:

- The historical data on the occurrence of cyanobacteria, in contrast to higher plants or animals, are usually very scarce and limited to very short time-scales; these limitations can be exacerbated by the potential taxonomic misidentification of species.
- Native ranges of cyanobacteria against their substantial dispersal capabilities.
- Natural barriers inhibiting expansion—the unfavourable environmental conditions in the new areas are usually the major one.
- Invasion stage—within the conceptual model of the invasion process presented in Fig. 4.1.

In summary, the discussion on a unified concept of invasion processes of organisms has not been completed yet. Major concerns are often associated with the subjective assessment of spatial and temporal scales of invasion, its potential ecological impacts, and species concepts. Therefore, when studying the expansion of cyanobacterial range, it must be indicated that each "invasion" should be considered individually and that we should not call all invasive species or populations of non-native cyanobacteria, even if one of those populations was found to be invasive somewhere else.

4.3 Occurrence of Invasive and Alien Cyanobacteria

An increasing number of non-native cyanobacteria have been documented in lakes and water reservoirs during recent decades. Among them, the most common species belong to the Nostocales, including *Cylindrospermopsis raciborskii*, *Chrysosporum ovalisporum* (*Aphanizomenon ovalisporum*), *Sphaerospermopsis aphanizomenoides* (*Aphanizomenon aphanizomenoides*), *Chrysosporum bergii* (*Anabaena bergii*), *Dolichospermum lemmermannii* (*Anabaena lemmermannii*), *Nodularia spumigena*, and *Raphidiopsis mediterranea*. Much less is known about invasive cyanobacteria from other taxonomic groups. The benthic filamentous, marine *Lyngbya majuscula* and planktonic, freshwater *Lyngbya wollei*, however, having broad natural distributions, are considered the most common, opportunistic invaders among the Oscillatoriales [13]. Also from this order, the expansion of the holarctic cyanobacterium *Planktothrix rubescens* toward southern parts of Europe has been observed [14]. Moreover, the common inhabitants of tropical regions, *Planktolyngbya microspira*, *Planktolyngbya circumcreta* (Pseudanabaenales), and *Cyanodictyon tropicale* (Synechococcales), have been considered non-native in European countries. Recently, *Synechococcus capitatus* (Chroococcales), common for cold waters in Scotland, Scandinavia, and northern Japan, also has been recorded in freshwater lakes in the Czech Republic and considered as an alien species [14].

4.3.1 Examples of the Expansion of Invasive and Alien Cyanobacteria

The most frequently reported invader is *C. raciborskii*, originally described as *Anabaena raciborskii* by Woloszynska from inland Java as a typical tropical species [15]. In Europe, *C. raciborskii* is commonly considered as an invasive species [16]; however, the lack of historical data in many cases prevents determination of whether this species was present before the first records and, thus, when the introduction of the species may have occurred. Nevertheless, it is clear that long-term phytoplankton studies and detailed monitoring programmes have shown a significant increase in the occurrence of *C. raciborskii* in comparison to its marginal occurrence in the past. For example, in Hungary and Bulgaria, this species is recently reported as a bloom-forming species, while one or two decades ago, *C. raciborskii* made a minor contribution to phytoplankton biomass [17–20]. In the Balkans, the appearance and dominance of this species were recently described in a temperate lowland river in Serbia [21]. In the Mediterranean regions, huge summer blooms of *C. raciborskii* were described in southern Italy (Lake Biviere di Gela, Sicilia) [22], while the presence or even blooms of this species were documented also in Greece (Lake Kastoria) [23, 24] and Spain (Lake Albufera) [25].

The invasion of *C. raciborskii* to the warm monomictic Lake Kinneret (Sea of Galilee, Israel) was first reported in 2000 as a minor component of the summer phytoplankton community but dominated the community in summer 2005 with biomass as high as 68 g (ww) m^{-2} [26]. The fact that the invasive strain is a non-cylindrospermopsin-producer suggests that it may be related to European populations, as opposed to Australian populations known to produce cylindrospermopsin or to Brazilian populations that have been reported to produce paralytic shellfish poisoning toxins [27].

In the 1960s, *C. raciborskii* was identified in Belarus and Ukraine [28–30], and in Poland during the 1970s in lakes that received heated water discharge from a power plant [31]. Currently, it frequently occurs also in parts of northern and central Europe including Germany and Poland [32, 33]. However, the growth and proliferation of *C. raciborskii* in these regions were rather limited to summer and have never been found to dominate the phytoplankton community. High biomass of *C. raciborskii* was recently also documented in western Russia in the large, hypertrophic Lake Nero [34].

Nevertheless, the expansion of this cyanobacterium toward subtropical and temperate zones seems to be a worldwide phenomenon as it spread out in Australia and New Zealand, South and North America, South and East Asia, the Middle East, and some countries in Africa [4].

Another example of an invasive cyanobacterium is *Nodularia spumigena*. In the past decade, this brackish/saltwater species appeared in freshwater lakes in Turkey. *N. spumigena* is very well known in Europe, especially in coastal waters of the Black Sea and the Baltic Sea, where it forms massive blooms. However, it was recorded in the freshwater Lake Iznik (Turkey) during the summer months of 2005 with densities up to 1.3×10^5 filaments L^{-1}. Although it did not cause a huge bloom, the species was found to produce nodularin [35]. Another episode occurred 5 years later in Hazar Lake in eastern Turkey, which is hundreds of kilometres from the Iznik lake. This time it was a surface bloom and the biomass reached 38.4 mg L^{-1} [36]. Although these two lakes are far from each other, both have a tectonic origin and high alkalinity. Moreover, there was also a heavy *Nodularia spumigena* bloom in a hyposaline lake (Lake Bafa) in 2011. Although the first record of this species in this lake was recorded in 1996 [37], the severity of the blooms has increased during recent years. In all cases, nodularin was detected and posed hazards to the ecosystem.

The absence of regular monitoring studies in the past makes it difficult to distinguish indisputably non-native species from species that were overlooked or present at minor densities (i.e. cryptogenic species). Nevertheless, it is undeniable that the expansion of the known geographical ranges colonized by cyanobacteria if fully or partly linked to increases in new observations. *A. ovalisporum*, *A. bergii*, and *R. mediterranea* are examples of cyanobacteria that are considered alien species in some regions. Among these the most common is *A. ovalisporum*, which was described in Lake Kucuk Cekmece in Turkey [38] and reported later in Australia [39], United States [40], and different Mediterranean countries [4, 41, 42] including the Iberian Peninsula, where it is known since 2005 to have produced blooms [43].

Recent studies have also identified this cyanobacterium in Poland [44]; however, its abundance was very low. *A. bergii* is another example of the expansion of species from tropical/subtropical latitudes to northern regions. It was reported and described for the first time in the plankton of Lake Aral and later documented in the Caspian Sea and Turkey [14]. Besides southern European countries, it is considered as an alien species in Slovakia, Germany, Poland, Czech Republic, and Lithuania [14, 32, 44, 45]. Similarly, the occurrence of *A. aphanizomenoides* was reported from northern parts of Europe including Germany and Poland [33, 46]. The contribution to phytoplankton biomass in these countries was generally low and these species never dominated the phytoplankton community.

Another recent case of geographical expansion by heterocytous cyanobacteria was documented in the large lake district south of the Alps. Extended surface blooms of *Dolichospermum lemmermannii* were recorded for the first time in Lake Garda (65 m above sea level) between 1990 and 1991 [47]. Since then, blooms of this species were documented in Lakes Iseo (second half of the 1990s), Maggiore (2005), and Como (2006) [48]. The blooms were recorded exclusively during the summer months as irregular stripes or layers of yellow-green colour developing exclusively in the first centimetres (around 0–2 cm) of the water column. During the blooms, populations of *Dolichospermum* could reach values $> 100{,}000$ cells mL^{-1}. Nevertheless, being strictly localised in the upper centimetres of the water column, the episodes give a false impression of cyanobacterial abundance. Actually, in Lake Garda the biovolume of *Dolichospermum* integrated to the upper 20 m was generally below 45 $mm^3\ m^{-3}$ [49]. For these reasons, the blooms in the deep southern subalpine lakes were referred as "oligotrophic blooms" [47]. The occurrence of measurable quantities of microcystins and anatoxin-a in Lake Garda were however linked to the development of other species, namely *Planktothrix rubescens* and *Tychonema bourrellyi*, respectively [50].

These cyanobacterial populations are very often at different stages of expansion. According to the proposed conceptual model, alien species may pass through a series of stages including: (1) Transport – pickup from native range and transport of individuals into new areas; (2) Establishment – release, survival and establishment of localized and rare populations within new range; and (3) Spread – growth, giving rise to dominant and widespread populations. In general, the term invasive should be restricted to those alien species able to expand their geographical range, with the development of dominant populations. For example, populations of *C. raciborskii* in Lake Kinneret or lakes in Hungary or Bulgaria where they dominate the phytoplankton community, should be considered invasive, while scarce populations in lakes in northern part of Europe (e.g. Poland) should be considered alien. Similarly, *A. ovalisporum* or *A. bergii* rarely are dominating phytoplankton communities in the lakes of northern Europe and thus should not be called invasive in relation to the above concepts.

4.4 Factors Enhancing the Expansion of Alien Cyanobacteria

4.4.1 Physiological Factors

Invasive cyanobacteria owe their success in colonizing new ecosystems to a combination of physiological characteristics. The invasion initiates with dispersion to new zones and is facilitated by human activities, migrating animals or birds, and wind. Obviously, aeolian transport is an efficient means to transfer free-living microorganisms including prokaryotes and protists [51]. Once dispersion has occurred, the invader needs a variety of traits that support its establishment and proliferation. Two distinctive traits of the Nostocales are: 1) The ability to form dormant cells (akinetes) that may survive long and extreme dispersion routes, serve as an overwintering form that survives unfavourable conditions and assure perennial germination and proliferation; 2) The ability to fix atmospheric nitrogen in the absence of combined inorganic nitrogen sources. This last trait opens invasion opportunities for members of the Nostocales in oligotrophic aquatic ecosystems, besides ecosystems temporarily limited in combined nitrogen, therefore extending the spectrum of ecosystems that they can potentially invade [4].

The establishment of invasive cyanobacteria in their new environment was further attributed to other physiological traits as summarized by Sukenik et al. [4]. These include (1) the persistence of vegetative forms under winter conditions providing additional sources for recruitment [52, 53], (2) high competitive ability [53], (3) high growth rate [52], (4) high resource use efficiency [53], (5) efficient response to high or fluctuating resource supply, and (6) an ability to use resources not used by the resident community [16].

In many cases, floating cyanobacterial populations create a thick scum that blocks light penetration through the water column and inhibits the growth of other species. Thus, under high cyanobacterial growth, the populations remain established for longer periods, minimizing competition by other algal groups [54]. Filaments floating at the water surface gain advantage over subsurface phytoplankton populations. They directly intercept CO_2 diffusing into the water from the atmosphere, thus minimizing inorganic carbon limitation of photosynthetic growth. Buoyant species contain UV-absorbing compounds including mycosporine-like amino acids and scytonemin that ensure survival under extremely high irradiance conditions [55].

4.4.2 Environmental Factors

Increased water temperatures and decreasing epilimnetic turbulence, mostly during the summer, in subtropical and temperate zones are considered major factors supporting the invasion of

C. raciborskii and other Nostocales [53, 56, 57]. Both *C. raciborskii* and *A. ovalisporum* are known to proliferate at temperatures higher than 20°C [58–60]; thus, a moderate elevation in water temperatures may open new ecological habitats for them.

Nostocales species are recently reported to reside in and dominate eutrophic as well as oligotrophic aquatic systems. Eutrophic conditions provide ample nutrients that can be easily and competitively used by opportunistic invaders [61]. Such eutrophic systems maintain high internal nutrient recycling processes that fortify primary production and biomass accumulation. The ability of many Nostocales to efficiently use low light allows their growth in turbid waters as well as at high biomass concentrations, but Nostocales species can survive and propagate also under oligotrophic conditions. This is achieved due to their various efficient phosphate acquisition capabilities [62–65] and nitrogen fixation, which seem to be more relevant under oligotrophic than under eutrophic conditions. Thus, in oligotrophic systems, phosphate rather than nitrogen availability controls the development of N_2 fixing populations and their growth. Consequently, management efforts to control eutrophication by reducing N loads, apparently effective in reducing blooms of *Microcystis* and other Chroococcales species, may actually support the growth of Nostocales which can easily outcompete native species in such N-limited ecosystems [4, 66]. Resistance to grazing [67], and the ability to control vertical migration preventing sedimentation in warmer, stratified waters [68] are further important factors that might enhance expansion of some cyanobacteria.

The appearance of *D. lemmermannii* in the largest lakes south of the Alps can be regarded as an example of proliferation not exclusively related to the trophic status but possibly also to temperature. Although the first blooms of this species observed in Lake Maggiore were documented just when the lake was at the end of a long-term oligotrophication process (e.g. [69]), the trophic status of Lakes Garda, Iseo, and Como ranged between oligotrophy and mesoeutrophy. Different strains of *D. lemmermannii* were characterized by high adaptability to a wide range of temperatures, although they are commonly found in cold environments [70]. Other strains of *D. lemmermannii* show higher temperature optima (between 19 and 26°C) [71], having, therefore, the capability to develop surface water blooms in stratified and stable water columns during calm weather. Salmaso et al. [48] showed that in Lakes Garda, Iseo, Como, and Maggiore, this species showed high and abrupt relative increases in abundance at water temperatures higher than 16–17°C, which coincided with the beginning of stratified conditions. It was speculated that the absence of *Dolichospermum* in Lake Lugano (the only deep and large lake where no blooms were detected) could be due to the lower water temperatures and higher mixing that characterized its epilimnion [48].

The successful establishment of an alien species in a new habitat depends both on its invasive potential and on the resilience of the native community [72]. The results of long-term phytoplankton studies in Bulgaria showed that non-native cyanobacterial species can appear occasionally at various sites but continue their development and cause blooms in waterbodies that have already been stressed and frequently disturbed (e.g. with drastic changes in hydrological regimen). This includes changes in water level, seawater ingression in combination with high temperatures, and hypertrophication [20].

4.5 Impact of Cyanobacterial Invasion on Ecosystem

The introduction of non-native species can radically alter existing ecosystems and is now ranked second only to habitat destruction in terms of potential ecological catastrophe [73]. Competitive exclusion of native taxa can often result in an overall decrease in biological diversity and seriously compromise the ability of the original population(s) to adapt to new selective pressures, thus increasing the chance of extinction, as reviewed by Booth et al. [74]. Although many studies have investigated how community characteristics such as diversity and disturbance relate to invisibility [72], the mechanisms underlying biotic resistance to introduced species are not well understood [75]. The same is valid for the exact traits and mechanisms that enable and enhance the impact of alien nostocalean species and other cyanobacteria on ecosystems. The problem is deepened by the fact that many morphological and physiological properties in general are common for native and invasive species. It is possible to suggest that the quantitative parameters of the traits of certain species (e.g. competitive potential) are of major importance, especially in cases when the ecological niches of the invaders and native species are similar or overlapping. Examples are provided not only by the already mentioned ability of cyanobacteria (and Nostocales especially) to fix nitrogen and to produce resting stages, but also by specific antagonistic relations, in which certain species deplete the relevant biogenes, or other resources and change them in a way which makes them inaccessible for other organisms. Cyanobacteria provide a classic illustration of this principle since they have well-developed systems for iron assimilation and, at the same time, excrete chelators, which actively bind it. In this way, the other algae become inhibited through secondary effects of starvation and cyanobacteria start to dominate [76].

Another trait commonly found among the Nostocales, but also in other cyanobacteria (e.g. *Microcystis*), is the ability to synthesize a suite of secondary metabolites, some of which were identified as toxic to humans and animals [77]. Many of these metabolites may affect community structure and function as allelochemicals that inhibit other phytoplankton species or deter and reduce grazing [78, 79]. Allelopathy was suggested as a beneficial trait of *C. raciborskii* that contributes to its stable dominance and geographic expansion [80]. A unique feature of *A. ovalisporum* to acquire phosphate under oligotrophic conditions was reported by Bar-Yosef et al. [65]. Excreted chemicals including the toxin cylindrospermopsin cause cells of some other phytoplankton species to make extracellular alkaline phosphatase, in turn producing inorganic phosphate that can be used by *A. ovalisporum* due to its high affinity to phosphate, thus contributing to its proliferation and domination.

The domination of invaded Nostocales can affect the entire ecosystem as native phytoplankton species become rare and displaced and higher trophic levels may be affected via their toxins or due to the loss of essential components of the food chain. *C. raciborskii* is known to affect the growth, reproduction, and feeding of zooplankton [81], and cylindrospermopsin has been shown to be accumulated in fish and freshwater mussels [82, 83].

In summary, the expansion of cyanobacterial species is similar to that of other examples of biological invasions, which are important parts of functioning of aquatic ecosystems. However, since a single unified concept or model of the invasion process has not yet been achieved by ecologists, we have proposed definitions of alien and invasive species of cyanobacteria here, whilst also providing criteria to distinguish the species at different stages of biological expansion. Recently, Sukenik et al. [84] indicated that alien and invasive cyanobacteria may have potentially significant impact on both aquatic and terrestrial ecosystems [84]. Former studies have revealed decreases in species diversity after invasion by alien species and alterations of ecosystem functioning due to the reproduction and feeding behaviour of organisms and because of the occurrence of newly introduced (after invasion) chemical compounds, including cyanotoxins associated with the massive occurrence of invasive cyanobacteria and accumulation of these toxins in food chains. Therefore, further studies on invasive cyanobacteria should enable us to better recognize and mitigate the threats associated with invasion processes.

References

[1] Blackburn,T.M., Pyšek, P., Bacher, S. *et al.* (2011) A proposed unified framework for biological invasions. *Trends in Ecology and Evolution* **26** (7), 333–339.

[2] Heger, T., Saul, W.C. and Trepl, L. (2013) What biological invasions "are" is a matter of perspective. *Journal for Nature Conservation* **21** (2), 93–96.

[3] Colautii, R.I. and Richardson, D.M. (2009) Subjectivity and flexibility in invasion terminology: Too much of a good thing? *Biological Invasions* **11** (6), 1225–1229.

[4] Sukenik, A., Hadas, O., Kaplan, A. and Quesada, A. (2012) Invasion of Nostocales (cyanobacteria) to subtropical and temperate freshwater lakes – physiological, regional, and global driving forces. *Frontiers in Microbiology* **3** (86), 1–9.

[5] Colautii, R.I. and MacIsaac, H.J. (2004) A neutral terminology to define "invasive" species. *Diversity and Distributions* **10** (2), 135–141.

[6] Executive Presidential Order. Executive Order 13112 of February 3 (1999) Invasive Species. *Federal Register* **64** (25), 6183–6186.

[7] Lockwood, J.L., Hoopes, M.F. and Marchetti, M.P. (eds.) (2013) *Invasive Ecology*, 2nd ed., Wiley-Blackwell.

[8] Krebs, C.J. (2009) *Ecology. The experimental analysis of distribution and abundance.* Benjamin Cummings, San Francisco.

[9] Komárek, J. and Mareš, J. (2011) An update to modern taxonomy of freshwater planktic heterocytous cyanobacteria. *Hydrobiologia* **698** (1), 327–351.

[10] Cohen, A.H. and Carlton, J.T. (1998) Accelerating invasion rate in a highly invaded estuary. *Science* **279** (5350), 555–558.

[11] Carlton, J.T. (1996) Biological invasions and cryptogenic species. *Ecology (Ecological Society of America)* **77**, 1653–1655.

[12] Burdick, A. (2006) *Out of Eden: An Odyssey of Ecological Invasion*, Farrar, Straus and Giroux.

[13] Paerl, H.W. and Huisman, J. (2009) Climate change: a catalyst for global expansion of harmful cyanobacterial blooms. *Environmental Microbiology Reports* **1** (1), 27–37.

[14] Kaštovský, J., Hauer, T., Mareš, J. *et al.* (2010) A review of the alien and expansive species of freshwater cyanobacteria and algae in the Czech Republic. *Biological Invasions* **12** (10), 3599–3625.

[15] Woloszynska, J. (1912) O glonach planktonowych niektórych jezior jazuanskich, z irwzględnieniem glonózu Sazvy. Das Phytoplankton einiger Javanian Seen mit Berücksichtigung des Sawa-Planktons. *Bulletin international de l'Académie des Sciences de Cracovie. Classe des Sciences Mathématiques et Naturelles. Série B: Sciences Naturelles* **6**, 649–709.

[16] Padisak, K. J (1997) *Cylindrospermopsis raciborskii* (Woloszynska) Seenayya et Subba Raju, an expanding, highly adaptative cyanobacterium: worldwide distribution and review of its ecology. *Archiv für Hydrobiologie, Supplement* **107**, 563–593.

[17] Antal, O., Karisztl-Gácsi, M., Farkas, A. *et al.* (2011) Screening the toxic potential of *Cylindrospermopsis raciborskii* strains isolated from Lake Balaton, Hungary. *Toxicon* **57** (6), 831–840.

[18] Stoyneva, M. (1995) Algal flora of the Danube River (Bulgarian sector) and adjoined water basins. V. Algal flora of the water bodies adjacent to the Lake of Srebarna. *Annuaire de l'Universite de Sofa"St. Klimnet Ohridski", Faculte de Biologie, Livre 2-Botanique* **88**, 5–19.

[19] Stoyneva, M.P (1998) Development of the phytoplankton of the shallow Srebarna Lake (North-Eastern Bulgaria) across the trophic gradient, in *Phytoplankton and trophic gradients* (eds. A. Cobelas, M.C.S. Reynolds, P. Sanchez-Castillo and J. Kristiansen), Kluwer Academic Publishers *Hydrobiologia* **369/370**, 259–367.

[20] Stoyneva, M.P. (2003) Steady-state phytoplankton assemblages in shallow Bulgarian wetlands, in *Phytoplankton and Equilibrium Concept: The Ecology of Steady-State Assemblages* (eds. N. Flores, L.J. Padisak and M. Dokulil), Kluwer Academic Publishers, *Hydrobiologia* **502** (1), 169–176.

[21] Karadžić, V., Simić, G.S., Natić, D. *et al.* (2013) Changes in the phytoplankton community and dominance of *Cylindrospermopsis raciborskii* (Wolosz.) Subba Raju in a temperate lowland river (Ponjavica, Serbia). *Hydrobiologia* **711** (1), 43–60.

[22] Barone, R., Castelli, G. and Naselli-Flores, L. (2010) Red sky at night cyanobacteria delight: The role of climate in structuring phytoplankton assemblage in a shallow, Mediterranean lake (Biviere di Gela, southeastern Sicily). *Hydrobiologia* **639** (1), 43–53.

[23] Moustaka-Gouni, M., Kormas, K.A. and Vardaka, E. *et al.* (2009) *Raphidiopsis mediterranea* Skuja represents non-heterocytous life-cycle stages of *Cylindrospermopsis raciborskii* (Woloszynska)

Seenayya et Subba Raju in Lake Kastoria (Greece), its type locality: Evidence by morphological and phylogenetic analysis. *Harmful Algae* **8** (6–8), 864–872.

[24] Katsiapi, M., Moustaka-Gouni, M., Vardaka, E. and Kormas, K. (2013) Different phytoplankton descriptors show asynchronous changes in a shallow urban lake (L. Kastoria, Greece) after sewage diversion. *Fundamental and Applied Limnology / Archiv für Hydrobiologie* **182** (3), 219–230.

[25] Romo, S., Soria, J., Fernández, F. *et al.* (2013) Water residence time and the dynamics of toxic cyanobacteria. *Freshwater Biology* **58** (3), 513–522.

[26] Hadas, O. Pinkas, R. Malinsky-Rushansky, N. *et al.* (2012) Appearance and establishment of diazotrophic cyanobacteria in Lake Kinneret, Israel. *Freshwater Biology* **57** (6), 1214–1227.

[27] Neilan, B.A., Saker, M.L., Fastner, J. *et al.* (2003) Phylogeography of the invasive cyanobacterium *Cylindrospermopsis raciborskii*. *Molecular Ecology* **12** (1), 133–140.

[28] Micheeva, T.M. (1968) On the occurrence of *Anabaenopsis raciborskii* (WOŁOSZ.) V. MILLER. *Acta Universitatis Carolinae Biologica* **1967**, 257–265.

[29] Vladimirova, K.S. and Danilova, L.E. (1968) Vodorosli Dunaja, zalivov Kilijskoj delty i pridunajskich vodojomov v predelach SSSR (Algae of Danube, Kilijska delta and adjacent to the Danube water bodies in the territory of Soviet Union). *Limnologische Berichte der X Jubiläumstagung Donauforschung* (Sofia) **10**, 141–168.

[30] Kondratjeva, N.V. (1968) *Viznachnik prisnovodnikh vodorostej Ukrainskoi RSS*, I. Chastina 2, sinozeleni vodorosti (The manual of freshwater algae of the Ukraine, Cyanophyta, part 2, Hormogoniophyceae). Naukova dumka, Kiev.

[31] Burchardt, L. (1977) Changes in phytoplankton composition of Lake Patnowskie, receiver of heated water and waste water from sugar factory (1972/73). *Seria Biologia* **8**, 1–119.

[32] Stüken, A., Rücker, J., Endrulat, T. *et al.* (2006) Distribution of three alien cyanobacterial species (Nostocales) in northeast Germany: *Cylindrospermopsis raciborskii*, *Anabaena bergii* and *Aphanizomenon aphanizomenoides*. *Phycologia* **45** (6), 696–703.

[33] Kokociński, M. and Soininen, J. (2012) Environmental factors related to the occurrence of *Cylindrospermopsis raciborskii* (Nostocales, Cyanophyta) at the north-eastern limit of its geographical range. *European Journal of Phycology* **47** (1), 12–21.

[34] Babanazarova, O.V., Kurmayer, R., Sidelev, S.I. *et al.* (2011) Phytoplankton structure and microcystin concentration in the highly eutrophic Nero Lake. *Water Resources* **38** (2), 223–231.

[35] Akcaalan, R., Mazur-Marzec, H., Zalewska, A. and Albay M. (2009) Phenotypic and toxicological characterization of toxic *Nodularia spumigena* from a freshwater lake in Turkey. *Harmful Algae* **8** (2), 273–278.

[36] Sen, B., Sonmez, F., Canpolat, O. *et al.* (2010) *Nodularia spumigena* threat the lake Hazar (Elazığ, Türkiye). 8th Int. Conference on Toxic Cyanobacteria, 29 August – 4 September, İstanbul, Turkey.

[37] Kazancı, N., Girgin, S. and Dügel, M. (2008) Research on the limnology of Bafa Lake in South-Western Turkey and climate change impact. *Review of Hydrobiology* **2** (1), 207–223.

[38] Fortii, A. (1911) *Diagnoses myxophycearum novarum. Atti e Memorie dell'Accademia di Agricoltura, Scienze e Lettere, Arti e Commercio di Verona, Series I*, **12**, 3–5.

[39] Shaw, G.R., Sukenik, A., Livne, A. *et al.* (1999) Blooms of the cylindrospermopsin containing cyanobacterium, *Aphanizomenon ovalisporum* (Forti), in newly constructed lakes, Queensland, Australia. *Environmental Toxicology* **14** (1), 167–177.

[40] Yilmaz, M., Philips, E.J., Szabo, N.J. and Badylak, S. (2008) A comparative study of Florida strains of *Cylindrospermopsis* and *Aphanizomenon* for cylindrospermopsin production. *Toxicon* **51**(1), 130–139.

[41] Barone, R., Castelli, G., Naselli-Flores, L. (2010) Red sky at night cyanobacteria delight: The role of climate in structuring phytoplankton assemblage in a shallow, Mediterranean lake (Biviere di Gela, southeastern Sicily). *Hydrobiologia* **639** (1), 43–53.

[42] Cirés, S., Wörmer, L., Agha, R. and Quesada, A. (2013) Overwintering populations of *Anabaena*, *Aphanizomenon* and *Microcystis* as potential inocula for summer blooms. *Journal of Plankton Research* **35** (6), 1254–1266.

[43] Quesada, A., Moreno, E., Carrasco, D. *et al.* (2006) Toxicity of *Aphanizomenon ovalisporum* (Cyanobacteria) in Spanish water reservoir. *European Journal of Phycology* **41**(1), 39–45.

[44] Kokociński, M., Mankiewicz-Boczek, J., Jurczak, T. *et al.* (2013) *Aphanizomenon gracile* (Nostocales), a cylindrospermopsin-producing cyanobacterium in Polish lakes. *Environmental Science and Pollution Research* **20** (8), 5243–5264.

[45] Koreiviene, J. and Kasperovičiene, J. (2011) Alien cyanobacteria *Anabaena bergii var. limnetica* Couté et Presig from Lithuania: some aspects of taxonomy, ecology and distribution. *Limnologica* **41** (4), 325–333.

[46] Rücker., J, Stüken, A., Nixdorf, B. *et al.* (2007) Concentrations of particulate and dissolved cylindrospermopsin in 21 *Aphanizomenon*-dominated temperate lakes. *Toxicon* **50** (6), 800–809.
[47] Salmaso, N. (2000) Factors affecting the seasonality and distribution of cyanobacteria and chlorophytes: A case study from the large lakes south of the Alps, with special reference to Lake Garda. *Hydrobiologia* **438** (1), 43–63.
[48] Salmaso, N., Buzzi, F., Garibaldi, L. *et al.* (2012) Effects of nutrient availability and temperature on phytoplankton development: A case study from large lakes south of the Alps. *Aquatic Sciences* **74** (1), 555–570.
[49] Salmaso, N. and Padisák, J. (2007) Morpho-functional groups and phytoplankton development in two deep lakes (Lake Garda, Italy and Lake Stechlin, Germany). *Hydrobiologia* **578** (1), 97–112.
[50] Shams, S., Capelli, C. Cerasino, L. *et al.* (2015) Anatoxin-a producing *Tychonema* (cyanobacteria) in European waterbodies. *Water Research* **69**, 68–79.
[51] Smith, D.J., Griffin, D.W. and Jaffe, D.A. (2011) The high life: transport of microbes in the atmosphere. *Eos Transactions* **92** (30), 249–250.
[52] Suikkanen, S., Kaartokallio, H., Hällfors, S. *et al.* (2010) Life cycle strategies of bloom-forming, filamentous cyanobacteria in the Baltic Sea. Deep Sea Res. Part 2 *Topical Studies in Oceanography* **57** (3), 199–209.
[53] Rücker, J., Tingwey, E., Wiedner, C. *et al.* (2009) Impact of the inoculum size on the population of Nostocales cyanobacteria in temperate lakes. *Journal of Plankton Research* **31** (10), 1151–1159.
[54] Calandrino, E.S. and Paerl, H. W. (2011) Determining the potential for the proliferation of the harmful cyanobacterium *Cylindrospermopsis raciborskii* in Currituck Sound, North Carolina. *Harmful Algae* **11**, 1–9.
[55] Paerl, H.W. and Paul, V.J. (2011) Climate change: links to global expansion of harmful cyanobacteria. *Water Research* **46** (5), 1349–1363.
[56] Briand, J.F., Leboulanger, C., Humbert, J.F. *et al.* (2004) *Cylindrospermopsis raciborskii* (Cyanobacteria) invasion at mid-latitudes: Selection, wide physiological tolerance, or global warming? *Journal of Phycology* **40** (2), 231–238 (2004).
[57] Wiedner, C., Rücker, J., Brüggemann, R. and Nixdorf, B. (2007) Climate change affects timing and size of populations of invasive cyanobacterium in temperate regions. *Oecologia* **152** (3), 473–484.
[58] Mehnert, G., Leunert, F., Cirés, S. *et al.* (2010) Competitiveness of invasive and native cyanobacteria from temperate freshwaters under various light and temperature conditions. *Journal of Plankton Research* **32** (7), 1009–1021.
[59] Hadas, O., Pinkas, R., Delphine, E. *et al.* (1999) Limnological and ecophysiological aspects of *Aphanizomenon ovalisporum* bloom in Lake Kinneret, Israel. *Journal of Plankton Research* **21** (8), 1439–1453.
[60] Hadas, O., Pinkas, R., Malinsky-Rushansky, N. *et al.* (2002) Physiological variables determined under laboratory conditions may explain the bloom of *Aphanizomenon ovalisporum* in Lake Kinneret. *European Journal of Phycology* **37** (2), 259–267.
[61] Piehler, M.F., Dyble, J., Moisander, P.H. *et al.* (2009) Interactions between nitrogen dynamics and the phytoplankton community in Lake George, Florida, USA. *Lake and Reservoir Management* **25** (1), 1–14.
[62] Isvánovics, V., Shafik, H.M., Présing and M., Juhos, S. (2000) Growth and phosphate uptake kinetics of the cyanobacterium, *Cylindrospermopsis raciborskii* (Cyanophyceae) in throughflow cultures. *Freshwater Biology* **43** (2), 257–275.
[63] Spröber, P., Shafik, H.M., Présing, M. *et al.* (2003) Nitrogen uptake and fixation in the cyanobacterium *Cylindrospermopsis raciborskii* under different nitrogen conditions. *Hydrobiologia* **506–509** (1–3), 169–174.
[64] Posselt, A.J., Burford, M.A. and Shaw, G. (2009) Pulses of phosphate promote dominance of the toxic cyanophyte *Cylindrospermopsis raciborskii* in a subtropical water reservoir. *Journal of Phycology* **45** (3), 540–546.
[65] Bar-Yosef, Y., Sukenik, A., Hadas, O. *et al.* (2010) Enslavement in the water body by toxic *Aphanizomenon ovalisporum*, inducing alkaline phosphatase in phytoplanktons. *Current Biology* **20** (17), 1557–1561.
[66] Schindler, D.W., Hecky, R.E., Findlay, D.L. *et al.* (2008) Eutrophication of lakes cannot be controlled by reducing nitrogen input: Results of a 37-year whole-ecosystem experiment. *Proceedings of the National Academy of Sciences U S A* **105**, 11254–11258.
[67] Reynolds, C.S. (2006) *The Ecology of Phytoplankton*. Cambridge University Press, Cambridge.

[68] Lürling M., Eshetu, F., Faassen E. J. *et al.* (2013) Comparison of cyanobacterial and green algal growth rates at different temperatures. *Freshwater Biology* **58** (3), 552–559.
[69] Morabito, G., Oggioni, A. and Panzani, P. (2003) Phytoplankton assemblage at equilibrium in large and deep subalpine lakes: A case study from Lago Maggiore (N. Italy). *Hydrobiologia* **502** (1), 37–48.
[70] Willén, E. (2003) Dominance patterns of planktonic algae in Swedish forest lakes. *Hydrobiologia* **502** (1–3), 315–324.
[71] Zapomělová, E., Řeháková, K., Jezberová, J. and Komárková, J. (2010) Polyphasic characterization of eight planktonic *Anabaena* strains (Cyanobacteria) with reference to the variability of 61 *Anabaena* populations observed in the field. *Hydrobiologia* **639** (1), 99–113.
[72] Lonsdale, W.M. (1999) Global patterns of plant invasions and the concept of invasibility. *Ecology* **80** (5), 1522–1536.
[73] Wilcove, D.S., Rothstein, D., Dubow, J. *et al.* (1998) Quantifying threats to imperiled species in the United States. *BioSciences* **48** (8), 607–615.
[74] Booth, D., Provan, J. and Maggs, C.A. (2007) Molecular approaches to the study of invasive seaweeds. *Botanica Marina* **50** (5/6), 285–296.
[75] Britton-Simmons, K.H. (2006) Functional group diversity, resource preemption and the genesis of invasion resistance in a community of marine algae. *Oikos* **113** (3), 395–401.
[76] Temniskova, D.T. and Stoyneva, M. (2011) *Algology*. Vol. 1, General Part, Vol. 2, Systematic Part, Pensoft, Sofia-Moscow.
[77] Stewart, I., Carmichael, W.W. and Backer, L.C. (2011) Toxic Cyanobacteria, in *Water and Sanitation-Related Diseases and the Environment* (ed. J.M.H. Selendy), John Wiley & Sons Ltd. Chichester, pp. 95–110.
[78] Paerl, H.W., Hall, N.S. and Calandrino, E.S. (2011) Controlling harmful cyanobacterial blooms in a world experiencing anthropogenic and climatic-induced change. *Science of the Total Environment* **409** (10), 1739–1745.
[79] Fastner, J., Rücker, J., Stücken, A. *et al.* (2007) Occurrence of the cyanobacterial toxin cylindrospermopsin in the Northeast Germany. *Environmental Toxicology* **22** (1), 26–32.
[80] Figueredo, C.C., Giani A. and Bird D.F. (2007) Does allelopathy contribute to *Cylindrospermopsis raciborskii* (cyanobacteria) bloom occurrence and geographic expansion? *Journal of Phycology* **43** (2), 256–265.
[81] Leonard, J. and Paerl, H.W (2005) Zooplankton community structure, micro-zooplankton grazing impact, and seston energy content in the St. Johns River system, Florida as influenced by the toxic cyanobacterium *Cylindrospermopsis raciborskii*. *Hydrobiologia* **537** (1), 89–97.
[82] Saker, M. L. and Eaglesham, G. K. (1999) The accumulation of cylindrospermopsin from the cyanobacterium *Cylindrospermopsis raciborskii* in tissues of the redclaw crayfish *Cherax quadricarinatus*. *Toxicon* **37** (7), 1065–1077.
[83] Saker, M.L., Metcalf, J.S., Codd, G.A. and Vasconcelos, V.M. (2004) Accumulation and depuration of the cyanobacterial toxin cylindrospermopsin in the freshwater mussel *Anodonta cygnea*. *Toxicon* **43** (2), 185–194.
[84] Sukenik, A., Quesada, A., and Salmaso, N. (2015) Global expansion of toxic and non-toxic cyanobacteria: effect on ecosystem functioning. *Biodiversity and Conservation* **24** (4), 889–908.

Section III

Sampling, Monitoring and Risk Management

5

Health and Safety During Sampling and in the Laboratory

Roberta Congestri[1], James S. Metcalf[2], Luca Lucentini[3], and Federica Nigro Di Gregorio[3]

[1]Department of Biology, University of Rome 'Tor Vergata,' Italy
[2]Institute for Ethnomedicine, Jackson, USA
[3]Department of Environment and Primary Prevention, Italian National Institute of Health, Rome, Italy

5.1 Introduction

This chapter presents occupational safety recommendations for personnel exposed to cyanobacteria and their toxins during sample collection (water, bloom material) and in the laboratory (use of purified cyanotoxins and toxic cyanobacteria, mass culture of isolates, routine extraction, isolation and purification of toxins, disposal of nuisance biomass). Basic guidelines are summarized in Fig. 5.1.

5.2 Sampling Safety

During the collection of water samples for cyanotoxin analyses, cell enumeration, identification, and biomass estimates, exposure to cyanotoxins, even at very low doses, can present potential health risks including skin rashes (contact dermatitis), upper respiratory irritation, and other effects from recurring low-level exposures (chronic exposures). Samplers should have training for cyanobacterial sampling in accordance with national and local requirements, including (but not limited to) safety protocols and appropriate use of personal protective equipment such as the selection and use of appropriate gloves to reduce skin exposure while sampling. Samplers are

Handbook of Cyanobacterial Monitoring and Cyanotoxin Analysis, First Edition.
Edited by Jussi Meriluoto, Lisa Spoof and Geoffrey A. Codd.

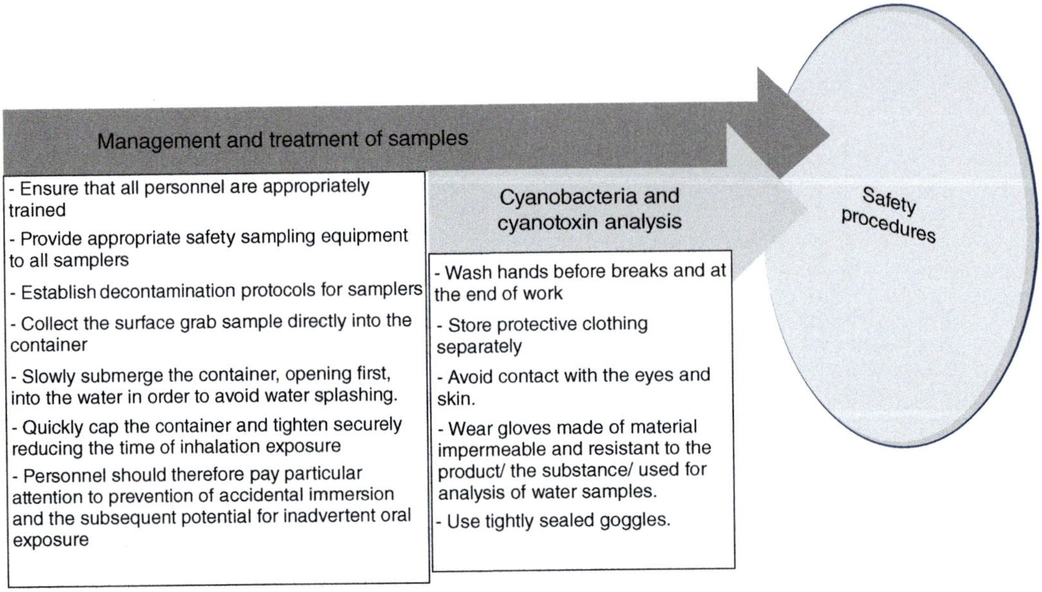

Figure 5.1 Safety recommendations during sampling and in the laboratory

encouraged to wear appropriate gloves throughout sample collection and processing and to avoid splashing or other indirect contact with the skin.

Skin and eye contact with dense surface accumulations should be avoided. If dense surface cyanobacteria are going to be sampled, arms should be covered by using gloves that extend to the shoulder. If contact with dense surface accumulations occurs, the affected area should be rinsed immediately with clean water. Personnel with a recent history of asthma or respiratory illness should take precautions during sampling activities to avoid inhalation of aerosols.

Moreover, unobtrusive and common-sense personal protective equipment such as waterproof boots may be used to harvest bloom biomass. Workers undertaking such activities should be familiar with the risks of accidental immersion, and care must be taken to ensure a secure footing.

5.3 Laboratory Safety

Good laboratory practice is an essential first step to protecting scientists from any potential health risks. Protective clothing, such as lab coats, aprons (if necessary), masks, gloves, and eyewear are the principal requirements for protection. For processing samples, care should be taken when lifting or manipulating large volumes of bloom and culture material, due to the weight and the potential for accidents. If lifting, care should be taken to prevent injury. When concentrating cyanobacterial samples, through buoyancy or centrifugation, any potential toxins present within the samples will also be concentrated and this should be taken into consideration. When purified cyanotoxins are purchased or created, knowledge of the local or national laws governing their storage and use should be followed as some may be subject to antiterrorism legislation [1].

5.4 Cyanotoxin Production and Application

Laboratory cultures grown for toxin production are likely to contain higher concentrations of specific cyanotoxins than are usually found in the field. Regular occupational handling of mass culture material would therefore be expected to present a greater human health risk should accidental exposures occur.

Oral, inhalation, dermal, and parenteral routes are all potential occupational exposure pathways during the various stages of cyanotoxin production and application. The dermal route is heightened with the use of organic solvents as vehicles. The international and intranational transports of living cyanobacteria for research purposes risks establishing feral cyanobacterial populations, so disinfection of culture equipment and the destruction of cells by autoclaving, incineration, and/or chlorination are recommended in order to prevent viable cyanobacteria from escaping research and production facilities and during transportation [2].

Investigation of toxicological or pharmacological properties using in vivo models may present specific risks when radiolabelled cyanotoxins are used, and the potential for occupational exposure via the cutaneous route is increased with the use of organic solvents as vehicles.

Also, the international and intranational deliveries of cyanobacteria for research purposes represent an environmental and occupational risk; thus, disinfection of culture equipment and destruction of cells by autoclaving, incineration, and/or chlorination are recommended.

5.5 Contamination due to Equipment, Glassware, and Accidents

Residues remaining on glassware after the preparation of crude extracts or purified products can be safely decontaminated by rinsing with a solvent such as acetone or ethanol, followed by a water rinse. Gloves provide only a temporary protection against material including cyanobacteria and cyanotoxin analytical standard solutions and may be permeable to some chemical reagents without visible deterioration, the use of appropriate gloves, such as nitrile gloves, for solvent exposure is recommended.

References

[1] Metcalf, J.S., Meriluoto, J.A.O. and Codd, G.A. (2006) Legal and security requirements for the air transportation of cyanotoxins and toxigenic cyanobacterial cells for legitimate research and analytical purposes. *Toxicology Letters* **163**, 85–90.

[2] Stewart, I., Carmichael, W.W., Sadler, R. *et al.* (2009) Occupational and environmental hazard assessments for the isolation, purification and toxicity testing of cyanobacterial toxins. *Environmental Health* **8**, 52 doi:10.1186/1476-069X-8-52.

6

Basic Guide to Detection and Monitoring of Potentially Toxic Cyanobacteria

Nico Salmaso[1], Cécile Bernard[2], Jean-François Humbert[3], Reyhan Akçaalan[4], Meriç Albay[4], Andreas Ballot[5], Arnaud Catherine[2], Jutta Fastner[6], Kerstin Häggqvist[7], Mária Horecká[8], Katarzyna Izydorczyk[9], Latife Köker[4], Jiří Komárek[10], Selma Maloufi[2], Joanna Mankiewicz-Boczek[9,11], James S. Metcalf[12], Antonio Quesada[13], Catherine Quiblier[2,14], and Claude Yéprémian[2]

[1] IASMA Research and Innovation Centre, Fondazione Edmund Mach-Istituto Agrario di S. Michele all'Adige, Trento, Italy

[2] UMR 7245 MCAM CNRS-MNHN, Muséum National d'Histoire Naturelle, and Département Sciences du Vivant, Université Paris Diderot, Paris, France

[3] Institute of Ecology and Environmental Sciences, UPMC, Paris, France

[4] Faculty of Fisheries, Istanbul University, Laleli-Istanbul, Turkey

[5] Norwegian Institute for Water Research (NIVA), Oslo, Norway

[6] Section Drinking Water Resources and Water Treatment, German Environment Agency, Berlin, Germany

[7] Biochemistry, Faculty of Science and Engineering, Åbo Akademi University, Turku, Finland

[8] Slovak Limnological Society, Slovak Academy of Sciences, Bratislava, Slovakia

[9] European Regional Centre for Ecohydrology of the Polish Academy of Sciences, Łódź, Poland

[10] Department of Plant Ecology, Institute of Botany, Academy of Sciences of the Czech Republic, Třeboň, Czech Republic

[11] Department of Applied Ecology, Faculty of Biology and Environmental Protection, University of Łódź, Poland

[12] Institute for Ethnomedicine, Jackson, USA

[13] Departamento de Biología, Universidad Autónoma de Madrid, Spain

[14] Département Sciences du Vivant, Université Paris Diderot, Paris, France

Handbook of Cyanobacterial Monitoring and Cyanotoxin Analysis, First Edition.
Edited by Jussi Meriluoto, Lisa Spoof and Geoffrey A. Codd.

6.1 Introduction

The design of monitoring programmes is the first crucial requirement in the study of the effects caused by the development of cyanobacteria. Monitoring includes both environmental sampling activities and laboratory analyses. The use of the most accurate, state-of-the-art analytical techniques is simply a waste of time if samples are not adequately collected. Most important, the inability to obtain a reliable estimation of cyanobacterial abundance and cyanotoxin concentrations can have critical consequences for risk assessment and health safeguarding plans. Cyanotoxin alarms can strongly decrease the economic and amenity value of freshwater ecosystems, while underestimation of the toxins can represent a threat for human health [1].

The description and implementation of monitoring programmes was the object of the seminal book on toxic cyanobacteria edited by Chorus and Bartram [2]. Successively, other user-friendly manuals have been published, addressed to scientists, water quality managers, and operators (among others, see [3–5]). In this chapter, we will critically examine the monitoring of cyanobacteria and cyanotoxins, highlighting the available state-of-the-art standard procedures and pointing toward new approaches fostered by the development of new technologies and Internet resources.

The design of a monitoring programme is determined by the general aim and specific objectives. Our general focus is on risk assessment and management. Specific goals include the sampling and measurement of cyanobacterial abundance (density and biovolumes) and cyanotoxin concentrations, as well as the identification and measurement of the principal environmental factors associated with the growth of cyanobacteria in drinking water sources and at recreational sites. The chapter will present and discuss the most up-to-date options currently available. Detailed descriptions of the laboratory methods will be presented in the respective standard operating procedures (SOPs).

Compared with other basic limnological investigations (see [6]), the approach used to study cyanobacteria and the causes of their development with a risk-oriented approach has many peculiarities. The temporal and spatial distribution of cyanobacteria and cyanotoxins is highly variable, depending on the characteristics of the species/strains, the physiographic and hydrological characteristics of the waterbodies, and meteorological and climatic conditions (Fig. 6.1). Cyanobacteria can have a homogeneous distribution throughout a waterbody. However, species with aerotopes (groups of gas vesicles), which are able to control their vertical position, can develop in delimited vertical layers and, during surface bloom formation, form irregular patches. Spatial distribution may change within days or even hours because of hydrological variability and the action of wind, which can cause surface blooms to accumulate along leeshores and in harbours and sheltered bays, making it difficult to assess the level of the associated risk in the framework of programmed strategies. The managing of unanticipated bloom events can require the adoption of flexible, reactive approaches [7]. Specific case studies of environmental sampling adapted to different lake types and different dominant cyanobacteria are reported in Chapter 7.

Taking into account these limitations, the ideal monitoring programme should be based on high-frequency observations in the whole waterbody. Insights into the application of new advanced real-time technologies and remote sensing are further explored in Chapters 8 and 9.

The majority of toxic and bloom-forming cyanobacteria are colonial or filamentous. The largest colonies and aggregated filaments of *Microcystis*, *Dolichospermum*, and *Aphanizomenon* can be easily observed by the naked eye, reaching dimensions of a few millimetres. Such colonies have a three-dimensional structure, making it difficult to proceed with microscopic examinations to estimate cell abundance. Similar difficulties are encountered when examining long and thin filamentous species, such as *Planktothrix*. Since many strains develop with toxic genotypes, a common conservative and precautionary approach is to assume toxicity, evaluating quantitatively

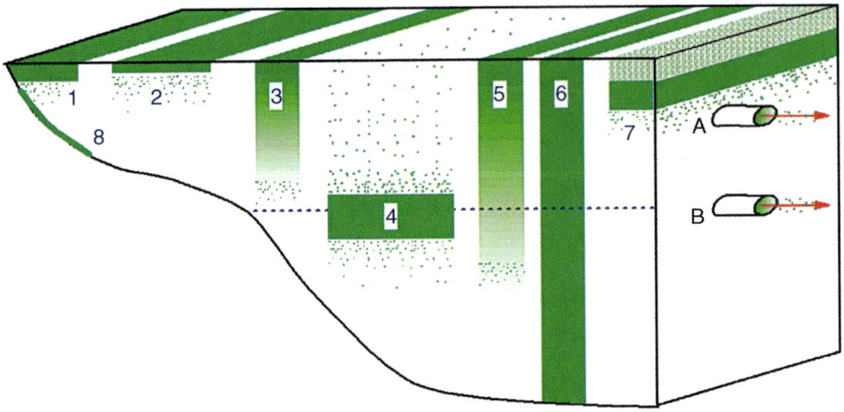

Figure 6.1 *Localization of cyanobacterial accumulations in a natural or artificial lake typical of temperate climates (modified from Lindholm et al. [1] and Codd et al. [7]). 1, 2, surface water blooms; 3, 5, 6, development along the water column, during summer, above (3) or extending over (5) the euphotic depth and during winter overturn (6); 7, below the ice; 4, metalimnetic blooms (e.g.* Planktothrix rubescens*); 8 cyanobacterial benthic mats. The horizontal dashed line indicates the lower limit of the euphotic zone, which corresponds to the layer that receives the 1% of the light incident on the surface of the lake. A and B indicate possible positioning of abstraction pipes for drinking water treatment and supply*

the presence of cyanobacteria rather than their toxins [7]. Therefore, a reliable evaluation of cyanobacterial abundance is the second crucial step in the study of the effects caused by the presence of cyanobacteria. These aspects are considered in the second part of this chapter and in SOPs 1–5.

6.2 Monitoring of Cyanobacteria: Sampling Strategies*

6.2.1 Selection of Variables

The collection and analysis of biological samples should be replicated by a simultaneous collection of samples for physical and chemical analyses and accompanied by in situ measurements at the time of sampling. Environmental variables should include all the main relevant drivers controlling the development of cyanobacteria in a specific waterbody, thereby also taking into account the specificity of the single sites.

6.2.1.1 Physical Variables

Cyanobacterial blooms may increase in warm and stable water columns of eutrophic waterbodies, and therefore water temperature should be measured along a vertical profile (at least every 1 m). At the same time, light availability is essential for the survival and development of photosynthetic organisms. The Secchi disk depth (z_{SD}; section 6.2.4.2) gives an objective value of transparency and can be used to roughly estimate the euphotic (z_{eu}) productive zone through the use of conversion

* *This section was prepared by* Jean-François Humbert, Reyhan Akçaalan, Meriç Albay, Cécile Bernard, Jutta Fastner, Kerstin Häggqvist, Katarzyna Izydorczyk, Latife Köker, Joanna Mankiewicz-Boczek, James S. Metcalf, Catherine Quiblier, and Nico Salmaso.

factors. Common values of the ratio $z_{eu}:z_{SD}=2.7$ have been generally used, along with conversion factors considered more suitable for humic (1.3) and turbid (3.3) lakes [8]. Nevertheless, more accurate light measurements require the use of underwater quantum sensors [9]. The ratio between the euphotic and mixing depths (z_{eu}/z_{mix}) can be used to estimate light availability in the surface mixed layer.

In reservoirs, knowledge of the hydraulic retention time is essential, as it is typically shorter than natural lakes. Long retention times usually increase the stability of the water column and the risk of cyanobacterial blooms [10]. In rivers, measurements of current velocity and water temperature are needed [11]. If nuisance in the river is related to cyanobacteria biofilms, substrate type and size should be noted. Conductivity should also be measured as a proxy for salinity.

6.2.1.2 *Chemical Variables*

Phosphorus and nitrogen are the main nutrients influencing the size and composition of cyanobacterial populations. Nutrient concentrations are more widely available than phytoplankton data for many lakes and rivers; therefore, they can be used (along with physical data) to estimate the potential development of cyanobacterial blooms. Phosphorus availability, which is usually determined by measuring the total fraction (TP; total phosphorus), is one of the main factors linked to the abundance of cyanobacteria. As estimated by Downing et al. [12], the risk of cyanobacterial dominance (>50% of total biovolume) is 0–10% when TP is between 0 and $30\,\mu g\,L^{-1}$, rising to about 40% between 30 and $70\,\mu g\,L^{-1}$, and increasing around 80% near $100\,\mu g\,L^{-1}$. Similarly, according to Carvalho et al. [13], the biggest increase in cyanobacterial biomass occurs in the TP range from about $20\,\mu g\,L^{-1}$ to $100\,\mu g\,L^{-1}$.

When fixed inorganic nitrogen concentrations exceed the assimilative capacity of phytoplankton, non–N_2 fixing taxa (e.g. *Microcystis* and *Planktothrix*) can predominate [14]. During fixed nitrogen deficit, diazotrophic species (e.g. heterocytous taxa such as *Aphanizomenon* and *Cylindrospermopsis*) can take advantage over other phytoplankton species [15]. In this regard, however, the use of N:P ratios to evaluate the potential development of cyanobacteria has been strongly criticised [16] and cannot be evaluated separately from the absolute concentrations of N and P.

pH gives a rough estimate of carbon availability, providing, along with dissolved oxygen concentrations, further indications to evaluate water quality and productivity. High pH values and carbon dioxide depletion are often associated with cyanobacterial blooms [17].

6.2.1.3 *Biological Variables*

Chlorophyll a (Chl a) is an important proxy for total phytoplankton biomass. Laboratory analyses of Chl a are comparatively easy and fast (see SOP 2). If a bloom is due to cyanotoxin-producing species (Table 6.1 and Appendix 2), Chl a can also provide an indication of the potential toxin production. Thus, during cyanobacterial dominance, Chl a may indicate when cyanotoxin analyses are needed [18]. However, during bloom senescence, Chl a declines, but the toxins can still present health risks. Quantitative determinations of cyanobacterial abundance and biovolume provide more reliable estimates of cyanobacterial biomass (section 6.3), but these analyses are time demanding and require taxonomic knowledge. For basic monitoring, recognition and quantification of dominant cyanobacterial species or genera in the sample may be sufficient (see SOP 1). Fresh, unpreserved phytoplankton samples, taken with a plankton net (e.g. 20-µm mesh size), should be inspected, and species preferably also photographed, helping in species recognition in the preserved samples.

Table 6.1 List of potentially toxin-producing cyanobacteria

Toxins	Genera
Microcystins	***Anabaena***, *Anabaenopsis, Annamia, Aphanizomenon**, *Aphanocapsa, Arthrospira, Calothrix*, ***Dolichospermum***, *Fischerella, Geitlerinema, Hapalosiphon*, ***Leptolyngbya***, *Limnothrix, Merismopedia*, ***Microcystis***, ***Nostoc***, *Oscillatoria*, ***Phormidium***, ***Planktothrix***, *Pseudanabaena, Radiocystis, Spirulina*, ***Synechococcus***, *Trichodesmium**, *Trichormus, Woronichinia**
Nodularins	***Nodularia***, *Nostoc*
Anatoxin-a	***Anabaena***, ***Aphanizomenon***, *Arthrospira*, ***Cuspidothrix***, *Cylindrospermum*, ***Dolichospermum***, *Microcoleus, Microcystis*, ***Oscillatoria***, ***Phormidium***, *Planktothrix**, *Pseudanabaena, Tychonema*
Anatoxin-a(S)	***Dolichospermum***
Cylindrospermopsins	*Anabaena*, ***Aphanizomenon***, ***Chrysosporum***, ***Cylindrospermopsis***, *Dolichospermum**, *Lyngbya, Oscillatoria*, ***Raphidiopsis***, *Umezakia*
Lyngbyatoxin	***Lyngbya****, ***Moorea****†
Saxitoxins	***Aphanizomenon***, ***Cuspidothrix***, *Cylindrospermopsis*, ***Dolichospermum***, *Hydrocoleum**, *Limnothrix* (?), *Lyngbya, Raphidiopsis, Scytonema, Trichodesmium**
Aplysiatoxin	*Lyngbya**, ***Moorea****†
Debromoaplysiatoxin	*Leibleinia**, *Lyngbya**
Ciguatoxin	*Trichodesmium**
Palytoxin	*Trichodesmium**

* Potentially toxic genera identified exclusively in environmental samples.
† *Moorea producens* has often been incorrectly cited in the literature as *Lyngbya majuscula* or *L. sordida* [60].
Not all of the species from the listed genera are always present with toxic genotypes. For a complete list of species and references, see Appendix 2. The genera mostly described as toxin-producers in the literature are indicated in bold.

6.2.2 Choice of Sampling Locations in Relation to the Typology of Waterbodies and to Bloom-Forming Cyanobacteria

The definition of sampling programmes is strictly dependent on the nature of the problem. A clear definition of objectives will determine the scales of time and space in which sampling should be performed. General introductions to sampling strategy specifically addressed to biologists and environmental scientists can be found in Green [19] and Thompson [20].

6.2.2.1 Standing Waters

A specific introduction to phytoplankton and cyanobacterial sampling can be found in Sournia [21] and Utkilen et al. [22]. These books provide the basics to evaluate statistically the number of replicates necessary to obtain specific levels of precision. Nevertheless, the underlying frequency distribution of phytoplankton is seldom known, while numerous studies of microdistribution showed that plankton is rarely, if ever, dispersed, according to Poisson (random) expectation [21]. Most often, phytoplankton and cyanobacteria have a temporally changing and aggregate distribution (overdispersed).

Number and Location of Sampling Points The identification of sampling points is determined by the aim of the investigation and by the characteristics of the waterbodies. In the assessment of hazards, cyanobacterial samples must be taken in areas where blooms or potential accumulations can affect human health (e.g. bathing zones) or livestock. Due to spatial variations, several sampling points are needed to obtain a good assessment of the overall cyanobacterial biomass, requiring monitoring schemes that must be individually defined and adapted for each lake. The number and location mainly depend on the surface and the shape of the waterbodies. For example, it has been shown in a small French pond (0.08 km^2) that three sampling points located around the lake were good enough to get an estimate of total cyanobacterial biomass. In this case, the most reliable estimations of biomass were obtained using sampling points located on the shore opposite to the prevailing wind direction over the lake [23].

In large lakes, many sampling points are necessary to obtain a good estimation of cyanobacterial biomass. Hillmer et al. [24] demonstrated that phytoplankton and cyanobacterial concentrations in large lakes can develop with horizontal patches ranging from a couple of kilometres to basin scale. As shown in Duan et al. [25], the early stages in the development of *Microcystis* blooms in Lake Taihu occur in northern bays of the lake before the spread to the centre and south of the lake. The dispersion of cyanobacteria in lakes is due to wind action and lake circulations. When remote sensing data are available, they can be used to target the areas where the cyanobacteria are growing. When no aerial data are available, it is recommended to perform a survey of the bays and gulfs, which are the most susceptible areas to cyanobacterial proliferations and accumulations, and to complete this survey by large transects on the lake. For round basins, two perpendicular transects extending from one shore to the other can be sufficient to obtain a good coverage, whereas for long basins, a transect covering the maximum length is needed. Fluorometric probes are suitable to perform wide spatial surveys (section 6.2.4.3). Indeed, except for surface blooms restricted to the very first centimetres of the water column (e.g. *Dolichospermum lemmermannii* in oligotrophic lakes; [26]), they can provide a real-time estimation of the cyanobacterial biomass, which can limit the number of water samples taken for further analyses [27, 28]. Besides transects, the addition of supplementary sampling points located near public bathing sites or near the intake of water treatment plants is recommended when waterbodies are used for recreational activities or for drinking water supply (Fig. 6.1, cases A, B). If the aim is to prevent health hazards in recreational sites, only the bathing areas can be surveyed and sampled, with preparedness to take further adequate action in the case of accumulations of cyanobacterial cells.

Sampling Depth Depending on the depth of the waterbodies and on the ecology of the different cyanobacterial species, the distribution of cyanobacteria in the water column can display large variations, thereby requiring different sampling strategies in the water column. When blooms are localized at the surface (scums), the maximum abundance of cyanobacteria and highest risk can be estimated by taking a sample where the accumulation of cyanobacteria is thickest [22]. In stratified waterbodies, the daily vertical migratory cycles observed in many Nostocales, which move up to near the surface by night and to depths during evening (e.g. *Dolichospermum*, *Aphanizomenon*; [29]), should be taken into account in the design of sampling procedures. Whatever the species, the use of fluorometric probes is of first interest to estimate the vertical distribution of populations. For example, such tools can provide a very good estimation of the vertical distribution and biomass peaks of *Planktothrix rubescens*, a red pigmented cyanobacterium blooming during summer in the metalimnic layers (10–30 m) of numerous deep alpine lakes [30]. When no probe is available, a discrete sampling of the water should be performed at several depths in the trophogenic layer.

Figure 6.2 Heterogeneity in the coverage of cyanobacterial biofilms in a French river (A, left panel, dark bands) and in a New Zealand river (B, right panel, dark area). Photo taken by J.F. Humbert, INRA

6.2.2.2 *Flowing Waters*

For lotic ecosystems, there is no published study describing a sampling strategy for the monitoring of benthic cyanobacteria at the scale of an entire river. Taking into account the knowledge available on the ecology of benthic cyanobacteria and the field campaigns recently performed in New Zealand and France, the number and the surface of areas characterized by high coverages of biomass greatly vary from one river to another. Consequently, the identification of these areas is a necessary preliminary step in the definition of the monitoring schemes in rivers.

As shown in Fig. 6.2, the coverage of cyanobacterial biofilms can display a great heterogeneity at a small spatial scale. It has been shown that the most important processes controlling the development of biofilms are access to light, flow velocities, water temperature, and the nature of the substrate [11].

6.2.3 Monitoring Frequency

6.2.3.1 *Visual Monitoring*

Visual monitoring of waterbodies is essential to detect the beginning of a cyanobacterial proliferation as early as possible to identify changes in water colour or surface accumulations and scums. This is especially relevant in the risk assessment of recreational waters. Closer inspection of the water in a transparent glass container may also reveal cyanobacterial colonies, bundles, and filaments, which rise to the surface if left undisturbed for 1 hour. The start of visual monitoring will depend on the climatic characteristics of the regions where waterbodies are located and on the ecology of the bloom-forming species. Usually, it will be sufficient to perform visual surveys every 2 weeks during the spring, increasing the frequency (to weekly or up to at least twice weekly) in the case of a notable increase in temperature during several days. In temperate latitudes, visual monitoring should be maintained throughout all of the summer and into the autumn as necessary.

6.2.3.2 *Water Sampling*

In order to minimize the cost of the survey, twice-monthly sampling could be the norm for monitoring, but only if it is complemented by regular visual inspection. In this case, changes in the appearance of the water (its colour or presence of scums) between two successive dates would lead to an immediate increase in the sampling frequency. If it is not possible to carry out visual monitoring, only a weekly sampling strategy can ensure a sufficient monitoring of cyanobacteria during a bloom as shown, for example, in Pobel et al. [23].

6.2.3.3 *Use of Probes for Continuous or High-Frequency Monitoring*

The efficiency of monitoring can be improved by the use of probes (physical, chemical, and fluorometric probes; section 6.2.4), which ensure continuous or high-frequency measurements (more than once a week). Their use is recommended to complement the visual monitoring and during the proliferation of cyanobacteria, especially in recreational areas or in the vicinity (same depth) of drinking water abstraction sites.

6.2.4 Equipment

6.2.4.1 *Water Sampling*

Sampling methods for cyanobacteria depend on the thickness of scums, the layers to be sampled, and the uses of the waterbodies [7]. Comparison of sampling methods for monitoring of cyanobacteria-dominated surface waters has been reviewed by Ahn et al. [31]. If sampling is conducted in bathing areas, samples can be taken from the shore with dipping equipment consisting of a beaker attached to a pole. Scums along the shoreline are sampled more easily with a wide-necked plastic or glass container. However, taking samples from a boat may ensure a greater efficiency and flexibility in the choice of sampling points. During a surface bloom, the inflow method can be used to take samples into a half-submerged wide-mouth bottle. In this case, the quantitative evaluation of areal density values can be improved using a surface bloom sampler (SUBS; [32]). Depth-integrated samples are recommended if a representative sample of the water column is desired. The samples can be collected by using a flexible hose-pipe sampler [33]. The length of the sample pipe should reflect the appropriate depth to which the cells are likely to be mixed. In the case of metalimnetic blooms, collection of water from specific depths can be carried out using 0.5- to 1.0-m sampling bottles (e.g. Niskin type). Plankton nets are preferred when large biomass is required or when only qualitative information is sufficient. The depth at which to deploy nets should be chosen according to the species present. If the species are buoyant (e.g. *Microcystis, Cylindrospermopsis, Dolichospermum*), the cells can be harvested in the upper layers, whereas if metalimnetic species are present (e.g. *Planktothrix rubescens*) nets should be lowered down to the deeper parts of the trophogenic layers (cf. Fig. 6.1). Nevertheless, plankton nets are not capable of retaining the small picocyanobacteria cells and, depending on the mesh size, other nanophytoplanktonic (<20 μm) or even smaller microphytoplanktonic organisms, leading to an overestimation of larger organisms (e.g. *Microcystis* colonies).

6.2.4.2 *Tools for the Physical and Chemical Characterization of Water*

The Secchi disk is a simple device to measure water transparency, which is affected by phytoplankton, other suspended organic and inorganic particles, and dissolved compounds. It is a 20- to 30-cm circular plate painted white or with black and white quarters. It is lowered into the water tied to a measuring rope, recording the depth at which it is no longer visible. A low Secchi disk depth value means higher turbidity. Although it is used as an indicator of phytoplankton abundance and general lake productivity, it is basically a rough proxy of optical properties, which should be also analyzed by using underwater quantum sensors [9].

Water temperature, pH, dissolved oxygen, and conductivity can be measured with separate or multiparametric probes. Specific care must be taken to control and calibrate instrumentation before sampling. The manufacturers' calibration procedures should be followed exactly.

Low-flow velocity is an important factor for proliferation of benthic cyanobacteria. To calculate the flow of rivers, the flow and cross-sectional area need to be determined. The velocity can be measured by using the float method and a current meter. Automated river flow systems can be installed for continuous measurements.

6.2.4.3 *Fluorometric Tools*

Spectrofluorometric methods provide reliable and rapid measurements and are unparalleled for achieving high spatial and temporal coverage of phytoplankton community occurrence and dynamics. They involve the measurement of the fluorescence emitted by Chl a after the excitation of antenna/accessory pigments. This approach has been used for decades to quantify total Chl a in water samples [34]. Subsequent work (e.g. [35]), together with improvements in light-emitting diode (LED) optics, have provided refinement of the method, which is now implemented in commercially available devices, including the bbe Moldaenke FluoroProbe®, for planktic phytoplankton, and the Benthotorch, for phytobenthos (bbe Moldaenke GmbH, Germany).These devices can measure total phytoplanktonic Chl a ($\mu g\,L^{-1}$ or $\mu g\ cm^{-2}$ Chl a) and parse it among four phytoplanktonic groups: "green" (Chlorophyta and Euglenophyta), "blue" (Cyanobacteria), "brown" (Bacillariophyta, Dinophyta, and Chrysophyta), and "red" (Cryptophyta) on the basis of their fluorescence emission spectra (see [28] for details). When used for cyanobacterial biomass quantification, these devices may be subjected to measurement bias due to, for instance, the occurrence of large colonies and/or phycoerythrin-containing cyanobacteria (which will be considered as a mixture of cyanobacteria and cryptophytes by the device) and cell physiology [28].

6.2.5 Storage and Transport

Samples can be collected separately into different containers for each analysis or collected into a single container to be subdivided in the laboratory. In both cases, waterproof-labelled sample containers should be prepared in advance. If the samples are subdivided in the field, different amounts of sample and different kind of containers are needed according to the analyses to be carried out. Different storage procedures of samples could be used for different kind of analysis.

6.2.5.1 *Conservation of Water for Chemical Analysis*

Water samples for nutrient analysis should be taken in 500- to 1000-mL polyethylene bottles. To avoid contamination in the determination of phosphorus compounds, sampling bottles should be prewashed with sulphuric acid ($4.5\,mol\,L^{-1}$) or hydrochloric acid. Nutrient analysis can be done from unfiltered (thereby including particulate matter, e.g. TP) or filtered samples (dissolved nutrients, e.g. soluble reactive phosphorus [SRP]). A 0.45-µm cellulose filter is typically used. If possible, filtration should be done in the field, especially for SRP. Samples should be kept in cold and dark conditions, and the analyses should to be done rapidly, preferably within 24 hours.

6.2.5.2 *Samples for Phytoplankton Counting and Identification*

Unpreserved plankton net or grab samples are indispensable to check taxonomic details using live cyanobacteria and to obtain individuals for the culturing of single strains. Most living and diluted algal samples can be preserved for a few days in incubators at 15–20°C and dim light (around 20–25 µE m^{-2} s^{-1}). For counting (see SOP 1), brown glass bottles (100–200 mL) can be used to preserve phytoplankton samples. In the field, after filling bottles with the samples, Lugol's iodine solution should be added immediately (1 mL Lugol's iodine solution:100-mL sample). For long-term storage (longer than 6 months), samples should be stored at 4°C, in the dark, with the bottle caps sealed with Parafilm®. In this case, transparent bottles can be used, facilitating the visual inspection of the presence of Lugol's solution in the sample, which should look dark yellow/light brown.

6.2.5.3 Filters for Chl a Quantification, Cyanotoxin Analysis, or/and DNA Analysis

Chlorophyll can be degraded very easily by sunlight; therefore, glass or plastic dark bottles of 1-L capacity are recommended. Depending on cyanobacterial population density, at least 1–2 L of sample for cyanotoxin analysis should be taken in amber glass bottles to avoid potential adsorption of cyanotoxins in plastic containers. For DNA analysis, 0.5- to 2-L samples should be taken in a precleaned, sterile plastic or glass bottle. Unpreserved samples should be transported to the laboratory as soon as possible in dark and cold conditions, preferably in a cool box. Water samples for Chl a, toxin, and DNA analysis should be filtered (Whatman GF/C filters) immediately after arrival at the laboratory or kept for the shortest time possible at 4°C. Once the cyanobacterial cells are on the filter, they become highly susceptible to degradation. Therefore, filters should be kept in a freezer (–20°C) and processed for extraction as soon as possible. For cyanotoxins and DNA analysis, samples can be also lyophilized for long-term storage.

6.3 Cyanobacterial Identification and Quantification*

6.3.1 Taxonomic Classification

Cyanobacteria are phylogenetically very ancient, prokaryotic gram-negative organisms, living in a wide range of ecosystems, from subaerophytic to all aquatic environments [36]. Morphologically, cyanobacteria are quite diverse, from small unicellular forms (picoplankton, see Chapter 3) to colonial, filamentous, and branching filamentous types. Some groups have specialized cells such as heterocytes (where N_2 fixation may take place) and akinetes (spore-like cells).

The nomenclature of cyanobacteria is ruled by both the International Code of Botanical Nomenclature [37] and the International Code of Nomenclature of Prokaryotes (ICNP, "The Bacteriological Code") [38]. The botanical classification is based on cell morphology, presence of envelopes or sheaths, cell division, branching, ecological features, and molecular phylogenetic approaches (e.g. [39, 40]). The bacteriological approach also considers morphological features, but with, in addition, the physiological, cytological, and biochemical properties of live, axenic cyanobacterial strains, and it is not based on the species concept [41]. Attempts to match the two systems are still under debate. Oren and Garrity [42] proposed to exclude cyanobacteria from the groups of organisms whose nomenclature is covered by the ICNP code. However, modern cyanobacterial taxonomic classification summarizes both of these concepts (a polyphasic approach), classifying the whole cyanobacterial diversity into more phylogenetically supported clusters (e.g. [43, 44]). As a result, several important changes have been proposed for freshwater planktic genera (see Appendix 1). For example, the previous genus *Anabaena*, which included both benthic

* *This section was prepared by* Cécile Bernard, Reyhan Akçaalan, Andreas Ballot, Arnaud Catherine, Jutta Fastner, Kerstin Häggqvist, Maria Horecka, Latife Köker, Jiri Komárek, Selma Maloufi, Antonio Quesada, Claude Yéprémian, and Nico Salmaso.

and planktic species, has been recently reviewed and divided into two genera: *Anabaena* and *Dolichospermum* [45]. The genus *Aphanizomenon* has also been revised and divided into several clusters. All of these newly defined genera are also determinable according to morphological characters. A brief description of the main planktic and potentially cyanotoxin-producing genera is given next. A list of recent nomenclatural changes is given in Appendix 1, and the toxins known to be potentially produced by these genera and species are listed in Table 6.1 and Appendix 2.

Microcystis Kützing ex Lemmermann 1907 (order Chroococcales)

The genus contains numerous morphospecies occurring in stagnant freshwaters and eutrophic reservoirs; few of them are restricted in tropical or temperate regions. *Microcystis* is a unicellular genus but forms irregular, up to macroscopic colonies and producing heavy water blooms. The main morphospecies from temperate zones are *M. aeruginosa* (type species), *M. botrys*, *M. flos-aquae*, *M. ichthyoblabe*, *M. novacekii*, *M. panniformis*, *M. smithii*, *M. viridis*, and *M. wesenbergii* [46].

Planktothrix Anagnostidis & Komárek 1988 (order Oscillatoriales).

Planktothrix is a genus of filamentous cyanobacteria with about 15–20 species. It belongs genetically to the family Microcoleaceae, to which also belong numerous species of the traditionally known genus *Phormidium*. All species are usually without sheaths, not branched, rarely forming small, and easy disintegrating fascicles (groups, in the mass developments). The trichomes (filaments) are more or less straight or slightly waved, isopolar, and free living. A few species cause water blooms, mainly in mesotrophic to eutrophic reservoirs (usually lakes), such as the most common species *Planktothrix agardhii*. *Planktothrix rubescens*, with reddish cells, occurs instead in lakes and reservoirs from northern to southern Europe [47].

Dolichospermum (Ralfs ex Bornet et Flahault) Wacklin et al. 2009 (order Nostocales)

This genus was previously classified as *Anabaena* Bory ex Bornet et Flahault 1886, which included both benthic and planktic species. It has been recently found that the species must be divided into genera according to molecular criteria: benthic *Anabaena* and planktic *Dolichospermum* [45]. The genus *Anabaena* now comprises only benthic and soil types, which never have gas vesicles in their cells (type species: *Anabaena oscillarioides*). Numerous planktic free-floating species mostly belong to the genus *Dolichospermum* and can create heavy water blooms and regulate their buoyancy in water reservoirs. The gas vesicles in their cells are gathered in groups (aerotopes) (type species: *Dolichospermum flos-aquae* (formerly *Anabaena flos-aquae*).

Aphanizomenon Morren ex Bornet et Flahault 1888 (order Nostocales)

This genus has been divided recently into several genera, which differ according to both phylogenetic and morphological criteria: typical *Aphanizomenon* sensu stricto (with the type species *Aph. flos-aquae*), *Cuspidothrix* (*C. issatschenkoi*), and *Chrysosporum* (*C. ovalisporum*). *Aph. flos-aquae* grows in nature in macroscopic elongated bundles, in contrast with the related genera that grow as simple filaments. Most species may produce intense water blooms and scums.

Cuspidothrix Rajaniemi et al. 2005 (order Nostocales)

This genus, which develops with solitary trichomes characterized by narrowed and pointed ends, is derived from the genus *Aphanizomenon* only. The most common species appears to be the type species, *C. issatschenkoi*, which occurs widely in central European eutrophic reservoirs.

Cylindrospermopsis Seenayya & Subba Raju 1972 (order Nostocales)

This planktic genus with about 12 species is of pantropical distribution [48]. The most common is the type species *C. raciborskii*, an invading species that has shown an expansion of its geographic range to temperate lakes and water reservoirs in recent decades [48; see Chapter 4)]. It forms heavy water blooms in temperate (Europe and North America) and tropical (South America) regions.

Nodularia Mertens ex Bornet & Flahault 1886 (order Nostocales)
This genus contains planktic (with aerotopes) and benthic species. The planktic species sometimes form heavy water blooms in brackish and slightly saline waters in distant regions throughout the world (in both temperate and subtropical zones). In Europe, it is extremely common over the large areas of the Baltic Sea; another important locality is the Caspian Sea. The most common is the planktic type species, *N. spumigena*.

6.3.2 Identification

6.3.2.1 Main Cyanotoxin-Producing Genera

The cyanobacteria that produce known cyanotoxins belong to more than 100 taxa (species or genera; see Appendix 2). The list of potentially toxin-producing genera is shown in Table 6.1. This is based on the examination of clonal cyanobacterial cultures and environmental blooms, where contamination by mixed populations cannot be excluded. In this last case, reports should be confirmed by analyses made on monocyanobacterial, possibly axenic, strains. The list is not exhaustive, and many other cyanotoxin-producers are expected to be identified in the near future [50]. Based on the detection of cyanotoxin biosynthesis–encoding genes, further species have been shown to be potentially toxic (see, e.g. [51, 52]).

As for other bacteria, the concept of species in cyanobacteria has no distinct boundaries, and the taxonomy and phylogeny of this group are under dynamic progress [40, 53–55]. This is the reason why the determination and identification to the species level of toxic cyanobacteria require scientific work and expertise in taxonomy. Cyanobacterial taxonomic identification can be carried out by microscopy, allowing recognition of the major genera and species forming blooms in specific waterbodies and geographical regions. This basic information is essential for the evaluation of the risks presented by potentially toxic cyanobacteria. For risk assessment and management the identification of cyanobacteria to genus level, considering potentially toxic taxa, may suffice because the current understanding of the cyanotoxin production indicates that cyanotoxin content varies widely at both subgenus and population levels. Further, identification at species level by microscopy might be extremely difficult for nonexperts, and in many cases this requires confirmation by genetic analyses (section 6.3.2.4).

Widely used taxonomic manuals and keys assisting in the determination of cyanobacterial genera or species include Komárek and Anagnostidis [56, 57] and Komárek [44]. These are supplemented by photo-guides (e.g. [58, 59]).

6.3.2.2 Examples of Principal Cyanobacterial Genera That Produce Blooms and Cyanotoxins

Examples of widespread genera producing toxic blooms are shown in Fig. 6.3.

6.3.2.3 Internet Resources

Modern molecular phylogenetic studies and the discovery of new phylogenetic relationships are putting taxonomy into a continuous state of change. Further, the mass digitization of biological information is radically revolutionizing algal systematics and the study of biodiversity. A list of algal names (including cyanobacteria) is publicly available and regularly updated on the AlgaeBase website [61]. This can be browsed at different taxonomic levels (e.g. species, genus, family) and *via* author names. Besides reporting current taxonomic status, the archive provides information on publications and key references, synonyms, geographic distribution, and, when available, photographs, links to NCBI nucleotide sequences, and culture collections. Though AlgaeBase is

Figure 6.3 *Cyanobacteria producing blooms and cyanotoxins. (A) Bloom of* Microcystis, *photograph taken by T. Lindholm, Finland. (B) Bloom of* Aphanizomenon, *C. Bernard, MNHN. (C) Bloom of* Planktothrix, *C. Bernard, MNHN. (D) Bloom of* Cylindrospermopsis, *M. Bouvy, IRD. (E) Bloom of* Dolichospermum, *N. Salmaso, FEM. (F) Bloom of* Microcystis, *B. Marie, MNHN. (G) Mats of* Phormidium, *J.F. Humbert, INRA. (H) Mats of* Phormidium, *J.F. Humbert, INRA. (I)* Planktothrix, *L. Brient, Université de Rennes. (J)* Cylindrospermopsis, *C. Bernard, MNHN. (K)* Aphanizomenon, *C. Duval, MNHN. (L)* Dolichospermum, *T. Lindholm, Finland.*

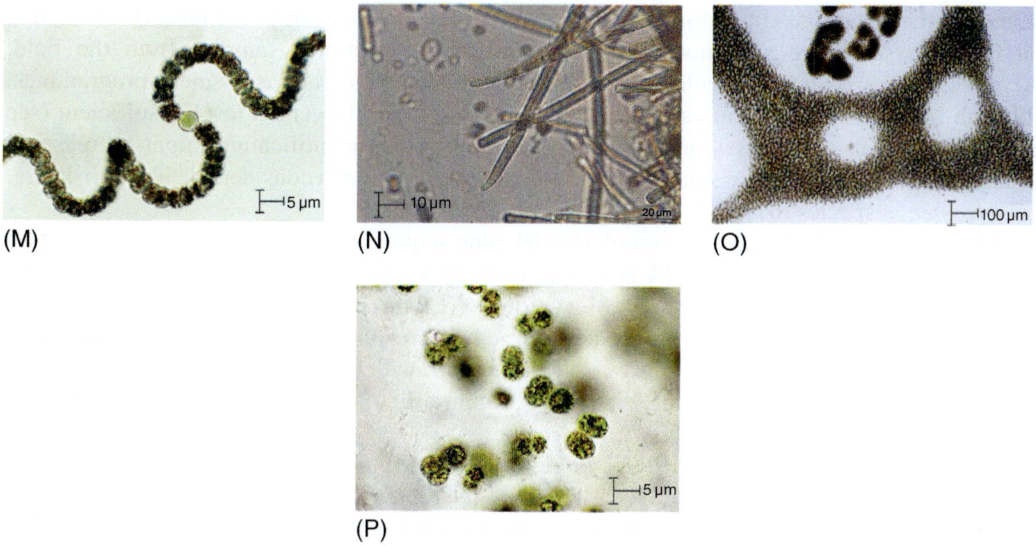

Figure 6.3 (*Continued*) *(M)* Dolichospermum, *A. Couté, MNHN. (N)* Phormidium, *I. Echenique, MNHN. (O)* Microcystis, *A. Couté, MNHN. (P)* Microcystis, *C. Bernard, MNHN*

the result of specific choices by the site managers, the frequent updates are linked to the most up-to-date publications. As highlighted by the authors, "AlgaeBase is purely meant as an aid to taxonomic studies and not a definitive source in its own right. You should always check the information included prior to use."

Another important online database dealing only with cyanobacterial genera and species is CyanoDB [62]. The site provides descriptions of genera, species lists, synonyms, relevant references, as well as a list of rejected and problematic genera. The taxonomy is updated according to Hoffman et al. [63] and newer works. Other Internet resources include images and more complete photographic atlases. A simple Internet search of the most common bloom-forming organisms usually provides tens of results. Nevertheless, particular care is necessary in the evaluation and reliability of information. Examples of atlases are given in www.fytoplankton.cz/fytoatlas.php and www-cyanosite.bio.purdue.edu.

6.3.2.4 Molecular Methods

The monitoring of cyanobacteria has so far been based on microscopic counting and the identification of single organisms and using analytical and bioanalytical methods to identify and quantify the toxins. The difficulty in identifying some of these microorganisms at the species or even genus level is a restriction on the use of these methods. In addition, cyanotoxin- and non–toxin-producing strains of the same species cannot be discriminated under the microscope (but see [64, 65]). Molecular methods represent complementary tools enabling detection and quantification of potentially toxic cyanobacteria species or strains [66].

DNA-based detection methods of the genes involved in cyanotoxin biosynthesis have become popular because of their potential specificity, sensitivity, speed, and ability to identify potentially toxin-producing strains of cyanobacteria prior to the mass production and potential release of toxins into the waterbody. In addition to these methods, DNA microarrays are powerful tools that can facilitate surveying and monitoring cyanobacteria in complex samples from the field. The effectiveness of the molecular approach is apparent in the risk assessment programmes. Although in some management strategies taxonomic resolution to genera level is sufficient (see section 6.3.2.1), in some cases precise cyanobacterial species identification might be relevant, because morphologically quite similar species may or not represent a considerable health risk (e.g. *Aphanizomenon* species [67]).

In PCR-based methods, several cyanobacterial gene sequences are used for species detection (e.g. *rpoC1*, *rbcLX*, *rpoB*, 16S rRNA gene fragment, rRNA 16S-23S ITS fragment) or identification of gene clusters involved in toxin biosynthesis (e.g. microcystins, saxitoxins, anatoxin-a, and cylindrospermopsins). To increase the accuracy and resolution of the molecular methods, multiplex polymerase chain reaction (PCR) can be used, which is able to identify a number of genes of the biosynthetic pathway, detecting potentially different toxin-producing strains simultaneously [68].

Real-time PCR (e.g. quantitative PCR or reverse transcriptase PCR) is used to quantify the cyanobacterial genes or target transcripts. This technique represents a robust and sensitive molecular method to evaluate the proportion of potential toxin-producing and non–toxin-producing genotypes in a population and bloom samples [69, 70].

Microarray technologies have been recently developed to achieve simultaneous detection and semiquantification of potential toxin-producing cyanobacteria in environmental samples. 16S rDNA oligonucleotide probes give a qualitative estimation of the presence or absence of the cyanobacterial genera but do not enable quantification. Microarray methods have high specificity and sensitivity, enabling reduction of the detection limit to 1 to 5 fmol of the PCR product. Their automation will be a key to real-time monitoring of freshwater ecosystems.

The low cost, high accuracy, and fast analysis of molecular methods is currently driving a revolution in which some management administrations are using these tools as a routine analytical procedure [71], which may provide a new dimension with a higher number of samples and sampling frequency. However, the limits of these techniques applied to field samples have to be defined because it was demonstrated that field samples with some biosynthesis genes detected by PCR amplification were not producing the toxins [69].

6.3.3 Population Density Estimation

Many excellent contributions on the theory and practice of phytoplankton counting have been published during the last 50 years (e.g. [21, 33, 72]). General procedures based on the use of the inverted light microscope and sedimentation chambers (Utermöhl method), including preservation and storage of samples, have been collected and described in a European standard [73], which should be consulted. Generally, these procedures can be used in phytoplankton research. However, their application to evaluate the population density of cyanobacteria requires great care and sometimes further appropriate pretreatments of the samples [74]. For multicellular taxa, the number of single constituent cells, as well as number of colonies and filaments, should be evaluated. Estimation of cell concentrations is essential for calculation of population biovolumes, which are computed from the volumes of the single cells (section 6.3.4.2). The number of individual colonies/filaments is a basic requirement for estimation of counting errors (section 6.3.3.2). Step-by-step procedures for the quantitative evaluation of cyanobacteria are reported in SOP 1. In this section, we will identify and briefly discuss a few critical points linked to the specific characteristics of cyanobacteria, and the implications for the statistical evaluation of counts.

6.3.3.1 Pretreatment of Samples

Ideally, in addition to samples preserved with acetic Lugol's solution, the examination of nonpreserved samples can greatly aid taxonomic identification. The classic Utermöhl counting method allows quantitative determination of different species, from single cells to large colonies. If cyanobacteria are too concentrated, the fixed samples should be properly diluted with Lugol-preserved tap water. If the cyanobacteria are too sparse, the sample can be concentrated by letting a large sample volume settle in combined plate chambers (e.g. 50 mL) for at least 4 h per 1-cm height of the chamber. If large and fast-sinking cyanobacteria are the only ones to be counted, even shorter periods may be sufficient [33]. High cylinders (≥100 mL) are not recommended because of the attachment of organisms to the wall and the formation of convection currents, which hinder cell sedimentation. Despite Lugol's fixation, buoyant species can have difficulties in fully settling. In such cases, the disruption of aerotopes can be obtained by a brief (about 10 seconds) ultrasonication or by application of a sudden hydrostatic pressure to the sample [72, 74]. Buoyant cyanobacteria can also be counted in a Sedgewick-Rafter cell (1 mm deep), or in a 0.1-mm-deep CYRUS I chamber [75].

For counting large and stable colonies or tangled filaments, such as several *Microcystis* and *Aphanizomenon* taxa, disintegration of the aggregates and counting of the individual cells are a more convenient option than counting cells in the intact colonies (section 6.3.3.2). An adequate disintegration of colonies to release the individual cells can be achieved after rapid (≤1 minute) ultrasonic treatment [76]. Other recommended methods include hot (80–90°C) alkaline hydrolysis, vortexing, TiO_2 treatment, and boiling [76–78]. Before routine use, the above methods should be checked to assure no substantial cell destruction of individual cells occurs.

6.3.3.2 Counting Methods

The more abundant taxa are counted at 400× in randomly selected optical fields or along transects (diameters of the chamber). Rare and large species are determined on larger sample fractions at lower magnification (100–200×). The precision of the counts depends on the number of the counted units (cells/colonies/filaments). The estimation of the number of units to count to obtain a certain precision generally assumes a nonaggregated, random distribution of individuals, which can be described by a Poisson distribution [21, 74]. Under this assumption, the precision, defined as ±95% confidence limits, is approximated by $2/\sqrt{N} \times 100$, where N is the number of units counted. For example, the count of 100 units has an error of ±20% (see Appendix 6.1). Before counting, the bottom of the chamber should be checked for uneven distribution of organisms and the presence of aggregates. In such cases, filling procedures should be verified and repeated. However, a random distribution of units is not common and cannot easily be obtained [21]. Procedures to evaluate random distributions are reported in Appendix 6.1. If a random distribution of units cannot be assessed, the estimate of the variation between different replicates represents a practical solution [79]. As a further cautionary note, the precision of a count does not represent the overall precision of a multilevel sampling programme ([21]; section 6.2.2).

If colonies are counted without pretreatment, their number (not the number of cells) is the determinant for evaluating the precision of the counting. In such cases, the number of individual cells can be roughly estimated relating cell concentrations to colony sizes [76], or by counting small fractions of the colonies [72]. As for thin and long filaments (e.g. *Planktothrix*), cell concentrations can be easily computed by dividing the total length of filaments (evaluated during counting) with the average length of the cells.

6.3.3.3 Digital Processing

Conventional counting procedures with use of an inverted light microscope are time demanding. Flow cytometers offer a faster, automated way of counting phytoplankton units. Individuals in the sample are moved in a narrow stream passing a light source. Detection is based on light scattering and fluorescence of the particles in the sample, and the signals are processed by computer. Phytoplankton groups can be identified based on their characteristic pigments. However, flow cytometers do not recognize different species, although some models are able to discriminate between particles of different sizes and shapes. Samples are best analysed fresh, as preservation with typical fixatives (formalin, Lugol's) will affect the characteristics used for identification and quantification. More detailed information on flow cytometry can be found in Dubelaar et al. [80]. Applications for real-time measurements in the field have been described in Pomati et al. [81].

6.3.4 Cyanobacterial Biomass Estimation

Chl a and other specific photosynthetic pigments (e.g. phycocyanin) are widely used as a proxy for cyanobacterial biomass. Techniques are available for measurements in the field, by using spectrofluorometric devices (section 6.2.4.3), and in the laboratory, by pigment quantification using spectrophotometry. Both are rapid and simple methods, but they are open to interference by Chl a contained in other (noncyanobacterial) phytoplankton cells. The quantification of cyanobacteria needs more detailed taxonomic resolution if population analysis or toxin content are to be predicted (see SOP 1).

6.3.4.1 Biomass Estimation Based on Pigment Quantification

Despite not providing a reliable estimate of phytoplankton [82], pigment quantification is widely used in routine analysis of water quality. Routine methods are based on spectrophotometry. Chl a quantification gives information on total phytoplankton biomass including cyanobacteria (see SOP 2), while phycocyanin quantification gives information on cyanobacterial biomass (see SOP 3). Other standard methods are based on high-performance liquid chromatography analysis of marker pigments, such as Chl a, phycocyanin, and carotenoids (see SOP 5).

6.3.4.2 Biomass Estimation by Microscopy

Cell volume is highly variable among cyanobacteria species (>25,000-fold; [83]). Thus, cell population densities can represent inappropriate estimates of cyanobacterial abundance. To overcome this issue, the cell biovolume of, at least, the dominant taxa (e.g. relative abundance >5%) should be measured. The biovolumes ($mm^3\,m^{-3} \times 10^3$) of any single species are calculated by multiplying densities (cells mL^{-1}) and specific biovolumes (μm^3) approximated to simple geometrical solids [73, 84]. It is advised to perform biovolume measurements on any given sample, as cyanobacteria cell size shows spatiotemporal variations in response to cell physiology and external environmental conditions (e.g. phase of the cell cycle, nutrient and light availability). Cell dimensions can be obtained using an optical microscope and a calibrated ocular micrometre (see SOP 1) or image analysis software. Semiautomated methods using epifluorescence image analysis are also available (see SOP 1). Examples of specific cell volume calculations are given in Olenina et al. [83], Druart and Rimet [85], and Kremer et al. [86].

Appendix 6.1 Testing Phytoplankton Distributions: χ^2 Test (Pearson Goodness-of-Fit Test)

To test if a given count follows a Poisson distribution, we will consider a real example based on the quantification of the small, monocellular species *Ankyra judayi* (a chlorophyte), which has typical dimensions of 35–40 μm × 2.4 μm (Table 6.A1). The sample was collected in Lake Garda on 6 May 2003, at 10-m depth. Individuals were counted using the Utermöhl method (10-mL sample) on 30 random optical fields at 400× (equivalent to 1.2% of the whole bottom area, see Rott et al. [79], for further details). The method can be applied to single cyanobacterial and phytoplankton units (i.e. single cells, entire filaments, and colonies).

Number of optical fields, $n = 30$
Total count, $\sum x = N = 89$
Mean, $\bar{x} = 2.97$
Variance, $s^2 = 2.86$

The mean (μ) of a Poisson function is equal to the variance (σ^2); thus, the comparison of the sample mean and variance provides a first indication of fit. In the analysed sample, $\bar{x}$ and s^2 are very close, suggesting a good approximation to a Poisson distribution.

The observed frequency distribution of the cells is compared to that expected from a Poisson distribution with the same mean and variance (Fig. 6.A1). To test if the counts are well described by a Poisson function, a χ^2 test for goodness of fit is commonly used. In the example, the test was computed using the packages `vcd` and `fitdistrplus` in the free R statistical software [87]. If the two packages are not installed, from the R console run this command:

```
install.packages("vcd")
install.packages("fitdistrplus")
```

In Fig. 6.A1, the fitted frequencies have been obtained with the script:

```
library(vcd)
goodfit(spec, type= "poisson", method= "ML")
```

where `spec` is a numeric vector of integer data values (Table 6.A1).

The χ^2 test has been computed using the package `fitdistrplus`. In `fitdistrplus`, the breaks of the cells used to compute the χ^2 statistic can be defined, helping to meet the assumptions of the test. More specifically, >80% of the expected counts should be >5 and all individual expected counts should be ≥1 [88]. The R code used to obtain the $\chi2$ test is:

```
library(fitdistrplus)
fit <- fitdist(spec, "pois")
gofstat(fit, chisqbreaks=c(1, 2, 3, 4))
```

Table 6.A1 *Number of a small monocellular species in 30 random optical fields*

Optical field	1	2	3	4	5	6	7	8	9	10	11	12	13	14	15	16	17	18	19	20	21	22	23	24	25	26	27	28	29	30
N. of individuals	5	4	5	5	4	4	1	0	1	2	2	0	5	3	5	3	3	2	3	4	4	3	3	3	7	1	1	1	3	2

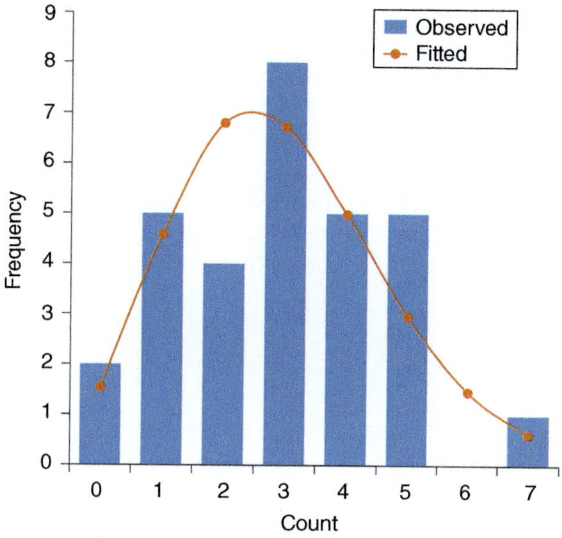

Figure 6.A1 *Count data vs Poisson distribution.*

where `chisqbreaks` is a numeric vector defining the breaks of the cells used to compute the χ^2 statistic. If not defined, the breaks are automatically computed from the raw data. This gives the following table, which reports the observed and expected counts, and the χ^2 statistic:

```
Chi-squared table:
       obscounts theocounts
<= 1   7.000000  6.125478
<= 2   4.000000  6.795506
<= 3   8.000000  6.720000
<= 4   5.000000  4.984000
>4     6.000000  5.375015
```

$\chi^2 = 1.59$
$df = 3$
$p = .66$

The null hypothesis H_0 (the data follow a Poisson distribution) is not rejected. The same conclusions are obtained with different commercial statistical software and accepting the default parameters in the package `fitdistrplus`. When present, slight differences depend on the way in which the breaks of the cells and the expected counts <5 are managed.

Under the assumption of a Poisson distribution, the precision, expressed as percentage of count, and defined as ±95% confidence limits, is given by:

$$precision\,(\%) = \left(t/\sqrt{N}\right) \times 100,$$

which can be approximated to ($t \approx 2$):

$$precision\,(\%) = \left(2/\sqrt{N}\right) \times 100\,[73]$$

Therefore, in Table 6.A1, the counting will result in approximately 95% confidence limits of ±21%. If the units are not randomly distributed, the precision estimate should be regarded only as a rough approximation. Precision and N are related by a nonlinear relationship. An increase of precision to ±10% requires a count of 400 units. In general, to obtain twice the accuracy, 4 times the number of organisms must be counted [33]. As a rule of thumb, counting 40–50 units of the dominant species (with 95% confidence limits of 32%–28%) was considered sufficient for random to moderately dispersed organisms [21, 79]. Increasing the number of counted fields attains little increase in precision. Therefore, if time and resources are available, it is more efficient to count replicates (subsamples) for increasing the accuracy of population density estimations in multilevel sampling programmes.

References

[1] Lindholm, T., Eriksson, J.E. and Meriluoto, J.A.O. (1989) Toxic cyanobacteria and water quality problems—Examples from a eutrophic lake on Åland, South West Finland. *Water Research* **23** (4), 481–486.

[2] Chorus, I. and Bartram, J. (eds.) (1999) *Toxic Cyanobacteria in Water: A Guide to Their Public Health Consequences, Monitoring and Management*, Published on behalf of UNESCO, WHO and UNEP by E&FN Spon, London.

[3] Meriluoto, J. and Codd, G.A. (eds.) (2005) *Cyanobacterial Monitoring and Cyanotoxin Analysis*, Åbo Akademi University Press.

[4] Graham, J.L., Loftin, K.A., Ziegler, A.C. and Meyer, M.T. (2008) Cyanobacteria in lakes and reservoirs: Toxin and taste-and-odor sampling guidelines (ver. 1.0), in *U.S. Geological Survey Techniques of Water-Resources Investigations*.

[5] Newcombe, G. (ed.) (2009) *International guidance manual for the management of toxic cyanobacteria*, Global Water Research Coalition, http://www.waterra.com.au/cyanobacteria-manual/.

[6] Wetzel, R.G. and Likens, G. (2000) *Limnological Analyses*, Springer.

[7] Codd, G.A., Chorus, I. and Burch, M. (1999) Design of monitoring programmes, in *Cyanobacteria in Water: A guide to their public health consequences, monitoring and management* (eds. Chorus, I. and Bartram, J.), Published on behalf of WHO by F & FN Spon, London, pp. 313–328.

[8] Kallf, J. (2002) *Limnology*, Prentice-Hall, Inc., Upper Saddle River, NJ.

[9] Kirk, J.T.O. (2011) *Light and photosynthesis in aquatic ecosystems*, Cambridge University Press, Cambridge.

[10] O'Neil, J.M., Davis, T.W., Burford, M.A. and Gobler, C.J. (2012) The rise of harmful cyanobacteria blooms: The potential roles of eutrophication and climate change. *Harmful Algae* **14**, 313–334.

[11] Quiblier, C., Wood, S., Echenique-Subiabre, I., et al. (2013) A review of current knowledge on toxic benthic freshwater cyanobacteria: Ecology, toxin production and risk management. *Water research* **47** (15), 5464–5479.

[12] Downing, J.A., Watson, S.B. and McCauley, E. (2011) Predicting Cyanobacteria dominance in lakes. *Canadian Journal of Fisheries and Aquatic Sciences* **58** (10), 1905–1908.

[13] Carvalho, L., McDonald, C., de Hoyos, C., et al. (2013) Sustaining recreational quality of European lakes: minimizing the health risks from algal blooms through phosphorus control. *Journal of Applied Ecology* **50** (2), 315–323.

[14] Dolman, A.M., Rücker, J., Pick, F.R., et al. (2012) Cyanobacteria and cyanotoxins: The influence of nitrogen versus phosphorus. *PloS One* **7** (6), e38757.

[15] Paerl, H.W. (2008) Nutrient and other environmental controls of harmful cyanobacterial blooms along the freshwater–marine continuum, *in Cyanobacterial Harmful Algal Blooms: State of the Science and Research Needs: Advances in Experimental Medicine and Biology*, vol. **619**, pp. 217–237.

[16] Oliver, R.L., Hamilton, D.P., Brookes, J.D. and Ganf, G.G. (2012) Physiology, blooms and prediction of planktonic cyanobacteria, in *Ecology of Cyanobacteria II: Their Diversity in Space and Time* (eds.Whitton, B.A.), Springer Netherlands, Dordrecht, pp. 155–194.

[17] Verspagen, J.M.H., Van de Waal, D.B., Finke, J.F., et al. (2014) Rising CO2 levels will intensify phytoplankton blooms in eutrophic and hypertrophic lakes. *PloS One* **9** (8), e104325.

[18] Rogalus, M.K. and Watzin, M.C. (2008) Evaluation of sampling and screening techniques for tiered monitoring of toxic cyanobacteria in lakes. *Harmful Algae* **7** (4), 504–514.
[19] Green, R.H. (1979) *Sampling Design and Statistical Methods for Environmental Biologists*, John Wiley & Sons.
[20] Thompson, K. (2012) *Sampling*, John Wiley & Sons, New Jersey.
[21] Sournia, A. (ed.) (1978) Phytoplankton manual, in *Monographs on Oceanographic Methodology*, vol. **6**, UNESCO, Paris, pp. 1–337.
[22] Utkilen, H., Fastner, J. and Bartram, J. (1999) Fieldwork: site inspection and sampling, in *Toxic Cyanobacteria in Water: A Guide to their Public Health Consequences, Monitoring and Management* (eds. Chorus, I. and Bartram, J.), published on behalf of WHO by F & FN Spon, London, pp. 329–345.
[23] Pobel, D., Robin, J. and Humbert, J.-F. (2011) Influence of sampling strategies on the monitoring of cyanobacteria in shallow lakes: lessons from a case study in France. *Water Research* **45** (3), 1005–1014.
[24] Hillmer, I., van Reenen, P., Imberger, J. and Zohary, T. (2008) Phytoplankton patchiness and their role in the modelled productivity of a large, seasonally stratified lake. *Ecological Modelling* **218** (1–2), 49–59.
[25] Duan, H., Ma, R., Xu, X., et al. (2009) Two-decade reconstruction of algal blooms in China's Lake Taihu. *Environmental Science & Technology* **43** (10), 3522–3528.
[26] Salmaso, N., Capelli, C., Shams, S. and Cerasino, L. (2015) Expansion of bloom-forming *Dolichospermum lemmermannii* (Nostocales, Cyanobacteria) to the deep lakes south of the Alps: Colonization patterns, driving forces and implications for water use. *Harmful Algae* **50**, 76–87.
[27] Brient, L., Lengronne, M., Bertrand, E., et al. (2008) A phycocyanin probe as a tool for monitoring cyanobacteria in freshwaterbodies. *Journal of Environmental Monitoring: JEM* **10** (2), 248–255.
[28] Catherine, A., Escoffier, N., Belhocine, A., et al. (2012) On the use of the FluoroProbe®, a phytoplankton quantification method based on fluorescence excitation spectra for large-scale surveys of lakes and reservoirs. *Water Research* **46** (6), 1771–1784.
[29] Reynolds, C.S. (2006) *The Ecology of Phytoplankton*, Cambridge University Press.
[30] Leboulanger, C., Dorigo, U., Jacquet, S., et al. (2002) Application of a submersible spectrofluorometer for rapid monitoring of freshwater cyanobacterial blooms: A case study. *Aquatic Microbial Ecology* **30**, 83–89.
[31] Ahn, C.-Y., Joung, S.-H., Park, C.-S., et al. (2007) Comparison of sampling and analytical methods for monitoring of cyanobacteria-dominated surface waters. *Hydrobiologia* **596** (1), 413–421.
[32] Bertoni, R. (2011) A Surface Bloom Sampler (SUBS) for cyanobacteria and algae. *Journal of Limnology* **70** (2), 313.
[33] Lund, J.W.G., Kipling, C. and Le Cren, E.D. (1958) The inverted microscope method of estimating algal numbers and the statistical basis of estimations by counting. *Hydrobiologia* **11** (2), 143–170.
[34] Yentsch, C.S.M. and Yentsch, C.S.M. (1979) Fluorescence spectral signatures: Characterization of phytoplankton by the use of excitation and emission spectra. *Journal of Marine Research* **3**, 471–483.
[35] Beutler, M., Wiltshire, K.H., Meyer, B., et al. (2002) A fluorometric method for the differentiation of algal populations in vivo and in situ. *Photosynthesis Research* **72** (1), 39–53.
[36] Whitton, B.A. (ed.) (2012) *Ecology of Cyanobacteria II Their Diversity in Space and Time*, Springer Dordrecht.
[37] McNeill, J., Barrie, F.R., Buck, W.R., et al. (eds.) (2012) *International code of nomenclature for algae, fungi and plants (Melbourne code): adopted by the Eighteenth International Botanical Congress Melbourne, Australia, July 2011*, Regnum Vegetabile, 154, Koeltz Scientific Books.
[38] Lapage, S., Sneath, P., Lessel, E., et al. (eds.) (1992) *International Code of Nomenclature of Bacteria*, American Society of Microbiology (ASM) Press, Washington (DC).
[39] Anagnostidis, K. and Komárek, J. (1988) Modern approach to the classification system of cyanophytes. 3 - Oscillatoriales. *Algological Studies/Archiv für Hydrobiologie, Supplement Volumes* **50–53**, 327–472.
[40] Komárek, J., Kaštovský, J., Mareš, J. and Johansen, J.R. (2014) Taxonomic classification of cyanoprokaryotes (cyanobacterial genera) 2014, using a polyphasic approach. *Preslia* **86**, 295–335.
[41] Castenholz, R.W. (2001) Phylum BX. Cyanobacteria. Oxygenic Photosynthetic Bacteria, in *Bergey's manual of Systematic Bacteriology: Volume One, The Archaea and the Deeply Branching and Phototrophic Bacteria, 2nd ed.*, Springer, New York, pp. 473–487.
[42] Oren, A. and Garrity, G.M. (2014) Proposal to change General Consideration 5 and Principle 2 of the International Code of Nomenclature of Prokaryotes. *International Journal of Systematic and Evolutionary Microbiology* **64** (Pt 1), 309–310.

[43] Zapomělová, E., Jezberová, J., Hrouzek, P., et al. (2010) Polyphasic characterization of three strains of *Anabaena reniformis* and *Aphanizomenon aphanizomenoides* (cyanobacteria) and their reclassification to *Sphaerospermum* gen. nov. (incl. *Anabaena kisseleviana*). *Journal of Phycology* **46** (2), 415–415.
[44] Komárek, J. (2013) Cyanoprokaryota. part 3: Heterocytous Genera, in *Süßwasserflora von Mitteleuropa, Band 19/3*, Springer Spektrum.
[45] Wacklin, P., Hoffmann, L. and Komárek, J. (2009) Nomenclatural validation of the genetically revised cyanobacterial genus *Dolichospermum* (Ralfs ex Bornet et Flahault) comb, nova. *Fottea* **9** (1), 59–64.
[46] Komárek, J. and Komárková, J. (2002) Review of the European *Microcystis*-morphospecies (Cyanoprokaryotes) from nature. *Czech Phycology, Olomouc* **2**, 1–24.
[47] Vareli, K., Briasoulis, E., Pilidis, G. and Sainis, I. (2009) Molecular confirmation of *Planktothrix rubescens* as the cause of intense, microcystin-Synthesizing cyanobacterial bloom in Lake Ziros, Greece. *Harmful Algae* **8** (3), 447–453.
[48] Komárek, J. and Komárková, J. (2003) Phenotype diversity of the cyanoprokaryotic genus *Cylindrospermopsis* (Nostocales); review 2002. *Czech Phycology, Olomouc* **3**, 1–30.
[49] Sukenik, A., Hadas, O., Kaplan, A. and Quesada, A. (2012) Invasion of nostocales (cyanobacteria) to subtropical and temperate freshwater lakes: Physiological, regional and global driving forces. *Frontiers in Microbiology* **3**, 86.
[50] Shams, S., Capelli, C., Cerasino, L., et al. (2015) Anatoxin-a producing *Tychonema* (Cyanobacteria) in European waterbodies. *Water Research* **69**, 68–79.
[51] Lopes, V.R., Ramos, V., Martins, A., et al. (2012) Phylogenetic, chemical and morphological diversity of cyanobacteria from Portuguese temperate estuaries. *Marine Environmental Research* **73**, 7–16.
[52] Mihali, T.K., Carmichael, W.W. and Neilan, B.A. (2011) A putative gene cluster from a *Lyngbya wollei* bloom that encodes paralytic shellfish toxin biosynthesis. *PloS One* **6** (2), e14657.
[53] Thomazeau, S., Houdan-Fourmont, A., Couté, A., et al. (2010) The contribution of sub-saharan african strains to the phylogeny of cyanobacteria: focusing on the Nostocaceae (Nostocales, Cyanobacteria). *Journal of Phycology* **46** (3), 564–579.
[54] Palinska, K.A. and Surosz, W. (2014) Taxonomy of cyanobacteria: a contribution to consensus approach. *Hydrobiologia* **740** (1), 1–11.
[55] Oren, A. (2015) Comments on "Taxonomy of cyanobacteria: a contribution to consensus approach", by K. A. Palinska and W. Surosz (Hydrobiologia 740: 1–11, 2014, doi:10.1007/s10750-014-1971-9). *Hydrobiologia* **758** (1), 1–2.
[56] Komárek, J. and Anagnostidis, K. (1999) Cyanoprokaryota Teil 1 (Chroococcales), in *Süßwasserflora von Mitteleuropa, Band 19*, Gustav Fischer Verlag, Stuttgart.
[57] Komárek, J. and Anagnostidis, K. (2005) Cyanoprokaryota. part 2: Oscillatoriales, in *Süßwasserflora von Mitteleuropa, Band 19/2.*, Springer Spektrum.
[58] Cronberg, G. and Annadotter, H. (1999) "Manual on aquatic cyanobacteria", *ISSHA and IOC of UNESCO, Copenhagen*, **106**.
[59] Cirés, G. and Quesada, A. (2011) *Catálogo de cianobacterias planctónicas potencialmente tóxicas de las aguas continentales españolas*, Ministerio de Medio Ambiente y Medio Rural y Marino.
[60] Engene, N., Rottacker, E.C., Kaštovský, J., et al. (2012) *Moorea producens* gen. nov., sp. nov. and *Moorea bouillonii* comb. nov., tropical marine cyanobacteria rich in bioactive secondary metabolites. *International Journal of Systematic and Evolutionary Microbiology* **62** (Pt 5), 1171–1178.
[61] Guiry, M.D. and Guiry, G.M. (2015) AlgaeBase. World-wide electronic publication. National University of Ireland, Galway."
[62] Komárek, J. and Hauer, T. (2013) "CyanoDB.cz - On-line database of cyanobacterial genera, World-wide electronic publication - Univ. of South Bohemia and Inst. of Botany AS CR."
[63] Hoffmann, L., Komárek, J. and Kaštovský, J. (2005) System of cyanoprokaryotes (cyanobacteria) – state in 2004. *Algological Studies* **117** (1), 95–115.
[64] Kokociński, M., Stefaniak, K., Izydorczyk, K., et al. (2011) Temporal variation in microcystin production by *Planktothrix agardhii* (Gomont) Anagnostidis and Komárek (Cyanobacteria, Oscillatoriales) in a temperate lake. *Annales de Limnologie - International Journal of Limnology* **47** (4), 363–371.
[65] Via-Ordorika, L., Fastner, J., Kurmayer, R., et al. (2004) Distribution of microcystin-producing and non-microcystin-producing *Microcystis* sp. in European freshwaterbodies: detection of microcystins and microcystin genes in individual colonies. *Systematic and Applied Microbiology* **27** (5), 592–602.

[66] Humbert, J.F., Quiblier, C. and Gugger, M. (2010) Molecular approaches for monitoring potentially toxic marine and freshwater phytoplankton species. *Analytical and Bioanalytical Chemistry* **397** (5), 1723–1732.

[67] Ballot, A., Fastner, J., Lentz, M. and Wiedner, C. (2010) First report of anatoxin-a-producing cyanobacterium *Aphanizomenon issatschenkoi* in northeastern Germany. *Toxicon* **56** (6), 964–971.

[68] Baker, L., Sendall, B.C., Gasser, R.B., et al. (2013) Rapid, multiplex-tandem PCR assay for automated detection and differentiation of toxigenic cyanobacterial blooms. *Molecular and Cellular Probes* **27** (5-6), 208–214.

[69] Al-Tebrineh, J., Gehringer, M.M., Akcaalan, R. and Neilan, B.A. (2011) A new quantitative PCR assay for the detection of hepatotoxigenic cyanobacteria. *Toxicon* **57** (4), 546–54.

[70] Campo, E., Lezcano, M.-Á., Agha, R., et al. (2013) First TaqMan assay to identify and quantify the cylindrospermopsin-producing cyanobacterium *Aphanizomenon ovalisporum* in Water. *Advances in Microbiology* **03** (05), 430–437.

[71] Chorus, I. (ed.) (2012) *Current Approaches to Cyanotoxin Risk Assessment, Risk Management and regulations in Different Countries (2012)*, Umweltbundesamt.

[72] Hötzel, G. and Croome, R. (1999) *A Phytoplankton Methods Manual for Australian Freshwaters | Land and Water Australia*, Land and Water Resources Research and Development Corporation, Canberra.

[73] EN-15204:2006 (2006) *Water quality - Guidance standard on the enumeration of phytoplankton using inverted microscopy (Utermöhl technique) | SIPE.*

[74] Lawton, L., Marsalek, B., Padisák, J. and Chorus, I. (1999) Chapter 12. Determination of cyanobacteria in the laboratory, in *Toxic Cyanobacteria in Water: A guide to their public health consequences, monitoring and management* (eds. Chorus, I. and Bartram, J.), published on behalf of WHO by F & FN Spon, London, pp. 347–367.

[75] Czech Technical Standard-757717 (2004) Water quality – Determination of planktonic cyanobacteria (in Czech).

[76] Reynolds, C.S. and Jaworski, G.H.M. (1978) Enumeration of natural *Microcystis* populations. *British Phycological Journal* **13** (3), 269–277.

[77] Box, J.D. (1981) Enumeration of cell concentrations in suspensions of colonial freshwater microalgae, with particular reference to *Microcystis aeruginosa*. *British Phycological Journal* **16** (2), 153–164.

[78] Joung, S.-H., Kim, C.-J., Ahn, C.-Y., et al. (2006) Simple method for a cell count of the colonial Cyanobacterium, *Microcystis* sp. *Journal of Microbiology (Seoul, Korea)* **44** (5), 562–565.

[79] Rott, E., Salmaso, N. and Hoehn, E. (2007) Quality control of Utermöhl-based phytoplankton counting and biovolume estimates—an easy task or a Gordian knot? *Hydrobiologia* **578** (1), 141–146.

[80] Dubelaar, G.B.J., Casotti, R., Tarran, G.A. and Biegala, I.C. (2007) Phytoplankton and their analysis by flow cytometry, in *Flow Cytometry with Plant Cells. Analysis of Genes, Chromosomes and Genomes* (eds. Doleel, J., Greilhuber, J. and Suda, J.), Wiley-VCH Verlag GmbH & Co. KGaA, Weinheim, Germany, pp. 287–322.

[81] Pomati, F., Kraft, N.J.B., Posch, T., et al. (2013) Individual cell based traits obtained by scanning flow-cytometry show selection by biotic and abiotic environmental factors during a phytoplankton spring bloom. *PloS One* **8** (8), e71677.

[82] Kruskopf, M. and Flynn, K.J. (2006) Chlorophyll content and fluorescence responses cannot be used to gauge reliably phytoplankton biomass, nutrient status or growth rate. *The New Phytologist* **169** (3), 525–536.

[83] Olenina, I., Hajdu, S., Edler, L., et al. (2006) Biovolumes and size-classes of phytoplankton in the Baltic Sea. *Helcom Baltic Sea Environment Proceedings* **106**, 1–144.

[84] Sun, J. and Liu, D. (2003) Geometric models for calculating cell biovolume and surface area for phytoplankton. *Journal of Plankton Research* **25** (11), 1331–1346.

[85] Druart, J.-C. and Rimet, F. (2008) *Protocoles d'analyse du phytoplancton de l'INRA: Prélèvement, dénombrement et biovolumes*, INRA-Thonon, Rapport SHL 283.

[86] Kremer, C.T., Gillette, J.P., Rudstam, L.G., et al. (2014) A compendium of cell and natural unit biovolumes for >1200 freshwater phytoplankton species. *Ecology* **95** (10), 2984.

[87] R Core Team (2016) *R: A language and environment for statistical computing.*, R Foundation for Statistical Computing, Vienna, Austria. URL https://www.R-project.org/.

[88] McHugh, M.L.M.L. (2013) The chi-square test of independence. *Biochemia Medica* **23** (2), 143–149.

7

Case Studies of Environmental Sampling, Detection, and Monitoring of Potentially Toxic Cyanobacteria

Kerstin Häggqvist[1], Reyhan Akçaalan[2], Isidora Echenique-Subiabre[3], Jutta Fastner[4], Mária Horecká[5], Jean-François Humbert[6], Katarzyna Izydorczyk[7], Tomasz Jurczak[8], Mikołaj Kokociński[9], Tore Lindholm[10], Joanna Mankiewicz-Boczek[7,8], Antonio Quesada[11], Catherine Quiblier[3,12], and Nico Salmaso[13]

[1] *Biochemistry, Faculty of Science and Engineering, Åbo Akademi University, Turku, Finland*
[2] *Faculty of Fisheries, Istanbul University, Laleli-Istanbul, Turkey*
[3] *UMR 7245 MCAM CNRS-MNHN, Muséum National d'Histoire Naturelle, Paris, France*
[4] *Section Drinking Water Resources and Water Treatment, German Environment Agency, Berlin, Germany*
[5] *Slovak Limnological Society, Slovak Academy of Sciences, Bratislava, Slovakia*
[6] *Institute of Ecology and Environmental Sciences, UPMC, Paris, France*
[7] *European Regional Centre for Ecohydrology of the Polish Academy of Sciences, Łódź, Poland*
[8] *Department of Applied Ecology, Faculty of Biology and Environmental Protection, University of Łódź, Poland*
[9] *Department of Hydrobiology, Adam Mickiewicz University, Poznań, Poland*
[10] *Environmental and Marine Biology, Faculty of Science and Engineering, Åbo Akademi University, Turku, Finland*
[11] *Departamento de Biología, Universidad Autónoma de Madrid, Spain*
[12] *Département Sciences du Vivant, Université Paris Diderot, Paris, France*
[13] *IASMA Research and Innovation Centre, Fondazione Edmund Mach-Istituto Agrario di S. Michele all'Adige, Trento, Italy*

Handbook of Cyanobacterial Monitoring and Cyanotoxin Analysis, First Edition.
Edited by Jussi Meriluoto, Lisa Spoof and Geoffrey A. Codd.

7.1 Introduction

The morphology of waterbodies and, consequently, their hydrodynamics, hydrology, and hydrochemistry can differ remarkably. Moreover, the conditions in a single waterbody may vary considerably in time. In turn, the associated organisms may display an array of different adaptions to the prevailing conditions. The sampling, detection, and monitoring of potentially toxic cyanobacteria are therefore in many cases not straightforward procedures. Nevertheless, good and purposeful techniques for sampling, detection, and monitoring of potentially toxic cyanobacteria are central requirements, especially in waterbodies used as drinking water supplies or for recreational purposes. In this chapter, case studies representing problems characteristic of six different water types are provided as examples on how cyanobacterial detection, sampling, and monitoring have been handled in these diverse waters.

7.2 Shallow Lakes

Shallow lakes (mean water depth 3 m) are characterised by a frequently mixed water column and dominance by macrophytes or algae/cyanobacterial blooms. These states can change from year to year and even within one season, though with higher trophic levels, often cyanobacteria dominate. Also with cyanobacterial dominance, the annual patterns may be variable due to changes in the mixing conditions and, thus, frequent monitoring of shallow lakes is especially important when they are used for bathing or for raw water abstraction.

Under eutrophic conditions, filamentous cyanobacteria (e.g. *Planktothrix*, *Aphanizomenon*) often dominate in frequently mixed waterbodies (see section 7.2.2). However, colonial *Microcystis* also can be abundant in some shallow lakes during calm and warm weather. Sampling must be adapted to these specific conditions. When the waterbody is constantly mixed and cyanobacteria are expected to be evenly distributed throughout the water column, only one sample from a defined depth (preferable 0–1 m) may be sufficient. When buoyant cyanobacteria such as *Microcystis* are present, preferably integrated samples, or in case of scum formation, surface samples to assess maximal toxin concentrations should be taken.

7.2.1 Variability in Microcystin Concentrations along the River Havel, Germany

The Havel is a slowly flowing river in Berlin with several river lakes, including the Wannsee, which has a mean water depth of around 5 m and a maximum depth of 9 m. High phosphorus concentrations lead to annual blooms of cyanobacteria, mainly *Microcystis* in summer and *Planktothrix agardhii* in autumn. *Microcystis* is trapped in the frequent bays of the river and proliferates in these areas in the summers. In addition to the calm bays, areas with a high flow velocity as well as quickly changing wind directions lead to a high temporal and spatial variability of cell and microcystin concentrations over several orders of magnitude in this river (0.1–25,000 $\mu g\ L^{-1}$ microcystin-LR equiv.). Rather constant relationships of microcystin–to–chlorophyll a ratios of about 0.1 indicate that the different microcystin concentrations are indeed due to accumulations of cells and not due to changing microcystin content of the cells [1]. These features give the river an interesting character reminiscent of both a slow-flowing river and a shallow lake, with subsequent influences on performed monitoring schemes.

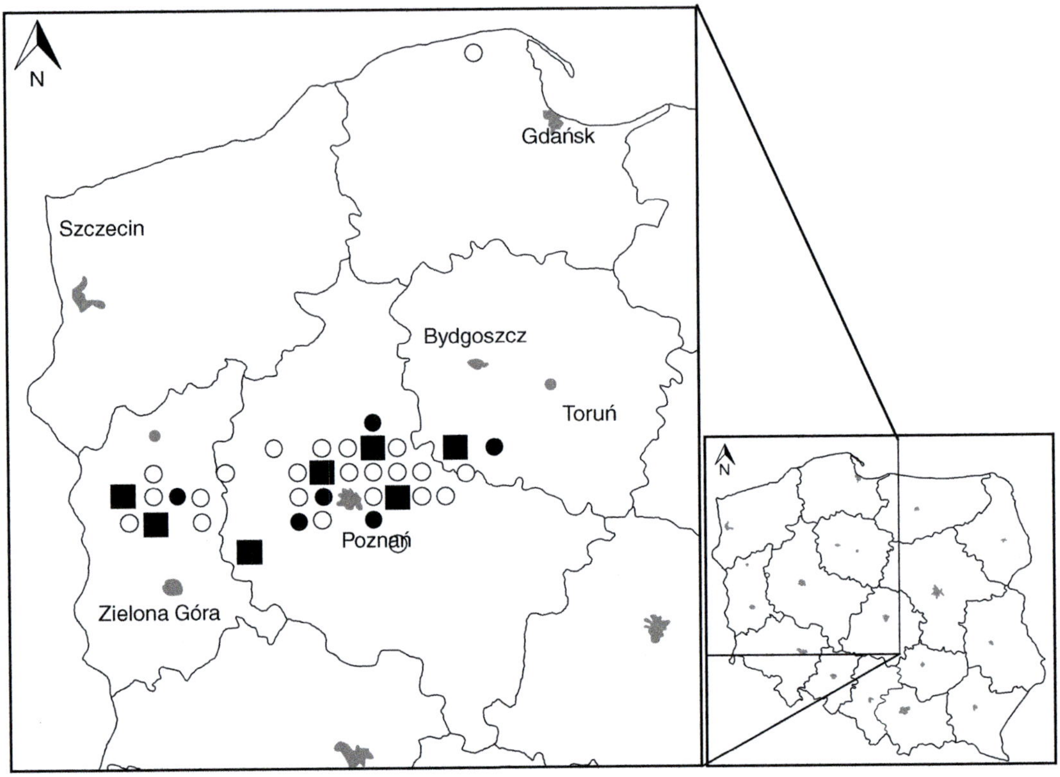

Figure 7.1 Distribution of cylindrospermopsin in different types of lakes in Poland. Black squares: cylindrospermopsin in shallow lakes, black dots: cylindrospermopsin in stratified lakes and white dots: lakes without detectable cylindrospermopsin. Adapted from [2]

7.2.2 Cylindrospermopsin in Shallow Lakes in Poland

A recent study demonstrated common occurrence of cylindrospermopsin in lakes of western Poland [2]. Particulate cylindrospermopsin was detected in 13 of 36 investigated lakes (Fig. 7.1) sampled twice during the beginning and at the end of summer 2010. Cylindrospermopsin was present in lakes differing in productivity and mixing. However, seven of these were classified as shallow lakes and accounted for almost 30% of all investigated shallow lakes. Cylindrospermopsin concentrations in these lakes did not differ significantly between study periods and ranged from trace levels to 0.78 µg L^{-1}.

The phytoplankton of the shallow lakes was dominated by cyanobacteria, including several potential cylindrospermopsin-producers, such as *Chrysosporum bergii* (*Anabaena bergii*), *Chrysosporum ovalisporum* (*Aphanizomenon ovalisporum*), *Aphanizomenon flos-aquae*, *Aphanizomenon gracile*, and *Cylindrospermopsis raciborskii*. However, only the last two species were abundant in the samples and only one isolated culture of *Aphanizomenon gracile* was found to be a cylindrospermopsin-producer. The majority of the cylindrospermopsin-positive shallow lakes were eutrophic and turbid. Statistical analyses did not reveal any strong environmental factors related to cylindrospermopsin production. A weak positive influence of phosphate, ammonium, and temperature was observed.

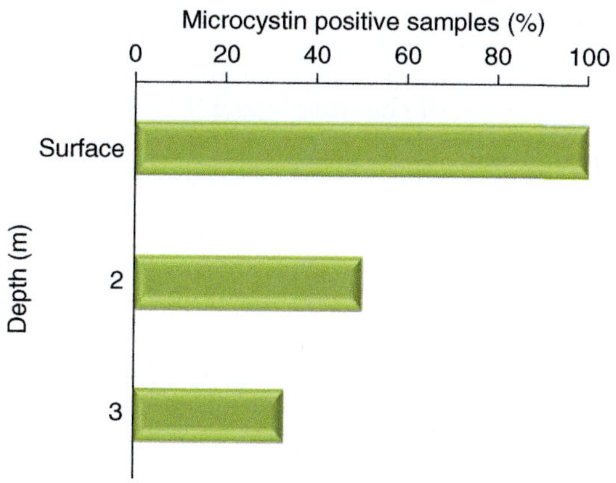

Figure 7.2 The proportion of microcystin-positive samples versus depth in Lake Taskisi ($N = 15$)

Many of the investigated lakes are used for recreation or as water supplies for agricultural farms, corroborating the importance of considering cylindrospermopsin in monitoring programmes. This is especially crucial as the toxin was shown to be present throughout the summer irrespective of lake conditions.

7.2.3 *Microcystis aeruginosa* and Microcystins in Lake Taskisi, Turkey

Lake Taskisi is a eutrophic and shallow lake (maximum depth 4.5 m). Samples taken from the surface, 2 m and 3 m revealed *Microcystis aeruginosa* as the dominant species in the summer months, also producing surface scums [3]. The cyanobacteria *Woronichinia compacta*, *Merismopedia tenuissima*, and *Anabaena* sp. contributed to phytoplankton biomass in low proportions. Microcystin-LR was the dominant cyanotoxin in the lake; however, microcystin-LW and -RR were also found in samples taken in July and August, respectively. Microcystin concentrations correlated significantly with the presence of *M. aeruginosa* according to depth. All samples from surface waters were microcystin-positive, whereas only 50% of the samples were microcystin-positive at 2 m and 33% at 3 m (Fig. 7.2). Thus, in this shallow lake, the highest microcystin concentrations were measured at the surface, posing high risks to those frequently using the lake for recreation and necessitating integrated and/or surface sampling to correctly estimate the toxin concentrations.

7.2.4 Natural Swimming Pools – Lakes and Dams in Slovakia

About 37 waterbodies, gravel-pit type lakes and dams, in which bathing is explicitly authorised, or in which bathing is not prohibited and traditionally practised by a large number of people, are monitored by the public health authorities in Slovakia [4]. Increasing water blooms were observed in about 10–12 of the localities during 2008–2012 (Table 7.1). In one lake in the Draždiak region (Bratislava), an interesting water bloom of *Dolichospermum lemmermannii* was observed in November 2011. During the bloom, the species rose repeatedly to the surface and sank to the bottom, indicating that sampling procedures comparable to those during the occurrence of *Planktothrix rubescens* in deep lakes (see section 7.3) sometimes need to be considered also in shallow lakes.

Table 7.1 Examples of water blooms in recreational lakes in Slovakia

Location	Toxin	Toxin concentration dry biomass^{-1} (μg g^{-1})	Dominant genera/species
Šaštín (lake)	Microcystin	220–1626	*Microcystis aeruginosa*, *Microcystis* spp., *Woronichinia naegeliana*, *Dolichospermum flos-aquae*, *Cuspidothrix issatschenkoi*, *Aphanizomenon gracile*
Nitrianske Rudno (dam)	Cylindrospermopsin	37.7	*Aphanizomenon flos-aquae*, *Microcystis* spp.
Ružiná (dam)	Microcystin	153–506	*Microcystis* spp., *Woronichinia naegeliana*
Senec (lake)	Microcystin	1037	*Microcystis aeruginosa*
Kanianka (dam)	Microcystin	4817	*Planktothrix rubescens* (bloom under the ice)

7.3 Deep Lakes

Lake Garda and other subalpine lakes are classic cases of environmental sampling in deep, oligotrophic environments. The dominant cyanobacteria in these lakes are mostly represented by *Planktothrix rubescens*, followed by a minor presence of *Planktolyngbya limnetica* and *Limnothrix* sp. [5]. *P. rubescens* is typically found in deep lakes in the temperate zone. In fact, it can develop huge populations in only weakly illuminated (2–20 μmol m^{-2} s^{-1}) metalimnetic layers [6]. At higher irradiances, the excess of carbohydrates produced by photosynthesis acts as ballast, causing sinking of filaments to the "neutral buoyancy depth," around the limit of the euphotic zone [7]. This behaviour can explain why an excessive increase in eutrophication, with a drastic decrease of light in the metalimnetic layers, causes a drastic reduction of *P. rubescens* in surface layers of deep lakes [8]. Therefore, the vertical positioning of this species should be taken carefully into account when defining sampling strategies. For example, in Lake Garda the euphotic depth in summer is generally around 10–25 m, and samples for chemistry and phytoplankton analyses are collected in three integrated layers (0–2, 9–11, 19–21 m) and at 60 m. The choice of samples is a compromise between the minimum level of knowledge that can be obtained by microscopic examinations, and the time/costs needed for the analyses. Nevertheless, the parallel use of fluorometric probes (phycocyanin detection and fluoroprobes) ensures the recording of a complementary set of information on the actual distribution and biomass of cyanobacteria developing along the water column [9]. In mesotrophic lakes, such as Lake Iseo (in the same lake district as Lake Garda), *P. rubescens* developed with massive autumn and winter surface blooms along the shores. When the focus is on risk assessment, reactive samplings should be organised, taking into account potential hazards connected with the use of the water resource (see Chapter 6).

At the other extreme, occasional surface blooms of *Dolichospermum lemmermannii* pose a serious challenge for proper evaluation and classification of problems. In many oligotrophic and oligomesotrophic large and deep lakes south of the Alps, this species appears regularly in summer, forming wide and unevenly distributed blooms strictly localised in the upper centimetres of the

(A) (B)

Figure 7.3 *(A) Blooms of* Dolichospermum lemmermannii *documented in the southeast basin of Lake Garda, Bardolino harbour, 27 August 2002. (B) Coiled filaments of* D. lemmermannii *(Lake Garda). The clusters of larger, green cells are akinetes. These specialised cells (around 15–20 μm) are able to remain dormant in the sediments for several decades, providing an inoculum for the development of new filaments. Source: "Salmaso N (2002) Fattori influenzanti lo sviluppo dei cianobatteri nei laghi profondi. Rapporti ISTISAN 02/9: 44-55"*

water column (Fig. 7.3, see Chapter 4). In such cases, the inflow method was used to take several samples into a half-submerged sampling bottle [10] (see Chapter 6). Considering the wide heterogeneity of the cyanobacterial patches, this approach provides information exclusively limited to the area of interest. As fluorometric probes have a poor efficiency in the measurement of thin surface blooms, the study of patches in large areas should be better evaluated using aerial and remote sensing techniques (see Chapter 9).

7.4 Reservoirs

Man-made water reservoirs typically serve as drinking water sources, for irrigation, industrial or cooling water supplies, for power generation, flood control, fishing, and/or recreation. Reservoirs constitute an intermediate aquatic system between flowing rivers and natural lakes, with particular features depending on the management of the hydrological balance. Therefore, in reservoirs, two main zones may be distinguished: (1) an upstream riverine zone, located in the inflow part of the reservoir, often narrow, shallow, channelized, and characterised by high inflow, and (2) the lacustrine zone at the reservoir end, a broad and deep lake-like basin with low flow rate. The two main zones are divided by a transitional zone [11].

Reservoirs are often characterised by high nutrient inputs from the catchment (point and diffuse sources), as they, with reference to their purposes, are often located in densely populated areas or areas with intense agriculture. They are frequently subjected to intense depth variations depending on the management strategies, triggering a remarkably variable water retention time and a peculiar shoreline with limited riparian vegetation. Contrary to lakes, the outlet of water in reservoirs can be variable and on many occasions the depth of water withdrawal can be adjusted according to requirements. For example, during summer and in the presence of cyanobacterial blooms, the water manager might withdraw hypolimnetic water, modifying the stratification dynamics and thus the ecological behaviour of the system (nutrients, photic depth, temperature, etc.).

The sampling design for water reservoirs is particularly important and should consider: (i), morphometry of the basins, (ii) expected hydraulic conditions, and (iii) hydrodynamics of the water. Additionally, if it is possible, sampling points should represent the two zones together with main tributaries and at each point integrated water samples should be collected.

7.4.1 The Eutrophic, *Microcystis*-Dominated Sulejow Reservoir, Poland

The shallow and eutrophic Sulejow Reservoir is situated in Central Poland, in the middle course of the River Pilica. At its full capacity, this reservoir covers an area of 22 km^2, its mean depth is 3.3 m and volume 75×10^6 m^3. Mean water retention time is 30 days, but during summer it increases up to 60 days. Total phosphorus concentration varies within the range 0.08–0.22 mg L^{-1}. The Sulejow Reservoir is important for flood control, power generation, fishing, and recreation. The toxic cyanobacteria in the reservoir have been extensively studied. The cyanobacteria are usually present from late spring (May–June) to early autumn (October) [12–15]. Multiannual studies, conducted regularly from 2001 to 2010, revealed that *Microcystis aeruginosa* was the dominant species of the bloom-forming cyanobacteria. Its average biomass was within the range 1–13 mg fresh biomass L^{-1} and reached a maximum of 70 mg L^{-1}. The average concentration of microcystins during the same monitoring period varied between 1–3 µg L^{-1} with a maximum of 10 µg L^{-1}.

The occurrence and amount of cyanobacteria were strictly dependent on water temperature (18–25°C), concentration of total phosphorus (>0.1 mg L^{-1}), and water retention time in the reservoir. A substantial shortening in water retention time (<20 days) resulted in a significant reduction in the concentration of cyanobacteria [12, 14, 15]. Research during 2009–2010 showed that the concentration and toxicity of microcystins were directly dependent on the quantity of toxic cyanobacterial genotypes. They regulated production and toxicity of microcystins at a similar mean level (1 µg L^{-1}), regardless of differences in cyanobacterial biomass (1–13 mg L^{-1}) [15].

These data illustrate that despite many years of domination by *Microcystis aeruginosa*, its biomass and toxicity is dynamically changing and dependent on a variety of environmental parameters. Therefore, only multifaceted monitoring, including analyses of physical, chemical, hydromorphological, and biological parameters supports a proper risk assessment and further actions aimed at improving the water quality.

7.4.2 Species, Morphospecies, and Toxin Variability in Santillana Reservoir, Spain

Santillana Reservoir on the Iberian Peninsula is a typical water reservoir with a maximum volume of about 0.08 km^3. It has two basins, a shallow basin (average depth 3 m) and a smaller deep basin (maximum depth 26 m). Cyanobacterial blooms usually begin in spring (April–May) with *Aphanizomenon* and *Dolichospermum* spp. but are later dominated by *Microcystis* spp. The *Microcystis* species follow a peculiar ecological succession, changing morphospecies and/or toxicity. This succession follows a periodicity [16] with bloom peaks every 3 weeks over the summer, reaching up to 250 µg L^{-1} chlorophyll a and over 70 µg L^{-1} microcystins. Fast changes in toxicity have been found to parallel shifts in genetic composition (strains) of the same morphospecies (*M. aeruginosa*).

This reservoir has been subjected to marked interannual variations apparently not linked to known environmental variables [17]. In 1999, it shifted from *M. aeruginosa*-dominated blooms over the summer to *Microcystis wesenbergii* in 2000 [17]. Sedimentation experiments demonstrated that *Microcystis* blooms sediment along the summer season although the highest sedimentation rate takes place in the autumn, at the end of the blooming period. The process was described as a continuous rain of toxic particles that accumulate at the sediment surface [18, 19].

These data illustrate how water reservoirs are complex systems in which cyanobacterial biomass and toxicity can change dramatically in very short periods of time and thus management requires considerable monitoring efforts.

7.5 Rivers

New Zealand is one of the countries with most experience in benthic cyanobacterial monitoring and sampling in rivers. Chapters are devoted to this issue in the New Zealand national guidelines for cyanobacteria surveillance [20]. The monitoring and sampling scheme is based on placing transects (40–60 m) in selected areas, including some riffles and runs. In each surveillance area, four transects are inspected with an underwater viewer. The transects are divided into five points, starting from the shore to a maximal depth of 0.6 m. For all the points on the transects, an estimation of the percentage coverage of cyanobacterial mats is performed. Samples for microscopic observations and cyanotoxin analyses can be collected by scraping the substrate at any of the sampling points. The alert-level framework proposed in the New Zealand guidelines is only based on the percentage coverage by the benthic cyanobacterial mats. It presents three levels (≤20%, 20–50%, and ≥50% coverage) with corresponding actions concerning: (i) frequency of the surveillance (fortnightly when ≤20% coverage and weekly when >20%); (ii) potential increase of the sampling areas (when percentage coverage is >20%); (iii) potential toxin analyses (when percentage coverage is >20%); and (iv) information to be given to the public and to public health units.

The Tarn and the Loue are two French rivers in which proliferations of benthic cyanobacteria producing anatoxin-a have caused the death of several dogs since 2000 [21, 22]. The two rivers are located in a karst landscape, the Loue River in the east part of France and the Tarn River in the southern part. During summer, these shallow rivers display a succession of calm and rapid areas. To better know where the most favourable areas for development of benthic cyanobacteria are and to implement monitoring programmes with the aim to limit the sanitary risks linked to the proliferation of these microorganisms, different sampling strategies and tools have been tested during the last two years in the rivers. In line with the New Zealand guidelines [20], the use of transects and an underwater viewer were tested (Fig. 7.4). This method allows a quick estimation of the coverage of potentially toxic cyanobacteria attached to substrate. In the two French rivers, areas appearing most favourable for development of benthic cyanobacteria, with percentage coverage varying from 5% to 80%, were identified using the method.

The use of a spectrofluorimetric probe (BenthoTorch) was also tested in the two French rivers. The probe allows a direct estimation of the biomass of chlorophytes, diatoms, and cyanobacteria. Results are expressed in units (μg) of chlorophyll a per unit of surface (cm^2). For the Loue River, a good overall correlation was obtained between the biomasses estimated by using the probe and those obtained by a spectrometric measurement of total chlorophyll a. For total chlorophyll a, pigment extraction was done on samples collected from the rivers at the same places where the measurements were performed with the probe. However, the biomasses estimated by using the probe were underestimated compared with those obtained from spectrometric measurements of the chlorophyll a. For the Tarn River, no correlation was found between the values provided by the probe and those obtained by spectrometric measurements for chlorophyll a. The main explanation for these differences in the efficiency of the probe is that biofilms of the Loue River were less thick and more diverse (a mix of diatoms, cyanobacteria and sometimes chlorophytes) than those of the Tarn River, which were strongly dominated by one cyanobacterium of the *Phormidium* genus. This cyanobacterium is red pigmented and probably requires a different calibration of the probe to be correctly detected.

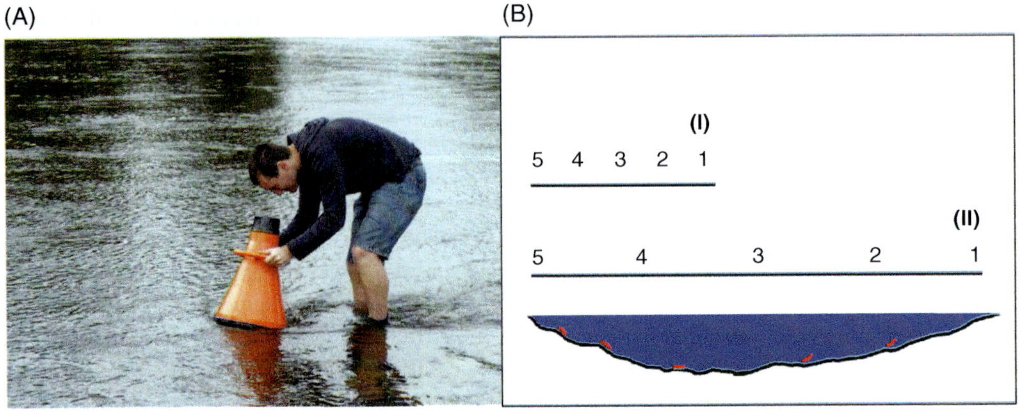

Figure 7.4 (A) Usage of an underwater viewer. If cyanobacterial mats are present, steps to protect the skin should be taken (see Chapter 5). (B) When monitoring cyanobacteria in rivers, transects are drawn to the standard maximum depth 0.6 m (I). In shallow rivers (<0.6 m), transects cover the whole river width (II). Five equidistant underwater views, starting from the deepest point, should be used when possible. Adapted from [20]

7.6 The Baltic Sea

The Baltic Sea is a brackish, semi-enclosed sea bordered by nine countries, with five more in the catchment area. The present Baltic Sea was formed through several freshwater and marine phases after the last Ice Age. Today it has limited inflow of sea water via the Kattegat. The Baltic Sea consists of several basins, 45–460 m deep. Salinity ranges from about 20‰ in the southern basins to almost freshwater in the eastern and northern areas. Major inflows of dense, saline water sink to the bottom and promote a stratified environment. In summer, the stratification is further stabilised by a thermocline. Vast areas of the bottom are oxygen-depleted.

The Baltic Sea is locally heavily eutrophicated and large blooms of cyanobacteria occur almost annually. Blooms in the open sea are usually dominated by *Nodularia spumigena*, *Aphanizomenon* and *Dolichospermum* (*Anabaena*). In the coastal areas *Microcystis*, *Planktothrix* and *Pseudanabaena* occur [23, 24]. Concentrations of <100–2400 μg g dry wt^{-1} of the hepatotoxin nodularin in *N. spumigena* blooms have been measured [25]. The highest reported concentration was 18,000 μg g dry wt^{-1} [26]. Microcystin-LR (2–4 μg g dry wt^{-1}) was also detected in the Baltic Sea [27].

Cyanobacteria have been noted in the Baltic Sea ever since phytoplankton recording started (>100 years ago), but the intensity of the blooms has probably increased during recent decades [23, 28]. The blooms are partly consequences of the high nutrient concentrations, which are due to internal phosphorus loading that continues to be a problem, despite a reduction in the external nutrient input [24]. Nitrogen fixation by the cyanobacteria further contributes to a vicious cycle of increasing nutrients [24]. In the open sea, cyanobacteria develop in June and by July–August they usually reach high densities. During 1993–2014, the biomass of *Aphanizomenon*, *Dolichospermum* (*Anabaena*) and *Nodularia* ranged between an average of 40 μg L^{-1} in the Bothnian Sea to a maximum of 7470 μg L^{-1} in the Gulf of Finland [29]. Phosphorus is an important bloom-regulating factor [24]. In the coastal areas, the blooms are variable and mainly related to local nutrient inputs, calm and warm weather [23, 24].

The Helsinki Commission (HELCOM) [30] was established in 1974 as an intergovernmental body consisting of the European Community and the nine countries bordering the Baltic Sea. The main objectives of HELCOM are to protect the Baltic Sea marine environment and to ensure a common environmental strategy among the participants, by for example common monitoring guidelines [31] and monitoring programmes (COMBINE). Monitoring is achieved in national programmes, by research vessels, satellites and automated sensors on commercial ships (ships-of-opportunity). In COMBINE, the Baltic Sea is divided into seven parts, each with regularly visited sampling stations (sampling dependent on environmental conditions) and high frequency sampling stations (sampled >12 times per year). The member states are responsible for monitoring their own catchments, territorial waters, and exclusive economic zones. The monitoring includes mandatory measurements of several hydrographical and chemical variables. Measurements relating to phytoplankton include chlorophyll a, primary production, species composition, abundance, and biovolume [31]. For chlorophyll a, sampling depths are standardised for the open sea (1, 5, 10, 15, and 20 m) and for coastal areas, 1 m, or an integrated sample from 1–10 m. One sample from the mixed surface layer is adequate during ship-of-opportunity and helicopter samplings. However, during cyanobacterial blooms chlorophyll a may be underestimated with ship-of-opportunity sampling, where 4 m is the usual sampling depth [32].

A hose sampler [33] is primarily recommended for the sampling. Phytoplankton is quantitatively determined in an integrated sample from 0–10 m, although in coastal areas 0–1 m may be sufficient. Alternatively, discrete samples from 0 to 1, 2.5, 5, 7.5, and 10 m may be pooled. It is also recommended to include a net sample of phytoplankton (0–20 m) for species identification. Preservation is done with Lugol's solution, quantification by the Utermöhl method, and the biovolumes are calculated according to Olenina et al. [34]. An alternative option to the time-demanding Utermöhl method is a semiquantitative method, where species are ranked into five classes based on their observed abundances. The monitoring results are published online by several HELCOM member states. An assessment system and quality assurance programme aim to ensure that the monitoring methods and results are comparable and reliable.

7.7 Waterbodies Used for Drinking Water Production

Groundwater and surface waters are main sources of drinking water, of which good quality and safety are essential. Eutrophication of surface waters (lakes/reservoirs) limits their use as drinking water sources due to the presence of potentially toxic cyanobacteria blooms. Water intake points in rivers downstream from eutrophic reservoirs are also exposed to the risk of cyanobacteria in the raw water. The provisional guideline value of microcystin-LR, one of the most commonly occurring cyanotoxins, in drinking water recommended by the World Health Organization (WHO) is 1 μg L^{-1} [35]. In raw water the concentration of microcystins can exceed 10 μg L^{-1} and in the scum of a bloom the concentration can be up to 50 μg L^{-1}.

Cyanobacteria and cyanotoxins can enter a water treatment plant in two forms, as intracellular toxins (bound within the cells of cyanobacteria) or as extracellular (dissolved) toxins. Some treatments (e.g. preoxidation, chemical coagulation) can break down cyanobacterial cells and release toxins into the water [36]. Conventional treatment processes can be ineffective for removing extracellular toxins [37, 38]. The effectiveness of water treatment processes in removal of cyanobacteria and cyanotoxins depends on a proper water intake system and management strategy in the treatment process. These should include: (i), permanent controls and changes of draw-off levels and/or place; (ii), chemical treatment such as powdered activated carbon, or oxidant dosing such as ozone during the treatment process; and (iii), changes in treatment technology (e.g. application of granular activated carbon beds, ozonation, or both). Additionally, control of raw water (cyanobacteria counting, chlorophyll a concentration and cyanotoxin analyses if possible) and optimisation of selected stages of the water treatment process (after coagulation, filtration and treated water) are necessary. Moreover, a proper reservoir management strategy, including measures against eutrophication, is important to minimize the risk of mass development of cyanobacteria in the raw water.

7.7.1 The Vertically Stratified Lake Sapanca, Turkey

Lake Sapanca is an oligomesotrophic lake with a maximum depth of 55 m. It supplies water for drinking water treatment for a nearby city with a population of 560,000. The lake is known for clear water, the euphotic depth can be 18 m in summer. It has a monomictic character and is stratified between May and November. These features enable the cyanobacterium *Planktothrix rubescens* to grow in the metalimnion of the lake [3]. Since its first record in 1997, *P. rubescens* has consistently formed annual blooms between 10 and 20 m during the stratification period of the lake. The intake level for drinking water coincides with exactly the same layer. This shows that it is essential to adjust the sampling strategy according to the cyanobacterial species when monitoring a lake. If the lake is deep enough to stratify and has low nutrient concentrations for cyanobacterial growth in the epilimnion, it is necessary to take samples over the whole depth to fully characterise the distribution of the species. If the thermal stratification pattern is known before the sampling, it can be used to identify representative depths for cyanobacterial detection. Moreover, management of water abstraction is very effective for minimising the risk of cyanobacteria presence in the raw water by choosing the appropriate depth for the withdrawal. For example, the vertical distribution of *P. rubescens* can be monitored using fluorometric tools to select the optimum position for taking the raw water.

7.7.2 Wind-Sheltered and Sensitive Lake Borgsjön, Finland

Since 1987, Lake Borgsjön has been a water supply for about 120 households in the parish of Sund on the Åland Islands (southwest Finland). The maximum depth of the lake is 14 m and the area is 18 ha. Its theoretical retention time is long, about 12 years, making the lake sensitive to changes. Borgsjön is sheltered by cliffs, steep hills, and forests. Some forest areas were clear cut in the 1980s, and other parts in the following decades. As a result, there have been signs of eutrophication. The Secchi depth of Borgsjön was formerly about 5–7 m, but in recent decades it has decreased to 2–4 m. Due to strong stratification, bottom anoxia develops in the summers. The use of the lake as a water supply increases the water level fluctuations, which affect the lake shores. The phytoplankton of the lake is usually dominated by cyanobacteria (especially different *Dolichospermum*/*Anabaena* spp.), with chrysophytes, cryptophytes, and dinoflagellates.

In autumn 2013, after problems in the local water works, phytoplankton and zooplankton occurred in the distributed water. Chrysophytes (*Uroglena* sp. and *Dinobryon* sp.) were abundant both in the lake and in the tap water, together with a *Dolichospermum* species (Fig. 7.5). Concentration of toxins (type of toxin not reported) in the tap water were below the detection limit (press release from the local environment and health authority 15.10.2013), and after 25 days the tap water could be used again. The *Dolichospermum* species in the lake probably live on the bottom, entering the free water during periods of low light or due to the autumn circulation. This is important to consider when selecting the depth for withdrawal of drinking water and during monitoring of the lake.

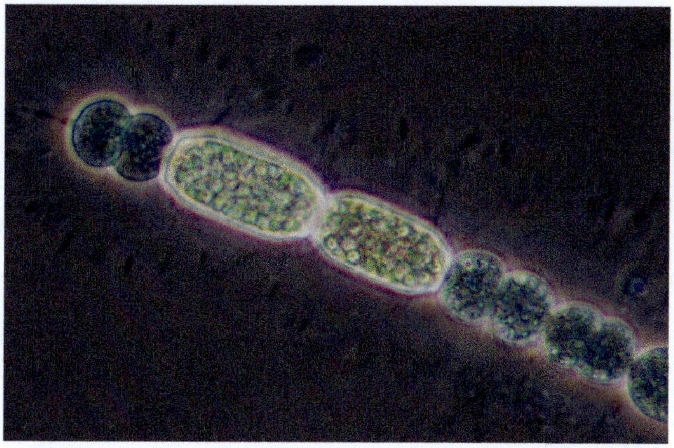

Figure 7.5 Dolichospermum *sp. from Lake Borgsjön, Åland Islands (SW Finland), 18 October 2013. The filament, with two akinetes, is surrounded by mucilage rich in bacteria (600× magnification). Photo: Tore Lindholm*

7.7.3 Management of Drinking Water in Sulejow, Zegrzynski, and Dobromierz Reservoirs and Pilica River, Poland

Sulejow Reservoir was for many years an important freshwater resource for the city of Lodz (characteristics of the reservoir and problems with the water quality were described in section 7.4.1). In 2003, to increase drinking water quality, the management strategy of the Sulejow-Lodz waterworks company included mixing of surface water and groundwater from seven deep wells located near the Sulejow Reservoir. This significantly reduced the problem of cyanobacteria and cyanotoxins (microcystins) in the water treatment process. Currently, only groundwater is used but Sulejow Reservoir remains an alternative source of drinking water for Lodz.

The Tomaszow-Lodz waterworks company supplies drinking water to the city of Tomaszow Mazowiecki from the Pilica River. The water intake point is located only a few kilometres downstream from the Sulejow Reservoir. Due to the eutrophication problems in Sulejow Reservoir, cyanobacteria were identified at the water intake point in the Pilica River. Microcystin concentrations were effectively reduced in the water treatment process by seasonal application of powdered activated carbon during the coagulation process.

Zegrzynski and Dobromierz reservoirs are drinking water supplies for the cities of Warsaw and Swiebodzice. To reduce the eutrophication symptoms in the Zegrzynski Reservoir, the water retention time was decreased from 30 days to 10 days in 2004. In the case of Dobromierz Reservoir, three different levels of water intake are used to reduce cyanobacterial occurrence in the raw water.

These measures to reduce potential cyanotoxin concentrations in drinking water demonstrate the importance of management strategies based on the ecohydrology concept, that is, regulation of both hydrological and ecological processes for the reduction of eutrophication [39] to maintain good drinking water quality.

References

[1] Fastner, J., Neumann, U., Wirsing, B. *et al.* (1999) Microcystins (hepatotoxic heptapeptides) in German fresh waterbodies. *Environmental Toxicology* **14** (1), 13–22.

[2] Kokociński, M., Mankiewicz-Boczek, J., Jurczak, T. *et al.* (2013) *Aphanizomenon gracile* (Nostocales), a cylindrospermopsin-producing cyanobacterium in Polish lakes. *Environmental Science and Pollution Research* **20** (8), 5243–5264.

[3] Albay, M., Akcaalan, R., Tufekci, H. *et al.* (2003) Depth profiles of cyanobacterial hepatotoxins (microcystins) in three Turkish freshwater lakes. *Hydrobiologia* **505** (1–3), 89–95.

[4] Horecká, M., Nagyová, V. and Košťálová, J. (2004) Sinicové vodné kvety na rekreačných vodách Slovenska. [Water blooms in recreational waters of Slovak Republic]. *Czech Phycology* **4**, 175–182.

[5] Salmaso, N. (2010) Long-term phytoplankton community changes in a deep subalpine lake: responses to nutrient availability and climatic fluctuations. *Freshwater Biology* **55** (4), 825–846.

[6] Walsby, A. and Schanz, F. (2002) Light-dependent growth rate determines changes in the population of *Planktothrix rubescens* over the annual cycle in Lake Zürich, Switzerland. *New Phytologist* **154** (3), 671–687.

[7] Walsby A.E., Schanz, F. and Schmid, M. (2006) The Burgundy-blood phenomenon: a model of buoyancy change explains autumnal waterblooms by *Planktothrix rubescens* in Lake Zürich. *New Phytologist* **169** (1), 109–122.

[8] Lampert, W., Sommer, U. and Haney, J.F. (1997) *Limnoecology: the Ecology of Lakes and Streams*, Oxford University Press, New York.

[9] Leboulanger, C., Dorigo, U., Jacquet, S. *et al.* (2002) Application of a submersible spectrofluorometer for rapid monitoring of freshwater cyanobacterial blooms: a case study. *Aquatic Microbial Ecology* **30** (1), 83–89.

[10] Salmaso, N. (2000) Factors affecting the seasonality and distribution of cyanobacteria and chlorophytes: a case study from the large lakes south of the Alps, with special reference to Lake Garda. *Hydrobiologia* **438** (1–3), 43–63.

[11] Rast, W. and Straškraba, M. (2000) Similarities, differences and importance. *Lakes and Reservoirs* **1**, 1–24.

[12] Tarczynska, M., Romanowska-Duda, Z., Jurczak, T. and Zalewski, M. (2001) Toxic cyanobacterial blooms in a drinking water reservoir - causes, consequences and management strategy. *Water Science and Technology: Water Supply* **1** (2), 237–246.

[13] Mankiewicz-Boczek, J., Urbaniak, M., Romanowska-Duda, Z. and Izydorczyk, K. (2006) Toxic cyanobacteria strains in lowland dam reservoir (Sulejow Res., Central Poland): amplification of *mcy* genes for detection and identification. *Polish Journal of Ecology* **54** (2), 171–180.

[14] Izydorczyk, K., Jurczak, T., Wojtal-Frankiewicz, A. *et al.* (2008) Influence of abiotic and biotic factors on microcystin content in *Microcystis aeruginosa* cells in a eutrophic temperate reservoir. *Journal of Plankton Research* **30** (4), 393–400.

[15] Gągała, I., Izydorczyk, K., Jurczak, T. *et al.* (2014) Role of environmental factors and toxic genotypes in the regulation of microcystins-producing cyanobacterial blooms. *Microbial Ecology* **67** (2), 465–479.

[16] Carrasco, D., Moreno, E., Sanchis, D. *et al.* (2006) Cyanobacterial abundance and microcystin occurrence in Mediterranean water reservoirs in Central Spain: microcystins in the Madrid area. *European Journal of Phycology* **41** (3), 281–291.

[17] Sanchis, D., Carrasco, D., Padilla, C. *et al.* (2002) Spatial and temporal heterogeneity in succession of cyanobacterial blooms in a Spanish reservoir. *International Journal of Limnology* **38** (3), 173–183.

[18] Wörmer, L., Cirés, S. and Quesada A. (2011) Importance of natural sedimentation in the fate of microcystins. *Chemosphere* **82** (8), 1141–1146.

[19] Cirés, S., Wörmer, L., Carrasco, D. and Quesada, A. (2013) Sedimentation patterns of toxin-producing *Microcystis* morphospecies in freshwater reservoirs. *Toxins* **5** (5), 939–957.

[20] Wood, S.A., Hamilton, D.P., Safi, K.A. and Williamson, W.M. (2009) New Zealand Guidelines for Cyanobacteria in Recreational Fresh Waters – Interim Guidelines. *Ministry for the Environment and Ministry of Health, Wellington.*

[21] Gugger, M., Lenoir, S., Berger, C. *et al.* (2005) First report in a river in France of the benthic cyanobacterium *Phormidium favosum* producing anatoxin-a associated with dog neurotoxicosis. *Toxicon* **45** (7), 919–928.

[22] Cadel-Six, S., Peyraud-Thomas, C., Brient, L. *et al.* (2007) Different genotypes of anatoxin-producing cyanobacteria coexist in the Tarn River, France. *Applied and Environmental Microbiology* **73** (23), 7605–7614.

[23] Finni, T., Kononen, K., Olsonen, R. and Wallström, K. (2001) The history of cyanobacterial blooms in the Baltic Sea. *Ambio* **30** (4), 172–178.

[24] Vahtera, E., Conley, D.J., Gustafsson, B.G. *et al.* (2007) Internal ecosystem feedbacks enhance nitrogen-fixing cyanobacteria blooms and complicate management in the Baltic Sea. *Ambio* **36** (2), 186–194.

[25] Sivonen, K., Kononen, K., Carmichael, W.W. *et al.* (1989) Occurrence of the hepatotoxic cyanobacterium *Nodularia spumigena* in the Baltic Sea and structure of the toxin. *Applied and Environmental Microbiology* **55** (8), 1990–1995.

[26] Kononen, K., Sivonen, K. and Lehtimäki, J. (1993) Toxicity of phytoplankton blooms in the Gulf of Finland and Gulf of Bothnia, Baltic Sea, in *Toxic Phytoplankton Blooms in the Sea*, 1st edn (eds T.J. Smayda and Y. Shimizu), Elsevier Scientific Publications, Amsterdam, pp. 269–274.

[27] Karlsson, K.M., Kankaanpää, H., Huttunen, M. and Meriluoto, J. (2005) First observation of microcystin-LR in pelagic cyanobacterial blooms in the northern Baltic Sea. *Harmful Algae* **4** (1), 163–166.

[28] Suikkanen, S., Pulina, S., Engström-Öst, J. *et al.* (2013) Climate change and eutrophication induced shifts in northern summer plankton communities. *PLOS One* **8** (6), 1–10.

[29] Wasmund, N., Busch, S., Gromisz, S. *et al.* (2014) *Cyanobacteria Biomass, HELCOM Baltic Sea Environment Fact Sheets*, http://www.helcom.fi/baltic-sea-trends/environment-fact-sheets/eutrophication/cyanobacteria-biomass/ (accessed 5 September 2016).

[30] HELCOM (2015) *Baltic Marine Environment Protection Commission,* http://www.helcom.fi (accessed 6 March 2015).

[31] HELCOM (2014) *Manual for Monitoring in the COMBINE Programme of HELCOM*, http://helcom.fi/helcom-at-work/publications/manuals-and-guidelines (accessed 6 March 2015).

[32] Majaneva, M., Autio, R., Huttunen, M. *et al.* (2009) Phytoplankton monitoring: the effect of sampling methods used during different stratification and bloom conditions in the Baltic Sea. *Boreal Environment Research* **14**, 313–322.

[33] Lindahl, O. (1986) A Dividable Hose for Phytoplankton Sampling. *Report of the ICES Working Group on Exceptional Algal Blooms. CM Documents – ICES Annex III.*

[34] Olenina, I., Hajdu, S., Edler, L. *et al.* (2006) Biovolumes and size-classes of phytoplankton in the Baltic Sea. *HELCOM Baltic Sea Environment Proceedings* **106**, 1–144.

[35] WHO (1996) Health Criteria and Other Supporting Information. *WHO Guidelines for Drinking-Water Quality, Volume 2, Second Edition.*

[36] Jurczak, T., Tarczynska, M., Izydorczyk, K. *et al.* (2005) Elimination of microcystins by water treatment processes – examples from Sulejow Reservoir, Poland. *Water Research* **39** (11), 2394–2406.

[37] Hitzfeld, B.C., Hoger, S.J. and Dietrich. D.R. (2000) Cyanobacterial toxins: removal during drinking water treatment, and human risk assessment. *Environmental Health Perspectives* **108** (Suppl 1), 113–122.

[38] Hoeger, S.J., Shaw, G., Hitzfeld, B.C. and Dietrich, D.R. (2004) Occurrence and elimination of cyanobacterial toxins in two Australian drinking water treatment plants. *Toxicon* **43** (6), 639–649.

[39] Zalewski, M. (2000) Ecohydrology – the scientific background to use ecosystem properties as management tools toward sustainability of water resources. *Ecological Engineering* **16** (1), 1–8.

8

New Tools for the Monitoring of Cyanobacteria in Freshwater Ecosystems

Jean-François Humbert[1] *and Andrea Törökné*[2]

[1]*Institute of Ecology and Environmental Sciences, UPMC, Paris, France*
[2]*Knot Ltd. Budapest, Hungary*

8.1 Introduction

A significant improvement to the traditional monitoring programmes for cyanobacteria based on discrete sampling of water would be the inclusion of continuous/semicontinuous and real-time analysis of chlorophyll-*a* and of other cyanobacterial pigments and of some physicochemical parameters in raw surface waters. Indeed, knowing that the spatial and temporal distribution of planktonic cyanobacteria in freshwater ecosystems can display short time-scale variations, such systems will permit significant improvements to the monitoring efficiency. These data are also very useful for the identification of the main factors and processes controlling the population dynamics and the spatial distribution of cyanobacteria. In this global framework, numerous fluorimetric tools have been developed over the past 30 years, for the detection and quantification of cyanobacteria and/or other photosynthetic microorganisms in freshwater ecosystems and in raw water in water treatment plants. Other technologies including holographic microscopy combined with fluorescence or flow cytometry are also being developed for the automatic detection and quantification of cyanobacteria in water. Some of these tools are just beginning to be integrated into buoys in combination with physicochemical probes, with the goal to implement real-time and continuous/semicontinuous monitoring systems for waterbodies.

Handbook of Cyanobacterial Monitoring and Cyanotoxin Analysis, First Edition.
Edited by Jussi Meriluoto, Lisa Spoof and Geoffrey A. Codd.

8.2 Use of Photosynthetic Pigments for the In Situ Quantification of Cyanobacteria and Other Phytoplankton in Water

In vivo fluoroscopy of intracellular photosynthetic pigments has permitted the development of several tools for the detection of microalgae and cyanobacteria in water. The first of them targeted only chlorophyll-*a*, which is present in all oxygenic photosynthetic microorganisms. More recently, other tools using a multispectral approach have been developed for the specific detection of cyanobacteria and for the quantification of the main phytoplankton groups. All of these tools require only basic training and call for much less time and skills than are required for microscopic identification and counting.

Chlorophyll-*a* and phycocyanin probes are sensors emitting light at only one wavelength. The former have been used for some time in marine and freshwater ecosystems for the monitoring of phytoplankton communities, and there are many articles showing that biomasses of phytoplankton communities can be estimated by using in vivo chlorophyll analyses (e.g. [1, 2]). The second type of probes targets phycocyanin, an accessory pigment shared by all cyanobacteria. These tools have been developed for the monitoring of these microorganisms in freshwater ecosystems and in water treatment plants. There have been many papers in recent years [e.g. 3–5] that show that in vivo fluorescence measured by the probes is significantly correlated with the biomass of cyanobacteria. Several dual-channel fluorescence sensors (chlorophyll/phycocyanin) also allow researchers to distinguish cyanobacterial blooms from other microalgal blooms.

More recently, the development of a spectrofluorimetric probe allows discrimination between the main phytoplanktonic groups (i.e. diatoms and dinoflagellates, cyanobacteria, green algae, cryptophytes) in mixed populations on the basis of the relative fluorescence intensity of chlorophyll-*a* at 680 nm (following sequential light excitation at different wavelengths). As for the other tools, many articles have shown that the spectrofluorimetric probe is very useful for the estimation of phytoplankton biomass in freshwater ecosystems and of the proportions of the main phytoplanktonic groups, including cyanobacteria (e.g. [6–8]).

Whilst the widespread efficiency of all these tools for the monitoring of cyanobacteria has been demonstrated, it is still critical to keep in mind that the data provided can be strongly impacted by various factors and processes that lead generally to an underestimation of the values. Among them, it has been shown that changes in the phycocyanin and chlorophyll-*a* contents of cells at different phases of their growth cycle affect fluorescent probe measurements [9]. Another problem with these sensors is due to the photochemical quenching of the cells when they are exposed to high irradiances. This photochemical quenching can induce an attenuation of the fluorescence signal (up to 40%) and, consequently, an underestimation of the cyanobacterial biomass. Among the bloom-forming cyanobacteria, some species, for example the red-coloured *Planktothrix rubescens*, display a specific pigment composition that can lead to differences in fluorescence patterns [6]. Consequently, specific calibration of the probe is necessary when this kind of species is dominant in an ecosystem. The same limitation in the use of a spectrofluorimetric probe due to pigment composition has been found for the monitoring of benthic cyanobacteria [10]. The cellular organization of some cyanobacterial species into colonies or bundles of filaments and the thickness of biofilms can also lead to severe underestimations of the abundance/biomass of cyanobacteria, probably because only the cyanobacterial cells located at the surface of the colonies, bundles of filaments, or biofilms are quantified by the probe [10]. Finally, water turbidity and organic matter concentrations are factors that can strongly influence the results provided by the in vivo measurement of the fluorescence.

Figure 8.1 Monitoring buoy for the survey of cyanobacteria in the reservoir of Grangent (France)

8.3 Integration of Physicochemical and Fluorescence Sensors in Buoys

Buoys now allow for real-time and long-term monitoring of freshwater ecosystems. They can provide data on water quality and meteorological parameters. Meteorology sensors can supply data on, for example, humidity, air temperature, wind speed and direction, barometric pressure, and liquid precipitation. Below the water surface, depending on the sensors, a continuous monitoring of temperature, dissolved oxygen, pressure, photosynthetically active radiation, and chlorophyll-*a* can be performed. Some of these buoys are able to perform measurements at only one fixed position in the water column, whereas some others can do so throughout vertical profiles (Fig. 8.1) [11].

These buoys are autonomous, and their data are transmitted automatically via GSM cellphone, satellite, or radio. In some buoys, an antifoulant device allows to protect the sensors against biofouling.

8.4 New Methods for Automatic Cell Counting in Water Samples

Cell counting under the microscope allows the direct assessment of microorganisms present in a cyanobacterial bloom, but this approach is very laborious and time-consuming and requires advanced knowledge in taxonomy for species identification. To overcome these limitations, several alternative methods have been developed, in particular with the objective to automate this process.

A microbiological monitoring device, WaterScope, has been developed using Digital Holographic Microscopes (DHM) at the Computer and Automation Research Institute of the Hungarian

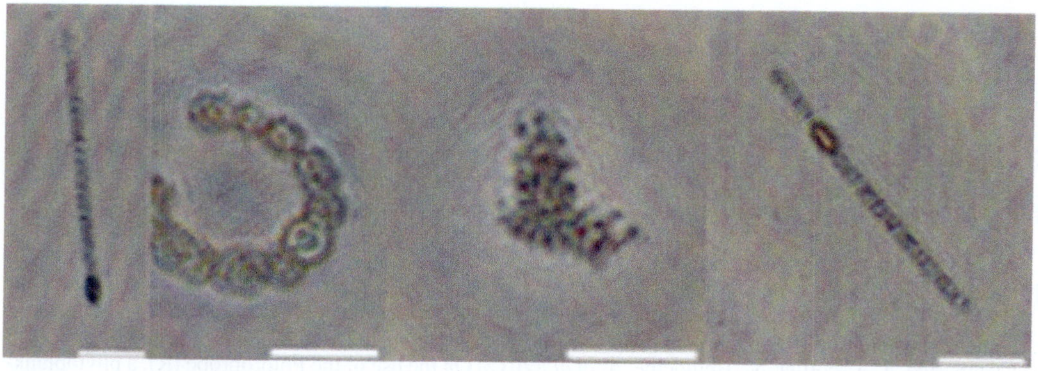

Figure 8.2 Cylindrospermopsis raciborskii, Dolichospermum (Anabaena) circinalis, Microcystis *sp., and* Aphanizomenon flos-aquae*: images made, recognized, and archived by WaterScope. White bars: 20 μm*

Academy of Sciences (MTA-SZTAKI) [12]. After an initial learning phase of the local microscopic flora and fauna, WaterScope can recognize the aquatic organisms by means of an advanced recognition system.

Reconstructing software is used to recall/render the images of the individual planes in different depths. Different species of organisms found in any depth can be identified and classified with the pattern recognition software (Fig. 8.2). The number and size of the recognised organisms can be determined and, thus, the biomass can be estimated. DHM may scan a volume of 1 mm^3 with a resolution of about 0.7 μm.

The role of WaterScope is to provide an early warning of when cyanobacteria develop in lakes, ponds, and reservoirs. The system includes a flow-through chamber giving real-time data on water quality. With this equipment, it may be possible to detect cyanobacteria as soon as they have started to spread in a waterbody. The prototype of WaterScope is under testing by the laboratory of the Trans Danubian Regional Water Works, where it was built in the source water pipe of Lake Balaton.

Similarly, it has been shown that flow cytometry can be a successful approach for phytoplankton monitoring [13]. Indeed, scanning flow cytometry allows researchers to monitor individual particles on the basis of various physiomorphological features such as size, cellular organization (colonies, filaments), pigment type and content, and pigment distribution within cells or colonies. For example, Pomati et al. [14] have obtained a similar description of the phytoplankton spring bloom in an alpine lake by flow cytometry and microscopy. Depending on the study, the richness of the phytoplankton communities estimated by flow cytometry and that estimated by microscopy were similar [14]. But, as highlighted by Becker et al. [15], if flow cytometry can be used to measure phytoplankton composition on the basis of biovolume and chlorophyll-*a* content, the application of this method needs careful calibration with microscopy and/or high-performance liquid chromatography pigment analysis.

References

[1] Berman, T. (1972) Profiles of chlorophyll concentration by the in vivo fluorescence: some limnological applications. *Limnology and Oceanography* **17**, 616–618.

[2] Powell, T.M., Cloern, J.E. and Huzzey, L.M. (1989) Spatial and temporal variability in South San Francisco Bay (USA). I. Horizontal distributions of salinity, suspended sediments, and phytoplankton biomass and productivity. *Estuarine, Coastal and Shelf Science* **28**, 583–597.

[3] Brient, L., Lengronne, M., Bertrand, E. *et al.* (2008) A phycocyanin probe as tool for monitoring cyanobacteria in freshwaterbodies. *Journal of Environmental Monitoring* **10**, 248–255.

[4] McQuaid, N., Zamyadi, A., Prevost, M. *et al.* (2011) Use of in vivo phycocyanin fluorescence to monitor potential microcystin-producing cyanobacterial biovolume in a drinking water source. *Journal of Environmental Monitoring* **13**, 455–463.

[5] Bastien, C., Cardin, R., Veilleux, E. *et al.* (2011) Performance evaluation of phycocyanin probes for the monitoring of cyanobacteria. *Journal of Environmental Monitoring* **13**, 101–118.

[6] Leboulanger, C., Dorigo, U., Jacquet, S. *et al.* (2002) Use of a submersible spectrofluorometer (FluoroProbe) for the survey of a toxic cyanobacteria, *Planktothrix rubescens*, in a large alpine lake. *Aquatic Microbial Ecology.* **30**, 83–89.

[7] Rolland, A., Rimet, F. and Jacquet, S. (2010) A 2-year survey of phytoplankton in the Marne reservoir (France): A case study to validate the use of an in situ spectrofluorometer by comparison with algal taxonomy and chlorophyll a measurements. *Knowledge and Management of Aquatic Ecosystems* **398**, 02.

[8] Catherine, A., Escoffier, N., Belhocine, A. *et al.* (2012) On the use of the Fluoroprobe (R), a phytoplankton quantification method based on fluorescence excitation spectra for large-scale surveys of lakes and reservoirs. *Water Research* **46**, 1771–1784.

[9] Escoffier, N., Bernard, C., Hamlaoui, S. *et al.* (2015) Quantifying phytoplankton communities using spectral fluorescence: the effects of species composition and physiological state. *Journal of Plankton Research* **37**, 233–247.

[10] Echenique, I., Dalle, C., Duval C. *et al.* (in revision) Application of a spectrofluorimetric tool (BBE BenthoTorch) for the monitoring of potentially toxic benthic cyanobacteria in rivers. In revision for *Environmental Science and Technology.*

[11] Le Vu, B., Vinçon-Leite, B., Lemaire, B.J. *et al.* (2011) High-frequency monitoring of phytoplankton dynamics within the European water framework directive: application to metalimnetic cyanobacteria. *Biogeochemistry.* **106**, 229–242.

[12] Orzó, L., Wittner, B. and Tőkés, S (2013) High speed water monitoring systems based on digital holographic microscopy. In *Computer Science and Information Technologies* (CSIT), pp 1–9.

[13] Rutten T.P.A., Sandee, B. and Hofman, A.R.T. (2005) Phytoplankton monitoring by high performance flow cytometry: A successful approach? *Cytometry Part A* **64A**, 16–26.

[14] Pomati, F., Kraft, N.J.B., Posch, T. *et al.* (2013) Individual cell based traits obtained by scanning flow-cytometry show selection by biotic and abiotic environmental factors during a phytoplankton spring bloom. *PLoS One* **8**, e71677.

[15] Becker, A., Meister, A. and Wilhelm, C. (2002) Flow cytometry discrimination of various phycobilin-containing phytoplankton groups in a hypertrophic reservoir. *Cytometry* **48**, 45–47.

9

Remote Sensing of Cyanobacterial Blooms in Inland, Coastal, and Ocean Waters

Peter D. Hunter[1], Mark W. Matthews[2], Tiit Kutser[3], and Andrew N. Tyler[1]

[1]*Biological and Environmental Sciences, School of Natural Sciences, University of Stirling, United Kingdom*

[2]*CyanoLakes (Pty) Ltd., Cape Town, South Africa*

[3]*Estonian Marine Institute (LOMI), University of Tartu, Tallinn, Estonia*

9.1 Introduction

Phytoplankton are one of the main colour-forming components in inland, coastal, and open-ocean waters. The absorption and scattering of light by phytoplankton cells, filaments, or colonies strongly influence water-leaving reflectance signals (i.e. the light exiting the water column) measured by remote sensing instruments whether deployed *in situ* or on aircraft or satellites. The interpretation of these water-leaving reflectance signals can thus be used to obtain information on the presence, abundance, and spatial distribution of phytoplankton blooms within sunlit near-surface waters (Fig. 9.1). Satellites in sun-synchronous orbits have the ability to capture such information over the entire globe and repeat this every few days. These observations are now used for monitoring the open ocean, shelf seas, and coastal zones on an operational basis, thereby providing near real-time information on phytoplankton dynamics and early warnings of harmful algal and cyanobacterial bloom (HAB) developments. The use of satellite observations for monitoring phytoplankton blooms in inland waters has developed in tandem with ocean colour remote sensing. However, their smaller size, greater complexity in optical properties, and the increased continentality of the overlying atmosphere pose additional challenges [1]. Nevertheless, significant progress has been made on the development of methods for the retrieval of information on phytoplankton biomass and the detection and monitoring of cyanobacterial blooms in lakes, reservoirs, and large rivers. Moreover, the

Handbook of Cyanobacterial Monitoring and Cyanotoxin Analysis, First Edition.
Edited by Jussi Meriluoto, Lisa Spoof and Geoffrey A. Codd.

Figure 9.1 *Extensive bloom of* Nodularia *in the Baltic Sea observed from the European Space Agency's Envisat MERIS sensor on 13 July 2005. Photograph: ESA*

imminent launch of new sensors and data products with improved spatial (pixel size), spectral (number of wavelength bands), and radiometric (signal-to-noise ratio) resolution has the potential to result in a step-change in our capacity to operationally monitor cyanobacteria blooms from space.

9.2 Bio-optical Properties of Marine and Inland Waters

The subsurface light field of natural waters is largely determined by a small number of colour-forming components including coloured dissolved organic matter (CDOM), mineral particles, and other nonliving organic detritus (nonalgal particles [NAP]) and phytoplankton. In the open ocean, the water-leaving reflectance signal is largely determined by the abundance, composition, and physiology of phytoplankton and related breakdown products—although in very clear waters, the signal may also be influenced by bubbles, viruses, and bacteria. This makes it relatively straightforward to map the abundance of phytoplankton in these waters using satellite remote sensing. By contrast, the optical properties of shelf-seas, coastal zones, and, in particular, inland waters are much more complex being influenced by non-covarying concentrations of phytoplankton, NAP, and CDOM. Accordingly, the analysis and interpretation of water-leaving reflectance signals over inland waters are more challenging and require different methodological approaches.

The magnitude and shape of the water-leaving reflectance signal are strongly influenced by wavelength-dependent absorption by phytoplankton pigments and scattering from cell structures. The main light-harvesting photosynthetic pigment in eukaryotic phytoplankton is chlorophyll a (Chl a). Chl a has *in vivo* absorption maxima at blue (ca. 440 nm) and red (ca. 675 nm) wavelengths, and these features can be observed as local, trough-like minima in the water-leaving reflectance signals measured by remote sensing. In addition to Chl a, all cyanobacteria (except prochlorophytes) also contain light-harvesting phycobilipigments in their photosynthetic machinery (mainly phycocyanin, phycoerythrin, and allophycocyanin). The phycobilipigments have very different absorption spectra to Chl a with single *in vivo* absorption maxima centred at approximately 560 nm (phycoerythrin), 620 nm (phycocyanin), and 650 nm (allophycocyanin). The phycobilipigments are found only in high concentrations in cyanobacteria (although they occur as minor pigments in some cryptophtes and rhodophytes), and as such they provide useful biomarkers for the presence and abundance of cyanobacteria in waterbodies. Similar to Chl a, in waters dominated by cyanobacteria the presence of phycobilipigments can be observed in water-leaving reflectance signals as local minima in proximity to the *in vivo* absorption maxima. These spectral features provide a route to the detection and quantification of cyanobacteria using remote sensing data (see Fig. 9.2).

9.3 Platforms and Sensors

Remote sensing data can be captured using sensors mounted on *in situ* platforms such as research vessels, ships of opportunity and fixed moorings, unmanned and manned aircraft, and Earth-observing satellites. Data from *in situ* sensors are most widely used for the development of algorithms or validation of airborne or satellite observations, although high-frequency data obtained from instrumented moorings or ships have significant potential for the early detection and tracking of cyanobacterial blooms (e.g. [2]). However, the real benefit of remote sensing for cyanobacterial bloom detection and monitoring comes from the ability to obtain data routinely over local, regional, and global scales using observations from aircraft and satellites. Data captured by aircraft have been used previously to support method development and for proof-of-concept studies [3–5], but the high acquisition costs means that such data are too expensive for routine monitoring. The recent emergence of unmanned aerial vehicles as a flexible, low-cost option for environmental monitoring now offers an alternative to the use of larger piloted aircraft [6], but such technology poses other challenges and is currently still some way from being used operationally for bloom monitoring. Thus satellites currently offer the greatest potential for operational monitoring of cyanobacteria blooms.

The potential of satellite data for water quality monitoring was first recognised during the 1970s following the launch of the inaugural Landsat-1 satellite, with several early studies demonstrating the potential of remote sensing for mapping the turbidity of lakes and reservoirs (e.g. [7, 8]). Latterly, far more advanced satellite sensors have been launched with improved spectral resolution

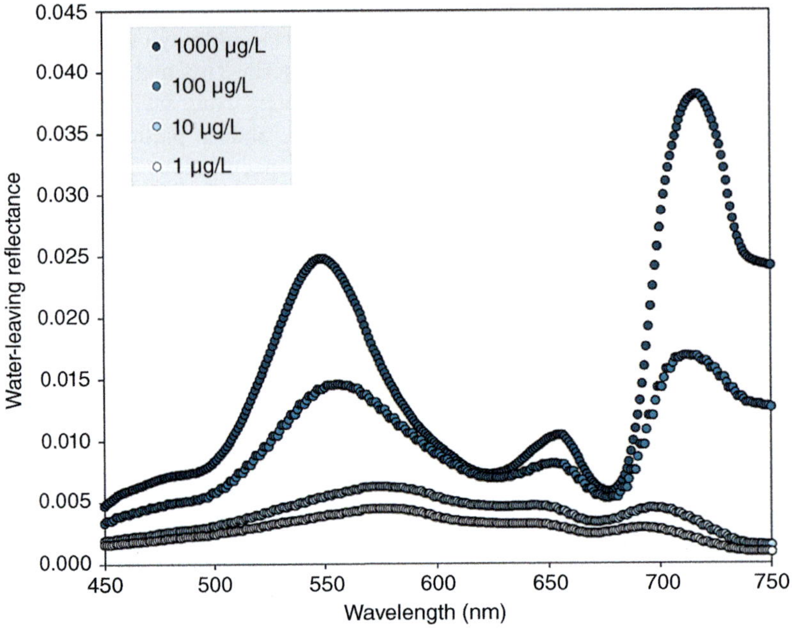

Figure 9.2 Water-leaving reflectance spectra measured from lakes with phycocyanin concentrations varying over 4 orders of magnitude (1–1000 µg L^{-1}). The measurements from lakes with high concentrations show local reflectance minima related to absorption by Chl a (~675 nm) and phycocyanin (620 nm) as well as prominent peaks (between 690 and 720 nm) caused light scattering by cyanobacterial cells. Data obtained from the LIMNADES database (http://www.globolakes.ac.uk/limnades)

(measuring more wavelengths), radiometric sensitivity (higher dynamic range and calibration improvements) and revisit (sampling) frequencies (Table 9.1). This has allowed information on water quality to be obtained globally at spatial resolutions of between 30 m and 1 km, with temporal frequencies between 3 and 16 days (cloud cover permitting). In particular, the dedicated ocean colour remote sensing missions such as CZCS (Coastal Zone Color Scanner), SeaWiFS (Sea-Viewing Wide Field-of-View Sensor), MODIS (Moderate Resolution Imaging Spectroradiometer), and MERIS (Medium Resolution Imaging Spectrometer) have been fundamental to the scientific advances made in the detection and tracking of phytoplankton blooms, including cyanobacteria, from space.

9.4 Overview of Approaches

Many different approaches have been used to detect and/or quantify cyanobacteria using remote sensing data. The methods currently available vary in their specificity and accuracy; some are based only on the recognition of high Chl a concentrations during the bloom season, while other approaches exploit the distinctive optical signatures of cyanobacterial blooms to distinguish them from other bloom-forming taxa and provide more reliable indications of presence and abundance. Kutser [9] developed an algorithm (i.e. numerical model relating the water-leaving reflectance

Table 9.1 *Satellite missions sensors (past and present) with potential for the detection and quantification of cyanoabacterial blooms in inland, coastal, and ocean waters*

Satellite	Sensor	Organisation	Resolution				Swath width (km)	Operational lifetime	Chlorophyll retrieval	Phycocyanin retrieval
			Spectral (nm) [bands]	Temporal (days)	Radiometric	Spatial (m)				
CZCS	Nimbus-7	NASA	433–12,500 [6]	2	8-bit	825	1556	1978–1986	Proven	No
SeaWiFS	WiFS	NASA	402–885 [8]	2	10-bit	1000	2801	1997–2010	Proven	No
Terra/Aqua	MODIS	NASA	405–14,385 [36]	2	12-bit	250–500	2330	1999–present	Proven	No
Envisat	MERIS	ESA	390–1040 [15]	3	16-bit	300	1150	2002–2012	Proven	Proven
ISS	HICO	NASA	300–1000 [87]	n/a	14-bit	100	45	2003–2015	Proven	Feasible
NPP and JPSS	VIIRS	NASA	402–11,800 [22]	1	12-bit	370/740	3000	2011–present	Proven	No
IRS-P4	OCM-2	ISRO	400–900 [8]	2	12-bit	360	1420	2009–present	Proven	Feasible
Landsat 5-7	MSS/TM/ETM+	USGS	435–2294 [9]	16	8-bit	30	185	1985–present	Feasible	No
Landsat-8	OLI	USGS	430–12,510 [11]	16	12-bit	30	185	2013–present	Feasible	No
EO-1	Hyperion	NASA	400–2500 [220]	16	12-bit	30	7.5	2000–present	Feasible	Feasible
EO-1	ALI	NASA	430–2350 [9]	16	11-bit	30	37	2000–present	Feasible	No
Sentinel-2a/b	MSI	ESA	443–2190 [12]	5	12-bit	10–60	290	Expected 2015	Feasible	No
Sentinel-3a/b	OLCI	ESA	400–1020 [21]	1	16-bit	300	1270	Expected 2015	Feasible	Feasible

The capability of each sensor for the specific detection and retrieval of Chl a and phycocyanin data is indicated as feasible (theoretically possible, in some cases supported by published proof-of-concept studies) or proven (established and supported by at least several published studies).

signal to phytoplankton pigment concentrations) to estimate the concentration of Chl a using data from NASA's ALI and Hyperion satellites and applied this to map the distribution of cyanobacterial blooms in the Gulf of Finland. Similarly, Alikas et al. [10] used the Maximum Chlorophyll Index algorithm to map cyanobacterial blooms in three large European lakes based on the recognition of high Chl a concentrations. This same type of approach has also been used to map blooms of *Trichodesmium* spp. in ocean waters [11, 12]. These approaches benefit from the fact it is possible to retrieve estimates of Chl a from a range of different satellite sensors with varying spatial and temporal resolutions. However, the reliability of such approaches for quantitative mapping of cyanobacterial blooms is clearly dependent on the attribution of the high Chl a concentrations to the presence of cyanobacteria. This is very much specific to the environment under consideration, and thus the applicability of such approaches is limited in situations where other taxa regularly form blooms or the phytoplankton community is not dominated by cyanobacteria.

Mass aggregations of buoyant cyanobacteria in surface waters can be identified through their distinct optical signatures, very bright near-infrared reflectance, and characteristically patchy distribution in satellite images. These features can be readily recognised in satellite images and, while quantitative estimates of abundance might be difficult to obtain, approaches aimed at identifying these features can provide a route to the detection and tracking of blooms over time. Kahru and Elmgren [13] developed an algorithm for the detection of near-surface accumulations and scums of cyanobacteria that could be readily adapted to data from sensors with different spatial and spectral resolutions. The algorithm was successfully applied to data from multiple satellite sensors to produce a 35-year time-series of cyanobacterial bloom dynamics in the Baltic Sea. Matthews et al. [14] presented an approach (expanded in [15]) for the optical detection of four cases: (i), eukaryotic phytoplankton blooms; (ii), cyanobacteria-dominant blooms; (iii), floating cyanobacterial scum; and (iv), floating vegetation. The identification of cyanobacteria-dominant blooms and scum was achieved used spectral features associated with phycocyanin absorption and a trough (in the 681-nm MERIS band) associated with the lack of discernible sun-induced Chl a fluorescence, which is distinctive for cyanobacterial blooms [16]. The quantification of cyanobacterial biomass in terms of Chl a was achieved using the maximum peak height algorithm by measuring the height of the peak associated with cyanobacterial backscattering near 710 nm. Therefore, biomass estimates for cyanobacterial blooms were obtained that were distinct from those for eukaryote blooms. The use of spectral flagging procedures for identifying cyanobacteria, as demonstrated in [14], is required a priori for obtaining biomass estimates from Chl a specific to cyanobacteria.

Methods based solely on the detection of high Chl a concentrations, or the presence of surface aggregations or scums, cannot provide a completely quantitative estimate of cyanobacterial abundance, particularly in waters with more mixed phytoplankton composition. However, more specific estimates of cyanobacteria abundance can be obtained through the use of algorithms for the detection and quantification of phycobilipigments. These approaches commonly target wavelengths in the water-leaving reflectance spectra that correspond to the *in vivo* absorption maxima of phycocyanin and phycoerythrin. Dekker [17] developed one of the first algorithms for the estimation of phycocyanin from remotely sensed data that used a baseline subtraction approach to estimate the depth of the reflectance trough near 620 nm related to phycocyanin absorption.

More recently, alternative algorithms have been developed for the estimation of phycocyanin, particularly for application to data from the European Space Agency's MERIS sensor. This includes algorithms developed using semiempirical methods [5, 18] and more analytical, physics-based approaches [19–21]. Most studies conducted thus far have developed and/or tested algorithms using *in situ* measurements of water-leaving reflectance but several have demonstrated their potential with airborne or satellite data. Riddick et al. [22] were able to estimate phycocyanin in Lake

Balaton using MERIS data to within ±12 mg m^{-3} (±54%) of the concentrations measured in directly in water samples using the algorithm developed previously [18]. Wheeler et al. [23] report comparable accuracies for the estimation of phycocyanin in Lake Champlain using MERIS data.

There are a number of factors that can impede the accurate estimation of phycocyanin by remote sensing. The presence of other secondary phytoplankton pigments that absorb strongly in the vicinity of the phycocyanin absorption maximum (e.g. chlorophyll b) can lead to errors in the estimation of phycocyanin [25]. The presence of surface aggregations and scums also affect the relationship between phycocyanin concentrations estimated from satellite data and the actual abundance of cyanobacteria due to the extremely shallow depth of signal penetration. Further, the use of phycocyanin as a quantitative indicator of cyanobacteria abundance is predicated on a strong relationship between pigment concentration and cyanobacterial cell numbers or biovolumne. However, the photosynthetic pigmentation of cyanobacteria is highly variable among different genera, and intracellular phycobilipigment quotas can vary markedly in response to factors such as photoacclimation or nutrient-related stress. These factors can weaken the quantitative relationship between phycocyanin concentrations and cyanobacterial abundance, leading some to suggest that Chl a might be a more robust biomass indicator in such situations [24].

Despite these issues, some studies have demonstrated strong correlations between phycocyanin concentrations estimated from remotely sensed data and cyanobacterial cell counts or biovolumes [5, 26, 27]. Further, some studies have also demonstrated strong correlation between remotely sensed estimates of phycocyanin with *in situ* measured cyanotoxin concentrations [5], although it is recognised such relationships are difficult to extrapolate in space or time. In situations where strong relationships exist between phycocyanin and cyanobacterial cell concentrations, it is possible to cautiously infer the possible risk to human health by relating remote sensing data to existing guidelines (e.g. World Health Organisation) regarding the health risks associated with cyanobacteria in waterbodies [28].

9.5 Case Study Examples

9.5.1 Mapping Cyanobacteria Blooms Using Airborne Remote Sensing

Hunter et al. (2010) used airborne remote sensing to map a cyanobacterial bloom in Esthwaite Water, a small, eutrophic shallow lake in the United Kingdom. The concentration of phycocyanin was mapped using data from a Specim Eagle airborne imaging spectrometer acquired on 26 April 2007 during a bloom of *Anabaena circinalis* (Fig. 9.3). The concentration of phycocyanin was estimated in this example using the algorithm developed by Simis et al. [18] and was validated against phycocyanin data obtained from the analysis of contemporaneous water samples in the laboratory by spectrophotometry after freeze-thaw extraction. In this study, the concentration of phycocyanin indicated by remote sensing was estimated to be within ±3.98 μg L^{-1} (±13.8%) of that measured *in situ,* with high surface concentrations clearly visible near the northwestern shoreline due to downwind accumulation of buoyant cells. The identification and mapping of near-shore cyanobacterial accumulations using the technique demonstrated here are of significant importance for reducing the health risks for recreational water users.

9.5.2 Mapping Cyanobacterial Blooms Using Satellite Remote Sensing

Matthews et al. [30] used the maximum peak height algorithm applied to MERIS time-series data collected between 2002 and 2012 for 50 South African lakes. This allows unique insights into the

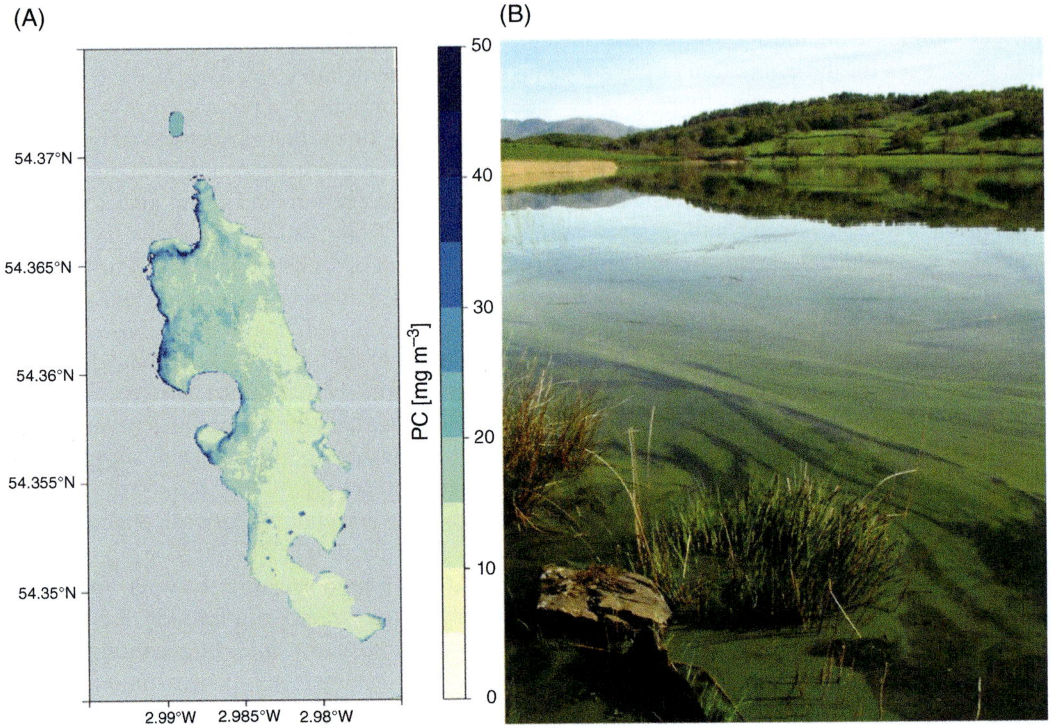

Figure 9.3 *(A) The concentration of phycocyanin in Esthwaite Water on 26 April 2007 as mapped using data from an airborne imaging spectrometer. (B) Photograph showing surface scums of* Anabaena circinalis *adjacent to the northwest shoreline at the time of the airborne data acquisition*

phenology of cyanobacterial blooms in lakes and the ability to monitor long-term changes caused by catchment changes and/or climatological drivers on continental scales. The example shown here, taken from Hartbeespoort Dam in South Africa, demonstrates the application for long-term monitoring of the seasonal cycle of Chl a, cyanobacterial blooms and surface scum (in this instance, defined as Chl a concentrations larger than 350 mg m^{-3}) in a small to medium-sized reservoir (Fig. 9.4). Cyanobacterial blooms occur regularly during the summer months, with clear phases corresponding to winter climates with reduced heat and solar radiation. The use of a medium resolution instrument with systematic global acquisition capabilities such as MERIS is, therefore, a powerful tool for routine, long-term monitoring of trophic status and cyanobacteria blooms in small to medium-sized inland waters.

9.6 Future Prospects

The next few years will see the launch of several satellite missions with potential for cyanobacterial bloom monitoring. Most notably, these include the European Space Agency's Copernicus Sentinel-2 MSI (Multispectral Imager) with a 10- to 60-m spatial resolution and 5-day global revisit cycle with two operational satellites and Sentinel-3 OLCI (Ocean Land Colour Imager) with a 300-m spatial resolution and daily revisit cycle globally with two satellites (http://www.esa.int/

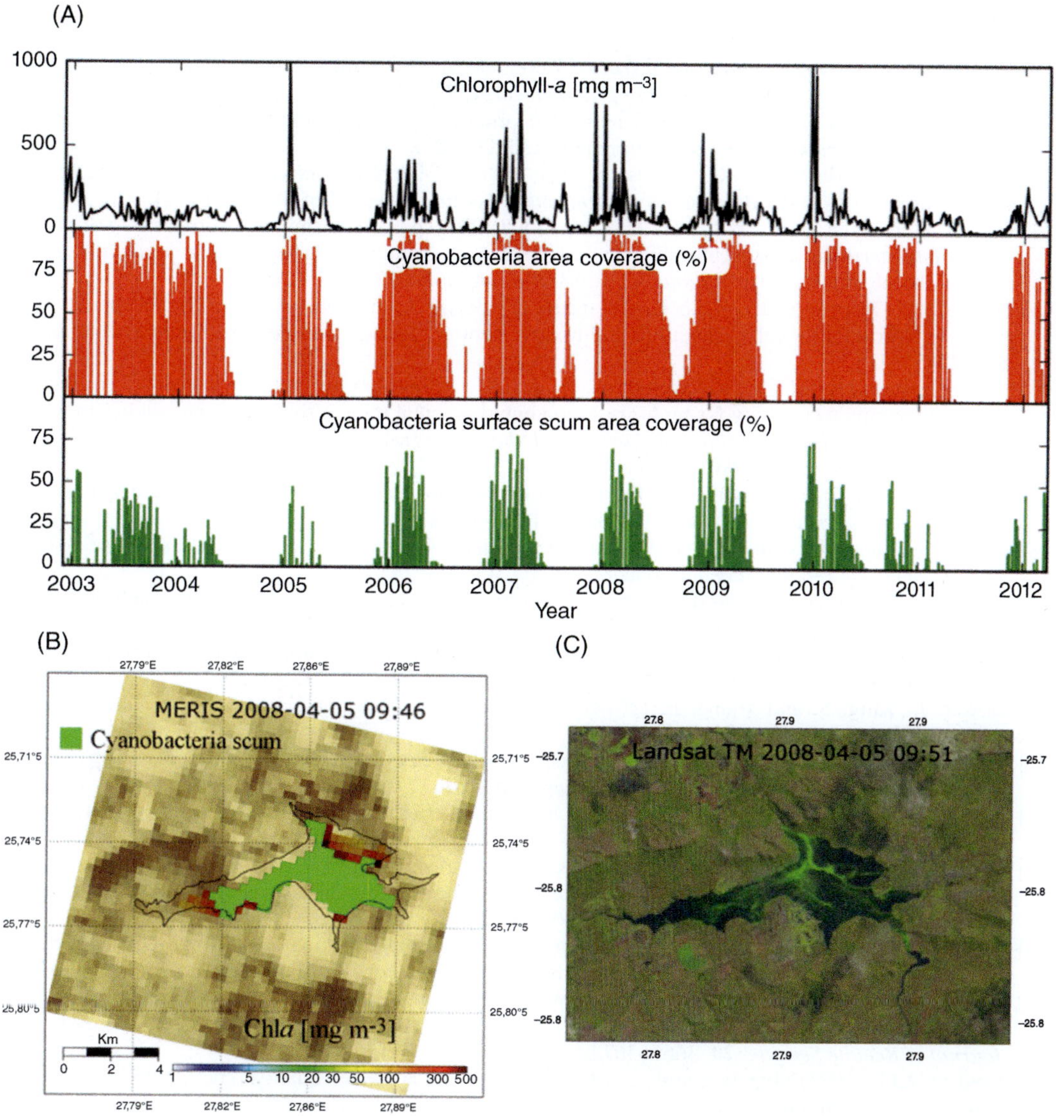

Figure 9.4 (A) MPH algorithm products derived from the MERIS 10-year archive for Hartbeespoort Dam, South Africa. (B) MERIS image showing Chl a concentration and surface scum mask. (C) Near-simultaneously acquired Landsat image showing cyanobacteria surface scum on the water surface. MERIS data provided by the European Space Agency. Landsat data available from the U.S. Geological Survey

Our_Activities/Observing_the_Earth/Copernicus). The latter has a strong MERIS heritage and includes the 620-, 665-, and 709-nm bands crucial to the estimation of Chl a and phycocyanin in inland and coastal waters. Significantly, the data from the Copernicus programme will be provided free of charge to end-users, greatly reducing the potential costs of operational satellite services for cyanobacterial bloom monitoring in inland, coastal, or ocean waters.

References

[1] Palmer, S.J.C., Kutser, T. and Hunter P.D. (2015) Remote sensing of inland waters: Challenges, progress and future directions. *Remote Sensing of Environment*, **157**, 1–8.

[2] Simis, S.G.H. and Olssen J. (2013) Unattended processing of shipborne hyperspectral reflectance measurements. *Remote Sensing of Environment*, **135**, 202–212.

[3] Szekielda, K.H., Marmorino, G., Maness, S. *et al.* (2007) Airborne hyperspectral imaging of cyanobacteria accumulations in the Potomac River. *Journal of Applied Remote Sensing*, **1**, 013544.

[4] Hunter, P.D., Tyler, A.N., Willby, N.J. *et al.* (2008) The spatial dynamics of vertical migration by *Microcystis aeruginosa* in a eutrophic shallow lake: A case study using high spatial resolution time-series airborne remote sensing. *Limnology and Oceanography*, **53**, 2391 – 2406.

[5] Hunter, P.D., Tyler, A.N., Carvalho, L. *et al.* (2010) Hyperspectral remote sensing of cyanobacterial pigments as indicators for cell populations and toxins in eutrophic lakes. *Remote Sensing of Environment*, **114**, 2705–2718.

[6] Honkavaara, E., Hakala, T., Kirjasniemi, J. *et al.* (2013) New light-weight stereosopic spectrometric airborne imaging technology for high-resolution environmental remote sensing – case studies in water quality mapping. *The International Archives of the Photogrammetry, Remote Sensing and Spatial Information Sciences*, **XL-1/W1**, 139–144.

[7] Strong, A.E. (1975) Surface chlorophyll distribution in Great Lakes as monitored from space by Landsat. *Transactions-American Geophysical Union*, **56**, 1003–1003.

[8] Gervin, J.C. and Marshall, M.L. (1977) Landsat investigation of water quality in Lake Okeechobee. *Photogrammetric Engineering and Remote Sensing*, **43**, 743–743.

[9] Kutser, T. (2004) Quantitative detection of chlorophyll in cyanobacterial blooms by satellite remote sensing. *Limnology and Oceanography*, **49**, 2179 – 2189.

[10] Alikas, K., Kangro, K. and Reinart, A. (2010) Detecting cyanobacterial blooms in large North European lakes using the Maximum Chlorophyll Index. *Oceaanologia*, **52**, 237–257.

[11] Gower, J., King, S. and Young, E. (2014) Global remote sensing of *Trichodesmium*. *International Journal of Remote Sensing*, **34**, 5459–5466.

[12] McKinna, L.I.W. (2015) Three decades of ocean-color remote-sensing *Trichodesmium* spp. in the World's oceans: A review. *Progress in Oceanography*, **131**, 177–199.

[13] Kahru, M. and Elmgren R. (2014) Multidecadal time series of satellite-detected accumulations of cyanobacteria in the Baltic Sea. *Biogeosciences*, **11**, 3619–3633.

[14] Matthews, M.W., Bernard, S. and Robertson, L. (2012) An algorithm for detecting trophic status (chlorophyll-a), cyanobacterial-dominance, surface scums and floating vegetation in inland and coastal waters. *Remote Sensing of Environment*, **124**, 637–652.

[15] Matthews, M.W. and Odermatt, D. (2015) Improved algorithm for routine monitoring of cyanobacteria and eutrophication in inland and near-coastal waters. *Remote Sensing of Environment*, **156**, 374–382.

[16] Wynne, T.T., Stumpf, R.P., Tomlinson, M.C., Warner, R.A., Tester, P.A., Dyble, J. and Fahnenstiel, G.L. (2008). Relating spectral shape to cyanobacterial blooms in the Laurentian Great Lakes. *International Journal of Remote Sensing*, **29**, 3665–3672.

[17] Dekker, A.G. (1993) Hyperspectral remote sensing of inland water quality. The Netherlands: Vrije University, Amsterdam.

[18] Simis, S.G.H., Peters, S.W.M. and Gons, H.J. (2005) Remote sensing of the cyanobacterial pigment phycocyanin in turbid inland water. *Limnology and Oceanography*, **50**, 237 – 245.

[19] Mishra, S. and Mishra D.R. (2014) A novel remote sensing algorithm to quantify phycocyanin in cyanobacterial algal blooms. *Environmental Research Letters*, **9**, 114003.

[20] Li, L., Li, L. and Song, K. (2015), Remote sensing of freshwater cyanobacteria: An extended IOP Inversion Model of Inland Waters (IIMIW) for partitioning absorption coefficient and estimating phycocyanin. *Remote Sensing of Environment*, **157**, 9–23.

[21] Qi, L., Hu, C., Duan, H. *et al.* (2014) A novel MERIS algorithm to derive cyanobacterial phycocyanin pigment concentrations in a eutrophic lake: Theoretical basis and practical considerations. *Remote Sensing of Environment*, **154**, 298–317.

[22] Riddick, C.A., Hunter, P.D., Tyler, A.N. *et al.* (in review) Evaluation of algorithms for retrieval of cyanobacterial pigments in highly turbid, optically complex waters using MERIS data. *Remote Sensing of Environment.*

[23] Wheeler, S.M., Morrissey, L.A., Levine, S.N. *et al.* (2012) Mapping cyanobacterial blooms in Lake Champlain's Missisquoi Bay using QuickBird and MERIS satellite data. *Journal of Great Lakes Research*, **38**, 68–75.

[24] Matthews, M.W. and Bernard, S. (2013). Characterizing the absorption properties for remote sensing of three small optically-diverse South African reservoirs. *Remote Sensing*, **5**, 4370–4404.

[25] Ruiz-Verdu, A., Simis, S.G.H., de Hoyos, C. *et al.* (2008) An evaluation of algorithms for the remote sensing of cyanobacterial biomass. *Remote Sensing of Environment*, **112**, 3996-4008.

[26] Wynne, T.T., Stumpf, R.P., Tomlinson, M.C. *et al.* (2010) Characterizing a cyanobacterial bloom in western Lake Erie using satellite imagery and meteorological data. *Limnology and Oceanography*, **55**, 2025–2036.

[27] Lunetta, R.S., Schaeffer, B.A., Stumpf, R.P. *et al.* (2015) Evaluation of cyanobacteria cell count detection derived from MERIS imagery across the eastern USA. *Remote Sensing of Environment*, **157**, 24–34.

[28] Hunter, P.D., Tyler, A.N., Gilvear, D.J. *et al.* (2009). Using remote sensing to aid the assessment of human health risks from blooms of potentially-toxic cyanobacteria. *Environmental Science & Technology*, **43**, 2627 – 2633.

[29] Matthews, M.W. and Bernard, S. (2013). Using a two-layered sphere model to investigate the impact of gas vacuoles on the inherent optical properties of Microcystis aeruginosa. *Biogeosciences* **10**, 8139–8157.

[30] Matthews, M.W. (2014). Eutrophication and cyanobacterial blooms in South African inland waters: 10 years of MERIS observations. *Remote Sensing of Environment* **155**, 161–177.

10

The Italian System for Cyanobacterial Risk Management in Drinking Water Chains

Luca Lucentini[1], Liliana La Sala[2], Rossella Colagrossi[2], and Roberta Congestri[3]

[1] *Department of Environment and Primary Prevention, Italian National Institute of Health, Rome, Italy*
[2] *Department of Prevention, Ministry of Health, Rome, Italy*
[3] *Department of Biology, University of Rome 'Tor Vergata,' Italy*

10.1 Introduction

Approximately 85% of the drinking water supplies in Italy rely on groundwater. However, the use of surface waters has been increasing because of the increase in water demand, with natural and artificial freshwater basins playing a crucial role in emergency water supply in big municipalities.

As in other countries, Italian surface waters have been affected by cyanobacterial blooms, and many regions have experienced abnormal proliferations of potentially toxic planktonic species. Cyanotoxins are relevant health hazards associated with these phenomena, since they may be transferred through the water supply chain and ultimately exert toxic effects on consumers.

The protection of consumers from the risks of cyanotoxin exposure is a concern of the Directive 98/83/CE, stating that water for human consumption should be free from any substances that, in numbers or concentrations, constitute a potential danger to human health. However, regular monitoring of cyanotoxins in drinking water is not expressly requested by the Directive or by its Italian national transposition, the Legislative Decree 31/2001. According to this national decree, the surveillance of cyanotoxins is the responsibility of the local health authorities, which are currently required to investigate, on a case-by-case basis, any health-related parameters of drinking water whose concentrations can be a potential threat to human health.

Handbook of Cyanobacterial Monitoring and Cyanotoxin Analysis, First Edition.
Edited by Jussi Meriluoto, Lisa Spoof and Geoffrey A. Codd.

In order to support local health and environmental authorities with the emerging cyanobacterial threat, the National Group for Cyanobacterial Risk Management in Water for Human Consumption (CyanoRisk National Group) was formed in 2010. The group joins multidisciplinary experts from the Ministry and National Institute of Health, National Research Council, universities, and research institutions along with local health and environmental regulatory agencies and water suppliers that have specific expertise in the study and management of cyanobacterial blooms and their impact on drinking water systems.

The effort of the group is primarily focused on an extensive analysis of cyanobacterial bloom development and toxicity with a review of international criteria, methods and regulatory standards. In parallel, a specific risk assessment of cyanobacterial outbreaks in Italian drinking water supply systems, including involved toxic species and cyanotoxins, mitigation, and response measures to potential human exposure events, was performed. This work is reported in "Cyanobacteria in Water for Human Consumption: Health Risk Assessment" [1] and represents the basis of the Italian guidelines for the management of risks related to cyanobacteria in water for human consumption [2]. The guidelines are conceived as a reference tool for health and environmental regulatory authorities, by describing the construction and implementation of a system based on the Alert Level Framework (ALF) and Water Safety Plan (WSP) principles of the World Health Organization (WHO) [3] and by establishing a parametric limit for cyanotoxin concentration in water for human consumption.

10.2 Risk Assessment of Toxic Cyanobacterial Outbreaks in Water for Human Consumption in Italy

10.2.1 Toxic Species and Associated Toxins

Cyanobacterial toxic blooms in Italian reservoirs are mainly ascribed to the genera *Aphanizomenon, Cylindrospermopsis*, *Dolichospermum* (ex *Anabaena*), *Microcystis,* and *Planktothrix.* Toxic species occurrence has been recorded in more than 60 basins in 13 of 20 different regions. The toxins produced include microcystins (MCs), cylindrospermospin (CYN), and anatoxin-a (ATX-a). MC is the prevalent toxin, while the occurrence of CYN and neurotoxins (ATX-a) is rare. Seven MC variants have been recorded: desmethyl-MC-RR, desmethyl-MC-LR, MC-RR, MC-YR, MC-LR, MC-LA, and MC-LW, and the simultaneous occurrence of two or more MC congeners in the same sample is common. The detection of CYN is commonly associated with the presence of *Chrysosporum* (ex *Aphanizomenon*) *ovalisporum* and *Cylindrospermopsis raciborskii*, which have been rarely recorded so far. ATX-a, produced by *Dolichospermum flos-aquae* and *D. lemmermanii*, has, so far, been found in only one case in Italy.

10.2.2 Water Supply System Vulnerability

The assessment of water supply vulnerability related to the risk of cyanotoxins in water for human consumption requires adequate training of involved operators. The long-term analysis of toxic cyanobacterial proliferation episodes and that of the effectiveness in the management consequently implemented has great relevance for the vulnerability assessment of reservoirs. Figure 10.1 describes the flow scheme, implemented for the Italian Guidelines for the protection of drinking water systems, based on the analysis of risk prevention and mitigation measures for cyanotoxin presence in water at the point of compliance.

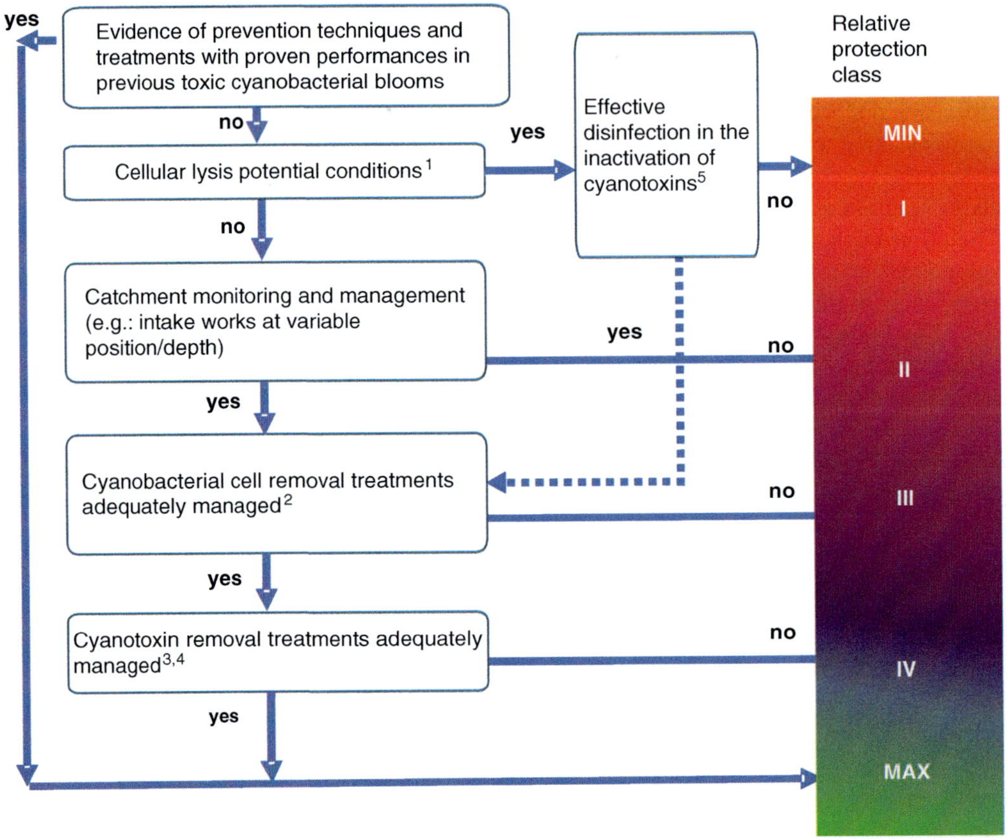

1. Events that can induce cellular lysis: algicide treatments, pre-oxidation, cyanobacterial population aging process, high pressure pumping.
2. Methods for removal of cyanobacteria cells: coagulation and sedimentation or flotation, sand filtration, ultrafiltration.
3. Methods for cyanotoxin removal: filtration on activated carbon, ultrafiltration, ozone.
4. Removal/inactivation system effectiveness depends on the technologies applied, plant size, water characteristics, cyanobacteria in inlet waters and toxins levels, maintenance, etc.
5. Disinfection system effectiveness, depends on technologies applied, plant size, water characteristics, entering flow of cyanobacteria and toxins, maintenance, etc.

Figure 10.1 Estimated protection level of the water supply system

10.3 Framework of Risk Management of Toxic Cyanobacterial Outbreaks in Water for Human Consumption

10.3.1 Risk Management Framework

Table 10.1 describes the risk assessment and management system planned by the CyanoRisk National Group as a regulatory strategy for the early warning and management of toxic cyanobacterial development and cyanotoxin presence within the entire water catchment-treatment-distribution system. The decision tree foresees different actions of monitoring and management, according to different estimated alert levels.

The approach is based on the international principles of ALF and WSP, depending on the estimated contamination risk level in raw water and taking into account the water monitoring and

Table 10.1 *Scheme of the risk assessment and management system planned by the CyanoRisk National Group*

Decision basis	Threshold definition and *expected risk*[1]	Recommended actions	Possible water use restrictions[2]
Detection level 0			
Detection of potentially toxic cyanobacteria during water monitoring[3,4]	Potentially toxic species[4]: 500–2,500 cell/mL or chlorophyll a: 1–2.5 µg/L • *Detection of reduced concentrations of cyanobacteria* • *No immediate health risks*	Deepen reservoir visual inspection Implement continuous monitoring, at least on a fortnightly basis, of cyanobacterial and chlorophyll *a* concentrations[10,11]	Use water restrictions[2] following the detection of toxin concentrations exceeding the maximum acceptable values in distributed waters[7,12]
Alert level 1			
Alert for possible health risk: cyanobacterial concentrations associated with a potential presence of cyanotoxins in the water (before the drinking water treatment chain) at a level equal to the maximum reference value[6,7]	*Planktothrix. rubescens:* 2,500 cells/mL [4] or other toxic species: 5,000 cells/mL[4] *Settlement and development of cyanobacterial population in water indicative, in the worst possible scenario*[6]*, of a potential presence of toxins at concentrations close to the maximum reference value (1.0 µg MC-LR/L*[7]*)*	Notify local health authority in case of recurring/systematic and adequately managed phenomena[5] Implement continuous monitoring on a fortnightly basis or better weekly basis, through cell counts[10,11] at least in ingoing and outgoing waters If the system protection level is considered inadequate[9], implement cyanotoxin analysis, on a weekly basis[9–11], in inlet waters, and, if necessary, in the outlet and/or in distribution[12] Optimize, as far as possible, mitigation measures in the whole water treatment chain[8,13] Ensure adequate water chlorination[14]	

(Continued)

Table 10.1 (*Continued*)

Decision basis	Threshold definition and *expected risk*[1]	Recommended actions	Possible water use restrictions[2]
High alert level 2 High alert level of possible health risk: cyanobacterial concentrations associated with a potential presence of cyanotoxins in the water (before the drinking water treatment chain) at a level equal to 10 times the maximum reference[6,7] value for waters for human consumption	*Planktothrix rubescens:* 25,000 cells/mL[4] or other toxic species: 50,000 cells/mL[4] *Settlement and development of cyanobacterial population in water indicative, in the worst possible scenario*[6]*, of a potential presence of toxins at concentrations) about 10-fold the maximum reference value (1.0 µg/L MC-LR*[7]*). Adequate prevention and treatment measures must be implemented so as to mitigate risks, otherwise use restrictions and emergency response plans together with an adequate training and communication*	Notify local health authority[5] Continuous monitoring on a weekly basis, or better twice a week, through cell counts[10–11] Cyanotoxin identification, on a weekly or better twice a week basis[11], in inlet and outlet waters from the water treatment system and in distribution[7,9–12] Optimize and/or enhance mitigation measures in the whole water treatment chain[9,14] Ensure an adequate chlorination[14] Organise emergency, information and communication plans	

[1] Threshold is the maximum acceptable level of toxin concentration in the reservoir water entering the water treatment system (as a preventative measure it is equal to the maximum acceptable concentration of cyanotoxin in the water at the consumer site), assessed from the cyanobacterial concentration in waters, according to a conservative assessment of the potential content of toxin per cyanobacterial cell (*toxin quota*); the expected risk is considered as the potential occurrence in the water distributed for human consumption of cyanotoxins at a concentration near or exceeding the maximum acceptable value.

[2] Restrictions of water uses must be considered, taking into account the general provisions of Directive 98/83/EC, concerning the need for considering the excess amount and potential risks for human health related to the risk scenario, as well as those risks that could derive from an interruption of supply or from a use-estriction of distributed waters. The possible use restrictions should therefore envisage the recourse to alternative supply and the urgent implementation of adequate treatments.

[3] The routine monitoring on the reservoir must be envisaged according to the regulation in force or to risk plans by the water supplier, based on a WSP approach.

[4] Values established on samples collected near the water intake site, or in raw waters entering the water supply system.

[5] If the risk related to cyanobacteria occurs in the water system on a regular or a systematic and/or prolonged basis (in general occurring in one or more months during which the alert level 1–2 persists), data on internal monitoring concerning the risk due to cyanobacteria and cyanotoxins, including monitoring data, can be provided by the operator through methods and timings to be agreed with the health authority on a case-by-case basis, according to contamination level, prevention and mitigation measures adopted by the treatment plant (according to plant size), technologies used for water treatment, level of systems maintenance and monitoring and to the experience gained by the water system. As a general rule, for water systems with multiannual experience in risk management using adequate and tested monitoring systems, a communication of data on a six-monthly or annual basis can be envisaged (summary of cyanobacterial concentration data and cyanotoxin concentrations in the inlet, outlet and distributed treated waters) together with the realized interventions, except for extraordinary cases.

[6] Maximum precautionary criteria by considering the entire cyanobacterial population as potential cyanotoxins-producers, assessing a high level of toxin produced per cell and assuming that all the of tcyanooxins produced are free in water, represent the most toxic varieties and are not removed during treatments.

[7] The parameter value established for microcystin-LR (MC-LR) is 1.0 µg/L, referred to the total toxin content (intra-plus extra-cellular) as the sum total of concentrations of the various congeneric elements of microcystin present in the sample, considered to be equivalent to microcystin-LR, on a concentration basis according to a conservative approach to health protection; for this reason, identifiable microcystin congeners must be detected using the best analytical capabilities available: at the very minimum, at least the congeners for which there are commercially available analytical standards must be detected and quantified.

[8] The analysis of cyanotoxins in water intended for human consumption is not subject to routine controls or checks which occur at pre-arranged intervals, but must only be carried out in cases identified following a specific risk assessment.

[9] As reference, a class I-II protection level indicated in Fig. 1, can be considered as inadequate, even if system size and functioning and maintenance conditions are appropriate. Moreover, water treatment is also required before they can be delivered for human consumption as stated in the Legislative Decree 152/2006 concerning pre-treatments based on water body classification.

[10] Cyanotoxin determinations must be referred to the overall content (intra-and extra-cellular) and the analytical method must be adequate for the purpose (see note 7).

[11] Monitoring frequency must be established according to the assessed risk level (first, cyanobacterial cell and cyanotoxin concentrations in raw intake waters and where necessary in outgoing and distributed, treated waters, with assessment of the reduction through the system during the monitoring and other operative indicators such as for example turbidity, TOC, *etc.*).

[12] To maintain monitoring frequency until two consecutive results state a reduced risk level, after that the frequency can be gradually decreased.

[13] As far as distributed (treated) water monitoring is concerned, it is required to take into consideration a proximal and a distal point of the network where the water is not mixed with water having other origins;

[14] Protection measures III-VI (Fig. 1) must be implemented for concentrations exceeding alert level 1.

[15] About 0.1–0.2 mg/L of residual chlorine, increasing, where possible, the contact time during disinfection; higher chlorine concentrations (e.g. 0.5 mg/L) can be recommended for short periods during an emergency, taking however into consideration that the monitoring of disinfection by-products can be critical due to the increase of organic substance charge caused by cyanobacterial and algal biomass in intake waters; it is therefore required to monitor the levels of disinfection by-products such as trihalomethanes and/or chlorite according to the techniques adopted for the possible pre-oxidation and post-disinfection processes.

mitigation measures adopted in the system. The application of the risk assessment and management schemes here described harmonizes with the proposed revision of Directive 98/83/EC on the quality of water intended for human consumption, with a clear orientation towards the WSP model, proposed in the WHO Guidelines and framed in a policy of sustainable management of water resources.

10.3.2 Emergency Response Plans

Emergency is the evidence of a health risk for consumers due to cyanotoxin concentrations exceeding the maximum acceptable values. The occurrence of a state of emergency can be rapid, even a few days. Therefore the management of an emergency, and the health, economic and social impact of these phenomena, depends on the promptness of intervention and the suitability of the action. The following aspects need to be considered:

- Existence of a well-defined emergency plan
- Coordinated and planned engagement of all the parties involved in the emergency stage as far as suitable procedures are concerned
- Preparation of operators, through adequate training

10.4 Risk Information and Communication

It is crucial that reference institutions design and implement communication plans according to strategically established methods, thus avoiding to adopt interventions only in case of "emergency." Local health authorities play a crucial role in planning, activation, and assessment of communication activities, by sharing contents and actions with other institutions and organizations at local, regional, and national levels. These include water suppliers, local administrations, operators of health institutions within the territory, institutions, associations at the regional and provincial levels, local media, scientific institutions, and other parties that can play, if organized, the role of communicative process leaders.

References

[1] Lucentini, L. and Ottaviani M. (eds.) (2011) Cyanobacteria in water for human consumption: Guidelines for risk management, *Rapporti Istisan* 11/35 pt 1. Italian National Institute of Health, Rome.
[2] Lucentini, L. and Ottaviani M. (eds.) (2011) Cyanobacteria in water for human consumption: Guidelines for risk management, *Rapporti Istisan* 11/35 pt 2. Italian National Institute of Health, Rome.
[3] World Health Organization (2011) *Guidelines for drinking-water quality*. 4th Edition. WHO, Geneva.

Section IV

Toxins and Bioactive/Noxious Compounds from Cyanobacteria

11

Microcystins and Nodularins

Arnaud Catherine[1], Cécile Bernard[1], Lisa Spoof[2], and Milena Bruno[3]

[1] *UMR 7245 MCAM CNRS-MNHN, Muséum National d'Histoire Naturelle, Paris, France*
[2] *Biochemistry, Faculty of Science and Engineering, Åbo Akademi University, Turku, Finland*
[3] *Department of Environment and Primary Prevention, Istituto Superiore di Sanità, Rome, Italy*

11.1 Chemical Characteristics and Diversity of Microcystins and Nodularins

Microcystins (MCs), a diverse family of cyclic heptapeptide hepatotoxins, are a group of frequently found toxins of cyanobacteria. Their general structure (Fig. 11.1A) is cyclo(-D-Ala-L-**X**-D-*erythro*-β-methylAsp(iso-linkage)-L-**Z**-Adda-D-Glu(iso-linkage)-*N*-methyldehydro-Ala), where Adda denotes the unique β-amino acid (2*S*,3*S*,8*S*,9*S*)-3-amino-9-methoxy-2,6,8-trimethyl-10-phenyldeca-4,6-dienoic acid [1–5]. The L-amino acid residues 2 (**X**) and 4 (**Z**) constitute the main variable part in MCs, and the corresponding MC name is indicated by a two-letter suffix [e.g. MC-LR contains leucine (L) and arginine (R) at positions 2 and 4, respectively]. The word iso-linkage indicates that the peptide bond is made through the carboxyl group in the side chain. However, all other MC residues are also variable to some extent (Appendix 3), these variations being then described by a prefix. Common variations include demethylations in residues 3 and 7 and naturally occurring changes in the Adda stereochemistry. The overall structure of MCs is relatively hydrophilic (polar), which originates from carboxylic acids at positions 3 and 6 and a frequent occurrence of arginine at positions 2 and 4. MCs are well soluble in aqueous methanol, although they tend to form aggregates at high concentrations ($mmol\,L^{-1}$). MCs are also soluble in neat water, but they are known to adsorb on some laboratory plastics if the methanol percentage of the solvent is low.

Nodularins (NODs) are cyclic pentapeptide hepatotoxins (Fig. 11.1B) with the general structure cyclo(-D-*erythro*-β-methylAsp(iso-linkage)-L-**Z**-Adda-D-Glu(iso-linkage)-2-(methylamino)-2(*Z*)-dehydrobutyric acid) [2, 6]. NODs share many structural and functional properties with MCs, the former lacking two amino acid residues and showing a substitution of the residue *N*-methyldehydroalanine with *N*-methyldehydrobutyrine. The second amino acid residue (**Z**) is

Handbook of Cyanobacterial Monitoring and Cyanotoxin Analysis, First Edition.
Edited by Jussi Meriluoto, Lisa Spoof and Geoffrey A. Codd.

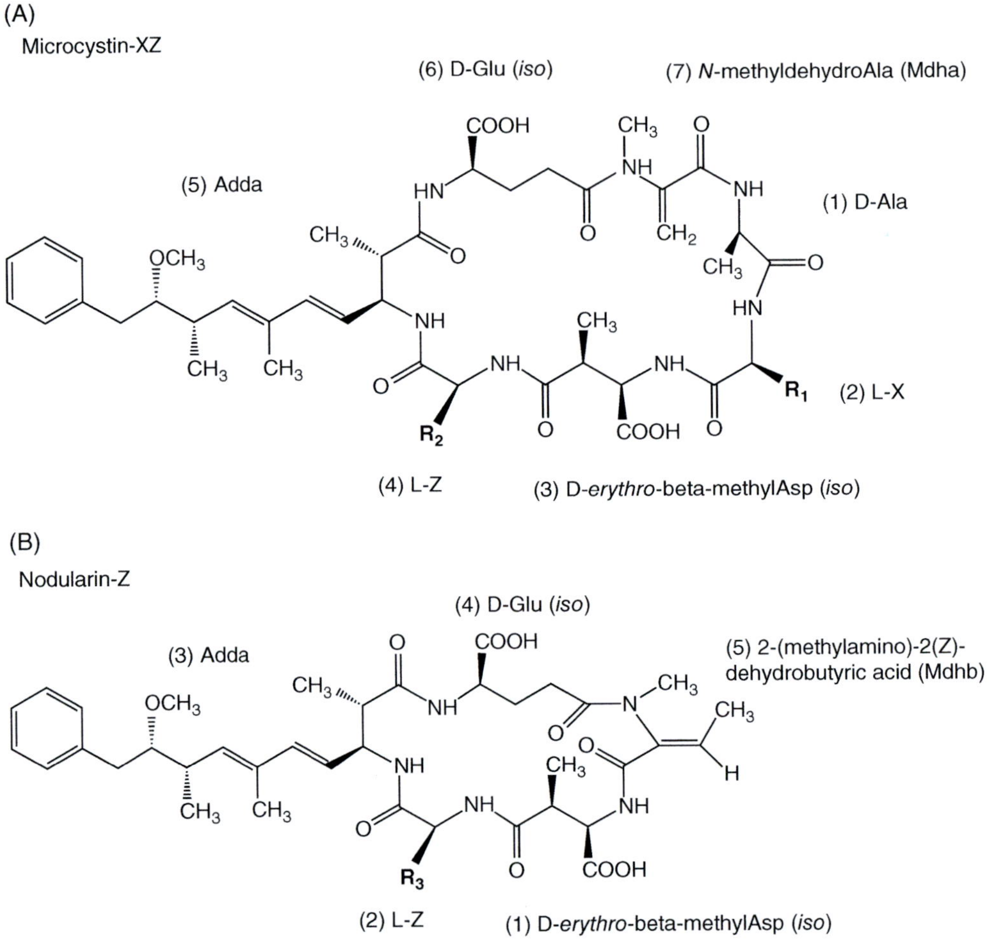

Figure 11.1 General chemical structures of microcystins (A) and nodularins (B)

L-Arg in the frequently occurring nodularin-R (also known as "nodularin" without a suffix). For identified NOD congeners, please see Appendix 3.

11.2 Biosynthesis and Genetics of MC and NOD Production

MCs and NODs are produced through nonribosomal peptide synthesis and exhibit biosynthetic routes that share many similarities. The gene clusters responsible for NOD (*nda*) and MC (*mcy*) biosynthesis are supposed to have a common evolutionary origin, the NOD gene cluster having probably evolved from an MC synthetase [7]. In natural cyanobacteria populations, MC-producing genotypes coexist with genotypes lacking some (or all) of the *mcy* gene cluster. Similarly, nontoxic strains of the potentially NOD-producing *Nodularia spumigena* have been reported [8].

MC biosynthesis genes were first characterized for *Microcystis aeruginosa* (strains PCC7806 and K139) [9, 10], *Planktothrix agardhii* (strain NIVA-CYA 126) [11], and *Anabaena* (present name *Dolichospermum*) sp. (strain 90) [12]. The biosynthesis pathway of MCs follows the same general course of reactions, despite some differences in domain specificity and/or operon organization. MCs are synthesized nonribosomally via a large enzyme complex, encoded by the *mcyA-J* gene cluster, comprising nonribosomal peptide synthetases (NRPSs), polyketide synthases (PKSs), and additional modifying enzymes (for a review, see [13]). NRPSs exhibit a modular structure and catalyse the formation of peptides through a thiotemplate mechanism. Each domain is responsible either for the activation, thiolation, modification, and condensation of one specific substrate amino acid. In the MC biosynthesis pathway, Adda is first synthesized by the activity of PKSs and then the remaining six amino acids are incorporated sequentially by peptide synthetases. Additional tailoring activities are implemented by specific domains of these large multifunctional proteins or by enzymes that are encoded by separate genes [14].

Ten genes encode for the biosynthetic enzymes complex in *Microcystis* and *Dolichospermum* (*Anabaena*) and nine genes in *Planktothrix* (see [15] for details on the organization of the MC and NOD gene clusters among the main MC- and NOD-producing genera). The *mcyA-B-C* genes encode NRPSs, while *mcyD* and *mcyE-G* encode, respectively, PKSs and NRPS-PKS hybrid enzymes. The *mcyJ* gene encodes an *O*-methyltransferase that transfers a methyl group to the side chain of the characteristic Adda moiety of MCs [11]. Two additional enzymes are required prior to the incorporation of D-*erythro*-β-methyl-aspartic acid: the 2-hydroxy acid dehydrogenase (*mcyI*) and the asparate racemase (*mcyF*). In *Planktothrix*, a distinct thioesterase enzyme is encoded by the gene *mcyT* (located at the 5' end of the gene cluster). The MC synthetase gene cluster in *Microcystis* and *Dolichospermum* (*Anabaena*) lacks *mcyT*-type genes, which might be linked to higher capacity for MC production in *Planktothrix* compared to other MC-producing genera [11, 16]. Finally, the *mcyH* gene encodes an ABC transporter, which is composed of a membrane domain and a cytosolic ATP-binding domain. It is considered as putatively involved in an active export of MCs, which may be responsible for the thylakoid localization of the toxin [17] or for the export of the toxin under certain growth conditions such as exposure to high and/or red light [18]. The operon and gene arrangement also exhibits variations among MC-producing genera [15]. Whereas *mcy* genes in *Microcystis* and *Dolichospermum* (*Anabaena*) are transcribed from a bidirectional promoter region (between *mcyA-D* and *mcyA-G*, respectively), MC genes in *Planktothrix* are mostly unidirectional (except for *mcyT*). *Dolichospermum* (*Anabaena*) is the only genus in which modules organization is in agreement with the different steps of MC biosynthesis route [19].

The NOD biosynthetic pathway was first characterized for a *Nodularia spumigena* NSOR10 strain [7]. The NOD genes are organized within a 48-kb cluster consisting of nine ORFs encoding NRPS, PKS, and tailoring enzymes. The main changes between MC and NOD biosynthesis cluster organization are (i) the deletion of the second module of *mcyA* and the first module of *mcyB*, (ii) the fusion of the remaining modules of *mcyA* and *mcyB* into a bimodular NRPS *ndaA*, and (iii) the substrate specificity of *ndaA* (L-Tyr instead of L-Ser) [7, 13].

MC synthetase genes are subjected to frequent recombination events and are further diversified by point mutations, leading to high MC variant diversity [20]. MC production was first considered a neutral trait [21]. Subsequently, it was shown that while the adenylation domains of *mcyB1* and *mcyC* are under purifying selection, a number of amino acid residues in the adenylation domain of these genes appear to be under positive selection [19]. The positive selection of residues involved in active site selectivity of these adenylation domains suggests that the production of novel MC variants might provide some competitive advantage.

The ecological significance of MCs production might be genus (or species) dependent. For instance, the *mcy* operon in *Planktothrix* exhibits a lower recombination rate and stronger purifying selection in comparison to other genera, which suggests stronger functional constraints on MC production in the *Planktothrix* genus [20]. Recombination events in *Microcystis* were shown to lead to two monophyletic main clades (A and B) and an intermediate outlier clade (group X), which could reflect some difference in the ecology of cyanobacteria forming the *Microcystis* genus. However, selection force is considered to be small in the creation of the clades A and B and might be the result of homology-dependent recombination and neutral genetic drift [22]. A recent study showed that the loss of MC production ability in the genus *Planktothrix* did not lead to phylogenetic diversification [23]. Diversification, thus, probably arose from adaptation to ecological factors not causally related to MC production. However, peptide dissimilarity among individual *Planktothrix* strains was found to increase with geographic distance, which suggests that microevolutionary processes and/or random drift generate secondary metabolite structural variations among local *Planktothrix* populations [22]. Further studies should seek to evaluate the relative importance of microevolutionary processes and random drift in generating peptide dissimilarity to better understand the ecological significance of secondary metabolite production (including MCs and NODs) in cyanobacteria.

11.3 Occurrence of MCs and NODs

MCs are the toxins most frequently produced by freshwater cyanobacteria (e.g. species from the genera *Anabaena*, *Aphanizomenon*, *Dolichospermum*, *Limnothrix*, *Microcystis*, *Nostoc*, *Phormidium*, and *Planktothrix*). To date, MC-producing cyanobacteria blooms have been described in 80 countries, which underlines the global distribution of MC-producing cyanobacteria (Fig. 11.2). Twenty-three cyanobacterial genera and at least 47 species are known or suspected to produce MCs (see Appendix 2). MC production was also recently documented in some marine cyanobacteria, including species from genera *Geitlerinema*, *Leptolyngbya*, *Roseofilum*, and *Synecococcus* [24–26], in cyanobacterial symbionts of terrestrial lichens belonging to the genus *Nostoc* [27], and in other terrestrial cyanobacteria, such as *Hapalosiphon hibernicus* [28].

In Europe, a number of medium- to large-scale studies on MC-producing cyanobacteria blooms have been published in the past decades (Table 11.1). MC concentrations usually range between 0.1 and 10 $\mu g\,L^{-1}$ in pelagic waters but can reach values as high as 2.8 $mg\,L^{-1}$ [29]. This value may often be exceeded when cyanobacterial scums are sampled.

Elevated MC concentrations may be found even with low cell abundances [30]. Thus, in case of toxin persistence after cell lysis, cell counts alone may not give a valuable indication on actual MC concentrations [31, 32].

While new case studies on cyanobacterial blooms are published every year, the lack of sampling design for selecting sites and of information on how the water samples were collected often does not allow evaluating the prevalence of cyanobacterial blooms in a given region and comparing bloom frequency among regions. In addition, very limited long-term regional dataset are available to assess variations in frequency of MC-producing blooms through time.

NODs have first been identified in *Nodularia spumigena*, a filamentous, heterocytous, nitrogen-fixing species that thrives in saline and brackish waters. The first report of a *N. spumigena* toxic bloom dates back to the late 1800s with reports of toxic events in Lake Alexandria, Australia [51]. In Europe, *N. spumigena* blooms are frequently observed in the Baltic Sea [52]. *N. spumigena* blooms have also been documented in estuaries and brackish lakes all over the world [53–56]. Recently, *N. spumigena* was also identified in a Turkish freshwater (salinity = 0.5 ppt) lake [57].

Figure 11.2 Countries where at least one ISI publication documented the occurrence of microcystin-producing cyanobacteria

[L-Har[2]]NOD production has also been identified in the benthic species *Nodularia sphaerocarpa* PCC7804 [58, 59]. Recently, *Nostoc* sp., a cyanobacterial symbiont of lichens, was also shown to produce high amount of NODs (up to 60 $\mu g\, g^{-1}$) in some lichen thalli [60].

11.4 Toxicological Effects and Associated Health Risk

11.4.1 Mechanisms of Toxicity

MCs and NODs are potent hepatotoxins acting as specific inhibitors of protein phosphatases (PPs) [61–64]. MCs are known to inhibit PPs of type 1, 2A, 3 (for MC-LA) [28], 4 and 5 [65], and, to a lesser extent, 2B [66], while NOD chemical targets are PPs of types 1, 2A, and 3 [67]. Contrary to MCs, NODs do not form a covalent bond with PPs [68]. PP dephosphorylate serine and threonine residues in animals and plants. The inhibition of PP1 and PP2A results in an increased phosphorylation of proteins in liver cells, affecting several processes such as metabolic pathways (e.g. glycogen synthesis), cell contractility, membrane transport, secretion, cell division, and gene transcription and translation [69].

MCs were shown to be potent tumour promoters [70–72], endocrine disruptors [73–77], and immunotoxicants [78]. MCs can induce oxidative DNA damage [79] and genotoxicity [80] and cause the activation of proto-oncogenes c-*jun*, c-*fos*, and c-*myc* [81].

NODs display stronger initiating and tumour promoting activities compared with MCs. This may originate from their smaller ring structure, allowing easier intake by hepatocytes [82].

Table 11.1 *Sampling surveys of MC occurrence in European countries*

Country	Region	Date	Number of samples	Number of sites	Sampling strategy	%>LoQ (s/S)	MC concentration (μg L^{-1}) % of samples			MC concentration range (μg L^{-1})	Reference
							0.1–1	>1–10	>10		
Belgium and Luxembourg	C. & S. Belgium	1997–2001	64	49	Cya	57/59	?	?	?	?	[33]
Bulgaria	National	2004	8	15	n.d.	100/40	?	?	?	?	[34]
Belgium	National	2007–2008	206	5	Cya	100/100	31%	48%	21%	0.1–251	[35]
Czech Rep.	National	2004	206	94	n.d.	70/85	?	?	?	0.1–37.0	[36]
Finland	Åland Islands	1999–2000	?	134	n.d.	?/13	33%	67%	0%	0.4–12.0	[37]
Finland	C. & S. Finland	2002	?	70	n.d.	?/83	?	?	?	0.1–3.3	[38]
France	Britany	2001	52	52	n.d.	73/73	16%	82%	12%	0.1–44.6	[39]
France	National	2002–2004	4273	442	n.d.	35%	88%	8%	4%	0.1–1600	[40]
France	Paris area	2006	50	50	RS	40/40	65%	25%	10%	0.1–25.4	A. Catherine (personal communication)
Germany	National	1995–1996	533	55	n.d.	72/?	70%		30%	0.1–366	[16]
Germany	Berlin area	1996–1998	129	15	n.d.	100/100	26%	36%	38%	0.1–119	[41]
Germany	S. Germany	1999	400	155	RS	?/26	?	?	?	11.0–560	[42]
Greece	National	1987–2000	?	33	n.d.	?/21	?	?	?	?	[43]
Greece	National	2005	13	13	n.d.	100/100	77%	23%	0%	0.2–3.7	[44]
Ireland	N. Ireland	2008	14	14	n.d.	100/100	?	?	?	?	[45]
Italy	National	1989–2006	66	28	n.d.	?/64	67%	23%		?	[46]
Poland	National	1996–2001	36	8	Cya	100/100	?	?	?	?	[47]
Portugal	National	1989–1992	12	9	Cya	100/100	?	?	?	?	[48]
Slovakia	National	1995–1998	12	7	Cya	75/57	?	?	?	?	[49]
Spain	Segura basin	2000–2001	32	8	n.d.	44/75	?	?	?	?	[50]
Netherlands	National	2009–2012	88	86	n.d.	77/88	19%	36%	27%	0.02–2800	[29]

C., central; N., northern; S., southern; n.d., not described; Cya, selective sampling based on cyanobacterial bloom occurrence; RS, representative sampling; %>LoQ, percentage of samples (s) or sites (S) displaying MC concentrations higher than the limit of quantification (LoQ).

Similar to MCs, NODs also induce the production of intracellular reactive oxygen species, causing peroxidation of lipids, proteins, and DNA [83].

MC-LR constitutes the first and the most-studied variant of the MC family and was considered as the most toxic one. However, a recent study identified MC-LW and MC-LF as having higher toxic effects on human hepatocytes [84]. This has led to recognize the need for a better understanding of congener-specific MC kinetics to assess chronic exposure risks [29].

The LD_{50} of MC-LR via intratracheal administration ($100\,\mu g\,kg^{-1}$) proved to be comparable to the LD_{50} following intraperitoneal injection ($100\,\mu g\,kg^{-1}$), leading to similar liver and pulmonary damages [85]. Daily subacute doses of MCs, intranasally administered to mice over 7-day periods, were shown to cause cumulative pathologic effects 2 times greater than those produced by the sum of MCs administered as a single dose [86], which may originate from bioaccumulation [87]. In experimental studies, NOD LD_{50} following intraperitoneal injection ranged from 30 to $70\,\mu g\,kg^{-1}$ [88].

11.4.2 Exposure Routes

Exposure to hepatotoxins originating from the consumption of contaminated water is known to have resulted in poisoning events [88]. In addition, MCs from contaminated lakes can percolate and contaminate groundwater. Contamination level of groundwater at water treatment plants pumping stations was shown to be proportional to the duration of toxic bloom events [32, 90]. However, only a few reports document poisoning from contaminated drinking water consumption, probably because the symptoms of poisoning caused by cyanobacterial hepatotoxins in drinking water resemble a number of other gastrointestinal illnesses linked to pathogenic bacteria [91].

Among other exposure routes, MC-producing cyanobacteria are sometimes found in food supplements, creating another source of possible human intoxication [92]. Then, MCs and NODs were also detected in a wide range of organisms. They enter aquatic animals through the ingestion of cyanobacterial cells or the direct uptake of the dissolved compounds. MC concentrations between $0.34\,\mu g\,kg^{-1}$ [93] and $36.4\,\mu g\,kg^{-1}$ [94] were measured in the muscle tissue of wild or farmed fish, indicating that even the consumption of contaminated fish muscle might constitute a threat for human health. Similarly, NOD was detected in clams, mussels [95] and fish (e.g. herring, stickleback, and salmon) [96].

11.4.3 Tolerable Daily Intake Guidelines

The toxic levels in water and other matrices are often the result of a sum of several coexisting toxins, sometimes belonging to different families [32, 97]. Consequently, the acute tolerable daily intake (TDI) limits for the various cyanotoxins have to be carefully considered and water use restriction guideline limits should be lowered when these substances coexist in water.

Following World Health Organization (WHO) recommendations, the MC-LR acute TDI for a 60-kg adult was set to $0.04\,\mu g\,kg$ body $wt^{-1}\,d^{-1}$ [98]. The WHO commission decided not to discuss a guideline for chronic effects [99], as reliable toxicological studies were not available [100]. Based on updated data, the U.S. Environmental Protection Agency (EPA) recently proposed acute and chronic TDI guidelines (0.006 and 0.003 MC-LR $\mu g\,kg$ body wt $^{-1}$ d^{-1}, respectively) [101]. Besides fixing an acute TDI level 7-fold lower than that of WHO, these guidelines set for the first time a chronic TDI level, recognizing a role for these toxins in giving rise to chronic human pathologies. In 2006, the International Agency for Research on Cancer (IARC) classified MC-LR as a possibly carcinogenic to humans (group 2B) [102]; however, available studies are inadequate to propose guidelines for carcinogen risk [101].

Recent studies on human MC daily intake and the high incidence of primary liver cancer (PLC) in the Chinese district of Haimen indicated values of 2.2–3.9 μg MC-LR per day per adult [103, 104] and 0.4–2.0 μg MC-LR per day per child [105]. These values are below or around the WHO limit but well above the U.S. EPA chronic TDI limit. Two cases of human intoxication occurred during toxic *Planktothrix agardhii* bloom episodes responsible for severe outbreaks of acute illness, one in Finland in 1989, where MCs were detected between 0.1 and 0.5 $\mu g L^{-1}$, and another in Sweden in 1994, with MCs up to 0.82 $\mu g L^{-1}$ [106]. According to these levels, a mean daily exposure should have ranged from 0.2 to 1 μg in the first case and reached 1.6 μg in the second case, both values being well below the WHO TDI limit but above the U.S. EPA acute limit. These findings lead us to consider the U.S. EPA limits for MCs as more adequate for a modern risk assessment. Indirect evidence also supports tumour promotion in human following chronic MC exposure in Florida [107] and in Serbia [108].

Considering the lack of exposure data regarding NODs, no TDI guidelines are available [83]. However, considering the higher toxicity of NODs compared with MCs [82], exposure guidelines would most probably have to be more conservative.

11.4.4 Impacts on Aquatic Ecosystems

Toxin-producing cyanobacteria have been linked to diverse fatal incidents involving terrestrial domestic and wild animals [109, 110]. In addition, various studies focused on evaluating potential deleterious effects of MCs and NODs on a diversity of aquatic organisms, including bacteria, aquatic plants, brown algae, phytoplankton, zooplankton, molluscs, amphibians, and fishes [111–114]. These organisms may be exposed to large quantities of toxins through cyanobacteria ingestion or, to a lesser extent, contact with dissolved toxins during bloom events [115]. Considering current debates on the putative ecological role of MCs and NODs, the potential impacts of these toxins on aquatic organisms and on aquatic ecosystem functioning are still under close scrutiny.

Bacteria do not appear to be strongly affected by exposure to MCs. *M. aeruginosa* extracts were shown to be nontoxic to *Bacillus subtilis, Escherichia coli, Pseudomonas hydrophila*, or *Staphylococcus aureus* [116], which may originate from the fact that MCs are inhibitors of eukaryotic protein phosphatases. On the contrary, many aquatic bacteria were shown to have the ability to use MCs as a substrate [117–119] and to have biodegradation activity. However, only a small fraction of bacteria are cultivable and experimental data may not reflect potential toxin-associated interactions between bacteria and cyanobacteria [120].

MCs are also known to bioaccumulate in aquatic plants [121–123]. MCs cause growth and photosynthesis inhibition, change of chlorophyll a/b ratio [124], and oxidative stress [125] at environmentally relevant concentrations. Similar effects were also found on planktonic autotrophs including growth inhibition [126–131], reduced photosynthesis activity [129, 130, 132], induction of oxidative stress [131, 133–135], and impacts on cell and chloroplast volumes and pigment content [136]. However, many studies were performed using extracts (and not pure toxins) at high MCs concentrations (>50 $\mu g L^{-1}$). In addition, results obtained from exposure of planktonic autotrophs to environmentally relevant concentrations showed deleterious effects only on a limited number of sensitive species [128, 133, 137]. These data suggest that allelopathy is not the main ecological function of MCs. Similar results have been reported for NOD impact on Baltic phytoplankton communities [138].

Cyanobacteria filaments and colonies are considered to be relatively inedible for grazers. In addition, they lack sterols and polyunsaturated fatty acids [139] and produce protease inhibitors [140].

Despite cyanobacteria constituting a poor-quality food, MCs and NODs were shown to accumulate in zooplankton [141–144]. However, it remains unclear to what extent these hepatotoxins act as an antiherbivore chemical defence. Indeed, recent meta-analysis showed large differences in species-specific response to cyanobacterial toxins exposure [145, 146]. In addition, recent research found that *Daphnia* spp. exposed to recurrent toxin-producing cyanobacteria blooms showed greater tolerance to cyanotoxins [147–149]. These results suggest that adaptation, selection of resistant genotypes or evolution may reduce the sensitivity of zooplankton to repeated exposure to MCs and NODs.

Finally, cyanobacterial hepatotoxins were also found on higher trophic levels (e.g., molluscs, amphibians, fishes, birds) with associated deleterious effects. More specifically, several large-scale fish death outbreaks have been associated to massive occurrence of cyanobacteria in waterbodies [150, 151]. The toxicity of MCs in fish depends on the balance between accumulation and metabolism, and the observed species-specific sensitivities have been interpreted as the result of anatomical, physiological, behavioural differences, and detoxification capacities *via* the glutathione-*S*-transferase pathway among the various fish orders [152, 153].

Despite toxin accumulation on all trophic levels, there is little support for biomagnification of MCs and NODs [115, 143, 154]. The lack of biomagnification probably originates from the degradation of the toxins by detoxification pathways or excretion and the high rate of biodegradation of MCs and NODs in natural waters (half-life = 4–18 days) [155].

11.5 Available Methods for the Analysis of MCs and NODs

Structural characterisation of MCs and NODs has mainly used ^{1}H and ^{13}C nuclear magnetic resonance spectroscopy (NMR), various mass spectrometry (MS) techniques, and the advanced Marfey [156] method, which assigns the stereoconfiguration of the amino acid residues. MCs have also been cocrystallized with a protein phosphatase and subjected to x-ray crystallography for three-dimensional structural determination [157].

Quantitative analyses of MCs and NODs can be based on physicochemical trace analyses, biochemical methods, and bioassays. In many cases it is beneficial to use a combination of methods: one method for screening and another one for trace analysis.

Biochemical methods are well suited to quick screening of MCs and NODs in water samples. Commercial enzyme-linked immunosorbent assays and protein phosphatase inhibition assays are available. Both methods provide semiquantitative estimates of toxin concentrations. Reported whole-organism bioassays for MCs/NODs have used either animals (mice and invertebrates such as *Artemia salina*) [158, 159] or plants (such as *Sinapis alba*) [160]. Biochemical methods and bioassays cannot identify individual variants of the toxins, and the quantitation of the toxins may be less accurate than with physicochemical methods [161].

Major congeners of MCs and NODs are fairly easily resolved and quantified by reversed-phase high-performance liquid chromatography (HPLC) [162, 163]. The high variability in toxin polarity makes gradient elution practically necessary. The toxins have a strong chromophore in the Adda residue, that is, the conjugated diene absorbing at 238 nm, making the toxins well amenable to UV absorbance detection [164]. The characteristic UV spectrum shape is indicative of MCs and NODs. While UV absorbance detection can detect subnanogram amounts of toxin on column, analysis of water samples by HPLC necessitates sample preconcentration by solid-phase extraction [165]. LC–liquid chromatography (MS)/MS analyses using electrospray ionisation and different mass analysers (including the triple quadrupole, ion trap, and various exact mass analysers) are becoming the gold standard in critical MC/NOD applications [166]. MS offers remarkable sensitivity and selectivity improvements over optical detection and characteristic mass fragments, produced by either in-source or by MS/MS fragmentation, can be used for the reliable identification of the toxins [166]. In order to gain sufficient confidence in identification and quantitation, the use of tandem MS is practically mandatory in analytical work concerning complex matrices such as animal tissues [165, 167]. Matrix-assisted laser desorption/ionization–time of flight MS can be used to obtain a quick, semiquantitative assessment of MC/NOD presence in a cyanobacterial sample [168]. The so-called MMPB method (for 2-methyl-3-methoxy-4-phenylbutyric acid method), which quantifies an oxidative derivative of Adda, can report a sum of MCs and NODs in the sample [169]. All the physicochemical methods are dependent on available reference materials as standards but, unfortunately, only a small fraction of all known MCs and NODs are commercially available.

Methods measuring free MC, based on prior solvent extraction of the analyte, may underestimate the total MC content because of covalent binding between the toxin and active site of protein phosphatases, with consequent inability of solvent to extract enzyme-bound toxins, which are slowly released and are toxicologically active [170, 171]. Conventional methods may thus underestimate total MCs, especially when dealing with tissue samples [172, 173]. Later chapters and SOPs in this handbook provide details about the methods available for the determination of MCs and NODs.

References

[1] Botes, D.P., Tuinman, A.A., Wessels, P.L. *et al.* (1984) The structure of cyanoginosin-LA, a cyclic heptapeptide toxin from the cyanobacterium *Microcystis aeruginosa. Journal of the Chemical Society, Perkin Transactions* **1**, 2311–2318.

[2] Rinehart, K.L., Harada, K., Namikoshi, M. *et al.* (1988) Nodularin, microcystin, and the configuration of Adda. *Journal of the American Chemical Society* **110** (25), 8557–8558.

[3] Namikoshi, M., Rinehart, K.L., Sakai, R. *et al.* (1992) Identification of 12 hepatotoxins from a Homer Lake bloom of the cyanobacteria *Microcystis aeruginosa*, *Microcystis viridis*, and *Microcystis wesenbergii*: nine new microcystins. *The Journal of Organic* Chemistry **57** (3), 866–872.

[4] Rinehart, K.L., Namikoshi, M. and Choi, B.W. (1994) Structure and biosynthesis of toxins from blue-green algae (cyanobacteria). *Journal of Applied Phycology* **6** 159–176.

[5] Spoof, L. and Meriluoto, J. (2005) Analysis of microcystins by high-performance liquid chromatography with photodiode-array detection, in *TOXIC: Cyanobacterial Monitoring and Cyanotoxin Analysis* (eds. J. Meriluoto and G.A. Codd), Åbo Akademi University Press, Turku, pp. 77–84.

[6] Namikoshi, M., Choi, B.W., Sakai, R. *et al.* (1994) New nodularins: a general method for structure assignment. *The Journal of Organic Chemistry* **59** (9), 2349–2357.

[7] Moffitt, M.C. and Neilan, B.A. (2004) Characterization of the nodularin synthetase gene cluster and proposed theory of the evolution of cyanobacterial hepatotoxins. *Applied and Environmental Microbiology* **70** (11), 6353–6362.

[8] Mazur-Marzec, H., Kaczkowska, M.J., Blaszczyk, A. *et al.* (2013) Diversity of peptides produced by *Nodularia spumigena* from various geographical regions. *Marine Drugs* **11** (1), 1–19.

[9] Dittmann, E., Neilan, B.A., Erhard, M. *et al.* (1997) Insertional mutagenesis of a peptide synthetase gene that is responsible for hepatotoxin production in the cyanobacterium *Microcystis aeruginosa* PCC 7806. *Molecular Microbiology* **26** (4), 779–787.

[10] Nishizawa, T., Asayama, M., Fujii, K. *et al.* (1999) Genetic analysis of the peptide synthetase genes for a cyclic heptapeptide microcystin in *Microcystis* spp. *The Journal of Biochemistry* **126** (3), 520–529.

[11] Christiansen, G., Fastner, J., Erhard, M. *et al.* (2003) Microcystin biosynthesis in *Planktothrix*: genes, evolution, and manipulation. *Journal of Bacteriology* **185** (2), 564–572.

[12] Rouhiainen, L., Vakkilainen, T., Siemer, B.L. *et al.* (2004) Genes coding for hepatotoxic heptapeptides (microcystins) in the cyanobacterium *Anabaena* strain 90. *Applied and Environmental Microbiology* **70** (2), 686–692.

[13] Dittmann, E., Fewer, D.P. and Neilan, B.A. (2013) Cyanobacterial toxins: biosynthetic routes and evolutionary roots. *FEMS Microbiology Reviews* **37** (1), 23–43.

[14] Tillett, D., Dittmann, E., Erhard, M. *et al.* (2000) Structural organization of microcystin biosynthesis in *Microcystis aeruginosa* PCC7806: an integrated peptide-polyketide synthetase system. *Chemistry & Biology* **7** (10), 753–764.

[15] Kalaitzis, J.A., Lauro, F.M. and Neilan, B.A. (2009) Mining cyanobacterial genomes for genes encoding complex biosynthetic pathways. *Natural Product Reports* **26** (11), 1447–1465.

[16] Fastner, J., Erhard, M., Carmichael, W.W. *et al.* (1999) Characterization and diversity of microcystins in natural blooms and strains of the genera *Microcystis* and *Planktothrix* from German freshwaters. *Archiv für Hydrobiologie* **145** (2), 147–163.

[17] Young, F.M., Thomson, C., Metcalf, J.S. *et al.* (2005) Immunogold localisation of microcystins in cryosectioned cells of *Microcystis*. *Journal of Structural Biology* **151** (2), 208–214.

[18] Kaebernick, M., Neilan, B.A., Börner, T. and Dittmann, E. (2000) Light and the transcriptional response of the microcystin biosynthesis gene cluster. *Applied and Environmental Microbiology* **66** (8), 3387–3392.

[19] Dittmann, E. and Börner, T. (2005) Genetic contributions to the risk assessment of microcystin in the environment. *Toxicology and Applied Pharmacology* **203** (3), 192–200.

[20] Tooming-Klunderud, A., Fewer, D.P., Rohrlack, T. *et al.* (2008) Evidence for positive selection acting on microcystin synthetase adenylation domains in three cyanobacterial genera. *BMC Evolutionary Biology* **8** 256.

[21] López-Rodas, V., Costas, E., Bañares E. *et al.* (2006) Analysis of polygenic traits of *Microcystis aeruginosa* (Cyanobacteria) strains by Restricted Maximum Likelihood (REML) procedures: 2. Microcystin net production, photosynthesis and respiration. *Phycologia* **45** (3), 243–248.

[22] Tanabe, Y., Sano, T., Kasai, F. and Watanabe, M.M. (2009) Recombination, cryptic clades and neutral molecular divergence of the microcystin synthetase (*mcy*) genes of toxic cyanobacterium *Microcystis aeruginosa*. *BMC Evolutionary Biology* **9**, 115.

[23] Kurmayer, R., Blom, J.F., Deng, L. and Pernthaler, J. (2015) Integrating phylogeny, geographic niche partitioning and secondary metabolite synthesis in bloom-forming *Planktothrix*. *The ISME Journal* **9** (4), 909–921.

[24] Richardson, L.L., Sekar, R., Myers, J.L. *et al.* (2007) The presence of the cyanobacterial toxin microcystin in black band disease of corals. *FEMS Microbiology Letters* **272** (2), 182–187.

[25] Gantar, M., Sekar, R. and Richardson, L.L. (2009) Cyanotoxins from black band disease of corals and from other coral reef environments. *Microbial Ecology* **58** (4), 856–864.

[26] Casamatta, D., Stanić, D., Gantar, M. and Richardson, L.L. (2012) Characterization of *Roseofilum reptotaenium* (Oscillatoriales, Cyanobacteria) gen. et sp. nov. isolated from Caribbean black band disease. *Phycologia* **51** (5), 489–499.

[27] Oksanen, I., Jokela, J., Fewer D.P. *et al.* (2004) Discovery of rare and highly toxic microcystins from lichen-associated cyanobacterium *Nostoc* sp. strain IO-102-I. *Applied and Environmental Microbiology* **70** (10), 5756–5763.

[28] Prinsep, M.R., Caplan, F.R., Moore, R.E. *et al.* (1992) Microcystin-LA from a blue-green alga belonging to the stigonematales. *Phytochemistry* **31** (4), 1247–1248.

[29] Faassen, E.J. and Lürling, M. (2013) Occurrence of the microcystins MC-LW and MC-LF in Dutch surface waters and their contribution to total microcystin toxicity. *Marine Drugs* **11** (7), 2643–2654.

[30] Wood, S.A., Briggs, L.R., Sprosen, J. *et al.* (2006) Changes in concentrations of microcystins in rainbow trout, freshwater mussels, and cyanobacteria in Lakes Rotoiti and Rotoehu. *Environmental Toxicology* **21** (3), 205–222.

[31] Conti, A.L., Guerrero, J.M. and Regueira, J.M. (2005) Levels of microcystins in two Argentinean reservoirs used for water supply and recreation: differences in the implementation of safe levels. *Environmental Toxicology* **20** (3), 263–269.

[32] Messineo, V., Mattei, D., Melchiorre, S. *et al.* (2006) Microcystin diversity in a *Planktothrix rubescens* population from Lake Albano (Central Italy). *Toxicon* **48** (2), 160–174.

[33] Willame, R., Jurczak, T., Iffly, J.-F. *et al.* (2005) Distribution of hepatotoxic cyanobacterial blooms in Belgium and Luxembourg. *Hydrobiologia* **551** (1), 99–117.

[34] Pavlova, V., Babica, P., Todorova, D. *et al.* (2006) Contamination of some reservoirs and lakes in Republic of Bulgaria by microcystins. *Acta Hydrochimica et Hydrobiologica* **34** (5), 437–441.

[35] Descy, J.-P., Pirlot, S., Verniers, L. *et al.* (2011) Cyanobacterial Blooms: Toxicity, Diversity, Modeling and Management 'B-BLOOMS2'. Belgian Science Policy, Brussels. http://www.belspo.be/belspo/ssd/science/Reports/B_BLOOMS%20FinalReport_ML.pdf

[36] Bláhová, L., Babica, P., Maršálková, E. *et al.* (2007) Concentrations and seasonal trends of extracellular microcystins in freshwaters of the Czech Republic: results of the national monitoring program. *CLEAN – Soil Air Water* **35** (4), 348–354.

[37] Lindholm, T., Vesterkvist, P., Spoof, L. *et al.* (2003) Microcystin occurrence in lakes in Åland, SW Finland. *Hydrobiologia* **505** (1-3), 129–138.

[38] Rantala, A., Rajaniemi-Wacklin, P., Lyra, C. *et al.* (2006) Detection of microcystin-producing cyanobacteria in Finnish lakes with genus-specific microcystin synthetase gene E (*mcyE*) PCR and associations with environmental factors. *Applied and Environmental Microbiology* **72** (9), 6101–6110.

[39] Brient, L., Vézie, C. and Bertru, G. (2001) Evaluation des Efflorescences à Cyanobactéries dans des Eaux de Cours d'Eau et Plans d'Eau Bretons. Direction Régionale de l'Environnement, Rennes. http://parlonstechneau.univ-rennes1.fr/2009/files/evaluation-des-efflorescences.pdf

[40] Agence Française de Sécurité Sanitaire des Aliments (2006) Rapport sur l'Evaluation des Risques Lies à la Présence de Cyanobactéries et de Leurs Toxines Dans les Eaux Destinées à l'Alimentation, à la Baignade et Autres Activités Récréatives. *Agence Française de Sécurité Sanitaire des Aliments*, Paris. http://www.ars.limousin.sante.fr/fileadmin/LIMOUSIN/Documents_ARS_Limousin/Votre_sante/environnement/eau/AFSSA_AFSSETcomplet.pdf

[41] Fromme, H., Köhler, A., Krause, R. and Führling, D. (2000) Occurrence of cyanobacterial toxins: microcystins and anatoxin-a in Berlin waterbodies with implications to human health and regulations. *Environmental Toxicology* **15** (2), 120–130.

[42] Frank, C.A.P. (2002) Microcystin-producing cyanobacteria in recreational waters in southwestern Germany. *Environmental Toxicology* **17** (4), 361–366.

[43] Cook, C.M., Vardaka, E. and Lanaras, T. (2004) Toxic Cyanobacteria in Greek freshwaters, 1987—2000: occurrence, toxicity, and impacts in the Mediterranean region. *Acta Hydrochimica et Hydrobiologica* **32** (2), 107–124.

[44] Papadimitriou, T., Kagalou, I., Bacopoulos, V. and Leonardos, I.D. (2010) Accumulation of microcystins in water and fish tissues: an estimation of risks associated with microcystins in most of the Greek Lakes. *Environmental Toxicology* **25** (4), 418–427.

[45] Mooney, K.M., Hamilton, J.T.G., Floyd, S.D. *et al.* (2011) Initial studies on the occurrence of cyanobacteria and microcystins in Irish lakes. *Environmental Toxicology* **26** (5), 566–570.

[46] Messineo, V. Bogialli, S., Melchiorre, S. *et al.* (2009) Cyanobacterial toxins in Italian freshwaters. *Limnologica - Ecology Management of Inland Waters* **39** (2), 95–106.

[47] Jurczak, T., Tarczyñska, M., Karlsson, K. and Meriluoto, J. (2004) Characterization and diversity of cyanobacterial hepatotoxins (microcystins) in blooms from Polish freshwaters Identified by liquid chromatography-electrospray ionisation mass spectrometry. *Chromatographia* **59** (9), 571–578.

[48] Vasconcelos, V.M., Sivonen, K., Evans, W.R. *et al.* (1996) Hepatotoxic microcystin diversity in cyanobacterial blooms collected in Portuguese freshwaters. *Water Research* **30** (10), 2377–2384.

[49] Maršálek, B., Bláha, L. and Hindák, F. (2000) Review of toxicity of cyanobacteria in Slovakia. *Biologia* **55** (6), 645–652.

[50] Aboal, M. and Puig, M.-A. (2005) Intracellular and dissolved microcystin in reservoirs of the river Segura basin, Murcia, SE Spain. *Toxicon* **45** (4), 509–518.

[51] Francis, G. (1878) Poisonous Australian lake. *Nature* **18** 11–12.

[52] Sivonen, K., Kononen, K., Carmichael, W.W. *et al.* (1989) Occurrence of the hepatotoxic cyanobacterium *Nodularia spumigena* in the Baltic Sea and structure of the toxin. *Applied and Environmental Microbiology* **55** (8), 1990–1995.

[53] Blackburn, S.I., McCausland, M.A., Bolch, C.J.S. *et al.* (1996) Effect of salinity on growth and toxin production in cultures of the bloom-forming cyanobacterium *Nodularia spumigena* from Australian waters. *Phycologia* **35** (6), 511–522.

[54] Del C. Pérez, M., Bonilla, S., De León, L. *et al.* (1999) A bloom of *Nodularia baltica-spumigena* group (Cyanobacteria) in a shallow coastal lagoon of Uruguay, South America. *Algological Studies - Archiv für Hydrobiologie Supplements* **93** 91–101.

[55] Falcón, L.I., Escobar-Briones, E. and Romero, D. (2002) Nitrogen fixation patterns displayed by cyanobacterial consortia in Alchichica crater-lake, Mexico, in *Advances in Mexican Limnology: Basic and Applied Aspects* (eds J. Alcocer and S.S.S. Sarma), Springer Netherlands, Dordrecht, pp. 71–78.

[56] Woodward, C.A. and Shulmeister, J.A. (2005) Holocene record of human induced and natural environmental changes from Lake Forsyth (Te Wairewa), New Zealand. *Journal of Paleolimnology* **34** (4), 481–501.
[57] Akcaalan, R., Mazur-Marzec, H., Zalewska, A. and Albay, M. (2009) Phenotypic and toxicological characterization of toxic *Nodularia spumigena* from a freshwater lake in Turkey. *Harmful Algae* **8** (2), 273–278.
[58] Lehtimäki, J., Sivonen, K., Luukkainen, R. and Niemelä, S.I. (1994) The effects of incubation time, temperature, light, salinity and phosphorus on growth and hepatotoxin production by *Nodularia* strains. *Archiv für Hydrobiologie* **130** (3), 269–282.
[59] Moffitt, M.C. and Neilan, B.A. (2001) On the presence of peptide synthetase and polyketide synthase genes in the cyanobacterial genus *Nodularia. FEMS Microbiology Letters* **196** (2), 207–214.
[60] Gehringer, M.M. Adler, L., Roberts, A.A. *et al.* (2012) Nodularin, a cyanobacterial toxin, is synthesized in planta by symbiotic *Nostoc* sp. *The ISME Journal* **6** (10), 1834–1847.
[61] Teixeira, M. da G., Costa, M. da C., de Carvalho, V.L. *et al.* (1993) Gastroenteritis epidemic in the area of the Itaparica Dam, Bahia, Brazil. *Bulletin of the Pan American Health Organization* **27** (3), 244–253.
[62] Codd, G. (1995) Cyanobacterial toxins: occurrence, properties and biological significance. *Water Science and Technology* **32** (4), 149–156.
[63] Jochimsen, E.M., Carmichael, W.W., An, J.S. *et al.* (1998) Liver failure and death after exposure to microcystins at a hemodialysis center in Brazil. *New England Journal of Medicine* **338** (13), 873–878.
[64] Gácsi, M. Antal, O., Vasas, G. *et al.* (2009) Comparative study of cyanotoxins affecting cytoskeletal and chromatin structures in CHO-K1 cells. *Toxicology In Vitro* **23** (4), 710–718.
[65] Hastie, C.J., Borthwick, E.B., Morrison, L.F. *et al.* (2005) Inhibition of several protein phosphatases by a non-covalently interacting microcystin and a novel cyanobacterial peptide, nostocyclin. *Biochimica Biophysica Acta* **1726** (2), 187–193.
[66] MacKintosh, C., Beattie, K.A., Klumpp, S. *et al.* (1990) Cyanobacterial microcystin-LR is a potent and specific inhibitor of protein phosphatases 1 and 2A from both mammals and higher plants. *FEBS Letters* **264** (2), 187–192.
[67] Honkanen, R.E., Codispoti, B.A., Tse, K. *et al.* (1994) Characterization of natural toxins with inhibitory activity against serine/threonine protein phosphatases. *Toxicon* **32** (3), 339–350.
[68] Annila, A. Lehtimäki, J., Mattila, K. *et al.* (1996) Solution structure of nodularin: an inhibitor of serine/threonine-specific protein phosphatases. *The Journal of Biological Chemistry* **271** (28), 16695–16702.
[69] Dawson, R. (1998) The toxicology of microcystins. *Toxicon* **36** (7), 953–962.
[70] Nishiwaki-Matsushima, R., Ohta, T., Nishiwaki, S. *et al.* (1992) Liver tumor promotion by the cyanobacterial cyclic peptide toxin microcystin-LR. *Journal of Cancer Research and Clinical Oncology* **118** (6), 420–424.
[71] Ito, E., Kondo, F., Terao, K. and Harada, K.-I. (1997) Neoplastic nodular formation in mouse liver induced by repeated intraperitoneal injections of microcystin-LR. *Toxicon* **35** (9), 1453–1457.
[72] Humpage, A.R., Fenech, M., Thomas, P. and Falconer, I.R. (2000) Micronucleus induction and chromosome loss in transformed human white cells indicate clastogenic and aneugenic action of the cyanobacterial toxin, cylindrospermopsin. *Mutation Research* **472** (1–2), 155–161.
[73] Rojas, M., Nuñez, M.T. and Zambrano, F. (1990) Inhibitory effect of a toxic peptide isolated from a waterbloom of *Microcystis* sp. (Cyanobacteria) on iron uptake by rabbit reticulocytes. *Toxicon* **28** (11), 1325–1332.
[74] Sayed, S.B., Whitehouse, B.J. and Jones, P.M. (1997) Phosphoserine/threonine phosphatases in the rat adrenal cortex: a role in the control of steroidogenesis? *Journal of Endocrinology* **154** (3), 449–458.
[75] Ford, S.L., Abayasekara, D.R., Persaud, S.J. and Jones, P.M. (1996) Role of phosphoprotein phosphatases in the corpus luteum: I. Identification and characterisation of serine/threonine phosphoprotein phosphatases in isolated rat luteal cells. *Journal of Endocrinology* **150** (2), 205–211.
[76] Hernandez, M. (2000) Modulation of human polymorphonuclear leukocyte adherence by cyanopeptide toxins. *Environmental Research* **84** (1), 64–68.
[77] Leiers, T., Bihlmayer, A., Ammon, H.P.T. and Wahl, M.A. (2000) $[Ca^{2+}]_{i}$ and insulin-stimulating effect of the non-membranepermeable phosphatase-inhibitor microcystin-LR in intact insulin-secreting cells (RINm5F). *British Journal of Pharmacology* **130** (6), 1406–1410.
[78] Lankoff, A., Carmichael, W.W., Grasman, K.A. and Yuan, M. (2004) The uptake kinetics and immunotoxic effects of microcystin-LR in human and chicken peripheral blood lymphocytes *in vitro. Toxicology* **204** (1), 23–40.

[79] Žegura, B., Sedmak, B. and Filipič, M. (2003) Microcystin-LR induces oxidative DNA damage in human hepatoma cell line HepG2. *Toxicon* **41** (1), 41–48.
[80] Bouaïcha, N., Maatouk, I., Plessis, M.J. and Périn, F. (2005) Genotoxic potential of Microcystin-LR and nodularin *in vitro* in primary cultured rat hepatocytes and *in vivo* in rat liver. *Environmental Toxicology* **20** (3), 341–347.
[81] Li, H., Xie, P., Li, G. *et al.* (2009) *In vivo* study on the effects of microcystin extracts on the expression profiles of proto-oncogenes (*c-fos*, *c-jun* and *c-myc*) in liver, kidney and testis of male Wistar rats injected i.v. with toxins. *Toxicon* **53** (1), 169–175.
[82] Ohta, T., Sueoka, E., Iida, N. *et al.* (1994) Nodularin, a potent inhibitor of protein phosphatases 1 and 2A, is a new environmental carcinogen in male F344 rat liver. *Cancer Research* **54** (24), 6402–6406.
[83] Žegura, B., Štraser, A. and Filipič, M. (2011) Genotoxicity and potential carcinogenicity of cyanobacterial toxins: a review. *Mutation Research* **727** (1–2), 16–41.
[84] Fischer, A., Hoeger, S.J., Stemmer, K. *et al.* (2010) The role of organic anion transporting polypeptides (OATPs/SLCOs) in the toxicity of different microcystin congeners *in vitro*: a comparison of primary human hepatocytes and OATP-transfected HEK293 cells. *Toxicology and Applied Pharmacology* **245** (1), 9–20.
[85] Ito, E., Kondo, F. and Harada, K.-I. (2001) Intratracheal administration of microcystin-LR, and its distribution. *Toxicon* **39** (2–3), 265–271.
[86] Fitzgeorge, R.B., Clark, I.A. and Keevil, C.W. (1994) Routes of intoxication, in *Detection Methods for Cyanobacterial Toxins*, (eds G.A. Codd, T.M. Jefferies, C.W. Keevil and E. Potter), The Royal Society of Chemistry, Cambridge, pp. 69–74.
[87] Solter, P.F., Wollenberg, G.K., Huang, X. *et al.* (1998) Prolonged sublethal exposure to the protein phosphatase inhibitor microcystin-LR results in multiple dose-dependent hepatotoxic effects. *Toxicological Sciences* **44** (1), 87–96.
[88] Chen, Y., Shen, D. and Fang, D. (2013) Nodularins in poisoning. *Clinica Chimica Acta* **425**, 18–29.
[89] Falconer, I.R., Beresford, A.M. and Runnegar, M.T. (1983) Evidence of liver damage by toxin from a bloom of the blue-green alga, *Microcystis aeruginosa*. The *Medical Journal of Australia* **1** (11), 511–514.
[90] Eynard, F. (2000) Risk of cyanobacterial toxins in Riga waters (Latvia). *Water Research* **34** (1), 2979–2988.
[91] Falconer, I.R. and Humpage, A.R. (2005) Health risk assessment of cyanobacterial (blue-green algal) toxins in drinking water. *International Journal of Environmental Research and Public Health* **2** (1), 43–50.
[92] Bruno, M., Fiori, M., Mattei, D. *et al.* (2006) ELISA and LC-MS/MS methods for determining cyanobacterial toxins in blue-green algae food supplements. *Natural Product Research* **20** (9), 827–834.
[93] Magalhães, V.F., Soares, R.M. and Azevedo, S.M. (2001) Microcystin contamination in fish from the Jacarepaguá Lagoon (Rio de Janeiro, Brazil): ecological implication and human health risk. *Toxicon* **39** (7), 1077–1085.
[94] Bruno, M., Melchiorre, S., Messineo, V. *et al.* (2009) Microcystin detection in contaminated fish from italian lakes using ELISA immunoassays and LC-MS/MS analysis, in *Handbook on Cyanobacteria* (eds. P.M. Gault and H.J. Marler), Nova Science Publishers Inc., New York, pp. 191–210.
[95] Sipiä, V.O., Kankaanpää, H.T., Pflugmacher, S. *et al.* (2002) Bioaccumulation and detoxication of nodularin in tissues of flounder (*Platichthys flesus*), mussels (*Mytilus edulis*, *Dreissena polymorpha*), and clams (*Macoma balthica*) from the Northern Baltic Sea. *Ecotoxicology and Environmental Safety* **53** (2), 305–311.
[96] Sipiä, V., Kankaanpää, H., Peltonen, H. *et al.* (2007) Transfer of nodularin to three-spined stickleback (*Gasterosteus aculeatus* L.), herring (*Clupea harengus* L.), and salmon (*Salmo salar* L.) in the northern Baltic Sea. *Ecotoxicology and Environmental Safety* **66** (3), 421–425.
[97] Messineo, V., Melchiorre, S., Di Corcia, A. *et al.* (2010) Seasonal succession of *Cylindrospermopsis raciborskii* and *Aphanizomenon ovalisporum* blooms with cylindrospermopsin occurrence in the volcanic Lake Albano, Central Italy. *Environmental Toxicology* **25** (1), 18–27.
[98] Chorus, I. and Bartram, J. (eds) (1999) *Toxic Cyanobacteria in Water: A Guide to Their Public Health Consequences, Monitoring, and Management*. E & FN Spon, New York.
[99] Carmichael, W. W. (2011) Peer Review of Cyanotoxin Toxicity Criteria and Health Based Water Concentrations to Protect Human Swimmers, Dogs and Cattle. State Water Resources Control Board - Division of Water Quality, Sacramento. http://www.swrcb.ca.gov/water_issues/programs/peer_review/docs/calif_cyanotoxins/carmichael_review_june2011.pdf

[100] Heinze, R. (1999) Toxicity of the cyanobacterial toxin microcystin-LR to rats after 28 days intake with the drinking water. *Environmental Toxicology* **14** (1), 57–60.

[101] US Environmental Protection Agency (2006). Toxicological Reviews of Cyanobacterial Toxins: Microcystins LR, RR, YR and LA. US Environmental Protection Agency, Cincinnati.

[102] International Agency for Research on Cancer (2010) Ingested Nitrate and Nitrite, and Cyanobacterial Peptide Toxins. World Health Organization, International Agency for Research on Cancer, Lyon. http://monographs.iarc.fr/ENG/Monographs/vol94/

[103] Chen, J., Xie, P., Li, L. and Xu, J. (2009) First identification of the hepatotoxic microcystins in the serum of a chronically exposed human population together with indication of hepatocellular damage. *Toxicological Sciences* **108** (1), 81–89.

[104] Zhang, H., Zhang, J. and Zhu, Y. (2009) Identification of microcystins in waters used for daily life by people who live on Tai Lake during a serious cyanobacteria dominated bloom with risk analysis to human health. *Environmental Toxicology* **24** (1), 82–86.

[105] Li, Y., Chen, J., Zhao, Q. *et al.* (2011) A cross-sectional investigation of chronic exposure to microcystin in relationship to childhood liver damage in the Three Gorges Reservoir Region, China. *Environmental Health Perspectives* **119** (10), 1483–1488.

[106] Annadoter, H., Cronberg, G., Lawton L. *et al.* (2001) An extensive outbreak of gastroenteritis associated with the toxic cyanobacterium *Planktothrix agardhii* (Oscillatoriales, Cyanophyceae) in Scania, South Sweden, in *Cyanotoxins. Occurrences, Causes, Consequences* (ed. I. Chorus), Springer, Berlin, pp. 200–208.

[107] Fleming, L.E., Rivero, C., Burns J. *et al.* (2002) Blue green algal (cyanobacterial) toxins, surface drinking water, and liver cancer in Florida. *Harmful Algae* **1** (2), 157–168.

[108] Svirčev, Z., Krstić, S., Miladinov-Mikov, M. *et al.* (2009) Freshwater cyanobacterial blooms and primary liver cancer epidemiological studies in Serbia. *Journal of Environmental Science and Health, Part C: Environmental Carcinogenesis and Ecotoxicology Reviews* **27** (1), 36–55.

[109] Duy, T.N., Lam, P.K.S., Shaw, G.R. and Connell, D.W. (2000) Toxicology and risk assessment of freshwater cyanobacterial (Blue-green algal) toxins in water, in *Reviews of Environmental Contamination and Toxicology*, vol. **163** (ed. G.W. Ware), Springer, New York, pp. 113–186.

[110] Briand, J.-F., Jacquet, S., Bernard, C. and Humbert, J.-F. (2003) Health hazards for terrestrial vertebrates from toxic cyanobacteria in surface water ecosystems. *Veterinary Research* **34** (4), 361–377.

[111] Wiegand, C. and Pflugmacher, S. (2005) Ecotoxicological effects of selected cyanobacterial secondary metabolites: a short review. *Toxicology and Applied Pharmacology* **203** (3), 201–218.

[112] Zurawell, R.W., Chen, H., Burke, J.M. and Prepas, E.E. (2005) Hepatotoxic cyanobacteria: a review of the biological importance of microcystins in freshwater environments. *Journal of Toxicology and Environmental Health, Part B: Critical Reviews* **8** (1), 1–37.

[113] Pflugmacher, S., Olin, M. and Kankaanpää, H. (2007) Nodularin induces oxidative stress in the Baltic Sea brown alga *Fucus vesiculosus* (Phaeophyceae). *Marine Environmental Research* **64** (2), 149–159.

[114] Giaramida, L., Manage, P.M., Edwards, C. *et al.* (2013) Bacterial communities' response to microcystins exposure and nutrient availability: linking degradation capacity to community structure. *International Biodeterioration & Biodegradation* **84**, 111–117.

[115] Ibelings, B.W. and Havens, K.E. (2008) Cyanobacterial toxins: a qualitative meta-analysis of concentrations, dosage and effects in freshwater, estuarine and marine biota, in *Cyanobacterial Harmful Algal Blooms: State of the Science and Research Needs*, vol. **619**, (ed. H.K. Hudnell), Springer, New York, pp. 675–732.

[116] Foxall, T.L. and Sasner, J.J. (1981) Effect of a hepatic toxin from the cyanophyte *Microcystis aeruginosa*, in *The Water Environment: Algal Toxins and Health* (ed. W.W. Carmichael), Plenum Press, New York, pp. 365–387.

[117] Jones, G. and Orr, P.T. (1994) Release and degradation of microcystin following algicide treatment of a *Microcystis aeruginosa* bloom in a recreational lake, as determined by HPLC and protein phosphatase inhibition assay. *Water Research* **28** (4), 871–876.

[118] Lam, A.K.-Y., Fedorak, P.M. and Prepas, E.E. (1995) Biotransformation of the cyanobacterial hepatotoxin microcystin-LR, as determined by HPLC and protein phosphatase bioassay. *Environmental Science and Technology* **29** (1), 242–246.

[119] Lahti, K. (1997) Persistence of cyanobacterial hepatotoxin, microcystin-LR in particulate material and dissolved in lake water. *Water Research* **31** (5), 1005–1012.

[120] Mazur-Marzec, H., Toruńska, A., Błońska, M.J. *et al.* (2009) Biodegradation of nodularin and effects of the toxin on bacterial isolates from the Gulf of Gdańsk. *Water Research* **43** (11), 2801–2810.

[121] Pflugmacher, S., Wiegand, C., Beattie, K.A. *et al.* (1998) Uptake of the cyanobacterial hepatotoxin microcystin-LR by aquatic macrophytes. *Journal of Applied Botany* **72** (5–6), 228–232.

[122] Pflugmacher, S., Codd, G.A. and Steinberg, C.E.W. (1999) Effects of the cyanobacterial toxin microcystin-LR on detoxication enzymes in aquatic plants. *Environmental Toxicology* **14** (1), 111–115.

[123] Pflugmacher, S. Wiegand, C., Beattie, K.A. *et al.* (2001) Uptake, effects, and metabolism of cyanobacterial toxins in the emergent reed plant *Phragmites australis* (cav.) trin. ex steud. *Environmental Toxicology and Chemistry* **20** (4), 846–852.

[124] Pflugmacher, S. (2002) Possible allelopathic effects of cyanotoxins, with reference to microcystin-LR, in aquatic ecosystems. *Environmental Toxicology* **17** (4), 407–413.

[125] Pflugmacher, S. (2004) Promotion of oxidative stress in the aquatic macrophyte *Ceratophyllum demersum* during biotransformation of the cyanobacterial toxin microcystin-LR. *Aquatic Toxicology* **70** (3), 169–178.

[126] Christoffersen, K. (1996) Ecological implications of cyanobacterial toxins in aquatic foodwebs. *Phycologia* **35** (6S), 42–50.

[127] Sedmak, B. and Kosi, G. (1998) The role of microcystins in heavy cyanobacterial bloom formation. *Journal of Plankton Research*, **20** (4), 691–708.

[128] Kearns, K.D. and Hunter, M.D. (2000) Green algal extracellular products regulate antialgal toxin production in a cyanobacterium. *Environmental Microbiology* **2** (3), 291–297.

[129] Singh, D.P., Tyagi, M.B., Kumar, A. *et al.* (2001) Antialgal activity of a hepatotoxin-producing cyanobacterium, *Microcystis aeruginosa. World Journal of Microbiology and Biotechnology* **17** (1), 15–22.

[130] Hu, Z., Liu, Y. and Li, D. (2004) Physiological and biochemical analyses of microcystin-RR toxicity to the cyanobacterium *Synechococcus elongatus. Environmental Toxicology* **19** (6), 571–577.

[131] Hu, Z., Liu, Y., Li, D. and Dauta, A. (2005) Growth and antioxidant system of the cyanobacterium *Synechococcus elongatus* in response to microcystin-RR. *Hydrobiologia* **534** (1–3), 23–29.

[132] Escoubas, J.M., Lomas, M., LaRoche, J. and Falkowski, P.G. (1995) Light intensity regulation of cab gene transcription is signaled by the redox state of the plastoquinone pool. *Proceedings of the National Academy of Sciences USA* **92** (22), 10237–10241.

[133] Pietsch, C., Wiegand, C., Amé, M.V. *et al.* (2001) The effects of a cyanobacterial crude extract on different aquatic organisms: evidence for cyanobacterial toxin modulating factors. *Environmental Toxicology* **16** (6), 535–542.

[134] Vardi, A., Schatz, D., Beeri, K. *et al.* (2002) Dinoflagellate-cyanobacterium communication may determine the composition of phytoplankton assemblage in a mesotrophic lake. *Current Biology* **12** (20), 1767–1772.

[135] Ou, D., Song, L., Gan, N. and Chen, W. (2005) Effects of microcystins on and toxin degradation by *Poterioochromonas* sp. *Environmental Toxicology* **20** (3), 373–380.

[136] Sedmak, B. and Elersek, T. (2005) Microcystins induce morphological and physiological changes in selected representative phytoplanktons. *Microbial Ecology* **50** (2), 298–305.

[137] Kearns, K.D. and Hunter, M.D. (2001) Toxin-producing *Anabaena flos-aquae* induces settling of *Chlamydomonas reinhardtii,* a competing motile alga. *Microbial Ecology* **42** (1), 80–86.

[138] Suikkanen, S. (2006) Allelopathy of Baltic Sea cyanobacteria: no evidence for the role of nodularin. *Journal of Plankton Research* **28** (6), 543–550.

[139] Martin-Creuzburg, D., Von Elert, E. and Hoffmann, K.H. (2008) Nutritional constraints at the cyanobacteria– *Daphnia magna* interface: the role of sterols. *Limnology and Oceanography* **53** (2), 456–468.

[140] Schwarzenberger, A., Zitt, A., Kroth, P. *et al.* (2010) Gene expression and activity of digestive proteases in *Daphnia*: effects of cyanobacterial protease inhibitors. *BMC Physiology* **10**, 6.
[141] Thostrup, L. and Christoffersen, K. (1999) Accumulation of microcystin in *Daphnia magna* feeding on toxic *Microcystis*. *Archiv für Hydrobiologie* **145** (4), 447–467.
[142] Karjalainen, M., Reinikainen, M., Spoof, L. *et al.* (2005) Trophic transfer of cyanobacterial toxins from zooplankton to planktivores: consequences for pike larvae and mysid shrimps. *Environmental Toxicology* **20** (3), 354–362.
[143] Karjalainen, M., Engström-Ost, J., Korpinen, S. *et al.* (2007) Ecosystem consequences of cyanobacteria in the northern Baltic Sea. *Ambio* **36** (2–3), 195–202.
[144] Oberhaus, L., Gelinas, M., Pinel-Alloul, B. *et al.* (2007) Grazing of two toxic *Planktothrix* species by *Daphnia pulicaria*: potential for bloom control and transfer of microcystins. *Journal of Plankton Research* **29** (10), 827–838.
[145] Wilson, A.E., Sarnelle, O. and Tillmanns, A.R. (2006) Effects of cyanobacterial toxicity and morphology on the population growth of freshwater zooplankton: meta-analyses of laboratory experiments. *Limnology and Oceanography* **51** (4), 1915–1924.
[146] Tillmanns, A.R., Wilson, A.E., Pick, F.R. and Sarnelle, O. (2008) Meta-analysis of cyanobacterial effects on zooplankton population growth rate: species-specific responses. *Archiv für Hydrobiologie* **171** (4), 285–295.
[147] Hairston, N.G., Lampert, W., Caceres, C.E. *et al.* (1999) Lake ecosystems: rapid evolution revealed by dormant eggs. *Nature* **401**, 446–446.
[148] Gustafsson, S. and Hansson, L.-A. (2004) Development of tolerance against toxic cyanobacteria in *Daphnia*. *Aquatic Ecology* **38** (1), 37–44.
[149] Sarnelle, O. and Wilson, A.E. (2005) Local adaptation of *Daphnia pulicaria* to toxic cyanobacteria. *Limnology and Oceanography* **50** (5), 1565–1570.
[150] Jewel, M.A.S., Affan, M.A. and Khan, S. (2003) Fish mortality due to cyanobacterial bloom in an aquaculture pond in Bangladesh. *Pakistan Journal of Biological Sciences* **6** (12), 1046–1050.
[151] Zimba, P.V., Camus, A., Allen, E.H. and Burkholder, J.M. (2006) Co-occurrence of white shrimp, *Litopenaeus vannamei*, mortalities and microcystin toxin in a southeastern USA shrimp facility. *Aquaculture* **261** (3), 1048–1055.
[152] Fischer, W.J. and Dietrich, D.R. (2000) Pathological and biochemical characterization of microcystin-induced hepatopancreas and kidney damage in Carp (*Cyprinus carpio*). *Toxicology and Applied Pharmacology* **164** (1), 73–81.
[153] Cazenave, J., Wunderlin, D.A., Bistoni, M., de Los, Á. *et al.* (2005) Uptake, tissue distribution and accumulation of microcystin-RR in *Corydoras paleatus*, *Jenynsia multidentata* and *Odontesthes bonariensis*. *Aquatic Toxicology* **75** (2), 178–190.
[154] Ibelings, B.W., Bruning, K., de Jonge, J. *et al.* (2005) Distribution of microcystins in a lake foodweb: no evidence for biomagnification. *Microbial Ecology* **49** (4), 487–500.
[155] Edwards, C., Graham, D., Fowler, N. and Lawton, L.A. (2008) Biodegradation of microcystins and nodularin in freshwaters. *Chemosphere* **73** (8), 1315–1321.
[156] Fujii, K., Ikai, Y., Mayumi, T. *et al.* (1997) A nonempirical method using LC/MS for determination of the absolute configuration of constituent amino acids in a peptide: elucidation of limitations of Marfey's method and its separation mechanisms. *Analytical Chemistry* **69** (16), 3346–3352.
[157] Barford, D. and Keller, J.C. (1994) Co-crystallization of the catalytic subunit of the serine/threonine specific protein phosphatase 1 from Human in complex with microcystin LR. *Journal of Molecular Biology* **235** (2), 763–766.
[158] Schantz, E.J., McFarren, E.F., Schafer, M.L. and Lewis, K.H. (1958) Purified shellfish poison for bioassay standardization. *Journal of the Association of Official Agricultural Chemists* **41**, 160–168.
[159] Kiviranta, J., Sivonen, K., Niemalä, S.I. and Huovinen K. (1991) Detection of toxicity of cyanobacteria by *Artemia salina* bioassay. *Environmental Toxicology and Water Quality* **6** (4), 423–436.
[160] Kós, P., Gorzó, G., Surányi, G. and Borbély, G. (1995) Simple and efficient method for isolation and measurement of cyanobacterial hepatotoxins by plant tests (*Sinapis alba* L.). *Analytical Biochemistry* **255** (1), 49–53.

[161] Sangolkar, L.N., Maske, S.S. and Chakrabarti, T. (2006) Methods for determining microcystins (peptide hepatotoxins) and microcystin-producing cyanobacteria. *Water Research* **40** (19), 3485–3496.

[162] Meriluoto, J. (1997) Chromatography of microcystins. *Analytica Chimica Acta* **352** (1–3), 277–298.

[163] ISO 20179:2005 (2005) Determination of microcystins – Method using solid phase extraction (SPE) and high performance liquid chromatography (HPLC) with ultraviolet (UV) detection. I*nternational Standards Organization*, Geneva.

[164] Lawton, L.A., Edwards, E. and Codd, G.A. (1994) Extractionand high-performance liquid chromatographic method for the determination of microcystins in raw and treated waters. *Analyst* **119** (7), 1525–1530.

[165] Pérez, S. and Aga, D.S. (2005) Recent advances in the sample preparation, liquid chromatography tandem mass spectrometric analysis and environmental fate of microcystins in water. *Trend in Analytical Chemistry* **24** (7), 658–670.

[166] Spoof, L., Vesterkvist, P., Lindholm, T. and Meriluoto, J. (2003) Screening of cyanobacterial hepatotoxins, microcystins and nodularin in environmental water samples by reversed-phase liquid chromatography-electrospray ionisation mass-spectrometry. *Journal of Chromatography A* **1020** (1), 105–119

[167] Li, W., Xie, P., Chen, J., Guo, X., *et al.* (2014) Quantitative liquid chromatography-tandem mass spectrometry method for determination of microcystin-RR and its glutathione and cysteine conjugates in fish plasma and bile. *Journal of Chromatography B* **963**, 113–118.

[168] Welker, M., Fastner, J., Erhard, M. and von Döhren, H. (2002) Applications of MALDI-TOF MS analysis in cyanotoxin research. *Environmental Toxicology* **17** (4), 367–374.

[169] Sano, T., Nohara, K., Shiraishi, F. and Kaya. K. (1992) A method for micro-determination of total microcystins content in waterblooms of cyanobacteria (blue-green algae). *International Journal of Analytical Chemistry* **49**, 163–170.

[170] Soares, R.M., Yuan, M., Servaites, J.C. *et al.* (2006) Sublethal exposure from microcystins to renal insufficiency patients in Rio de Janeiro, Brazil. *Environmental Toxicology* **21** (2), 95–103.

[171] Smith, J.L., Schulz, K.L., Zimba, P.V. and Boyer, G.L. (2010) Possible mechanism for the foodweb transfer of covalently bound microcystins. *Ecotoxicology and Environmental Safety* **73** (5), 757–761.

[172] Peng, L., Liu, Y., Chen, W. *et al.* (2010) Health risks associated with consumption of microcystin-contaminated fish and shellfish in three Chinese lakes: significance for freshwater aquacultures. *Ecotoxicology and Environmental Safety* **73** (7), 1804–1811.

[173] Berry, J.P., Lee, E., Walton, K. *et al.* (2011) Bioaccumulation of microcystins by fish associated with a persistent cyanobacterial bloom in Lago de Patzcuaro (Michoacan, Mexico). *Environmental Toxicology and Chemistry* **30** (7), 1621–1628.

12

Cylindrospermopsin and Congeners

Mikołaj Kokociński[1], Ana Maria Cameán[2], Shmuel Carmeli[3], Remedios Guzmán-Guillén[2], Ángeles Jos[2], Joanna Mankiewicz-Boczek[4,5], James S. Metcalf[6], Isabel Maria Moreno[2], Ana Isabel Prieto[2], and Assaf Sukenik[7]

[1]*Department of Hydrobiology, Adam Mickiewicz University, Poznań, Poland*
[2]*Area of Toxicology, Faculty of Pharmacy, University of Sevilla, Spain*
[3]*Raymond and Beverly Sackler School of Chemistry, Tel Aviv University, Tel Aviv, Israel*
[4]*European Regional Centre for Ecohydrology of the Polish Academy of Sciences, Łódź, Poland*
[5]*Department of Applied Ecology, Faculty of Biology and Environmental Protection, University of Łódź, Poland*
[6]*Institute for Ethnomedicine, Jackson, USA*
[7]*Kinneret Limnological Laboratory, Israel Oceanographic & Limnological Research, Migdal, Israel*

12.1 Chemical Characteristics of Cylindrospermopsin and Congeners

Cylindrospermopsins (CYNs) are guanidine alkaloid hepatotoxins, produced by a number of cyanobacterial genera (see later) [1]. Currently, five analogues of CYN are known, namely CYN, 7-epi-CYN, and 7-deoxy-CYN (Fig. 12.1) and the two recently characterized congeners, 7-deoxydesulfo-CYN and 7-deoxydesulfo-12-acetyl-CYN [2]. The structures of CYN and 7-epi-CYN have repeatedly gone through name changes, with each at some point being called CYN. At the present, 7-(*S*)-CYN is considered to be the principal toxin produced by *Cylindrospermopsis raciborskii*. Based on reevaluation of absolute configuration at C-7 of the isolated CYN, revised synthesis [3, 4], and asymmetric synthesis [5], the absolute configuration of all of the asymmetric centres of CYN, 7-epi-CYN, and 7-deoxy-CYN were established. It was thus concluded that the names remain the same as assigned but the stereochemistry is opposite to that of Ohtani et al. [6] and Banker et al. [7].

Handbook of Cyanobacterial Monitoring and Cyanotoxin Analysis, First Edition.
Edited by Jussi Meriluoto, Lisa Spoof and Geoffrey A. Codd.

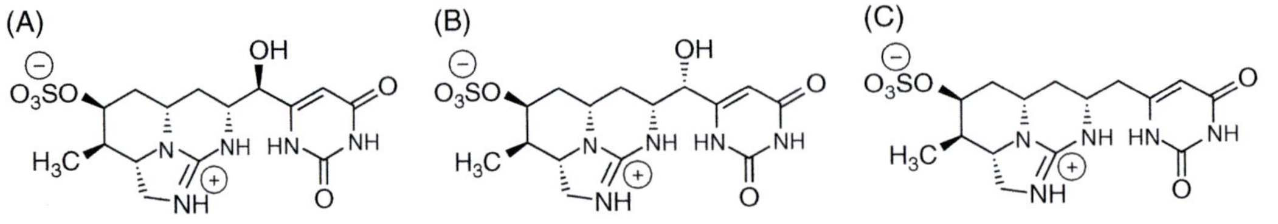

Figure 12.1 *The structures of cylindrospermopsin (A), 7-epicylindrospermopsin (B), and 7-deoxycylindrospermopsin (C)*

12.2 Genes Involved in CYN Biosynthesis

Polyketide synthases (PKSs) and nonribosomal peptide synthetases (NRPSs) are enzymes involved in secondary metabolite biosynthesis in cyanobacteria. Putative genes encoding PKS and NRPS, characteristic of CYN synthesis, were first identified in an Australian strain of the cyanobacterium *C. raciborskii* by Schembri et al. [8]. Then, Shalev-Alon et al. [9] identified a novel gene (*aoaA*) encoding aminotransferase (AoaA) involved in CYN synthesis *Chrysosporum ovalisporum* (*Aphanizomenon ovalisporum*). The complete gene cluster (*cyr*) involved in synthesis of CYN in *C. raciborskii* AWT205 was described by Mihali et al. [10]. The homologous cluster of *cyr* genes, but with a different gene organization, was found in *Oscillatoria sp.* strain PCC 6506, although the toxin was still produced [11].

According to Mihali et al. [10], the biosynthetic pathway starts with formation of the carbon skeleton involved in the synthesis of guanidinoacetate by the amidinotransferase CyrA (AoaA homolog), which is then activated by the adenylation domain of an NRPS/PKS hybrid, CyrB (AoaB homolog). This is followed by five polyketide extensions involving NRPS, CyrC-F. The formation of the carbon skeleton is completed by generation of the uracil ring (CyrG–H). Then, in order to complete the biosynthesis, tailoring reactions occur and involve sulfation (CyrJ) and hydroxylation (CyrI). Mihali et al. [10] revealed that the *cyrJ* gene was present only in CYN-producing strains, supporting the involvement of this gene in the biosynthesis of CYN. The *cyrJ* gene was therefore suggested as a genetic marker to detect CYN-producing cyanobacterial strains. Subsequently, nontoxic strains of *C. raciborskii* and *A. ovalisporum* isolated from Lake Kinneret (Israel) were characterized to lack *cyrJ/aoaJ* gene cluster [12, 13].

The knowledge about genes involved in CYN synthesis is applicable in environmental research, as the detection of toxigenic genotypes indicated the potential toxicity of cyanobacteria, acting as an effective, early alert to a possible health risk. A study of 36 Polish lakes supported the notion that *cyrJ* and *cyrA* genes can be used as suitable qualitative genetic markers for the determination of toxigenic genotypes in environmental and cultivable samples and therefore can be used to assess the potentiality of cyanobacteria to produce CYN [14].

12.3 CYN Producers and Distribution

The first identified CYN producer was *C. raciborskii* isolated from a drinking water supply reservoirs in Palm Island, northern Queensland, Australia [6], following an outbreak of hepatoenteritis in the island population. CYN-producing strains of this cyanobacterium have also been isolated in New Zealand [15], Thailand [16], and Japan [17, 18]. Concurrently, non–CYN-producing strains of this *C. raciborskii* were isolated and characterized in South, Central, and North America, in

Africa, and in Europe [19]. Harada et al. [20] reported that *Umezakia natans* produced CYN. Initially, that toxic strain was allocated within the Stigonematales order, but recently this species was reclassified into the Nostocales [21]. More recently, CYN has been detected worldwide as a product of *Chrysosporum ovalisporum* (*Aphanizomenon ovalisporum*) in Israel, Australia, Spain, USA, and Italy [7, 22–25]. Other CYN producers such as *Chrysosporum bergii* (*Anabaena bergii*), *Lyngbya wollei*, and *Raphidiopsis mediterranea* were reported in Australia [8, 26, 27], and *Raphidiopsis curvata* was reported in China [28]. In Europe, in addition to *A. ovalisporum*, several other CYN-producing species were identified. These include *Anabaena lapponica* in Finland [29], *Aphanizomenon flos-aquae* in Germany [30], and *Aphanizomenon gracile* in Germany and Poland [14, 31]. In addition, *A. flos-aquae* var. *klebahnii* (name used in the original report) in Czech Republic and *Anabaena planctonica* in Portugal and France were reported as potential producers of CYN [32, 33]. Recently, *Oscillatoria* sp. strain PCC 6506 and several benthic strains isolated from the Tarn River in France were found capable of synthesising CYN [11]. The variety of the CYN producers clearly indicates that the production is not species specific and that the list of potential CYN producers may remain incomplete. Moreover, it must be emphasizeded that for most of the known CYN-producing species, both CYN-producing and nonproducing strains have been observed. The naturally occurring congeners of CYN—deoxy-CYN and 7-epi-CYN—were found to be produced by *C. raciborskii*, *R. curvata*, *R. mediterranea*, *A. ovalisporum*, and *Oscillatoria sp.* [11, 27, 28, 34, 35]. The main CYN producers, *C. raciborskii* and *A. ovalisporum*, are considered invaders in many lakes and water reservoirs of subtropical and temperate areas [36, 37]. The routes and global and regional conditions that support their invasion are discussed in Chapter 4.

The effect of environmental conditions on CYN production has been only partly resolved. Laboratory experiments have shown effects of temperature, nitrogen source, phosphate availability, and light intensity on CYN production [19, 38]. It must be indicated, however, that the influence of environmental conditions on CYN production may vary among different producing species and strains, especially those isolated from different geographical regions. Interestingly, optimal conditions for CYN production may not be optimal for growth of the CYN producer. For example, it was found that the highest concentrations of CYN in *C. raciborskii* cultures occurred when they were grown in the absence of fixed-nitrogen source, while the lowest when supplied with ammonium, whereas growth rates were the highest in the presence of 2 mmol L^{-1} ammonium [39]. Similarly, the highest concentration of CYN was observed in temperatures much lower than those found as optimal for growth but under much higher light intensities than optimal light conditions for growth [40]. Furthermore, a decrease in CYN cell content of *A. ovalisporum* under phosphate limitation has been shown [41].

12.4 Toxicity of CYN

12.4.1 Mechanism of Toxicity

CYN is a potent inhibitor of protein synthesis. Terao et al. [42] described the liver as its main target, with four consecutive phases of pathological changes: protein synthesis inhibition, membrane proliferation, fat droplet accumulation, and cell death. Runnegar et al. [43, 44] found that CYN exposure led to a decrease in the content of reduced glutathione (GSH) in rat hepatocytes. This reduction was attributed to the inhibition of GSH synthesis rather than to an increase in its consumption [45]. Norris et al. [46] suggested that activation of CYN by cytochrome P450 is of primary importance in its mechanism of action. Activation of CYN by *cyt-p450* results in higher toxicity; thus, it is considered as a progenotoxic substance [47]. Moreover, oxidative stress may play a significant role in CYN toxicity *in vitro* [48] and *in vivo* [49]. CYN may also inhibit progesterone production in human granulose cells, indicating its potential role as an endocrine disruptor [50]. CYN inhibited in a noncompetitive manner the *in vitro* activity of uridine monophosphate (UMP) synthase complex (responsible for the conversion of orotic acid to UMP) in a cell-free liver extract from mice. However, exposure of mice to subacute concentrations of CYN only slightly affected the *in vitro* activity of UMP synthase [51]. Irregular metabolic responses such as abnormal red blood cell formation, increases in haematocrit, and deviated cholesterol metabolism were reported in mice exposed to subacute concentrations of CYN [51, 52].

Overall, CYN appears as a wide-range toxic molecule. It affects various organs and metabolic pathways, either directly or after some metabolic modification. The lack of a specific target for CYN hinders further efforts to understand its potent toxicity and to define acceptable thresholds of exposures.

12.4.2 Human Intoxication

A well-documented incident of human toxicity associated with CYN is the Palm Island event reported in 1979, when more than 100 children of Aboriginal families had to be hospitalized with symptoms of hepatoenteritis [6, 53]. The outbreak occurred a few days after the sole source of reticulated water supply on the island (Solomon Dam) had been dosed with copper sulphate to control a dense algal bloom. All people affected declared having drunk from the infected water at some point. A 4-year postevent ecological study of the reservoir showed that accumulation of cyanobacterial blooms in the water body was a common event [54]. One of the prominent components, which was proved to be highly toxic, was *C. raciborskii*, a producer of CYN [6, 53]. In another serious incident of human intoxication, at a dialysis centre in Caruaru (Brazil) in 1996, 100 patients developed acute liver failure, and of these, 76 died. Analyses of the clinic's water treatment system, serum, and liver tissue of the patients led to the conclusion that despite the occurrence of two types of cyanobacterial toxins, microcystins (MCs) and CYN, only MCs were the major contributing factor to the death of dialysis patients [55].

People exposed to direct ingestion of CYN (the Palm Island incident) developed symptoms such as anorexia, vomiting, headache, hepatomegaly, initial constipation followed by bloody diarrhoea, and dehydration. Skin and eye irritation, diarrhoea, and vomiting may occur by exposure to cyanobacteria during recreational water activities [56], but no direct reference to CYN was reported.

12.4.3 Effects on Animals and Ecosystems

Unlike other cyanotoxins, studies about the effects of CYN on animals are much more limited. It has the potential to affect a wide variety of aquatic and semiaquatic plant and animal species, due to its cytotoxic mechanism. Further, due to CYN bioaccumulation and trophic transfer, its environmental damage may include largely terrestrial organisms. The deaths of cattle that drank from a farm dam's water bloom of *C. raciborskii* was attributed to high CYN concentration as clearly demonstrated by liquid chromatography–mass spectrometry analysis [57].

Rodents are among the most used experimental models, followed by different species of fish, crustaceans, and molluscs. Studies carried out on mice reported alterations mainly in liver and kidney, although other organs can be involved (stomach, lungs, small intestines, thymus, and spleen). Both oral and intraperitoneal CYN administration revealed scopic histopathological lesions ending in cellular necrosis. CYN is also characterized as a potential carcinogen, as seen by neoplastic processes and tumours. Alterations in body weight have also been observed [58].

Several laboratory experiments have been reported in diverse animal species (amphibians, bivalves, crustaceans, snails, and fish) with different degrees of injury. Amphibians (*Bufo marinus*) exposed to whole-cell extracts and live *C. raciborskii* cells revealed necrotic tissues (liver and kidney), moderate mortality, changes in the normal tadpole behaviour, and lower relative growth rate, predominantly with live cells [59]. Neither of these effects were shown in some studies carried out in the freshwater gastropod *Melanoides tuberculata*; nevertheless, high CYN concentration was demonstrated to decrease hatchlings [50]. Studies on fish revealed embryo malformations using zebrafish exposed to CYN cell extracts. More recent studies of *C. raciborskii* and *A. ovalisporum* revealed that the teratogenicity in zebrafish embryo model is due to the occurrence polymethoxyalkenes (PMAs) in CYN-producing and -nonproducing strains [60]. Moreover, there are dose-dependent histopathological effects mainly in liver and kidney of tilapia, influenced by exposure route (gavage and intraperitoneal injection) [61]. Noteworthy are the effects on the oxidative stress status, as measured by lipid peroxidation, DNA oxidation, reduced-oxidized glutathione ratio, and changes in the activity of different enzymes such as glutathione-*S*-transferase, glutathione peroxidase, superoxide dismutase, catalase, and γ-glutamyl-cysteine synthetase [49].

Field studies demonstrated the presence of CYN in aquatic species including bivalves (*Alathyria pertexta pertexta*), crustaceans (Redclaw crayfish *Cherax quadricarinatus*), snails (*Pomecea patula catemacensis*), and fish (*Salmo trutta*, *Oreochromis aureus*, and other species of finfish) harvested from dams, ponds, and natural lakes [47, 50, 62]. Anderson et al. [63] studied CYN accumulation in bivalves, and demonstrated that the mussel *A. pertexta pertexta*, accumulated amounts of 130–560 μg CYN kg fresh weight $(fw)^{-1}$. In crustaceans, CYN has been detected in the red claw crayfish *C. quadricarinatus*, collected from a reservoir with less than 0.8 μg CYN L^{-1}, in both hepatopancreas (0.9 μg g dry weight $[dw]^{-1}$) and muscle (4.3 μg g dw^{-1}). In fish, Messineo et al. [25] demonstrated for the first time that CYN was present in salmonids from a deep natural lake (2.7 ng g fw^{-1} in viscera and 0.8 ng g fw^{-1} in muscle from *Salmo trutta*). More recently, Berry et al. [64] performed a study on CYN accumulation in fish, particularly in muscle, showing values between 0.09 and 1.26 μg CYN kg fw^{-1} in different fish species. Plants exposed to CYN (via irrigation) accumulate the toxin in their tissues, including edible parts (leaves, roots, and fruits). Brassica species (*Brassica oleracea* var. *sabellica*, *Brassica juncea*, and *Sinapis alba*) exposed to CYN showed significant CYN uptake, with CYN levels ranging from 10% to 21% in the leaves compared to the CYN exposure concentration applied to the roots (18–35 μg L^{-1}) [65]. Crop plants irrigated with CYN-containing water may represent a significant source of this toxin within the food chain and CYN toxicity in different species. The isolated toxin from *A. ovalisporum* produced growth inhibition of etiolated *Sinapsis alba* seedlings [66]. CYN produced inhibition of tobacco (*Nicotiana tabacum*) pollen germination at concentrations greater than 5 μg/ml [67] and decreases in root and shoot elongation in the aquatic macrophyte *Phragmites australis* at concentrations between 0.5 and 40 μg mL^{-1} [68]. Short-term plant exposure to toxic *A. ovalisporum* cyanobacterial extracts containing CYN has shown to increase oxidative stress (by alteration of glutathione-*S*-transferase and reduced-oxidized glutathione ratio activities), with longer exposure periods leading to tissue necrosis [69]. A recent study reported growth inhibition and *Microcystis aeruginosa* cell necrosis at CYN concentrations between 10 and 50 μg L^{-1} and inhibitory effect on the excretion of microcystin-LR at CYN concentrations between 1 and 50 μg L^{-1} [70]. Moreover, the combined exposure to CYN and MC-LR at low concentrations (1 μg L^{-1}) induced a significant enhancement in the fresh weight of lettuce leaves and at the proteomic level it increased the abundance of a high number of proteins. In contrast, other proteins exhibited a decreased abundance or were absent in the gels after the simultaneous exposure to 10 and 100 μg L^{-1} MC-LR/CYN, and a significant decrease in the fresh weight was observed [71].

12.5 The Biological Role of CYN

The main thought regarding the biological role of CYN is related to direct competitive advantage, through either grazing defence or allelopathy. The pressure of resource competition, light, and nutrients, from other primary producers, presents a threat for the cyanobacterial population. Thus, toxins may act as allelopathic compounds [72, 73]. A recent study has suggested that the presence of CYN causes the secretion of alkaline phosphatase (APase) by other phytoplankton, thus increasing the amount of inorganic phosphate (Pi) available to the CYN producer, and allowing it to outcompete other species in environments with limited inorganic phosphate. Thus, CYN is used to "enslave" other organisms to fulfill the Pi demands of the "toxic" strain [74]. This proposed role was also supported by field results with enzyme-labelled fluorescence (ELF-APase), a strong correlation between the abundance of CYN-producing *A. ovalisporum* and high APase activity in various phytoplankton species but not in *Aphanizomenon* [74, 75]. Another earlier study has also reported the increase in APase during the presence of *A. ovalisporum* [76].

12.6 Degradation of CYN

Cylindrospermopsin is usually produced at concentration of 1 and 10 μg L^{-1} [77] however much higher concentrations of 42.3 and 126 μg L^{-1} for particulate and dissolved concentration respectively were reported for example from two lakes in central Italy [78]. Generally, higher concentrations of the extracellular fraction of CYN are observed due to slow degradation of this toxin in the water [79], possible membrane permeability, and its high water solubility [80]. Moreover, release of this toxin into water varies among cyanobacterial species. CYN was demonstrated to be stable to a wide range of heat, light, and pH conditions. Effective removal of CYN from drinking water resources can be achieved by flocculation, sand filtration, and chlorination or ozonation. Photocatalytic degradation with titanium dioxide and UV irradiation were also demonstrated as effective procedures for CYN elimination [78]. In natural ecosystems, photodegradation and biodegradation are two major mechanisms for removal of CYN. Photodegradation is effective however, when high concentrations of plant pigments are present [79], while the knowledge on biodegradation is still very scarce and opposing results were achieved in a few studies. Effective CYN degradation was also observed in sediments under oxic conditions and higher temperature [81].

12.7 Available Methods for Determining CYN in Waters

Unlike many other cyanotoxins, relatively little work has been done on CYN detection methods for its analysis in water samples used for drinking water or recreational purposes. Compared to microcystins and saxitoxins, CYN is often present in the extracellular (dissolved) form, reaching up to 90% of the total toxin; taking this into account, it is necessary to develop validated confirmation methods for dissolved CYN determination in real water samples. These methods include mouse bioassay and enzyme-linked immunosorbent assay for fast screening, liquid chromatography, and capillary electrophoresis. ELISA permits CYN quantification with a sensitive level of detection and from several types of samples; however, it is nonselective for CYN analogues and cross-reactivity may exist. Moreover, currently, molecular methods are widely applied (polymerase chain reaction and characterization of DNA and RNA unique to CYN-producing cyanobacteria) [50].

Regarding extraction procedures, sample acidification with 1% formic acid and addition of sodium chloride has been proved useful to obtain a robust recovery of dissolved CYN in environmental samples [82], as well as freeze-thaw cycles (for cell disruption and toxin release) combined with ultrasonication and stirring, for intracellular CYN. Due to hydrophilic nature of CYN, some authors reported that it cannot be concentrated from water samples with solid phase extraction cartridges such as C18, but graphitized carbon has been used successfully. Furthermore, the use of a solution of methanol: dichloromethane as eluent is strongly recommended, allowing increased bed weights while guaranteeing optimum recovery in small elution volumes [82, 83]. Kubo et al. [84] used a double system of cartridges, which permits the isolation of both CYN and deoxy-CYN from culture samples. Methods based on reversed-phase high-performance (LC-DAD) have been proposed for analyzing CYN in real water samples and lyophilized cyanobacterial cells and despite being less expensive, it does not adequately detect trace quantities of CYN, not offering unequivocal and definite identification. Liquid chromatography–tandem mass spectrometry was proved to be ideal to determine trace amounts of CYN in water samples due to its sensitivity and specificity [85], and they have been established as a standard method for this purpose. An optimized and validated new method has been described for determining dissolved CYN in waters based on solid phase extraction with graphitized carbon cartridges and quantification by liquid chromatography–tandem mass spectrometry was proved to be ideal to determine trace amounts of CYN in water samples due [83]. Additional information on analytical aspects can be found in Chapters 21, 23, 24, and 25 and SOPs 15, 16, and 19.

References

[1] Metcalf, J.S. and Codd, G.A. (2012) Cyanotoxins, in *Ecology of cyanobacteria II: their diversity in time and space* (ed. B.A. Whitton), Springer, Dordrecht, Heidelberg, New York, London, pp. 651–673.

[2] Wimmer, K.M., Strangman, W.K. and Wright, J.L.C. (2014) 7-Deoxy-desulfo-cylindrospermopsin and 7-deoxy-desulfo-12-acetylcylindrospermopsin: Two new cylindrospermopsin analogs isolated from a Thai strain of *Cylindrospermopsis raciborskii*. *Harmful Algae* **37**, 203–206.

[3] White, J.D., Hansen, J.D. (2002) Asymmetric synthesis of epicylindrospermopsin via intramolecular nitrone cycloaddition. Assignment of absolute configuration. *Journal of the American Chemical Society* **124** (18), 4950–4951.

[4] White, J. and Hansen, J.D. (2005) Total synthesis of (–)-7epicylindrospermopsin, a toxic metabolite of the freshwater cyanobacterium *Aphanizomenon ovalisporum*, and assignment of its absolute configuration. *Journal of Organic Chemistry* **70** (6), 1963–1977.

[5] Looper, R.E., Runnegar, M.T.C. and Williams, R.M. (2006) Syntheses of the cylindrospermopsin alkaloids and their toxicological evaluation. *Tetrahedron* **62** (18), 4549–4562.

[6] Ohtani, I., Moore, R.E. and Runnegar, M.T.C. (1992) Cylindrospermopsin: a potent hepatotoxin from the blue-green alga *Cylindrospermopsis raciborskii*. *Journal of the American Chemical Society* **114** (20), 7941–7942.

[7] Banker, R., Carmeli, S., Hadas, O. *et al.* (1997) Identification of cylindrospermopsin in *Aphanizomenon ovalisporum* (Cyanophyceae) isolated from Lake Kinneret. *Israeli Journal of Phycology* **33** (4), 613–616.

[8] Schembri, M.A., Neilan, B.A. and Saint, C.P. (2001) Identification of genes implicated in toxin production in the cyanobacterium *Cylindrospermopsis raciborskii*. *Environmental Toxicology* **16** (5), 413–421.

[9] Shalev-Alon, G., Sukenik, A., Livnah, O. *et al.* (2002) A novel gene encoding amidinotransferase in the cylindrospermopsin producing cyanobacterium *Aphanizomenon ovalisporum*. *FEMS Microbiology Letters* **209** (1), 87–91.

[10] Mihali, T.K., Kellmann, R., Muenchhoff, J. *et al.* (2008) Characterization of the gene cluster responsible for cylindrospermopsin biosynthesis. *Applied and Environmental Microbiology* **74** (3), 716–722.

[11] Mazmouz, R., Chapuis-Hugon, F., Mann, S. *et al.* (2010) Biosynthesis of cylindrospermopsin and 7-epicylindrospermopsin in *Oscillatoria sp.* strain PCC 6506: identification of the *cyr* gene cluster and toxin analysis. *Applied and Environmental Microbiology* **76** (15), 4943–4949.

[12] Alster, A., Kaplan-Levy, R., Sukenik, A. and Zohary, T. (2010) Morphology and phylogeny of non-toxic invasive *Cylindrospermopsis raciborskii* from a Mediterranean Lake. *Hydrobiologia* **639** (1), 115–128.

[13] Ballot, A., Ramm, G., Rundberget, T. *et al.* (2011) Occurence of non-cylindrospermopsin-producing *Aphanizomenon ovalisporum* and *Anabaena bergii* in Lake Kinneret (Israel). *Journal of Plankton Research* **33** (11), 1736–1746.

[14] Kokociński, M., Mankiewicz-Boczek, J., Jurczak, T. *et al.* (2013) *Aphanizomenon gracile* (Nostocales), a cylindrospermopsin-producing cyanobacterium in Polish lakes. *Environmental Science and Pollution Research* **20** (8), 5243–5264.

[15] Wood, S.A. and Stirling, D.J. (2003) First identification of the cylindrospermopsin-producing cyanobacterium *Cylindrospermopsis raciborskii* in New Zealand. *New Zealand Journal of Marine and Freshwater Research* **37** (4), 821–828.

[16] Li, R., Carmichael, W.W., Brittain, J.E. *et al.* (2001) Isolation and identification of the cyanotoxin cylindrospermopsin and deoxycylindrospermopsin from a Thailand strain of *Cylindrospermosis raciborskii* (Cyanobacteria). *Toxicon* **39** (7), 973–980.

[17] Chonudomkul, D., Yongmanitchai, W., Theeragool, G. *et al.* (2004) Morphology, genetic diversity, temperature tolerance and toxicity of *Cylindrospermopsis raciborskii* (Nostocales, Cyanobacteria) strains from Thailand and Japan. *FEMS Microbiology Ecology* **48** (3), 345–355.

[18] Zarenezhad, S., Sano, T., Watanabe, M.M. and Kawachi, M. (2012) Evidence of the existence of toxic form of *Cylindrospermopsis raciborskii* (Nostocales, Cyanobacteria) in Japan. *Phycological Research* **60** (2), 98–104.

[19] Burford, M. and Davis, T.W. (2011) Physical and chemical processes promoting dominance of the toxic cyanobacterium *Cylindrospermopsis raciborskii. Chinese Journal of Oceanology and Limnology* **29** (4), 883–891.

[20] Harada, K.I., Ohtani, I., Iwamato, K. *et al.* (1994) Isolation of cylindrospermopsin from cyanobacterium *Umezakia natans* and its screening method. *Toxicon* **32** (1), 73–84.

[21] Niiyama, Y., Tuji, A. and Tsujimura, S. (2011) *Umezakia natans* M. Watan. does not belong to Stigonemataceae but Nostocaceae. *Fottea* **11** (1), 163–169.

[22] Shaw, G.R., Sukenik, A., Livne, A. *et al.* (1999) Blooms of cylindrospermosin containing cyanobacterium *Aphanizomenon ovalisporum* (Forti), in newly constructed lakes, Queensland, Australia. *Environmental Toxicology* **14** (1), 167–177.

[23] Quesada, A., Moreno, E., Carrasco, D. *et al.* (2006) Toxicity of *Aphanizomenon ovalisporum* (Cyanobacteria) in Spanish water reservoir. *European Journal of Phycology* **41** (1), 39–45.

[24] Yilmaz, M., Philips, E.J., Szabo, N.J. and Badylak, S. (2008) A comparative study of Florida strains of *Cylindrospermopsis* and *Aphanizomenon* for cylindrospermopsin production. *Toxicon* **51** (1), 130–139.

[25] Messineo, V., Bogialli, S., Melchiorre, S. *et al.* (2010) Seasonal succession of *Cylindrospermopsis raciborskii* and *Aphanizomenon ovalisporum* blooms with cylindrospermopsin occurence in the volcanic Lake Albano, Central Italy. *Environmental Toxicology* **25** (1), 18–27.

[26] Seifert, M., McGregor, G., Eaglesham, G. *et al.* (2007) First evidence for the production of cylindrospermopsin and deoxy-cylindrospermopsin by the freshwater benthic cyanobacterium *Lyngbya wollei* (Farlow ex Gomont) Speziale and Dyck. *Harmful Algae* **6** (1), 73–80.

[27] Mc Gregor, G.B., Sendall, B.C., Hunt, L.T. and Eaglesham, G.K. (2011) Report of the cyanotoxins cylindrospermopsin and deoxy-cylindrospermopsin from *Raphidiopsis mediterranea* Skuja (Cyanobacteria/Nostocales). *Harmful Algae* **10** (4), 402–410.

[28] Li, R., Carmichael, W.W., Brittain, S. *et al.* (2001) First report of the cynotoxins cylindrospermopsin and deoxycylindrospermopsin from *Raphidiopsis curvata* (Cyanobacteria). *Journal of Phycology* **37** (6), 1121–1126.

[29] Spoof, L., Berg, K.A., Rapala, J. *et al.* (2006) First observation of cylindrospermopsin in *Anabaena lapponica* isolated from the boreal environment (Finland). *Environmental Toxicology* **21** (6), 552–560.

[30] Preussel, K., Stüken, A., Wiedner, C. *et al.* (2006) First report on cylindrospermopsin producing *Aphanizomenon flos-aquae* (Cyanobacteria) isolated from two German lakes. *Toxicon* **47** (2), 156–162.

[31] Stüken, A., Campbell, R.J., Quesada, A. *et al.* (2009) Genetic and morphologic characterization of four putative cylindrospermopsin producing species of the cyanobacterial genera *Anabaena* and *Aphanizomenon*. *Journal of Plankton Research* **31** (5), 465–480.

[32] Blahova, L., Oravec, M., Marsalek, B. *et al.* (2009) The first occurrence of the cyanobacterial alkaloid toxin cylindrospermopsin in the Czech Republic as determined by immunochemical and LC/MS methods. *Toxicon* **53** (5), 519–524.

[33] Brient, L., Lengronne, M., Bormans, M. and Fastner, J. (2009) First occurence of cylindrospermopsin in freshwater in France. *Environmental Toxicology* **24** (4), 415–420.

[34] Norris, R.L., Eaglesham, G., Pierens, G. *et al.* (1999) Deoxycylindrospermopsin, an analog of cylindrospermopsin from *Cylindrospermopsis raciborskii*. *Environmental Toxicology* **14** (1), 163–165.

[35] Banker, R., Teltsch, B., Sukenik, A. and Carmeli, S. (2000) 7-Epicylindrospermopsin, a toxic minor metabolite of the cyanobacterium *Aphanizomenon ovalisporum* from Lake Kinneret, Israel. *Journal of Natural Products* **63**, 387–389.

[36] Wiedner, C., Rücker, J., Brüggemann, R. and Nixdorf, B. (2007) Climate change affects timing and size of populations of invasive cyanobacterium in temperate regions. *Oecologia* **152** (3), 473–484.

[37] Sukenik, A., Hadas, O., Kaplan, A. and Quesada, A. (2012) Invasion of Nostocales (cyanobacteria) to subtropical and temperate freshwater lakes—physiological, regional, and global driving forces. *Frontiers in Microbiology* **3**, 86.

[38] Neilan, B.A., Pearson, L.A., Muenchhoff, J. *et al.* (2013) Environmental conditions that influence toxin biosynthesis in cyanobacteria. *Environmental Microbiology* **15** (5), 1239–1253.

[39] Saker, M.L. and Neilan, B.A. (2001) Variable diazotrophies, morphologies and toxicities of genetically similar isolates of *Cylindrospermopsis raciborskii* (Nostocales, Cyanophyceae) from northern Australia. *Applied and Environmental Microbiology* **67** (4), 1839–1845.

[40] Dyble, J., Tester, P.A. and Litaker, R.W. (2006) Effects of light on cylindrospermopsin production in the cyanobacterial HAB species *Cylindrospermopsis raciborskii*. *African Journal of Marine Science* **28** (2), 309–312.

[41] Bacsi, I., Vasas, G., Suranyi, G. *et al.* (2006) Alteration of cylindrospermopsin production in sulfate- or phosphate starved cyanobacterium *Aphanizomenon ovalisporum*. *FEMS Microbiology Letters* **269** (2), 303–310.

[42] Terao, K., Ohmori, S., Igarashi, K. *et al.* (1994) Electron microscopic studies on experimental poisoning in mice induced by cylindrospermopsin isolated from blue-green alga *Umezakia natans*. *Toxicon* **32** (7), 833–843.

[43] Runnegar, M.T., Kong, S., Zhong, Y.Z., Ge, J.L and Lu, S.C. (1994) The role of glutathione in the toxicity of a novel cyanobacterial alkaloid cylindrospermopsin in cultured rat hepatocytes. *Biochemical and Biophysical Research Communications* **201** (1), 235–241.

[44] Runnegar, M.T., Xie, C., Snider, B.B. *et al.* (2002) In vitro hepatotoxicity of the cyanobacterial alkaloid cylindrospermopsin and related synthetic analogues. *Toxicological Sciences* **67** (1), 81–87.

[45] Runnegar, M.T., Kong, S., Zhong, Y.Z. and Lu, S.C. (1995) Inhibition of reduced glutathione synthesis by cyanobacterial alkaloid cylindrospermopsin in cultured rat hepatocytes. *Biochemical Pharmacology* **49** (2), 219–225.

[46] Norris, R.L., Seawright, A.A., Shaw, G.R. *et al.* (2002) Hepatic xenobiotic metabolism of cylindrospermopsin in vivo in the mouse. *Toxicon* **40** (4), 471–476.

[47] Kinnear, S. (2010) Cylindrospermopsin: A decade of progress on bioaccumulation research. *Marine Drugs* **8** (3), 542–564.

[48] Gutiérrez-Praena, D., Pichardo, S., Jos, Á. and Cameán, M.A. (2011) Toxicity and glutathione implication in the effects observed by exposure of the liver fish cell line PLHC-1 to pure cylindrospermopsin. *Ecotoxicology and Environmental Safety* **74** (6), 1567–1572.

[49] Gutiérrez-Praena, D., Jos, A., Pichardo, S. and Cameán, A.M. (2011) Oxidative stress responses in tilapia (*Oreochromis niloticus*) exposed to a single dose of pure cylindrospermopsin under laboratory conditions: Influence of exposure route and time of sacrifice. *Aquatic Toxicology* **105** (1–2), 100–106.

[50] Moreira, C., Azevedo, J., Antunes, A. and Vasconcelos, V. (2012) Cylindrospermopsin: occurrence, methods of detection and toxicology. *Journal of Applied Microbiology* **114** (3), 605–620.

[51] Reisner, M., Carmeli, S., Werman, M. and Sukenik, A. (2004) The cyanobacterial toxin cylindrospermopsin inhibits pyrymidine nuceotide synthesis and alters cholesterol distribution in mice. *Toxicological Sciences* **82** (2), 620–627.

[52] Sukenik, A., Reisner, M., Carmeli, S. and Werman, M. (2006) Oral toxicity of the cyanobacterial toxin cylindrospermopsin in mice: long-term exposure to low doses. *Environmental Toxicology* **21** (6), 575–582.
[53] Griffiths, D.J. and Saker, M.L. (2003) The Palm Island mystery disease 20 years on: a review of research on the cyanotoxin cylindrospermopsin. *Environmental Toxicology* **18** (2), 78–93.
[54] Hawkins, P.R., Runnegar, M.T., Jackson, A. and Falconer, I. (1985) Severe hepatotoxicity caused by the tropical cyanobacterium (blue-green alga) *Cylindrospermopsis raciborskii* (Woloszynska) Seenaya and Subba Raju isolated from domestic water supply reservoir. *Applied and Environmental Microbiology* **50** (5), 1292–1295.
[55] Azevedo, S.M., Carmichael, W.W., Jochimsen, E.M. *et al.* (2002). Human intoxication by microcystins during renal dialysis treatment in Caruaru-Brazil. *Toxicology* **181–182**, 441–446.
[56] Pilotto, L.S., Douglas, R.M., Burch, M.D. *et al.* (1997) Health effects of exposure to cyanobacteria (blue-green algae) during recreational water-related activities. *Australian and New Zealand Journal of Public Health* **21** (6), 562–566.
[57] Saker, M.L., Thomas, A.D. and Norton, J.H. (1999) Cattle mortality attributed to the toxic cyanobacterium *Cylindrospermopsis raciborskii* in an outback region of north Queensland. *Environmental Toxicology* **14** (1), 179–182.
[58] Falconer, I.R. and Humpage, A.R. (2001) Preliminary evidence for in vivo tumour initiation by oral administration of extracts of the blue green alga *Cylindrospermopsis raciborskii* containing the toxin cylindrospermopsin. *Environmental Toxicology* **16** (2), 192–195.
[59] White, S.H., Dulvenvoorden, L.J., Fabbro, L.D. and Eaglesham, G.K. (2007) Mortality and toxin bioaccumulation in *Bufo marinus* following exposure to *Cylindrospermopsis raciborskii* cell extracts and live cultures. *Environmental Pollution* **147** (1), 158–167.
[60] Jaja-Chimedza, A., Gantar, M., Gibbs, P.D.L., Schmale, M.C. and Berry, J.P. (2012) Ploymethoxy-1-alkenes from *Aphanizomenon ovalysporum* inhibit vertebrate development in zebrafish (*Danio rerio*) embryo model. *Marine Drugs* **10** (10), 2322–2336.
[61] Gutiérrez-Praena, D., Jos, Á., Pichardo, S. *et al.* (2012) Time-dependent histopathological changes induced in Tilapia (*Oreochromis niloticus*) after acute exposure to pure cylindrospermopsin by oral and intraperitoneal route. *Ecotoxicology and Environmental Safety* **76** (2), 102–113.
[62] Gutiérrez-Praena, D., Jos, Á., Pichardo, S. *et al.* (2013) Presence and bioaccumulation of microcystins and cylindrospermopsin in food and the effectiveness of some cooking techniques at decreasing their concentrations: a review. *Food and Chemical Toxicology* **53**, 139–152.
[63] Anderson, L., Fabbro, L.D. and Cowden, K. (2003) Assessment of blue-green algal toxins in Barramundi, Red Clay and Mussels from Awoonga Dam. Central Queensland University, Gladstone, Australia.
[64] Berry, J.P., Jaja-Chimedza, A., Davalos-Lind, L. and Lind, O. (2012) Apparent bioaccumulation of cylindrospermopsin and paralytic shellfish toxins by finfish in Lake Catemaco (Veracruz, Mexico). *Food Additives and Contaminants* **29** (3), 314–321.
[65] Kittler, K., Schreiner, M., Krumbein, A. *et al.* (2012) Uptake of the cyanobacterial toxin cylindrospermopsin in *Brassica* vegetables. *Food Chemistry* **133** (3), 875–879.
[66] Vasas, G., Gáspár, A., Surányi, G. *et al.* (2002) Capillary electrophoretic assay and purification of cylindrospermopsin, a cyanobacterial toxin from *Aphanizomenon ovalisporum*, by plant test (blue-green Sinapis test). *Analitycal Biochemistry* **302** (1), 95–103.
[67] Metcalf, J.S., Barakate, A. and Codd, G.A. (2004) Inhibition of plant protein synthesis by the cyanobacterial hepatotoxin, cylindrospermopsin. *FEMS Microbiology Letters* **235** (1), 125–129.
[68] Beyer, D., Surányi, G., Vasas, G., Roszik, J., Erdodi, F. *et al.* (2009) Cylindrospermopsin induces alterations of root histology and microtubule organization in common reed (*Phragmites australis*) plantlets cultured in vitro. *Toxicon* **54** (4), 440–449.
[69] Prieto, A., Campos, A., Cameán, A.M. and Vasconcelos, V. (2011) Effects on growth and oxidative stress status of rice plants (*Oryza sativa*) exposed to two extracts of toxin-producing cyanobacteria (*Aphanizomenon ovalisporum* and *Microcystis aeruginosa*). *Ecotoxicology and Environmental Safety* **74** (7), 1973–1980.
[70] Rzymski, P., Poniedziałek, B., Kokociński, M. *et al.* (2014) Interspecific allelopathy in cyanobacteria: Cylindrospermopsin and *Cylindrospermopsis raciborskii* effect on the growth and metabolism of *Microcystis aeruginosa*. *Harmful Algae* **35**, 1–8.
[71] Freitas, M., Campos, A., Azevedo, J. *et al.* (2015) Lettuce (*Lactuca sativa* L.) leaf-proteome profiles after exposure to cylindrospermopsin and a microcystin-LR/cylindrospermopsin mixture: a concentration-dependent response. *Phytochemistry* **110**, 91–103.

[72] Holland, A. and Kinnear, S. (2013) Interpreting the possible ecological role(s) of cyanotoxins: compounds for competitive advantage and/or physiological aide? *Marine Drugs* **11** (7), 2239–2258.
[73] Kaplan, A., Harel, M., Kaplan-Levy, R.N. *et al.* (2012) The languages spoken in the water body (or the biological role of cyanobacterial toxins). *Frontiers in Microbiology* **3**, 138.
[74] Bar-Yosef, Y., Sukenik, A., Hadas, O. *et al.* (2010) Enslavement in the water body by toxic *Aphanizomenon ovalisporum*, inducing alkaline phosphatase in phytoplanktons. *Current Biology* **20** (17), 1557–1561.
[75] Bar-Yosef, Y., Murik, A., Sukenik, A. *et al.* (2012) Multiannual variations in phytoplankton populations: what distinguished the blooms of *Aphanizomenon ovalisporum* in Lake Kinneret in 2010 from 2009? *Environmental Microbiology Reports* **4** (5), 498–503.
[76] Hadas, O., Pinkas, R., Delphine, E. *et al.* (1999) Limnological and ecophysiological aspects of *Aphanizomenon ovalisporum* bloom in Lake Kinneret, Israeli. *Journal of Plankton Research* **21** (8), 1439–1453.
[77] Falconer, I.R. and Humpage, A.R. (2006) Cyanobacterial (blue-green) algal toxins in water supplies: cylindrospermopsins. *Environmental Toxicology* **21** (4), 299–304.
[78] De la Cruz, A., Hiskia, A., Kaloudis, T. *et al.* (2013) A review on cylindrospermopsin: the global occurrence, detection, toxicity and degradation of potent cyanotoxin. *Environmental Sciences: Processes & Impacts* **15**, 1979–2003.
[79] Chiswell, R.K., Shaw, G.R., Eaglesham, G. *et al.* (1999) Stability of cylindrospermopsin, the toxin from the cyanobacterium, *Cylindrospermopsis raciborskii*: effect of pH, temperature, and sunlight on decomposition. *Environmental Toxicology* **14** (1), 155–161.
[80] Sivonen, K. and Jones, G. (1999) Cyanobacterial toxins, in *Toxic cyanobacteria in water, a quide to their public health consequences, monitoring and management* (eds. I. Chorus and J. Bartman), E and FN Spoon, London, pp. 41–411.
[81] Klitzke, S. and Fastner, J. (2012) Cylindrospermopsin degradation in sediments – the role of temperature, redox conditions, and dissolved organic carbon. *Water Research* **46** (5), 1549–1555.
[82] Wörmer, L., Carrasco, D., Cirés, S. and Quesada, A. (2009) Advances in solid phase extraction of the cyanobacterial toxin cylindrospermopsin. *Limnology and Oceanography: Methods* **7**, 568–575.
[83] Guzmán-Guillén, R., Prieto, A. I., González, A. G. *et al.* (2012) Cylindrospermopsin determination in water by LC–MS/MS: Optimization and validation of the method and application to real samples. *Environmental Toxicology and Chemistry* **31** (10), 2233–2238.
[84] Kubo, T., Sano, T., Hosoya, K. *et al.* (2005) A new simply and effective fractionation method for cylindrospermopsin analyses. *Toxicon* **46** (1), 104–107.
[85] Eaglesham, G., Norris, K.R., Shaw, G.R. *et al.* (1999) Use of HPLC-MS/MS to monitor cylindrospermopsin, a blue-green algal toxin, for public health purposes. *Environmental Toxicology* **14** (1), 151–154.

13

Anatoxin-a, Homoanatoxin-a, and Natural Analogues

Milena Bruno[1], Olivier Ploux[2], James S. Metcalf[3], Annick Mejean[2], Barbara Pawlik-Skowronska[4], and Ambrose Furey[5]

[1]Department of Environment and Primary Prevention, Istituto Superiore di Sanità, Rome, Italy
[2]LIED UMR 8236 CNRS, University Paris Diderot, and Chimie ParisTech, Paris, France
[3]Institute for Ethnomedicine, Jackson, USA
[4]Department of Hydrobiology, University of Life Sciences in Lublin, Poland
[5]Mass Spectrometry Research Centre, Department of Physical Sciences, Cork Institute of Technology (CIT), Cork, Ireland

13.1 Introduction

Anatoxin-a (ANTX) and homoanatoxin-a (HANTX) are potent neurotoxins, produced by a number of cyanobacterial species. ANTXs contaminate lakes and rivers and have been associated with a number of animal fatalities, including cattle and dogs. Originally, ANTX was called the very fast death factor (VFDF) due to the rapidity of death for animals who consumed toxic cyanobacteria.

13.2 Chemical Structure, Synthesis, and Reactivity

The structures of ANTX and HANTX are represented in Figure 13.1. Other ANTX derivatives have also been identified in samples at varying concentrations. ANTX and HANTX undergo rapid chemical degradation in nature; the main degradation products are dihydro- and epoxy analogues (Fig. 13.1) [1–3]. In 2003, Namikoshi et al. identified a new HANTX derivative, 4-hydroxyhomoanatoxin-a (4-hydroxyHANTX), from the cyanobacteria *Raphidiopsis mediterranea* [4, 5].

Handbook of Cyanobacterial Monitoring and Cyanotoxin Analysis, First Edition.
Edited by Jussi Meriluoto, Lisa Spoof and Geoffrey A. Codd.

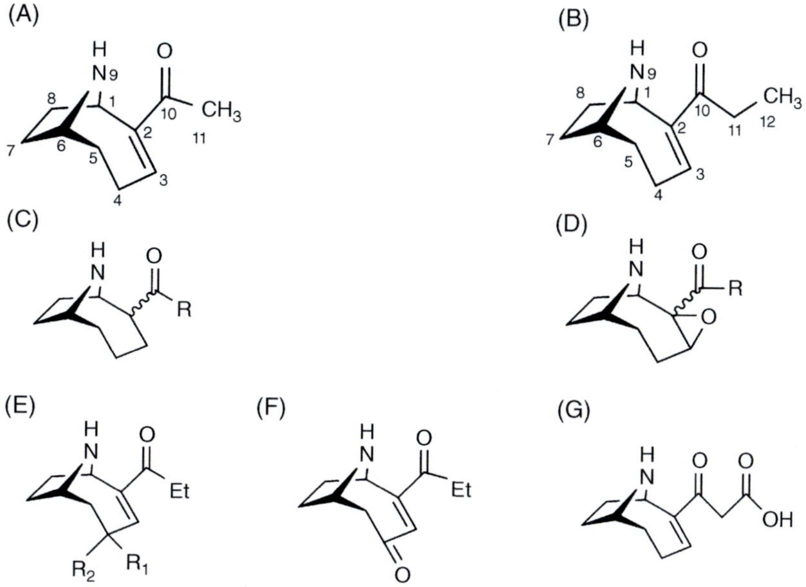

Figure 13.1 *Structure of (A) anatoxin-a (ANTX) (B) homoanatoxin-a (HANTX), and their degradation products: (C) dihydroanatoxin-a (R = Me; two diastereoisomers:* cis *or* trans*) and dihydrohomoanatoxin-a (R = Et; two diastereoisomers:* cis *or* trans*); (D) 2,3-epoxyanatoxin-a (R = Me; two diastereoisomers) and 2,3-epoxyhomoanatoxin-a (R = Et; two diastereoisomers); (E) (4R)-4-hydroxyhomoanatoxin-a (R_1 = H, R_2 = OH) and (4S)-4-hydroxyhomoanatoxin-a (R_1 = OH, R_2 = H); (F) 4-oxohomoanatoxin-a, and (G) 11-carboxyanatoxin-a. There are two diastereoisomers (*cis *or* trans*) for each dihydroderivative and epoxyderivative, depending on the side chain orientation*

In 2007, Selwood et al. discovered another ANTX derivative, 11-carboxyl anatoxin-a, in a strain of *Cuspidothrix issatschenkoi* (CAWBG02) which was cultured for ANTX production and was then acidified. It is believed that this compound is a biosynthetic precursor in the production of ANTX [6]. This discovery supported previous findings by Hemscheidt et al., following radiolabelled studies, that 11-carboxyl intermediates were involved in the biosynthesis of the toxin [7]. The known degradation products of ANTX and HANTX are not believed to be significantly toxic [3, 5, 8, 9]; however, it is believed that dihydroanatoxin-a (dihydroANTX) as a mixture of diastereoisomers, retains about 10% of the toxicity of ANTX [10–12].

Many different syntheses of ANTX or HANTX have been reported, and the early synthetic works have been reviewed previously [13, 14]. More recent synthetic approaches have been proposed including the short and efficient synthesis developed by Parsons et al. [15]. Also, the *cis*- and *trans*-dihydroderivatives of ANTX and HANTX have been recently prepared [16].

The major chemical properties of ANTX are summarized in Table 13.1. ANTX has two reactive functions (an amine and a ketone), but modification of the amine has been widely used to derivatize this alkaloid, mainly to facilitate its detection. The *N*-acetyl, and *N*-(7-nitro-2,1,3-benzoxadiazole)-4-anatoxin-a have been used in analytical studies (see Section 13.6). ANTX is rapidly (half-life = 100 min) degraded by sunlight [3]. It is also oxidized very rapidly by ozone, for instance. This oxidation might be potentially used for water treatment.

Table 13.1 Summary of the essential molecular properties of ANTX

Systematic name	(1*R*,6*R*)-1-(9-azabicyclo[4.2.1]non-2-en-2-yl)ethanone
Formula and exact mass	$C_{10}H_{15}NO$; 165.11536 Da
pKa	9.36 ± 0.06
Optical rotation	$[\alpha]_D = +43.2$ (c 0.676, ethanol)
UV absorption	$\lambda_{max} = 227$ nm, $\varepsilon = 10\,700$ cm mol L^{-1}
NMR data	See Koskinen and Rappoport, 1985 [17]

13.3 Biosynthesis of ANTX, HANTX, and dihydroANTX

The cluster of genes responsible for the biosynthesis of ANTX and HANTX, thus called *ana*, was first identified and sequenced in *Oscillatoria* sp. PCC6506 [18]. Subsequently, the *ana* cluster was sequenced in three other cyanobacteria strains. Based on the bioinformatic analysis of the *ana* cluster and on *in vitro* studies on the biosynthetic enzymes, a biosynthetic route for ANTX and HANTX, starting from proline, has been proposed [19]. Lately, it has been found that *Cylindrospermum stagnale* sp. PCC 7417, produced dihydroANTX [20]. Interestingly, the *ana* cluster found in this cyanobacterium possesses an extra gene coding for a reductase that is likely to be involved in the reduction of the double bond of ANTX or of a biosynthetic precursor [21].

13.4 Occurrence and Producing Strains

Distribution of ANTX and/or HANTX-producing cyanobacteria is worldwide (see Appendix 2). Planktonic filamentous cyanobacteria such as *Dolichospermum* spp. (syn. *Anabaena*), *Aphanizomenon* spp., *Cuspidothrix issatschenkoi* (syn. *Aphanizomenon issatschenkoi*), *Planktothrix rubescens*, *Oscillatoria* sp., *Arthrospira fusiformis*, and several benthic *Oscillatoria* spp., *Phormidium favosum*, etc., are able to produce the neurotoxins, and were identified both in freshwaters and brackish/marine waters in almost all geographical regions. So far, they have been reported from lakes, artificial water bodies, rivers, and marine reservoirs in 10 European countries, North and South America, Asia, Africa, Australia, New Zealand, and New Caledonia. Different strains of cyanobacteria with various levels of neurotoxin production occur and may coexist in the same water reservoir [22, 23]. For example, only 25% of samples collected in German lakes and reservoirs were AN positive [24]. However, in Finnish lakes [25] and Portuguese natural and artificial freshwater reservoirs [26], about 60% of cyanobacteria strains were capable of ANTX production, and the range of its concentration in the cyanobacteria biomass was very broad (0.06–24.6 μg g dry wt^{-1}). In *Cuspidothrix issatschenkoi* (New Zealand strain CAWBG02), even 1683 μg ANTX g dry wt^{-1} has been recently reported [27]. ANTX was first reported in *Dolichospermum flos-aquae* NRC-44h [28] and HANTX in oscillatoriales cyanobacteria (e.g. *Phormidium formosum* (syn. *Oscillatoria formosa*; [29]). The earliest reports of ANTX occurrence in Poland concerned the brackish coastal waters of the Baltic sea [30] and the shallow hypereutrophic Zemborzycki dam reservoir near Lublin city, where perennial water blooms (formed by a mixture of different proportions of *Dolichospermum flos-aquae, D. circinale, D. spiroides, D. planctonicum, Aph. flos-aquae,* and *Aph. gracile* together with *Planktothrix agardhii* and *Microcystis* spp.) had been developing [31, 32]. The concentration of ANTX was very variable and correlated positively with the density of *Dolichospermum* spp. but not with the density of

Aph. flos-aquae. Seasonal changes in species composition of cyanobacteria assemblages are common; hence, neurotoxin-producing species/strains may be dominant for only several days or weeks [32]. ANTX (up to 1.4 mg L^{-1} of cyanobacterial scum) was also detected in newly constructed shallow dam reservoirs with blooms of *Dolichospermum flos-aquae* (syn. *Anabaena flos-aquae*) or *Aph. flos-aquae* [33] and in a hypereutrophic lake in Polesie region (eastern Poland) where a positive correlation between intracellular ANTX concentration (0.4–3.5 μg L^{-1}) and the biomass of *D. lemmermanni, D. flos-aquae* and *C. issatschenkoi* was noted [34].

13.5 Toxicity and Pharmacology

13.5.1 Mechanism of Toxicity

ANTX and its analogue, HANTX, are nicotinic agonists and have been shown to bind to nicotinic acetylcholine receptors. ANTX is an extremely potent toxin, with an LD_{50} of 250 μg kg^{-1} [28]. ANTX binds to the acetylcholine receptors and remains there, due to the inability of acetylcholine esterase to remove the toxin from the acetylcholine site. Consequently, the neuron continuously propagates neuronal impulses, which ultimately result in nerve depolarisation through movement of positively charged ions across the receptor, in addition to desensitivity of the receptor, blocking neurotransmission [43, 44]. ANTX has been shown to affect muscle-type acetylcholine receptors ($\alpha 1_2\beta\gamma\delta$; [43]) and chicken neuronal type acetylcholine receptors (α482 and α7; [45]). At sufficiently high concentrations, multiple neurons will be affected, such that paralysis results, with death due to respiratory arrest. As nicotinic acetylcholine receptors are known to be affected by ANTX, and as these nicotinic receptors are known to affect the release of dopamine, the effect of ANTX was studied on the release of dopamine [46]. It was observed that ANTX affected the release of dopamine, with an increase in the presence of ANTX in a dose-dependent manner [46].

Additional toxicity mechanisms have also been proposed for ANTX. During perfusion of various doses of ANTX, blood pressure, heart rate, and gas exchange were affected [47]. Acidosis, hypoxia, and respiratory arrest were recorded, potentially explaining some of the symptoms observed with ANTX poisoning [44, 47].

13.5.2 Animal Poisonings

There have been a number of reports of wild and domestic animals poisoned by ANTX. These include cattle, dogs and birds, with microcystin (MC)- and ANTX-containing cyanobacterial mats being considered responsible for cattle deaths in Alpine lakes [48, 49]. Dogs have been shown to be particularly susceptible to ANTX, largely through consumption of benthic mats, composed primarily of *Phormidium*, and such deaths have been documented in Scotland [50], New Zealand [51], and France [52]. This susceptibility of dogs to ANTX poisoning by benthic cyanobacteria is considered to be due to two factors. Firstly, such benthic *Phormidium* mats are common producers of taste and odour compounds such as geosmin, and dogs appear to be attracted to such smells [53]. Further, dogs clean and groom their fur through licking, and if they have been swimming or wading in waters where the cyanobacteria adhere to their fur, they can thereby ingest the toxin.

Birds deaths have also been attributed to cyanobacterial toxins [49]. In particular, the lesser Flamingo (*Phoeniconaias minor*), which consumes cyanobacteria (*Arthrospira*) as a major part of their diet, are known to succumb to ANTX poisoning. During such intoxication events, it was shown that MCs and ANTX were present within tissues of the lesser Flamingo [54], ANTX was shown to be present also within the Flamingo feathers post mortem [55]. A further indication of

ANTX intoxication is the observation of opisthotonos. Consequently, when birds suffer from ANTX poisoning, the muscle at the back of the neck contracts, so that the head is thrown backward and, in the case of lesser Flamingo, the head and neck will lie along the back of affected birds.

13.6 Analytical Methodologies

13.6.1 Available Methods for Determination and Quantification

Table 13.2 summarizes the major published analytical methods for the detection and quantification of ANTX and its derivatives. GC-MS was widely used in the past to detect ANTX, usually as its more volatile *N*-acetyl derivative [35]. More recently, LC-MS–based methods seem to be now accepted as the best methodologies. Further, with the advent of more sophisticated LC-MS apparatus using MS^2 detection, the detection/quantification of ANTX without derivatization makes for faster sample preparation [37–41]. However, it should be emphasized that pure standards are needed for accurate quantification. Very recently, a colorimetric receptor-binding assay for ANTX has been released on the market by Abraxis. This assay is based on the competition of the binding of the analyte to an immobilized nicotinic acetylcholine receptor (nAChR), with biotinylated α-bungarotoxin. The remaining α-bungarotoxin is detected by horseradish peroxidase (HRP) linked to streptavidin. It is too early to give an accurate comparison of this new technique with the LC-MS techniques. However, it should be mentioned that the receptor-binding assay will detect any potent agonist of the nAChR and will not specifically determine ANTX.

13.6.2 Methods for the Detection and Quantitation of AN

The mouse bioassay has been applied extensively to determine the presence of ANTX in samples or extracts [56, 57]. The test is observational: mice or rats are injected intraperitoneally (i.p.) and the presence of ANTX is based on identifying a classic set of symptoms after injection of the purified or extracted toxin [29, 58]. In practice, the method has many drawbacks: it is not quantitative, it cannot detect the presence of co-occurring toxins, and it s regarded by many as cruel when other instrumental methods are available (see Table 13.2). Marc et al. [59] attempted to refine the synthesis of an analogue of ANTX for monoclonal antibody production, which may in the future be used as the basis for a bioassay kit for determining ANTX in samples.

GC-MS methods have been applied to the analysis of ANTX and HANTX; the method is somewhat laborious; involving the conversion of the target compounds into their *N*-acetyl derivatives to increase volatility [1, 2, 28, 35, 50, 60].

LC with UV detection has also been applied to the analysis of ANTX and HANTX, taking advantage of the absorbance arising from the α,β-unsaturated ketone moiety on these compounds [50, 61–64]. However, ANTX and HANTX undergo rapid degradation to their dihydro- and epoxy-products [1, 63], destroying the α,β-unsaturated ketone in the process, thus removing the chromophore and hence excluding them from UV detection [8].

LC-FLD has also been used to determine ANTX and HANTX in samples, along with their characteristic degradation products. James et al. [8] derivatised the targets with 4-fluoro-7-nitro-2,1,3-benzoadiazole (NBD-F) prior to LC-FLD analysis; the method was sensitive and reproducible, and since unreacted NBD-F does not fluoresce strongly, very clean chromatography ensues. Namera et al. [65] devised an LC-FLD method using solid-phase microextraction (SPME) for ANTX and associated compounds that incorporated on-fibre derivatisation of targets prior to instrumental analysis. Rawn et al. [66] also developed an SPME-LC-FLD method using

Table 13.2 *Selected published analytical methods for the detection and quantification of ANTX, HANTX and their derivatives*

Method	Extraction (matrix)	Derivatization	LOD	LOQ	Comments	Reference
GC-MS	Methanol/HCl (cyanobacteria or water)	Formation of *N*-acetyl- ANTX and preconcentration on Sep-Pack silica cartridge	100 ng L^{-1}	n.r.	Good selectivity	[35]
HPLC-UV	50 mmol L^{-1} acetic acid (cyanobacteria)	No derivatization	20 ng on column	n.r.	Low sensitivity (estimated from the calibration curve) and low selectivity	[2]
HPLC-FLD	HCl (1 mol L^{-1}; 100 μL) in methanol (10 mL) and weak cation exchange SPE (cyanobacteria or water)	Derivatization with 4-fluoro-7-nitro-2,1, 3-benzoxadiazole	10 ng L^{-1}	n.r.	High sensitivity. Determination of ANTX, HANTX and their natural derivatives. See also Azevedo *et al.* [35] for improvements.	[8]
LC-MS	NH_4OH, pH 10m and reverse-phase SPE (water)	Online derivatization using fluorenylmethyloxycarbonyl chloride	2.1 ng L^{-1}	15.2 ng L^{-1}	High sensitivity. Automated process.	[37]
LC-MS^n	Water, pH 7, purification with weak cation exchange SPE (water)	No derivatization	600 ng L^{-1}	2500 ng L^{-1}	Mass fragmentation pathways for ANTX, HANTX and their natural derivatives. Low sensitivity.	[38]
LC-MS^2	Filtration (water)	No derivatization	8 ng L^{-1}	13 ng L^{-1}	High sensitivity.	[39]
LC-MS^2	50% acetic acid in methanol (cyanobacteria)	No derivatization	0.7 ng	1.15 ng	Very high sensitivity. Analysis of ANTX and derivatives.	[40]
LC-MS^2	Aqueous NH_4OH, pH 10.5, porous graphitic carbon SPE (water)	No derivatization	0.65 ng L^{-1}	1.96 ng L^{-1}	Very high sensitivity. Use of d_5-phenylalanine as internal standard.	[41]

FLD, fluorescence detection; GC, gas chromatography; HPLC, high-performance liquid chromatography; LC, liquid chromatography; LOD, limit of detection; LOQ, limit of quantification; MS, mass spectrometry; n.r., not reported; SPE, solid-phase extraction. biomass of *D. lemmermanni*, *D. flos-aquae*, and *C. issatschenkoi* was noted [42].

NBD-F derivatisation and found the low volumes associated with SPME compromised the sensitivity of the method. The problem with NBD-F derivatisation methods, however, is that the tag is reactive to most primary amines; so that interfering amines in the matrix can lead to false positive results.

LC-MS/MS offers perhaps the best analytical approach to the determination of ANTX, HANTX and their degradation products and analogues [38–40]. Further, the method can be applied to a range of sample types. However, the method is hindered by the presence of the amino acid Phe, which shares a nominal mass with ANTX and possesses a similar chromatographic retention time [66]. Once known, though, this co-elution possibility can be easily resolved [41]. For this reason, the identification of unique fragmentation ions in the MS/MS spectrum is essential for the unequivocal identification of the toxin group. Fragmentation data obtained using the high mass accuracy facility on a quadrupole-time-of-flight (qTOF-MS) and multiple fragmentation steps (MS^n; where $n = 1–4$) on a quadrupole ion trap (QIT MS) have identified characteristic fragments for this group of toxins [68]. Another caveat when using LC-MS/MS is that care must be taken to prevent ion suppression (or enhancement) of the MS signal, which can be caused by interferants in the matrix. Furey et al. [69] reviewed the causes of ion suppression and the approaches to quantifying and/or eliminating the phenomenon for routine analysis.

Several comprehensive reviews of the analytical approaches for determining ANTX and associated compounds have been published to date and should be consulted by interested parties [5, 70].

References

[1] Smith, R.A. and Lewis, D. (1987) A rapid analysis of water for anatoxin-a, the unstable toxic alkaloid from *Anabaena flos-aquae*, the stable non-toxic alkaloids left after bioreduction and a related amine which may be nature's precursor to anatoxin-a. *Vet. Hum. Toxicol.*, **29** (2), 153–154.

[2] Harada, K., Kimura, Y., Ogawa, K. *et al.* (1989) A new procedure for the analysis and purification of naturally occurring anatoxin-a from the blue-green alga *Anabaena Flos-Aquae*. *Toxicon*, **27** (12), 1289–1296.

[3] Stevens, D.K. and Krieger, R.I. (1991) Stability studies on the cyanobacterial nicotinic alkaloid anatoxin-a. *Toxicon*, **29** (2), 167–179.

[4] Namikoshi, M., Murakami, T., Watanabe, M.F. *et al.* (2003) Simultaneous production of homoanatoxin-a, anatoxin-a, and a new non-toxic 4-hydroxyhomoanatoxin-a by the cyanobacterium *Raphidiopsis mediterranea* Skuja. *Toxicon*, **42** (5), 533–538.

[5] Osswald, J., Rellan, S., Gago, A. and Vasconcelos, V. (2007) Toxicology and detection methods of the alkaloid neurotoxin produced by cyanobacteria, anatoxin-a. *Environment International*, **33** (8), 1070–1089.

[6] Selwood, A.I., Holland, P.T., Wood, S.A., Smith, K.F., McNabb, P.S. (2007) Production of anatoxin-a and a novel biosynthetic precursor by the cyanobacterium *Aphanizomenon issatschenkoi*. *Environ. Sci. Technol.*, **41** (2), 506–510.

[7] Hemscheidt, T., Rapala, J., Sivonen, K. and Skulberg, O.M. (1995) Biosynthesis of anatoxin-a in *Anabaena flos-aquae* and homoanatoxin-a in *Oscillatoria formosa*. *J. Chem. Soc. Chem. Commun.*, **13**, 1361–1362.

[8] James, K.J., Furey, A., Sherlock, I.R. *et al.* (1998) Sensitive determination of anatoxin-a, homoanatoxin-a and their degradation products by liquid chromatography with fluorimetric detection. *J. Chromatogr. A*, **798** (1-2), 147–157.

[9] James, K.J., Crowley, J., Duphard, J. *et al.* (2007) Anatoxin-a and analogues: discovery, distribution and toxicology, in *Phycotoxins: Chemistry and Biochemistry* (ed. L. Botana), Blackwell Publishing: UK, pp. 141–158.

[10] Wonnacott, S., Jackman, S., Swanson, K.L. *et al.* (1991) Nicotinic pharmacology of anatoxin analogs. II. Side-chain structure-activity-relationships at neuronal nicotinic ligand binding sites. *J. Pharmacol. Expt. Therapeut.*, **259**, 387–391.

[11] Bates, H.A. and Rapoport, H. (1979) Synthesis of anatoxin-a via intra-molecular cyclization of iminium salts. *J Am Chem Soc.*, **101**, 1259–1265.
[12] Kellman, R., Ploux, O. and Neilan, B. (2013) Neurotoxic alkaloids from cyanobacteria, in *Handbook of Natural Products (ISBN: 978-3-642-22143-9)* (eds K.G. Ramawat and J.M. Merillon), Springer-Verlag Berlin Heidelberg, pp. 39-83, DOI: 10.1007/978-3-642-22144-6_47.
[13] Mansell, H.L. (1996) Synthetic approaches to anatoxin-a. *Tetrahedron*, **52** (17), 6025–6061.
[14] Arbella, N.A., O'Callaghan, K., Furey, A. and James, K.J. (2007) Chemistry of cyanobacterial neurotoxins-anatoxin-a: synthetic approaches, in *Phycotoxins: Chemistry and Biochemistry* (ed. L. Botana), Blackwell Publishing Inc.: USA, pp. 119–140.
[15] Parsons, P.J., Camp, N.P., Edwards, N. and Sumoreeah, L.R. (2000) Synthesis of (±)-anatoxin-a and analogues. *Tetrahedron*, **56** (2), 309–315.
[16] Mann, S., Cohen, M., Chapuis-Hugon, F. *et al.* (2012) Synthesis, configuration assignment, and simultaneous quantification by liquid chromatography coupled to tandem mass spectrometry, of dihydroanatoxin-a and dihydrohomoanatoxin-a together with the parent toxins, in axenic cyanobacterial strains and in environmental samples. *Toxicon*, **60** (8), 1404–1414.
[17] Koskinen, A.M.P. and Rapoport, H. (1985) Synthetic and conformational studies on anatoxin-a: a potent acetylcholine agonist. *J. Medicinal Chem*, **28**, 1301–1309.
[18] Méjean, A., Mann, S., Maldiney, T. *et al.* (2009) Evidence that biosynthesis of the neurotoxic alkaloids anatoxin-a and homoanatoxin-a in the cyanobacterium *Oscillatoria PCC 6506* occurs on a modular polyketide synthase initiated by L-proline. *J Am Chem Soc.*, **131** (22), 7512–7523.
[19] Mejean, A., Mann, S., Vassiliadis, G. *et al.* (2010) *In vitro* reconstitution of the first steps of anatoxin-a biosynthesis in *Oscillatoria PCC 6506*: from free L-proline to acyl carrier protein bound dehydroproline. *Biochemistry*, **49** (1), 103–113.
[20] Mejean, A., Paci, G., Gautier, V. and Ploux, O. (2014) Biosynthesis of anatoxin-a and analogues (anatoxins) in cyanobacteria. *Toxicon*, **91**, 15–22.
[21] Shih, P.M., Wu, D., Latifi, A. *et al.* (2013) Improving the coverage of the cyanobacterial phylum using diversity-driven genome sequencing. *Proceedings of the National Academy of Sciences of the United States of America* **110** (3), 1053–1058.
[22] Araóz, R., Nghiêm, H.O., Rippka, R. *et al.* (2005) Neurotoxins in axenic oscillatorian cyanobacteria: coexistence of anatoxin-alpha and homoanatoxin-alpha determined by ligand-binding assay and GC/MS. *Microbiology*, **151** (Pt 4), 1263–1273.
[23] Kotut, K., Ballot, A. and Krienit, L. (2006) Toxic cyanobacteria and their toxins in standing waters of Kenya: implications for water resources use. *Journal of Water Health*, **4**, 233–245.
[24] Bumke-Vogt, C., Mailahn, W. and Chorus, I. (1999) Anatoxin-a and neurotoxic cyanobacteria in German lakes and reservoirs. *Environ. Toxicol.*, **14**, 117–125.
[25] Sivonen, K., Himberg, K., Luukkainen, R. *et al.* (1989) Preliminary characterization of neurotoxic cyanobacteria blooms and strains from Finland. *Toxic. Ass.*, **4** (3), 339–352.
[26] Osswald, J., Rellán, S., Gago-Martinez, A. and Vasconcelos, V. (2009) Production of anatoxin-a by cyanobacterial strains isolated from Portuguese water systems. *Ecotoxicology*, **18** (8), 1110–1115.
[27] Gagnon, A. and Pick, F.R. (2012) Effect of nitrogen on cellular production and release of the neurotoxin anatoxin-a in a nitrogen-fixing cyanobacterium. *Frontiers in Microbiology*, **3** (211), 1–7.
[28] Devlin, J.P., Edwards, O.E., Gorham, P.R. *et al.* (1977) Anatoxin-a, a toxic alkaloid from Anabaena flos-aquae NRC-44h. *Can. J. Chem.*, **55**, 1367–1371.
[29] Skulberg, O.M., Skulberg, R., Carmichael, W.W. *et al.* (1992) Investigations of a neurotoxic oscillatorialean strain (Cyanophyceae) and its toxin. Isolation and characterisation of homoanatoxin-a. *Environ. Toxic. Chem.*, **11** (3), 321–329.
[30] Mazur, H.M. and Plinski, M. (2003) *Nodularia spumigena* blooms and the occurrence of hepatotoxin in the Gulf of Gdańsk. *Oceanologia*, **45** (2), 305–316.
[31] Pawlik-Skowronska, B., Skowronski, T., Pirszel, J. and Adamczyk, A. (2004) Relationship between cyanobacterial bloom composition and anatoxin-a and microcystin occurrence in the eutrophic dam reservoir (SE Poland). *Polish J. Ecol.*, **52** (4), 479–490.
[32] Pawlik-Skowronska, B., Kalinowska, R. and Skowronski, T. (2013) Cyanotoxin diversity and food web bioaccumulation in a reservoir with decreasing phosphorus concentrations and perennial cyanobacterial blooms. *Harmful Algae*, **28**, 118–125.
[33] Pawlik-Skowronska, B. and Toporowska, M. (2001) Blooms of toxin-producing cyanobacteria – a real threat in small dam reservoirs at the beginning of their operation. *International Journal of Oceanography and Hydrobiology*, **40** (4), 30–37.

[34] Pawlik-Skowronska, B., Toporowska, M. and Rechulicz J. (2012) Simultaneous accumulation of anatoxin-a and microcystins in three fish species indigenous to lakes affected by cyanobacterial blooms. *Oceanological and Hydrobiological Studies*, **41** (4), 53–65.
[35] Himberg, K. (1989) Determination of anatoxin-a. the neurotoxin of *Anabaena flos-aquae* cyanobacterium, in algae and water by gas chromatosraphy-mass spectrometry. *Journal of Chromatography*, **481** (0), 358–362.
[36] Azevedo, J., Osswald, J., Guilhermino, L. and Vasconcelos, V. (2011) Development and validation of an SPE-HPLC-FL method for the determination of anatoxin-a in water and trout (*Oncorhincus mykiss*). *Analytical Letters*, **44** (8), 1431–1441.
[37] Takino, M., Daishima, S., Yamaguchi, K. (1999) Analysis of anatoxin-a in freshwaters by automated on-line derivatization-liquid chromatography-electrospray mass spectrometry. *Journal of Chromatography*, **862** (2), 191–197.
[38] Furey, A., Crowley, J., Lehane, M. and James, K.J. (2003) Liquid chromatography with electrospray ion-trap mass spectrometry for the determination of anatoxins in cyanobacteria and drinking water. *Rapid Commun Mass Spectrom.*, **17**, 583–588.
[39] Bogialli, S., Bruno, M., Curini, R. *et al.* (2006) Simple and rapid determination of anatoxin-a in lake water and fish muscle tissue by liquid-chromatography-tandem mass spectrometry. *Journal of Chromatography A*, **1122** (1–2),180–185.
[40] Hiller, S., Krock, B., Cembella, A. and Luckas, B. (2007) Rapid detection of cyanobacterial toxins in precursor ion mode by liquid chromatography tandem mass spectrometry. *Journal of Mass Spectrometry* **42** (9), 1238–1250.
[41] Dimitrakopoulos, I.K., Kaloudis, T.S., Hiskia, A.E. *et al.* (2010) Development of a fast and selective method for the sensitive determination of anatoxin-a in lake waters using liquid chromatography-tandem mass spectrometry and phenylalanine-d5 as internal standard. *Anal Bioanal Chem.*, **397** (6), 2245–2252.
[42] Pawlik-Skowronska, B., Toporowska, M. and Rechulicz, J. (2012) Simultaneous accumulation of anatoxin-a and microcystins in three fish species indigenous to lakes affected by cyanobacterial blooms. *Oceanological and Hydrobiological Studies*, **41**(4), 53–65.
[43] Spivak, C.E., Witkop, B. and Albuquerque, E.X. (1980) Anatoxin-a: a novel, potent agonist at the nicotinic receptor. *Mol. Pharmacol.*, **18**, 384–394.
[44] Araóz, R., Molgó, J. and Tandeau de Marsac, N. (2010) Neurotoxic cyanobacterial toxins. *Toxicon*, **56** (5), 813–28.
[45] Thomas, P., Stephens, M., Wilkie, G. *et al.* (1993) (+)-Anatoxin-a is a potent agonist at neuronal nicotinic acetylcholine receptors. *J. Neurochem.*, **60** (6), 2308–2311.
[46] Soliakov, L., Gallagher, T. and Wonnacott, S. (1995) Anatoxin-a evoked [H-3] dopamine release from rat stiatial synaptosomes. *Neropharmacol.*, **34**, 1535–1541.
[47] Adeyemo, O.M. and Siren, A.L. (1992) Cardio-respiratory changes and mortality in the conscious rat induced by (+)- and (+/-)-anatoxin-a. *Toxicon*, **30**, 899–905.
[48] Mez, K., Beattie, K., Codd, G. *et al.* (1997) Identification of a microcystin in benthic cyanobacteria linked to cattle deaths on alpine pastures in Swtzerland. *European Journal of Phycology*, **32** (2), 111–117.
[49] Codd, G.A., Lindsay, J., Young, F.M. *et al.* (2005) Harmful cyanobacteria. *Aquatic Ecology Series*, **3**, 1–23.
[50] Edwards, C., Beattie, K.A., Scrimgeour, C.M. and G.A. Codd (1992) Identification of anatoxin-a in benthic cyanobacteria (blue-green algae) and in associated dog poisonings at Loch Insh, Scotland. *Toxicon*, **30** (10), 1165–1175.
[51] Wood, S.A., Selwood, A.I., Rueckert, A. *et al.* (2007) First report of homoanatoxin-a and associated dog neurotoxicosis in New Zealand. *Toxicon*, **50** (2), 292–301.
[52] Cadel-Six, S., Peyraud-Thomas, C., Brient, L. *et al.* (2007) Different genotypes of anatoxin-producing cyanobacteria coexist in the Tarn River, France. *Applied and Environmental Microbiology*, **73** (23), 7605–7614.
[53] Codd, G.A., Edwards, C., Beattie, K.A. *et al.* (1992) Fatal attraction to cyanobacteria? *Nature*, **359**, 110–111.
[54] Krienitz, L., Ballot, A., Kotut, K. *et al.* (2003) Contribution of hot spring cyanobacteria to the mysterious deaths of lesser Flamingos at Lake Bogoria, Kenya. *Fems Microbiology Ecology*, **43** (2), 141–148.
[55] Metcalf, J.S., Morrison, L.F., Krienitz, L. *et al.* (2006) Analysis of the cyanotoxins anatoxin-a and microcystins in lesser Flamingo feathers. *Toxicological and Environmental Chemistry*, **88** (1), 159–167.

[56] Gorham, P.R., McLachlan, J., Hammer, U.T. and Kim, W.K. (1964) Isolation and culture of toxic strains of Anabaena flos-aquae (Lyngb.) de Breb. *Mitt. internat. Verein. Limnol.*, **15**, 796–804.
[57] Carmichael, W.W. and Gorham, P.R. (1974) An improved method for obtaining axenic clones of planktonic blue-green algae. *J. Phycol.* **10**, 238–241.
[58] Carmichael, W.W., Biggs D.F. and Gorham, P.R. (1975) Toxicology and pharmacological action of Anabaena flos-aquae toxin. *Science*, **187**, 542–544.
[59] Marc, M., Outurquin, F., Renard, P.-Y. *et al.* (2009) Synthesis of a (+)-anatoxin-a analogue for monoclonal antibodies production. *Tetrahedron Letters*, **50** (31), 4554–4557.
[60] James, K.J., Furey, A., Sherlock, I.R. *et al.* (1998) The analysis of the toxic and non-toxic anatoxins from cyanobacteria, in *Harmful Algae* (eds. B.B. Reguera, J. Blanco, M.L. Fernandez, T. Wyatt), Xunta de Galicia and Intergovernmental Oceanographic Commission of UNESCO: Vigo, pp. 525–528.
[61] Astrachan, N.B. and Archer, B.G. (1981) Simplified monitoring of anatoxin-a by reverse phase high performance liquid chromatography and sub-acute effects of anatoxin-a in rats., in *The Water Environment: Algal Toxins and Health*, W.W. Carmicheal, Editor. Plenum Press: New York, pp. 437–446.
[62] Wong, H.S. and Hindin, E. (1982) Detecting an algal toxin by high-pressure liquid chromatography. *J. Am. Water Works Assoc.*, **74**, 528–529.
[63] Harada, K.-I. *et al.* (1993) Liquid chromatography/mass spectrometric detection of anatoxin-a, a neurotoxin from cyanobacteria. *Tetrahedron*, **49** (41), 9251–9260.
[64] Zotou, A., Jefferies, T.M., Brough, P.A. and Gallagher, T. (1993) Determination of anatoxin-a and homoanatoxin in blue-green algal extracts by high-performance liquid chromatography and gas chromatography-mass spectrometry. *Analyst*, **118** (7), 753–758.
[65] Namera, A., So, A. and Pawliszyn, J. (2002) Analysis of anatoxin-a in aqueous samples by solid-phase microextraction coupled to high-performance liquid chromatography with fluorescence detection and on-fiber derivatization. *Journal of Chromatography A*, **963** (1-2), 295–302.
[66] Rawn, D.F.K., Lau, B.P., Niedzwiadek, B. and Lawrence, J.F. (2005) Improved method for the determination of anatoxin-a and two of its metabolites in blue-green algae using liquid chromatography with fluorescence detection. *Journal of AOAC International*, **88** (6), 1741–1747.
[67] Furey, A., Crowley, J., Hamilton, B. *et al.* (2005) Strategies to avoid the mis-identification of anatoxin-a using mass spectrometry in the forensic investigation of acute neurotoxic poisoning. *Journal of Chromatography A*, **1082** (1), 91–97.
[68] James, K.J., Crowley, J., Hamilton, B. *et al.* (2005) Anatoxins and degradation products, determined using hybrid quadrupole time-of-flight and quadrupole ion-trap mass spectrometry: forensic investigations of cyanobacterial neurotoxin poisoning. *Rapid Commun. Mass Spectrom.*, **19** (9), 1167–1175.
[69] Furey, A., Moriarty, M., Bane, V. *et al.* (2013) Ion suppression; a critical review on causes, evaluation, prevention and applications. *Talanta*, **115**, 104–122.
[70] James, K.J., Dauphard, J., Crowley, J. and Furey, A. (2008) Cyanobacterial neurotoxins, anatoxin-a and analogues: detection and analysis, in *Seafood and Freshwater Toxins: Pharmacology, Physiology and Detection* (ed. L. Botana), CRC Press Taylor and Francis Group, pp. 809–822.

14

Saxitoxin and Analogues

Andreas Ballot[1], *Cécile Bernard*[2], *and Jutta Fastner*[3]

[1]*Norwegian Institute for Water Research (NIVA), Oslo, Norway*
[2]*UMR 7245 MCAM CNRS-MNHN, Muséum National d'Histoire Naturelle, Paris, France*
[3]*Section Drinking Water Resources and Water Treatment, German Environment Agency, Berlin, Germany*

14.1 Introduction

Saxitoxins (STXs), also called paralytic shellfish poisoning (PSP) toxins or paralytic shellfish toxins (PSTs), are potent neurotoxic alkaloids, produced by marine eukaryotic dinoflagellates and freshwater prokaryotic cyanobacteria. They represent a remarkable example of toxins that are produced by organisms belonging to different kingdoms of life [1]. In freshwater environments, STXs are mainly associated with filamentous cyanobacteria of various genera, including *Aphanizomenon*, *Dolichospermum* (*Anabaena*), *Lyngbya*, *Cylindrospermopsis*, *Raphidiopsis*, *Scytonema*, *Geitlerinema*, *Cylindrospermum*, and *Phormidium* [2–5] (for details, see Appendix 2). STX production has also been suggested for bacteria associated with dinoflagellates, but this remains controversial [6–9].

Saxitoxin (STX) was the first saxitoxin analogue that was identified [10–12]. Saxitoxin and its analogues (STXs) are composed of a 3,4-perhydropurine tricyclic system and have two guanidinium groups [13]. The STX molecule can be substituted at various positions (Fig. 14.1). Currently, 57 STXs are described [3].

Depending on their chemical structures, the STXs can be classified into several groups including carbamoyl (C) toxins, decarbamoyl (dc) toxins, *N*-sulphocarbamoyl (G) toxins, gonyautoxins (GTX), and deoxydecarbamoyl (LW) toxins. A detailed overview of the 57 known STXs found in different sources (e.g. cyanobacteria, dinoflagellates, pufferfish, xanthid crabs, and frogs) is given by Wiese et al. [3].

Handbook of Cyanobacterial Monitoring and Cyanotoxin Analysis, First Edition.
Edited by Jussi Meriluoto, Lisa Spoof and Geoffrey A. Codd.

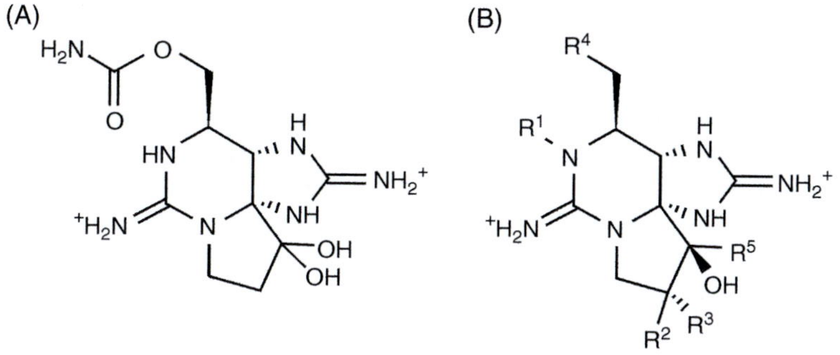

Figure 14.1 *General structures of (A) STX and (B) STX analogues*

14.2 Toxicity of STXs

STXs are potent neurotoxins. They block voltage-gated sodium ion channels in neuronal cells, acting on the extracellular sides through interaction with a neurotoxin receptor known as site 1 [14, 15]. Recently, STX has been discovered to also bind to potassium and calcium channels [16].

Ingestion of STXs by humans leads to syndromes known as paralytic shellfish poisoning (PSP). In severe cases, PSP causes death through respiratory failure [17]. From the marine environment, around 2000 cases of human poisoning through shellfish and fish consumption are reported annually with a mortality rate of approximately 15% [18]. An antidote or detoxification pathway is currently not known for PSP.

The most potent STXs are C toxins, including saxitoxin (STX) and neosaxitoxin (NEO). In mice, STX has an intravenous LD_{50} of 3.4 μg kg body wt^{-1}, an intraperitoneally LD_{50} of 10 μg kg body wt^{-1}, and an oral LD_{50} of 263 μg kg body wt^{-1} [19, 20].

STXs accumulate throughout aquatic foodwebs in marine environments. Traditional vectors for STXs are filter-feeding molluscs like clams, mussels, and oysters, but they are also transferred by nontraditional vectors like fish, crabs, or snails to terrestrial biota including humans causing PSP incidents [21]. No reported cases of human intoxication or transfer throughout foodwebs related to STXs from freshwater environments are known [22].

14.3 Occurrence

Compared with microcystins, there are relatively few studies on the occurrence of STXs in freshwaters. STXs have been reported worldwide with frequencies between 10% and 80% either on a per-lake or per-sample basis from Australia, New Zealand, North and South America, Asia, and Europe [23–30].

In samples from Europe, United States, Asia, and Australia, the detection of STXs has mostly been associated with the presence of *Aphanizomenon* and *Dolichospermum* (*Anabaena*) species (e.g. [23, 24, 28, 31]). In Denmark and Finland, STX occurrence was strongly associated with *Dolichospermum lemmermannii (Anabaena lemmermannii)*. Confirmed producers are *Dolichospermum circinale (Anabaena circinalis)* in Australia and *Aphanizomenon gracile* in Portugal, Germany, France, and Spain [23–25, 32–35]. Other identified STX producers are *Cylindrospermopsis raciborskii* and *Raphidiopsis brookii* in Brazil [36, 37]. STXs have also been detected in benthic (i.e. attached to surfaces) cyanobacteria such as *Lyngbya wollei* in, for example, Canada and Florida; *Geitlerinema*, *Cylindrospermum*, and *Phormidium* in Brazil; and *Scytonema* in New Zealand [4, 5, 38, 39].

Observed concentrations of STXs are generally quite low, with concentrations<< 1 to a few μg L^{-1} reported from Germany, United States, Brazil, and France; however, concentrations in the range of 10–1000 μg L^{-1} have been reported on occasion (e.g. in Spain, Finland) [25, 26, 28, 29, 34, 40]. Though high biomass of potential STX producers may result in high toxin concentrations, such as 1070 μg L^{-1} STX-equivalents measured during a bloom of *D. lemmermannii* (*A. lemmermannii*) in Finland [25], toxin concentrations can be low when STX-producing genotypes are less abundant in the population [41]. It is possible that some studies have underestimated the STX concentration since the share of extracellular STXs can be up to 40% in cultures and up to almost 100% in field samples [40, 42].

For benthic cyanobacteria, but also for some planktonic samples, STX concentrations were reported on a per-weight basis with findings in the range of about 3.5–19 μg g dry wt^{-1} in *Scytonema* samples from New Zealand [4], <100–290 μg g dry wt^{-1} in *L. wollei* samples from the United States and Canada [38, 39], between a few μg g^{-1} to around 1000 μg g^{-1} in Danish and Finnish samples [23, 25], and 2500–4500 μg g dry wt^{-1} in *D. circinale* (*A. circinalis*)–dominated samples from Australia [24].

Cyanobacteria may produce different STX analogues with a broad variety of profiles. Some similarities exist between European and Asian samples and strains of *Aphanizomenon* (*Aph. gracile*) showing various profiles with predominantly STX, NEO, dcSTX, and GTXs, while in *D. lemmermannii* (*A. lemmermannii*)–containing samples from Denmark and Finland, mainly or solely STX has been detected [23, 25, 33]. In contrast, *D. circinale* (*A. circinalis*) from Australia produces predominantly C toxins and GTXs with comparable profiles among samples and strains [24]. In *L. wollei*, dcSTX and dcGTX have been found in addition to *Lyngbya* toxins (LWT 1–6) [38].

14.4 Genetics and Biosynthesis

The gene cluster for STX biosynthesis (*sxt*), recently identified in strains of *Dolichospermum circinale* (*Anabaena circinalis*), *Aphanizomenon* sp., *Cylindrospermopsis raciborskii,* and *Raphidiopsis brookii*, as well as in a mat sample of *Lyngbya wollei*, provides new insight into the biosynthesis pathway of STX and its analogues [43–47]. A set of 14 core genes is common to all known *sxt* gene clusters, while another set of genes varies between *sxt* clusters of different strains [1].

The *Dolichospermum* (*Anabaena*) and *Aphanizomenon sxt* gene clusters seem to be more similar in gene content and organization than the *sxt* gene cluster in *C. raciborskii* [45, 46]. Although the *L. wollei sxt* gene cluster is highly similar in gene content to the *sxt* gene cluster of other cyanobacteria, it contains several unique genes [46]. According to Murray et al. [48], it is likely that a common ancient Nostocales ancestor of the *C. raciborskii*/*R. brookii* and the *Dolichospermum* (*Anabaena*)/*Aphanizomenon* clade possessed the *sxt* gene cluster. The identified *sxt* gene clusters are characterized by a mosaic structure. Genes have apparently been transposed as mobile small cassettes, leading to different gene arrangements in the known STX biosynthesis gene clusters [43, 44]. Multiple horizontal gene transfer events from different bacterial and cyanobacterial sources were proposed to explain the origin of parts of the STX biosynthetic machinery [49]. However, further evidence suggests that the *sxt* gene cluster has evolved through gene loss and rearrangements, as well as recombination and positive selection resulting in extraordinarily conserved biosynthetic machinery through stabilizing selection [48]. The coexistence of STX-producing and nonproducing strains in various cyanobacterial species can be explained by the repeated loss of the whole or parts of the *sxt* gene cluster over time [49].

These *sxt* gene clusters range in size from 25.7 kb (*R. brookii*) to 36 kb (*L. wollei*), and the presence or absence of *sxt* genes and their organization reflect the specific toxin profile expressed by each strain. The biosynthesis pathway of STXs starts with arginine, acetyl-CoA, *S*-adenosyl methionine (SAM), and carbamoylphosphate as precursors. A polyketide synthase-like enzyme, SxtA, is assumed to perform the first steps of STX biosynthesis. This is followed by transamidination, cyclization, epoxidation, aldehyde reduction, and finally carbamoylation and hydroxylation reactions, which may result in the STX parent molecule. Then, several tailoring enzymes produce decarbamoylated, *N*-1-hydroxylated (e.g. NEO), or sulphated analogues of STX (for review, see [1]).

14.5 Detection Methods

Methods for the detection and quantification of STXs include physicochemical methods (TLC, HPLC-FLD, LC-MS/MS), toxicity-based methods (e.g. mouse bioassay, cell-based assays), and assays using macromolecular or chemosensors (e.g. ELISA). A detailed review of analytical tools and bioassays for STXs is given by Humpage et al. [50].

Until recently, the mouse bioassay has been the official method for the detection of STXs in food and was also used in early studies on cyanotoxins. Its major advantage, despite the justified ethical concerns, is that it detects all (including unknown) STXs in a sample and that the results reflect the toxicity of the sample. In comparison, physicochemical methods face problems such as the diverse chemical properties of STXs (i.e. charge state), the possibility of chemical and biological transformations, new analogues not yet identified, and lack of certified standards. Nevertheless, after 30 years and numerous studies (for details, see Humpage et al. [50]), the analysis of STXs by LC with postcolumn oxidation and fluorescence detection is now accepted as the official method in food chemistry [51]. The method has been proved to be comparable to the mouse bioassay though being much more sensitive and has been applied to freshwater samples containing cyanobacteria in the past.

In the last decade, LC-MS has been successfully adapted for the analysis of STXs, often using hydrophilic interaction liquid chromatography (HILIC) for separation and the selective reaction monitoring mode to obtain high selectivity and sensitivity.

In order to replace the mouse bioassay, cell assays based on the mode of action of STXs (i.e. sodium channel blocking) have been developed, such as the neuroblastoma cell assay. Toxicities determined with cell assays have been found to be comparable to the mouse bioassay, while the cell assays have a greater sensitivity (e.g. [52]). Disadvantages are their sensitivity to all substances inhibiting sodium channels and the long performing times, though more rapid end points are under validation.

Other methods use molecular or chemical sensors that have the affinity to bind molecules with structural characteristics of STXs. ELISA is the easiest to use and most widespread method using antibodies as sensor molecules. Due to the highly variable cross-reactivity with the different STX analogues, ELISA is not considered to be quantitative and it is recommended for use as a screening tool [50]. Recent developments also use optical biosensor technologies for the analysis of STXs [53].

14.6 Guidance Values or National Regulations or Recommendations for Managing STXs

No official guideline value from the World Health Organization exists (as it does for microcystin). Some countries (Australia, Brazil, New Zealand, France) recommend 3 μg L^{-1} (as STX-equivalents) for drinking water [54].

References

[1] Dittmann, E., Fewer, D.P. and Neilan, B.A. (2013) Cyanobacterial toxins: biosynthetic routes and evolutionary roots. *FEMS Microbiology Reviews*, **37** (1), 23–43.

[2] Pearson, L., Mihali, T., Moffitt, M. *et al.* (2010) On the chemistry, toxicology and genetics of the cyanobacterial toxins, microcystin, nodularin, saxitoxin and cylindrospermopsin. *Marine Drugs*, **8** (5), 1650–1680.

[3] Wiese, M., D'Agostino, P.M., Mihali, T.K. *et al.* (2010) Neurotoxic alkaloids: saxitoxin and its analogs. *Marine Drugs*, **8** (7), 2185–2211.

[4] Smith, F.M.J., Wood, S.A., Wilks, T. *et al.* (2012) Survey of *Scytonema* (cyanobacteria) and associated saxitoxins in the littoral zone of recreational lakes in Canterbury, New Zealand. *Phycologia*, **51** (5), 542–551.

[5] Borges, H.L.F., Branco, L.H.Z., Martins, M.D. *et al.* (2015) Cyanotoxin production and phylogeny of benthic cyanobacterial strains isolated from the northeast of Brazil. *Harmful Algae*, **43**, 46–57.

[6] Martins, C.A., Alvito, P., Tavares, M.J. *et al.* (2003) Reevaluation of production of paralytic shellfish toxin by bacteria associated with dinoflagellates of the Portuguese coast. *Applied and Environmental Microbiology*, **69** (9), 5693–5698.

[7] Prol, M.J., Guisande, C., Barreiro, A. *et al.* (2009) Evaluation of the production of paralytic shellfish poisoning toxins by extracellular bacteria isolated from the toxic dinoflagellate *Alexandrium minutum*. *Canadian Journal of Microbiology*, **55** (8), 943–954.

[8] Kodama, M. (2010) Paralytic shellfish poisoning toxins: Biochemistry and origin. *Aqua-Bioscience Monographs*, **3** (1), 1–38.

[9] Stüken, A., Orr, R.J.S., Kellmann, R. *et al.* (2011) Discovery of nuclear-encoded genes for the neurotoxin saxitoxin in dinoflagellates. *PLoS ONE*, **6** (5), e20096.

[10] Schantz, E.J., Mold, J.D., Stanger, D.W. *et al.* (1957) Paralytic shellfish poison. VI. A procedure for the isolation and purification of the poison from toxic clam and mussel tissues. *Journal of the American Chemical Society*, **79** (19), 5230–5235.

[11] Bordner, J., Thiessen, W.E., Bates, H.A. and Rapoport, H. (1975) Structure of a crystalline derivative of saxitoxin. Structure of saxitoxin. *Journal of the American Chemical Society*, **97** (21), 6008–6012.

[12] Schantz, E.J., Ghazarossian, V.E., Schnoes, H.K. *et al.* (1975) Structure of saxitoxin. *Journal of the American Chemical Society*, **97** (5), 1238–1239.

[13] Shimuzu, Y. (2000) Chemistry and mechanisms of action, in *Seafood and freshwater toxins; Pharmacology, physiology and detection* (ed. L.M. Botana), Marcel Dekker, New York, pp. 176–201.

[14] Lipkind, G.M. and Fozzard, H.A. (1994) A structural model of the tetrodotoxin and saxitoxin binding site of the Na^+ channel. *Biophysical Journal*, **66**, 1–13.

[15] Cestèle, S. and Catterall, W.A. (2000) Molecular mechanisms of neurotoxin action on voltage-gated sodium channels. *Biochimie*, **82** (9 – 10), 883–892.

[16] Llewellyn, L.E. (2006) Saxitoxin, a toxic marine natural product that targets a multitude of receptors. *Natural Product Reports*, **23** (2), 200–222.

[17] Kelly, B.J. and Luce, J.M. (1991) The diagnosis and management of neuromuscular diseases causing respiratory failure. *Chest*, **99** (6), 1485–1494.

[18] Hallegraeff, G.M. (1993) A review of harmful algal blooms and their apparent global increase. *Phycologia*, **32** (2), 79–99.

[19] Wiberg, G.S. and Stephenson, N.R. (1960) Toxicologic studies on paralytic shellfish poison. *Toxicology and Applied Pharmacology*, **2** (6), 607–615.

[20] IPCS (1984) International Programme on Chemical Safety. Aquatic (Marine and Freshwater) Biotoxins. *Environmental Health Criteria* **37**. World Health Organization (WHO). http://www.inchem.org/documents/ehc/ehc/ehc37.htm (accessed 27 March 2015).

[21] Deeds, J.R., Landsberg, J.H., Etheridge, SM. *et al.* (2008) Non-traditional vectors for paralytic shellfish poisoning. *Marine Drugs*, **6** (2), 308–348.

[22] van Apeldoorn, M.E., van Egmond, H.P., Speijers, G.J. and Bakker, G.J. (2007) Toxins of cyanobacteria. *Molecular Nutrition & Food Research*, **51** (1), 7–60.

[23] Kaas, H. and Henriksen, P. (2000) Saxitoxins (PSP toxins) in Danish lakes. *Water Research*, **34** (7), 2089–2097.

[24] Velzeboer, R.M.A., Baker, P.D., Rositano, J. *et al.* (2000) Geographical patterns of occurrence and composition of saxitoxins in the cyanobacterial genus *Anabaena* (Nostocales, Cyanophyta) in Australia. *Phycologia*, **39** (5), 395–407.

[25] Rapala, J., Robertson, A., Negri, A.P. *et al.* (2005) First report of saxitoxin in Finnish lakes and possible associated effects on human health. *Environmental Toxicology*, **20** (3), 331–340.

[26] Costa, I.A.S., Azevedo, S.M.F.O., Senna, P.A.C. *et al.* (2006) Occurrence of toxin-producing cyanobacteria blooms in a Brazilian semiarid reservoir. *Brazilian Journal of Biology*, **66** (1b), 211–219.

[27] Wood, S.A., Holland, P.T., Stirling, D.J. *et al.* (2006) Survey of cyanotoxins in New Zealand water bodies between 2001 and 2004. *New Zealand Journal of Marine and Freshwater Research*, **40** (4), 585–597.

[28] Graham, J.L., Loftin, K.A., Meyer, M.T. and Ziegler, A.C. (2010) Cyanotoxin mixtures and taste-and-odor compounds in cyanobacterial blooms from the midwestern United States. *Environmental Science & Technology*, **44** (19), 7361–7368.

[29] Wörmer, L., Cirés, S., Agha, R. *et al.* (2011) First detection of cyanobacterial PSP (paralytic shellfish poisoning) toxins in Spanish freshwaters. *Toxicon*, **57** (6), 918–921.

[30] Dolman, A.M., Rücker, J., Pick, F.R. *et al.* (2012) Cyanobacteria and cyanotoxins: The influence of nitrogen versus phosphorus. *PLoS ONE*, **7** (6), e38757.

[31] Liu, Y., Chen, W., Li, D. *et al.* (2006) First report of aphantoxins in China—waterblooms of toxigenic *Aphanizomenon flos-aquae* in Lake Dianchi. *Ecotoxicology and Environmental Safety*, **65** (1), 84–92.

[32] Pereira, P., Li, R., Carmichael, W.W. *et al.* (2004) Taxonomy and production of paralytic shellfish toxins by the freshwater cyanobacterium *Aphanizomenon gracile* LMECYA40. *European Journal of Phycology*, **39** (4), 361–368.

[33] Ballot, A., Fastner, J. and Wiedner, C. (2010) Paralytic shellfish poisoning toxin producing cyanobacterium *Aphanizomenon gracile* in northeast Germany. *Applied and Environmental Microbiology*, **76** (4), 1173–1180.

[34] Ledreux, A., Thomazeau, S., Catherine, A. *et al.* (2010) Evidence for saxitoxins production by the cyanobacterium *Aphanizomenon gracile* in a French recreational water body. *Harmful Algae*, **10** (1), 88–97.

[35] Cirés, S., Wörmer, L., Ballot, A. *et al.* (2014) Phylogeography of cylindrospermopsin and paralytic shellfish toxin-producing Nostocales cyanobacteria from Mediterranean Europe (Spain). *Applied and Environmental Microbiology*, **80** (4), 1359–1370.

[36] Molica, R., Onodera, H., García, C. *et al.* (2002) Toxins in the freshwater cyanobacterium *Cylindrospermopsis raciborskii* (Cyanophyceae) isolated from Tabocas reservoir in Caruaru, Brazil, including demonstration of a new saxitoxin analogue. *Phycologia*, **41** (6), 606–611.

[37] Yunes, J.S., De La Rocha, S., Giroldo, D. *et al.* (2009) Release of carbohydrates and proteins by a subtropical strain of *Raphidiopsis brookii* (cyanobacteria) able to produce saxitoxin at three nitrate concentrations. *Journal of Phycology*, **45** (3), 585–591.

[38] Foss, A.J., Phlips, E.J., Yilmaz, M. and Chapman, A. (2012) Characterization of paralytic shellfish toxins from *Lyngbya wollei* dominated mats collected from two Florida springs. *Harmful Algae*, **16**, 98–107.

[39] Lajeunesse, A., Segura, P.A., Gélinas, M. *et al.* (2012) Detection and confirmation of saxitoxin analogues in freshwater benthic *Lyngbya wollei* algae collected in the St. Lawrence River (Canada) by liquid chromatography-tandem mass spectrometry. *Journal of Chromatography A*, **1219**, 93–103.

[40] Fastner, J. (2011) Cyanobacterial hepato- and neurotoxins in the studied lakes. NOSTOTOX Final report, pp. 23-32. http://www.kompetenz-wasser.de/index.php?id=382&type=0&jumpurl=..%2Ffileadmin%2Fuser_upload%2Fpdf%2Fforschung%2FNOSTOTOX%2FNostotox_final_web.pdf.

[41] Al-Tebrineh, J., Merrick, C., Ryan, D. *et al.* (2012) Community composition, toxigenicity, and environmental conditions during a cyanobacterial bloom occurring along 1,100 kilometers of the Murray River. *Applied and Environmental Microbiology*, **78** (1), 263–272.

[42] Negri, A.P., Jones, G.J., Blackburn, S.I. *et al.* (1997) Effect of culture and bloom development and of sample storage on paralytic shellfish poisons in the cyanobacterium *Anabaena circinalis*. *Journal of Phycology*, **33** (1), 26–35.

[43] Mihali, T.K., Kellmann, R. and Neilan, B.A. (2009) Characterisation of the paralytic shellfish toxin biosynthesis gene clusters in *Anabaena circinalis* AWQC131C and *Aphanizomenon* sp. NH-5. *BMC Biochemistry*, **10**:8, doi:10.1186/1471-2091-10-8.

[44] Kellmann, R., Mihali, T.K., Jeon, Y.J. *et al.* (2008) Biosynthetic intermediate analysis and functional homology reveal a saxitoxin gene cluster in cyanobacteria. *Applied and Environmental Microbiology*, **74** (13), 4044–4053.

[45] Stucken, K., John, U., Cembella, A. *et al.* (2010) The smallest known genomes of multicellular and toxic cyanobacteria: Comparison, minimal gene sets for linked traits and the evolutionary implications. *PLoS ONE*, **5** (2), e9235.

[46] Mihali, T.K., Carmichael, W.W. and Neilan, B.A. (2011) A putative gene cluster from a *Lyngbya wollei* bloom that encodes paralytic shellfish toxin biosynthesis. *PLoS ONE*, **6**, e14657.

[47] Carmichael, W.W., Evans, W.R., Yin, Q.Q. *et al.* (1997) Evidence for paralytic shellfish poisons in the freshwater cyanobacterium *Lyngbya wollei* (Farlow ex Gomont) comb. nov. *Applied and Environmental Microbiology*, **63** (8), 3104–3110.

[48] Murray, S.A., Mihali, T.K. and Neilan, B.A. (2011) Extraordinary conservation, gene loss, and positive selection in the evolution of an ancient neurotoxin. *Molecular Biology and Evolution*, **28** (3), 1173–1182.

[49] Moustafa, A., Loram, J.E., Hackett, J.D. *et al.* (2009) Origin of saxitoxin biosynthetic genes in cyanobacteria. *PLoS ONE*, **4** (6), e5758.

[50] Humpage, A.R., Magalhaes, V.F. and Froscio, S.M. (2010) Comparison of analytical tools and biological assays for detection of paralytic shellfish poisoning toxins. *Analytical and Bioanalytical Chemistry*, **397** (5), 1655–1671.

[51] AOAC (2011) Official Method 2011.02. Paralytic Shellfish Toxins in Mussels, Clams, Oysters, and Scallops. Post-Column Oxidation (PCOX) Method. First Action 2011. Final Action 2013. Gaithersburg, MD, USA: AOAC International.

[52] Humpage, A.R., Ledreux, A., Fanok, S. *et al.* (2007) Application of the neuroblastoma assay for paralytic shellfish poisons to neurotoxic freshwater cyanobacteria. Interlaboratory calibration and comparison with HPLC, LC-MS/MS, and Jellett Mist Alert™ for PSPs. *Environmental Toxicology and Chemistry*, **26** (7), 1512–1519.

[53] Campbell, K., Rawn, D.F.K., Niedzwiadek, B. and Elliott, C.T. (2011) Paralytic shellfish poisoning (PSP) toxin binders for optical biosensor technology: problems and possibilities for the future: a review. *Food Additives & Contaminants: Part A*, **28** (6), 711–725.

[54] Chorus, I. (2012) Current approaches to cyanotoxin risk assessment, risk management and regulations in different countries. Federal Environment Agency (Umweltbundesamt) **63**, http://www.uba.de/publikationen/current-approaches-to-cyanotoxin-risk-assessment.

15

Anatoxin-a(*S*)

James S. Metcalf[1] *and Milena Bruno*[2]

[1] *Institute for Ethnomedicine, Jackson, USA*

[2] *Department of Environment and Primary Prevention, Istituto Superiore di Sanità, Rome, Italy*

15.1 Chemical Structure of Anatoxin-a(*S*)

Anatoxin-a(*S*) is a phosphorylated cyclic *N*-hydroxyguanidine (Fig. 15.1) produced by strains of *Anabaena*. It is an organophosphorous compound, with the structure (5*S*)-2-amino-1-((hydroxymethoxyphosphinyl)oxy)-*N,N*-dimethyl-4,5-dihydro-1*H*-imidazole-5-methanamin. Currently, it is the only naturally occurring organophosphate molecule to exist and, in common with pesticides and insecticides, anatoxin-a(*S*) also inhibits acetylcholine esterases (AChEs). Although the total synthesis of anatoxin-a(*S*) has not been elucidated, the side chain of anatoxin-a(*S*) can be chemically synthesized using a 2,3-diamino acid as an intermediate [1].

15.2 Biosynthesis

Anatoxin-a(*S*) was originally isolated from a strain of *Anabaena* (NRC 525-17; [2]). Through a procedure of extraction, Sephadex, and high-performance liquid chromatography, anatoxin-a(*S*) was isolated [2]. Using this strain, Moore et al. [3] examined the origin of the carbons of this neurotoxin. Carbon molecules 2, 4, 5, and 6 were found to be derived from the guanido carbons C-5, C-4, and C-3 from L-arginine. Although no biosynthetic enzymes have yet been elucidated, erythro-4-hydroxy-L-arginine, present within 525-17, was considered to be a precursor in anatoxin-a(*S*) biosynthesis [3], and (2*S*,4*S*)-4-hydroxyarginine was found to be an intermediate in anatoxin-a(*S*) biosynthesis, originating from L-arginine [4]. Although the biosynthesis of

Handbook of Cyanobacterial Monitoring and Cyanotoxin Analysis, First Edition.
Edited by Jussi Meriluoto, Lisa Spoof and Geoffrey A. Codd.

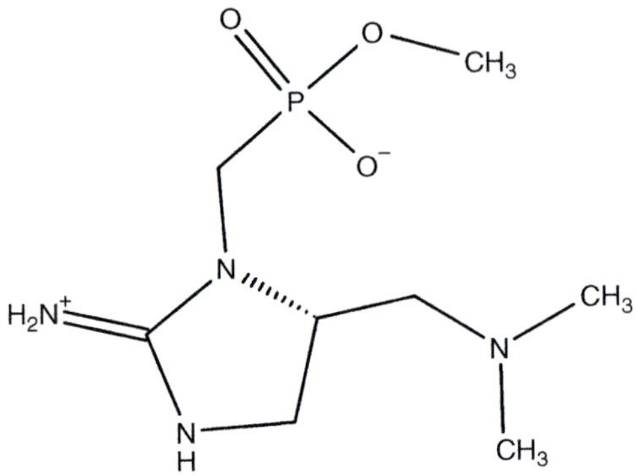

Figure 15.1 Structure of anatoxin-a(S)

Table 15.1 Anatoxin-a(S)–producing species and strains

Species	Strains	References
Anabaena flos-aquae	NRC 525-17	[7, 13]
Anabaena lemmermannii		[9, 14]
Anabaena spiroides	ITEP-024; ITEP-025; ITEP-026	[10]

anatoxin-a(*S*) is relatively unknown, when detected, either in strains or blooms of *Anabaena* (now termed *Dolichospermum*), microcystins are also generally known to co-occur with this neurotoxic organophosphate [5, 6].

15.3 Occurrence and Producing Strains

Anatoxin-a(*S*) has been identified in Canada [7], the United States [8], Denmark [9], and Brazil [10]. The presence of anatoxin-a(*S*) was also indicated by AChE inhibition assay in desert crust material from Qatar [11, 12].

Anatoxin-a(*S*) is produced by several species of the genus *Anabaena*: *A. flos-aquae*, *A. lemmermanni,* and *A. spiroides.* Several anatoxin-a(*S*)–producing strains have been isolated from species in bloom material (Table 15.1).

15.4 Toxicology and Pharmacology

15.4.1 Mechanism of Toxicity

Anatoxin-a(*S*) is a naturally occurring organophosphate that irreversibly inhibits AChE, similar to organophosphorus and carbamate insecticides [2]. The inhibition of AChE results in the neurotransmitter acetylcholine not being hydrolyzed at the synapse and, as such, the postsynaptic membrane cannot be repolarized and the nerve influx is blocked. This toxin is structurally unrelated to the other neurotoxic alkaloids produced by cyanobacteria, acting in a similar mechanism as the insecticides paraoxon, physostigmine, pyridostigmine, and sarin [15, 16]. Anatoxin-a(*S*) is unable to cross the blood–brain barrier, being a peripheral cholinesterase inhibitor [17].

15.4.2 Animal Poisonings

Symptoms in mammals are hypersalivation, lacrimation, fasciculation, urinary incontinence, convulsions, and death by respiratory arrest [18, 19]. The toxin kills mice in a few minutes (2–30 min) after intraperitoneal injection with an LD_{50} of 20–40 μg kg^{-1} in mice and 20 μg kg^{-1} in rats [2, 15, 20]. Inhibition of AChE activity was shown in experimental studies on fish (*Odontesthes argentinensis*) and crab (*Callinectes sapidus*) by aqueous extracts of *Anabaena spiroides* [21]. Recent experiments on *Daphnia similis* showed effects of an anatoxin-a(*S*)–producing strain (ITEP-024) on survivorship and somatic growth, with intact cell samples more effective than aqueous extracts [22]. Animal poisonings have been reported for dogs, ducks, swine, and birds [6, 8, 17].

15.5 Analytical Methods for Determination and Quantification

As anatoxin-a(*S*) is a naturally occurring organophosphate, methods to determine such compounds can be used in the analysis of anatoxin-a(*S*). During the original isolation of anatoxin-a(*S*), Mahmood and Carmichael [2] used a colorimetric AChE inhibition assay to monitor inhibitory activity. This assay has been used to measure and infer the presence of anatoxin-a(*S*) in strains and environmental samples of cyanobacteria [2, 7, 12].

Analysing blooms from Danish lakes, Henriksen et al. [6] detected up to 3.3 mg anatoxin-a(*S*) g phytoplankton dry wt^{-1} in blooms of *Anabaena lemmermanii* during 1993 and 1994, associated with bird kills. Analysis of tissues from the dead birds resulted in detectable anatoxin-a(*S*) concentrations between 2 and 89.7 μg kg body wt^{-1} [6]. When strains of *Anabaena lemmermanii* isolated from the 1993 bloom were tested, anatoxin-a(*S*) concentrations equivalent to between 29 and 743 μg g^{-1} were detected. Although *Anabaena* appears to be the only known cyanobacterial producer of anatoxin-a(*S*), Metcalf et al. [11], whilst studying cryptogamic crusts from Qatar, composed mainly of *Microcoleus* and Chroococcalean cyanobacteria, found inhibition of AChE in this material.

As the AChE inhibition assay is useful for the detection of anatoxin-a(*S*), other organophosphorous compounds may inhibit this enzyme. Although the enzyme assay is often laboratory based, Villatte et al. [23] used electric eel AChE to produce a disposable amperometric sensor to detect anatoxin-a(*S*). In order to overcome the possibility of insecticides inhibiting the assay, they used oxime reactivation to discriminate between insecticides and anatoxin-a(*S*) [23]. A different approach was taken by Devic et al. [24], who created mutants of AChE, to try and discriminate between anatoxin-a(S) and other organophosphorus compounds. They successfully produced an AChE inhibition assay that could selectively detect anatoxin-a(*S*) and applied this to anatoxin-a(*S*) in environmental cyanobacterial material [24].

Although enzyme inhibition assays are useful for measuring anatoxin-a(*S*), very few studies have successfully developed analytical methods to detect this neurotoxin, largely because of the lack of a chromophore on this organophosphorus molecule. However, Dörr et al. [25] examined a number of liquid chromatography (LC)–mass spectrometry (MS/MS) methods to see if anatoxin-a(*S*) could be successfully detected. Of the methods tested, a hydrophobic interaction liquid chromatography method showed the best sensitivity and retention for use. Using transitions of *m/z* 253>58 (quantitation ion), 253>159, 235>98, and 235>96, anatoxin-a(*S*) could be reliably detected [25]. Further, LC-MS/MS should be used to confirm anatoxin-a(*S*) in environmental samples and especially in animal or human poisoning events.

References

[1] Moura, S. and Pinto, E. (2010) Synthesis of cyclic guanidine intermediates of anatoxin-a(S) in both racemic and enantiomerically pure forms. *Synthetic Letters* **6**, 967–969.

[2] Mahmood, N.A. and Carmichael, W.W. (1986) The pharmacology of anatoxin-a(s), a neurotoxin produced by the fresh-water cyanobacterium *Anabaena flos-aquae* NRC 525-17. *Toxicon* **24** (5), 425–434.

[3] Moore, B.S., Ohtani, I., de Koning, C.B. *et al.* (1992) Biosynthesis of anatoxin-a(S). Origin of the carbons. *Tetrahedron Letters* **33** (44), 6595–6598.

[4] Hemscheidt, T., Burgoyne, D.L. and Moore, R.E. (1995) Biosynthesis of anatoxin-a(S). (2S,4S)-4-hydroxyarginine as an intermediate. *Journal of the Chemical Society Chemical Communications* **2**, 205–206.

[5] Harada, K.I., Ogawa, K., Kimura, Y. *et al.* (1991) Microcystins from *Anabaena flos-aquae* NRC 525-17. *Chemical Research in Toxicology* **4** (5), 535–540.

[6] Henriksen, P., Carmichael, W.W., An, J. *et al.* (1997) Detection of an anatoxin-a(s)-like anticholinesterase in natural blooms and cultures of cyanobacteria/blue-green algae from Danish lakes and in the stomach contents of poisoned birds. *Toxicon* **35** (6), 901–913.

[7] Carmichael, W.W., Gorham, P.R. and Biggs, D.F. (1977) 2 Laboratory case studies on oral toxicity to calves of freshwater cyanophyte (blue-green-alga) *Anabaena flos-aquae* NRC-44-1. *Canadian Veterinary Journal* **18** (3), 71–75.

[8] Mahmood, N.A., Carmichael, W.W. and Pfahler, D. (1988) Anticholinesterase poisonings in dogs from a cyanobacterial (blue-green-algae) bloom dominated by *Anabaena flos-aquae. American Journal of Veterinary Research* **49** (4), 500–503.

[9] Onodera, H., Oshima, Y., Henriksen, P. and Yasumoto, T. (1997) Confirmation of anatoxin-a(s), in the cyanobacterium *Anabaena lemmermannii*, as the cause of bird kills in Danish lakes. *Toxicon* **35** (11), 1645–1648.

[10] Molica, R.J.R., Oliveira, E.J.A., Carvalho, P.V. V.C. *et al.* (2005) Occurrence of saxitoxins and an anatoxin-a(S)-like anticholinesterase in a Brazilian drinking water supply. *Harmful Algae* **4** (4), 743–753.

[11] Metcalf, J.S., Richer, R., Cox P.A. *et al.* (2012) Cyanotoxins in desert environments may present a risk to human health. *Science of the Total Environment* **421-422**, 118–123.

[12] Chatziefthimiou, A.D., Richer, R., Rowles, H. *et al.* (2014) Cyanotoxins as a potential cause of dog poisonings in desert environments. *Veterinary Record* **174** (19), 484–485.

[13] Carmichael, W.W. and Gorham, P.R. (1978) Anatoxins from clones of *Anabaena flos-aquae* isolated from lakes of western Canada. *Mitteilungen Internationale Vereinigung für Theoretische und Angewandte Limnologie* **21**, 285–295.

[14] Landsberg, J.H. (2002) Anatoxins, In: R. R. Stickney, ed., *The Effects of Harmful Algal Blooms on Aquatic Organisms. Reviews in Fisheries Sciences* **10** (2), 241–243. doi:10.1080/20026491051695

[15] Cook, W.O., Beasley, V.R., Dahlem, A.M. *et al.* (1988) Comparison of effects of anatoxin-a(s) and paraoxon, physostigmine and pyridostigmine on mouse-brain cholinesterase activity. *Toxicon* **26** (8), 750–753.

[16] Pita, R., Anadon, A. and Martinez-Larranaga, M.R. (2003) Neurotoxins with anticholinesterase activity and their possible use as warfare agents. *Medicina Clinica-Barcelona* **121** (13), 511–517.

[17] Cook, W.O., Dellinger, J.A., Singh, S.S. *et al.* (1989) Regional brain cholinesterase activity in rats injected intraperitoneally with anatoxin-a(s) or paraoxon. *Toxicology Letters* **49** (1), 29–34.

[18] Codd, G.A., Bell, S.G., Kaya, K. *et al.* (1999) Cyanobacterial toxins, exposure routes and human health. *European Journal of Phycology* **34** (4), 405–415.

[19] Chorus I. and Bartram J. (1999) *Toxic Cyanobacteria in Water: A Guide to Their Public Health Consequences, Monitoring and Management.* E and FN Spon, London.

[20] Falconer I.R. (1996) Potential impact on human health of toxic cyanobacteria. *Phycologia* **35** (1), 6–11.

[21] Monserrat, J.M., Yunes, J.S. and Bianchini, A. (2001) Effects of *Anabaena spiroides* (Cyanobacteria) aqueous extracts on the acetylcholinesterase activity of aquatic species. *Environmental Toxicology and Chemistry* **20** (6), 1228–1235.

[22] De Abreu, F.Q. and Da Silva Ferrão-Filho, A. (2013) Effects of an anatoxin-a(S)-producing strain of *Anabaena spiroides* (Cyanobacteria) on the survivorship and somatic growth of two *Daphnia similis* clones. *Journal of Environmental Protection* **4** (6A), 12–18.

[23] Villatte, F., Schulze, H., Schmid, R. *et al.* (2002) A disposable acetylcholinesterase-based electrode biosensor to detect anatoxin-a(S) in water. *Analytical and Bioanalytical Chemistry* **372** (2), 322–326.
[24] Devic, E., Li, D., Dauta, A., Henriksen, P. *et al.* (2002) Detection of anatoxin-a(S) in environmental samples of cyanobacteria by using a biosensor with engineered acetylcholinesterases. *Applied and Environmental Microbiology* **68** (8), 4102–4106.
[25] Dörr, F.A., Rodriguez, V., Molica, R. *et al.* (2010) Methods for detection of anatoxin-a(S) by liqud chromatography coupled to electrospray ionization-tandem mass spectrometry. *Toxicon* **55** (1), 92–99.

16

β-*N*-Methylamino-L-Alanine and (*S*)-2,4-Diaminobutyric Acid

Olivier Ploux[1], *Audrey Combes*[2], *Johan Eriksson*[3], *and James S. Metcalf*[4]

[1] *LIED UMR 8236 CNRS, University Paris Diderot, and Chimie ParisTech, Paris, France*
[2] *LSABM, UMR CBI 8231 CNRS/ESPCI-ParisTech, ESPCI, Paris, France*
[3] *Department of Botany, Stockholm University, Sweden*
[4] *Institute for Ethnomedicine, Jackson, USA*

16.1 Historical Overview

Interest in β-*N*-methylamino-L-alanine (BMAA) and its isomers emerged through studies investigating a high incidence of a neurological disease called *Lytico bodig* and later termed amyotrophic lateral sclerosis (ALS)/parkinsonism dementia complex (ALS/PDC) among Chamorro villagers on the island of Guam. ALS/PDC was of scientific merit due largely to the fact that the disease included components of ALS, Alzheimer disease, and Parkinson disease, and the incidence was 50–100 times greater than elsewhere in the world. Further, those with ALS/PDC also manifested various clinical aspects of the three diseases, a finding which suggested that ALS, Alzheimer's, and Parkinson's may have some biological connection [1].

Investigations concerning the various potential causes of ALS/PDC were performed, with diet found to be the only statistical correlate with disease, with no known genetic causes on Guam [2]. Further, people who married into the Chamorro way of life were equally at risk of developing ALS/PDC, and Chamorros who left Guam and emigrated often developed ALS 10 to 15 years later. Studying the Chamorro diet led to clues concerning the potential etiology of the disease, and a cycad (*Cycas micronesica*) was examined as a possible source of toxins. A number of different toxic compounds were isolated from the cycad, including methazoxymethanol, sterol glycosides, and neurotoxic amino acids. Of these compounds, BMAA was investigated due to its neurotoxicity

Handbook of Cyanobacterial Monitoring and Cyanotoxin Analysis, First Edition.
Edited by Jussi Meriluoto, Lisa Spoof and Geoffrey A. Codd.

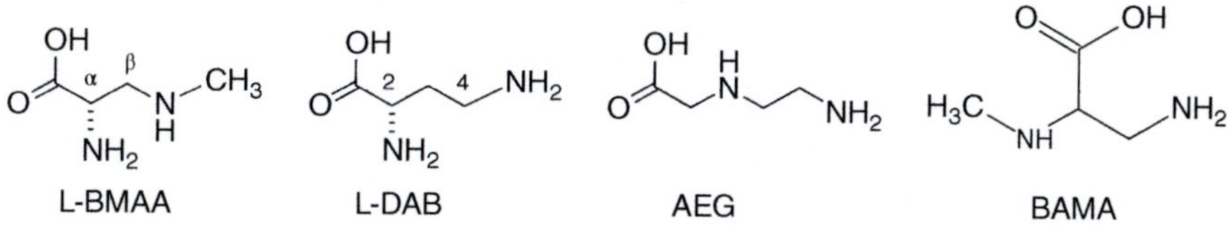

Figure 16.1 Structures of L-BMAA, L-DAB, AEG, and BAMA

in chicks and other organisms. Blinded analysis of Chamorro and control human brains found that the Chamorro brains contained BMAA, whereas the control brains largely did not. In addition, a number of Canadian Alzheimer brains were included, and these were also found to contain BMAA when the blind concerning sample identification was broken [3].

The finding of BMAA in Alzheimer's and Chamorro brains from people who died of ALS/PDC led to the wider consideration that BMAA may be important with respect to human neurodegenerative disease. A study replicating the earlier work of analysing Alzheimer's brains also included ALS and Huntington disease brains, the latter being a purely genetic disease. As all samples were blinded to the scientists prior to analysis, when the analytical blind was broken, in comparison to controls and Huntington brains, which were largely BMAA free, the ALS, Parkinson's, and Alzheimer's brains all contained differing concentrations of BMAA [4].

16.2 Structure, Synthesis, and Molecular Properties

Figure 16.1 shows the chemical structure of (*S*)-2-amino-3-methylaminopropanoic acid (L-BMAA) and that of its natural isomers (*S*)-2,4-diaminobutyric acid (L-DAB), *N*-(2-aminoethyl)glycine (AEG), and β-amino-*N*-methylalanine (BAMA). There are, in fact, other possible isomers of BMAA, molecules that share the same stoichiometric molecular formula ($C_4H_{10}N_2O_2$), and hence the same molecular mass (118.074228 U for the exact mass). However, this chapter will be restricted to the BMAA isomers that have been identified in biological samples. While AEG is achiral, BMAA, BAMA, and DAB are chiral molecules, and there are thus two possible enantiomers. It has been shown that cycads, and likely cyanobacteria, produce L-BMAA [5], while L- and D-DAB are produced by bacteria, as constituents of the peptidoglycan cell wall [6] or of secondary metabolites [7]. Several syntheses of L- or D-BMAA have been proposed [5, 8], as well as syntheses of labelled BMAA [9]. The pKa values of BMAA have been determined (2.1, 6.6, and 9.8) and its pI is 8.1 [5, 10, 11]. The amino functions of BMAA can be modified by specific reagents for derivatization (see Analytical Methods). BMAA also forms carbamates by reacting with carbonates [10] and forms chelates with divalent metal cations [11]. Although not discovered in cyanobacteria, BAMA has been detected in molluscs [12].

16.3 Neurotoxicity

When L-BMAA was originally isolated from cycad seeds [5, 13], toxicity experiments were performed in chicks, with the animals found to have clonus convulsions [14]. Further experiments with animals found that in all systems tested, BMAA showed toxicity (reviewed in [15]). Work by Spencer et al. [16] with Macaque monkeys dosed with BMAA found that at high doses, adverse neurological effects were observed.

Although in vivo studies showed that toxic effects could be observed with BMAA, *in vitro* studies were required to determine a molecular mechanism. Work by Weiss and colleagues showed that BMAA could potentiate neurotoxic insults with mixed motor neurons, and they determined that AMPA/kainate receptors were molecular targets for BMAA [17]. Only when bicarbonate was added did they elucidate that BMAA was capable of forming carbamates with a very close structural resemblance to glutamate, potentially explaining the ability of BMAA to bind to glutamate receptors and be toxic to motor neurons. Neurotoxicity work by Lobner and colleagues showed that BMAA was also capable of affecting the glutamate transporter system X_C and that the neurotoxicity of BMAA can be potentiated by other neurotoxic insults [18].

16.4 Methods for Identification and Quantification

Conventionally, trichloroacetic acid (TCA, 0.1 mol L^{-1} in water) is used to extract free BMAA and to precipitate proteins in samples (cyanobacteria, cycad seeds, or tissues). The proteins in the precipitate are then hydrolyzed (6 mol L^{-1} HCl, 110°C, overnight), releasing the bound fraction of BMAA [3, 19]. The physicochemical properties of BMAA and the absence of chromophore or fluorophore groups complicate its analysis in native form (no retention in reverse-phase chromatography and no possible UV or fluorescence detection). Cox and collaborators in 2003 [20] developed and validated an HPLC-fluorescence method generally applied for amino acid analysis [21]. This analytical procedure included (i) a derivatization step for free and/or bound BMAA extracts with 6-aminoquinolyl-*N*-hydroxysuccinimidyl carbamate (AQC) that reacts with primary and secondary amines, (ii) separation by conventional reverse phase (C18) to resolve BMAA from isomers, and (iii) detection by fluorescence (λ excitation = 250 nm, λ emission = 395 nm) or mass spectrometry at *m/z* 459 (M + H) or 230 (M + 2H) [20]. The high specificity of the detection of AQC-derivatized BMAA by mass spectrometry in multiple reaction monitoring (MRM) mode [21] has now largely replaced fluorescence detection, measuring BMAA concentrations in a wide range of samples, including cyanobacteria, shellfish, crustaceans, fish, and sharks, and has now been validated by the AOAC [22]. However, it has been reported that the fluorescence method could possibly lead to an overestimation of BMAA concentrations in complex samples [23], and the LC-MS/MS method could result in an underestimation of BMAA concentration due to adduct formation [24]. Recent application of hydrophilic interaction liquid chromatography (HILIC) designed for the separation of polar compounds, enabled the separation of BMAA and DAB without derivatization. This separation mode coupled with detection by mass spectrometry used in MRM mode allows identification and quantification [23, 25, 26]. However, extraction of free BMAA with TCA can lead to very complex samples, which generate a strong matrix effect during analysis by HILIC LC-MS/MS (decrease of the signal and increase in the limit of quantification of the method). To help avoid this bias, samples can be cleaned-up with solid-phase extraction cartridges (SPE) prior to analysis by LC-MS/MS [25, 26]. This analytical procedure allowed the identification and the quantification of underivatized BMAA in complex samples at trace amounts (cyanobacteria, river water, biofilms) [26].

16.5 Occurrence in Cyanobacteria, Plants, and Animals

BMAA was initially thought to be produced by a cycad (*Cycas micronesica*) on the island of Guam. Later research by Cox *et al.* [27] showed that a symbiotic cyanobacterium produced BMAA in the coralloid root of the cycad. The BMAA was then transferred to the cycad and accumulated in the seeds of the plant. Flying foxes that fed on the seeds contained high concentrations of BMAA and were also found to be a vector for transfer of BMAA to humans.

BMAA has been reported to be produced by the majority of cyanobacteria tested (e.g. [19]), although some research groups have reported an inability to identify this toxin in cyanobacteria, often when using HILIC methods (e.g. [28]). Recent findings have shown that eukaryotic organisms are also capable of producing BMAA, as both diatoms and dinoflagellates are considered to produce BMAA [29, 30]. Studies on bioaccumulation in both brackish and temperate waters, as well as marine tropical waters showed a similar trend [31, 32]. The highest concentrations were found in bottom or near-bottom dwelling species, suggesting a slightly different accumulation pathway than the classic bioaccumulation pattern, with the highest concentrations in bottom or near-bottom feeding species, such as molluscs, crabs, flatfish, and shrimp. In a study of Florida bay, very high concentrations, up to 7 mg g dry wt^{-1} of BMAA were detected in blue crabs (*Callinectes sapidus*).

In blue mussels from the Swedish west coast, high concentrations of BMAA have been found, but, surprisingly, no BMAA has been detected in blue mussels from the Baltic Sea, where heavy cyanobacterial blooms occur annually, indicating a different feeding strategy or metabolic pathway for BMAA in this brackish ecosystem. On the Swedish west coast, the dominating phytoplankton blooms are various diatom species, recently shown to also be able to produce BMAA [29], suggesting a different primary producer of BMAA than cyanobacteria in this marine ecosystem.

References

[1] Bradley, G. and Mash, D.C. (2009) Beyond Guam: The cyanobacteria/BMAA hypothesis of the cause of ALS and other neurodegenerative diseases. *Amyotrophic Lateral Sclerosis* **10** (S2), 7–20.

[2] Reed, D., Labarthe, D., Chen, K.M. *et al.* (1987) A cohort study of amyotrophic lateral sclerosis and parkinsonism-dementia on Guam and Rota. *American Journal of Epidemiology* **125** (1), 92–100.

[3] Murch, S.J., Cox, P.A., Banack, S.A. *et al.* (2004) Occurrence of β-methylamino-L-alanine (BMAA) in ALS/PDC patients from Guam. *Acta Neurologica Scandinavica* **110** (4), 267–269.

[4] Pablo, J., Banack, S.A., Cox, P.A. *et al.* (2009) Cyanobacterial neurotoxin BMAA in ALS and Alzheimer's disease. *Acta Neurologica Scandinavica* **120** (4), 216–225.

[5] Vega, A., Bell, E.A. and Nunn, P. B. (1968) The preparation of L- and D-α-amino-β-methylaminopropionic acids and the identification of the compound isolated from Cycas circinalis as the L-isomer. *Phytochemistry* **7** (10), 1885–1887.

[6] Sasaki, J., Chijimatsu, M. and Suzuki, K. (1998) Taxonomic significance of 2,4-diaminobutyric acid isomers in the cell wall peptidoglycan of actinomycetes and reclassification of Clavibacter toxicus as Rathayibacter toxicus comb. nov. *International Journal of Systematic Bacteriology* **48**, 403–410.

[7] Gross, H. and Loper, J.E. (2009) Genomics of secondary metabolite production by *Pseudomonas* spp. *Natural Product Reports* **26** (11), 1408–1446.

[8] Moura, S. and Pinto, E. (2007) One-pot synthesis of N-Cbz-l-BMAA and derivatives from N-Cbz-l-serine. *Tetrahedron Letters* **48** (13), 2325–2327.

[9] Hu, Y. and Ziffer, H. (1990) Synthesis and optical resolution of the neurotoxin 2-amino-3-([15N]-methylamino)propanoic acid (BMAA). *Journal of Labelled Compounds and Radiopharmaceuticals* **28** (5), 581–586.

[10] Nunn, P.B. and O'Brien, P. (1989) The interaction of β-N-methylamino-L-alanine with bicarbonate: an 1H-NMR study. *FEBS Letters* **251** (1–2), 31–35.

[11] Nunn, P.B., O'Brien, P., Pettit, L.D. *et al.* (1989) Complexes of zinc, copper, and nickel with the nonprotein amino acid L-α-amino-β-methylaminopropionic acid: A naturally occurring neurotoxin. *Journal of Inorganic Biochemistry* **37** (2), 175–183.

[12] Jiang, L., Aigret, B., De Borggraeve, W.M. *et al.* (2012) Selective LC-MS/MS method for the identification of BMAA from its isomers in biological samples. *Analytical and Bioanalytical Chemistry* **403** (6), 1719–1730.

[13] Vega, A. and Bell, E.A. (1967) α-Amino-β-methylaminopropionic acid, a new amino acid from seeds of *Cycas circinalis*. *Phytochemistry* **6** (5), 759–762.

[14] Polsky, F.I., Nunn, P.B. and Bell, E.A. (1972) Distribution and toxicity of α-amino-β-methylaminopropionic acid. *Federation Proceedings* **31** (5), 1473–1475.

[15] Karamyan, V.T. and Speth, R.C. (2008) Animal models of BMAA neurotoxicity: a critical review. *Life Sciences* **82** (5), 233–246.

[16] Spencer, P.S., Nunn, P.B., Hugon, J. *et al.* (1987) Guam amyotrophic lateral sclerosis-parkinsonism-dementia linked to a plant excitant neurotoxin. *Science* **237** (4814), 517–522.

[17] Weiss, J.H., Koh, J-Y. and Choi, D.W. (1989) Neurotoxicity of β-*N*-methylamino-L-alanine (BMAA) and β-*N*-oxalylamino-L-alanine (BOAA) on cultured cortical neurons. *Brain Research* **497** (1), 64–71.

[18] Liu, X., Rush, T., Zapata, J. *et al.* (2009) β-*N*-Methylamino-L-alanine induces oxidative stress and glutamate release through action on system X_c^-. *Experimental Neurology* **217** (2), 429–433.

[19] Cox, P.A., Banack, S.A., Murch, S.J. *et al.* (2005) Diverse taxa of cyanobacteria produce β-N-methylamino-l-alanine, a neurotoxic amino acid. *Proceedings of the National Academy of Sciences of the USA* **102** (14), 5074–5078.

[20] Banack, S.A. and Cox, P.A. (2003) Biomagnification of cycad neurotoxins in flying foxes: implications for ALS-PDC in Guam. *Neurology* **61** (3), 387–389.

[21] Cohen, S.A. and Michaud, D.P. (1993) Synthesis of a fluorescent derivatizing reagent, 6-aminoquinolyl-N-hydroxysuccinimidyl carbamate, and its application for the analysis of hydrolysate amino acids via high-performance liquid chromatography. *Analytical Biochemistry* **211** (2), 279–287.

[22] Glover, W.B., Baker, T.C., Murch, S.J., Brown, P. (2015) Determination of β-N-methylamino-L-alanine, N-(2-aminoethyl)glycine, and 2,4-diaminobutyric acid in food products containing cyanobacteria by ultra-performance liquid chromatography and tandem mass spectrometry: single-laboratory validation. *Journal of the AOAC International* **98** (6), 1559–1565.

[23] Faassen, E.J., Gillissen, F. and Lürling, M. (2012) A comparative study on three analytical methods for the determination of the neurotoxin BMAA in cyanobacteria. *PLoS ONE* **7** (5), e36667.

[24] Glover, W.B., Liberto, C.M., McNeill, W.S. *et al.* (2012) Reactivity of β-methylamino-L-alanine in complex sample matrixes complicating detection and quantification by mass spectrometry. *Analytical Chemistry* **84** (18), 7946–7953.

[25] Spácil, Z., Eriksson, J., Jonasson, S. *et al.* (2010) Analytical protocol for identification of BMAA and DAB in biological samples. *Analyst* **135** (1), 127–132.

[26] Combes, A., El Abdellaoui, S., Sarazin, C. *et al.* (2013) Validation of the analytical procedure for the determination of the neurotoxin β-N-methylamino-l-alanine in complex environmental samples. *Analytica Chimica Acta* **771** (1), 42–49.

[27] Cox, P.A., Banack, S.A. and Murch, S.J. (2003) Biomagnification of cyanobacterial neurotoxins and neurodegenerative disease among the Chamorro people of Guam. *Proceedings of the National Academy of Sciences of the USA* **100** (23), 13380–13383.

[28] Li, A., Tian, Z., Yu, R., Banack, S.A., Wang. Z. (2010) Detection of the neurotoxin BMAA within cyanobacteria isolated from freshwater in China. *Toxicon* **55** (5), 947–953.

[29] Jiang, L., Eriksson, J., Lage, S. *et al.* (2014) Diatoms: A novel source for the neurotoxin BMAA in aquatic environments. *PLoS One* doi:10.1371/journal.pone.0084578.

[30] Lage, S., Costa, P.R., Molta, T. *et al.* (2014) BMAA in shellfish from two Portuguese transitional water bodies suggests the marine dinoflagellate *Gymnodinium catenatum* as a potential BMAA source. *Aquatic Toxicology* **152**, 131–138.

[31] Jonasson, S., Eriksson, J., Berntzon, L. *et al.* (2010) Transfer of a cyanobacterial neurotoxin within a temperate aquatic ecosystem suggests pathways for human exposure. *Proceedings of the National Academy of Sciences of the USA* **107** (20), 9252–9257.

[32] Brand, L.E., Pablo, J., Compton, A. *et al.* (2010) Cyanobacterial blooms and the occurrence of the neurotoxin beta-N-methylamino-L-alanine (BMAA) in South Florida aquatic food webs. *Harmful Algae* **9** (6), 620–635.

17

Lipopolysaccharide Endotoxins

Sílvia Monteiro[1], Ricardo Santos[1], Luděk Bláha[2], and Geoffrey A. Codd[3,4]

[1]Instituto Superior Técnico, Laboratorio de Análises, Lisboa, Portugal
[2]Faculty of Science, RECETOX, Masaryk University, Brno, Czech Republic
[3]Biological and Environmental Sciences, University of Stirling, Scotland, United Kingdom
[4]School of the Environment, Flinders University, Adelaide, Australia

17.1 Lipopolysaccharide Endotoxins: Structure

Lipopolysaccharides (LPS) are toxins and irritants from the outer membrane of Gram-negative bacteria including cyanobacteria. LPS are responsible for a wide range of outcomes in humans, including sepsis and septic shock. Cyanobacterial LPS are a particular class of LPS and have been implicated in several illnesses, from gastrointestinal illness to allergy and other respiratory diseases, fever, and skin rashes. However, relatively little information exists on the real implication of cyanobacterial LPS on human health, and further studies in this field are required [1–3].

LPS are endotoxins, a term given by Richard Pfeiffer in the late nineteenth century. He discovered that heat-inactivated lysates of *Vibrio cholerae* induced a pathophysiological event in guinea pigs [4]. The original chemical and structural analyses of these molecules revealed the presence of a polysaccharide and a lipid, the material being thus named "lipopolysaccharide."

The basic chemical structure of LPS is similar in most organisms. The structure consists of four covalently linked segments: a surface carbohydrate polymer (O-specific chain), a central oligosaccharide with an inner and outer region, and an acylated glycolipid (lipid A) (Fig. 17.1). Lipid A is the most conserved region and docks the LPS onto the outer membrane of the producer-cell, while the O-specific chain is the most variable region and is the core region for serological specificity [5].

Several studies on LPS using polysaccharide-deficient mutants and chemically synthesised lipid A have proved that lipid A is the endotoxic component of LPS [5–8]. Alterations in the nature and

Handbook of Cyanobacterial Monitoring and Cyanotoxin Analysis, First Edition.
Edited by Jussi Meriluoto, Lisa Spoof and Geoffrey A. Codd.

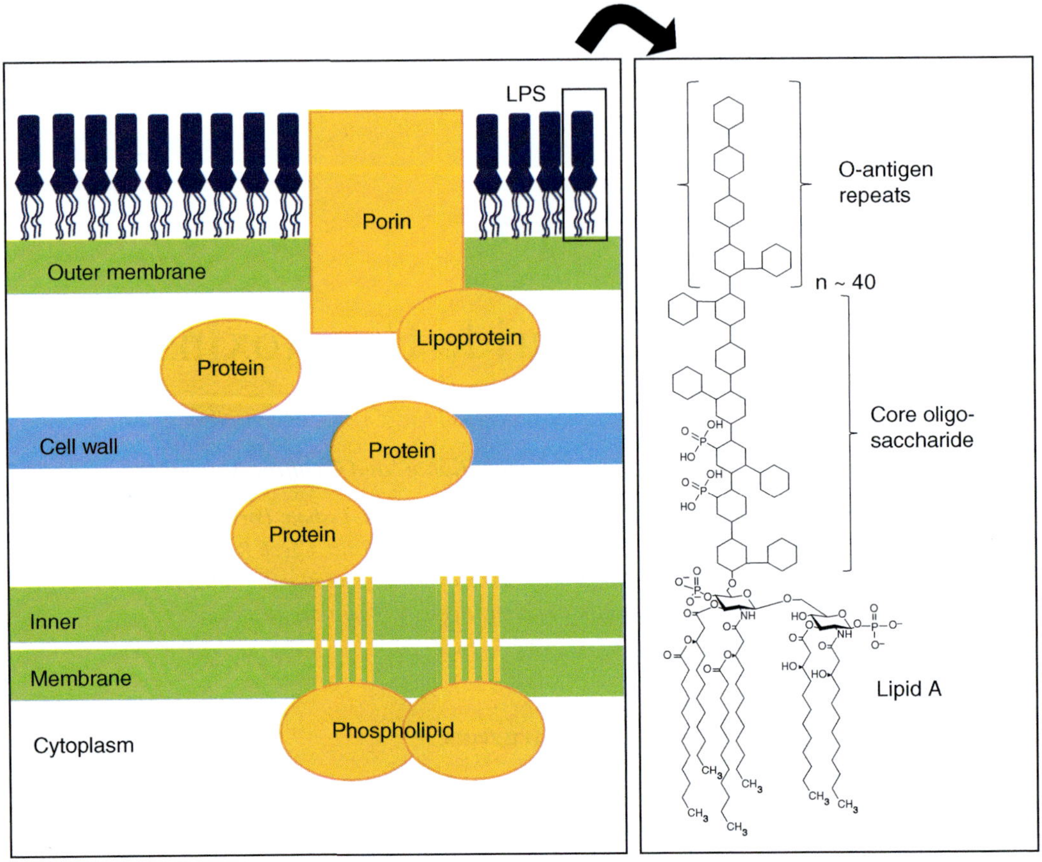

Figure 17.1 Location in bacterial cell membrane and general structure of LPS

location of acyl groups or in the hydrophilic backbone of lipid A can create partial or total loss of biological activity, even though the basic structure remains the same for phylogenetically diverse organisms [4, 6, 9–11].

Analyses of the supramolecular structure of lipid A might help to determine the biological activity of LPS [12]. This structure is a result of its monomeric conformation, which in turn is a function of the primary molecular structure [11–13]. The molecular conformation and binding capacity of LPS to serum proteins are highly influenced by the quantity of negative charges on the disaccharide structure [11, 14]. A highly negatively charged disaccharide is indispensable for agonistic and antagonistic properties, and the substitution of these negative charges may lead to a partial or total loss of biological activity [11, 15]. Seydel et al. [15] suggested that the same molecular conformational rule may be applied to lipid A structures and their subsequent biological activity. They discussed that lipid A molecules acquiring a conical/concave shape are formed of hexagonal or cubic supramolecular structures and as a result have high endotoxic potential, while lipid A structures with cylindrical conformations have a lamellar structure and hence are either inactive or LPS antagonists [13]. An example of these findings is lipid A from the bacteria *Rubrivivax gelatinosus* and *Chromobacterium violaceum*. The first has a high endotoxic activity and adopts a hexagonal supramolecular conformation, whereas the second, known for its low endotoxic activity, adopts cylindrical shapes with lamellar aggregate structures [12, 13, 16]. Thus, a conical conformation of the lipid A monomer is crucial for the endotoxicity of LPS [12]. Most of the publications on the toxicity of cyanobacterial LPS have concluded that these molecules are less toxic than the LPS from heterotrophic bacteria.

However, one study addressing the ability of isolated cyanobacterial LPS to decrease the activity of glutathione *S*-transferases (GSTs) in embryos of zebrafish (*Danio rerio*) suggests opposite results [17]. In this study, cyanobacterial LPS had a greater impact on the decrease of microsomal and soluble GSTs in vivo, compared with LPS from *Escherichia coli* and *Salmonella typhimurium*. The authors concluded that the reduction of GST may cause a decrease in the use of glutathione in conjugation reactions, affecting the capacity of the organisms to detoxify microcystins. However, the reduction of GSTs in zebrafish may also be indicative of an anti-inflammatory response stronger than that observed with LPS from heterotrophic bacteria or a consequence of the reduction of glutathione levels, as intracellular glutathione secretion is higher on the onset of apoptosis [18].

17.2 Occurrence of LPS Endotoxins

There are few studies on the specific presence of cyanobacterial endotoxins in natural waters, partly because it is not straightforward to determine if the LPS present in a sample is from cyanobacteria or from other Gram-negative bacteria. Since the late 1980s, there has been a move to report LPS endotoxin concentrations (or, more accurately, endotoxin activity) in endotoxin units (EU) rather than in units of weight (typically ng mL^{-1}). This form of reporting reflects the fact that endotoxin potency depends on the genus of producer-bacteria and, within species, on the specific lot or batch from which the reference endotoxin is isolated. Much of the literature available, however, does not use the newer terminology.

Rapala et al. [19] reported LPS concentrations between 20 and 38,000 EU mL^{-1} in Finnish surface waters experiencing cyanobacterial blooms. For groundwater, a Danish study [20] found concentrations ranging from 1 to 200 EU mL^{-1}. On the other hand, in surface water, Di Luzio and Friedmann [21] reported ranges between 1 and 400 μg mL^{-1} in several locations of Alabama and of the Mississippi River. For saltwater, Watson et al. [22] reported values between 0.04 and 17.8 ng mL^{-1}.

There are only a few studies in which the concentrations of LPS along water treatment systems have been measured. Haas et al. [23] measured the total LPS concentration in raw water (8.8 ng mL^{-1}), following prechlorination (3.5 ng mL^{-1}), post rapid sand filtration (3.5 ng mL^{-1}), and post granular activated carbon (3.4 ng mL^{-1}), in finished water (3.7 ng mL^{-1}), as well as in distribution systems (4.7–7.5 ng mL^{-1}). A similar study was conducted in Alberta, Canada, by Huck et al. [24], where the LPS concentrations ranged between 2.91 ng mL^{-1} in raw water and 0.33 ng mL^{-1} after anthracite-sand filter postchlorination. The Finnish survey by Rapala et al. [19] measured endotoxin concentrations at nine different drinking water treatment plants. Endotoxin concentrations ranged from 18 to 356 EU/mL^{-1} at the plant intakes. The drinking water treatment processes, which varied in complexity, removed from 59% to 97% of the endotoxin present in the raw water. Concentrations in the distribution system were very low, ranging from 14 to 32 EU mL^{-1}. In general, the endotoxin levels are usually low in treated waters, even if the concentration in raw water is high. It was proved that when a cyanobacterial bloom is in progress, the LPS levels rise dramatically, even to concentrations that could harm human health.

17.3 Toxic Effects of LPS Endotoxins

Symptoms of endotoxin exposure in humans are general and include fever, diarrhoea, and vomiting [25]; hypotension (i.e. a decrease in systolic blood pressure by 30 mm Hg or more [26]); and shock, intravascular coagulation, and death [27]. The latter symptoms are only exhibited at highly elevated concentrations. It is now commonly accepted that endotoxins are the primary triggers of septic shock. Sepsis and septic shocks arising from exposure to LPS from heterotrophic bacteria are responsible for an estimated 175,000 to 200,000 deaths per year in the United States and have been increasing in frequency since the 1950s [28–31].

Exposure to LPS induces a systemic inflammation, a complicated cellular and humoral response in the host [9, 33–35]. Endotoxins prompt several different reactions, releasing cytokines and arachidonic acid. Noncellular or humoral pathways include complement and coagulation cascades. If the endotoxins continue to stimulate the host system and the host cannot cope with this stimulation, it may lead to a prolonged inflammatory response that may endanger or ultimately kill the host. Further studies have shown several biological effects of cyanobacterial LPS, including the enhancement of toxicity of heavy metals [36] and toxicity and biotransformation on the conformation of microcystins [37, 38]. The observed effects of cyanobacterial LPS on glutathione-based detoxification pathways may explain previously observed interactive effects of LPS with other toxicants. In particular, cyanobacteria produce a wide array of toxic or otherwise bioactive metabolites, including several recognized toxins (e.g. microcystins, cylindrospermopsins, anatoxin-a, saxitoxins). In the case of the widespread microcystin hepatotoxins, for example, detoxification has been clearly linked to GST conjugation of GSH to the toxin [17, 18].

As a result of the effects of LPS on generic detoxification pathways, these molecules may not only exert direct toxic effects on susceptible organisms but also indirectly augment the toxicity of other known toxins.

Since the first reports by Pfeiffer in 1892 and Centanni in 1894, endotoxins have mainly been associated with toxic effects [39]. However, LPS were recently shown to have beneficial effects in several diseases such as allergies, diabetes, multiple sclerosis, and inflammatory bowel disease (Fig. 17.2).

Naturally occurring LPS may help to balance the host's immune system by keeping pathogens away and by circumventing the possibility of invasion by saprophytes/commensals, which are a greater cause of sepsis. Additionally, it has been shown that the immune system relies on several products including LPS for normal development [40]. Studies on children have shown that the development of atopic illnesses correlates negatively with of exposure to LPS [41]. Several models using mice have shown that treatment with LPS can help to decrease some of the allergic inflammation and related symptoms that the control mice showed [42, 43]. It has also been demonstrated that pretreatment of animals with LPS increases the recovery of myocardial functions following an ischemia/reperfusion heart injury [42, 43]. Mouihate et al. [44] stated that LPS exposure during early infancy reformulates the neuroimmune axis, which may be beneficial to the rest of human life. Such strengthened regulation of improper immune system activation in early childhood could help in reducing symptoms during acute systemic inflammatory reaction (e.g. sepsis/septic shock) or critical allergic diseases (e.g. asthma attack) during adult life.

Thus, LPS endotoxins, as with many bioactive molecules, may have both harmful and beneficial actions [45]. Endotoxins are therefore under investigation for the development of new drugs for the prevention and therapy of several human illnesses.

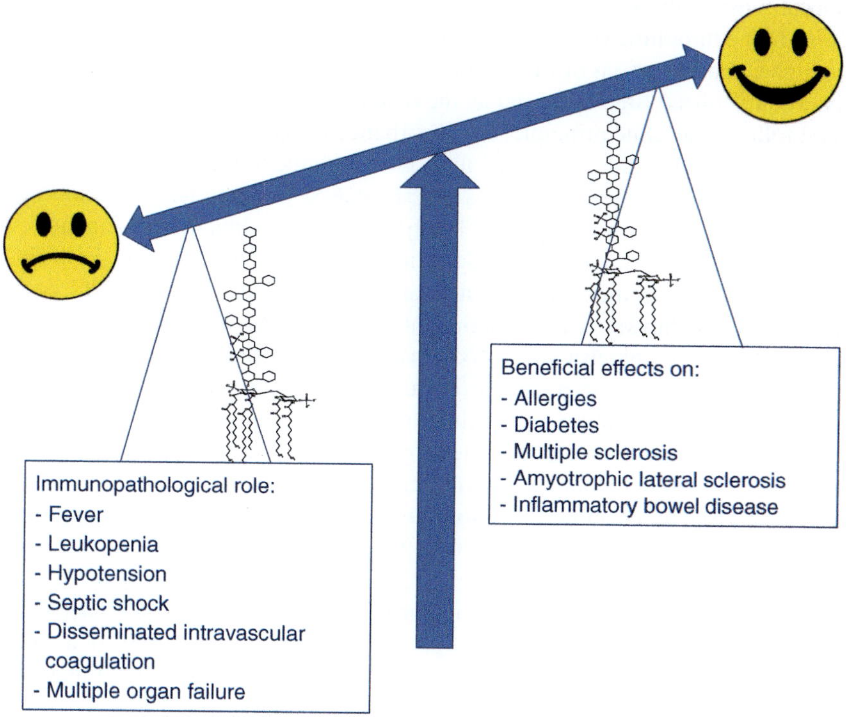

Figure 17.2 *Equilibrium between beneficial and detrimental effects of LPS endotoxins*

17.4 Methods for Determination of LPS Endotoxins

The presence of LPS is usually analysed wth the Limulus Amebocyte Lysate (LAL) test, which uses an aqueous extract obtained after lysis of blood cells (amoebocytes) of the horseshoe crab. The LAL test is based on a reaction between LPS and a clottable protein within amoebocytes. When LPS come in contact with LAL, they initiate a series of enzymatic reactions that result in the activation of a pathway for the production of at least three serine protease zymogens (factor C, factor B, and a proclotting enzyme). This pathway promotes amoebocyte coagulogen to form an insoluble coagulin gel. Serine proteases cleave peptide bonds in proteins, in which serine serves as the nucleophilic amino acid at the active site. In humans, they are responsible for coordinating various physiological functions, including digestion, immune response, blood coagulation and reproduction. Significantly, the blood coagulation reaction is similar in humans and the horseshoe crab [46].

Within the LAL test, there are several different options for the detection of LPS. The simplest method is the gel clot method. In this method, LPS activate a proenzyme in the LAL. The amount of active enzyme generated is dependent on the initial concentration of LPS. The active enzyme cleaves a clotting protein and the cleaved products aggregate to form a clot. The sensitivity limit of the gel clot method is 0.03 EU mL^{-1}. This is the least expensive method for LPS detection as no special instruments are needed.

Turbidimetric (end point or kinetic) testing is based on the development of turbidity due to the formation of insoluble coagulin in the LAL. The rate at which the turbidity increases is related to the LPS concentration in the sample. The samples are compared to a standard curve with known concentrations. In the turbidimetric end point method, absorbance is measured after a fixed period of time, with higher concentrations of LPS in the samples and standards resulting in higher turbidity. In this assay, a standardized time is crucial as the reaction cannot be stopped. The kinetic turbidimetric method follows the same principle, but absorbance measurements are made throughout the entire test. The test compares the time taken for a given sample to reach a specific optical density threshold, with the standard curve and gives a result based on this. The detection range of the turbidimetric test is very wide, from 0.001 to 100 EU mL^{-1} depending on the standard curve used.

In chromogenic (end point and kinetic) detection methods, LPS catalyze the activation of a proenzyme in the LAL. The initial rate of activation is determined by the LPS concentration. The activated enzyme catalyses the generation of *p*-nitroaniline from the colourless substrate. This release is measured photometrically at 405 nm continuously throughout the incubation period. The concentration of endotoxin in a sample is calculated by comparing its reaction time to the reaction time of solutions containing known amounts of endotoxin standard.

The PyroGene test is also used for the detection of endotoxins. In this, factor C (i.e. the first component of the enzymatic cascade when LPS comes in contact with the LAL) is activated by LPS binding. The created active moiety acts to cleave a synthetic substrate, resulting in the generation of a fluorogenic compound. The sensitivity of the test can be as low as 0.005 EU mL^{-1} and go up to 5.0 EU mL^{-1}.

Other approaches for the detection of LPS include the EndoLISA. This is a microplate test that first captures the LPS using a high affinity receptor coating and then uses a similar strategy as the PyroGene test (reaction of factor C with a fluorescent substrate). The detection range of this test is from 0.05 to 500 EU mL^{-1}.

References

[1] Weckesser, J. and Drews, G. (1979) Lipopolysaccharides of photosynthetic prokaryotes. *Annual Review of Microbiology*, **33**, 215–239.

[2] Codd, G.A., Bell, S.G., Kaya, K. *et al.* (1999) Cyanobacterial toxins, exposure routes and human health. *European Journal of Phycology*, **34** (4), 405–415.

[3] Steffensen, D., Burch, M., Nicholson, B. *et al.* (1999) Management of toxic blue-green algae (cyanobacteria) in Australia. *Environmental Toxicology*, **14** (1), 183–195.

[4] Schletter, J., Heine, H., Ulmer, A.J. and Rietschel, E.T. (1995) Molecular mechanisms of endotoxin activity. *Archives of Microbiology*, **164**, 383–389.

[5] Morrison, D.C. (1983) Bacterial endotoxins and pathogenesis. *Reviews of Infectious Diseases*, **5** (Suppl 4), S733–S747.

[6] Alexander, C. and Rietschel, E.T. (2001) Bacterial lipopolysaccharides and innate immunity. *Journal of Endotoxin Research*, **7** (3), 167–202.

[7] Ulevitch, R.J. and Tobias, P.S. (1999) Recognition of Gram-negative bacteria and endotoxin by innate immune system. *Current Opinion in Immunology*, **11** (1), 19–22.

[8] Galanos, C., Lüderitz, O., Rietschel, E.T. *et al.* (1985) Synthetic and natural *Escherichia coli* free lipid A express identical endotoxic activities. *European Journal of Biochemistry*, **148** (1), 1–5.

[9] Rietschel, E.T., Brade, H., Holst, O. *et al.* (eds) (1996) Bacterial endotoxin: Chemical construction, biological recognition, host response, and immunological detoxification, in *Pathology of Septic Shock* (eds. E.T. Rietschel and H. Wagner), Springer Berlin Heidelberg, *Current Topics in Microbiology and Immunology*, **216**, 39–81.

[10] Erwin, A.L., Mandrell, R.E. and Munford, R.S. (1991) Enzymatically deacylated *Neisseria* lipopolysaccharide (LPS) inhibits murine splenocyte mitogenesis induced by LPS. *Infection and Immunity*, **59** (6), 1881–1887.

[11] Wiese, A., Brandenburg, K., Ulmer, A.J. *et al.* (1999) The dual role of lipopolysaccharide as effector and target molecule. *Biological Chemistry*, **380** (7–8), 767–784.
[12] Seydel, U., Schroom, A.B., Blunk, R. and Brandenburg, K. (2000) Chemical structure, molecular conformation, and bioactivity of endotoxins. *Chemical Immunology*, **74**, 5–24.
[13] Schromm, A.B., Brandenburg, K., Loppnow, H. *et al.* (2000) Biological activities of lipopolysaccharides are determined by the shape of their lipid A portion. *European Journal of Biochemistry*, **267** (7), 2008–2013.
[14] Schromm, A.B., Brandenburg, K., Loppnow, H. *et al.* (1998) The charge of endotoxin molecules influences their conformation and IL-6-inducing capacity. *Journal of Immunology*, **161** (10), 5464–5471.
[15] Seydel, U., Scheel, O., Muller, M. *et al.* (2001) A K+ channel is involved in LPS signaling. *Journal of Endotoxin Research*, **7** (3), 243–247.
[16] Brandenburg, K., Mayer, H., Koch, M.H. *et al.* (1993) Influence of the supramolecular structure of free lipid A on its biological activity. *European Journal of Biochemistry*, **218** (2), 555–563.
[17] Best, J.H., Pflugmacher, S., Wiegand, C. *et al.* (2002) Effects of enteric bacterial and cyanobacterial lipopolysaccharides, and of microcytin-LR, on glutathione S-transferase activities in zebrafish (*Danio rerio*). *Aquatic Toxicology*, **60** (3-4), 223–231.
[18] van den Dobbelsteen, D.J., Nobel, C.S.I., Schlegel, J. *et al.* (1996) Rapid and specific efflux of reduced gluthatione during apoptosis induced by anti-Fas/APO-I antibody. *Journal of Biological Chemistry*, **271** (26), 15420–15427.
[19] Rapala, J., Lahti, K., Räsänen, L.A. *et al.* (2002) Endotoxins associated with cyanobacteria and their removal during drinking water treatment. *Water Research*, **36** (10), 2627–2635.
[20] Korsholm, E. and Søgaard, H. (1988) An evaluation of direct microscopical counts and endotoxin measurements as alternatives for total plate counts. *Water Research*, **22** (6), 783–788.
[21] Di Luzio, N.R. and Friedmann, T.J. (1973) Bacterial endotoxins in the environment. *Nature* (London), **244** (5410), 49–51.
[22] Watson, S.W., Novitsky, T.J., Quinby, H.L. and Valois F.W. (1977) Determination of bacterial number and biomass in the marine environment. *Applied and Environmental Microbiology*, **33** (4), 940–946.
[23] Haas, C.N., Meyer, M.A., Paller, M.S. and Zapkin, M.A. (1983) The utility of endotoxins as a surrogate indicator in potable water microbiology. *Water Research*, **17** (7), 803–807.
[24] Huck, P.M. (1998) Design of Biological Processes for Organics Control, American Water Works Association, USA.
[25] Madigan, M.T., Martinko, J.M. and Parker, J. (2000) Brock Biology of Microorganisms. 9th ed. Prentice Hall, Upper Saddle River, N.J. pp. 74–75, 793–794.
[26] Hindman, S.H., Favero, M.S., Carson, L.A. *et al.* (1975) Pyrogenic reactions during haemodialysis caused by extramural endotoxins. *Lancet*, **2** (7938), 732–734.
[27] Braude, A.L. (1982) Microbiology, (Eds. C.E. Davis and J. Fierer.), W.B. Saunders Company, Toronto, Canada. pp. 63–73, 796–800.
[28] American College of Chest Physicians (ACCP)/Society of Critical Care Medicine. (1992) Definitions for sepsis and organ failure and guidelines for the use of innovative therapies in sepsis. *Critical Care Medicine*, **20**, 864–874.
[29] Glauser, M.P., Zanetti, G., Baumgartner, J.D., Cohen, J. (1991) Septic shock: pathogenesis. *Lancet*, **338** (8769), 732–736.
[30] Hoffman, W., Natanson, C. (1993) Endotoxin in septic shock. *Anesthesia and Analgesia*, **77** (3), 613–624.
[31] Ulevitch, R.J., Tobias, P.S. (1995) Receptor-dependent mechanisms of cell stimulation by bacterial endotoxin. *Annual Review of Immunology*, **13**, 437–457.
[32] Bone, R. (1991) The pathogenesis of sepsis. *Annals of Internal Medicine*, **115** (6), 457–469.
[33] Bone, R.C., Balk, R.A., Cerra, F.B. *et al.* (1992) Definitions for sepsis and organ failure and guidelines for the use of innovative therapies in sepsis. The ACCP/SCCM Consensus Conference Committee. American College of Chest Physicians/Society of Critical Care Medicine. *Chest*, **101** (6), 1644–1655.
[34] Bone, R., Balk, R., Fein, A. *et al.* (1995) A second large controlled clinical study of E5, a monoclonal antibody to endotoxin: results of a prospective, multicenter, randomized, controlled trial. The E5 Sepsis Study Group. *Critical Care Medicine*, **23** (6), 994–1006.
[35] Remick, D.G. (1995) Applied molecular biology of sepsis. *Journal of Critical Care*, **10** (4), 198–212.
[36] Notch, E.G., Miniutti, D.M., Berry, J.P. and Mayer, G.D. (2011) Cyanobacterial LPS potentiates cadmium toxicity in zebrafish (*Danio rerio*) embryos. *Environmental Toxicology*, **26** (5), 498–505.

[37] Lindsay, J., Metcalf, J.S. and Codd, G.A. (2006) Protection against the toxicity of microcystin-LR and cylindrospermopsin in *Artemia salina* and *Daphnia* spp. by pre-treatment with cyanobacterial lipopolysaccharide (LPS). *Toxicon*, **48** (8), 995–1001.

[38] Jaja-Chimedza, A., Gantar, M., Mayer, G.D. *et al.* (2012) Effects of cyanobacterial lipopolysaccharides from *Microcystis* on glutathione-based detoxification pathways in the zebrafish (*Danio rerio*) embryo. *Toxins*, **4** (6), 390–404.

[39] Williams, K.L. (2007) Historical and emerging themes, in *Endotoxins, Pyrogens, LAL Testing and Depyrogenation*, (ed. K.L. Williams), Eli Lilly & Company, New York, NY, USA.

[40] Miller, S.I., Ernst, R.K. and Bader, M.W. (2005) LPS, TLR4 and infectious disease diversity. *Nature Reviews in Microbiology*, **3** (1), 36–46.

[41] Srinivasan, N. (2010) Telling apart friend from foe: discriminating between commensals and pathogens at mucosal sites. *Innate Immunity*, **16** (6), 391–404.

[42] Round, J.L., O'Connell, R.M. and Mazmanian, S.K. (2010) Coordination of tolerogenic immune responses by the commensal microbiota. *Journal of Autoimmunity*, **34** (3), J220–J225.

[43] Aumeunier, A., Grela, F., Ramadan, A. *et al.* (2010) Systemic toll-like receptor stimulation suppresses experimental allergic asthma and autoimmune diabetes in NOD mice. *PLoS One*, **5** (7), e11484.

[44] Mouihate, A., Galic, M.A., Ellis, S.L. *et al.* (2010) Early life activation of Toll-like receptor 4 reprograms neural anti-inflammatory pathways. *Journal of Neuroscience*, **30** (23), 7975–7983.

[45] Matera, G., Liberto, M.C. and Joosten, A.B. (2008) The Janus face of *Bartonella quintana* recognition by toll-like receptors (TLRs): a review. *European Cytokine Network*, **19** (3), 113–118.

[46] Cooper, J.F. (2004) Microbiological contamination control in parenteral manufacturing, *in Microbial Contamination Control in Parenteral Manufacturing* (ed. K. Williams), New York: Marcel Dekker, pp. 531–540.

18

Cyanobacterial Retinoids

Kunimitsu Kaya and Tomoharu Sano

Center for Environmental Measurement and Analyses, National Institute for Environmental Studies, Tsukuba, Japan

18.1 Introduction

"Retinoid" is a general term for retinol (vitamin A) metabolites and retinoid receptor–activating compounds. Retinoic acid (RA) is a typical component of retinoids. RA, a low-molecular-weight lipophilic metabolite of retinol, is the most potent natural retinoid. All-*trans* RA (ATRA) is a major metabolite that undergoes isomerization and metabolism in vivo, yielding several analogues including 5,6-epoxy RA and 4-oxo-RA [1–13]. These metabolites act by binding to retinoic acid receptors (RARs) and regulate proliferation, differentiation, apoptosis, cytokine production, gut mucosal immunity and the malignant transformation of cells by RA binding with receptors, gene expression, and the synthesis of functional peptides and proteins [14–18]. However, exogenous retinoids are known to cause various types of malformations.

Whilst there has been a focus on the relationship between deformation and retinoids [19–22], cyanobacterial retinoids had not been identified until recently.

Retinoid-like activity was detected in the biomass of several cyanobacterial species using an in vitro yeast two-hybrid assay, and a novel RA analogue, 7-hydroxy retinoic acid (7-OH RA), was identified [23]. Subsequently, 4-oxo-RA [24] and retinol as vitamin A [25] were detected in natural cyanobacterial blooms of Taihu Lake, China.

Handbook of Cyanobacterial Monitoring and Cyanotoxin Analysis, First Edition.
Edited by Jussi Meriluoto, Lisa Spoof and Geoffrey A. Codd.

18.2 Detection of Retinoids Produced by Cyanobacteria

RA, its analogues, and RAR-activating compounds are known to cause various types of malformations in zebrafish (*Danio rerio*) embryos including yolk sac, heart oedema, and brain and tail malformations [26, 27]. When extracts and exudates of *Microcystis aeruginosa, Dolichospermum flos-aquae, Cylindrospermopsis raciborskii,* and *Aphanizomenon klebahnii* were exposed to zebrafish embryos, remarkable events of bent tail, undeveloped eyes, or neural tube malformations were observed [28–32]. Also, deformed frogs have been observed in natural environments [20, 21].

For screening of cyanobacterial retinoids, a yeast two-hybrid retinoid activity assay method using yeast cells (*Saccharomyces cerevisae* T190) expressing the human receptor RAR, or RXR and co-activator TIF2, has been used. The recombinant cells were adapted to a chemiluminescent reporter gene method (for β-galactosidase) by using a 96-well culture plate [33]. This method has the merit of detection of total retinoid including unknown retinoid. The results of cyanobacterial retinoid content determination are shown [23]. Table 18.1 shows examples of contents of RAR-activating compounds in selected cyanobacteria.

The reporter gene assay is also effective for the detection of retinoids [28]. The murine embryonic carcinoma cell line P19 transfected with a luciferase reporter pRAREβ2-TK-luc plasmid (P19/A15 clone 34) is used. Retinoids and/or retinoid-like compounds in exudates from cultured microalgae were examined, and showed that retinoid-like activity was detected from cyanobacteria and one green alga (*Desmodesms quadricaudus*) [28].

At present, LC-MS/MS is the most effective method for the detection and identification of RA and its known analogues [24, 25]. However, in the case of unknown retinoid, in vivo or in vitro assay, fractionation and isolation of retinoids, and identification using spectrometry methods such as LC-MS/MS are necessary [23].

Table 18.1 *RAR-activating compounds in cyanobacterial cells*

Cyanobacteria	Agonistic activity (ng equivalent of ATRA per g of lyophilized cells)
Cylindrospermopsis raciborskii AWT205	5000
Gloeocapsa decorticans NIES-931	210
Microcystis aeruginosa NIES-44	2500
Microcystis aeruginosa NIES-1222	6000
Nostoc sp. DUN 901	2600
Oscillatoria limnetica NIES-36	3000
Phormidium tenue NIES-512	1100
Planktothrix agardhii NIES-204	2500
Planktothrix rubescens NIES-610	2000
Raphidiopsis sp. TEG-1	20,000
Spirulina platensis NIES-45	1500
Spirulina platensis NIES-46	660
Spirulina sp. (commercial product)	3400

A yeast two-hybrid RA activity assay using yeast cells (*Saccharomyces cerevisiae* T190) into which human RAR and coactivator TIF2 had been introduced was used. The agonistic activity is expressed as nanogram (ng) equivalent of all-*trans* RA (ATRA) per gram (g) of lyophilized cells. Table taken from Kaya, K., Shiraishi, F., Tatarazako, N., Sano, T. (2014) Does 7-hydroxy retinoic acid in cells of cyanobacteria act as ecotoxin? *International Conference of Asian Environmental Chemistry,* Nov. 24–26, 2014, pp. 163–166, Bangkok, Thailand. Japan Society for Environmental Chemistry. With permission of Japan Society for Environmental Chemistry.

18.3 Chemistry and Analysis of Retinoids

RA is a kind of isoprenoid (such as vitamins E and K), which consists of moieties of a β-ionone ring and isoprene units. RA is biosynthesized from β-carotene (Fig. 18.1). At the first step, β-carotene is cleaved between C-15 and C-16 of the molecule, to form retinol (vitamin A, alcohol type). Retinol is further converted to RA via retinal (aldehyde type). RA is isomerized to the *cis*-form and/or formed to the oxo-form such as 11-*cis* RA and 4-oxo-RA. Also, 7-OH RA is formed from RA by the hydroxylation at C-7. RA contains unstable conjugated double bonds under UV-A and oxygen. Under these oxidative conditions, RA changes to inactive forms. However, 7-hydroxy RA is more stable under UV-A than ATRA [23]. Probably, the stability is closely related to *keto-enol* tautomerism and *cis-trans* isomerization [35].

RA and its known analogues have been analysed using LC-MS/MS with reverse-phase columns and gradient-solvent systems [24, 25, 36–38] with detection of the molecular ions and/or typical fragment ions. For the determination, quality assurance and quality control are the most important factors. The validation parameters of instrument and method for retinoids (retinol, retinal, retinoic acid, and β-carotene) have been presented in detail [25].

In the case of 7-OH RA as a new analogue, 7-OH RA showed four peaks due to *keto-enol* tautomerism and *cis-trans* isomerization in LC chromatograms with an isocratic solvent system [23, 35]. Improved methods for extraction and determination of all *trans*- RA and 7-OH RA

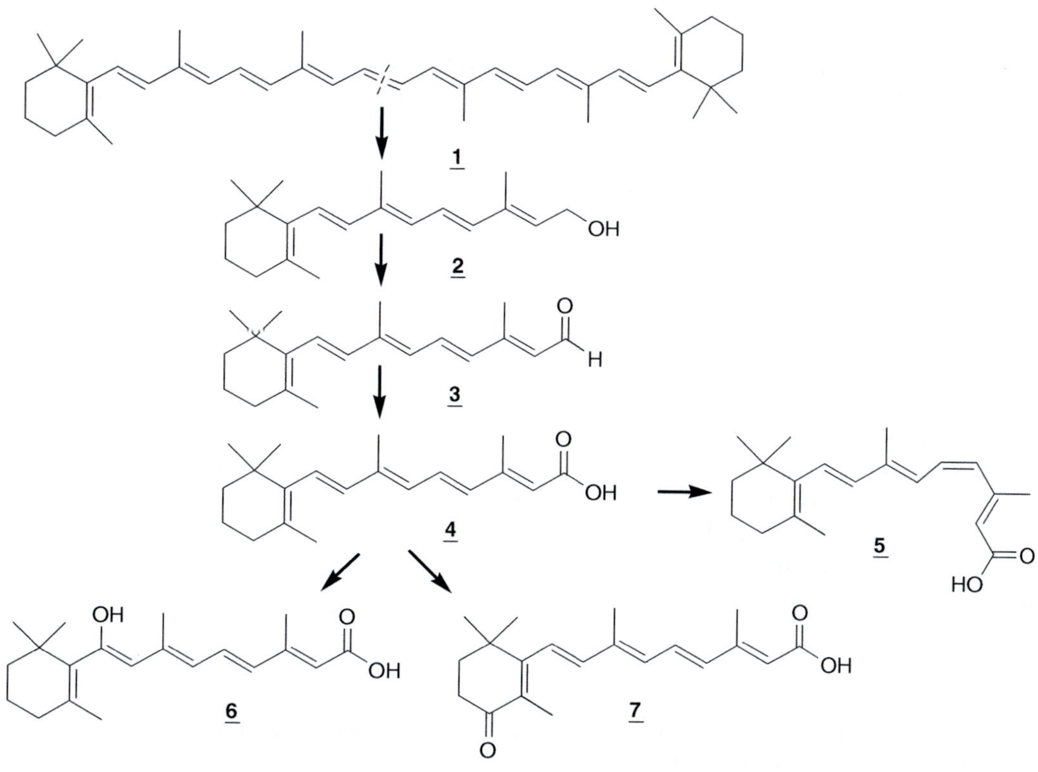

Figure 18.1 Biosynthesis of retinoic acid and its analogues. 1, β-Carotene; 2, retinol (vitamin A); 3, retinal; 4, retinoic acid; 5, 11-cis-retinoic acid; 6, 7-hydroxyl retinoic acid; 7, 4-oxo-retinoic acid

have been presented [39]. The extraction method included saponification of alcohol esters of RA, and fractions of anion-exchange cartridges. The analysis method was based on a reverse-phase column and a linear gradient of methanol (25–95% for 9.5 min) containing 0.1% v/v acetic acid at 50°C. MS analysis was in the negative MRM mode, and *m/z* for 7-OH RA and all-*trans* RA were 315.00 > 118.95 and 299.00 > 256.05, respectively. The four peaks of 7-OH RA appeared between 8.5 and 9.7 min, and the peak of ATRA occurred at 11.7 min. These methods are suitable for routine analysis.

18.4 Malformations by Cyanobacterial Retinoids

Science News reported a serious ecotoxicological problem involving retinoid concentrations in Taihu Lake, China (up to 20 ng L^{-1}), and an ecologist commented that "there's not enough information yet to link cyanobacteria to high rates of amphibian deformities" [40]. Malformations by cyanobacterial retinoids involving fish and frogs have been observed [3, 20–22, 27, 28, 30, 31]. Teratogenicity of cyanobacterial exudates was examined using zebrafish embryos and was concluded to be likely related to retinoids [28]. ATRA exposure of zebrafish embryos caused deformation of the tail tip, spine, and mouth and inhibition of hatching. At low concentrations (0.3–1.3 μg L^{-1}), tail tip deformation have been observed, and hatching inhibition has been observed at the high concentration [28, 35]. From these results, it is concluded that some cyanobacteria produce and release retinoids and/or retinoid-like compounds into the aquatic environment at the concentrations equivalent to those causing teratogenicity in fish and other vertebrate species.

Under the conditions of ATRA exposure to zebrafish embryos, incident light conditions, especially of UV-A, affect the activity of retinoid, the activity of ATRA in water was decreased to 30% of the original level by irradiation with fluorescent light (UV-A, 1.0 μW cm^{-2}) for 60 min [23]. The causes of the decrease were mainly oxidation by UV-A and dissolved oxygen. However, dissolved oxygen is also necessary for the development of embryos. For teratogenic experiments of retinoid exposures, at least UV-A should be omitted from the light sources. An ideal condition is darkness.

18.5 Concluding Remarks

Cyanobacterial waterblooms can accumulate on water surfaces, and cyanobacterial cell products including cyanotoxins may then be released into the water after cell decomposition [41]. Cyanobacteria in the water surface layers are usually exposed to strong sunlight including UV-A and the water pH is generally around pH 9 [42]. Under these conditions, retinoids are easy converted to inactive forms [23]. However, cyanobacteria produce many UV protectants including scytonemin [43–45] and mycosporine-like amino acids [46, 47], which act as photo-oxidation blockers. Therefore, cyanobacterial retinoids may be rather stable in certain conditions. The influence of retinoids and retinoid receptor–activating compounds on aquatic vertebrates in waters affected by cyanobacterial blooms should be investigated further.

References

[1] Heyman, R. A., Mangelsdorf, D. J., Dyck, J. A. *et al.* (1992) 9-*cis* Retinoic acid is a high affinity ligand for the retinoid-X receptor. *Cell*, **68**, 397–406 .

[2] Levin, A. A., Sturzenbecker, L. J., Kazmer, S. *et al.* (1992) 9-*cis* Retinoic acid stereoisomer binds and activates the nucleoreceptor RXR-alpha. *Nature*, **355**, 359–361.

[3] Kraft, J. C., Schuh, T., Juckau, M., Kimelman, D. (1994) The retinoid-X receptor-ligand, 9-*cis* retinoic acid, is a potential regulator of early *Xenopus* development. *Proceedings of the National Academy of Science USA*, **91**, 3067–3071.
[4] McComick, A. M., Napoli, J. L., Schnoes, H. K., Deluca, H. F. (1978) Isolation and identification of 5,6-epoxyretinoic acid-biologically-active metabolite of retinoic acid. *Biochemistry*, **17**, 4085–4090.
[5] Frolik, C. A., Roberts, A. B., Tavela, T. E. *et al.* (1979) Isolation and identification of 4-hydroxyretinoic acid and 4-oxoretinoic acid- in *vitro* metabolites of all-*trans*-retinoic acid in hamster trachea and liver. *Biochemistry*, **18**, 2092–2097.
[6] White, J. A., Guo, Y-D., Baetz, K. *et al.* (1996) Identification of retinoic acid-inducible all *trans*-retinoic acid 4-hydroxylase. *Journal of Biological Chemistry*, **271**, 29922–29927.
[7] Fujii, H., Sato, T., Kaneko, S. *et al.* (1997) Metabolic inactivation of retinoic acid by a novel P-450 differentially expressed in developing mouse embryos. *EMBO Journal*, **16**, 4163–4173.
[8] White, J. A., Beckett-Jones, B., Guo, Y-D. *et al.* (1997) Cytochromes P-450. *Journal of Biological Chemistry* **372**, 18538–18541.
[9] Ray, W., Bai, G., Ya, M., Gottlieb, D. I. (1997) CYP26, a novel mammalian Cytochrome P-450, is induced by retinoic acid and defines a new family. *Journal of Biological Chemistry*, **272**, 18702–18708.
[10] Abu-Abed, S. S., Beckett, R. R., Chiba, H. *et al.* (1997) Mouse P-450RAI (CYP26) expression and retinoic acid-inducible retinoic acid metabolism in F9 cells are regulated by retinoic acid receptor γ and retinoid X receptor α. *Proceedings of the National Academy of Sciences USA*, **77**, 3230–3223.
[11] Zile, M. H., Schnoes, H. K., Detuca, H. F. (1980) Characterization of retinyl β-glucuronide as minor metabolite of retinoic acid in bile. *Proceedings of the National Academy of Sciences USA*, **77**, 3230–3233.
[12] Zile, M. H., Inborn, R. C., Detuca, H. F. (1982) Metabolites of all-*trans* retinoic acid in bile-identification of all-*trans* retinoyl and 13-*cis* retinyl glucuronides, *Journal of Biological Chemistry.* **257**, 3537–3543.
[13] Zile, M. H., Inborn, R. C., Detuca, H. F. (1982) Metabolism *in vivo* of all-*trans* retinoic acid-biosynthesis of 13-*cis* retinoic acid and all-trans retinyl and 13-*cis* retinyl glucronides in the intestinal-mucosa of the rat. *Journal of Biological Chemistry*, **257**, 3544–3550.
[14] Fisher, G. J., Voorhees, J. J. (1996) Molecular mechanisms of retinoid actions in skin. *FASEB Journal*, **10**, 1002–1013.
[15] Napoli, J. L. (1996) Retinoic acid biosynthesis and metabolism. *FEMS Journal*, **10**,993–1001.
[16] Grenier, E., Maupas, F. S., Beaulieu, J. F. *et al.* (2007) Effect of retinoic acid on cell cell proliferation and differentiation as well as on lipid synthesis, lipoprotein secretion, and apolipoprotein biogenesis. *American Journal of Physiology, Gastrointestinal and Liver Physiology*, **293**, G1179–G1189.
[17] Mora, J. R., Iwata, M., Eksteen, B. *et al.* (2006) Generation of gut-homing IgA-secreting B cells by internal dendritic cells. *Science*, **314**, 1157–1160.
[18] Iwata, M., Eshima, Y., Kagechika, H. (2003) Retinoic acids exert direct effects on T cells to suppress T(h)1 development and enhance T(h)2 development via retinoic acid receptors. *International Immunology*, **15**, 1017–1025.
[19] Blaustein, A. R., Johnson, P. T. J. (2003) The complexity of deformed amphibians. *Frontiers in Ecology and the Environment*, **1**, 87–94.
[20] Gardiner, D. M., Hoppe, D. M. (1999) Environmentally induced limb malformations in mink frogs (*Rana septentrionalis*). *Journal of Experimental Zoology*, **284**, 207–216.
[21] Gardiner, D., Ndayibagira, A., Ggun, F., Blumberg, B. (2003) Deformed frogs and environmental retinoids. *Pure and Applied Chemistry*, **75**, 2263–2273.
[22] Skelly, D. K., Benard, M. F. (2010) Mystery unsolved: missing limbs in deformed amphibians. *Journal of Experimental Zoology. Part B, Molecular and Developmental Evolution*, **314**, 179–181.
[23] Kaya, K., Shiraishi, F., Uchida, H., Sano, T. (2011) A novel retinoic acid analogue, 7-hydroxy retinoic acid, isolated from cyanobacteria. *Biochimca et Biophysica Acta*, **1810**, 414–419.
[24] Wu, X., Jiang, J., Wan, Y., Glesy, J. P., Hu, J. (2012) Cyanobacteria blooms produce teratogenic retinoic acids. *Proceedings of the National Academy of Sciences*, **109**, 9477–9482.
[25] Wu, X., Jiang, J., Hu, J. (2013) Determination and occurrence of retinoids in a eutrophic lake (Taihu Lake, China): Cyanobactera blooms producing teratogenic retinal. *Environmental Science and Technology*, **47**, 807–814.
[26] Herrmann, K. (1995) Teratogenic effects of retinoic acid and related substances on the early development of zebrafish (*Brachydamio rerio)* as assessed by a novel scoring system. *Toxicology In Vitro*, **9**, 267–283.
[27] Parng, C., Roy, N. M., Ton, C., Lin, Y., McGrath, P. (2007) Neurotoxicity assessment using zebrafish. *Journal of Pharmacological and Toxicological Methods*, **55**, 103–112.

[28] Jonas, A., Buranova, V., Scholz, S., Fetter, E., Novakova, K. (2014) Retinoid-like activity and teratogenic effects of cyanobacteria exudates. *Aquatic Toxicology*, **155**, 283–290.

[29] Acs, A., Kovács, A. W., Csepregi, J. Z. *et al.* (2013) Ecotoxicological evaluation of *Cylindrospermopsis raciborskii* from Lake Balaton (Hungary) employing a battery of bioassays and chemical screening. *Toxicon*, **70C**, 98–106.

[30] Berry, J. P., Gibbs, P. D. L., Schmake, M. C., Saker, M. L. (2009) Toxicity of cylindrospermopsin and other apparent metabolites from *Cylindrospermopsis raciborskii* and *Aphanizomenon ovalisporum*, to the zebrafish (*Danio rerio*) embryo. *Toxicon*, **53**, 289–299.

[31] Ghazali, I, E. l., Saqrane, S., Cavalho, A. P. *et al.* (2009) Compensatory growth induced in zebrafish larvae after pre-exposure to a *Microcystis aeruginosa* natural bloom extract containing microcystins. *International Journal of Molecular Science*, **10**, 133–146.

[32] Oberemm, A., Becker, J., Codd, G. A., Steinberg, C. (1999) Effects of cyanobacterial toxins and aqueous crude extracts of cyanobacteria on the development of fish and amphibians. *Environmental Toxicology*, **14**, 77–88.

[33] Kamata, K., Shiraishi, F., Nishikawa, J., Yonemoto, J., Shiraishi, H. (2008) Screening and detection of the *in vitro* agonistic activity of xenobiotics on the retinoic receptor. *Toxicology In Vitro*, **22**, 1050–1061.

[34] Novák, J., Beníšek, M., Pachermík, J. *et al.* (2007) Interference of contaminated sediment extracts and environmental pollutants with retinoid signaling. *Environmental Toxicology and Chemistry*, **26**, 1591–1599.

[35] Kaya, K., Shiraishi, F., Tatarazako, N., Sano, T. (2014) *Does 7-hydroxy retinoic acid in cells of cyanobacteria act as ecotoxin?* International Conference of Asian Environmental Chemistry, Nov. 24–26, 2014, pp. 163–166, Bangkok, Thailand. Japan Society for Environmental Chemistry.

[36] MacCaffery, P., Evans, J., Kou, I.O. *et al.* (2002) Retinoid quantification by HPLC/MS^n. *Journal of Lipid Research*, **43**, 1143–1149.

[37] Ruhl, R. (2006) Method to determine 4-oxo-retinoid acid, retinoic acids and retinol in serum and cell extracts by liquid chromatography/diode-array detection atmospheric pressure. *Rapid Communications in Mass Spectrometry*, **20**, 2497–2504.

[38] Gundersen, T. E., Bastani, N. E., Blomhoff, R. (2007) Quantitative high-throughput determination of endogenous retinoids in human plasma using triple-stage liquid chromatograph/tandem mass spectrometry, *Rapid Communications in Mass Spectrometry*, **21**, 1176–1186.

[39] Sano, T., Kaya, K. (2014) *LC-MS/MS analysis of 7-hydroxy retinoic acid in cyanobacterial cells.* International Conference of Asian Environmental Chemistry, Nov. 24–26, 2014, pp. 290, Bangkok, Thailand. Japan Society for Environmental Chemistry.

[40] Milius, S. (2012) Blue-green algae release chemical suspected in some amphibian deformation: Retinoic acid levels high in waterways rich in cyanobacterial blooms, Science News (Web edition) http://www.scientificamerican.com/ (accessed 30 May 2012).

[41] Codd, G. A., Azevedo, S. M. F. O., Bagchi, S. N. *et al.* (eds) (2005) Cyanonet: A global network for cyanobacterial bloom and toxin risk management, *UNESCO technical documents in hydrology*, **76**, pp. 138. UNESCO's workshops, Paris, France.

[42] Kaya, K., Liu, D. Y., Shen, U. W., Xiao, B. D., Sano, T. (2005) Selective control of toxic *Microcystis* water blooms using lysine and malonic acid: an enclosure experiment. *Environmental Toxicology*, **20**, 170–179.

[43] Proteau, P. J., Grewick, W. H., Garcia-Pichel, F., Castenholz, R. W. (1993) The structure of scytonemin, an ultraviolet sunscreen pigment from the sheaths of cyanobacteria. *Experientia*, **49**, 825–829.

[44] Bultel-Poncé, V., Felix-Theodore, F., Sarthon, C., Ponge, J. F., Bodo, B. (2004) New pigments from the terrestrial cyanobacterium *Scytonema* sp. collected on the Mitaraka Inselberg, French Guyana. *Journal of Natural Products*, **67**, 678–681.

[45] Dillon, J. G., Tatsumi, C. M., Tandingan, P. G., Castenholz, R. W. (2002) Effect of environmental factors on the synthesis of sytonemin, a UV-screening pigment, in a cyanobacterium (*Chroococcidiopsis* sp.). *Archives of Microbiology*, **177**, 322–331.

[46] Cockell, C. S. (1998) The biological effects of high ultraviolet radiation on early Earth–a theoretical evaluation. *Journal of Theoretical Biology*, **193**, 717–731.

[47] Sinha, R. P., Häder, D. E. P. (2008) UV-Protectants in cyanobacteria. *Plant Science*, **174**, 278–289.

19

Other Cyanobacterial Bioactive Substances

Tina Elersek[1], Luděk Bláha[2], Hanna Mazur-Marzec[3], Wido Schmidt[4], and Shmuel Carmeli[5]

[1]Department of Genetic Toxicology and Cancer Biology, National Institute of Biology, Ljubljana, Slovenia
[2]Faculty of Science, RECETOX, Masaryk University, Brno, Czech Republic
[3]Institute of Oceanography, University of Gdansk, Gdynia, Poland
[4]DVGW Technologiezentrum Wasser TZW, Branch Office Dresden, Germany
[5]Raymond and Beverly Sackler School of Chemistry, Tel Aviv University, Tel Aviv, Israel

19.1 Introduction

Cyanobacteria are able to produce a very heterogeneous group of metabolites, which are recognised as biologically active (bioactive) substances. The chemical and structural variety of the metabolites reflects in very different biological activities (Table 19.1). The most prominent bioactive substances are peptides, retinoids, alkaloids, lactones, and phospholipids [1, 2]. Some of the compounds can be beneficial to mammals, such as a group of cyanobactins that possess antimalarial and antitumor activities [3, 4].

It is beyond the scope of this handbook to describe all the cyanobacterial bioactive substances, so mainly cyanobacterial metabolites with indication for mammalian toxicity were chosen for this chapter. Also the definition of the terms "toxicity" and "toxin" should be reconsidered, since there are many open questions regarding, for example, the biological level where toxicity is demonstrated: molecules, tissues, individuals, mammals, humans, aquatic organisms, or the whole aquatic ecosystem. In this chapter, we have used the terms "hepatotoxin" and "neurotoxin," referring to well-described mammalian toxins (e.g. microcystins, cylindrospermopsin, saxitoxins, anatoxin-a), although the same substance can be both hepatotoxic and neurotoxic, such as microcystin-LR [5].

The harmful effects of cyanobacteria cannot be attributed only to the known cyanotoxins; it was demonstrated that other bioactive substances also posess indication for mammalian toxicity. This indication can be mainly seen as inhibition of enzymes that take part in key metabolic pathways, skin irritation, action on nervous system, cytotoxic action, signalling disruption, influence

Handbook of Cyanobacterial Monitoring and Cyanotoxin Analysis, First Edition.
Edited by Jussi Meriluoto, Lisa Spoof and Geoffrey A. Codd.

Table 19.1 *Cyanobacteria produce a wide array of bioactive metabolites*

Cyanobacterial metabolite	Biological activity (indicating mammalian toxicity)
Aeruginosins (spumigins) [16–18]	Serine protease inhibitors
Anabaenopeptins [19–24]	Weak relaxation of norepinephrine-induced contraction of rat aortic preparations Inhibition of leukocyte elastase Inhibition of pancreatic elastase Impair the metabolic activity of normal human astrocytes Inhibition of protein phosphatases Inhibition of carboxypeptidase A
Biogenic amines [25, 26]	Blood pressure–increasing (e.g. tyramine) Blood pressure–reducing (e.g. histamine) Action on nervous system as psychoactive substance (e.g. norepinephrine)
Depsipeptides[1]/cyanopeptolins[1]/ micropeptins[1] [14, 19, 27]	Protease inhibitor and/or cytotoxic activities (selective inhibition of leukocyte elastase, pancreatic elastase and chymotrypsin) Negative influence on metabolism of normal human astrocytes Oscillapeptin J: inhibitor of trypsin and chymotrypsin is toxic to zooplankton grazers *Eudiaptomus* sp. and *Daphnia* sp., but not to mice (up to 1000 μg kg body wt^{-1})
Estrogenic compounds, endocrine disruptors, and novel tumour promoters [28–30]	Modulations of cytoplasmic (nuclear) receptors such as estrogen or androgen receptor and disruption of hormone-controlled homeostasis Reproduction disorders in both males and females, tumour development (e.g. breast cancer, prostate), pathophysiological changes Inhibition of intercellular communication via gap junctions, activation of mitogen-activated protein kinases (MAPK) Tumour promotion (stimulatory growth effects in normal stem-like cells)
Lyngbyatoxins[2] [31]	Strong cytotoxicity, signalling disruption (protein kinase C activation) Skin irritation and dermatitis Tumour-promotional effects in skin Erosion of the gastrointestinal tract and inflammation upon oral administration
Microginins [32, 33]	Amino protease inhibitors
Microviridins [34–36]	Serine protease inhibitors (elastase, chymotrypsin)
Oscillatorins [37]	Chymotrypsin inhibitor
Retinoic acids [2]	Teratogenic effects (deformed amphibians)

[1] Depsipeptides are peptides that contain esteric bond in their backbone, cyanopeptolins or the more used name micropeptins are the same and they inhibit serine proteases (both trypsin and chymotrypsin types).
[2] Found only in saline habitats.

on receptors, hormonal disruption, reproductive disorders, tumour promotion, etc. (Table 19.1). Also, the ecological impact of diverse cyanobacterial metabolites should not be neglected. These natural substances also influence the physiology of cyanobacteria themselves and may have a strong impact on their blooming capacity [6, 7]. It was reported that depsipeptides can trigger cyanophages with an ability to display lysogenic life cycle, which cause rapid lysis of cyanobacterial bloom [8]. Further, the presence of cyanobacterial metabolites has strong negative impact on phytoplankton biodiversity [9, 10]. Some cyanobacterial bioactive substances (e.g. anabaenopeptins and anabaenopeptilides) have been found in tissues of various aquatic animals [11]. There is also the possibility of synergistic interactions between different hepatotoxins and neurotoxins and other cyanobacterial metabolic products released in the water. Crude extracts of cyanobacteria are sometimes more toxic to vertebrates and invertebrates than the respective concentrations of pure hepatotoxins contained in the extract [12, 13]. Production of microcystins (and other well-characterized toxins) is often accompanied by the production of other bioactive substances, but since the advanced analytical procedures are not standardised, the occurrence in the environment of various other cyanobacterial metabolites is not properly evaluated.

The toxicological studies of nonhepatotoxic and nonneurotoxic cyanobacterial bioactive substances with the whole-animal systems are limited [14]. To date, we have found no record on health impact or on any operational or economic consequences regarding their occurrence in the environment. Nevertheless, without further studies, it is impossible to predict whether there is no health impact or simply the health impact is unstudied or unknown. Although the abundance of many bioactive substances in the environment has been preliminary assessed and some knowledge indicating mammalian toxicity has already been accumulated (Table 19.1), currently no guideline is set in the European Union. For all the reasons described here, the scale of threat associated with the occurrence of these substances for ecosystem and mammals should be properly evaluated in the near future. Nevertheless, some documents [15] pay attention to cyanobacterial cells as such and they do not focus to any particular bioactive substance (e.g. microcystin).

The most widely studied cyanobacterial bioactive substances are peptides, which can be classified according to (a) biosynthetic pathway (ribosomal, non-ribosomal, etc.), (b) chemical structure (linear, cyclic, depsipeptides, etc.), and (c) primary target organ or system in mammals (hepatotoxic, nonhepatotoxic, etc.).

Among peptides, the thiazole and oxazole containing cyclic peptides (Fig. 19.1) have been studied in great detail. Cyclic peptides involving heterocyclic modifications (usually called cyclamides) were

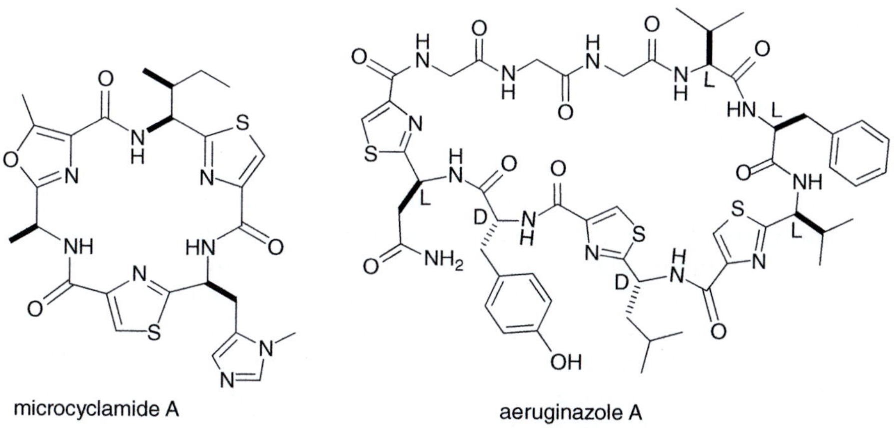

Figure 19.1 Structure of microcyclamide A and aeruginazole A

initially isolated from several marine organisms, which acquired compounds through a cyanobacterial diet (dolastatin E [38]), or from cyanobacterial symbionts (trisoxazoline, bistratamides [39]). Cyclic peptides were also characterized from marine species (venturamides [40]), cultured terrestrial cyanobacteria (raocyclamides, tenuecyclamides [41, 42]) and recently from bloom-forming strains of cyanobacteria such as *Microcystis* and *Nostoc* (microcyclamide [43–46], aerucyclamides [47], and banyascyclamides [48]). The concurrent isolation of the same cyclic hexapeptide from the cyanobacterium *Westiellopsis prolifica* (westiellamide [49]) and the ascidian *Lissoclinum bistratum* (trioxazolline [39]) marked the beginning of research efforts to reveal the biosynthesis of azoline and azole containing cyclic peptides in marine invertebrates and cyanobacteria. Prinsep et al. [49] suggested that these modified cyclic peptides are synthesised in the ascidian by its obligate prokaryotic symbiont, *Prochloron* sp. This suggestion was verified through the research of Donia et al. [50], who showed that the patellamides and related cyclic peptides indeed are synthesised by the symbiotic *Prochloron* spp., unexpectedly, through a ribosomal pathway. Recently, the same authors have shown that tenuecyclamides are synthesised by the nonsymbiotic cyanobacterium *Nostoc carneum*/*N. spongiaeforme* (current name *N. carneum*/former name *N. spongiaeforme* used in the cited reference) by a similar ribosomal pathway [51]. The cyclic octapeptides, patellamides, and patellins are produced by *Prochloron* spp. symbionts in ascidians and the cyclic hexapeptides are produced both by *Prochloron* spp. and nonsymbiotic strains of cyanobacteria. However, azoline and azole containing cyclic peptides with larger cycles have been isolated from marine cyanobacteria. One of these is kororamide, which is a cyclic nonapeptide that was isolated from a Palauan collection of *Lyngbya majuscula* [52] and the other, trichamide, whose structure was predicted from genome mining of *Trichodesmium erythraeum* [53]. The first thiazole-containing cyclic dodecapeptide, aeruginazole A, which was isolated from a waterbloom material of a *Microcystis* sp. collected in a water reservoir near Kfar-Yehoshua in Israel, was reported by the Raveh and Carmeli [44]. Since the isolation of aeruginazole A, several other related metabolites (aeruginazoles DA1497, DA1372, DA1338, DA1304, and DA1274 [54]) were isolated from *Microcystis* spp. blooms. Their presence was detected by MALDI-MS in many samples of bloom material from Israeli waterbodies (Carmeli, unpublished).

Peptides are most widely studied, but there are also other compound classes produced by cyanobacteria (e.g. biogenic amines, alkaloid structures, derivatives of lipids, etc.). The following paragraphs try to briefly summarize some of the known and studied groups of cyanobacterial bioactive compounds, which are of human health relevance.

19.2 Aeruginosins and Spumigins

Aeruginosins and spumigins are linear tripeptides and tetrapeptides. They are biosynthesized in cyanobacteria from amino acids or amino acid precursors by non–ribosomal peptide type enzyme assembly [17, 55]. The aeruginosins and spumigins usually contain phenyl lactic acid (Pla), or its *p*-hydroxy derivative (Hpla), at the *N*-terminus and an arginine derivative (if any) at the *C*-terminus (Fig. 19.2). Further, Hpla derivatives with mono chloro or bromo substituents at the *meta*-position, or di-Cl, di-Br, and Cl, Br substituents at the *m,m′* positions and an *O*-sulphate derivative were characterized. Various L- and D-amino acids may occupy the second position (Leu or Ile, Hty, Tyr, Hph, or Phe). The structure of the two classes of peptides (aeruginosins and spumigins) differs mainly with the respect to the residue at position 3, which in the spumigins is occupied by a proline (L-Pro) or MePro (i.e. spumigin A, Fig. 19.2) and in the aeruginosins by a 2-carboxy-6-hydroxy-octahydroindole (Choi) residue (Fig. 19.2). Four natural Choi stereoisomers have been

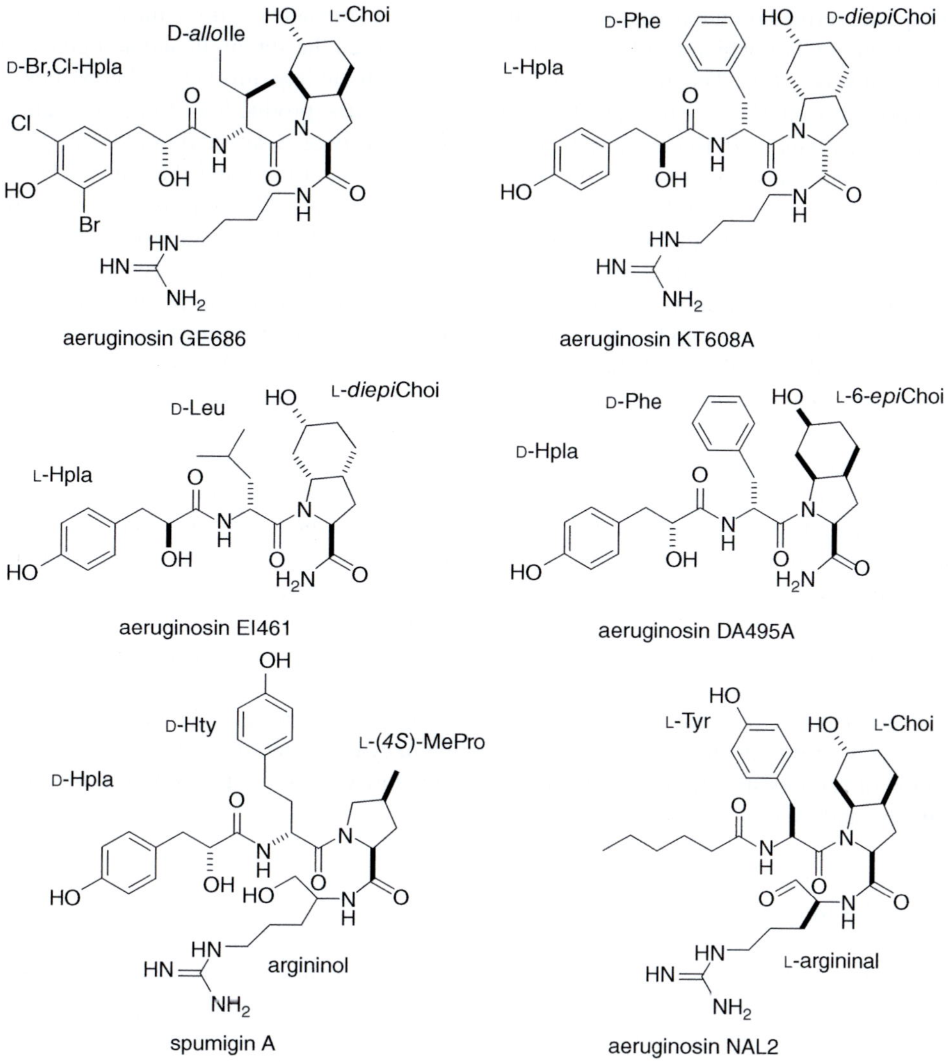

Figure 19.2 *Structure of some aeruginosins and spumigin A*

characterized in the literature to date: (i) the most abundant isomer, which is incorporated into 33 isolated compounds, is 2*S*,3a*S*,6*R*,7a*S*-Choi (L-Choi, *i.e.* aeruginosin GE686, Fig. 19.2, [34]); (ii) a single compound containing the 2*S*,3a*R*,6*R*,7a*R*-Choi isomer (L-3a,7a-*diepi*-Choi, *i.e.* aeruginosin EI461, Fig. 19.2, [56]) was the second to be identified; (iii) four compounds containing the 2*R*,3a*R*,6*R*,7a*R*-Choi isomer (D-3a,7a-*diepi*-Choi, i.e. aeruginosin KT608A, Fig. 19.2, [57]) were isolated from Lake Kinneret, Israel bloom material and, (iv) two compounds containing the 2*S*,3a*S*,6*S*,7a*S*-Choi isomer (L-6-*epi*-Choi, i.e. aeruginosin DA495A, Fig. 19.2, [58]) were recently characterized. The 6-OH of Choi is substituted in some of the aeruginosins by a sulphate or sugar unit. In *Planktothrix/Oscillatoria agardhii* [35] and *Nodularia spumigena* [59] the aeruginosins were glycosylated with the pentose sugar xylose. Recently, aeruginosins with short fatty acid chains at the *N*-terminus were identified in *Nodularia spumigena* [59]. The same species produces hybrid peptides termed pseudoaeruginosins containing short fatty acid chains at the *N*-terminus and MePro instead of Choi, which result from cooperation between the spumigin and aeruginosin biosynthetic pathways [60]. The presence of aeruginosins (including microcin SF608) has been mainly reported from *Microcystis* and *Planktothrix,* whereas the production of the spumigins was attributed to *N. spumigena*. Recently, however, spumigins were also detected in *Dolichospermum compactum*/*Anabaena compacta* [61]. Related compounds, suomilide and the banyasides, were isolated from *N. spumigena* [62] and *Nostoc* sp. [63], respectively. The approximate mass range of these linear peptides is from 461 Da (aeruginosin EI461) to 803 Da (aeruginosin 205A/B).

Both classes of these linear peptides attracted significant attention from synthetic chemists due to their modified peptide nature and their capacity to inhibit important proteolytic processes [16, 17]. Some aeruginosins and spumigins exhibit inhibitory activity against trypsin-type serine proteases (that hydrolyse the amide bond next to arginine or lysine). Weak activity of aeruginosins against papain, a cysteine protease, was also observed [16, 17]. As proteases are involved in many important physiological processes, inhibitory activity of aeruginosins and spumigins against these enzymes may have indirect adverse effects on the exposed organisms. For example, the inhibition of trypsin and chymotrypsin, the most important digestive enzyme in animals, reduces food digestion and lowers its nutritional quality. Experiments with *Daphnia magna* [64], showed that microcin SF608, an aeruginosin variant, lowered the cellular concentration of microsomal and soluble glutathione-*S*-transferase, the enzyme involved in detoxification processes in plants and animals. Due to inhibitory activity against enzymes involved in the blood coagulation cascade, some aeruginosin analogues are also considered to be promising therapeutic agents [16].

Spumigins and aeruginosins are often analysed by LC-MS/MS systems [17, 56, 65]. However, for complete structure elucidation of this group of compounds thorough NMR analysis is needed [18, 56–58, 66]. Aeruginosin and spumigin production in cyanobacterial strains can be revealed by application of primers specific for genes involved in the biosynthesis of the peptides (*aer* and *spu*, respectively) and PCR techniques [17].

19.3 Anabaenopeptins

Anabaenopeptins (APs) are a structurally diverse group of nonribosomal cyclic peptides commonly detected in cyanobacterial blooms. Anabaenopeptin type compounds were also classified as ferintoic acids, lyngbyaureidamides, oscillamides, nodulapeptins, schizopeptins, and melophluspeptins. Anabaenopeptins are composed of a single amino acid residue side chain (most frequently Arg, Ile, Tyr, or Phe), which is attached to the cyclic pentapeptide through an ureido bridge formed with the α-amine of D-lysine. This amino acid residue occupies position 1 and is the only nonvariable component in the AP structure (Fig. 19.3). At position 2, Val and Ile are most common. Position 3 is occupied by homo-variants of tyrosine (Hty) or phenylalanine (Hph). In position 4, *N*-methylated amino acid residues (e.g. *N*-MeAla, *N*-MeHty, *N*-MeHph or *N*-MeGly) are incorporated. The carboxylic end of the anabaenopeptins is usually occupied by Phe, Tyr, Hty, Ile/Leu, or Hph, although *O*-Ac-Ser appears in nodulapeptins.

Depending on the organization of the gene cluster involved in the biosynthesis of APs, and on the substrate promiscuity of the A domain, individual strains of cyanobacteria can produce peptides with only one (e.g. from *Nostoc* and *Nodularia*) or with different amino acids (e.g. from *Dolichospermum* and *Planktothrix*) in the exocyclic position. The mass weights of known APs range from 759 Da (AP I) to 956 Da (oscillamide C). So far, APs have been identified in cyanobacteria from the genera *Dolichospermum*/*Anabaena* [67], *Chrysosporum*/*Aphanizomenon* [22],

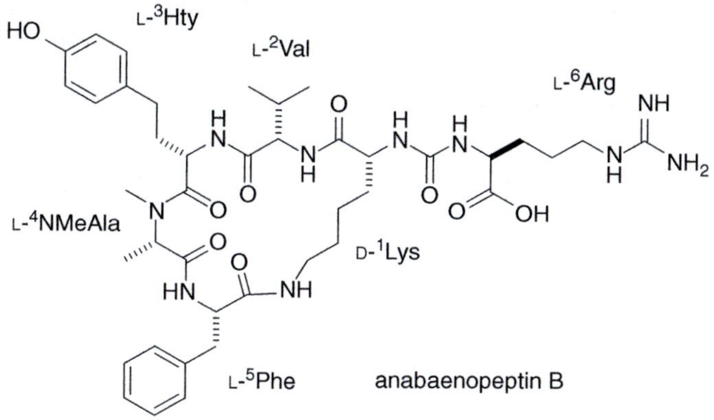

Figure 19.3 Structure of anabaenopeptin B

Planktothrix/Oscillatoria [21–24], *Microcystis* [66, 68], *Nodularia* [65, 67], *Nostoc* [67], *Lyngbya/Planktolyngbia* [69], and *Schizothrix* [70]. Several anabaenopeptin type compounds have been isolated from marine sponges [71].

Anabaenopeptin F, oscillamides B and C inhibit the activity of protein phosphatases (PP). Positively charged Arg in position 1 and the hydrophobic residue of *N*-MeHty in position 5 seem to play significant role in this activity [24]. Anabaenopeptins were found to also be active towards proteolytic enzymes such as trypsin (AP MM85, schizopeptin 791), chymotrypsin (AP MM823, AP MM850, oscillamide Y), elastase (AP MM823, AP MM850), and carboxypeptidase A (AP G, H, T, I, J,915) [22, 23, 72]. It was suggested that the type of the side chain residue has significant effect on the biological activity of the compounds. Anabaenopeptin B and AP 906 showed relaxing activity of norepinephrine-induced contraction of rat aortic preparations [20]. Anabaenopeptin B and AP F, the most commonly occurring APs, were shown to completely suppress the growth of several *M. aeruginosa* strains through induction of the lytic cycle in lysogenic cyanobacteria [8, 73]. As this process leads to cyanobacterial cell disruption, its ecological implications might be significant. A potential role of APs in the interaction with chytrid fungi infecting cyanobacteria was also documented [74].

Different mass spectrometer configurations have been used in analyses of APs [65, 67, 75]. The planar structure of the APs might be elucidated based on the fragmentation pattern obtained in the ion product mode (see SOP 25). When sufficient amount of pure AP is available (several milligrams), the application of NMR is possible [22, 68] and allows identification of the absolute configuration of the amino acid residues, including stereoisomers such as Ile and *allo*Ile [58].

19.4 Biogenic Amines

Cell-bound as well as extracellular organic material from living organisms contain complex mixtures of proteins, peptides, amino acids, and amines. Among these, the so-called biogenic amines (BA) are more and more in focus. These compounds are transformation and degradation products of amino acids, such as formed by decarboxylation. One of the most well-known is histamine, a known vasodilatant involved, for example, in allergic reactions, formed from the amino acid histidin. There are some regulations for the occurrence of histamine in food, such as in cheese and wine in several countries. Additionally, from a hygienic point of view, the degradation and transformation products of the nonproteinogenic amino acid ornithine (polyamines putrescine and cadaverine) are of significant relevance. The chemical structure of biogenic amines is quite variable. Aromatic structures like tryptamines or tyramines and aliphatic polyamines (spermidine, putrescine, and cadaverine) are the most significant ones.

The formation and occurrence of biogenic amines in cyanobacteria have been proved in recent years by several authors [76, 77]. These authors discussed biosynthesis and transport aspects of the polyamines putrescine, spermidine, and spermine in genus *Synechocystis*. During the life cycle of this cyanobacterium, the level of BA can vary significantly. In waters containing algal and cyanobacterial cells, which are treated for human consumption, decarboxylation of the protein- and peptide-like amino acids, in the frame of fouling and degradation processes, is one of the most relevant sources of biogenic amine formation. Because of their organoleptic properties (taste and odour), this process can be relevant for hygienic quality of drinking as well as bathing water. It is known that the precursor substances of biogenic amines, the amino acids, cannot be efficiently removed during many steps of the water treatment processes [78–81]. Therefore, traces of BA can remain in the drinking water, especially when it is obtained from treated surface water containing high amounts of cyanobacteria. The concentration of putrescine and cadaverine in drinking water covers the range up to 2 $\mu g\,L^{-1}$ per substance [83].

Published data describe “food toxicity” mainly due to monoamines (e.g. histamine, tyramine). Meanwhile, some amines, especially diamines and polyamines, are potential indicators of food spoilage. Amines act primarily on (a) the vascular system as blood pressure–increasing (e.g. tyramine) or blood pressure–reducing (e.g. histamine) and (b) nervous system as psychoactive substances (e.g. norepinephrine). Feeding studies on animals have given variable results (enhanced growth in some cases and toxicity in others) [25, 26].

The analytical procedure for detection of BA-traces in water has not been well examined. The method of choice is derivatization, such as with FMOC-Cl in water phase followed by liquid chromatography and mass spectrometric detection. Because of the low specificity of low-molecular amine-like substances, confirmation of positive results by high-resolution mass spectroscopy is urgent. This makes the analysis of such compounds time consuming and expensive [81, 82]. Using this technique, the putrescine and cadaverine in concentration of few hundred micrograms per mm^3 of biovolume can be measured. This level was found in cyanobacteria species like *Microcystis aeruginosa* and *Planktothrix rubescens* cultivated under laboratory conditions.

19.5 Depsipeptides

A depsipeptide is a peptide in which one or more of the amide (–CONHR–) bonds are replaced by ester (COOR) bonds. Cyanobacteria are capable of biosynthesizing both depsipeptides and cyclic depsipeptides, which present high structural variability and diverse biological activities. Some depsipeptides isolated from terrestrial and marine environments exhibit promising therapeutic properties as potential anticancer agents. For example, the cryptophycins [84], exhibit potent tumour-selective cytotoxicity [85]; cryptophycin-1 has an IC_{50} value of 20 $pmol\,L^{-1}$ against human ovarian carcinoma and has shown excellent activity against solid tumours implanted in mice, including a drug-resistant tumour [86]. Further, the cyclic depsipeptides majusculamide C and dolastatins 11 and 12 also show promising cytotoxic activity by arresting cells at cytokinesis by stimulating microfilament hyperpolymerization in a dose- and time-dependent manner [87, 88].

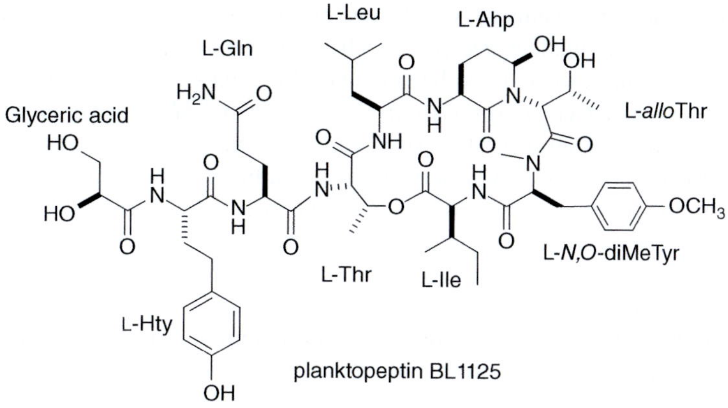

Figure 19.4 *Structure of planktopeptin BL1125*

The micropeptins are the most diverse group of cyclic depsipeptides in cyanobacteria and so far more than 142 variants have been identified [89]. They are characterized by a lactone ring consisting of six amino acid residues, one of which is the unique 3-amino-6-hydroxy-2-piperidone (Ahp), and a side chain of one to three acid units (e.g. planktopeptin BL1125 is shown on Fig. 19.4). They are regularly found in cyanobacterial bloom forming species [90] in the freshwater environment (e.g. *Microcystis* sp., *Dolichospermum/Anabaena* sp. and *Planktothrix rubescens*) but occasionally isolated from the marine environment as well (e.g. *Lyngbya/Planktolyngbia* sp., *Schizothrix* sp., and *Dichothrix utahensis* [91]).

The micropeptins are serine proteases inhibitors. The amino acid at the fifth position from the carboxylic terminus selects for the type of serine proteases that will be inhibited by these compounds. Basic amino acids (i.e. arginine and lysine) at this position select for inhibition of trypsin like serine protease, while aliphatic, aromatic and other neutral amino acids select for inhibition of chymotrypsin like serine proteases. Planktopeptin BL1125, is a potent chymotrypsin and elastase inhibitor, with an inhibition constant in the nM range [92] and was shown to influence the morphology and physiology of algae and cyanobacteria [93], and other micropeptins have been shown to alter gene expression in the cyanobacteria themselves [83].

The micropeptins usually give strong ions under ESI and MALDI ionization techniques. However, identification of their sequence by MS/MS techniques is rather limited due to their structure that include one to three residues in the side chain which present non-conventional fragmentation pattern. Relatively limited methods for detection of these compounds are often based on HPLC/MS [73]; see SOP 25.

19.6 Endocrine Disruptors and Novel Tumour Promoters

A number of recent studies indicate the presence of diverse bioactivities in freshwater cyanobacterial strains, which cannot be explained by the presence of known cyanotoxins or other identified cyanobacterial metabolites. It is therefore clear that cyanobacteria produce novel metabolites, although their structures remain to be resolved.

One of important toxicity mechanisms, which have recently been documented in cyanobacteria, is modulation of intracellular (or nuclear) receptors such as estrogen receptor or androgen receptor [30]. For example, the estrogenic effects have been evoked by cyanobacterial strains of *Planktothrix agardhii* (the strongest bioactivity), *Aphanizomenon gracile* and *Microcystis aeruginosa,* as well as in natural bloom samples dominated by these cyanobacteria. Nonphysiological modulations of nuclear receptors by xenobiotics (either activation or inhibition) are known to adversely affect hormonal signalling in all vertebrates causing endocrine disruption. The health outcomes include for example reproduction and developmental disorders, stimulation of growth of hormone-sensitive tumours, immunomodulations, and many others. The properties of the causative agents, i.e. cyanobacterial or algal phytoestrogens, are under investigation.

Other recent studies showed production of novel potential tumour promoters acting via disruption of intercellular communication [28, 29]. The studies showed that these fast acting inhibitors of gap-junctional communication are simple (most likely lipid derived) metabolites, and their bioactivities have been detected in several natural blooms as well as laboratory cultures of *Microcystis aeruginosa*, *Aphanizomenon flos-aquae*, and *Cylindrospermopsis raciborskii*. Nevertheless, *in vivo* toxicological relevance of these novel compounds remains to be investigated further.

19.7 Lyngbyatoxins and Other Toxins Produced by *Lyngbya majuscula*

More than 70 different bioactive molecules [31] have been isolated from *Lyngbya majuscula* or, more specifically, from *Moorea producens,* a newly separated species formed by tropical strains of *L. majuscula* [94, 95]. They include structurally diverse compounds such as peptides, alkaloids, phenolic bis-lactones, fatty acid derivatives, and others. The most widely studied are lyngbyatoxins A, B, and C and aplysiatoxin isomers (Fig. 19.5). Antillatoxin A (Fig. 19.5) is a structurally novel lipopeptide, isolated from *L. majuscula*, that activates voltage-gated sodium channels leading to sodium influx in cerebellar granule neurons and cerebrocortical neurons after 8 to 9 days' exposure in vitro [94].

M. producens is a marine benthic cyanobacterium with a global distribution throughout the tropics and subtropics in water to 30-m depth. Production of toxins by this species has been documented in isolates collected worldwide. The most important toxicological symptoms associated with lyngbyatoxins and aplysiatoxins but also several other toxins from *M. producens* include skin, eye, and respiratory irritation. The first causative reports linking *L. majuscula* with an epidemic of acute contact dermatitis in Hawaii are from the 1950s followed by a number of reports from other

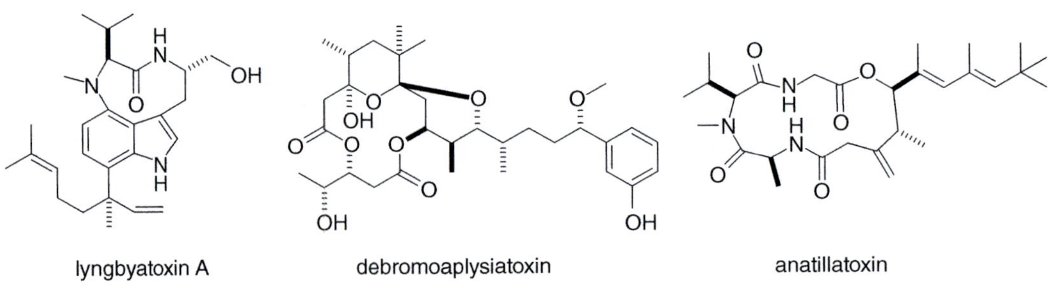

Figure 19.5 Structure of Lyngbya derived toxins

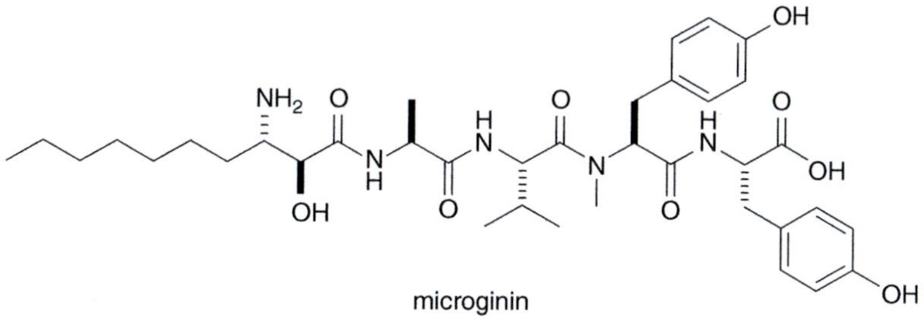

Figure 19.6 Structure of microginin

tropical localities. The commonly reported irritation effects include acute vesicular dermatitis with superficial desquamation and epidermis oedema with infiltrated chronic and acute inflammatory cells. The most important mechanism of these toxins is fast and strong disruption of normal cellular signalling via activation of protein kinase C.

Determination methods for the two major toxins are usually based on HPLC with mass spectrometry [31].

19.8 Microginins

Microginins represent a 25-member group of linear, nonribosomal peptides isolated from various bloom forming cyanobacteria strains, primarily the cosmopolitan species *Microcystis aeruginosa* [33]. They are composed of α-hydroxy-β-amino decanoic or octanoic acid, which is occasionally chlorinated at its terminal methyl group and is extended by three to five additional amino acids of which two are usually *N*-methylated (Fig. 19.6) [18, 33, 96, 97].

The microginins inhibit zinc metalloproteases such as angiotensin-converting enzyme and aminoproteinases, by binding of their α-hydroxy-β-amino residue to the zinc at the active cite of the enzyme. As such, they represent important drug candidates to treat hypertension and associated diseases such as chronic heart failure and diabetic nephropathy [98].

The relatively limited methods for microginin detection are often based on HPLC/MS. In addition, molecular DNA analyses of the microginin synthetase gene cluster can be used for detection of new microginin variants or microginin-like structures [98].

19.9 Microviridins

Microviridins are a group of 15 ribosomally synthesized three-cyclic depsipeptides (Fig. 19.7) [72, 99–101] produced by different genera of cyanobacteria [36]. Microviridins form one of the most intriguing classes of peptides, since they feature a cage-like structure (e.g. microviridin B). They are composed of 12 to 14 amino acids: an acetylated amino acid at the *N*-terminus, followed by one to three variable amino acids in the side chain, a conserved Thr at the fourth position, a variable amino acid at the fifth position, conserved sequences at positions 6 to 12 (^{6}Lys-^{7}Tyr-^{8}Pro-^{9}Ser-^{10}Asp-^{11}Trp(Phe)-^{12}Glu), an acidic amino acid (Asp, Glu, or hydroxyl-Asp) at the thirteen position and Tyr at the carboxylic terminus. The highly unusual three-cyclic architecture (three interconnected rings of amino acids) results from ester bonds between the side chain

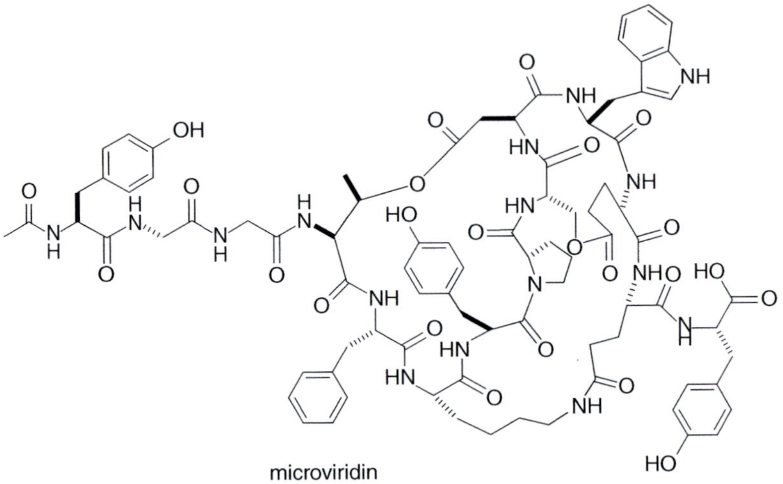

Figure 19.7 Structure of microviridin

hydroxyl of 4Thr and the side chain carboxyl of 10Asp and the side chain hydroxyl of 9Ser and side chain carboxyl of 12Glu and an amide bond between the side chains amine of 6Lys and the carboxyl of the acidic residue at position 13.

Main known producers of microviridins comes from the cosmopolitan genera *Microcystis*, *Planktothrix*, and *Nostoc*. Recent studies have revealed a unique biosynthetic mechanism for microviridins in *Microcystis* and the filamentous cyanobacterial genus *Planktothrix* [36, 99].

Most of the microviridin variants show inhibitory activities against serine-type proteases, most notably against elastase, which is a target enzyme in the treatment of lung emphysema [102]. One of the peptide isoforms, microviridin J, has been shown to inhibit the molting process of *Daphnia*, leading to death of the animals [103].

Available methods for microviridin detection are very limited. The microviridins give strong ions under ESI and MALDI ionization conditions. However, identification of their sequence by MS/MS techniques is challenging due to their structure that includes a tricyclic core and a side-chain of one to four residues, which present nonconventional fragmentation patterns. Molecular approaches are based on PCR, using two *mdn* biosynthetic gene clusters from *Microcystis* species [36].

References

[1] Singh, S., Kate, B.N. and Banerjee, U.C. (2005) bioactive compounds from cyanobacteria and microalgae: an overview. *Critical Reviews in Biotechnology*, **25** (3), 73–95.

[2] Wu, X., Jiang, J., Wan, Y. *et al.* (2012) Cyanobacterial blooms produce teratogenic retinoic acids. *Proceedings of the National Academy of Sciences of the United States of America*, **109** (24), 9477–9482.

[3] Linington, R.G., Gonzalez, J., Urena, L.D. *et al.* (2007) Venturamides A and B: antimalarial constituents of Panamanian marine cyanobacterium *Oscillatoria* sp. *Journal of Natural Products*, **70** (3), 397–401.

[4] Williams, D.E., Moore, R.E. and Paul, V.J. (1989) The structure of ulithiacyclamide B. Antitumor evaluation of cyclic peptides and macrolides from *Lissoclinum patella. Journal of Natural Products*, **52** (4), 732–739.

[5] Fischer, W.J., Altheimera, S., Cattori, V. *et al.* (2005) Organic anion transporting polypeptides expressed in liver and brain mediate uptake of microcystin. *Toxicology and Applied Pharmacology*, **203** (3), 257–263.

[6] Schatz, D., Keren, Y., Vardi, A. *et al.* (2007) Towards clarification of the biological role of microcystins, a family of cyanobacterial toxins. *Environmental Microbiology*, **9** (4), 965–970.
[7] Sukenik, A., Eshkol, R., Livne, A. *et al.* (2002) Inhibition of growth and photosynthesis of the dinoflagellate *Peridinium gatunense* by *Microcystis* sp. (cyanobacteria): a novel allelopathic mechanism. *Limnology and Oceanography*, **47** (6), 1656–1663.
[8] Sedmak, B., Carmeli, S. and Eleršek, T. (2008) "Non-toxic" cyclic peptides induce lysis of cyanobacteria - An effective cell population density control mechanism in cyanobacterial blooms. *Microbial Ecology*, **56** (2), 201–209.
[9] Sedmak, B. and Kosi, G. (2002) Harmful cyanobacterial blooms in Slovenia–bloom types and microcystin producers. *Acta Biologica Slovenica*, **45**, 17–30.
[10] Sedmak, B. and Kosi, G. (1998) The role of microcystins in heavy cyanobacterial bloom formation. *Journal of Plankton Research*, **20** (4), 691–708.
[11] Gkelis, S., Lanaras, T. and Sivonen, K. (2006) The presence of microcystins and other cyanobacterial bioactive peptides in aquatic fauna collected from Greek freshwaters. *Aquatic Toxicology*, **78** (1), 32–41.
[12] Jonas, A., Scholz, S., Fetter, E. *et al.* (2015) Endocrine, teratogenic and neurotoxic effects of cyanobacteria detected by cellular *in vitro* and zebrafish embryos assays. *Chemosphere*, **120**, 321–327.
[13] Šuput, D., Milutinovič, A., Serša, I. and Sedmak, B. (2002) Chronic exposure to cyanobacterial lyophilizate reveals stronger effects than exposure to pure microcystins – a MRI study. *Radiology and Oncology*, **36**, 165–167.
[14] Blom, J.F., Baumann, H.I., Codd, G.A. and Jüttner, F. (2006) Sensitivity and adaptation of aquatic organisms to oscillapeptin J and [D-Asp^3, [E]-Dhb^7]microcystin-RR. *Archiv fur Hydrobiologie*, **167** (1-4), 547–559.
[15] Council Directive 76/160/EEC (2003) Bathing water directive–concerning the quality of bathing water as amended by Council Directive 91/692/EEC (further amended by Council Regulation 1882/2003/EC), and Council Regulation 807/2003/EC.
[16] Ersmark, K., Del Valle, J.R. and Hanessian, S. (2008) Chemistry and biology of the aeruginosin family of serine protease inhibitors. *Angewandte Chemie International Edition*, **47** (7), 1202–1223.
[17] Fewer, D.P., Jokela, J., Rouhiainen, L. *et al.* (2009) The non-ribosomal assembly and frequent occurrence of the protease inhibitors spumigins in the bloom-forming cyanobacterium *Nodularia spumigena. Molecular Microbiology*, **73** (5), 924–937.
[18] Ishida, K., Okita, Y., Matsuda, H. *et al.* (1999) Aeruginosins, protease inhibitors from the cyanobacterium *Microcystis aeruginosa. Tetrahedron*, **55** (36), 10971–10988.
[19] Bubik, A., Sedmak, B., Novinec, M. *et al.* (2008) Cytotoxic and peptidase inhibitory activities of selected non-hepatotoxic cyclic peptides from cyanobacteria. *The Journal of Biological Chemistry*, **389** (10), 1339–1346.
[20] Harada, K.I., Mayumi, T., Shimada, T. *et al.* (1993) Occurrence of four depsipeptides, aeruginopeptins, together with microcystins from toxic cyanobacteria. *Tetrahedron Letters*, **34** (38), 6091–6094.
[21] Itou, Y., Suzuki, S., Ishida, K. and Murakami, M. (1999) Anabaenopeptins G and H, potent carboxypeptidase A inhibitors from the cyanobacterium *Oscillatoria agardhii* NIES-595. *Bioorganic and Medicinal Chemistry Letters*, **9** (9), 1243–1246.
[22] Murakami, M., Suzuki, S., Itou, Y. *et al.* (2000) New anabaenopeptins, potent carboxypeptidase-A inhibitors from the cyanobacterium *Aphanizomenon flos-aquae*. *Journal of Natural Products*, **63** (9), 1280–1282.
[23] Sano, T. and Kaya, K. (1995) Oscillamide Y, a chymotrypsin inhibitor from toxic *Oscillatoria agardhii. Tetrahedron Letters*, **36** (33), 5933–5936.
[24] Sano, T., Usui, T., Ueda, K. *et al.* (2001) Isolation of new protein phosphatase inhibitors from two cyanobacteria species, *Planktothrix* sp. *Journal of Natural Products*, **64** (8), 1052–1055.
[25] Meat and Livestock Australia (2001) Biogenic amines in meat meal (ISBN: 1740367022), pp. 1–4.
[26] European Food Safety Authority (2011) EFSA Scientific Opinion on risk based control of biogenic amine formation in fermented foods, *EFSA Panel on Biological Hazards, EFSA Journal*, **2393**; 9 (10).
[27] Petit, G.R., Kamano, Y., Dufresn, H.C. *et al.* (1989) Isolation and structure of the cytostatic depsipeptide dolastatin 13 from the sea hare *Dolabella auricularia. Journal of American Chemical Society*, **111** (13), 5015–5017.
[28] Bláha, L., Babica, P., Hilscherová, K. *et al.* (2010) Inhibition of gap-junctional intercellular communication and activation of mitogen-activated protein kinases by cyanobacterial extracts - indications of novel tumor promoting cyanotoxins. *Toxicon*, **55** (1), 126–134.

[29] Nováková, K., Kohoutek, J., Adamovský, O. *et al.* (2013) Novel metabolites in cyanobacterium *Cylindrospermopsis raciborskii* with potencies to inhibit gap junctional intercellular communication. *Journal of Hazardous Materials*, **262**, 571–579.

[30] Stepankova, T., Ambrozova, L., Blaha, L. *et al.* (2011) In vitro modulation of intracellular receptor signaling and cytotoxicity induced by extracts of cyanobacteria, complex water blooms and their fractions. *Aquatic Toxicology*, **105** (3-4), 497–507.

[31] Osborne, N.J.T., Webb, P.M. and Shaw, G.R. (2001) The toxins of *Lyngbya majuscula* and their human and ecological health effects. *Environment International*, **27** (5), 381–392.

[32] Neumann, U., Forchert, A., Flury, T. and Weckesser, J. (1997) Microginin FR1, a linear peptide from a water bloom of *Microcystis* species. *FEMS Microbiology Letters*, **153** (2), 475–478.

[33] Okino, T., Matsuda, H., Murakami, M. and Yamaguchi, K. (1993) Microginin, an angiotensin-converting enzyme inhibitor from the blue-green alga *Microcystis aeruginosa*. *Tetrahedron Letters*, **34** (3), 501–504.

[34] Murakami, M., Ishida, K., Okino,T. *et al.* (1995) Aeruginosins 98-A and B, trypsyn inhibitors from the blue-green alga *Microcystis aeruginosa* (NIES-98). *Tetrahedron Letters*, **36** (16), 2785–2788.

[35] Shin, H.J., Murakami, M., Matsuda, H. *et al.* (1995) Oscillapeptin, an elastase and chymotrypsin inhibitor from the cyanobacterium *Oscillatoria agardhii* (NIES-204). *Tetrahedron Letters*, **36** (29), 5235–5238.

[36] Ziemert, N., Ishida, K., Liaimer, A. *et al.* (2008) Ribosomal synthesis of tricyclic depsipeptides in bloom-forming cyanobacteria. *Angewandte Chemie International Edition*, **47** (40), 7756–7759.

[37] Sano, T. and Kaya, K. (1996) Oscillatorin, A chymotrypsin inhibitor from toxic *Oscillatoria agardhii*. *Tetrahedron Letters*, **37** (38), 6873–6876.

[38] Ojika, M., Nemoto, T., Nakamura, M. and Yamada, K. (1995) Dolastatin E, a new cyclic hexapeptide isolated from the sea hare *Dolabella auriculria*. *Tetrahedron Letters*, **36** (28), 5057–5058.

[39] Hambley, T.W., Hawkins, C.J., Lavin, M.F. *et al.* (1992) Cycloxazoline: a cytotoxic cyclic hexapeptide from the ascidian *Lissoclinum bistratum*. *Tetrahedron*, **48** (2), 341–348.

[40] Linington, R.G., Gonzàles, J., Ureña, L.D. *et al.* (2007) Venturamides A and B: antimalarial constituents of the Panamanian marine cyanobacterium *Oscillatoria* sp. *Journal of Natural Products*, **70** (3), 397–401.

[41] Admi, V., Afek, U. and Carmeli, S. (1996) Raocyclamide A and B; Novel cyclic hexapeptides isolated from the cyanobacterium *Oscillatoria raoi*. *Journal of Natural Products*, **59**, 396–399.

[42] Banker, R. and Carmeli, S. (1999) Inhibitors of serine proteases from a waterbloom of the cyanobacterium *Microcystis* sp. *Tetrahedron*, **55** (35), 10835–10844.

[43] Ishida, K., Nakagawa, H. and Murakami, M. (2000) Microcyclamide, a cytotoxic cyclic hexapeptide from the cyanobacterium *Microcystis aeruginosa*. *Journal of Natural Products*, **63** (9), 1315–1317.

[44] Raveh, A. and Carmeli, S. (2010) Aeruginazole A, a novel thiazole containing cycliopeptide from the cyanobacterium *Microcystis* sp. *Organic Chemistry Journal*, **12** (15), 3536–3539.

[45] Raveh, A., Moshe, S., Evron, Z. *et al.* (2010) Novel thiazole and oxazole containing cyclic hexapeptides from a waterbloom of the cyanobacterium *Microcystis aeruginosa*. *Tetrahedron*, **66**, 2705–2712.

[46] Zafrir-Ilan, E. and Carmeli, S. (2010) Two new microcyclamides from a water bloom of the cyanobacterium *Microcystis* sp. *Tetrahedron Letters*, **51** (50), 6602–6604.

[47] Portmann, C., Blom, J.F., Gademann, K. and Jüttner, F. (2008) Aerucyclamides A and B: isolation and synthesis of toxic ribosomal heterocyclic peptides from the cyanobacterium *Microcystis aeruginosa* PCC 7806. *Journal of Natural Products* **71**, 1193–1196.

[48] Ploutno, A. and Carmeli, S., (2002) Modified peptides from a water bloom of the cyanobacterium *Nostoc* sp. *Tetrahedron*, **58** (50), 9949–9957.

[49] Prinsep, M.R., Moore, R.E., Levine, I.A. *et al.* (1992) Westielamide, a bistratamide-related cyclic peptide from the blue-green alga *Westillopsis prolifica*. *Journal of Natural Products*, **55**, 140–142.

[50] Donia, M.S., Hathaway, B.J., Sudek, S. *et al.* (2006) Natural combinatorial peptide libraries in cyanobacterial symbionts of marine ascidians. *Nature Chemical Biology*, **2** (12), 729–735.

[51] Donia, M.S., Ravel, J. and Schmidt, E.W. (2008) A global assembly line for cyanobactins. *Nature Chemical Biology*, **4** (6), 341–343.

[52] Mitchell, S.S., Faulkner, D.J., Rubins, K. and Bushman, F.D. (2000) Dolastatin 3 and two novel cyclic peptides from a Palauan collection of *Lyngbya majascula*. *Journal of Natural Products*, **63** (2), 279–282.

[53] Sudek, S., Haygood, M.G., Youssef, D.T. and Schmidt, E.W. (2006) Structure of trichamide, a cyclic peptide from the bloom-forming cyanobacterium *Trichodesmium erythraeum*, predicted from the genome sequence. *Applied and Environmental Microbiology*, **72**, 4382–4387.

[54] Adiv, S., Ahronov-Nadborny, R. and Carmeli, S. (2012) New aeruginazoles, a group of thiazole-containing cyclic peptides from *Microcystis aeruginosa* blooms. *Tetrahedron*, **68** (5), 1376–1383.

[55] Ishida, K., Christiansen, G., Yoshida, W.Y. *et al.* (2007) Biosynthesis and structure of aeruginoside 126A and 126B, cyanobacterial peptide glycosides bearing a 2-carboxy-6-hydroxyoctahydroindole moiety. *Chemistry and Biology*, **14** (5), 565–576.

[56] Valls, N., Vallribera, M., Carmeli, S. and Bonjoch, J. (2003) Synthesis of both the putative and revised structure of aeruginosin EI461 bearing a new bicyclic α-amino acid. *Organic Chemistry Letters*, **5** (4), 447–450.

[57] Lifshits, M. and Carmeli, S. (2012) Metabolites of a *Microcystis aeruginosa* bloom material from Lake Kinneret, Israel. *Journal of Natural Products*, **75** (2), 209–219.

[58] Adiv, S. and Carmeli, S. (2013) Protease inhibitors from *Microcystis aeruginosa* bloom material collected from the Dalton Reservoir, Israel. *Journal of Natural Products*, **76** (12), 2307–2315.

[59] Fewer, D.P., Jokela. J., Paukku, E. *et al.* (2013) New structural variants of aeruginosin produced by the toxic bloom forming cyanobacterium *Nodularia spumigena. PLoS ONE*, **8** (9), e73618.

[60] Liu, L., Budnjo, A., Jokela, J. *et al.* (2015) Pseudoaeruginosins, nonribosomal peptides in *Nudularia* spumigena. *ACS Chemical Biology*, **10** (3), 725–733.

[61] Anas, A.R., Kisugi, T., Umezawa, T. *et al.* (2012) Thrombin inhibitors from the freshwater cyanobacterium *Anabaena compacta. Journal of Natural Products*, **75** (9), 1546–1552.

[62] Fujii, K., Sivonen, K., Adachi, K. *et al.* (1997) Comparative study of toxic and non-toxic cyanobacterial products: a novel glycoside, suomilide from non-toxic *Nudularia spumigena* HKVV. *Tetrahedron Letters*, **38** (31), 5529–5532.

[63] Ploutno, A. and Carmeli, S. (2005) Banyasin A and Banyasides A and B, three novel modified peptides from a water bloom of the cyanobacterium *Nostoc* sp. *Tetrahedron*, **61** (3), 575–583.

[64] Wiegand, C., Peuthert, A., Pflugmacher, S. and Carmeli, S. (2002) Effects of microcin SF608 and microcystin-LR, two cyanotobacterial compounds produced by *Microcystis* sp., on aquatic organisms. *Environmental Toxicology*, **17** (4), 400–406.

[65] Mazur-Marzec, H., Kaczkowska, M.J., Blaszczyk, A. *et al.* (2013) Diversity of peptides produced by *Nodularia spumigena* from various geographical regions. *Marine Drugs*, **11** (1), 1–19.

[66] Welker, M., Maršálek, B., Šejnohová, L. and von Döhren, H. (2006) Detection and identification of oligopeptides in *Microcystis* (Cyanobacteria) colonies: Toward an understanding of metabolic diversity. *Peptides*, **27** (9), 2090–2103.

[67] Rouhiainen, L., Jokela, J., Fewer, D.P. *et al.* (2010) Two alternative starter modules for the non-ribosomal biosynthesis of specific anabaenopeptin variants in *Anabaena* (Cyanobacteria). *Chem. Biol.*, **17** (3), 265–273.

[68] Gesner-Apter, S. and Carmeli. S. (2009) Protease inhibitors from a water bloom of the cyanobacterium *Microcystis aeruginosa. J. Nat. Prod.*, **72** (8), 1429–1436.

[69] Zi, J., Lantvit, D.D., Swanson, S.M. and Orjala, J. (2012) Lyngbyaureidamides A and B, two anabaenopeptins from the cultured freshwater cyanobacterium *Lyngbya* sp. (SAG 36.91). *Phytochemistry*, **74**, 173–177.

[70] Reshef, V. and Carmeli, S. (2002) Schizopeptin 791, a new anabaenopeptin-like cyclic peptide from the cyanobacterium *Schizothrix* sp. *Journal of Natural Products*, **65**, 1187–1189.

[71] Plaza, A., Keffer, J.L., Lloyd, J.R., Colin, P.L., Bewley, C.A. (2010) Paltolides A-C, anabaenopeptin-type peptides from the Palau sponge *Theonella swinhoei. Journal of Natural Products*, **73**, 485–488.

[72] Ishitsuka, M.O., Kusumi, T., Kakisawa, H. *et al.* (1990) Microviridin. A novel tricyclic depsipeptide from the toxic cyanobacterium *Microcystis viridis. Journal of American Chemical Society*, **112** (22), 8180–8182.

[73] Sedmak, B., Eleršek, T., Grach-Pogrebinsky, O. *et al.* (2008) Ecotoxicologically relevant cyclic peptides from cyanobacterial bloom (*Planktothrix rubescens*) - a threat to human and environmental health. *Radiology and Oncology*, **42** (2), 102–113.

[74] Sønstebø, J.H. and Rohrlack, T. (2011) Possible implication of chytrid parasitism for population subdivision in freshwater cyanobacteria of the genus *Planktothrix. Applied and Environmental Microbiology*, **77** (4), 1344–1351.

[75] Erhard, M., von Döhren, H. and Jungblut, P.R. (1999) Rapid identification of the new anabaenopeptin G from *Planktothrix agardhii* HUB 011 using matrix-assisted laser desorption/ionization time-of-flight mass spectrometry. *Rapid Communications in Mass Spectrometry*, **13** (5), 337–343.

[76] Incharoensakdi, A., Jantaro, S., Raksajit, W. *et al.* (2010) Polyamines in Cyanobacteria: Biosynthesis, transport and abiotic stress response, in *Current Research, Technology and Education Topics in Applied Microbiology and Microbial Biotechnology* (ed. A. Mendez-Vilas), Formatex, Spain, pp. 23–32.

[77] Raksajit, W., Mäenpää, P. and Incharoensakdi, A. (2006) Putrescine transport in a cyanobacterium *Synechocystis* sp PCC 6803. *Journal of Biochemistry and Molecular Biology*, **39** (4), 394–399.

[78] Chinn, R. and Barrett, S.E. (2000) Occurrence of amino acids in two drinking water sources, in *Natural organic matter and disinfection by-products characterization and control in drinking water* (Eds S.E. Barrett, S.W. Krasner and G.L. Amy), American Chemical Society, Washington, pp. 96–108.

[79] Chiu, C.A., Westerhoff, P. and Ghosh, A. (2012) GAC removal of organic nitrogen and other DBP precursors. *Journal of American Water Works Association*, **104** (7), 406–415.

[80] Dotson, A. and Westerhoff, P. (2009) Occurrence and removal of amino acids during drinking water treatment. *Journal of American Water Works Association*, **101** (9), 101–115.

[81] Grübel, A. (2013) *Avoidance of organoleptically contamination of drinking water – role of free amino acids during chlorination*. Dissertation. Technical University Dresden.

[82] Callejón, R.M., Troncoso, A.M. and Morales, M.L. (2010) Determination of amino acids in grape-derived products: A review. *Talanta*, **81** (4–5), 1143–1152.

[83] Schmidt, W. (2013) *Aminosäurebürtige Transformations- und Abbauprodukte im Roh- und Trinkwasser*, Studie, DVGW-TZW Karlsruhe, Außenstelle Dresden.

[84] Subbaraju, G.V., Galakoti, T., Patterson, G.M.L. *et al.* (1997) Three new cryptophycins from *Nostoc* sp. GSV 22460. *Journal of Natural Products*, **60** (3), 302–305.

[85] Smith, C.D., Zhang, X., Mooberry, S.L. *et al.* (1994) Cryptophycin: a new antimicrotubule agent active against drug-resistant cells. *Cancer Ressearch*, **54** (14), 3779–3784.

[86] Golakoti, T., Ogino, J., Heltzel, C.E. *et al.* (1995) Structure determination, conformational analysis, chemical stability studies and antitumor evaluation of cryptophycins. Isolation of 18 new analogs from *Nostoc* sp. strain GSV 224. *Journal of American Chemical Society*, **117** (49), 12030–12049.

[87] Bia, R., Verdier-Panard, P., Gangwar, S. *et al.* (2001) Dolastatin 11, a marine depsipeptide, arrests cells at cytokinesis and induces hyperpolymerization of purified actin. *Molecular Pharmacology*, **59** (3), 462–469.

[88] Harrigan, G.G., Yoshida, W.Y., Moore, R.E. *et al.* (1998) Isolation, structure determination, and biological-activity of dolastatin-12 and lyngbyastatin-I from *Lyngbya-Majuscula Schizothrix-Calcicola* cyanobacterial assemblages. *Journal of Natural Products*, **61** (10), 1221–1225.

[89] Bladt, T.T., Kalifa-Aviv, S., Larsen, T.O. and Carmeli, S. (2014) Micropeptins from *Microcystis* sp. collected in Kabul Reservoir, Israel. *Tetrahedron*, **70** (4), 936–943.

[90] Welker, M., Brunke, M., Preussel, K. *et al.* (2004) Diversity and distribution of *Microcystis* (Cyanobacteria) oligopeptide chemotypes from natural communities studied by single-colony mass spectrometry. *Microbiology*, **150** (Pt6), 1785–1796.

[91] Gunasekera, S.P. Miller, M.W., Kwan, J.C., Luesch, H. and Paul, V.J. (2010) Molassamide, a depsipeptide serine protease inhibitor from the marine cyanobacterium *Dichothrix utahensis. The American Chemical Society and American Society of Pharmacognosy*, **73** (3), 459–462.

[92] Grach-Pogrebinsky, O., Sedmak, B. and Carmeli, S. (2003) Protease inhibitors from a Slovenian Lake Bled toxic waterbloom of the cyanobacterium *Planktothrix rubescens*. *Tetrahedron*, **59** (42), 8329–8336.

[93] Sedmak, B. and Eleršek, T. (2006) Microcystins induce morphological and physiological changes in selected representative phytoplanktons. *Microbial Ecology*, **50** (2), 298–305.

[94] Engene, N., Rottacker, E.C., Kaštovský, J. *et al.* (2012) *Moorea producens* gen. nov., sp. nov. and *Moorea bouillonii* comb. nov., tropical marine cyanobacteria rich in bioactive secondary metabolites. *International Journal of Systematic and Evolutionary Microbiology*, **62** (5), 1171–1178.

[95] Jabba, S.V., Prakash, A., Dravid S.M. *et al.* (2010) Antillatoxin, a novel lipopeptide, enhances neurite outgrowth in immature cerebrocortical neurons through activation of voltage-gated sodium channels. *Journal of Pharmacology and Experimental Therapeutics*, **332** (3), 698–709.

[96] Bannage, M.E., Burke, A.J., Davies, S.G. and Goodwin, C.J. (1994) Asymmetric synthesis of [2S,3R]-3-amino-2-hydroxydecanoic acid – the unknown amino acid component of microginin. *Tetrahedron Asymmetry*, **5** (2), 203–206.

[97] Lifshits, M., Zafrir-Ilan, E., Raveh, A. and Carmeli, S. (2011) Protease inhibitors from three fishpond water blooms of the cyanobacterium *Microcystis* sp. *Tetrahedron*, **67** (22), 4017–4024.

[98] Kramer, D. (2010) Microginin producing proteins and nucleic acids encoding a microginin gene cluster as well as methods for creating novel microginins. US Patent 20110034680 A1, filed Oct. 22, 2010 and issued Feb.10, 2011.

[99] Philmus, B., Christiansen, G., Yoshida, W.Y. and Hemscheidt, T.K. (2008) Post-translational modification in microviridin biosynthesis. *ChemBioChem*, **9** (18), 3066–3073.

[100] Reshef, V. and Carmeli, S. (2006) New microviridins from a water bloom of the cyanobacterium *Microcystis aeruginosa*. *Tetrahedron*, **62** (31), 7361–7369.
[101] Rohrlack, T., Christoffersen, K., Hansen, P.E. *et al.* (2003) Isolation, characterization, and quantitative analysis of microviridin J, a new *Microcystis* metabolite toxic to *Daphnia*. *Journal of Chemical Ecology*, **29** (8), 1757–1770.
[102] Shapiro, S.D. (2002) Proteinases in chronic obstructive pulmonary disease. *Biochemical Society Transactions*, **30** (2), 98–102.
[103] Rohrlack, T., Christoffersen, K., Kaebernick, M. and Neilan, B.A. (2004) Cyanobacterial protease inhibitor microviridin J causes a lethal molting disruption in *Daphnia pulicaria*. *Applied and Environmental Microbiology*, **70** (8), 5047–5050.

20

Taste and Odour Compounds Produced by Cyanobacteria

Triantafyllos Kaloudis[1], *Theodoros M. Triantis*[2], *and Anastasia Hiskia*[2]

[1] *Water Quality Control, Athens Water Supply and Sewerage Company - EYDAP SA, Athens, Greece*
[2] *Institute of Nanoscience & Nanotechnology, National Center for Scientific Research "DEMOKRITOS," Athens, Greece*

20.1 Cyanobacterial Taste and Odour Compounds in Water Resources

During growth and decay of blooms cyanobacteria synthesise and release a large number of metabolites that are harmful to humans, animals, and ecosystems. Those "noxious algal metabolites" (NAM) generally belong to two groups: cyanotoxins and taste and odour (T&O) compounds [1]. NAM have serious negative impacts, for public health, drinking water supplies, aquaculture, and tourism. Although T&O compounds are generally nontoxic and could be regarded as second priority compared with cyanotoxins, their occurrence in water can have big negative socioeconomic impacts but can also have a diagnostic value for water quality problems [2].

A large number of T&O compounds that fall into several chemical groups are produced by algae (cyanobacteria or eukaryotic taxa). Those include terpenoids such as geosmin (GSM), 2-methylisoborneol (MIB), α- and β-ionones, β-cyclocitral, pyrazines such as 2,6-dimethyl pyrazine, 2-isobutyl-3-methoxy pyrazine, sulphurous compounds such as dimethyl trisulfide, aldehydes, and amines [2].

By far, the most frequent and intense episodes of T&O in source and water supplies are related to GSM and MIB (Fig. 20.1), which give strong "earthy" and "musty" odours, respectively. Earthy/musty off-odours are the second frequent cause of consumer complaints, after "chlorine smell" [3, 4]. GSM (4*S*,4a*S*,8a*R*-4,8a-dimethyl-1,2,3,4,5,6,7,8-octahydronaphthalen-4a-ol), whose name in Greek means "earth odour," is a bicyclic sesquiterpenic compound that presents an extremely intense muddy/earthy smell with an odour threshold of $4\,\mathrm{ng\,L^{-1}}$ in water [5].

Handbook of Cyanobacterial Monitoring and Cyanotoxin Analysis, First Edition.
Edited by Jussi Meriluoto, Lisa Spoof and Geoffrey A. Codd.

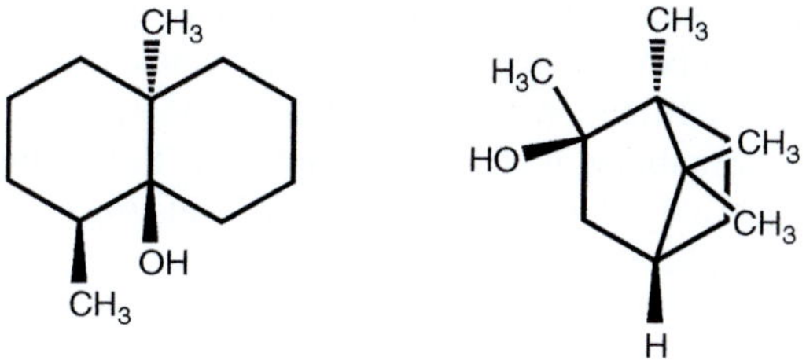

Figure 20.1 Chemical structures of geosmin (left) and 2-methylisoborneol (right)

MIB (1,6,7,7-tetramethylbicyclo[2.2.1]heptan-6-ol) is a bicyclic terpenoid with a strong musty odour that can be detected by human nose at concentrations above 6 ng L^{-1} [5]. GSM and MIB are produced by actinomycetes as well; however, the main source of GSM and MIB in water seems to be cyanobacterial metabolism [2], while the importance of the role of actinomycetes in off-odour problems in water is still under investigation [6].

GSM and MIB can serve as diagnostic signals but are not indicative of cyanotoxin-producing cyanobacteria [7]. Other T&O compounds, such as dimethyl sulphides (septic odour) or β-cyclocitral (tobacco/moldy odour), can be better indicators as noxious cyanobacteria are major sources of those compounds [8].

20.2 Analytical Methods for Taste and Odour Compounds

Detection and identification of T&O compounds produced by cyanobacteria can be carried out by sensory evaluation, instrumental chemical analysis techniques, or a combination [9]. Among sensory evaluation methods, flavour profile analysis (FPA) that was originally developed for testing of drinking water [10] is the most useful for forensic investigations of T&O incidents. FPA determines both the odour/flavour characteristics of the sample and the intensities of those characteristics. Results of FPA studies in water have enabled the development of the water "taste and odour wheel (TOW)" that is divided in circular sectors to present the primary taste and odour categories and the relative descriptions given by trained panels. Additionally, TOW links those descriptions with chemical compounds that have been confirmed as causative agents for the characteristic T&O. When FPA is used in combination with TOW, panelists can provide further guidance to chemical analysis laboratories on which compounds or groups of compounds should be targeted. TOW contains sectors (e.g. earthy/musty/mouldy, grassy/hay/woody, decaying/septic, fishy, etc.) where cyanobacterial metabolites are referred to as causative agents.

Threshold odour number (TON) and threshold flavour number (TFN) are sensory methods used widely by the drinking water industry to assess consumers' perceptions of water. In TON and TFN, the water samples are successively diluted with pure water and compared with a reference water. TON and TFN numbers represent the highest dilutions where odour and flavour are still perceptible. The main disadvantage of those methods is that they do not provide descriptions of the perceived odours/flavours and thus they cannot be used to determine possible causes. European Standard EN 1622:2006 describes procedures for the determination of TON and TFN in water [11], while Standard Methods for the Examination of Water and Wastewater [12] include sensory methods for determination of TON, flavour threshold test (FTT), flavour rating assessment (FRA), and flavour profile analysis (FPA).

Considering chemical analysis of T&O, gas chromatography (GC) is the method of choice since most of T&O compounds are volatile or semivolatile compounds. GC is usually coupled to mass spectrometry (MS) so that identification of compounds is possible in targeted or nontargeted analysis. GC-MS techniques, combined with suitable methods of sampling and sample preconcentration, can provide high selectivity, detection at very low levels ($ng\,L^{-1}$), and confirmation of the identity of compounds based on structural data.

Solid-phase microextraction (SPME) is a simple and efficient sample preparation method that can be used in combination with GC-MS for analysis of cyanobacteria T&O compounds. The concept of SPME technique is based on absorption of targeted analytes on a fused-silica fiber coated with a polymeric stationary phase, followed by thermal desorption of analytes directly into a GC injector [13]. SPME can be used in headspace mode (HS-SPME), where the fibre is exposed to the headspace of the sample. HS-SPME has many advantages in the analysis of volatile and semivolatile compounds as it is simple, it achieves high enrichment of analytes and generally less interferences. SPME techniques have been widely used in environmental analysis [14, 15]. SPME can be fully automated, with use of advanced autosamplers, making the technique suitable for routine, high-throughput monitoring. A number of methods for the determination of cyanobacterial T&O compounds (mostly geosmin and MIB) have been reported in the literature [e.g. 16–19] and an SPME procedure for geosmin and MIB is included in Standard Methods for the Examination of Water and Wastewater [20]. SOP 27 describes a procedure for the determination of geosmin and MIB in water by HS-SPME-GC/MS with LODs of $1\,ng\,L^{-1}$. Further improvements in sensitivity can be achieved with combination of SPME with MS/MS techniques. For example, Machado et al. (2011) reported a method for simultaneous analysis of personal-care products and earthy-musty compounds using an SPME-GC-MS/MS (ion trap) technique, achieving subnanogram-per-litre levels of detection for GSM [21].

Solid phase extraction (SPE) can be also used for sample enrichment in T&O analysis of water. Wright et al. (2014) have reported an SPE-GC-MS/MS (ion trap) method for analysis of T&O compounds including GSM, MIB, 2-isobutyl-3-methoxypyrazine (IBM) and 2-isopropyl-3-methoxypyrazine (IPM), with a 2000-enrichment factor, achieving LODs of less than $6\,ng\,L^{-1}$. However, SPE has limitations when low boiling-point volatile compounds (e.g. dimethyl sulphide) are included in the scope of analysis [22].

Purge and trap (P&T) is another solventless technique that can be used for sample extraction and preconcentration of T&O compounds. Lloyd et al. (1998) compared SPME with P&T for the analysis of GSM and MIB and obtained similar results in terms of reproducibility and sensitivity [23]. A limitation of P&T technique in the past was the unavailability of automated samplers that could handle multiple samples. Nowadays, P&T autosamplers are available and the technique is capable of high-throughput analysis. For example, Deng et al. (2011) reported a method for simultaneous determination of eight T&O compounds in water using automated P&T coupled to GC-MS, with very low LODs (e.g. $0.08\,ng\,L^{-1}$ for geosmin) [24]. Salemi et al. (2006) also reported automated P&T–GC-MS procedures for determination of 2-isopropyl-3-methoxypyrazine, 2-isobutyl-3-methoxypyrazine, MIB, and GSM in water at low $ng\,L^{-1}$ levels [25].

Closed-loop stripping analysis (CLSA) developed by Grob and Zurcer (1976) [26] has been widely used for extraction of T&O compounds from water, especially where extremely low detection limits are required [27, 28]. CLSA is based on the continuous purging of a water sample with an inert gas in a closed-loop recirculating system, where compounds are stripped in a small amount of an adsorbent (e.g. charcoal). Compounds are then extracted with a small volume of a solvent (e.g. 20–50 μL) and the extract is analysed, usually by GC-MS. Hayes and Burch (1988) have used CLSA to identify odourous algal compounds in water supply reservoirs in Australia, where they detected GSM, MIB, sesquiterpenes, β-cyclocitral, 3-methyl-1-butanol, 6-methyl-5-hepten-2-one, and the compound *trans*-2,*cis*-6-nonadienal, which was responsible for characteristic cucumber odour [29]. Malleret et al. (2001) developed a modified CLSA technique combined with large-volume GC-MS injection to analyse eight target T&O compounds with sensitivity of analysis at the pg L^{-1} level, which is equal to or better than olfactory sensitivity [28].

Stir-bar sorptive extraction (SBSE), introduced in 1999, is a new method for extraction of organic compounds from water samples [30]. SBSE is based on sorptive extraction of compounds into a polymer coating on a magnetic stirring rod and subsequent thermal desorption into a GC injector system that is specially designed for SBSE. Benanou et al. (2004) have used SBSE combined with GC-MS to determine off-flavour compounds in water, among others GSM and MIB with LODs 0.5 ng/L and 1 ng/L respectively [31]. Ochiai et al. (2013) developed a multiple-SBSE technique were two stir-bars coated with PDMS and ethyleneglycol-modified silicone were used simultaneously in extraction and desorption, in combination with GC-MS, in order to capture compounds with a wide range of polarities [32]. Using the SBSE principle, Tondelier et al. (2008) developed an "on-tap" passive sampling device equipped with seven SBSE stir bars coated with PDMS. The sampler, in combination with GC-MS analysis, was able to detect odorous compounds at subnanogram-per-litre levels [33].

GC-olfactometry (GC-O) (also called sensory chromatography) is a hybrid technique that uses human assessors to detect and evaluate compounds eluting from a GC column after separation [34]. GC-O can be used in combination with MS detection, resulting in simultaneous chemical and sensory analysis of a sample. Configurations of GC-O/MS devices for solving taste and odour episodes in water have been described in literature (e.g. [35]). GC-O and GC-O/MS methods have been successfully used in identifying the causative compounds of off-odour incidents in water (e.g. [36, 37]).

References

[1] Watson, S.B., Ridal, J. and Boyer, G.L. (2008) Taste and odour and cyanobacterial toxins: impairment, prediction, and management in the Great Lakes. *Canadian Journal of Fisheries and Aquatic Sciences*, **65** (8), 1779–1796.

[2] Watson, S.B. (2004) Aquatic taste and odor: a primary signal of drinking-water integrity. *Journal of Toxicology and Environmental Health, Part A: Current Issues*, **67** (20–22), 1779–1795.

[3] McGuire, M.J. (1995) Off-flavor as the consumer's measure of drinking water safety. *Water Science and Technology*, **31** (11), 1–8.

[4] Suffet, I.H., Corado, A., Chou, D., McGuire, M.J.M. and Butterworth, S. (1996) AWWA taste and odor survey. *Journal of the American Water Works Association*, **88** (4), 168–180.

[5] Young, W.F., Horth, H., Crane, R., Ogden, T. and Arnott, M. (1996) Taste and odor threshold concentrations of potential potable water contaminants. *Water Research*, **30** (2), 331–340.

[6] Zeitlin, B. and Watson, S.B. (2006) Actinomycetes in relation to taste and odour in drinking water: myths, tenets and truths. *Water Research*, **40** (9), 1741–1753.

[7] Watson, S.B. (2003) Cyanobacterial and eukaryotic algal odour compounds: signals or by-products? A review of their biological activity. *Phycologia*, **42** (4), 332–350.

[8] Jüttner, F. (1984) Dynamics of the volatile organic substances associated with cyanobacteria and algae in a eutrophic shallow lake. *Applied and Environmental Microbiology*, **47** (4), 814–820.

[9] Suffet, I.H., Schweitzer, L. and Khiari, D. (2004) Olfactory and chemical analysis of taste and odor episodes in drinking water supplies. *Reviews in Environmental Science & Biotechnology*, **3** (1), 3–13.

[10] Krasner, S.W., McGuire, M.J. and Ferguson, V.B. (1985) Tastes and odors: the Flavor Profile method. *Journal of the American Water Works Association*, **77** (3), 34–39.

[11] EN 1622:2006. (2006) *Water quality. Determination of the threshold odour number (TON) and threshold flavour number (TFN)*, CEN European Committee of Standardization.

[12] APHA, AWWA, WEF (2012) *Standard Methods for the Examination of Water and Wastewater*, 22nd ed. APHA, Washington, D.C., USA.

[13] Arthur, C. and Pawliszyn, J. (1990) Solid phase microextraction with thermal desorption using fused silica optical fibers. *Analytical Chemistry*, **62** (19), 2145–2148.

[14] Vas, G. and Vekey, K. (2004) Solid-phase microextraction: a powerful sample preparation tool prior to mass spectrometric analysis. *Journal of Mass Spectrometry*, **39** (3), 233–254.

[15] Lambropoulou, D., Konstantinou, I. and Albanis, T. (2007) Recent developments in headspace microextraction techniques for the analysis of environmental contaminants in different matrices. *Journal of Chromatography A*, **1152** (1–2), 70–96.

[16] Lloyd, S.W., Lea, J.M., Zimba, P.V. and Grimm, C.C. (1998) Rapid analysis of geosmin and 2-methylisoborneol in water using solid phase micro extraction procedures. *Water Research*, **32** (7), 2140–2146.

[17] Zimmerman, L.R., Ziegler, A.C. and Thurman, E.M. (2002) *Method of Analysis and Quality-Assurance Practices by U.S. Geological Survey Organic Geochemistry Research Group—Determination of Geosmin and Methylisoborneol in Water Using Solid-Phase Microextraction and Gas Chromatography/Mass Spectrometry*. Open-Fie Report 02-337, available from http://ks.water.usgs.gov (accessed 30 April 2015).

[18] Saito, K., Okamura, K. and Kataoka, H. (2008) Determination of musty odorants, 2-methylisoborneol and geosmin, in environmental water by headspace solid-phase microextraction and gas chromatography-mass spectrometry. *Journal of Chromatography A*, **1186** (1–2), 434–437.

[19] Yuan, B., Li, F., Xua, D., and Fub, M.L. (2013) Comparison of two methods for the determination of geosmin and 2-methylisoborneol in algae samples by stable isotope dilution assay through purge-and-trap or headspace solid-phase microextraction combined with GC/MS. *Analytical Methods*, **5** (7), 1739–1746.

[20] Greenberg, A., *et al.* (2005) Method 6040 D, in *Standard Methods for the Examination of Water and Wastewater*, 21st edn, American Public Health Association (APHA), American Water Works Association (AWWA) & Water Environment Federation (WEF), USA.

[21] Machado, S., Gonalves, C., Cunha, E., Guimarães, A. and Alpendurada, M.F. (2011) New developments in the analysis of fragrances and earthy–musty compounds in water by solid-phase microextraction (metal alloy fibre) coupled with gas chromatography–(tandem) mass spectrometry. *Talanta*, **84** (4), 1133–1140.

[22] Wright, E., Daurie, H. and Gagnon, G.A. (2014) Development and validation of an SPE GC-MS/MS taste and odour method for analysis in surface water. *International Journal of Environmental Analytical Chemistry*, **94** (13), 1302–1316.

[23] Lloyd, S.W., Lea, J.M., Zimba, P.V. and Grimm. C.C. (1998) Rapid analysis of geosmin and 2-methylisoborneol in water using solid phase microextraction procedures. *Water Research*, **32** (7), 2140–2146.

[24] Deng, X., Liang, G., Chen, J., Qi, M. and Xie, P. (2011) Simultaneous determination of eight common odors in natural water body using automatic purge and trap coupled to gas chromatography with mass spectrometry. *Journal of Chromatography A*, **1218** (24), 3791–3798.

[25] Salemi, A., Lacorte, S., Bagheri, H. and Barcelo, D. (2006) Automated trace determination of earthy-musty odorous compounds in water samples by on-line purge-and-trap–gas chromatography–mass spectrometry. *Journal of Chromatography A*, **1136** (2), 170–175.

[26] Grob, K. and Zurcher, F. (1976) Stripping of trace organic substances from water – Equipment and procedure. *Journal of Chromatography A*, **117** (2), 285–294.

[27] Izaguirre, G., Hwang, C.J., Krasner, S.W. and McGuire, M.J. (1982) Geosmin and 2-methylisoborneol from cyanobacteria in three water supply systems. *Applied and Environmental Microbiology*, **43** (3), 708–714.

[28] Malleret, L., Bruchet, A. and Hennion, M.-C. (2001) Picogram determination of "earthy-musty" odorous compounds in water using modified closed loop stripping analysis and large volume injection GC/MS. *Analytical Chemistry*, **73** (7), 1485–1490.

[29] Hayes, K. and Burch, M. (1989) Odourous compounds associated with algal blooms in South Australian water. *Water Research*, **23** (1), 115–121.

[30] David, F. and Sandra, P. (2007) Stir bar sorptive extraction for trace analysis. *Journal of Chromatography A*, **1152** (1–2), 54–69.

[31] Benanou, D., Acobas, F. and de Roubin, M.R. (2004) Optimization of stir bar sorptive extraction applied to the determination of odorous compounds in drinking water. *Water Science and Technology*, **49** (9), 161–170.

[32] Ochiai, N., Sasamoto, K., Ieda, T., David, F. and Sandra, P. (2013) Multi-stir bar sorptive extraction for analysis of odor compounds in aqueous samples. *Journal of Chromatography A*, **1315**, 70–79.
[33] Tondelier, C., Benanou, D., Thouvenot, T. and Genin, A. (2008) Novel on-tap passive sampling for odorous compounds in drinking water. *Water Science and Technology: Water Supply*, **8** (2), 197–205.
[34] Delahunty, C.M., Eyres, G. and Dufour, J.-P. (2006) Gas chromatography-olfactometry. *Journal of Separation Science*, **29** (14), 2107–2125.
[35] Hochereau, C. and Bruchet, A. (2004) Design and application of a GC-SNIFF/MS system for solving taste and odour episodes in drinking water. *Water Science and Technology*, **49** (9), 81–87.
[36] Khiari, D., Suffet, I.H. and Barrett, S.E. (1995) The determination of compounds causing fishy/swampy odors in drinking water supplies. *Water Science and Technology*, **31** (11), 105–112.
[37] Bruchet, A., Hochereau, C. and Campos, C. (2007) An acute taste and odour episode solved by olfactory GC-MS. *Water Science and Technology*, **55** (5), 223–230.

Section V

Screening and Trace Analysis of Cyanotoxins

21

Determination of Cyanotoxins by High-Performance Liquid Chromatography with Photodiode Array

Anastasia Hiskia[1], *Lisa Spoof*[2], *Triantafyllos Kaloudis*[3], *and Jussi Meriluoto*[2,4]

[1]*Institute of Nanoscience & Nanotechnology, National Center for Scientific Research "DEMOKRITOS," Athens, Greece*

[2]*Biochemistry, Faculty of Science and Engineering, Åbo Akademi University, Turku, Finland*

[3]*Water Quality Control, Athens Water Supply and Sewerage Company - EYDAP SA, Athens, Greece*

[4]*Laboratory for Paleoenvironmental Reconstruction, Faculty of Sciences, University of Novi Sad, Serbia*

21.1 Introduction: Application of High-Performance Liquid Chromatography for Different Classes of Cyanotoxins

High-performance liquid chromatography (HPLC) is a separation technique where a dissolved mixture of analytes (sample) is separated into individual components while passing through a column filled with adsorbent (stationary phase). The transport of analytes through the column is facilitated by a flow of liquid (mobile phase). HPLC uses polarity, charge, and, more rarely, size differences between the analytes to achieve separation. Analytes that have a UV chromophore are detectable by UV absorbance detection (e.g. microcystins, nodularins, cylindrospermopsin, anatoxin-a). Ionisable compounds (practically all cyanotoxins) can be detected by mass spectrometric detection. Some cyanotoxins are amenable to fluorimetric detection after derivatization: this applies to the neurotoxic anatoxin-a, saxitoxins, and β-*N*-methylamino-L-alanine (BMAA).

Handbook of Cyanobacterial Monitoring and Cyanotoxin Analysis, First Edition.
Edited by Jussi Meriluoto, Lisa Spoof and Geoffrey A. Codd.

HPLC is an efficient analytical tool for many cyanotoxins, and it can also be used for preparative purposes. Most HPLC separations of cyanotoxins use reversed-phase (RP) chromatography, which has been the usual mode of separation for microcystins, nodularins, cylindrospermopsin, derivatised anatoxin-a, and derivatised BMAA. Saxitoxins can be separated by RP HPLC if ion-pairing reagents are used to increase their retention. Some very hydrophilic cyanotoxins such as underivatised anatoxin-a, cylindrospermopsin, saxitoxins, and BMAA can benefit from being analysed by the HILIC technique (i.e. hydrophilic interaction liquid chromatography).

21.2 HPLC of Microcystins and Nodularins

The chromatography of microcystins and nodularins has been reviewed extensively [1]. Separations of cyanobacterial peptide toxins necessitate high resolution and good selectivity as the number of toxin analogues is high and coelution is likely. Generally speaking, acidic mobile phases are able to resolve more microcystins and nodularins than are mobile phases at neutral pH.

RP HPLC in acidic conditions has become the standard technique in analyses of microcystins and nodularins [2–4]. Mobile phases using acetonitrile as the organic phase give superior resolution and selectivity for most separations of peptides [5]. The elution order of some common microcystins in acidic water/acetonitrile mobile phases and RP chromatography is 3-demethylmicrocystin-RR, microcystin-RR, nodularin-R, microcystin-YR, -LR, -FR, -LA, -LY, -LW, and -LF [1, 2]. The different polarities of microcystin and nodularin variants require the use of gradient elution instead of isocratic conditions. In RP chromatography of microcystins and nodularins the gradients of acidic water/acetonitrile should span about 25–70% acetonitrile.

Trifluoroacetic acid (TFA) is often used to acidify the pH of the mobile phase. The typical percentages of TFA in HPLC solvents are 0.05% or 0.1%. The acidic pH of the mobile phase protonates the carboxyl groups in microcystins and nodularins, which results in stronger retention in RP chromatography. TFA also acts as an ion-pairing reagent with residual silanol groups on the silica surface and improves the peak shape of analytes carrying basic groups. In LC-MS work the use of formic acid (FA) is preferred because TFA causes ion suppression and interferes with spray formation. Usually 0.1% FA is added to the mobile phase.

Preparative chromatography of microcystins and nodularins often involves a neutral mobile phase with aqueous ammonium acetate ($1\,g\,L^{-1}$) and acetonitrile as the first HPLC step. Further purification can be accomplished by applying other solvent systems of different selectivity, such as mobile phases based on methanol with different buffers and pH values [6] or aqueous TFA-acetonitrile. By using a mobile phase of methanol-sodium sulphate in RP chromatography [6(*Z*)-Adda]microcystins could be resolved from the [6(*E*)-Adda]microcystins [7].

The traditional analytical HPLC-DAD protocols have used long RP columns (typically 25 cm length with 5 μm particles or 15 cm length with 3 μm particles) in order to achieve sufficient resolution of microcystin analogues. The gradients have typically lasted up to 45 min followed by re-equilibration of the column. The sample throughput in these conditions is only one sample per hour. While the use of established slow methods may be justified in some cases, new methods should use more modern technology.

Recent technological developments in HPLC enable fast analysis times without sacrificing chromatographic efficiency. A reasonable compromise that guarantees high sample throughput on traditional HPLC instrumentation is to employ columns with sub–3 μm particle size [8, 9]; an

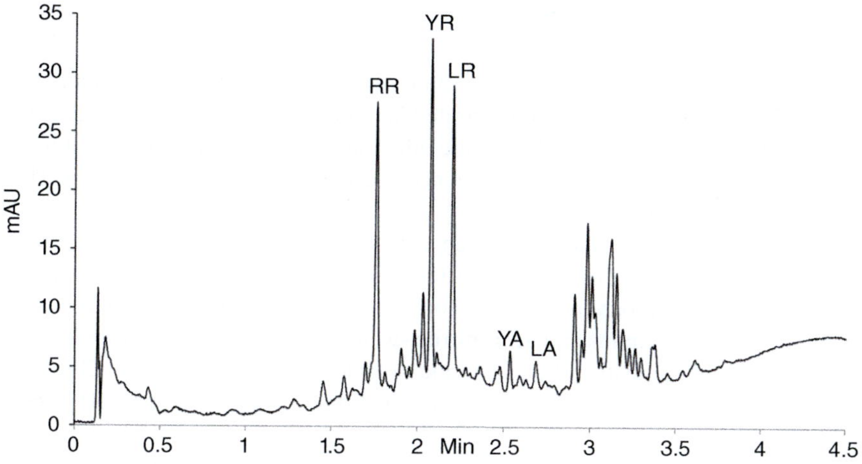

Figure 21.1 Trace of cyanobacterial extract prepared from bloom material collected from Sulejow reservoir, Poland. Five microcystins (-RR, -YR, -LR, -YA, and -LA) were identified. Column Supelco Ascentis Express C18, 2.7 µm fused core particles, 50 mm × 2.1 m. Mobile phase A: 0.05% TFA in water, B: 0.05% TFA in acetonitrile, gradient 20–70% B over 3 min, flow rate 1.1 mL min^{-1}. Agilent 1100 series HPLC instrumentation with quaternary pump, UV absorbance detection at 238 nm. The experiment has been performed under experimental conditions fully described [9]

illustrative chromatogram is presented in Fig. 21.1. Another possibility is to use monolithic silica technology [10]. Monolithic HPLC columns have a bimodal pore structure; macroporosity ("flow channels") combined with the large surface area composed by the mesopores allows very high flow rates without sacrificing resolution [11]. The highest flow rates are, however, not compatible with LC-MS and require reduction of flow, such as by splitting.

Ultra-performance liquid chromatography (UPLC, Waters trademark) and similar techniques represent another leap forward in cyanobacterial toxin analyses. Very fast analyses can be achieved by using novel sub-2 µm stationary phases and instrumentation that can withstand the ultra-high pressure, up to 1000 bar, caused by such small particles [12]. With the development of UPLC or ultra-high-performance liquid chromatography (UHPLC, a generic term) the sensitivity, resolution and speed of separations have been much improved compared with traditional HPLC. UPLC and corresponding UHPLC techniques have the potential to become the preferred choice in both LC-DAD and LC-ESI-MS/MS analysis of microcystins and nodularins [9, 13–19].

Fairly sensitive UV detection of the toxins (nanogram levels on column) is possible because microcystins and nodularins possess a conjugated diene in the Adda residue, a strong chromophore absorbing at 238 nm. Additional absorbance is given, for example, by the α,β-unsaturated carbonyl group in Mdha/Mdhb residue. The usual microcystin and nodularin spectra, those of microcystin-LR and nodularin-R, have a local maximum at 238 nm. The spectra of tryptophan-containing microcystins have an additional maximum at 222 nm, and the spectra of tyrosine-containing microcystins are clearly flatter in the region of 230–240 nm. Most microcystins and nodularins can be recognised as microcystins or nodularins by the typical UV spectrum even when there is no standard with a retention time match. Mass spectrometric detection will be covered in Chapter 23.

21.3 HPLC of Anatoxins

Anatoxin-a and anatoxin-a(*S*) are the most important neurotoxic compounds from an environmental health standpoint [20]. Due to an α,β-unsaturated keto group, anatoxin-a can be detected by UV with absorption maximum at 227 nm [21]. On the other hand, anatoxin-a(S) does not have a useful chromophore in its structure and as a consequence it cannot be determined by the HPLC-PDA technique.

The first HPLC-UV methods that have been developed for the determination of anatoxin-a without derivatization used a C_{18} column and buffers with methanol or acetonitrile as mobile phase [21, 22]. However, these methods have very poor sensitivity. More recently, anatoxin-a analysis has been reported with C_{18} columns and acetonitrile–water or methanol–water (acidified with TFA or HCOOH) as mobile phase using either isocratic or gradient elution [23–26]. Chromatographic separation of anatoxin-a with hydrophilic interaction liquid chromatography (HILIC) has also been reported [27].

A serious concern in the analysis of anatoxin-a is the possible coelution with the naturally occurred amino acid phenylalanine, and misidentification of anatoxin-a has been reported [28]. Efficient separation of the two compounds can be achieved using a C_{18} column and methanol–water with 0.1% HCOOH as mobile phase with isocratic elution [26] (see SOP 18).

21.4 HPLC of Cylindrospermopsin

Cylindrospermopsin can be analysed by RP HPLC and detected by PDA/UV detector (UV absorbance at 262 nm [29]) or MS [30]. Chromatographic analysis of cylindrospermopsin can be performed with a C_{18} column and methanol–water as mobile phase using either isocratic or gradient elution [31–33]. Acidified mobile phases (e.g. 0.1% TFA) have also been applied [32, 34]. HPLC-PDA determination of cylindrospermopsin in environmental samples is affected by the matrix background, and occasionally matrix constituents interfere with cylindrospermopsin determination [32]. For this reason, a clean-up step of samples prior to HPLC-PDA analysis is often necessary [34–36]. Trace quantities of cylindrospermopsin are best analysed by LC-MS/MS, which provides better sensitivity and selectivity.

HILIC coupled with electrospray mass spectrometry (HILIC-MS) has also been used for the determination of several cyanotoxins including cylindrospermopsin. For example, an isocratic method uses an HILIC TSK gel-amide column and buffered acetonitrile–water elution [27].

The chromatographic separation of cylindrospermopsin and its analogue 7-deoxy-cylindrospermopsin can be carried out with both RP chromatography [37, 38] and HILIC [27]. Separation is performed using acidified acetonitrile–water [37] or methanol–water [38] with gradient elution. When detection is with MS, FA is preferred for the acidification of the mobile phase.

21.5 Advantages and Disadvantages of HPLC-PDA

Table 21.1 summarizes the advantages and disadvantages of the HPLC-PDA technique compared to other available methods for determination of cyanotoxins.

Table 21.1 Advantages and Disadvantages of HPLC-PDA

Advantages	Disadvantages
Reasonable capital and running costs (compared with LC-MS)	Depends on availability of standard compounds
Fairly good sensitivity (nanogram levels on column with analytes containing a strong chromophore)	Risk for misidentification due to co-eluting substances
Continuous technical progress leading to higher sensitivity, resolution, and selectivity	Does not measure toxicity
Possible to target different classes of analytes in the same run	Does not provide definite identification (compared with LC-MS)
Can be automated	
No ethical issues (comparing to in vivo bioassays)	

References

[1] Meriluoto, J. (1997) Chromatography of microcystins. *Analytica Chimica Acta*, **352** (1–3), 277–298.

[2] Lawton, L.A., Edwards, C. and Codd, G.A. (1994) Extraction and high-performance liquid chromatographic method for the determination of microcystins in raw and treated waters. *Analyst*, **119** (7), 1525–1530.

[3] ISO 20179:2005. (2005) Water quality - Determination of microcystins - Method using solid phase extraction (SPE) and high performance liquid chromatography (HPLC) with ultraviolet (UV) detection. International Standards Organization, Technical Committee ISO TC147/SC2.

[4] Meriluoto, J. and Spoof, L. (2005) Analysis of Microcystins by high-performance liquid chromatography with photodiode-array detection, *in TOXIC–Cyanobacterial Monitoring and Cyanotoxin Analysi*, (eds J. Meriluoto and G.A. Codd), Åbo Akademi University Press, Åbo, pp. 77–84.

[5] Guo, D., Mant, C.T., Taneja, A.K. *et al.* (1986) Prediction of peptide retention times in reversed-phase high-performance liquid chromatography. *I. Determination of retention coefficients of amino acid residues of model synthetic peptides. Journal of Chromatography A*, **359** (1), 499–518.

[6] Harada, K., Ogawa, K., Matsuura, K. *et al.* (1991) Isolation of two toxic heptapeptide microcystins from an axenic strain of *Microcystis aeruginosa*, K-139. *Toxicon*, **29** (4–5), 479–489.

[7] Harada, K., Matsuura, K., Suzuki, M. *et al.* (1990) Isolation and characterization of the minor components associated with microcystins LR and RR in the cyanobacterium (blue-green algae). *Toxicon*, **28** (1), 55–64.

[8] Neffling, M.R., Spoof, L. and Meriluoto, J. (2009) Rapid LC-MS detection of cyanobacterial hepatotoxins microcystins and nodularins–Comparison of columns. *Analytica Chimica Acta*, **653** (2), 234–241.

[9] Spoof, L., Neffling, M.-R. and Meriluoto, J. (2010) Fast separation of microcystins and nodularins narrow-bore reversed-phase columns coupled to a conventional HPLC system. *Toxicon*, **55** (5), 954–964.

[10] Spoof, L. and Meriluoto, J. (2002) Rapid separation of microcystins and nodularin using a monolithic silica C_{18} column. *Journal of Chromatography A*, **947** (2), 237–245.

[11] Cabrera, K., Lubda, D,. Eggenweiler, H.-M. *et al.* (2000) A new monolithic-type HPLC column for fast separations. *Journal of High Resolution Chromatography*, **23** (1), 93–99.

[12] Guillarme, D., Nguyen, D.T., Rudaz, S. and Veuthey, J.L. (2007) Recent developments in liquid chromatography-impact on qualitative and quantitative performance. *Journal of Chromatography A*, **1149** (1), 20–29.

[13] Wang, J., Pang, X., Ge, F. and Ma, Z. (2007) An ultra-performance liquid chromatography-tandem mass spectrometry method for determination of microcystins occurrence in surface water in Zhejiang province, China. *Toxicon*, **49** (8), 1120–1128.

[14] Xu, W., Zhang, T., Cai, Z. *et al.* (2008) Development of ultra-performance liquid chromaotgraphy-electrospray ionization tandem triple quadrupole mass spectrometry for determination of seven microcystins in water samples. *Analytica Chimica Acta*, **626** (1), 28–36.

[15] Ortelli, D., Edder, P., Cognard, E. and Jan, P. (2008) Fast screening and quantitation of microcystins in microalgae dietary supplement products and water by liquid chromatography coupled to time of flight mass spectrometry. *Analytica Chimica Acta*, **617** (1–2), 230–237.

[16] Spoof, L., Neffling, M.R. and Meriluoto, J. (2009) Separation of microcystins and nodularins by ultra performance liquid chromatography. *Journal of Chromatography B*, **877** (30), 3822–3830.

[17] Oehrle, S., Southwell, B. and Westrick, J. (2010) Detection of various freshwater cyanobacterial toxins using ultra-performance liquid chromatography tandem mass spectrometry. *Toxicon*, **55** (5), 965–972.

[18] Li, W., Duan, J., Niu, C. *et al.* (2011) Determination of microcystin-LR in drinking water using UPLC tandem mass spectrometry–matrix effects and measurement. *Journal of Chromatographic Science*, **49** (9), 665–670.

[19] Beltran, E., Ibanez, M., Sancho, J.V. and Hernandez, F. (2012) Determination of six microcystins and nodularin in surface and drinking waters by on-line solid phase extraction-ultra high pressure liquid chromatography tandem mass spectrometry. *Journal of Chromatography A*, **1266**, 61–68.

[20] Carmichael, W.W. (1988) Toxins of freshwater algae, in *Handbook of Natural Product Toxins* (ed. A.T. Tu), Marcel Dekker, New York, pp. 121–147.

[21] Edwards, C., Beattie, K.A., Scrimgeour, C.M. and Codd, G.A. (1992) Identification of anatoxin-A in benthic cyanobacteria (blue-green algae) and in associated dog poisonings at Loch Insh, Scotland. *Toxicon*, **30** (10), 1165–1175.

[22] Harada, K., Kimura, Y., Ogawa, K. *et al.* (1989) A new procedure for the analysis and purification of naturally occurring anatoxin-a from the blue-green alga *Anabaena flos-aquae*. *Toxicon*, **27** (12), 1289–1296.

[23] Furey, A., Crowley, J., Lehane, M. and James, K.J. (2003) Liquid chromatography with electrospray ion-trap mass spectrometry for the determination of anatoxins in cyanobacteria and drinking water. *Rapid Communications in Mass Spectrometry*, **17** (6), 583–588.

[24] Metcalf, J.S. and Cood, G.A. (2005) Analysis of anatoxin-a by high-performance liquid chromatography with photodiode-array detection, in *TOXIC–Cyanobacterial Monitoring and Cyanotoxin Analysis*, (eds J. Meriluoto and G.A. Codd), Åbo Akademi University Press, Åbo, pp. 129–132.

[25] Bogialli, S., Bruno, M., Curini, R. *et al.* (2006) Simple and rapid determination of anatoxin-a in lake water and fish muscle tissue by liquid-chromatography-tandem mass spectrometry. *Journal of Chromatography A*, **1122** (1–2), 180–185.

[26] Dimitrakopoulos, I., Kaloudis, T., Hiskia, A. *et al.* (2010) Development of a fast and selective method for the sensitive determination of anatoxin-a in lake waters using liquid chromatography–tandem mass spectrometry and phenylalanine-d5 as internal standard. *Analytical and Bioanalytical Chemistry*, **397** (6), 2245–2252.

[27] Dell'Aversano, C., Eaglesham, G.K. and Quilliam, M.A. (2004) Analysis of cyanobacterial toxins by hydrophilic interaction liquid chromatography–mass spectrometry. *Journal of Chromatography A*, **1028** (1), 155–164.

[28] Furey, A., Crowley, J., Hamilton, B. *et al.* (2005) Strategies to avoid the mis-identification of anatoxin-a using mass spectrometry in the forensic investigation of acute neurotoxic poisoning. *Journal of Chromatography A*, **1082** (1), 91–97.

[29] Ohtani, I., Moore, R.E. and Runnegar, M.T.C. (1992) Cylindrospermopsin: a potent hepatotoxin from the blue-green alga Cylindrospermopsis raciborskii. *Journal of the American Chemical Society*, **114** (20), 7941–7942.

[30] De la Cruz, A.A., Hiskia, A., Kaloudis, T. *et al.* (2013) A review on cylindrospermopsin: the global occurrence, detection, toxicity and degradation of a potent cyanotoxin. *Environmental Science: Processes & Impacts*, **15** (11), 1979–2003.

[31] Harada, K.-I., Ohtani, I., Iwamoto, K. *et al.* (1994) Isolation of cylindrospermopsin from a cyanobacterium *Umezakia natans* and its screening method. *Toxicon*, **32** (1), 73–84.

[32] Welker, M., Bickel, H. and Fastner, J. (2002) HPLC-PDA detection of cylindrospermopsin—opportunities and limits. *Water Research*, **36** (18), 4659–4666.

[33] Metcalf, J.S. and Cood, G.A. (2005) Analysis of cylindrospermopsin by high-performance liquid chromatography with photodiode-array detection, in *TOXIC–Cyanobacterial Monitoring and Cyanotoxin Analysis* (eds J. Meriluoto and G.A. Codd), Åbo Akademi University Press, Åbo, pp. 133–136.

[34] Törökné, A., Asztalos, M., Bánkiné, M. *et al.* (2004) Interlaboratory comparison trial on cylindrospermopsin measurement. *Analytical biochemistry*, **332** (2), 280–284.

[35] Metcalf, J.S., Beattie, K.A., Saker, M.L. and Codd, G.A. (2002) Effects of organic solvents on the high performance liquid chromatographic analysis of the cyanobacterial toxin cylindrospermopsin and its recovery from environmental eutrophic waters by solid phase extraction. *FEMS Microbiology Letters*, **216** (2), 159–164.

[36] Wormer, L., Carrasco, D., Cirés, S. and Quesada, A. (2009) Advances in solid phase extraction of the cyanobacterial toxin cylindrospermopsin. *Limnology and Oceanography Methods*, **7** (7), 568–575.

[37] Li, R., Carmichael, W.W., Brittain, S. *et al.* (2001) Isolation and identification of the cyanotoxin cylindrospermopsin and deoxy-cylindrospermopsin from a Thailand strain of *Cylindrospermopsis raciborskii* (Cyanobacteria). *Toxicon*, **39** (7), 973–980.

[38] Graham, J.L., Loftin, K.A., Meyer, M.T. and Ziegler, A.C. (2010) Cyanotoxin mixtures and taste-and-odor compounds in cyanobacterial blooms from the midwestern United States. *Environmental Science and Technology*, **44** (19), 7361–7368.

22

Determination of Cyanotoxins by High-Performance Liquid Chromatography with Fluorescence Derivatization

James S. Metcalf[1] *and Paulo Baptista Pereira*[2]

[1] *Institute for Ethnomedicine, Jackson, USA*
[2] *Instituto Nacional de Saúde Dr Ricardo Jorge (INSA), Lisboa, Portugal*

22.1 Principle of the Technique and Why It Is Used for Cyanotoxins

Although the majority of cyanotoxins have chromophores, certain compounds have no natural absorbance properties and may require chemical modification in order that they can be detected. Furthermore, with the detection of very small molecules (~100 Da), chemical modification may increase the size and retention, and improve the chances of detecting the compound of interest with confidence. In certain cases, where the toxicity of the compound is extremely high, such as with the saxitoxins, chemical modification may increase the detectability of the compound, and thus the detection limits, so that there is a greater chance of protecting human health through guidelines and regulation. Of the chemical modifications that are possible, the addition of a fluorescent chemical tag has a number of advantages. Firstly, they can increase the mass of the molecule, so that methods such as mass spectrometry can identify the unique molecule, and secondly, the fluorescent chromophore can be detected by optical detectors. The fluorescence detector works by exciting the compound with a specific wavelength of light. Once excited and at a higher energy level, the electrons then drop down to a lower energy level. By dropping down to a lower energy level, the subsequent loss of energy is emitted as a second specific wavelength of light. The detector is

Handbook of Cyanobacterial Monitoring and Cyanotoxin Analysis, First Edition.
Edited by Jussi Meriluoto, Lisa Spoof and Geoffrey A. Codd.

able to produce light of a certain wavelength, which excites the compound, and also detect the emitted light at a specific wavelength, based on the derivatisation compound used.

22.2 Types of Reactions for Analysing Paralytic Shellfish Toxins Using High-Performance Liquid Chromatography with Fluorescence Derivatization

Paralytic shellfish toxins (PSTs) are neurotoxic alkaloids that share a common tricyclic structure based on a 3,4,6-trialkyltetrahydropurine skeleton. PST analogs can be divided into major categories based on substituent side chains. Actually, more than 30 structural variants are known to occur in nature, and evidence is still accumulating for the presence of other as yet uncharacterized toxins [1]. The diversity of structures makes the accurate detection and quantification of the PST a substantial challenge. Given that each variant can exhibit different toxicities, the accurate determination of variants is critical when assessing risk.

PSTs do not exhibit a strong UV absorbance or native fluorescence. However, the oxidation of the tricyclic ring system with appropriate oxidant agents, such as with acid, cleaves the propino ring to give a fluorescent aromatic aminopurine derivative (Fig. 22.1). The reaction is the basis for the detection and quantification of PST by high-performance liquid chromatography with fluorescence derivatization (HPLC-FLD). PST-oxidized derivatives have UV absorption maxima at 252 and 335 nm and emission maxima at 390 nm.

There are two major HPLC-FLD approaches for PST analysis: (1) the precolumn oxidation approach (based on Lawrence's method), which involves the oxidation of toxins prior to the chromatographic separation of the derivatised compounds [2], and (2) the postcolumn oxidation method (based on Oshima's method), which separates the toxins on the column and then oxidizes the separated toxins by way of a flow injection process before fluorescence detection [3]. Like all methods, each has its own advantages and limitations.

A major disadvantage with Oshima's postcolumn oxidation method is that anionic and cationic ion-paring reagents are required to obtain the adsorption and separation of PST analogues on a reverse-phase HPLC system by means of their different polarity and ionic charge. This amounts to three separate isocratic analyses being required to separate the various PSTs (STXs, GTXs, and C1–C4 toxins), rendering the method both time consuming and costly for multiple sample analyses. Moreover, the postcolumn oxidation method requires the capability to deliver reagents by flow injection for oxidation reactions between the chromatographic column and the fluorescence detector. Therefore, laboratories that do not possess such postcolumn delivery systems may consider the method unfavourable based on instrumentation requirements. Another drawback is that the method uses oxidizing reagents that are unstable over time and slight changes in pH, temperature, or delivery flow rate may affect the analysis considerably if not monitored carefully by an experienced technician. That said, the method is still regarded as the most satisfactory because it enables analysis of individual toxins with accurate results, being the only method capable of separating PST epimers (such as GTX1/4, GTX2/3, C1/2, and C3/4) and quantifying them individually. The Oshima method has given very consistent results for the analysis of PSTs from cyanobacteria.

The Lawrence precolumn oxidation method, on the other hand, uses a very simple reverse-phase chromatographic system that is easy to set up and found to be reliable and reproducible over many years of use in shellfish monitoring settings. It uses a binary gradient that does not change regardless of the PST being evaluated. No postcolumn system is required, and prechromatographic oxidation reactions are performed by adding proper volumes of sample and reagent in a vial before LC injection. The method has been the subject of a successful interlaboratory study [4] and

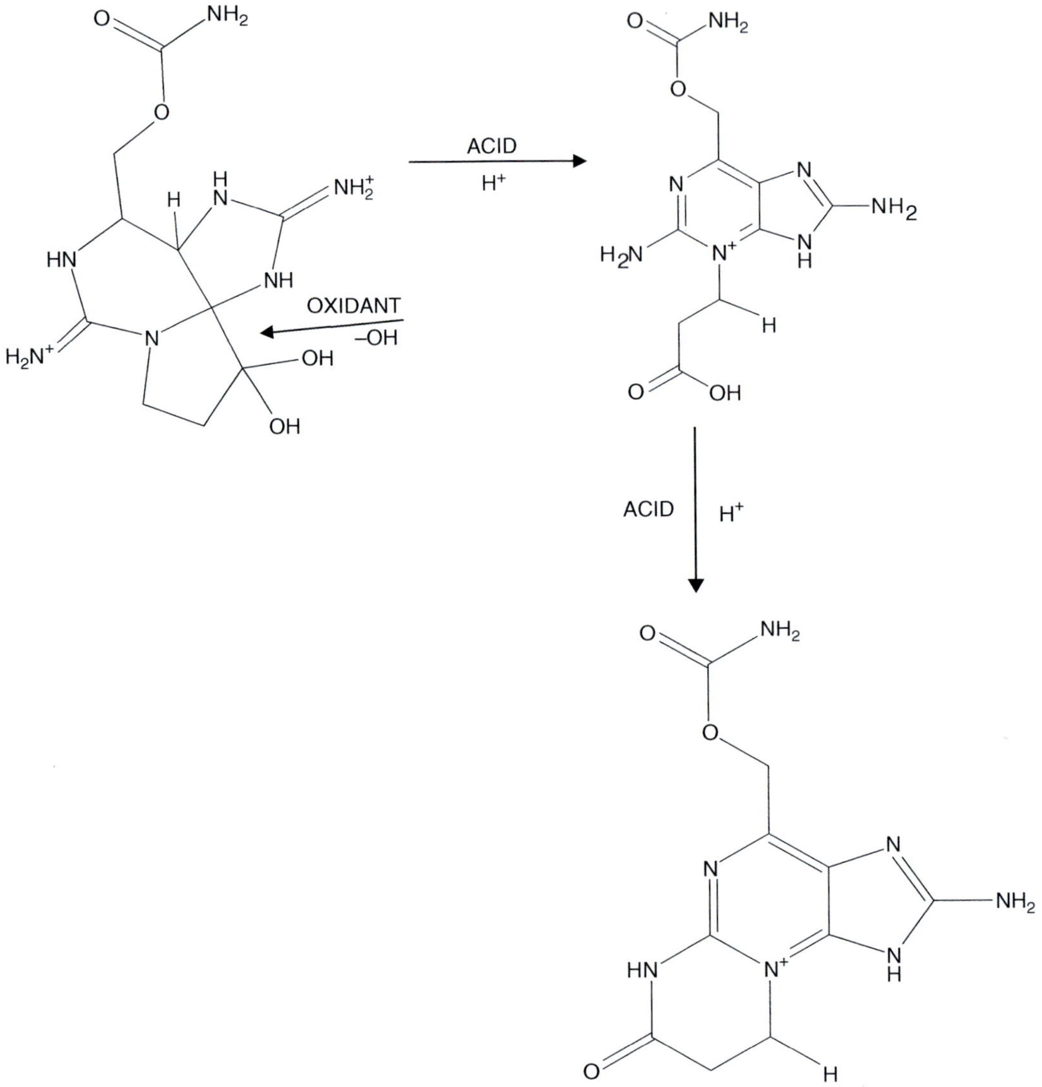

Figure 22.1 Chemical oxidation of the saxitoxin tricyclic ring system to give a fluorescent aromatic aminopurine derivative

collaborative study [5] and has been accepted by the AOAC as the first analytical alternative to the mouse bioassay for the quantitative determination of PSTs in shellfish (AOAC Official method 2005.06). The major inconvenience when applying the precolumn derivatization method is that some PSTs have multiple reaction products and different toxins may have the same oxidation reaction product. This is partially addressed by oxidizing the toxins using two oxidant reagents: peroxide oxidation is used for the quantification of STX, dcSTX, GTX2/3, GTX5, C1/C2, dcGTX2/3, and periodate oxidation is used for the quantification of GTX1/4 and neoSTX. If necessary, a COOH fractionation step is incorporated to further separate the toxins (this is particularly important for samples containing *N*-1-hydroxylated toxins such as NeoSTX, GTX1/4, GTX6, and C3/C4).

The analysis of different fractions using both oxidation solutions allows for a very clear distinction of exactly which toxins are present with excellent chromatographic performance. It is notable that for screening purposes only one aliquot (C18 with peroxide) is required for analysis. However, for samples containing significant amounts of toxins, total PST quantification usually requires the analysis of three aliquots (C18 with peroxide, COOH fractions 2 and 3 with periodate) with a fourth aliquot (COOH fraction 1 with peroxide) being necessary when C3/4 toxins are present. The process involves a rigorous and time-consuming sample preparation besides extensive data analysis. It should also be noted that the method cannot distinguish isomeric forms that may exhibit different toxicities (e.g. GTX1/4, GTX2/3, C1/2). Foss et al. [6] noted that the prechromatographic oxidation method failed to resolve the individual constituents of the *Lyngbya wollei*–derived PST profile as the oxidation products of dcneoSTX coeluted with the oxidation products of both dcSTX and STX. From a safety perspective, this is not a major issue as the relative toxicity of dcneoSTX is similar to that of dcSTX and less than that of neoSTX. In this situation, the worst case scenario would be an overestimation of total toxicity, further protecting the consumer.

During the past decade, alternative HPLC-FLD methods have been proposed, mostly to address the short comings of both the Lawrence and Oshima methods. Turner et al. [7] refined the Lawrence method to include additional toxins (dcneoSTX and dcGTX2/3) and to improve derivative stability and analytical sensitivity. Rourke et al. [8], on the other hand, made important modifications to Oshima's method, decreasing the number of injections needed to perform a complete analysis by combining two isocratic mobile phases into a binary step gradient (for the analysis of both STXs and GTXs). Rourke's new postcolumn oxidation method has undergone single [9] and collaborative laboratory validation [10] under the auspices of the AOAC and has recently been accepted as the official method for PST determination in shellfish (AOAC Official method 2011.02). To our knowledge, however, there is still no information regarding the application of this method to cyanobacteria. It is important to note that all the methodologies cited above have been developed and optimized for shellfish analysis. As the accurate determination of PSTs from cyanobacteria depends greatly on the methods applied during sample preservation, extraction, and analysis, methodological approaches validated for marine shellfish may not prove appropriate for cyanobacteria.

Regardless of the method used, it is common practice to also analyse samples without oxidation to distinguish PSTs from other fluorescent compounds that may co-elute with toxins. If after removing the oxidizing reagent the peak of interest changes in the same way as the standard, then this response may be used to further confirm the identity of this peak. Of course this step is much simpler with the precolumn method since no changes in instrumental setup are required. When using the postcolumn method it is necessary to wash out the postcolumn reagent with water and run each sample again under the new system setup, making this task a more tedious one.

The oxidation of the PST can also be done electrochemically with the use of a coulometric electrochemical cell. The first HPLC coupled electrochemical oxidation system for the analysis of PSTs was proposed by Boyer and Goddard [11]. The method was intended to be an alternative to the postcolumn chemical method, eliminating the problem of unstable reagents associated with the postcolumn reaction system. Electrochemical cells require a simpler instrumental setup and are less sensitive to changes in flow rate than postcolumn reaction systems. They also allow for the rapid identification of naturally fluorescent impurities, just by turning off the applied potential to the oxidizing cell and checking for changes in signal behaviour. As with the chemical oxidation system, excessive oxidation to multiple products often causes the electrochemical system to be less sensitive to the *N*-1-hydroxy compounds, such as neoSTX, GTX1 and GTX4. However, electrochemical oxidation is relatively easy to setup, being compatible with most of the mobile phases described for reversed phase chromatography.

One of the most interesting features is that, by eliminating the use of periodic acid, the electrochemical oxidation can be made compatible with parallel application of fluorescence and MS detection. Unfortunately, the mobile phase components used in reversed phase chromatography (such as ion-pair formers, phosphate buffers, etc.) prevent effective LC/MS coupling. To address this problem, Jaime et al. [12] described a method in which PSTs are separated on anion and cation exchange columns connected in series, using aqueous ammonium acetate as mobile phase. As the mobile phase does not contain ion-pair reagents or phosphate buffer, the method allows combining of both electrochemical FLD and MS detection. Such an FLD/MS system not only allows for the unambiguous determination of PST but also is well suited for the analysis and certification of standards.

22.3 Types of Reactions for Analysing β-*N*-Methylamino-L-Alanine and Isomers by HPLC-FLD

The cyanobacterial neurotoxin, BMAA is a small (118-Da) amino acid that is difficult to analyse underivatised. In addition, two further structural isomers of BMAA, 2,4-diaminobutyric acid (DAB) and *N*-(2-aminoethyl)glycine (AEG), have also been reported in cyanobacteria [13, 14]. Although HILIC methods have been published for the analysis of this compound, this latter method is considered unsuitable for BMAA analysis [15]. Consequently methods are required to modify this amino acid so that it can be more easily detected, as the derivatised molecule is much larger than the native amino acid. A variety of fluorescent derivatives of BMAA have been used with LC and GC separation, although the two most common are chloroformates (e.g. Ezfaast [13, 16]) and carbamates such as 6-aminoquinolyl-*N*-hydroxysuccinimidyl carbamate (AQC [13, 14]). The consequences of using such derivatisations is that the size of BMAA is increased from 118 to 332 or 458, when using Ezfaast or AQC, respectively.

22.4 Need for Confirmatory Techniques with HPLC-FLD

Although HPLC-FLD is an excellent tool for the analysis of known compounds, confirmatory methods such as mass spectrometry are essential to verify that the peak of interest is the compound which is being analysed. This is largely because with HPLC-FLD, retention time is most frequently the only factor used to determine the compound and the possibility exists that of the thousands of compounds that may be present and possibly derivatised, some may have retention times coincident with those of toxins. Therefore, using mass spectrometry to analyse a proportion of, or all of the samples, can be extremely useful to confirm the presence of the compound(s) of interest.

References

[1] Wiese, M., D'Agostino, M.P., Mihali, T.K. *et al.* (2010) Neurotoxic alkaloids: saxitoxins and its analogs. *Marine Drugs*, **8** (7), 2185–2211.

[2] Lawrence, J.F., Niedzwiadek, B. and Menard, C. (2001) Quantitative determination of paralytic shellfish poisoning toxins in shellfish by using prechromatographic oxidation and liquid chromatography with fluorescence detection: collaborative study. *Journal of the AOAC International*, **88** (6), 1714–1732.

[3] Oshima, Y. (1995) Post column derivatization liquid chromatographic method for paralytic shellfish toxins. *Journal of the AOAC International*, **78** (2), 528–532.
[4] Lawrence, J.F., Niedzwiadek, B. and Menard, C. (2004) Quantitative determination of paralytic shellfish poisoning toxins in shellfish using prechromatographic oxidation and liquid chromatography with fluorescence detection: Interlaboratory study. *Journal of the AOAC International*, **87** (1), 83–100.
[5] Lawrence, J.F., Niedzwiadek, B. and Menard, C. (2005) Quantitative determination of paralytic shellfish poisoning toxins in shellfish using prechromatographic oxidation and liquid chromatography with fluorescence detection: collaborative study. *Journal of the AOAC International*, **88** (6), 1714–1732.
[6] Foss, A.J., Phlips, E.J., Yilmaz, M. and Chapman, A. (2012) Characterization of paralytic sellfish toxins from *Lyngbya wollei* dominated mats collected from two Florida springs. *Harmful Algae*, **16**, 98–107.
[7] Turner, A.D., Norton, D.M., Hatfield, R.G. *et al.* (2009) Refinement and extension of AOAC method 2005.06 to include additional toxins in mussels: single-laboratory validation. *Journal of the AOAC International*, **92** (1), 190–200.
[8] Rourke, W.A., Murphy, C.J., Pitcher, G. *et al.* (2008) Rapid postcolumn methodology for determination of paralytic shellfish toxins in shellfish tissue. *Journal of the AOAC International*, **91** (3), 589–597.
[9] van de Riet, J.M., Gibbs, R.S., Chou, F.W. *et al.* (2009) Liquid chromatographic post-column oxidation method for analysis of paralytic shellfish toxins in mussels, clams, scallops and oysters: single laboratory validation. *Journal of the AOAC International*, **92** (6), 1690–1704.
[10] van de Riet, J.M., Gibbs, R.S., Muggah, P.M. *et al.* (2011) Liquid chromatography post-column oxidation (PCOX) method for the determination of paralytic shellfish toxins in mussels, clams, oysters, and scallops: collaborative study. *Journal of the AOAC International*, **94** (4), 1154–1176.
[11] Boyer, G.L. and Goddard, G.D. (1999) High performance liquid chromatography coupled with post-column electrochemical oxidation for the detection of PSP toxins. *Natural Toxins*, **7** (6), 353–359.
[12] Jaime, E., Hummert, C., Hess, P. and Lucas, B. (2001) Determination of paralytic shellfish poisoning toxins by high-performance ion-exchange chromatography. *Journal of Chromatography A*, **929** (1–2), 43–49.
[13] Banack, S.A., Downing, T.G., Spacil, Z. *et al.* (2010) Distinguishing the cyanobacterial neurotoxin β-*N*-methylamino-L-alanine (BMAA) from its structural isomer 2,4-diaminobutyric acid (2,4-DAB). *Toxicon*, **56** (6), 868–879.
[14] Banack, S.A., Metcalf, J.S., Spacil, Z. *et al.* (2011) Distinguishing the cyanobacterial neurotoxin β-*N*-methylamino-L-alanine (BMAA) from other diamino acids. *Toxicon*, **57** (5), 730–738.
[15] Cohen, S.A. (2012) Analytical techniques for the detection of α-amino-β-methylaminopropionic acid. *Analyst*, **137** (9), 1991–2005.
[16] Esterhuizen, M. and Downing, T.G. (2008) β-*N*-Methylamino-L-alanine (BMAA) in novel South African cyanobacterial isolates. *Ecotoxicology and Environmental Safety*, **71** (2), 309–313.

23

Liquid Chromatography–Mass Spectrometry

Josep Caixach[1], Cintia Flores[1], Lisa Spoof[2], Jussi Meriluoto[2,3], Wido Schmidt[4], Hanna Mazur-Marzec[5], Anastasia Hiskia[6], Triantafyllos Kaloudis[7], and Ambrose Furey[8]

[1]*Mass Spectrometry Laboratory/Organic Pollutants, IDAEA-CSIC, Barcelona, Spain*
[2]*Biochemistry, Faculty of Science and Engineering, Åbo Akademi University, Turku, Finland*
[3]*Laboratory for Paleoenvironmental Reconstruction, Faculty of Sciences, University of Novi Sad, Serbia*
[4]*DVGW Technologiezentrum Wasser TZW, Branch Office Dresden, Germany*
[5]*Institute of Oceanography, University of Gdansk, Gdynia, Poland*
[6]*Institute of Nanoscience & Nanotechnology, National Center for Scientific Research "DEMOKRITOS", Athens, Greece*
[7]*Water Quality Control, Athens Water Supply and Sewerage Company – EYDAP SA, Athens, Greece*
[8]*Mass Spectrometry Research Centre, Department of Physical Sciences, Cork Institute of Technology (CIT), Cork, Ireland*

23.1 Introduction

Mass spectrometry (MS) is a valuable tool to determine the amount and elemental make-up of a compound and to obtain certain structural information of a target analyte and compounds related to the target analyte [1, 2]. It can provide this information about a compound using trace quantities of the substance even in the presence of complex matrices [3, 4].

Handbook of Cyanobacterial Monitoring and Cyanotoxin Analysis, First Edition.
Edited by Jussi Meriluoto, Lisa Spoof and Geoffrey A. Codd.

A mass spectrometer consists of three main components:

1. The ion source (electron ionisation (EI), chemical ionisation (CI), fast atom bombardment (FAB), matrix-assisted laser desorption/ionisation (MALDI), atmospheric pressure chemical ionisation (APCI), electrospray ionisation (ESI), atmospheric pressure photoionisation (APPI))
2. The mass analyser (quadrupole ion trap (QIT), triple quadrupole (QqQ), quadrupole time of flight (Q-TOF), Fourier transform ion cyclotron resonance (FT-ICR), orbitrap, magnetic sector)
3. The detector (electron multipliers, Faraday cups, photographic plate, scintillation counter, channel electron multipliers, multichannel detector, conversion dynodes and cryogenic detectors)

MS is at its simplest based on the principle of weighing a charged analyte. The mass spectrometer produces ions that are separated according to their mass-to-charge (*m/z*) ratio and that are subsequently detected. The resulting mass spectrum is a graph of the relative abundance of the measured ions as a function of their *m/z* ratio. The mass spectrometer is designed to 'see' any ionisable compound. MS exploits the properties of ions in a vacuum which allows them to be stratified according to their mass and their number of charges and then transported along the mass spectrometer to the detector [3, 4]. A typical MS can provide the following information for a target molecule:

- The *m/z* ratio of the compound, which allows the nominal mass to be assigned
- The relative abundance of the target analyte
- The separation of isotopes (because isotopes have different masses, they can be distinguished by the MS)
- A tandem mass spectrometry (MS/MS), which can produce a fragmentation spectrum of the target compound that can aid in the structural characterisation of the target molecule (and closely related analogues)

The significant advantage of MS lies in its unparalleled selectivity and sensitivity, where it can provide reliable identification and quantitation at low concentration levels (ppb and/or ppt) of target analyte with acceptable accuracy and precision [3, 4]. The sensitivity of individual MS instruments depends on the following factors:

- The properties of the target analyte (i.e. the *m/z* value of the ion)
- The amount of sample used
- The properties of the calibration compound used on the instrument
- The resolving power of the instrument
- The ionisation efficiency and geometry of the ion source
- The scan time of the mass analyser (the higher the scan time, the lower the sensitivity, as there are less ions of a particular *m/z* value reaching the detector)
- The number of scan events used (i.e. the number of selected reaction monitoring (SRM) full scan and/or data-dependent scans used in the instrument method as these will affect the number of data points acquired for each scan)
- The ion current arriving at the detector [3, 4]

Mass spectrometers are generally used in tandem with separation techniques like liquid chromatography (LC) and gas chromatography (GC) [3, 4]. LC is the most commonly used technique as GC coupled to MS is limited to the analysis of compounds that are not thermally labile and that are volatile. Indeed, some cyanotoxins like microcystins are large molecules and have low volatility;

therefore GC separation requires more complex sample preparation usually including derivatisation. The advantages of using an LC system with MS include the following:

1. There is less emphasis on the need for complete chromatographic resolution of the analytes as the MS instrument can further filter components according to their *m/z* value [3, 4].
2. Number one thus facilitates shorter analysis run times.
3. The LC separation reduces considerably the possibility of the co-elution of isobaric compounds; in cases where isobaric compounds do co-elute, MS^n fragmentation studies can usually discriminate between them [3, 4].
4. LC separation can reduce ion suppression effects by separating matrix components from the analytes.
5. Chromatographic separation of components can improve the post-analysis assessment of a sample where untargeted compounds can be studied retrospectively in a full-scan mass spectrum [5].
6. LC separation can incorporate online sample clean-up, such as online solid-phase extraction (SPE), which can further enhance the sensitivity of the instrument.

Disadvantages associated with LC coupling to MS include the following:

1. An interface (ion source) between the two instruments is necessary to ionise target analytes from the LC column before introduction into the mass analyser. This ion source often necessitates lengthy optimisation of key parameters to obtain the best ionisation efficiency for analytes leaving the mobile phase [6].
2. LC-MS methods use a higher flow rate than infusion methods (e.g. ~200 $\mu L\ min^{-1}$ compared with <20 $\mu L\ min^{-1}$, respectively), and the use of aqueous mobile phases means that high temperatures may be required for desolvation. This can sometimes cause in-source fragmentation, resulting in reduced analyte signal and consequently reductions in sensitivity [6].
3. Some additives to LC mobile phases which are used to improve the chromatography can cause ion suppression, for example, trifluoroacetic acid (TFA).
4. While the use of ultrahigh-performance liquid chromatography (UHPLC)/rapid resolution LC (RRLC) can provide better resolution and shorter analysis time, they have a major disadvantage when used with MS. Namely, the peak widths in these runs are often too narrow to monitor a sufficient number of data points per peak in the MS, which means that only a small number of transitions may be monitored and this can limit accurate quantitation. Increased flow rate (>0.6 $mL\ min^{-1}$) can affect analyte ionisation leading to higher limit of detections (LODs) and limit of quantifications (LOQs) than with a method developed at a lower flow rate (~0.2 $mL\ min^{-1}$).

23.2 Ion Sources

To obtain good MS data, the analyte of interest must be converted into a form suitable to be analysed. In order for this to occur, the analyte must be ionised prior to its introduction into the mass analyser portion of the mass spectrometer. There are many ionisation techniques available such as:

- EI
- CI
- FAB
- MALDI
- APCI
- ESI
- APPI

EI and CI are used by GC/MS methodologies [7–12]. GC methods are used for the determination of total microcystins based on their oxidation, which splits the Adda side chain to produce the 3-methoxy-2-methyl-4-phenylbutyric acid (MMPB) derivative. The MMPB derivative can then be analysed either by GC/FID or GC/MS (as its methyl ester) or by HPLC with fluorescence detection [9, 12]. GC/MS offers total toxin content, namely, free and covalently bound microcystins in biological matrices. The limitation of this method is that it requires tedious sample preparation including extraction, clean-up, oxidation and post-treatment in order to eliminate the derivatisation reagents used before analysis by GC or HPLC. Cyanotoxins are therefore mostly detected by LC-MS. FAB, MALDI and ESI are the ionisation techniques commonly applied to the analysis of large biomolecules including proteins and peptides by LC-MS [1, 2].

23.2.1 Electron Ionisation (EI)

In **EI**, a filament heated to high temperature produces thermal electrons; these are then accelerated towards an anode by establishing a potential difference. The high-energy electrons collide with neutral molecules in the ion source, violently ejecting valence electrons, a process represented by the following reaction:

$$M + e^- \rightarrow M^{+\cdot} + 2e^-$$

Because the high-energy electrons deposit excess internal energy in the ions during the ionisation process, the initial molecular radical ion $M^{+\cdot}$ can fragment into smaller ions. This often results in little to no precursor ion being observed. Large libraries of EI mass spectra are used for fast identification of unknown analytes [6]. This is possible because the ionisation energy is kept constant at 70 eV and the de Broglie wavelength of electrons approximately matches the length of the bonds of organic compounds, resulting in reproducible spectra [13].

23.2.2 Chemical Ionisation (CI)

CI uses ion–molecule reactions to produce ions from the analyte. An ionisation plasma is formed through a number of reactions. Firstly, a high reagent gas pressure results in ion–molecule reactions between the reagent gas ions and reagent gas neutrals. Due to the excess of this reactant gas in the source, the accelerated electrons collide first with the reaction gas. The key event is the formation of a radical cation (usually) of the reagent gas (e.g. CH_4 gives rise to CH_4^{+}). This radical then reacts with other gas molecules (see Fig. 23.1). These complex reactions will finally achieve a relatively steady-state composition of reaction ions which can finally ionise the analyte by ion–molecule reactions [6].

$M + [BH]^+ \rightarrow [M+H]^+ + B$	proton transfer
$M + X^+ \rightarrow [M+X]^+$	electrophilic addition
$M + X^+ \rightarrow [M\text{-}A]^+ + AX$	anion abstraction
$M + X^{+*} \rightarrow M^{+*} + X$	charge exchange

Figure 23.1 *Typical reactions that lead to CI protonation*

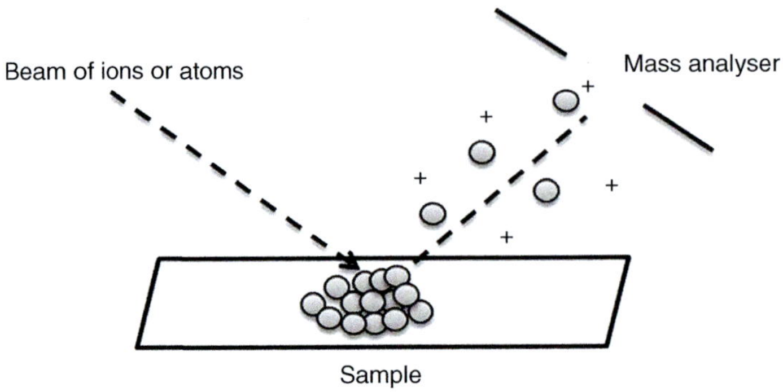

Figure 23.2 Schematic representation of fast atom bombardment [3, 4]

23.2.3 Fast Atom Bombardment (FAB)

In **FAB**, the analyte is dissolved in a non-volatile liquid matrix like glycerol or *m*-nitrobenzyl alcohol (about 1 μL) and is placed on a target platform. The target is bombarded with a fast atom beam that induces a shock wave that ejects ions and molecules from the solution. Cluster ions from the liquid matrix are also desorbed and produce a chemical background that varies with the matrix used. Ions are accelerated by a potential difference towards the analyser. This is a rapid, simple technique that is relatively tolerant of variations in sampling, and it is applicable to a large variety of compounds (see Fig. 23.2). However, its detection limits are defined by high chemical background, which may interfere with the identification of low molecular weight compounds [6]. It is important to note that the pioneering works of Sivonen *et al.* [14] in the 1990s on the structural elucidation of cyanotoxins were performed by FAB and high-resolution mass spectrometry (HRMS) using magnetic sector analysers [3, 4].

23.2.4 Matrix-Assisted Laser Desorption/Ionisation (MALDI)

In **MALDI**, the analyte is dissolved in a solution containing an excess of a matrix that has a chromophore which absorbs at the laser wavelength. The sample is dried and a small quantity of this ‘solid solution’ is placed on the laser target. The matrix absorbs the energy from the laser pulse and produces a plasma that results in the rapid heating of the crystals, causing vaporisation and ionisation of the analyte. The possibility of reanalysis is an advantage, in comparison to spray techniques (all API techniques), where the eluate is completely consumed in nebulisation. MALDI relies on the absorption of the laser energy by the absorbing matrix, allowing for increased efficiency of energy transfer from the laser to the analyte [3, 4, 6].

23.2.5 Atmospheric Pressure Ionisation (API)

Most routine LC-MS-based methods rely on this type of ionisation interface. In fact API is often regarded as the ionisation source of choice. This method can be further divided into specific types of API sources including ESI, APCI and APPI.

The API source has three main purposes:

- Removal of the mobile phase from the LC eluate
- Transfer of the analytes from the high-pressure (~100–200 bar) environment of the LC to the gas phase of the MS which functions at a high vacuum of around 10^{-6} Torr
- Ionisation of the target analytes

A combination of thermal and pneumatic (use of pressurised gas) means that API is used to desolvate the ions as they enter the ion source. This method gives improved ionisation efficiency up to 10^4 times greater than with a reduced pressure chemical ionisation source [15]. As the atmospheric pressure source is coupled with a very-low-pressure mass analyser, an important requirement for the focusing lenses/multipole lenses is that they have very small openings between both compartments. In addition, the use of several differential stages of high-capacity pumps to maintain the low pressure of the vacuum region of the mass spectrometer is necessary. A high-temperature transfer tube or the application of a heated dry nitrogen countercurrent on gas curtain is also necessary to eliminate the formation of cluster ions [3, 4, 6].

23.2.6 Atmospheric Pressure Chemical Ionisation (APCI)

The LC eluent is passed through a pneumatic nebuliser where the droplets are both generated and desolvated. The spray then passes through a heated region where the LC eluent droplet vapour is evaporated. The neutral species produced are then passed through a corona discharge. This corona discharge occurs when the electric field at the tip of the electrode is sufficiently high to ionise the gas surrounding it, but insufficiently high to cause a spark. The analyte is then ionised by CI, with the vaporised solvent acting as the reagent gas. Ionisation occurs when a reagent ion collides with an analyte molecule (see Fig. 23.3). APPI operates on a similar principal to APCI; however, ions are generated via a UV source instead of the corona discharge voltage that is used in APCI [3, 4].

23.2.7 Electrospray Ionisation (ESI)

Electrospray ionisation–mass spectrometry (ESI-MS) was introduced in the 1980s and increased in popularity when Fenn *et al.* [16, 17] illustrated that multiply charged ions were obtained from proteins (i.e. low *m/z* value due to high number of charges on the molecule), therefore allowing the determination of these high molecular weight compounds with low mass range mass spectrometers. ESI sources not only work efficiently for infusion methods but also provide an excellent interface for coupling to LC instruments.

The principle of ESI is the formation of ions, so it works well with polar (moderately polar) molecules. The analytes may be already ionised in the eluent (e.g. through dissociation), before

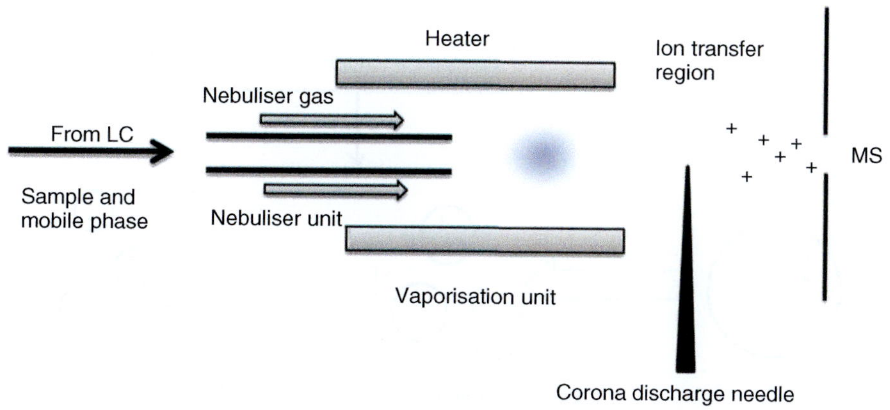

Figure 23.3 *Schematic representation of atmospheric pressure chemical ionisation [3, 4]*

they reach the ion source. The ESI source works by converting ions in solution into ions in the gas phase. This is achieved by applying a voltage to a capillary tube while the sampling cone is held at low voltage. When a low flow rate is used, the potential difference between the two is sufficient to create a spray of charged droplets, but when high flow rates (50–1000 μL min^{-1}) are used, a nitrogen flow (~8 L min^{-1}) is required to maintain a stable spray. A heating device is also used to help the desolvation process. A droplet containing the ionised analytes will form at the tip of the capillary, and an accumulation of charges will form at the surface of the droplet due to the electric field applied which will be positive or negative. As the solvent evaporates, the size of the droplets reduces and the density of charges on the droplet surface increases. This results in many-like charges accumulating closer and closer together until repulsion forces increase to a level where the droplet explodes (coulomb explosion), releasing smaller charged droplets. This process of desolvation continues during the droplet transfer from the spray needle to the heated capillary where the remaining solvent molecules are completely removed. The charge from the droplets is transferred to the analytes. The ions are then passed through the skimmer region and into the mass analyser region of the mass spectrometer (see Fig. 23.4).

Advances in ESI include the development of orthogonal ESI where the ESI spray is oriented at a 90° angle to the entrance of the mass spectrometer. This allows for greater sensitivity because non-charged matrix components are prevented from contaminating the entrance port (e.g. heated capillary) of the MS.

One type of ESI sources that are of interest is the TurboIonSpray ion source. The TurboIonSpray uses a heater gas to enhance the desolvation of the spray, yet it is mild enough to be used with labile compounds such as peptides, proteins and pharmaceuticals. This source can accommodate high

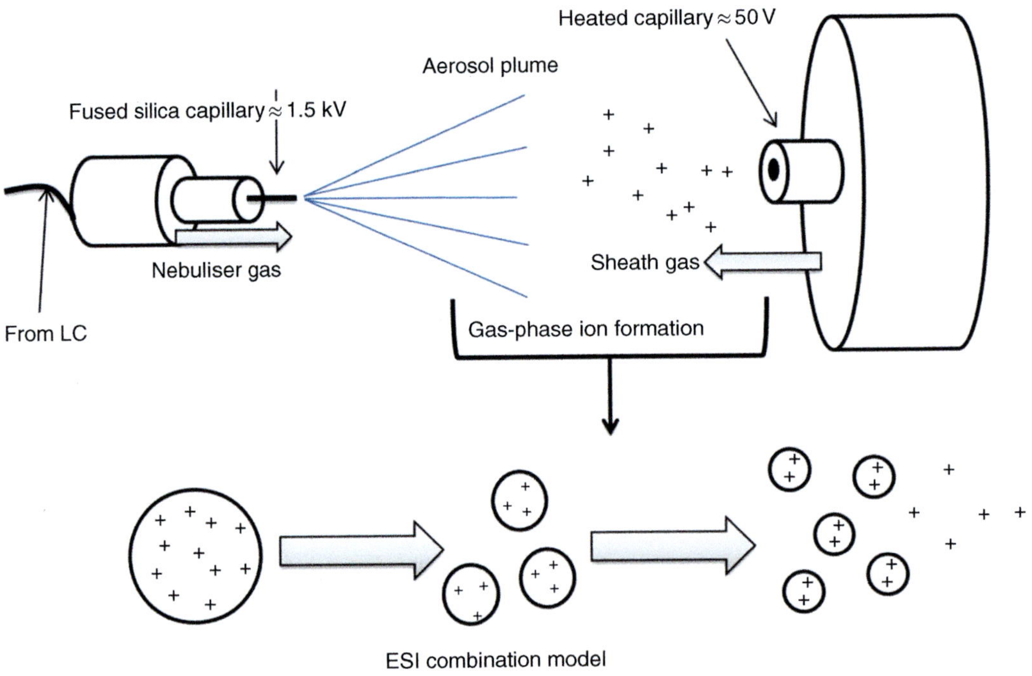

Figure 23.4 Schematic representation of electrospray ionisation [3, 4]

flow rates up to 1 mL min^{-1} and provides greater sensitivity in comparison with conventional ESI. It can vaporise 100% organic and 100% aqueous solvents. The sample is sprayed to the side of the orifice through the region of heated gas. A potential is applied to the orifice plate to direct either positive or negative ions into the MS entrance, leaving unwanted ions and neutral entities to gather on the curtain plate [3, 4].

The ESI source is more susceptible to ion suppression than other API interfaces [18]. This is because when desolvation of the ions occurs, the charged molecules are desorbed from the surface of the droplets. Sensitivity is usually higher for compounds that have properties that allow them to concentrate at the surface of the droplet, as compared with other compounds that do not possess this property (e.g. lipophilic compounds). When a mixture of compounds is analysed, especially samples with complex matrices, the presence of these types of molecules at the surface of the droplets can mask the presence of molecules that are more soluble in the bulk of the droplet, leading to an increase in suppression of the ionisation process (see section 23.7) of the analyte of interest [3, 4, 18].

23.3 Types of Mass Analysers

Each mass analyser has its own individual capabilities (MS/MS, MS^n, high resolution), applications (routine analysis, structure elucidation, high mass accuracy), benefits and limitations with regard to sensitivity, selectivity and resolution based on its specific design. The choice of mass analyser is based upon the analyte information and quality of result required from the analysis.

23.3.1 Quadrupole Ion Trap (QIT) Mass Spectrometer

The QIT, or ion trap, is a three-dimensional quadrupole. The ions generated in the ionisation region are focused through a skimmer-and-octapole transmission system into the trap (see Fig. 23.5). The trap consists of a ring electrode and two end-cap electrodes orientated above and below. The ions are stabilised in the trap by applying a radio-frequency (RF) voltage on the ring electrode. To give the maximum efficiency, the ions must be focussed near the centre where the trapping fields are closest to the ideal and less distorted than at the extremities. This is achieved by introducing a damping gas (helium) that cools injected ions, thus dampening down their oscillations until they

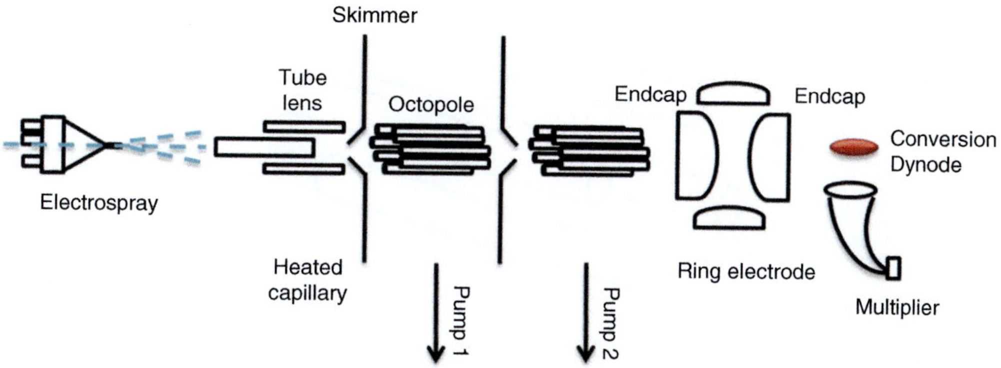

Figure 23.5 *Schematic diagram of an ion trap with an ESI source (Finnigan LCQ) [3, 4]*

stabilise and allowing for maximum resolution and sensitivity. By ramping the RF voltage or applying supplementary voltages on the end-cap electrodes or by a combination of both, it is possible to destabilise the ions and eject them progressively from the trap. Alternatively one ion of a given *m/z* value can be kept in the trap and then ejected to observe it specifically or fragmented by inducing vibrations to observe the product ions. This is an example of MS/MS experiments. This operation can be repeated several times to progressively fragment selected ions, which is MS^n. The QIT is a scanning device that is very useful in elucidating the fragmentation pattern of a molecule which can assist in identifying the structural characteristics of the compound [3, 4].

The ion trap is limited in that it can only observe *m/z* values greater than 30% of the selected precursor ion. This is known as the 'one third rule', and it is caused by the destabilisation of smaller fragment ions in the ion trap. The ion trap also suffers from space charge effects as it can only hold a limited amount of ions before the performance of the ion trap is reduced; it overcomes this by counting the number of ions and limiting the amount that enter the trap. This can become an issue when an analyte is present with a large population of background ions [3, 4].

23.3.2 Quadrupole and Triple Quadrupole Mass Spectrometry

A quadrupole mass spectrometer is a mass filter consisting of four parallel metal rods arranged as in the Figure 23.6. Two opposite rods have an applied potential of $(U+V\cos(\omega t))$, and the other two rods have a potential of $-(U+V\cos(\omega t))$, where U is a DC voltage and $V\cos(\omega t)$ is an RF voltage with an MHz frequency $=\omega/2\pi$ and an amplitude of Vo. Potentials on the two sets are out of phase by 180°. Quadrupole fields cause no acceleration along the z-axis. Quadrupole mass analysers operate in a similar fashion to the ion trap in that oscillating electric field is employed to select ions. The stability of ions in this oscillating electric field results in the ions being either carried along the ion path or colliding with the rods of the quadrupole and ejected out of the transmitted ion path. Only ions of a certain *m/z* ratio will therefore reach the detector for a given set of voltages. In quadrupole instruments a continuous beam of ions is generated (see Fig. 23.6), unlike in ion trap or TOF mass analysers in which pulses of ions are sent towards the detector [3, 4].

When coupled to LC, single quadrupole instruments are primarily used to perform selected ion monitoring (SIM) experiments. These methods are quite selective for applications in the pharmaceutical industry or for routine quality analysis, but when biological or complex matrices are

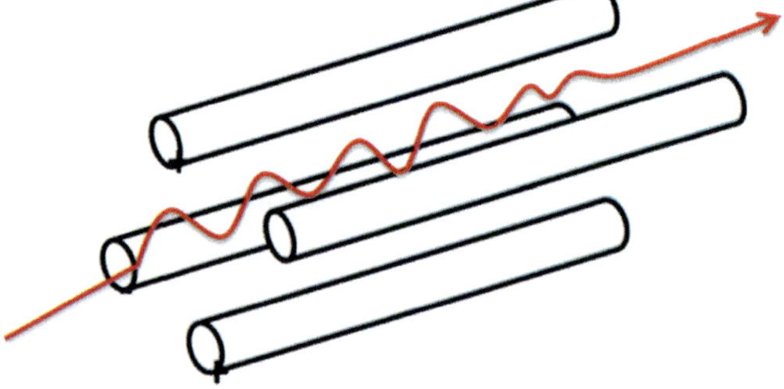

Figure 23.6 The four rods of the quadrupole act as two pairs to generate an electric field that focuses ions of a compatible m/z *through the assembly. All other ions collide with the quadrupole and out of the flight path [3, 4]*

involved, these methods can suffer from interference. Limited fragmentation information can be obtained from single quadrupole instruments.

A QqQ mass spectrometer is a tandem mass spectrometer consisting of three quadrupoles placed in tandem. The first (Q1) and third (Q3) quadrupoles serve as mass filters using a RF. The second quadrupole acts as a cell for collision-induced dissociation (CID) in the presence of an inert gas such as argon (Ar), helium (He) or nitrogen (N_2). This configuration is often abbreviated to QqQ.

QqQ instruments offer excellent versatility in the number of scan types in which they can operate. The most commonly utilised scan types for quantitation purposes on QqQ instruments are SRM or multiple reaction monitoring (MRM) scan events due to their high sensitivity, specificity and selectivity. These types of experiments are used in analytical method applications when coupled to LC. In essence, the QqQ mass spectrometer acts as a compound-specific detector with high sensitivity [19]. In QqQ instruments, three quadrupole mass filters are aligned such that ions can be manipulated in different ways through the three sections. The first quadrupole termed Q1 is used to select a certain discrete ion or range of ions depending on the scan type. The selected ions will be directed through this cell and into the second quadrupole region Q2, but all other ions will fall out of the paths directed by the electric field. Q2 is a collision cell and does not filter ions according to mass in the same way as Q1. In Q2 CID occurs, providing product ions of the precursor ion selected in Q1. The resultant product ions are channelled into Q3. Q3 acts in a similar fashion to Q1, whereby it can select a range of ions or a particular ion of a preselected *m/z* value (see Fig. 23.7).

QqQ instruments can also perform full MS scans across a desired mass range. Furthermore by selecting a precursor ion in Q1 and subjecting this to collision-induced fragmentation, a product ion MS/MS spectrum can be obtained. This type of scan is useful when structural elucidation data is needed or when choosing MRM Q1/Q3 ion pairs to develop a quantitative MS/MS method. A similar scan called a precursor ion scan is performed when a specific mass is filtered in Q3. In this type of scan, the electric field applied in Q1 is varied to allow ions across a mass range into the collision cell Q2. Ions from Q1 are fragmented and those that produce the preselected product ion(s) are sampled and correlated to their specific precursor ion. This data provides information as to the origin of a product ion and what ions it is formed from. Finally, a neutral loss scan is used

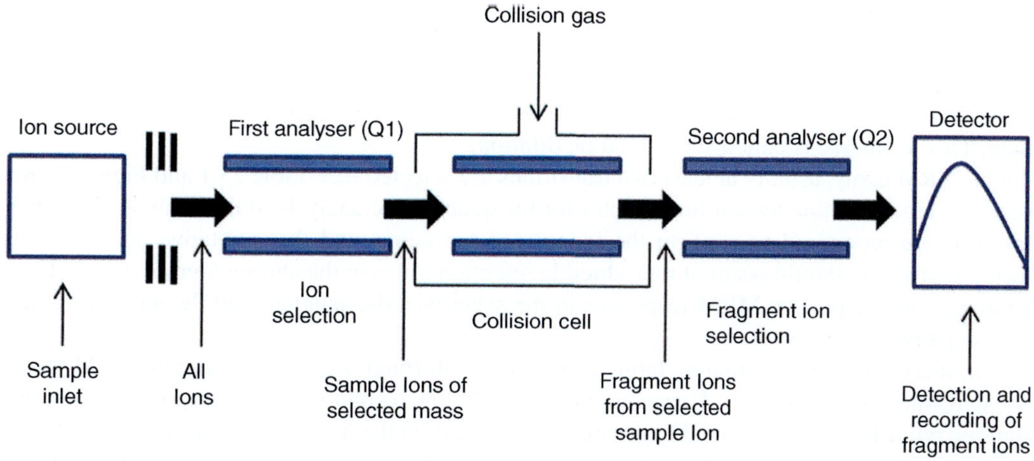

Figure 23.7 *Schematic representation of a triple quadrupole mass spectrometer*

for the screening of several compounds for common functionalities (e.g. glucuronides, sulphates, GSH adducts). Q1 scans a wide mass range. Ions are fragmented into product ions in Q2. Q3 nearly scans the same mass range as Q1; the mass range is lowered by an amount equivalent to the mass of a neutral fragment.

Pseudo-MS^3 scans are also possible on QqQ instruments. In pseudo-MS^3 scans, the source conditions are altered to partially fragment the analyte ions (in-source fragmentation) to a desired product ion. Q1 selects this product ion which is in turn fragmented in Q2. Selecting a third product ion at Q3 allows MS^3 spectra or LC-MS^3 data to be obtained. This type of method is useful in that the extra stage of MS increases the specificity of the method and can reduce noise in the resulting spectra and chromatograms.

QqQ instruments are frequently applied to analytical screening methods. They are ideal for this because they offer analyte selectivity and sensitivity. The requirement for only minimal sample preparation is an advantage of using triple-stage quadrupole MS due to the separation offered by both the chromatography and high analyte selectivity. Linear ranges for analytes have been shown to extend over several orders of magnitude and down to the ppt level. This reduces the need for sample preconcentration techniques such as SPE. A further advantage is the lack of necessity for complete chromatographic resolution in liquid chromatography–tandem mass spectrometry (LC-MS/MS) data once the analytes can be differentiated by their MS scans [4, 6].

The sensitivity of MRM assays can be increased by a prudent choice of transition ions. The signal has to be intense (excellent signal-to-noise (S/N) ratio), specific and selective. In some cases, the most intense ion from the CID spectrum may not be optimal for quantitative analyses in the MRM mode. First of all, the selected product ion should be stable and give reproducible results from scan to scan. It should also give a signal response with the best S/N ratio. Optimisation of parameters specific to each precursor → product ion transition pair must be performed.

During the optimisation procedure, similar conditions to the conditions used during the MRM assay are often applied. Flow injection analysis (FIA) can be performed to find the optimum conditions.

The collision energy (CE) is one of the parameters that have a significant effect on the intensity of the transition. When the CE is too high, the selected product ion can be lost due to disintegration. On the other hand, a low CE may result in the precursor ion not being induced to fragment to the expected product ions. The response of the detector can be disturbed by ion suppression or by impurities in the ion source or by changes in the collision cell pressure. In order to increase the reliability of the quantitative analyses and perform accurate measurement of analyte concentrations, the application of an internal standard (IS) is recommended. The use of the same compound, but labelled with a stable isotope analogue, which has very similar chemical properties, is the best option. This technique is called stable isotope dilution.

In the MRM assay, usually at least two transitions are selected (see Table 23.1 and Figure 23.8). One is used to determine the calibration plot for the quantitative analysis of the analyte. The other transitions are monitored to confirm the identity of the compound. In qualitative analyses, the selected transitions should occur at reproducible retention times in the chromatogram. In addition, the ratios of the respective MRM responses in the spectra of the standard and the analysed compound must be similar [3, 4].

In summary, LC-MS quantitative target analysis is performed principally using the SRM mode using QqQ mass spectrometers. However, the relatively low resolution of quadrupoles renders the technique prone to interference and may cause inaccuracies in the quantitation due to matrix effects (ME) in complex samples, and this increases the risk of having false positives. This subject has been particularly addressed by Kaufmann *et al.* [24].

Table 23.1 Selected transitions used in the detection of cyanobacterial toxin in MRM mode

Compound	Precursor ion, *m/z*	Product ion, *m/z*	CE, V	Reference
Microcystin-LR (MC-LR)	995.40	135.20 Q	65	[20]
		213.10	64	
Microcystin-RR (MC-RR)	519.8	135.20 Q	36	
		105.00	49	
dm-Microcystin-RR (dmMC-RR)	512.8	135.20 Q	34	
		103.05	55	
Microcystin-YR (MC-YR)	1045.45	135.20 Q	60	
		213.10	55	
Microcystin-WR (MC-WR)	1068.45	135.20 Q	65	
		213.10	64	
Microcystin-LA (MC-LA)	910.35	135.20 Q	60	
		213.10	52	
Microcystin-LY (MC-LY)	1002.40	135.20 Q	59	
		446.15	39	
Microcystin-LW (MC-LW)	1025.40	135.20 Q	62	
		213.10	59	
Microcystin-LF (MC-LF)	986.40	135.20 Q	60	
		375.20	40	
Nodularin (NOD)	825.35	135.20 Q	60	
		227.25	55	
Anatoxin-a (ANTX-a)	166	149 Q	25	[21]
		131	25	
		107	25	
		91	25	
Cylindrospermopsin (CYN)	416	194 Q	40	[22]
		176	40	
β-*N*-methylamino-L-alanine (BMAA)	119	102 Q	15	[23]
		101	10	
		88	20	
		74	25	

The quantitative transitions are marked with Q.

23.3.3 Time-of-Flight (TOF) Mass Spectrometry

TOF instruments combine the flexibility of quadrupole MS with the high-quality data obtained from a high-resolution detector [25]. TOF mass spectrometers offer high mass accuracy data that can deliver information on the molecular composition of selected target ions in the mass spectrum [26, 27].

The essential principle of TOF MS is to produce a population of ions moving in the same direction and having a distribution of masses that are accelerated under a (more or less) constant kinetic energy; this will result in the velocity of each ion being inversely proportional to the square root of the *m/z* value of that particular ion. To clarify this, consider a situation in which ions, under the influence of an external electric field, begin their acceleration from rest at the same time and from the same spatial plane normal to the acceleration vector; their arrival times at a target plane (parallel to the plane of origin) will be distributed according to the square root of *m/z* [28].

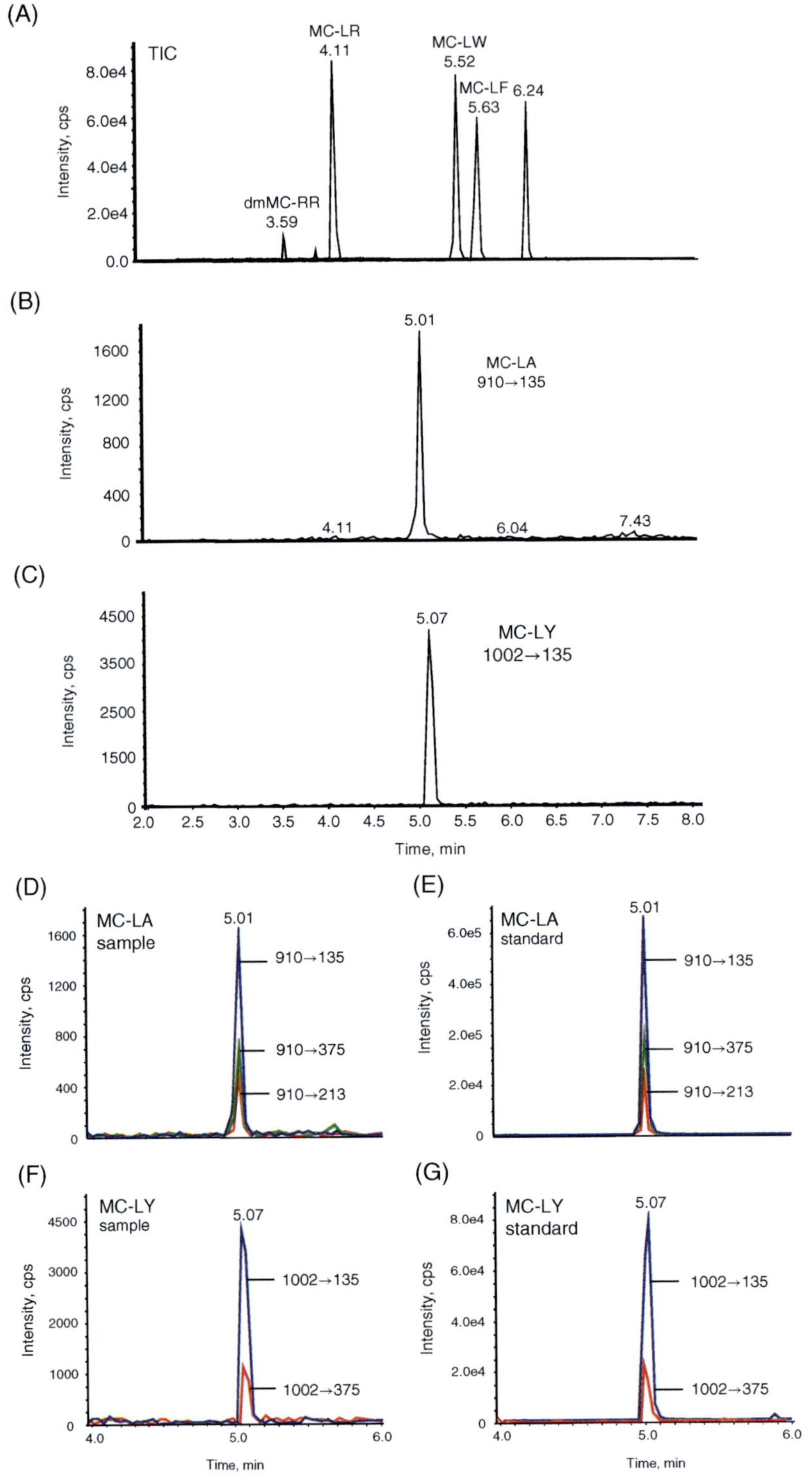

Figure 23.8 An example of MRM chromatogram for microcystins from cyanobacteria bloom sample (A) and extracted chromatograms for MC-LA (m/z 910>135) (B) and MC-LY (m/z 1002>135) (C). MRM transitions for MC-LA and MC-LY in bloom sample (D, F) and in the standard solution of the two microcystins (E, G)

The orthogonal acceleration QqTOF instrument has a quadrupole assembly in front of the high mass accuracy TOF detector. It is similar to a QqQ instrument in that the ion source and Q1/Q2 sections of the mass spectrometer work in the same way. Where a QqQ instrument has a Q3 quadrupole mass filter, the TOF instrument has a TOF mass analyser and detector (Fig. 23.9). The advantage of this arrangement is that MS/MS scans are possible, with the added benefit of high mass resolution. This type of data is invaluable in helping with structural elucidation work, as the molecular composition can be determined from the accurate masses of ions observed [3, 4].

23.3.4 The Orbitrap Mass Spectrometer

In the orbitrap analyser, injected ions cycle around a central electrode; simultaneously the ions oscillate along the horizontal axis [29, 30]. The orbitrap mass analyser consists of two uniquely designed central and outer electrodes (see Fig. 23.10). The main components of the orbitrap are an outer electrode that is shaped like a barrel and an inner spindle-like electrode placed along the horizontal axis of the trap. Each electrode is connected to a separate voltage supply [29, 30]. The outer electrode is divided into two main parts, one for excitation and the other for ion detection. The space between the inner and outer electrodes forms a measurement region that is linked to a

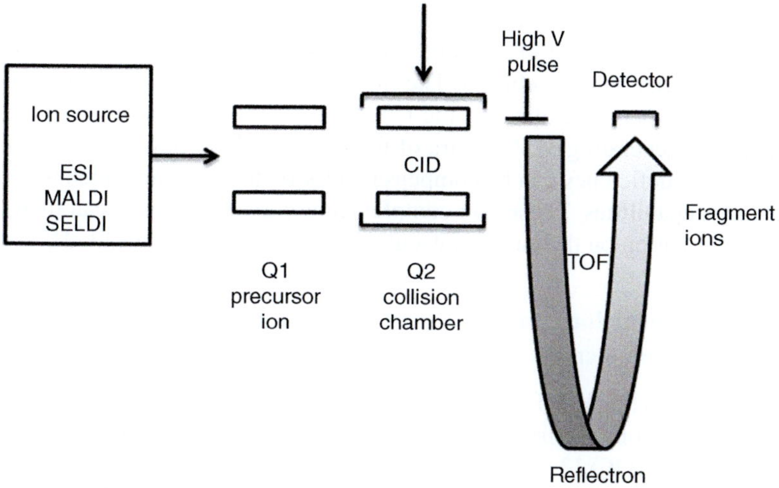

Figure 23.9 A QqTOF mass spectrometer has mass filtering capabilities in Q1 combined with the high-resolution TOF mass analysers. Ions are accelerated orthogonally to the flight path and bent towards the detector in durations related to their mass [3, 4]

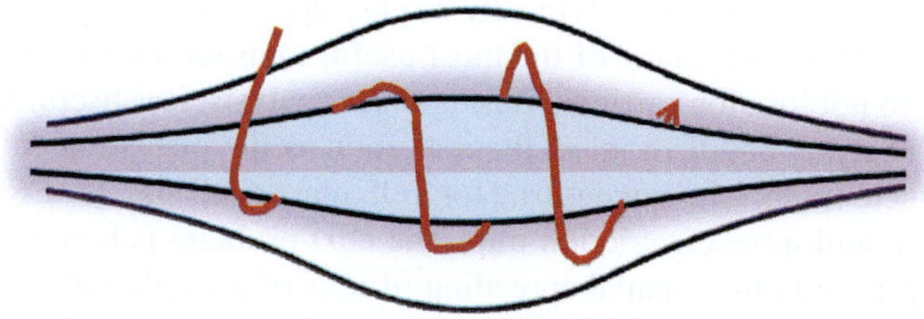

Figure 23.10 Schematic representation of the orbitrap mass analyser [3, 4]

vacuum pump to achieve a vacuum in the region of less than or equal to 10^{-8} Torr. The orbitrap mass analyser is preceded by an external trapping device which traps ions in an RF-only gas filled curved quadrupole (called the C-trap in the case of the LTQ orbitrap). For optimal signal response and high resolution, ions of each mass must be injected into the orbitrap in short, compressed 'packets' [31]. To inject the ions into the orbitrap from an external ion source, the magnitude of the electric field is first of all reduced. As the ion packets are injected into the space between the electrodes at an angle which is at a tangent to the orbitrap equator, the voltage is then ramped to increase the electric field on the inner electrode. As a result of the voltage ramping, the ions are compressed towards the inner electrode until they achieve the desired orbit within the trap. The voltage ramping is then terminated, and the field becomes static and detection of the ions can begin. Each ion packet comprises a legion of ions of differing velocities spread over a certain space volume; consequently the ions move with different rotational frequencies but with identical axial frequencies. As a result ions of specific *m/z* ratio spread into rings, rather like helices, that oscillate along the inner spindle-like electrode. The entire ring then oscillates along the central electrode harmonically with a period proportional to $(m/z)^{1/2}$, and this produces an image current on the split outer orbitrap electrodes. This signal is detected and undergoes Fourier transform to obtain a frequency-domain signal which is then converted to an *m/z* spectrum [32]. Linearity of detection combined with high fidelity of frequency determination accounts for the high resolution and mass accuracy achievable by the instrument [32].

The LTQ hybrid MS is a hybrid instrument which has two different but complementary mass analysers: a linear ion trap and an orbitrap. The mass analysers can be used independently or in conjunction with each other depending on the type of data desired. The linear trap can perform CID and is capable of generating MS^n spectra of the target analyte; however it can only measure the nominal mass. This deficiency can be compensated for by the orbitrap which we have seen has high mass accuracy capabilities. These two complementary mass analysers can therefore provide a rich variety of information on the target molecule and related compounds [3, 4].

23.3.5 The Hybrid Linear Ion Trap–Fourier Transform Ion Cyclotron Resonance Mass Spectrometer

Ion cyclotron resonance (ICR) and orbitrap mass analysers are based on measurements of RF rather than on ion deflection (electric/magnetic sectors), ion stability (quadrupole mass analyser, QIT) or time of transit (TOF) [3, 4, 33]. ICR and orbitrap frequencies are independent of ion energy, and the induced signal is linearly proportional to the ion motional amplitude. For such linear response systems, it is possible to obtain a frequency-domain spectrum by Fourier transformation of the time-domain signal for an initially spatially coherent ion packet. In ICR, initially spatially incoherent ions of a given *m/z* become spatially coherent following resonant RF electric field excitation.

Both ICR and orbitrap mass analysers typically require approximately 1 s for data acquisition, storage and processing. However, many of the most useful ionisation sources generate ions continuously. Therefore, to not lose ions arriving during mass analysis, it is useful to accumulate ions external to the mass analyser and then inject them as soon as the preceding data acquisition/processing cycle is complete (in fact the same is true for TOF mass analysis). In ICR, ions are conveniently accumulated in a multipole electric ion trap. The next problem is how to eject ions quickly from the external trap to minimise spatial spreading of ions of a single *m/z*. In ICR, ion ejection may be synchronised by applying 10–30 V to tilted wires placed between the rods of a multipole electric ion trap.

For FT-ICR, ions moving in a spatially uniform static magnetic field, B, rotate at a cyclotron frequency, ν_c (Hz), where

$$\nu_c = \frac{ezB}{2\pi m},$$

in which 'e' is the elementary charge. At room temperature, typical ion cyclotron orbital radii are at the submillimetre level; moreover, ions of a given *m/z* rotate with random phase. Thus, to generate a detectable signal, it is necessary to resonantly excite the ions with an oscillating or rotating electric field to yield a spatially coherent packet of ions of a given *m/z*. The motion of this ion packet gives rise to a time-domain signal consisting of the difference in current induced on a pair of opposed electrodes. This signal is digitised and subjected to discrete fast Fourier transformation to yield a spectrum of ion cyclotron frequencies, which may then be converted to a spectrum of *m/z*.

The hybrid linear ion trap–Fourier transform ion cyclotron resonance mass spectrometer (LT-FT-ICR) is an attractive alternative [34], compared with the TOF MS and LC-MS/MS, as routine analytical tool. Several reviews on the application of MS in the environmental and food analysis have been published [34, 35].

Finally, all the current recommended MS terminologies (e.g. nominal mass, accurate mass, exact mass, RDB, nitrogen rule, isotope ratio, SRM, transition ratio, etc.) are included in the IUPAC 2013 guidelines [36].

23.4 Application of LC-MS in Cyanotoxin Analyses

Recent reviews on methods and approaches used for the detection of cyanotoxins in environmental samples have been published [37–40].

Most cyanobacterial toxins are non-volatile, relatively hydrophilic and easily ionisable and thus successfully analysed by LC-MS.

FAB, MALDI and ESI are the ionisation sources most commonly applied to the analyses of cyanobacterial toxins. Specifically, positive ESI LC-MS has been applied to the separation and detection of cyanotoxins as a standard approach for quantitative purposes. Additionally, negative ESI mode has recently been applied to the analysis of certain cyanotoxins [41–43]. The ESI-MS has traditionally been understood as a concentration-dependent device. The best chromatographic resolution is achieved by miniaturising the chromatography in terms of column diameter and stationary phase particle size. Higher response and LOD are achieved using capillary columns. The small column diameters necessitate strictly zero-dead-volume capillary connections in order not to lose the high chromatographic efficiency. Furthermore, column clogging by particulates tends to be a more serious problem with narrow-bore and capillary columns, and great attention has to be paid to sample and mobile phase preparation and filtration (0.2 μm) to overcome the problem. Most mass spectrometers also perform better at lower mobile phase flow rate. The lowest mobile phase flow rates are achieved using specially designed interfaces including the nano-ESI ionisation sources.

Nowadays, four major trends can be taken into account within the cyanotoxin analysis by MS techniques: (i) the rapid analysis with minimum treatment of sample, (ii) the development of multi-class toxin methods [44–48] and the analysis of new toxins present in cyanobacteria or water samples [49, 50], (iii) its application to the structural characterisation of toxins and their degradation products mainly by HRMS and (iv) the introduction of quantification methods based on isotope dilution using IS [51]. A stable isotope-labelled standard has often been recommended as the ideal choice of IS [52, 53].

In order to have faster methods for toxin analysis, UHPLC [54–56], *online* SPE [57–59] and MALDI-TOF have been implemented in recent works. The current trend in routine environmental and biomedical LC–MS is to use short (typically 3–5 cm) narrow-bore columns with sub-3 μm particle size. As the small particle size results in a high chromatographic efficiency, the run times can be shortened without compromising resolution by reducing column length. However, with sub-2 μm particles and with the optimal linear velocity of the mobile phase, there is often a problem with high backpressure in the LC system. Thus the LC system has to produce/withstand pressures approaching or even exceeding 1000 bars. Such UPLC (Waters trademark) and UHPLC (a generic term for ultrahigh-performance liquid chromatography) systems are nowadays the preferred solvent delivery equipment in the rapid LC-MS of cyanobacterial toxins.

MALDI-TOF MS has been employed for simple and rapid screening over a wide molecular mass range and for the preliminary identification of cyanobacteria biotoxins [60]. MALDI-TOF is a very attractive approach for many analytical applications because it is a high-resolution and sensitive technique that requires little sample handling, has low sample consumption and allows accurate identification of toxin congeners simultaneously without the need for preceding separation (see Fig. 23.11). Unfortunately, MALDI-MS quantification is still troublesome. The poor spot-to-spot and shot-to-shot reproducibility of the analyte signal intensities makes it necessary to use an IS for quantitation [61]. Additionally MALDI cannot be coupled with LC to separate sample components.

Although the need for baseline-resolved chromatographic separation in LC-MS is not as critical as with optical detection, LC-MS analyses also benefit from well-resolved narrow peaks. Good

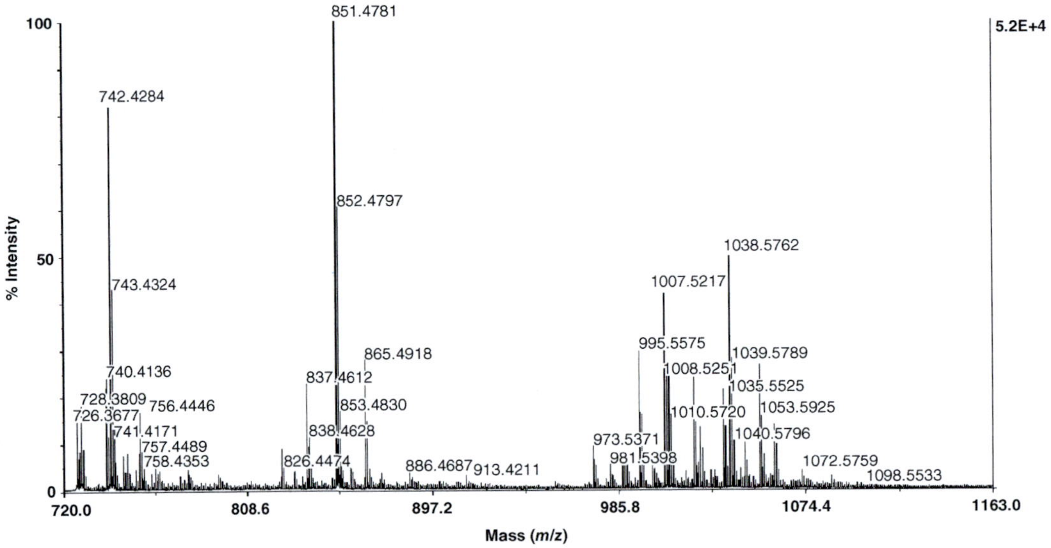

Figure 23.11 *Positive MALDI-TOF full-scan spectrum of a cyanobacterial* bloom *sample*

chromatography gives high analyte concentrations across the peaks, resulting in high S/N ratios. High resolution helps to minimise the ion suppression (which causes anomalies in quantification) and ensures the avoidance of co-elutants while also providing enough time for the instrument to do proper MS/MS data acquisition of the precursor ions.

Mainly three types of MS methods are employed when monitoring environmental toxins:

1. Those that give nominal masses and perform MS/MS (QqQ or QqLIT) [62–65]
2. Those that allow elucidation of fragmentation sequences by multiple tandem mass (MS^n) analysis (IT, QqLIT, LTQ orbitrap or LT-FT-ICR) [66, 67]
3. Those that are capable of obtaining accurate mass measurements by MS or MS/MS (TOF, QqTOF, orbitrap, LT-FT-ICR) [45, 68–73]

These three types of MS experiments are considered as complementary to each other in order to obtain the maximum information on target analytes. Although accurate mass measurements and MS/MS and MS^n studies are ideal to allow identification of unknown contaminants, QqQ instruments are better suited for the routine quantitative analysis of target contaminants. Several studies describe the identification of cyanotoxins by use of data recorded with different MS techniques (MS/MS and MS^n experiments and accurate mass measurements) [70, 72, 74, 75]. MS techniques have clearly demonstrated their potential for tentative structural identification of cyanotoxins, but final proof of identity can only be provided by the use of the MS data in combination with data recorded by NMR spectroscopic analyses. Some additional support of the proposed structure may be obtained by thiol derivatisation, chiral amino acid analyses, ultraviolet–visible spectroscopy and deuterium exchange experiments [75, 76].

23.5 Overview of Quantitation: Cyanobacterial Toxins

Quantitation of an analyte requires the construction of a calibration plot in which known concentrations of a reference substance (standards) are plotted against observed signal responses. Based on this plot, the signal response of an analysed sample can be converted to the corresponding analyte concentration.

The development of an LC-MS assay is usually initiated with an *external standard* approach where the standards and the samples are analysed in separate runs. In this case the standards are dilutions of the authentic target compound. In order to get good quantitation using external standards, the chromatographic and mass spectrometric conditions must remain constant during the sample series. Sample matrix often influences the ionisation efficiency of target ions which may be either suppressed (more common) or enhanced by the matrix. The successful use of external standards in LC-MS thus necessitates the use of matrix-matched samples and standards. It is typical to run external standards at the beginning and at the end of a sample sequence as well as to embed them throughout the sequence.

As is well known, the method based on isotope dilution using IS is the most reliable quantitation approach for MS analysis [51]. For example, nodularin is an IS widely used in microcystin analysis. However, an ideal IS in LC-MS is the target compound labelled with stable isotopes such as deuterium or ^{15}N [52, 53]; often however these are not commercially available.

Fluctuations in the injection volume, retention time, ionisation efficiency and detector performance can be checked by adding an IS to the samples just before the injection. The IS should be sufficiently different from the analyte in order to identify it in the mass spectrum but still related in behaviour, and it must be absent from the non-spiked samples. The IS must either be chromatographically resolved from the analyte (when using optical detection) or differ from the analyte by several mass units (MS detection). The expected signal of the IS must be consistent in all the spiked samples. The identification and quantitation of unknowns are thus based on relative retention times and relative response factors, respectively, established by analysing known concentrations of the target compound spiked with the IS.

A *surrogate standard* is similar to an IS but added to the sample at the beginning of sample workup. A surrogate standard has to mimic the properties of the target compound during sample preparation. The surrogate standard not only covers the purposes of an IS but also gives an idea of the target compound recovery during sample preparation. It is important to note that the definition of internal and surrogate standards varies [77].

23.5.1 Preparing of Standards for LC-MS Applications

With references to other chapters in this book, there are a number of parameters that should be considered when preparing cyanobacterial standards for LC-MS analysis. These include consulting the relevant material safety data sheets (MSDS) that outline the standard purity, stability and solubility in various solvent compositions. Ideally certified standards with corresponding deuterated IS should be sourced; however these are not always available. Therefore analytical standards (with and without supporting information) are the only alternative.

Choosing the correct solvent, mobile phase and mobile phase buffer(s) are also important parameters to optimise to ensure that efficient and optimised analyte ionisation and chromatographic separation are achieved for the range of targeted cyanobacterial toxins.

23.5.2 An Approach to the Tuning of Target Analytes (0.1–1 μg mL^{-1})

Considerations include the following:

- Preparation of the standards in an appropriate solvent (ensuring efficient ionisation).
- Option #1: Infusion of the individual standards or a mixture of standards directly into the MS (10 μL min^{-1} of a 1 μg mL^{-1} solution) (automatic/semi-automatically, manual calibration). Follow manufacture guidelines – SOPs.
- Option #2: Post-column infusion experiment. A T junction is placed between the LC system (with or without a column) and the MS source. A blank matrix extract is injected (10 μL) via the autosampler into a mobile phase at flow rate of 0.19 mL min^{-1}, and the mixture of standards is delivered via a syringe pump under a flow rate of 0.01 mL min^{-1}. Total flow is 200 μL min^{-1} (mimic the expected final chromatography flow rate). The response of the standard mixture is monitored continuously (in both full-scan mode and SRM) to produce a profile of the ME with the chromatographic region of the standards. This was repeated using all blank (target analyte-free) matrix extracts (cyanobacteria, tissues, raw and purified water) for the suite of samples typically tested on the LC-MS method.
- Option #3: Chromatographic injection (10 μL) of standards into the mobile phase to get chromatographic peak areas, applying (short) gradients – with a solvent mixture of acetonitrile/water over time. Plot of peak areas versus tune parameter (different transmission, focusing and declustering voltages, CE, etc.) to get optimised and robust parameter values.

LC-MS operator and training manuals should always be consulted during the process of tuning analytes.

23.6 Ion Suppression/Enhancement Considerations

LC-MS techniques provide unprecedented analytical capabilities. In fact LC-MS has often been marketed as the panacea for many troublesome (complex) analytes. However, in recent years users have come to realise that LC-MS can be susceptible to interferences from the sample matrix components that may adversely hinder analysis; this of course applies to the analysis of the cyanobacterial toxins [78]. Matrix constituents can cause 'ion suppression' or 'ion enhancement' collectively called 'matrix effects' in LC-MS methods that can compromise both the reproducibility and the accuracy of results [79].

23.6.1 Matrix Effects

ME interfere with the response of the analyte in proportion to analyte concentration. ME occur on account of reactions between the analyte(s) and some constituent of the matrix, resulting in a species that can suppress or enhance signal responses in the MS [80].

By far the most commonly encountered ME observed in LC-MS methods is ion suppression which usually occurs in the early stages of ionisation in the MS [81]. The main causes include:

a. Competition between matrix components and analyte ions which co-elute in the vaporised solution at the MS ionisation interface [81, 82]
b. Matrix components that compete for available charge
c. Matrix components that bind to the analyte or that induce the analyte to co-precipitate
d. Analyte ions that are neutralised in the gas-phase acid/base reactions
e. Additives to the mobile phase
f. Equipment design [83, 84]

Common ion suppressors include ionic species, very polar molecules and organic compounds. Some factors potentiate ion suppression, for example, high concentration, high mass, basicity and co-elution with the analyte. Ion suppression may be caused by external sources, that is, agents introduced during the sample collection and preparation steps [79].

The effects of ion suppression are as follows:

- The detection capability can be impaired due to the decrease in the analyte signal. This results in a poorer LOD for the analyte, a lower S/N ratio and a reduced linear range.
- Reproducibility can be compromised because the degree of ion suppression may vary from sample to sample, due to slight differences in the sample matrix.
- Severe ion suppression can mask the analyte leading to non-detection.
- Conversely, ion suppression that masks an IS signal can result in an overestimation of analyte concentration [79].

It is imperative that LC-MS method validation must address ME.

Where SIM or SRM is used, only the ions of a preselected *m/z* ratio appear in the mass spectra, and matrix components (unless isobaric) are absent from the spectra. However, components in the matrix constituents may have already diminished the ionisation efficiency of the target compound [78, 85, 86]. Therefore, during LC-, SIM- or SRM-MS method development, it is important to acquire a full-scan MS spectrum to identify potential ME, noting the region of the chromatogram where they elute.

23.6.2 Detecting and Evaluating of Ion Suppression

There are several strategies to investigate, quantitate and counteract ion suppression in LC-MS:

i. ME may be identified in a sample by comparison of the signal of the analyte standard spiked at a known concentration into a matrix-matched blank with that of the standard spiked into neat mobile phase or solvent. If the matrix-matched sample gives a lower response than the sample in solvent, this may indicate ion suppression. Similarly, calibration curves prepared from calibrants in solvent and in analyte-free matrix extracts can be compared. The calibration curve of a matrix that is free of ion suppression should be superimposable on the plot of the standard in solvent [87].

ii. An established method for detecting ion suppression is through a post-column infusion experiment, an approach which was first proposed by Bonfiglio *et al.* [88, 89]. A standard solution, containing the target, is continuously infused into the column effluent through a T-piece [18, 90]. A blank matrix sample is then injected into the system (while the analyte is continuously infused), and the blank sample is separated on the LC column. A drop in the baseline signal (indicating suppression of signal) in parts of the chromatogram highlights the presence of culprit matrix components [79, 91–94]. Post-column infusion can also be used to prevent ion suppression. This is achieved by infusing a buffer post-LC to adjust the pH to improve ionisation efficiency by promoting the formation of alternative adduct ions.

iii. A third method for detecting ion suppression is a variation of (i). It involves comparing peak areas of the analyte spiked into a matched matrix before extraction, into the sample extract after extraction (matrix-matched standard) and into neat solvent. Using this method the recovery and ME associated with the sample can be quantified [95]. In Matuszewski *et al.*, the 'absolute' ME was calculated using the formula

$$\text{ME}\ (\%) = \frac{\text{X}}{\text{Y}} \times 100$$

where X = peak area for analyte in solvent and Y = peak area of standards spiked into matched matrix after extraction.

Recovery (RE) can be calculated using

$$\text{RE}\ (\%) = \frac{\text{Z}}{\text{Y}} \times 100$$

where Z = peak area of standards spiked before extraction.

Overall process efficiency (PE) can then be calculated by

$$\text{PE}\ (\%) = \frac{\text{Z}}{\text{X}} \times 100 = \frac{\text{ME} \times \text{RE}}{100}$$

Any difference in the analyte response of sample extracts that were spiked with standard after extraction but before evaporation and reconstitution in the mobile phase will be due to the sample matrix [96].

Once ion suppression is identified in a sample matrix, how can it be reduced or removed?

- Selective extraction [82, 85, 95, 97]
- Effective sample clean-up after extraction [97, 98]
- Re-optimisation of the chromatographic method to attain full separation of ion-suppressing (or ion-enhancing) components from the analytes of interest [86, 99–101]

Sometimes, these approaches are not satisfactory as they could lead to analyte losses (a and b) or very long chromatographic runs (c). Other approaches include using:

- External calibration with matrix-matched samples
- Standard addition [80, 90]
- IS calibration [51–53,102]
- Sample dilution which can provide a very simple way to reduce the amount of coexisting matrix components in a sample

Sample clean-up to remove ME often includes SPE [97, 98, 103], liquid–liquid extraction (LLE) [97] and protein precipitation (PP) [104]. These techniques may involve a compromise between the removal of ion-suppressing agents from the matrix and significant losses of the sample and can be laborious and time consuming to develop. In addition care must be taken not to add ion-suppressing agents during sample preparation and clean-up.

Another option in reducing ME may be to change the ionisation polarity from positive to negative ESI [3, 95]. For some applications, different source designs influence ion suppression according to the general rule: Z-spray < orthogonal spray < linear spray (ranked in order of reducing ion suppression effects) [105]. It was found by Stahnke *et al.* that a Jet Stream ESI (Agilent Technologies Inc.) interface which deploys superheated sheath gas to desolvate ions delivered higher sensitivity than other ESI interfaces [106]. Changing the ionisation interface, for example, from ESI to APCI, may help [107]. Another ionisation source called APPI has been shown to reduce ion suppression significantly [108].

The use of nano-LC has also been found to reduce ME. In a study by Gangl *et al.* [109], reducing the LC flow rate entering the ESI source to the nano-litre per minute range led to increased desolvation, ionisation and ion transfer efficiency compared with conventional LC.

In conclusion, the removal/reduction of ME is a trial-and-error process [18]; it is however a process that needs to be incorporated into the validation protocol for all LC-MS methods designed for the testing of cyanobacterial toxins in water, bloom and tissue samples [78].

23.7 High-Resolution Mass Spectrometry (HRMS)

Although LC-MS on a QqQ MS operating in SRM is a method widely used in the literature for quantitative analysis of biotoxins, the HRMS approach provides at least comparable specificity, accuracy, linearity and sensitivity [110]. However, the relatively low resolution of quadrupoles (typically unit resolution) renders the technique prone to matrix interference from ions of similar mass in complex samples. Also, significant method development time is generally required, as SRM transitions are typically optimised for each analyte. Additionally, due to the targeted nature of SRM, only known toxins specified in the method will be detected.

The recent advances in MS instrumentation, especially those related to HRMS – Q-TOF, TOF-TOF and orbitrap – have created a paradigm shift in LC-MS analysis. The availability of sophisticated MS instruments (e.g. Q-TOF, QqLIT and Q-orbitrap) enables the combination of both features: it performs accurate mass measurements, and on the other hand, it offers high-sensitivity MS/MS and multiple MS (MS^n) experiments. As a result, the current trend in MS is an integrated quantitative and qualitative bioanalysis approach [35, 111, 112].

This new generation of MS technologies should be able to achieve resolution greater than 20 000 (full width at half maximum (FWHM)) that is feasible in the working mass range, to obtain an extended dynamic range and to be effectively coupled to current LC systems, such as fast LC and UHPLC. It should be noted that the great specificity due to mass accuracy of HRMS allows elemental formulas to be assigned which can aid with the assignment of tentative structures of analytes and identification of new toxins or toxin analogues.

The inherent quantitative strategy of HRMS uses full-scan acquisition mode with reconstructed ion currents for very narrow extraction windows (2–5 ppm around the theoretical/measured *m/z* value). In contrast, GC-HRMS analysis by magnetic sector analyser and electron ionisation have to be limited in mass range and usually use SIM for quantitation purposes. It is worth mentioning that GC-HRMS works by a resolution of 10000 (10%) for routine analysis and employs continuous calibration through the use of a lock mass to preserve the accuracy by in situ recalibration.

LC-HRMS has been applied to the characterisation of microcystins combined with positive ESI using various mass spectrometers including FT-ICR, Q-TOF and orbitrap [67, 68, 70, 73]. Recently a few studies have been published where negative ESI mode has been used for the analyses of these cyanotoxins [41].

In particular, orbitrap operating in full-scan and high-energy collision dissociation (HCD) acquisition modes is able to provide simultaneously the MS and product ion full-scan spectra with high mass accuracy. As an example Figure 23.12 shows the total and reconstructed chromatograms

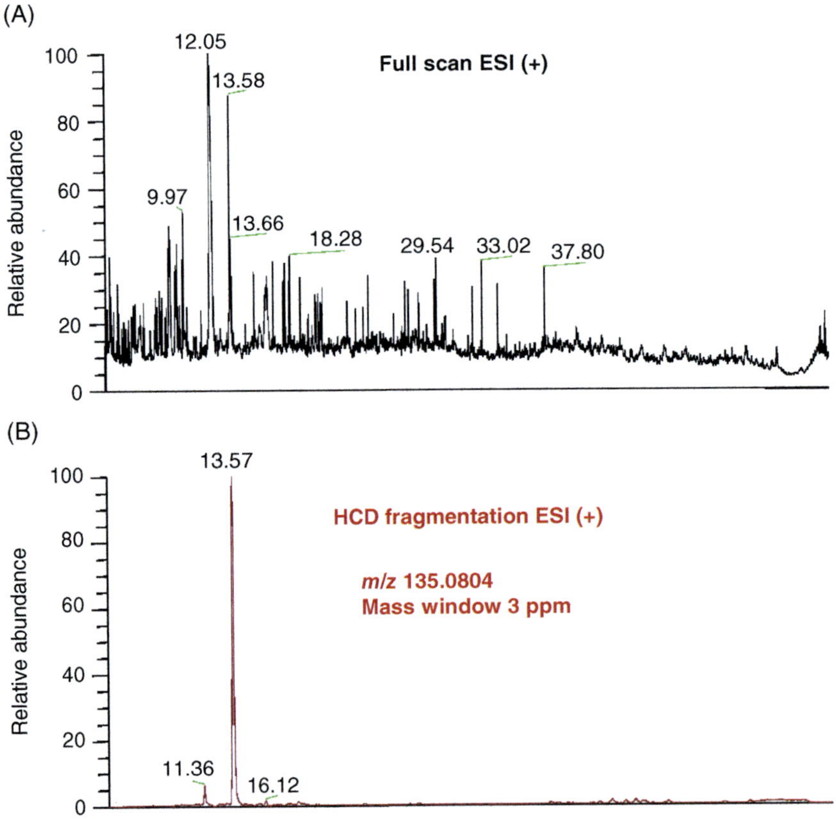

Figure 23.12 Chromatogram of LC/ESI-HRMS analysis by Orbitrap-Exactive HCD of intracellular fraction from Microcystis aeruginosa *bloom extract, Galicia reservoir, Spain, in positive and negative ionisation modes in (A) and (C) full-scan mode acquisitions and (B) and (D) HCD fragmentation mode, monitoring the diagnostic fragment ions*

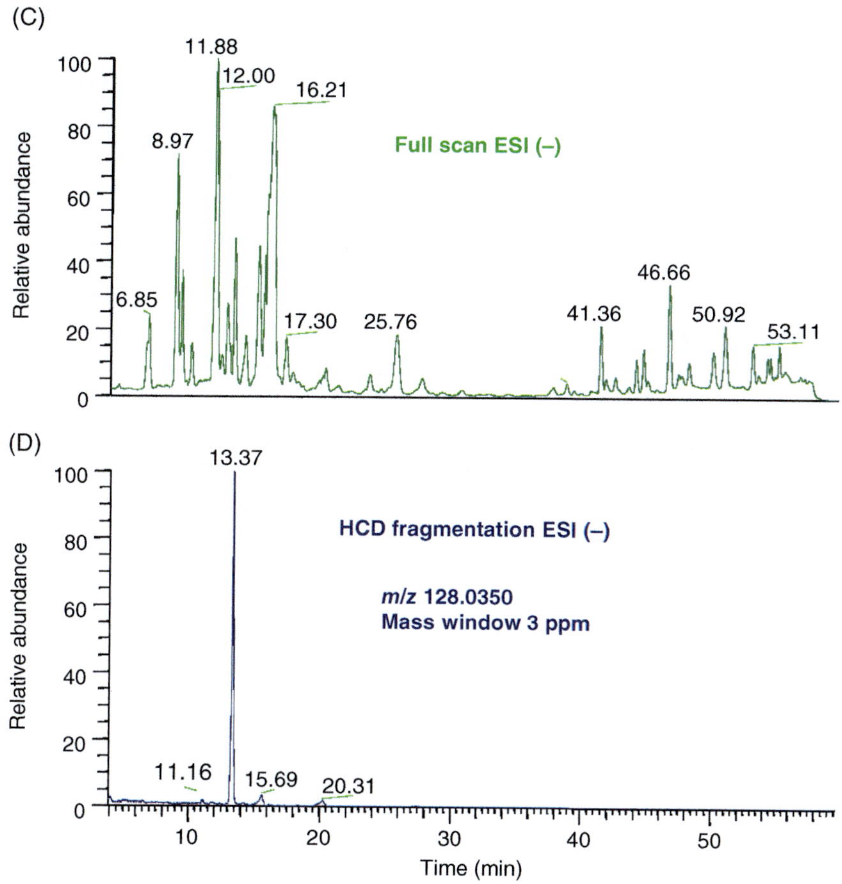

Figure 23.12 (*Continued*)

of a 'real' water sample by LC/ESI-HRMS analysis with an Orbitrap-Exactive HCD mass spectrometer in full-scan and HCD fragmentation acquisition modes with both ionisation modes (positive and negative). Two characteristic fragments, specifically the most selective and intense ions of *m/z* 135.0804 and 128.0353 in positive and negative ionisation modes, respectively (see Figures 23.12 and 23.13), are used in the reconstructed ion chromatogram for identification and quantitation purposes. Figure 23.13 shows the diagnostic ions in both ESI modes by HRMS and their accurate masses. Worth mentioning is the presence of ions with a nominal mass *m/z* 135 which may be significant in complex samples (blooms). As a consequence, these interferences may give false-positive results. Therefore, accurate mass measurements of the characteristic Adda fragment with a delimited extracted window (2–5 ppm) can provide reliable results.

Orbitrap also provides the product full-scan ion spectra with high mass accuracy. This increases the specificity due to the availability of product ions, which can be used for quantitation and confirmation. As an example, Figure 23.14 shows the reconstructed ion chromatogram of *m/z* 128.0350 (2 ppm of window) and the product ion spectra of a *Planktothrix rubescens* bloom extract from a Galicia reservoir, Spain, (A) and (C), respectively; and a mixture of individual commercial standards microcystins (demethylated microcystin MC-RR, MC-LR,

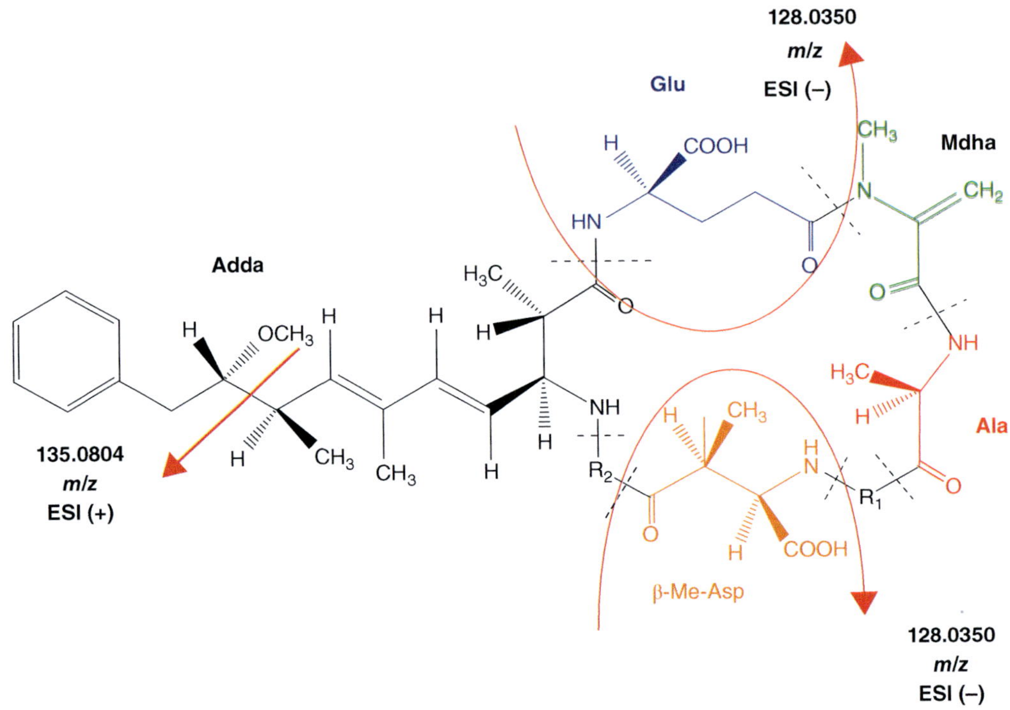

Figure 23.13 General chemical structure of microcystins with two main diagnostic ions identified in both ionisation modes: m/z *135.0804 (Adda) and* m/z *128.0353 (isobaric glutamic acid/β-Me-Asp) for positive ESI and negative ESI modes, respectively*

MC-YR, MC-LR, MC-WR and nodularin), (B) and (D), respectively. Demethylated microcystin-RR (MC-dmRR) and demethylated microcystin-LR (MC-dmLR) are clearly identified in the Galicia bloom extract. As an example, the matching product ion spectra of MC-dmLR from a bloom sample and a standard mixture are shown in Figure 23.12C and D, respectively. The data generated by LC-HRMS and HCD fragmentation (including retention time, accurate mass of precursor and product ions, ion ratio, MS and MS/MS spectra, isotope pattern, etc.) gives excellent linearity (5 orders) and good sensitivity. The interferences originating from the matrix are resolved with a resolution of 23 000.

In conclusion, HRMS is emerging as an attractive approach for combining the qualitative and quantitative analysis together, minimising the effect of the matrix and also reducing the inaccuracies commonly associated with quantitative LC-MS/MS [24].

23.8 MS Experiments for the Detection of Unknown Cyanotoxins

New and emerging MS technologies have opened up new vistas of possibilities in the field of toxin discovery. High-resolution and high mass accuracy technologies like TOF MS, orbitrap and FT-ICR MS and hybrid instruments such as the LTQ orbitrap MS [2, 24, 33, 113, 114], together with the ability to obtain multigenerational fragment ions, can facilitate the identification of not only targeted analytes but also related or even new compounds [2, 35, 66, 67]. These 'new' compounds (or previously undetected compounds) can include members of new toxin classes, unknown members of an established toxin family and minor analogues of well-known toxins and/or toxin metabolites, bioconversion products or toxin degradation products [45, 60, 61, 67–69, 71, 73]. Many of the new MS instruments allow for archived data to be subjected to a retrospective investigation to determine previously unknown but pertinent compounds.

The analytical basis for identifying new compounds is related to establishing accurate mass data which can provide unequivocal elemental composition information. Furthermore, many of the previously mentioned LC-MS technologies can provide fragmentation spectra that allow characteristic fragmentation ions to be identified (e.g. for many the presence of an ion at *m/z* 135.0804 in the positive ion MS/MS spectrum can indicate that the analyst has found a member of the microcystin family (see Fig. 23.15)). In other cases isotopic patterns can assist in the exploration of new analogues [32, 33, 36, 113–116]. If a potential analyte is found, its status as a toxin or compound of interest is confirmed by many assays outside of the realm of MS analysis.

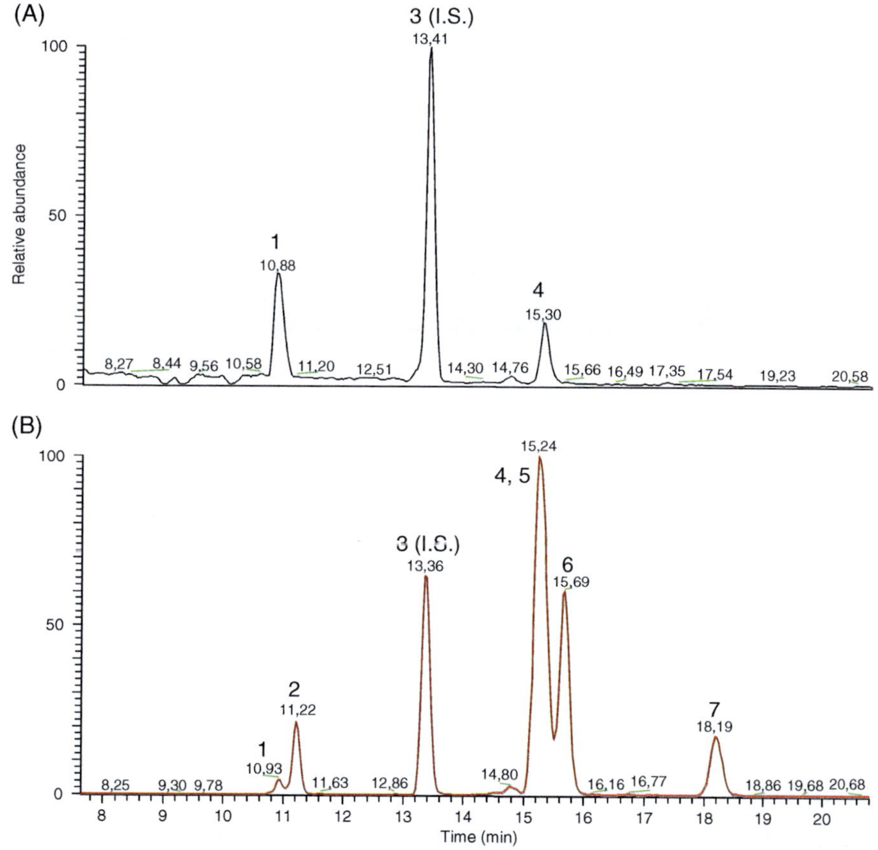

Figure 23.14 *Comparison of* m/z *128.0350 reconstructed ion chromatogram and product ion spectra of signal 1 from* Planktothrix rubescens *bloom extract from Galicia reservoir, Spain, (A) and (C), respectively; and a mixture of individual commercial standards microcystins, (B) and (D), respectively. Nodularin is used as internal standard (IS). (1, demethylated microcystin-RR; 2, microcystin-RR; 3, nodularin; 4, demethylated microcystin-LR; 5, microcystin-YR; 6, microcystin-LR; 7, microcystin-WR.)*

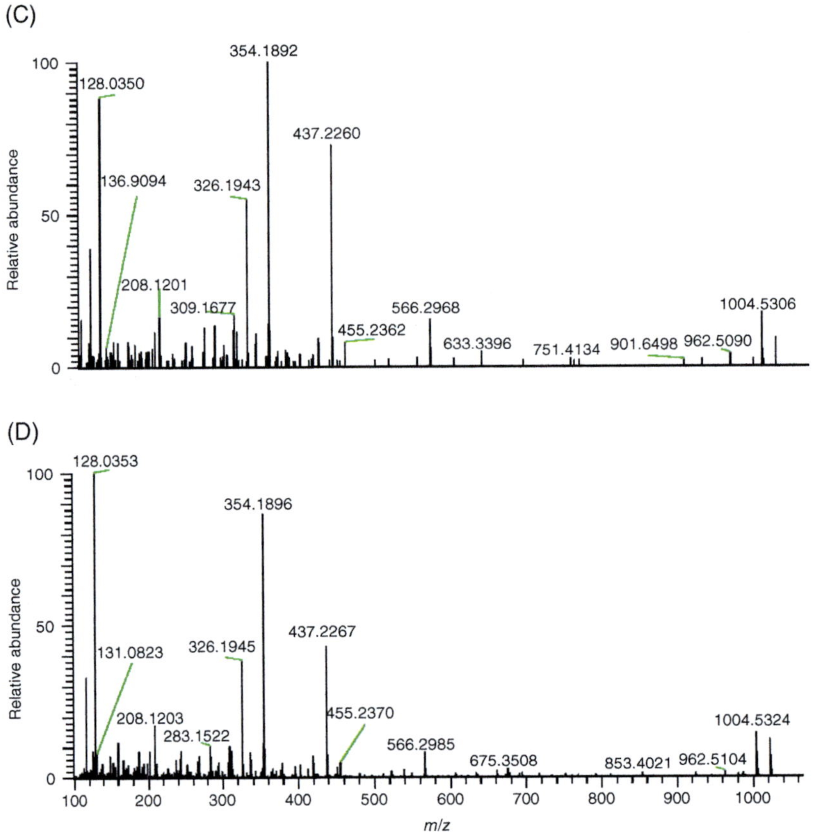

Figure 23.14 (*Continued*)

The 'new' compound will generally be isolated by advance chromatographic approaches and will be subjected to many analytical investigations, including UV and IR analysis as well as H^{1-} NMR analysis, if there is sufficient quantity of isolate. Further toxicological examinations involving *in vivo* and *in vitro* bioassay testing are likely to ensue to further classify and characterise the new 'toxin'. It is important to acknowledge the limitations of MS and concede that the identification of new and interesting compounds come from a selection of different analytical approaches [24, 72, 74, 117, 118].

23.8.1 Interpretation of Mass Spectra: Based on Examples of Different Cyanotoxins and Diagnostic Ions

Characterisation of cyanotoxin structure is usually based on the fragmentation spectrum. Every chemical structure or structural type has a specific fingerprint pattern in its fragmentation spectrum. The collected spectrum can be compared with the spectrum of a cyanotoxin standard collected under the same LC-MS/MS conditions. Optionally, the available published data, software or spectrum libraries can be used to identify the compound. When the exact *m/z* value of the molecular ion (with an accuracy of <10 ppm) is determined, the list of possible molecular formulas can be proposed, with a reduced risk of misinterpretation of the spectrum [67].

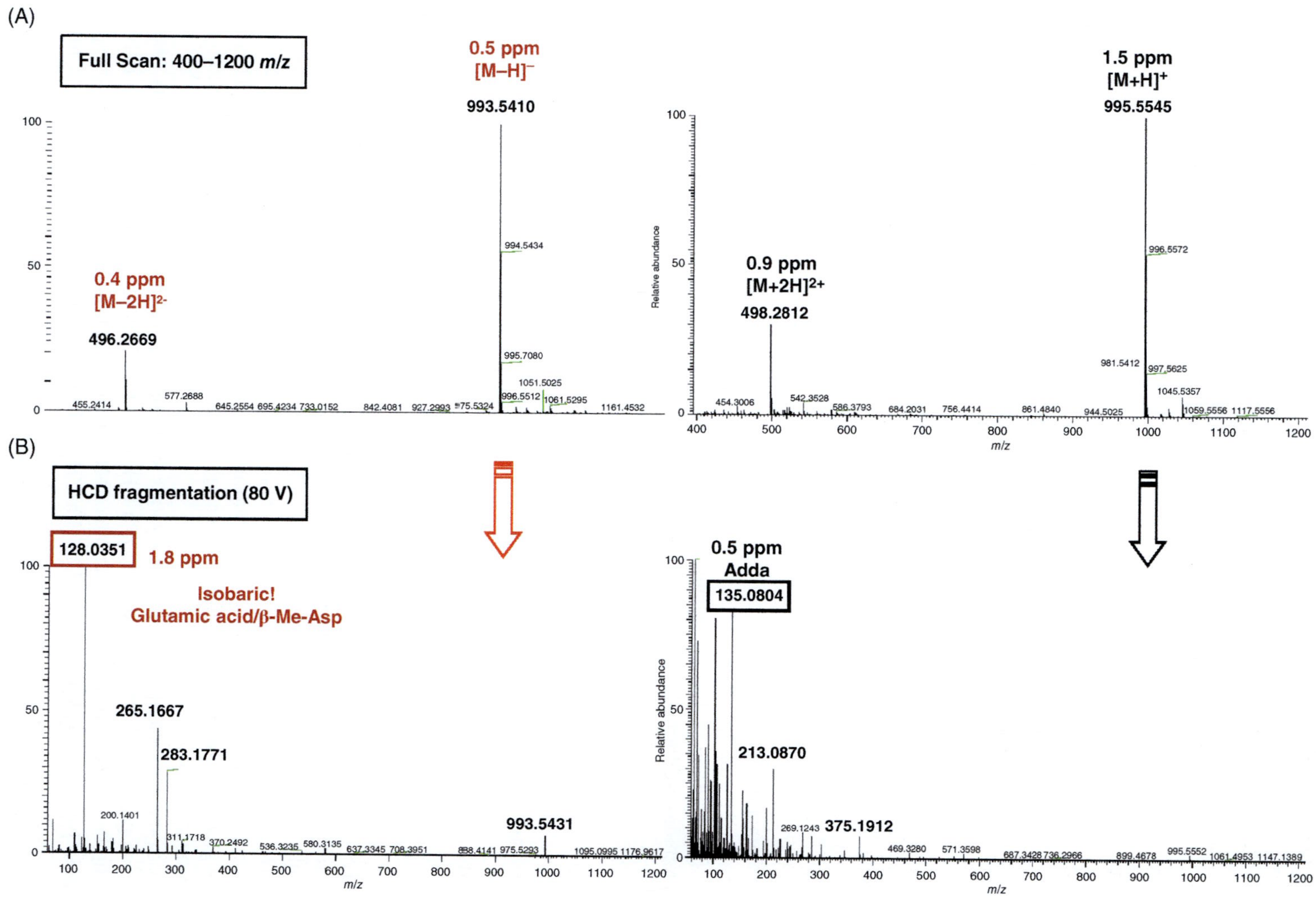

Figure 23.15 Spectra of microcystin-LR by orbitrap analyser in both ESI modes, negative and positive: (A) full-scan mode and (B) HCD fragmentation mode

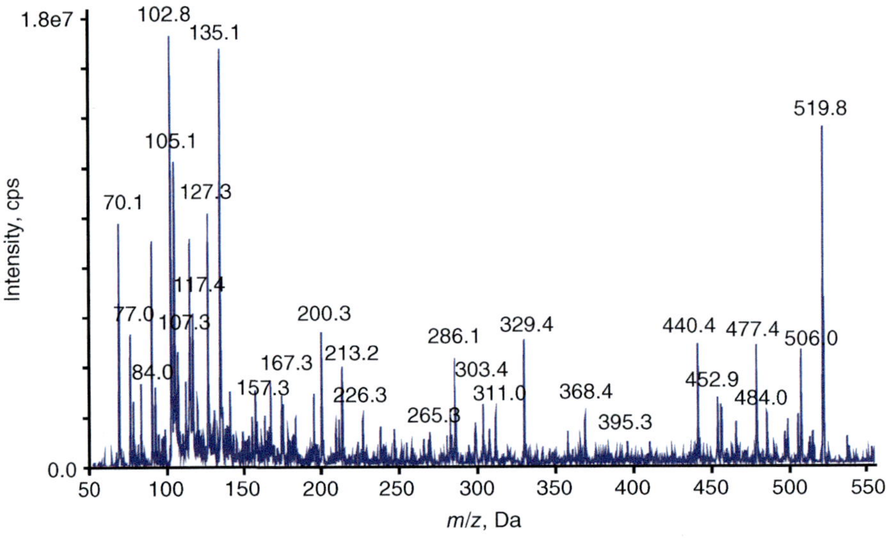

Figure 23.16 Collision-induced dissociation spectrum of microcystin-RR (collision energy (CE) 40 eV)

The determination of the elemental composition of cyanotoxins and their structure elucidation is also supported by the application of several rules of spectrum interpretation, including the nitrogen rule [119]. In the analyses of cyanotoxins, mostly the positive ionisation mode is used, and protonated ions are formed.

23.8.2 Microcystins

Characterisation of the microcystin structure includes the determination of amino acid composition and elucidation of their sequence. The presence of specific amino acid can be indicated by the low mass immonium ions with the general structure $XCH{=}NH^{2+}$, where X stands for the amino acid side chain (e.g. at *m/z* 86 for Leu or *m/z* 120 for Phe). However, these ions are not always observed in the CID spectra.

The amino acid composition of microcystins, especially the presence of a guanidine group containing Arg (R), has significant effect on the fragmentation spectrum [120, 121]. In the case of MC-RR variants with two Arg residues (in positions 2 and 4), the doubly charged ions $[M+2H]^{2+}$ dominate when ESI-MS is used (see Fig. 23.16). These ions (*m/z* 500–550) fragment more easily than the singly protonated pseudomolecular ions of the same peptide (see Fig. 23.15). When only one Arg is present in MC structure, both doubly and singly charged ions are produced, and the $[M+H]^+$ ion is one of the most abundant in the spectrum. In MCs without Arg (e.g. MC-LA, MC-LF and MC-LW), the methoxy group in Adda is protonated (see Fig. 23.17). These peptides give only singly charged ions and tend to fragment at a lower CE than Arg-containing microcystins.

In the fragmentation spectra of microcystins with an unmodified Adda functionality, the intense ion peak at *m/z* 135 representing [Ph-CH_2-CH(OCH_3)] is usually observed, especially when a higher CE is used. The fragmentation of MCs with an Adda containing a hydroxyl group [DMAdda[5]] gives a weak ion at *m/z* 121 [Ph-CH_2-CH(OH)], while in the case of MCs with

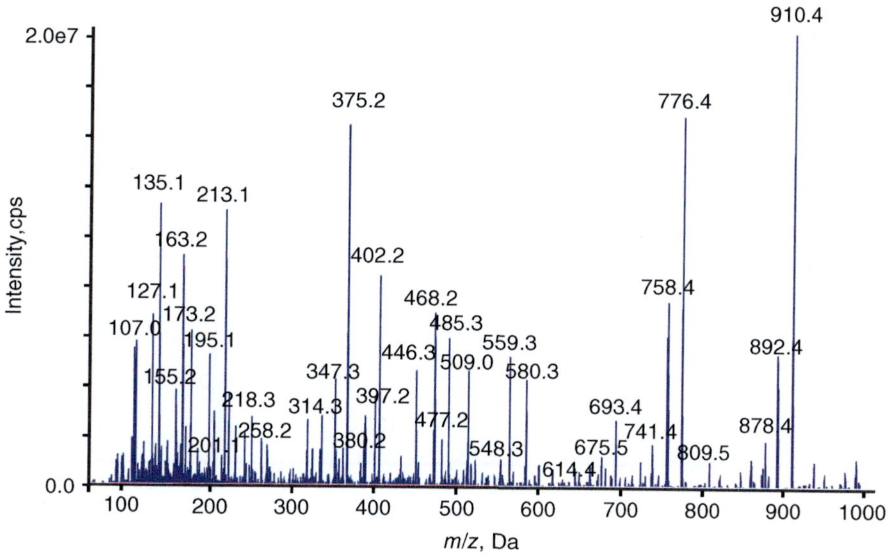

Figure 23.17 Collision-induced dissociation spectrum of microcystin-LA (collision energy (CE) 50 eV)

an acetoxy group [ADMAdda[5]], ion peaks at *m/z* 265 [Ph-CH_2-CH-$OCOCH_3$] and *m/z* 627 [Arg-ADMAdda-Glu + H]$^+$ are observed [122].

Also the ion [M + H-Adda]$^+$ generated by a loss of the Adda fragment and the cleavage of the C—N bond between Adda and Arg (or another residue) are the characteristic features of the spectra of MCs and nodularin. Further fragmentation of [M + H-Adda]$^+$ gives a series of ions containing a unit corresponding to the formula $C_{11}H_{14}O$. For example, in the CID spectrum of MC-LR, the following ions can occur in the MS/MS spectrum [$C_{11}H_{14}O$-Glu-Mdha-Ala + H]$^+$ (*m/z* 446), [$C_{11}H_{14}O$-Glu-Mdha + H]$^+$ (*m/z* 375), [$C_{11}H_{14}O$-Glu + H]$^+$ (*m/z* 292) and [$C_{11}H_{14}O$ + H]$^+$ (*m/z* 163). In the spectra of non-Arg-containing MCs, the intensities of these ion peaks are higher than in the spectra of microcystins with Arg. The fragmentation spectrum of MCs containing Arg incorporates a typical dissociation product at *m/z* 70, Arg side chain ([C_4H_8N]$^+$) (see Fig. 23.16).

The other ions that provide information on the sequence of amino acids in the structure of microcystins are *b*- and *y*-type ions. They are formed by a fission of the cyclic structure and subsequent cleavage of the peptide bonds. The difference between the *m/z* values of the precursor ion peak and the resultant product ion peak corresponds to the mass of the liberated amino acid residue. In the spectra of MCs without Arg, the *b*-ions usually dominate and can contain up to six residues. It was observed [120, 122, 123] that under MS/MS conditions, Arg-containing microcystins always give *b*-ions with Arg in their structure. The *b*-type ions frequently lose CO (−28 Da) and give a-type ion NH_3 (−17 Da). The loss of H_2O (−18 Da) or NH_3 (−17 Da) is also quite common.

The additional valuable diagnostic ions in the spectra of MCs and NOD occur at *m/z* 213 or 227 and correspond to [Glu-Mdha/dhb + H]$^+$ and [Glu-Mdhb + H]$^+$, respectively.

Interpretation of MCs and NOD fragmentation spectra can be complicated by the presence of amino acids which have the same molecular weight (Gly and MeAsp; dhb and MeDha) and form isobaric ions.

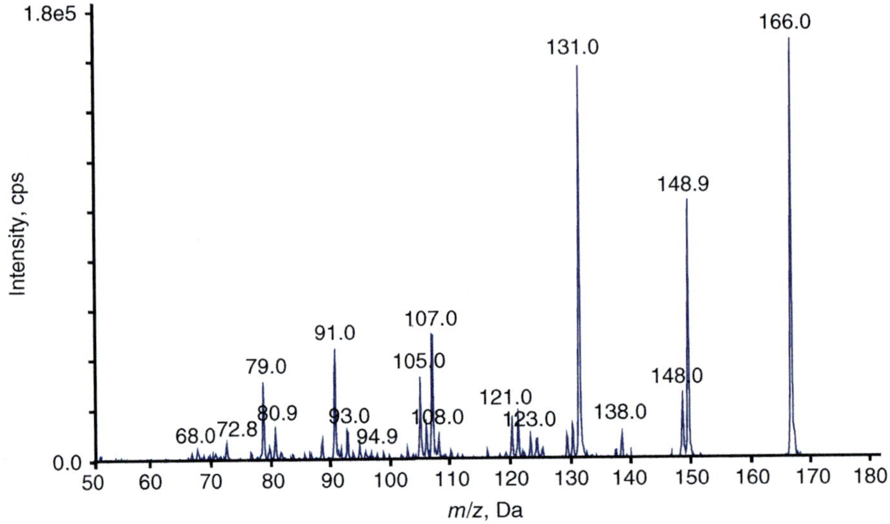

Figure 23.18 *Collision-induced fragmentation spectrum of anatoxin-a*

23.8.3 Anatoxin-a

In the CID spectrum of anatoxin-a (ANTX), the abundant product ions at *m/z* 149 and *m/z* 131, generated by the loss of ammonia $[M+H-NH_3]^+$ and subsequent loss of water $[M+H-NH_3-H_2O]^+$, are present (Fig. 23.18). Other characteristic ions occur in the spectrum at *m/z* 107 $[M+H-NH_3-COCH_2R]^+$ and *m/z* 91 $[C_7H_7]^+$. Two other product ions, *m/z* 43 and *m/z* 56, are assigned to $[CH_2{=}C{=}OH]^+$ and $[CH_3CH{=}C{=}OH]^+$, respectively. The molecular ion and some of the product ions of ANTX are characterised by similar *m/z* values to the ions generated by Phe. For proper identification of the toxin structure, the presence of an ion at *m/z* 149 $[M+H-NH_3]^+$ in the spectrum of ANTX and an ion at *m/z* 120 in the spectrum of Phe $[M+H-NH_3-H_2O]$ is crucial [124]. The accurate determination of the *m/z* value of the pseudomolecular ion $[M+H]^+$ also allows us to distinguish ANTX (166.12) from phenylalanine (166.07).

23.8.4 Cylindrospermopsin

The CID spectrum of cylindrospermopsin ($[M+H]^+=416$), collected in positive mode, is characterised by the presence of ion at *m/z* 336 $[M-80]^+$ which is generated by the cleavage of the sulphonic group [125] (see Fig. 23.19). Water loss from this ion gives an ion at *m/z* 318 $[M-80-H_2O]^+$. Another diagnostic ion present in the MS/MS spectrum of CYN occurs at *m/z* 274 and is assigned to the fragment ion formed by the loss of the [6-(2-hydroxy-4-oxo-3-hydropyrimidyl)]hydroxymethinyl part of CYN ($[M-C_5H_6N_2O_3+H]^+$) ion. Subsequent loss of SO_3 and H_2O from this fragment gives ion peaks at *m/z* 194 and *m/z* 176, respectively.

23.8.5 BMAA

The most important diagnostic ion in the CID spectrum of β-*N*-methylamino-L-alanine (BMAA) pseudomolecular ion (*m/z* 119.07) occurs at *m/z* 88 $[M+H–CH_3–NH_2]^+$ (Fig. 23.20). This ion is helpful in distinguishing between BMAA and 2,4-diaminobutyric acid (BAD). These two compounds can be also distinguished by the *m/z* 101.3 $[M+H-H_2O]^+$ ion which occurs as an intensive

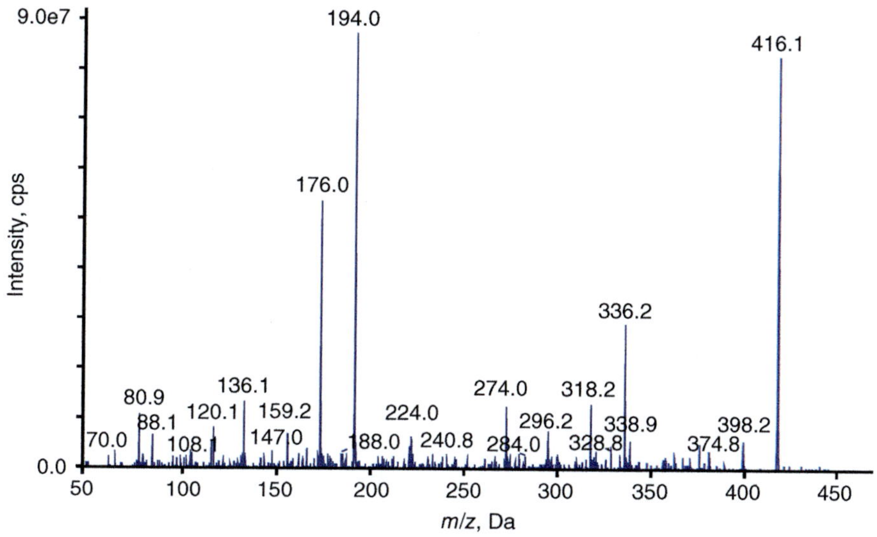

Figure 23.19 *Collision-induced fragmentation spectrum of cylindrospermopsin*

ion in the spectrum of DAB. Other fragment ions are observed at *m/z* 102 $[M+H-NH_3]^+$ and *m/z* 76 $[M+H\text{-}CH_2\text{-}N\text{-}CH_3]^+$. Loss of carboxylic group from BMAA generates an ion at *m/z* 73 $[M+H-HCOOH]^+$. To increase the sensitivity of detection and the reliability of BMAA identification, the derivatisation with a fluorescent reagent 6-aminoquinolyl-*N*-hydroxysuccinimidyl carbamate (AQC) [126, 127] or with propyl chloroformate [128] was proposed.

23.9 Performance Criteria of LC-MS Methods for Identification and Quantification of Cyanotoxins

Although cyanotoxins are currently not included in the list of priority substances in the field of water policy (Directive 2013/39/EU) [129] or the list of parameters concerning quality of water for human consumption (Council Directive 98/83/EC) [130], performance criteria for the analysis of cyanotoxins with LC-MS-based techniques should be generally in accordance with the criteria established for other organic pollutants included in regulations.

General performance criteria and technical specifications of analytical methods for monitoring water status according to Directive 2000/60/EC are described in Commission Directive 2009/90/EC [131]. This directive dictates that analytical methods should be validated and documented according to EN ISO 17025, uncertainty of measurement should be below 50% (estimated at the level as set out in the relevant environmental quality standard), and the LOQ should be below 30% of the relevant environmental quality standard. Regarding water intended for human consumption, Directive 98/83 EC establishes method performance criteria for trueness, precision and LOD. For most organic compounds included in this directive (e.g. pesticides, PAHs), all three of the aforementioned criteria are set to maximum 25% of the parametric value (regulatory limit).

More specific performance criteria, especially for LC-MS methods, can be found in the Commission Decision 2002/657/EC [132], although this decision primarily refers to the analysis

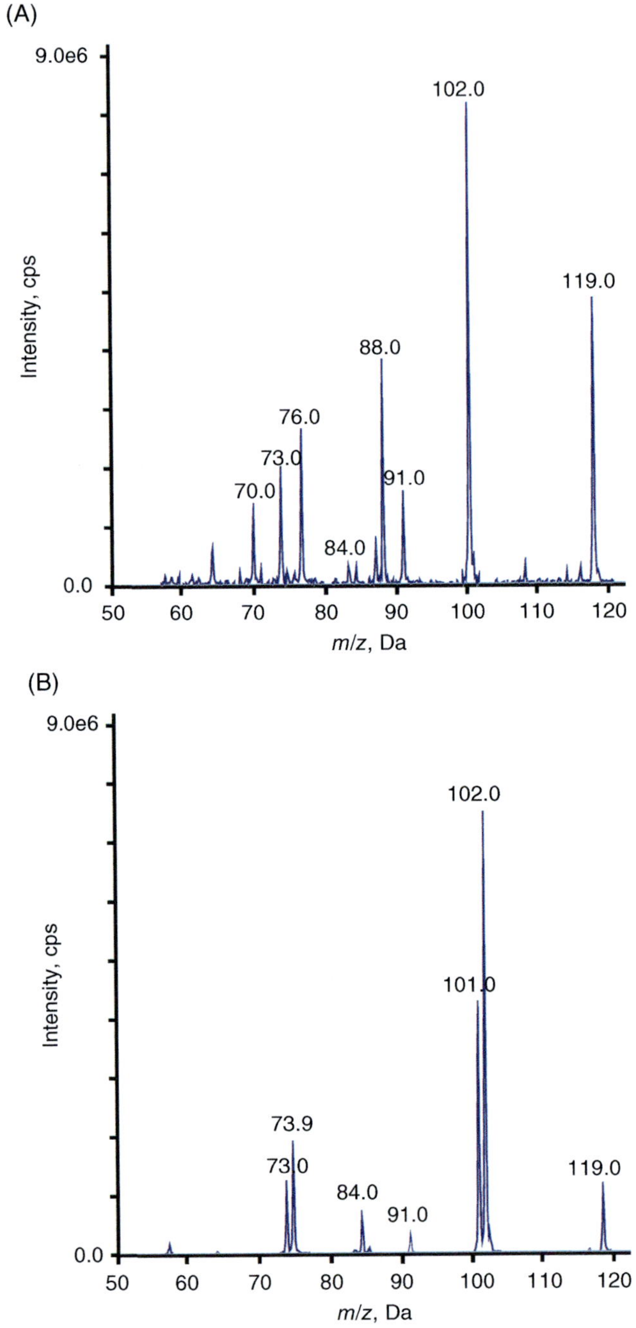

Figure 23.20 Collision-induced fragmentation spectra of BMAA (A) and DBA (B)

of residues in animal products. In this decision LC-MS methods are listed as 'suitable confirmatory methods for organic residues or contaminants', including 'MS-MS^n techniques, or other suitable MS or MS-MS^n techniques in combination with appropriate ionization modes'. According to this decision, confirmation requires at least 3 or 4 (depending on the analyte) 'identification points' for techniques that do not record the full mass spectra. Depending on the technique used, identification points earned per ion monitored are 1 (LC-MS^n, precursor ion), 1.5 (LC-MS^n, transition products), 2 ($HRMS^n$, precursor ion) and 2.5 ($HRMS^n$, transition ions). For example, an LC-MS/MS method that monitors one precursor and two daughters earns four identification points. Criteria for relative ion intensities are also established, for example, for LC-MS^n maximum permitted tolerance is ±20% when the relative intensity of the base peak is greater than 50%. Criteria for precision, trueness and detection capability are also described. For HRMS, the resolution should be greater than 10000 for the entire mass range at 10% valley.

Requirements for identification and quantification of analytes with LC-MS techniques are also included in the SANCO 'Guidance documents on analytical quality control and method validation procedures for pesticide residues analysis in food and feed' (latest version SANCO/12571/2013) [133]. The minimum number of required diagnostic ions is three for single-stage unit-mass resolution MS (e.g. quadrupole, ion trap, TOF), two for single-stage high-resolution/high mass accuracy MS (e.g. TOF, orbitrap) and 2 product ions for MS/MS (e.g. QqQ, ion trap, hybrid MS). The recommended maximum tolerance of ion ratios for LC-MS and LC-MS^n is ±30%. General method performance acceptability criteria are mean recovery 70–120% with associated repeatability $RSD_r < 20\%$. According to SANCO guidelines, LOQ is defined as the lowest validated concentration level that meets validation criteria.

Laboratories that carry out analysis of cyanotoxins with LC-MS techniques should use the aforementioned identification–confirmation–quantification criteria as rough guidelines for their in-house validation studies. Validation and routine quality control results should be documented, and any major deviations from the respective criteria should be justified. In any case that the criteria cannot be met, analysis should be carried out with the best available techniques with acceptable performance parameters according to the intended use.

References

[1] Gross, J.H. (2011) *Mass Spectrometry: A Textbook*, Springer-Verlag, Ltd, Berlin.

[2] Ham, B.M. (2008) *Even Electron Mass Spectrometry with Biomolecule Applications*, John Wiley & Sons, Ltd, Chichester.

[3] Greaver, J. and Roboz, J. (2014) *Mass Spectrometry for the Novice*, CRC Press Taylor & Francis Group, Boca Raton.

[4] Watson, J.T. and Sparkman, O.D. (2009) *Introduction to Mass Spectrometry: Instrumentation, Applications, and Strategies for Data Interpretation*, 4th edn, John Wiley & Sons, Ltd, Chichester.

[5] van den Ouweland, J.M. and Kema, I.P. (2012) The role of liquid chromatography-tandem mass spectrometry in the clinical laboratory. *Journal of Chromatography B*, **883–884**, 18–32.

[6] de Hoffmann, E. and Stroobant, V. (2007) *Mass Spectrometry: Principles and Applications*, 3rd edn, John Wiley & Sons, Ltd, Chichester.

[7] Sano, T., Nohara, K., Shiraishi, F. and Kaya, K. (1992) A method for microdetermination of total microcystin content in waterblooms of cyanobacteria (blue-green algae). *International Journal of Environmental Analytical Chemistry*, **49**, 163–170.

[8] Tanaka, Y., Takenaka, S., Matsuo, H. *et al.* (1993) Levels of microcystin in Japanese lakes. *Toxicological & Environmental Chemistry*, **39**, 21–27.

[9] Tsuji, K., Masui, H., Uemura, H. *et al.* (2001) Analysis of microcystins in sediments using MMPB method. *Toxicon*, **39**, 687–692.

[10] Harada, K.I., Fujiki, H., Shimada, T. *et al.* Two cyclic peptides, anabaenopeptins, a third group of bioactive compounds from the cyanobacterium *Anabaena flos-aquae* NRC 525-17. *Tetrahedron Letters*, **36** (9), 1511–1514.

[11] Harada, K., Oshikata, M., Shimada, T. *et al.* (1997) High-performance liquid chromatographic separation of microcystins derivatized with a highly fluorescent dienophile. *Natural Toxins*, **5**, 201–207.

[12] Suchy, P. and Berry, J. (2012) Detection of total microcystin in fish tissues based on lemieux oxidation and recovery of 2-methyl-3-methoxy-4-phenylbutanoic acid (MMPB) by solid-phase microextraction gas chromatography-mass spectrometry (SPME-GC/MS). *International Journal of Environmental Analytical Chemistry*, **92** (12), 1443–1456.

[13] de Broglie, L. (1970) The reinterpretation of wave mechanics. *Foundations of Physics*, **1** (1), 5–15.

[14] Sivonen, K., Namikoshi, M., Evans, W.R. *et al.* (1992) Three new microcystins, cyclic heptapeptide hepatotoxins from Nostoc sp. strain 152. *Chemical Research in Toxicology*, **5**, 464–469.

[15] Zhang, T., Watson, D.G., Azike, C. *et al.* (2007) Determination of vancomycin in serum by liquid chromatography-high resolution full scan mass spectrometry. *Journal of Chromatography B, Analytical Technologies in the Biomedical and Life Sciences*, **857** (2), 352–356.

[16] Fenn, J.B., Mann, M., Meng, C.K. *et al.* (1989) Electrospray ionization for mass spectrometry of large biomolecules. *Science*, **246** (4926), 64–71.

[17] Fenn, J.B., Mann, M. and Men, C.K. (1990) Electrospray ionization—principles and practice. *Mass Spectrometry Reviews*, **9**, 37–70.

[18] Furey, A., Moriarty, M., Vaishali, B. *et al.* (2013) Ion suppression; a critical review on causes, evaluation, prevention and applications. *Talanta*, **115**, 104–122.

[19] Hoffmann, E. (1996) Tandem mass spectrometry: a primer. *Journal of Mass Spectrometry*, **31**, 129–137.

[20] Yao, J., Hao, H., Huo, Y. *et al.* (2013) *Determination of Microcystins in Drinking Water by Ultra High Performance Liquid Chromatography/Triple Quadrupole Mass Spectrometry. Shimadzu Excellence*, Poster presented at the 61st ASMS Conference on Mass Spectrometry and Allied Topics, Minneapolis, June 9–13, 2013.

[21] Bogialli, S., Bruno, M., Curini, R. *et al.* (2006) Simple and rapid determination of anatoxin-a in lake water and fish muscle tissue by liquid-chromatography-tandem mass spectrometry. *Journal of Chromatography A*, **1122** (1–2), 180–185.

[22] Spoof, L., Berg, K.A., Rapala, J. *et al.* (2006) First observation of cylindrospermopsin in *Anabaena lapponica* isolated from the boreal environment (Finland). *Environmental Toxicology*, **21** (6), 552–650.

[23] Li, A., Fan, H., Ma, F. *et al.* (2012) Elucidation of matrix effects and performance of solid-phase extraction for LC-MS/MS analysis of β-*N*-methylamino-L-alanine (BMAA) and 2,4-diaminobutyric acid (DAB) neurotoxins in cyanobacteria. *Analyst*, **137**, 1210–1229.

[24] Kaufmann, A., Butcher, P., Maden, K. *et al.* (2010) Comprehensive comparison of liquid chromatography selectivity as provided by two types of liquid chromatography detectors (high resolution mass spectrometry and tandem mass spectrometry): "Where is the crossover point?" *Analytica Chimica Acta*, **673**, 60–72.

[25] Chernushevich, I.V. (2000) Duty cycle improvement for a quadrupole-time-of-flight mass spectrometer and its use for precursor ion scans. *European Journal of Mass Spectrometry*, **6**, 471–479.

[26] Chernushevich, I.V., Loboda, A.V. and Thomson, B.A. (2001) An introduction to quadrupole–time-of-flight mass spectrometry. *Journal of Mass Spectrometry*, **36**, 849–865.

[27] Lacorte, S. and Fernandez-Alba, A.R. (2006) Time-of-flight mass spectrometry applied to the liquid chromatographic analysis of pesticides in water and food. *Mass Spectrometry Reviews*, **25**, 866–880.

[28] Guilhaus, M. and Guilhaus, M. (1995) Principles and instrumentation in time-of-flight mass spectrometry. *Journal of Mass Spectrometry*, **30**, 1519–1532.

[29] Makarov, A. (2000) A Electrostatic axially-harmonic orbital trapping. A novel high-performance technique of mass analysis. *Analytical Chemistry*, **72**, 1156–1162.

[30] Hu, Q., Noll, R.J., Li, H. *et al.* (2005) The Orbitrap: a new mass spectrometer. *Journal of Mass Spectrometry*, **40** (4), 430–443.

[31] Perry, R., Cooks, G. and Noll, R. (2008) Orbitrap mass spectrometry: instrumentation, ion motion and applications. *Mass Spectrometry Reviews*, **27** (6), 661–699.

[32] Makarov, A. and Scigelova, M. (2010) Coupling liquid chromatography to Orbitrap mass spectrometry. *Journal of Chromatography A*, **1217**, 3938–3945.

[33] Marshall, A.G. and Hendrickson, C.L. (2008) High-resolution mass spectrometers. *Annual Review of Analytical Chemistry*, **1**, 579–599.

[34] Forcisi, S., Moritz, F., Kanawati, B. *et al.* (2013) Liquid chromatography–mass spectrometry in metabolomics research: mass analyzers in ultra high pressure liquid chromatography coupling. *Journal of Chromatography A*, **1292**, 51–56.
[35] Holcapek, M., Jiràsko, R. and Lísa, M. (2012) Recent developments in liquid chromatography-mass spectrometry and related techniques. *Journal of Chromatography A*, **1259**, 3–15.
[36] Murray, K.K., Boyd, R.K., Eberlin, M.N. *et al.* (2013) Definitions of terms relating to mass spectrometry (IUPAC Recommendations 2013). *Pure and Applied Chemistry*, **85**, 1515–1609.
[37] Merel, S., Walker, D., Chicana, R. *et al.* (2013) State of knowledge and concerns on cyanobacterial blooms and cyanotoxins. *Environment International*, **59**, 303–327.
[38] Kaushik, R. and Balasubramanian, R. (2013) Methods and approaches used for detection of cyanotoxins in environmental samples: a review. *Critical Reviews in Environmental Science and Technology*, **43**, 1349–1383.
[39] Lawton, L.A. and Edwards, C. (2008) Conventional laboratory methods for cyanotoxins, in *Cyanobacterial Harmful Algal Blooms: State of the Science and Research Needs*, (ed. K.H. Hudnell), Springer, Ltd, New York, pp. 513–537.
[40] Msagati, T.A., Siame, B.A. and Shushu D.D. (2006) Evaluation of methods for the isolation, detection and quantification of cyanobacterial hepatotoxins. *Aquatic Toxicology*, **78** (4), 382–397.
[41] Dörr, F.A, Oliveira-Silva, D, Lopes, N.P. *et al.* (2011) Dissociation of deprotonated microcystin variants by collision-induced dissociation following electrospray ionization. *Rapid Communications in Mass Spectrometry*, **25**, 1981–1992.
[42] Gambaro, A., Barbaro, E., Zangrando, R. and Barbante, C. (2012) Simultaneous quantification of microcystins and nodularin in aerosol samples using high-performance liquid chromatography/negative electrospray ionization tandem mass spectrometry. *Rapid Communications in Mass Spectrometry*, **26**, 1497–1506.
[43] Rodrigues, M.A., Reis, M.P. and Mateus, M.C. (2013) Liquid chromatography/negative electrospray ionization ion trap MS2 mass spectrometry application for the determination of microcystins occurrence in Southern Portugal water reservoirs. *Toxicon*, **74**, 8–18.
[44] Dahlmann, J., Budakowski, W.R. and Luckas, B. (2003) Liquid chromatography–electrospray ionisation–mass spectrometry based method for the simultaneous determination of algal and cyanobacterial toxins in phytoplankton from marine waters and lakes followed by tentative structural elucidation of microcystins. *Journal of Chromatography A*, **994** (1–2), 45–57.
[45] Blay, P., Hui, J.P.M., Chang, J. and Melanson, J.E. (2011) Screening for multiple classes of marine biotoxins by liquid chromatography–high-resolution mass spectrometry. *Analytical and Bioanalytical Chemistry*, **400**, 577–585.
[46] Yen, H.K., Lin, T.F. and Liao, P.C. (2011) Simultaneous detection of nine cyanotoxins in drinking water using dual solid-phase extraction and liquid chromatography–mass spectrometry. *Toxicon*, **58** (2), 209–218.
[47] Chen, J., Yan, T., Xu, J. *et al.* (2012) Simultaneous determination of toxins in algae and water samples by high-performance liquid chromatography with triple quadrupole mass spectrometry. *Journal of Separation Science*, **35**, 1094–1101.
[48] Herrmann, A., Rosén, J., Jansson, D. and Hellenäs, K.-E. (2012) Evaluation of a generic multi-analyte method for detection of >100 representative compounds correlated to emergency events in 19 food types by ultrahigh-pressure liquid chromatography–tandem mass spectrometry. *Journal of Chromatography A*, **1235**, 115–124.
[49] Sivonen, K., Leikoski, N., Fewer, D.P. and Jokela, J. (2010) Cyanobactins-ribosomal cyclic peptides produced by cyanobacteria. *Applied Microbiology and Biotechnology*, **86**, 1213–1225.
[50] Cohen, S.A. (2012) Analytical techniques for the detection of α-amino-β-methylaminopropionic acid. *Analyst*, **137**, 1991–2005.
[51] Kikuchi, S., Kubo, T. and Kaya, K. (2007) Cylindrospermopsin determination using 2-[4-(2-hydroxyethyl)-1-piperazinyl]ethanesulfonic acid (HEPES) as the internal standard. *Analytica Chimica Acta*, **583** (1), 124–127.
[52] Dimitrakopoulos, I.K., Kaloudis, T.S., Hiskia, A.E. *et al.* (2010) Development of a fast and selective method for the sensitive determination of anatoxin-a in lake waters using liquid chromatography–tandem mass spectrometry and phenylalanine-d5 as internal standard. *Analytical and Bioanalytical Chemistry*, **397** (6), 2245–2252.
[53] Sano, T., Takagi, H., Nagano, K. *et al.* (2011) Accurate LC–MS analyses for microcystins using per-15N-labeled microcystins. *Analytical and Bioanalytical Chemistry*, **399**, 2511–2516.

[54] Neffling, M.-R., Spoof, L. and Meriluoto, J. (2009) Rapid LC–MS detection of cyanobacterial hepatotoxins microcystins and nodularins—comparison of columns. *Analytica Chimica Acta*, **653** (2), 234–241.

[55] Oehrle, S.A., Southwell, B. and Westrick, J. (2010) Detection of various freshwater cyanobacterial toxins using ultra-performance liquid chromatography tandem mass spectrometry. *Toxicon*, **55** (5), 965–972.

[56] Jiang, L., Johnston, E., Åberg, K.M. *et al.* (2013) Strategy for quantifying trace levels of BMAA in cyanobacteria by LC/MS/MS. *Analytical and Bioanalytical Chemistry*, **405**, 1283–1292.

[57] Rivasseau, C., Vanhoenacker, G., Sandra, P. and Hennion, M.C. (2000) Online solid-phase extraction in microcolumn-liquid chromatography coupled to UV or MS detection: application to the analysis of cyanobacterial toxins. *Journal of Microcolumn Separations*, **12** (5), 323–332.

[58] Shan, Y., Shi, X., Dou, A. *et al.* (2011) A fully automated system with on-line micro solid-phase extraction combined with capillary liquid chromatography–tandem mass spectrometry for high throughput analysis of microcystins and nodularin-R in tap water and lake water. *Journal of Chromatography A*, **1218**, 1743–1748.

[59] Beltrán, E., Ibáñez, M., Sancho, J.V. and Hernández, F. (2012) Determination of six microcystins and nodularin in surface and drinking waters by on-line solid phase extraction-ultra high pressure liquid chromatography tandem mass spectrometry. *Journal of Chromatography A*, **1266**, 61–68.

[60] Welker, M., Fastner, J., Erhard, M. and von Dohren, H. (2002) Applications of MALDI-TOF MS analysis in cyanotoxin research. *Environmental Toxicology*, **17** (4), 367–374.

[61] Howard, K.L. and Boyer, G.L. (2007) Quantitative analysis of cyanobacterial toxins by matrix-assisted laser desorption ionization mass spectrometry. *Analytical Chemistry*, **79**, 5980–5986.

[62] Spoof, L., Vesterkvist, P., Lindholm, T. and Meriluoto, J. (2003) Screening for cyanobacterial hepatotoxins, microcystins and nodularin in environmental water samples by reversed-phase liquid chromatography–electrospray ionisation mass spectrometry. *Journal of Chromatography A*, **1020** (1), 105–119.

[63] Zhang, L.F., Ping, X.F., Yang, Z.G. (2004) Determination of microcystin-LR in surface water using high-performance liquid chromatography/tandem electrospray ionization mass detector. *Talanta*, **62** (1), 193–200.

[64] Karlsson, K.M., Spoof, L.E. and Meriluoto, J.A. (2005) Quantitative LC–ESI-MS analyses of microcystins and nodularin-R in animal tissue–matrix effects and method validation. *Environmental Toxicology*, **20**, 381–389.

[65] Faassen, E.J., Gillissen, F. and Lürling, M. (2012) A comparative study on three analytical methods for the determination of the neurotoxin BMAA in cyanobacteria. *PLoS ONE*, **7** (5), art no e36667.

[66] Zweigenbaum, J.A., Henion, J.D., Beattie, K.A. *et al.* (2000) Direct analysis of microcystins by microbore liquid chromatography–electrospray ionization ion-trap tandem mass spectrometry. *Journal of Pharmaceutical and Biomedical Analysis*, **23** (4), 723–733.

[67] Diehnelt, C.W., Peterman, S.M. and Budde, W.L. (2005) Liquid chromatography–tandem mass spectrometry and accurate measurements of cyclic peptide cyanobacteria toxins. *Trends in Analytical Chemistry*, **24**, 622–634.

[68] Diehnelt, C.W., Dugam, N.R., Peterman, S.M. and Budde, W.L. (2006) Identification of microcystin toxins from a strain of *Microcystis aeroginosa* into a hybrid linear Ion trap-Fourier ion cyclotron resonance mass spectrometer. *Analytical Chemistry*, **78**, 501–512.

[69] Merel, S., LeBot, B., Clément, M. *et al.* (2009) MS identification of microcystin-LR chlorination by-products. *Chemosphere*, **74**, 832–839.

[70] Ferranti, P., Fabbrocino, S., Nasi, A. *et al.* (2009) Liquid chromatography coupled to quadruple time-of-flight tandem mass spectrometry for microcystin analysis in freshwaters: method performances and characterisation of a novel variant of microcystin-RR. *Rapid Communications in Mass Spectrometry*, **23**, 1328–1336.

[71] Merel, S., Clement, M. and Mourot, A. *et al.* (2010) Characterization of cylindrospermopsin chlorination. *Science of the Total Environment*, **408** (16), 3433–3442.

[72] Ferranti, P., Nasi, A., Bruno, M. *et al.* (2011) A peptidomic approach for monitoring and characterizing peptide cyanotoxins produced in Italian lakes by matrix-assisted laser desorption/ionisation and quadrupole time-of-flight mass spectrometry. *Rapid Communications in Mass Spectrometry*, **25**, 1173–1183.

[73] Mil'man, B.L., Russkikh, Y.V., Nekrasova, L.V. and Zhakovskaya, Z.A. (2011) An approach to the mass spectrometry identification of cyanobacterial peptides. The case of demethylmicrocystin-LR. *Journal of Analytical Chemistry*, **66**, 1423–1431.

[74] James, K.J., Crowley, J., Hamilton, B. *et al.* (2005) Anatoxins and degradation products, determined using hybrid quadrupole time-of-flight and quadrupole ion-trap mass spectrometry: forensic investigations of cyanobacterial neurotoxin poisoning. *Rapid Communications in Mass Spectrometry*, **19**, 1167–1175.

[75] Krüger, T., Christian, B. and Luckas, B. (2009) Development of an analytical method for the unambiguous structure elucidation of cyclic peptides with special appliance for hepatotoxic desmethylated microcystins. *Toxicon*, **54**, 302–312.

[76] Miles, C.O., Sandvik, M., Nonga, H.E. *et al.* (2012) Thiol derivatization for LC–MS identification of microcystins in complex matrices. *Environmental Science & Technology*, **46**, 8937–8944.

[77] Smith, J.L. and Boyer, G.L. (2009) Standardization of microcystin extraction from fish tissues: a novel internal standard as a surrogate for polar and non-polar variants. *Toxicon*, **53**, 238–245.

[78] Neffling, M.-R., Lance, E. and Meriluoto, J. (2010) Detection of free and covalently bound microcystins in animal tissues by liquid chromatography–tandem mass spectrometry. *Environmental Pollution*, **158** (3), 948–952.

[79] Antignac, J.-P., de Wasch, K., Monteau F. *et al.* (2005) The ion suppression phenomenon in liquid chromatography–mass spectrometry and its consequences in the field of residue analysis. *Analytica Chimica Acta*, **529** (1–2), 129–136.

[80] Chin, C., Zhang, Z.P. and Karnes, H.T. (2004) A study of matrix effects on an LC/MS/MS assay for olanzapine and desmethyl olanzapine. *Journal of Pharmaceutical and Biomedical Analysis*, **35** (5), 1149–1167.

[81] Manini, P., Andreoli, R. and Niessen, W.M.A. (2004) Liquid chromatography–mass spectrometry in occupational toxicology: a novel approach to the study of biotransformation of industrial chemicals. *Journal of Chromatography A*, **1058** (1–2), 21–37.

[82] Freitas, L.G., Gotz, C.W., Ruff, M. *et al.* (2004) Quantification of the new triketone herbicides, sulcotrione and mesotrione, and other important herbicides and metabolites, at the ng/l level in surface waters using liquid chromatography–tandem mass spectrometry. *Journal of Chromatography A*, **1028** (2), 277–286.

[83] Gomes R.L., Avcioglu E., Scrimshaw M.D. and Lester J.N. (2004) Steroid-estrogen determination in sediment and sewage sludge: a critique of sample preparation and chromatographic/mass spectrometry considerations, incorporating a case study in method development. *Trends in Analytical Chemistry*, **23** (10–11), 737–744.

[84] Mortier, K.A., Verstraete, A.G., Zhang, G.-F. and Lambert, W.E. (2004) Enhanced method performance due to a shorter chromatographic run-time in a liquid chromatography–tandem mass spectrometry assay for paclitaxel. *Journal of Chromatography A*, **1041** (1–2), 235–238.

[85] Villagrasa, M., Guillamon, M., Eljarrat, E. and Barcelo, D. (2007) Matrix effect in liquid chromatography–electrospray ionization mass spectrometry analysis of benzoxazinoid derivatives in plant material. *Journal of Chromatography A*, **1157** (1–2), 108–114.

[86] Wang, S., Cyronak, M. and Yang, E. (2007) Does a stable isotopically labeled internal standard always correct analyte response?: a matrix effect study on a LC/MS/MS method for the determination of carvedilol enantiomers in human plasma. *Journal of Pharmaceutical and Biomedical Analysis*, **43** (2), 701–707.

[87] Gros, M., Petrovic, M. and Barcelo D. (2006) Development of a multi-residue analytical methodology based on liquid chromatography–tandem mass spectrometry (LC–MS/MS) for screening and trace level determination of pharmaceuticals in surface and wastewaters. *Talanta*, **70** (4), 678–690.

[88] Bonfiglio, R., King, R., Olah, T. and Merkle, K. (1999) The effects of sample preparation method on the variability of the electrospray ionization response for model drug compounds. *Rapid Communications in Mass Spectrometry*, **13** (12), 1175–1185.

[89] King, R., Bonfiglio, R., Fernandez-Metzler, C. *et al.* (2000) Mechanistic investigation of ionization suppression in electrospray ionization. *Journal of the American Society for Mass Spectrometry*, **11** (11), 942–950.

[90] Allis, O., Dauphard, J., Hamilton B. *et al.* (2007) Liquid chromatography–tandem mass spectrometry application, for the determination of extracellular hepatotoxins in Irish lake and drinking waters. *Analytical Chemistry*, **79** (9), 3436–3447.

[91] Benijts, T., Dams, R., Lambert, W. and De Leenheer, A. (2004) Countering matrix effects in environmental liquid chromatography–` electrospray ionization tandem mass spectrometry water analysis for endocrine disrupting chemicals. *Journal of Chromatography A*, **1029** (1–2), 153–159.

[92] Kall, M.A., Fu, I., Dige, T. *et al.* (2007) Development and validation of a selective and sensitive bioanalytical procedure for the quantitative determination of gaboxadol in human plasma employing mixed mode solid phase extraction and hydrophilic interaction liquid chromatography with tandem mass spectroscopic detection. *Journal of Chromatography B*, **858** (1–2), 168–176.

[93] Choi, B.K., Hercules, D.M. and Gusev, A.I. (2001) LC–MS/MS signal suppression effects in the analysis of pesticides in complex environmental matrices. *Fresenius Journal of Analytical Chemistry*, **369** (3–4), 370–377.

[94] Choi, B.K., Hercules, D.M. and Gusev, A.I. (2001) Effect of liquid chromatography separation of complex matrices on liquid chromatography–tandem mass spectrometry signal suppression. *Journal of Chromatography A*, **907** (1–2), 337–342.

[95] Matuszewski, B.K., Constanzer, M.L. and Chavez-Eng, C.M. (2003) Strategies for the assessment of matrix effect in quantitative bioanalytical methods based on HPLC–MS/MS. *Analytical Chemistry*, **75** (13), 3019–3030.

[96] Matuszewski, B.K., Constanzer, M.L. and Chavez-Eng, C.M. (1998) Matrix effect in quantitative LC/MS/MS analyses of biological fluids: a method for determination of finasteride in human plasma at picogram per milliliter concentrations. *Analytical Chemistry*, **70** (5), 882–889.

[97] Lawton, L.A., Edwards, C. and Codd, G.A. (1994) Extraction and high-performance liquid chromatographic method for the determination of microcystins in raw and treated waters. *Analyst*, **119**, 1525–1530.

[98] Karlsson, K., Sipia, V., Kankaanpaa, H. and Meriluoto, J. (2003) Mass spectrometric detection of nodularin and desmethylnodularin in mussels and flounders. *Journal of Chromatography B*, **784** (2), 243–253.

[99] Stuber, M. and Reemtsma, T. (2004) Evaluation of three calibration methods to compensate matrix effects in environmental analysis with LC–ESI-MS. *Analytical and Bioanalytical Chemistry*, **378** (4), 910–916.

[100] Annesley, T.M. (2003) Ion suppression in mass spectrometry. *Clinical Chemistry*, **49** (7), 1041–1044.

[101] Buhrman, D.L., Price, P.I. and Rudewicz, P.J. (1996) Quantitation of SR 27417 in human plasma using electrospray liquid chromatography–tandem mass spectrometry: a study of ion suppression. *Journal of the American Society for Mass Spectrometry*, **7** (11), 1099–1105.

[102] Sangster, T., Spence, M., Sinclair, P. *et al.* (2004) Unexpected observation of ion suppression in a liquid chromatography/atmospheric pressure chemical ionization mass spectrometric bioanalytical method. *Rapid Communications in Mass Spectrometry*, **18** (12), 1361–1364.

[103] Furey, A., Crowley, J., Lehane, M. and James, K.J. (2003) Liquid chromatography with electrospray ion-trap mass spectrometry for the determination of anatoxins in cyanobacteria and drinking water. *Rapid Communications in Mass Spectrometry*, **17**, 583–588.

[104] Souverain, S., Rudaz, S. and Veuthey, J.L. (2004) Protein precipitation for the analysis of a drug cocktail in plasma by LC–ESI-MS. *Journal of Pharmaceutical and Biomedical Analysis*, **35** (4), 913–920.

[105] Holcapek, M., Volna, K., Jandera, P. *et al.* (2004) Effects of ion-pairing reagents on the electrospray signal suppression of sulphonated dyes and intermediates. *Journal of Mass Spectrometry*, **39**, 43–50.

[106] Stahnke, H., Kittlaus, S., Kempe, G. *et al.* (2012) The influence of electrospray ion source design on matrix effects. *Journal of Mass Spectrometry*, **47**, 875–884.

[107] Larger, P.J., Breda, M., Fraier, D. *et al.* (2005) Ion-suppression effects in liquid chromatography–tandem mass spectrometry due to a formulation agent, a case study in drug discovery bioanalysis. *Journal of Pharmaceutical and Biomedical Analysis*, **39** (1–2), 206–216.

[108] Ghosh, C., Shinde, C.P. and Chakraborty, B.S. (2012) Influence of ionization source design on matrix effects during LC–ESI-MS/MS analysis. *Journal of Chromatography B*, **893–894**, 193–200.

[109] Gangl, E.T., Annan, M., Spooner, N. and Vouros, P. (2001) Reduction of signal suppression effects in ESI-MS using a nanosplitting device. *Analytical Chemistry*, **73** (23), 5635–5644.

[110] Gómez-Canela, C., Cortés-Francisco, N., Ventura, F. *et al.* (2013) Liquid chromatography coupled to tandem mass spectrometry and high resolution mass spectrometry as analytical tools to characterize multi-class cytostatic compounds. *Journal of Chromatography A*, **1276**, 78–94.

[111] Hopfgartner, G. and Varesio, E. (2005) New approaches for quantitative analysis in biological fluids using mass spectrometric detection. *Trends in Analytical Chemistry*, **24**, 583–589.

[112] Flores, C., Cortés-Francisco, N. and Caixach, J. (2013) *Capabilities of High Resolution Mass Spectrometry for the Identification of Microcystins*. Oral presentation at the Marine and Freshwater Toxins Analysis, fourth Joint Symposium and AOAC Task Force Meeting, Baiona, Spain, 5–9 May 2013.

[113] Cortés-Francisco, N., Flores, C., Moyano, E. and Caixach J. (2011) Accurate mass measurements and ultrahigh-resolution: evaluation of different mass spectrometers for daily routine analysis of small molecules in negative electrospray ionization mode. *Analytical and Bioanalytical Chemistry*, **400**, 3595–3606.

[114] Herniman, J.M., Langley, G.J., Bristow, W.T. and O'Connor, G. (2005) The validation of exact mass measurements for small molecules using FT-ICRMS for improved confidence in the selection of elemental formulas. *Journal of the American Society for Mass Spectrometry*, **16**, 1100–1108.

[115] Blom, K. (2001) Estimating the precision of exact mass measurement on an orthogonal time-of-flight mass spectrometer. *Analytical Chemistry*, **73**, 715–719.

[116] Kind, T. and Fiehn, O. (2007) Seven Golden Rules for heuristic filtering of molecular formulas obtained by accurate mass spectrometry. *BMC Bioinformatics*, **8**, 105.
[117] Yuan, M., Namikoshi, M., Otsuki, A. *et al.* (1999) Electrospray ionization mass spectrometric analysis of microcystins, cyclic heptapeptide hepatotoxins: modulation of charge states and $[M+H]^+$ to $[M+Na]^+$ ratio. *Journal of the American Society for Mass Spectrometry*, **10**, 1138–1151.
[118] Brunnée, C. (1987) The ideal mass analyzer: fact or fiction? *International Journal of Mass Spectrometry and Ion Processes*, **76**, 125–237.
[119] Holcapek, M., Jirásko, R. and Lísa, M. (2010) Basic rules for the interpretation of atmospheric pressure ionization mass spectra of small molecules. *Journal of Chromatography A*, **1217**, 3908–3921.
[120] Yuan, M., Namikoshi, M., Otsuki, A. *et al.* (1999) Low-energy collisionally activated decomposition and structural characterization of cyclic heptapeptide microcystins by electrospray ionization mass spectrometry. *Journal of Mass Spectrometry*, **34** (1), 33–43.
[121] Kubwabo, C., Vais, N. and Benoit, F.M. (2005) Characterization of microcystins using in-source collision-induced dissociation. *Rapid Communications in Mass Spectrometry*, **19**, 597–604.
[122] Yuan, M., Namikoshi, M., Otsuki, A. and Sivonen, K. (1998) Effect of amino acid side-chain on fragmentation of cyclic peptide ions: differences of electrospray ionization collision-induced decomposition mass spectra of toxic heptapeptide microcystins containing ADMAdda instead of Adda. *European Journal of Mass Spectrometry*, **4** (4), 287–298.
[123] Mayumi, T., Kato, H., Imanishi, S. *et al.* (2006) Structural characterization of microcystins by LC–MS/MS under ion trap conditions. *The Journal of Antibiotics*, **59**, 710–719.
[124] Furey, A., Crowley, J., Hamilton, B. *et al.* (2005) Strategies to avoid the mis-identification of anatoxin-a using mass spectrometry in the forensic investigation of acute neurotoxic poisoning. *Journal of Chromatography A*, **1082** (1), 91–97.
[125] Guzmán-Guillén, R., Prieto, A.I., González, A.G. *et al.* (2012) Cylindrospermopsin determination in water by LC–MS/MS: optimization and validation of the method and application to real samples. *Environmental Toxicology and Chemistry*, **31** (10), 2233–2238.
[126] Krüger, T., Mönch, B., Oppenhäuser, S. and Luckas, B. (2010) LC–MS/MS determination of the isomeric neurotoxins BMAA (beta-*N*-methylamino-L-alanine) and DAB (2,4-diaminobutyric acid) in cyanobacteria and seeds of *Cycas revoluta* and *Lathyrus latifolius*. *Toxicon*, **55** (2–3), 547–557.
[127] Spáčil, Z., Eriksson, J., Jonasson, S. *et al.* (2010) Analytical protocol for identification of BMAA and DAB in biological samples. *Analyst*, **135**, 127–132.
[128] Esterhuizen-Londt, M., Downing, S. and Downing, T.G. (2011) Improved sensitivity using liquid chromatography–mass spectrometry (LC–MS) for detection of propyl chloroformate derivatised β-*N*-methylamino-l-alanine (BMAA) in cyanobacteria. *Water SA*, **37**, 133–138.
[129] Council Directive 2013/39/EU (2013) European Parliament and of the Council, amending Directives 2000/60/EC and 2008/105/EC as regards priority substances in the field of water policy. *Official Journal of the European Union*, **L226**, 1–17.
[130] Council Directive 98/83/EC (1998) On the quality of water intended for human consumption. *Official Journal of the European Communities*, **L330**, 32–54.
[131] Commission Directive 2009/90/EC (2009) Laying down, pursuant to Directive 2000/60/EC of the European Parliament and of the Council, technical specifications for chemical analysis and monitoring of water status. *Official Journal of the European Union*, **L201**, 36–38.
[132] Anon. Commission Decision 2002/657/EC (2002) Implementing Council Directive 96/23/EC concerning the performance of analytical methods and the interpretation of results. *Official Journal of the European Communities*, **L221**, 8–36.
[133] Anon. Guidance document on analytical quality control and validation procedures for pesticide residues analysis in food and feed, SANCO/12571/2013, 19 November 2013; European Commission, Health and Consumer Protection Directorate-General, Brussels, Belgium.

24

Capillary Electrophoresis of Cyanobacterial Toxins

Gábor Vasas

Department of Botany, University of Debrecen, Hungary

24.1 Basic Theory and Introduction of Capillary Electrophoresis

Capillary electrophoresis (CE) is an analytical technique that separates ions based on their electrophoretic mobility with the use of an applied voltage. It is used predominantly because of the short analysis times and the high resolution it provides. It is a useful technique because there is a wide range of detection methods available.

The principal advantages include high efficiency (N = 100,000 to 1 million), short analysis times, simultaneous separation of anions and cations, small sample volumes (picolitres to nanolitres), low consumption of mobile phase, inexpensive columns, an being extremely versatile (with several separation methods).

There are six main types of capillary electroseparation available: capillary zone electrophoresis (CZE), capillary gel electrophoresis (CGE), micellar electrokinetic capillary chromatography (MECC), capillary electrochromatography (CEC), capillary isoelectric focusing (CIEF), and capillary isotachophoresis (CITP).

CE has demonstrated tremendous potential for a wide range of applications, from small molecules that include inorganic ions, organic acids, amino acids, peptides, drugs, nucleosides, nucleotides, vitamins, steroids, and carbohydrates and various natural products, to larger molecules, such as hormones, proteins, nucleic acids, and even living cells [1].

Handbook of Cyanobacterial Monitoring and Cyanotoxin Analysis, First Edition.
Edited by Jussi Meriluoto, Lisa Spoof and Geoffrey A. Codd.

24.2 Selection of Separation Methods

A major challenge in the analysis of cyanobacterial toxins is not only the high variability of the cyanotoxins but also the presence of the toxins in complex biological matrices, such as cyanobacterial cells, bloom materials, environmental and drinking waters, and tissue materials [2].

CZE separation methods are attracting interest because of their capability of generating rapid, high resolution separations of complex polar molecules. CZE is a form of free electrophoresis conducted in small diameter capillaries. Separations approaching 10^6 theoretical plates in less than 20 min have been demonstrated. Separation selectivity can be manipulated by changing the electrophoretic medium: typically, the pH and buffer composition. This separation method is recommended for the analysis of cyanobacterial toxins in protein-free model solutions, drinking waters, and purity testing, as well as in qualitative and quantitative analyses of toxin standards [1, 2].

MECC is a technique where, in order to separate compounds with very similar electrophoretic mobilities, a surfactant (like SDS) may be added to the buffer electrolyte at or above its critical micelle concentration. This technique has been extensively used for a wide range of sample types. MECC is an attractive alternative or a complementary technique to high-performance liquid chromatography (HPLC) because the solutes partition themselves between the micellar and the aqueous phase during the normal course of electrophoresis in a fashion analogous to reverse-phase HPLC. Solute partitioning between the micellar phase and aqueous phase is similar to reverse-phase HPLC except for the fact that the surfactant in MECC is a pseudo-stationary phase. This method makes it possible to separate complex biological samples such as intracellular extracts and bloom materials. Moreover, the technique is capable of separating nonpolar, uncharged cyanobacterial toxins [1, 2].

ITP is a nonlinear electrophoretic technique used in the separation of a variety of ionic compounds, ranging from small to large molecules. Unlike "linear" zone electrophoresis, ITP forms self-sharpening, adjacent zones of substantially pure solutes whose concentrations often exceed several milligrams per millilitre. In ITP, a multianalyte sample is usually introduced between the leading electrolyte (LE, containing leading ion) and the terminating electrolyte (TE, containing terminating ion) where the leading ion, the terminating ion, and the sample components must have the same charge and the sample ions must have lower electrophoretic mobilities than the leading ion but higher electrophoretic mobilities than the terminating ion. After the application of a fixed electric current, sample components move forward behind the leading ion and in front of the terminating ion and form discrete, contiguous zones according to their electrophoretic mobility order. Then, following a brief transient period where the discrete solute zones are formed, this ITP "stack" assumes a fixed concentration profile with a constant velocity moving in the direction of the leader. The isotachophoretic mode can provide a significant increase in terms of detection limits for cyanobacterial toxins in naturally contaminated samples [1–3].

24.3 Detection Methods

Although the use of small-diameter capillaries in CE effectively reduces zone-broadening effects from heat-induced convection, small sample sizes are required, which demands a sensitive detection method [1]. The sensitivity, particularly in the case of UV absorbance detection in conventional CE, is approximately two orders of magnitude less than in HPLC, because of the short light path (50-µm capillary diameter) [1–3]. On-column detection in CE has been accomplished using a variety of methods including UV absorption, conductivity, and fluorescence. The use of laser-induced fluorescence detection has also been reported for CE. The isotachophoresis preconcentration procedure combined with MS/MS enables access to valuable structural information and facilitates the identification of low levels of isomeric toxins [1, 3].

24.4 CE Methods of Cyanobacterial Toxins

24.4.1 PSP Toxins

The detection of PSP toxins has provided a considerable challenge to the development of analytical techniques. The low volatility and highly polar nature of these compounds and the lack of useful UV chromophores absorbing above 220 nm are some of the difficulties that prevent their separation and detection by conventional techniques [2]. CZE methods combined with either UV or MS detection were described for the analysis of native PSP toxins in extracts from toxic cyanobacteria and dinoflagellates. The high resolution capability of CZE provides high separation efficiencies for PSP toxins ($N > 150{,}000$) and enables separation of closely related isomers such as the epimeric GTXs. A study reports results on the use of CITP/CZE, with both UV and MS detection, for the analysis of PSP toxins in shellfish tissues [2, 3]. MS detection provided enhanced selectivity compared to UV detection, and meaningful spectra were generally obtained for toxins present at levels as low as 100 nmol L^{-1} in extracts of contaminated scallop liver tissues [3].

24.4.2 Microcystins

The microcystin (MC) toxins have very closely related structures and mass-to-charge ratios, and as a consequence, the small difference in their electrophoretic mobilities does not allow them to be fully resolved from each other. MECC was proved to be an efficient approach for the analysis of microcystins [4, 5]. MECC combines the high efficiency of CE with the separation mechanism of chromatography because the micelles of surfactant act as a pseudo-stationary phase. Conditions of 120 mmol L^{-1} borate with 20 mmol L^{-1} SDS at pH 9.5 provided the overall best response. The optimisation of parameters was based on peak areas, migration times, efficiency, effective mobility, and capacity factor of the toxins and the current developed in the capillary. Online preconcentration based on field-amplified sample stacking (FASS) was used for increasing the sensitivity achieved in the MEKC mode and contributed to an efficiency increase of the electrophoretic separation [2, 5]. These conditions were applied to the analysis of lyophilised material from a toxic bloom of cyanobacteria in freshwater (Fig. 24.1). The method allowed the determination of specific MCs in naturally contaminated samples [4, 8].

CE-DAD (capillary electrophoresis with diode array detector), LIF (laser-induced fluorescence) detector and CE–ESI-MS (capillary electrophoresis–electrospray ionization-mass spectrometry) have been applied for the analysis of MCs [2, 4, 5,].

24.4.3 Anatoxin-a

Capillary electrophoresis of anatoxin-a was first investigated by Jefferies et al. in 1994 [6]. The applicability of CZE and MEKC methods was investigated for the simultaneous determination of the three most frequently occurring cyanotoxins (anatoxin-a, CYN, MC-LR) by Vasas et al [4]. The developed methods provide suitable separation of the cyanotoxins present in model solutions and also in extracts of water bloom samples and in crude extracts of cyanobacterial strains. In CZE separation, while the alkaloid anatoxin-a exists only in cationic form, the cyclic peptide

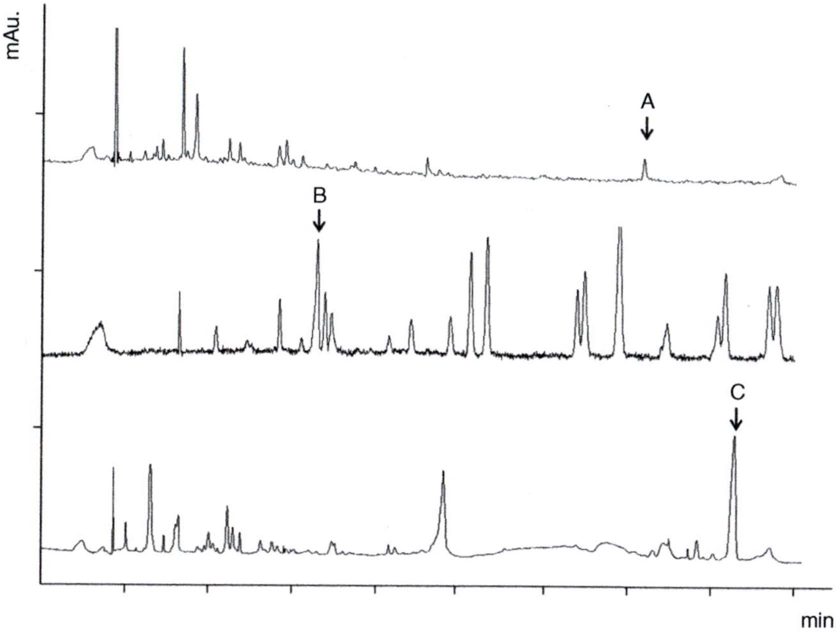

Figure 24.1 *Capillary electrophoresis of homoanatoxin-a (A) from* Oscillatoria sp. *PCC 6506 strain, cylindrospermopsin (B) from* Chrysosporum ovalisporum *BGSD-423 [D-Asp^3, Mdha^7] MC–RR* (C) from* Planktothrix rubescens *dominated bloom sample. Separation conditions: capillary: 80 cm, 50 µm i.d., buffer electrolyte: 25 mmol L^{-1} borate and 75 mmol L^{-1} SDS, pH 9.3, applied voltage: +25 kV, detection: UV absorption at 227 nm (A), 275 nm (B) 238 nm (C). *D-Aspartic acid in position 3,* N*-methyl-dehydroalanine in position 7*

cyanotoxins can be present in both cationic and anionic forms depending on the pH of the background electrolyte. Below pH 5.5 all three (cationic) components migrate before the electro-osmotic flow (EOF). Above pH 6.1 the net charge of MC-LR and CYN alters to negative and the EOF transports these "slow" components to the cathode. The optimal pH for the separation of the investigated cyanotoxins was found to be pH 9.3, where a quick separation and adequate resolution could be achieved and anatoxin-a migrated before the electro-osmotic flow, separately from the other components. To improve the separation, the influence of variable concentrations of SDS in the borate buffer was studied [4, 6]. The migration times of the anatoxin-a increased significantly with the SDS concentration (mostly in the range of 0–50 mmol L^{-1}). The behavior of the compounds is probably a consequence of their different hydrophobicities. Strong interaction could be expected between the micelles and anatoxin-a, increasing the migration times [4].

24.4.4 Cylindrospermopsin (CYN)

Capillary electrophoresis of CYN was investigated by Vasas et al. in 2002 and 2004 [4, 7]. The applicability of CZE and MEKC methods was investigated for the determination of the zwitterionic toxin. Below pH 5.5, many components migrate before the electro-osmotic flow (EOF); above pH 6.1, the net charge of CYN alters to negative and the EOF transports these "slow" components to the cathode. The optimal buffer concentration was found to be 25 mmol L^{-1}. The SDS content of the electrolyte had practically no effect on the migration of CYN, because significant interaction could not be supposed between the micelles and the polar molecule. The proposed CZE electrolyte contained 25 mmol L^{-1} sodium tetraborate (pH 9.1) and the MECC electrolyte contained the same with the addition of 100 mmol L^{-1} SDS [4]. In the case of MECC, the SDS micelles migrate with the sample proteins, causing the proteins to be eluted after the toxin peaks of interest. Separation of compounds was followed at 270 nm, since the elevated alkaline pH shifted the absorption maximum of CYN toward 270 nm (Fig. 24.1) [7].

24.5 Future Perspectives

CE methods have not been developed for some cyanobacterial toxins, like β-*N*-methylamino-L-alanine and the LPS fractions of cyanobacteria. Validated CE methods for cyanobacterial toxins are also missing. Online preconcentration techniques are possible in the CE analysis of cyanobacterial toxins, and thus, they need to be developed to enhance the sensitivity of the analysis. Due to the simplicity and speed of CE, it can be a routine qualitative and quantitative high-throughput screening technique in the future.

References

[1] Landers, J.P. (ed.) (1994) *Handbook of Capillary Electrophoresis*. CRC Press, Boca Raton.

[2] Gago-Martínez, A., Leão, J.M., Piñeiro, N. *et al.* (2003) An application of capillary electrophoresis for the analysis of algal toxins from the aquatic environment. *International Journal of Environmental Analytical Chemistry* **83**, 443–456.

[3] Wu, Y., Ho, A.Y., Qian, P.Y. *et al.* (2006) Determination of paralytic shellfish toxins in dinoflagellate *Alexandrium tamarense* by using isotachophoresis/capillary electrophoresis. *Journal of Separation Science*, **29**, (3), 399–404.

[4] Vasas, G., Gáspár, A., Páger, C. *et al.* (2004) Analysis of cyanobacterial toxins (anatoxin-a, cylindrospermopsin, microcystin-LR) by capillary electrophoresis. *Electrophoresis*, **25** (1), 108–115.

[5] Birungi, G., Li, S.F.Y. (2009) Determination of cyanobacterial cyclic peptide hepatotoxins in drinking water using CE. *Electrophoresis*, **30** (15), 2737–2742.

[6] Jefferies, T.M., Brammer, G., Zotou, A. *et al.* (1994) Determinations of anatoxin-a, homoanatoxin and propylanatoxin in cyanobacterial extracts by HPLC, GC-mass spectrometry and capillary electrophoresis. *Special Publications of the Royal Society of Chemistry* **149**, 34–39.

[7] Vasas, G., Gáspár, A., Surányi, G. *et al.* (2002) Capillary electrophoretic assay and purification of cylindrospermopsin, a cyanobacterial toxin from *Aphanizomenon ovalisporum* by plant test (blue-green Sinapis Test). *Analytical Biochemistry*, **302** (1), 95–103.

[8] Vasas, G., Szydlowska, D., Gáspár, A. *et al.* (2007) Determination of microcystins in environmental samples using capillary electrophoresis. *Journal of Biochemical and Biophysical Methods*, **66** (3), 87 – 97.

25

Immunoassays and Other Antibody Applications

James S. Metcalf[1] *and Geoffrey A. Codd*[2,3]

[1] *Institute for Ethnomedicine, Jackson, USA*
[2] *Biological and Environmental Sciences, University of Stirling, Scotland, United Kingdom*
[3] *School of the Environment, Flinders University, Adelaide, Australia*

25.1 Introduction

Antibodies are being increasingly applied in cyanotoxin detection and quantification in the laboratory and field. They can be used with samples of lysed or permeabilised cyanobacterial cells, raw and treated water samples, and (with some limitations) extracts of animal and plant tissues that may have been exposed to or have taken up cyanotoxins. Beginning with the production of monoclonal antibodies versus microcystin-LA (then called cyanoginosin-LA), the specificity of the antigen–antibody binding relationship in immunology has been exploited for cyanotoxin immunoassay for almost 30 years [1]. Monoclonal and polyclonal antibodies have been produced versus several purified complete cyanotoxin molecules and versus the microcystin fragment, ADDA. Increasing numbers of commercial immunoassay kits are becoming available in traditional microtitre plate formats and immunodiagnostic strips, and vigorous research is in progress into the application of antibodies to cyanotoxins in a range of biosensors. Essential requirements to verify performance specificity are discussed. Aims are also under way to enable low-technology/low-resource test methods to be performed at the sampling site, with rapid response or real-time data provision. Further applications of the antibodies for cyanotoxin localisation and quantification in individual cyanobacterial cells, colonies, and filaments are available.

Handbook of Cyanobacterial Monitoring and Cyanotoxin Analysis, First Edition.
Edited by Jussi Meriluoto, Lisa Spoof and Geoffrey A. Codd.

25.2 Production of Antibodies versus Cyanotoxins

The production of antibodies against large molecules, such as proteins, is usually a straightforward process, as the compound (the antigen) is large enough to be recognised by the mammalian immune system. This means that a large pool of polyclonal antibodies, or a number of monoclonal cell lines, can be produced and evaluated. However, in the case of small molecules, such as the majority of cyanotoxins, these compounds are not sufficiently large for the mammalian immune system to recognise. Such molecules are termed haptens, and special techniques are required to produce antibodies against these molecules. Generally, in order to produce antibodies against haptens, the small molecule of interest is attached to a larger protein, which is recognized by the mammalian immune system. Antibodies are then produced against the entire complex, including the cyanotoxins, which are attached to the outside surface of the carrier protein. The antibodies can be subsequently collected and often can be used without need for further purification. However, in some cases, affinity column purification may be needed to isolate the cyanotoxin antibodies which can then be used in various applications.

A number of cyanotoxins have been used for the production of antibodies, including microcystins [1] and saxitoxins [2], with commercial kits being produced for the immunoassay of these cyanotoxin families and also for cylindrospermopsin, the latter being used in a study in comparison with LC-MS [3]. In the majority of published studies, conjugation of the cyanotoxin to a carrier protein is carried out, and the antibodies subsequently produced. Antibodies were produced against both saxitoxin and neosaxitoxin in an attempt to provide some resolution between different variants in the saxitoxin group of neurotoxins [2]. However, due to the close structural similarities between saxitoxin variants, limited immunological discrimination occurred between certain variants and additional confirmatory methods are needed to determine which saxitoxin variants are present when uncharacterised samples are being tested by ELISA. Antibodies against microcystins have generally been made against the entire microcystin molecule, with published antibodies generated against microcystin-LA [4], microcystin-LR [5, 6], and microcystin-RR [7], versus nodularin [8]. The isolated ADDA fragment of the microcystins and nodularins has also been used to generate antibodies using a modified hapten [9] and is used in ELISA to detect these toxins.

25.3 Applications of Cyanotoxin Antibodies

As with any other antibodies, those that recognise cyanotoxins have the same potential for application in a wide variety of systems. The greatest use of cyanotoxin antibodies has occurred with the microcystin antibodies. The majority of systems have used the antibodies to detect free microcystins, detoxication products,and degradation products by ELISA. Such ELISA systems, although initially laboratory based, are being increasingly applied in the field and, with necessary precautions and controls, can provide valuable environmental and ecotoxicological information. Increasingly, the development of more rapid antibody tests, such as with lateral flow test kits [e.g. 10–12], has allowed the analysis of microcystins to go out of the laboratory. Such screening methods have proved very useful in providing rapid information concerning the potential concentration of microcystins in the field. As with all analytical methods for cyanotoxins, appropriate checks and controls should be carried out when using immunoassays for cyanotoxins in the laboratory and field. For example, with microcystin antibody-based methods, these should include the production of false positives by solvents including methanol, and the cross-reactivity between microcystin antibodies and the variants and detoxication products of microcystins. For example

[13], the cross reaction between microcystin-LR antibodies with the commonly produced products of microcystin-detoxication in animal and plant tissues (GSH-MC-LR, Cys-Gly-MC-LR, and Cys-MC-LR) can lead to overestimation of the actual concentration of the microcystin content of the sample and thus of toxicity.

Other applications of antibodies have included the production of immunoaffinity columns, which, although similar to clean-up methods, such as solid-phase extraction, allow the production of extremely clean extracts for analysis, as only compounds which bind microcystins should be retained by the columns [14].

25.4 Cyanotoxin Localisation and Quantification Using Antibodies

The localisation of pathogens and toxins at organ, tissue, and single cell levels in the producer-organism and in exposed animals and plants is widely possible via immunohistochemistry. In principle, it should be possible to use this method to localise all cyanotoxins in animal and plant tissues if the respective cyanotoxin antibodies are available. Hitherto, however, this only appears to have been carried out for microcystin localisation using antibodies to microcystin-LR and to the microcystin side chain, ADDA. After fixing and sectioning the tissues and exposing them to the cyanotoxin antibodies, the sites of the bound antibodies are visualised by chromogenic or fluorescence methods. For example, intestinal lining, spleen, liver, and individual hepatocytes are clearly labelled from fish exposed to microcystin-LR by gavage [15]. Immunohistochemistry has potential to contribute to the understanding of cyanotoxin pathology in animal and human intoxications using appropriate tissue fixation methods. However, in applications of microcystin antibodies to human, animal, and plant tissues in which toxin uptake has occurred, positive results should not be limited to infer the presence of the authentic toxins only [13].

Microcystin antibodies have also been successfully used to localise and quantify microcystins in single colonies and filaments of cyanobacteria. After micropipetting from laboratory cultures and environmental samples, serially washing, measuring, then lysing the colonies and filaments, their microcystin quotas can be measured and related directly to cell number, colony or filament size, and biovolume [16, 17]. Avoiding the use of alcohol-based cell dehydration and consequent microcystin-extraction, immunogold-electron microscopy is an effective tool for the localisation of cyanotoxins within the cyanobacterial cell. This procedure has shown that about half of the total intracellular microcystin pool is associated with the thylakoids, with lesser portions associated with polyphosphate bodies and in apparently free form in the cyanobacterial cytoplasm [16].

25.5 Other Cyanotoxin Antibody-Related Technologies

Success in producing recombinant antibody fragments in *Escherichia coli* was achieved over 10 years ago for the quantitative detection of a range of microcystins (see [1]), although this approach does not appear to have been more widely applied. Meanwhile, conventional monoclonal antibodies versus ADDA are being incorporated into modern formats including a multiwalled carbon nanotube-based electrochemical biosensor with performance characteristics suitable for microcystin analysis in water [18]. The ultimate value of further applications, whether involving currently available cyanotoxin and cyanotoxin-fragment antibodies, or future antibodies to cyanotoxins, will be determined to a large degree by the specificities of the antibodies, which need to be rigorously tested.

Analogous to the use of antibodies, aptamers are being developed for cyanotoxin analysis. This involves the selection of oligomer sequences from large numbers of randomly generated DNA and RNA fragments that show microcystin-binding properties [1, 19]. Electrochemical impedance-based biosensors involving aptamers offer prospects for cyanotoxin analysis in raw and treated waters and further applications of aptamers can be expected.

References

[1] Metcalf, J.S. and Codd, G.A. (2003) Analysis of cyanobacterial toxins by immunological methods. *Chemical Research in Toxicology* **16**, 103–112.

[2] Chu, F.S., Huang, S. and Hall, S. (1992) Production and characterisation of antibodies against neosaxitoxin. *Journal of the AOAC International* **75**, 341–345.

[3] Blahova, L., Oravec, M., Marsalek, B. *et al.* (2009) The first occurrence of the cyanobacterial alkaloid toxin cylindrospermopsin in the Czech Republic as determined by immunochemical and LC/MS methods. *Toxicon* **53**, 519–524.

[4] Kfir, R., Johannsen, E. and Botes, D.P. (1986) Monoclonal antibody specific for cyanoginosin-LA: Preparation and characterization. *Toxicon* **24**, 543–552.

[5] Chu, F.S., Huang, X., Wei, R.D. *et al.* (1989) Production and characterization of antibodies against microcystin. *Applied and Environmental Microbiology* **55**, 1928–1933.

[6] Metcalf, J.S., Bell, S.G. and Codd, G.A. (2000) Production of novel polyclonal antibodies against the cyanobacterial toxin microcystin-LR and their application for the detection and quantification of microcystins and nodularin. *Water Research* **34**, 2761–2769.

[7] Young, F.M., Metcalf, J.S., Meriluoto, J.A.O. *et al.* (2006) Production of antibodies against microcystin-RR for the assessment of purified microcystins and cyanobacterial environmental samples. *Toxicon* **48**, 295–306.

[8] Mikhailov, A., Harmala-Brasken, A-S., Polosukhina, E. *et al.* (2001) Production and specificity of monoclonal antibodies against nodularin conjugated through *N*-methyldehydrobutyrine. *Toxicon* **39**, 1453–1459.

[9] Fischer, W.J., Garthwaite, I., Miles, C.O. *et al.* (2001) Congener-independent immunoassay for microcystins and nodularins. *Environmental Science and Technology* **35**, 4849–4856.

[10] Tippkotter, N., Stuckmann, H., Kroll, S. *et al.* (2009) A semi-quantitative dipstick assay for microcystin. *Analytical and Bioanalytical Chemistry* **394**, 863–869.

[11] Lawton, L.A., Chambers, H., Edwards, C., Nwaopara, A.A., Healy, M. (2010) Rapid detection of microcystins in water. *Toxicon* **55**, 973–978.

[12] Metcalf, J.S. and Codd, G.A. (2012) Cyanotoxins: In (Whitton, B.A., ed.) Ecology of Cyanobacteria II: Their Diversity in Time and Space, Springer, Dordrecht, Heidelberg, New York, London, pp. 651–675.

[13] Metcalf, J.S., Beattie, K.A., Ressler, J. *et al.* (2002) Cross-reactivity and performance assessment of four microcystin immunoassays with detoxication products of the cyanobacterial toxin, microcystin-LR. *Journal of Water Supply: Research and Technology – AQUA* **51**, 145–151.

[14] Aranda-Rodriguez, R., Kubwabo, C. and Benoit, F.M. (2003) Extraction of 15 microcystins using immunoaffinity columns. *Toxicon* **42**, 587–599.

[15] Djediat, C., Malécot, M., de Luze, A. *et al.* (2010) Localization of microcystin-LR in medaka fish tissues after cyanotoxin gavage. *Toxicon* **55**, 531–535.

[16] Young, F.M., Morrison, L.F., James, J. *et al.* (2008) Quantification and localisation of microcystins in colonies of a laboratory strain of *Microcystis* (Cyanobacteria) using immunological methods. *European Journal of Phycology* **43**, 217–225.

[17] Akcaalan, R., Young, F.M., Metcalf, J.S. *et al.* (2006) Microcystin analysis in single filaments of *Planktothrix* spp. in laboratory cultures and environmental blooms. *Water Research* **40**, 1583–1590.

[18] Han, C., Doepke, A., Cho, W. *et al.* (2013) A multiwalled-carbon-nanotube-based biosensor for monitoring microcystin-LR in source of drinking water supplies. *Advances in Functional Materials* **23**, 1807–1816.

[19] Lin, Z., Huang, H., Xu, Y. *et al.* (2013) Determination of microcystin-LR in water by a label-free aptamer-based electrochemical impedance biosensor. *Talanta* **103**, 371–374.

26

Protein Phosphatase Inhibition Assays

James S. Metcalf[1], Anastasia Hiskia[2], and Triantafyllos Kaloudis[3]

[1] *Institute for Ethnomedicine, Jackson, USA*
[2] *Institute of Nanoscience & Nanotechnology, National Center for Scientific Research "DEMOKRITOS," Athens, Greece*
[3] *Water Quality Control, Athens Water Supply and Sewerage Company - EYDAP SA, Athens, Greece*

26.1 Background and Molecular Mechanism of Protein Phosphatase Inhibition

Through in vitro investigations, microcystins (MCs) were found to inhibit protein phosphatases [1]. Protein phosphatases are essential enzymes that regulate cell cycle events and dephosphorylate proteins, in opposition to protein kinases that phosphorylate proteins. Microcystins were found to inhibit PP1 and PP2A, in addition to some phosphoprotein phosphatases [1, 2]. When present, the interaction of protein phosphatase with the methyldehydroalanine moiety of MCs showed that MCs irreversibly bound to cysteine 273 of the enzyme [3], whereas nodularin (NOD), which contains methyldehydrobutyrine, did not bind to this amino acid residue but could sit in the active site and still inhibit enzyme activity. This difference led to the idea that MCs and NODs could be tumour promoters and carcinogens, respectively. This is largely due to the fact that protein phosphatases are essential in order that cell division can proceed and the divided cell is a correct copy that will function properly. If protein phosphatases are affected, then when cell division occurs, the new cell may have defects, and in some cases, have cancer-producing potential, depending on whether the enzyme is irreversibly (MCs) or reversibly inhibited (NODs).

Handbook of Cyanobacterial Monitoring and Cyanotoxin Analysis, First Edition.
Edited by Jussi Meriluoto, Lisa Spoof and Geoffrey A. Codd.

26.2 Classes of Compounds that Inhibit Protein Phosphatases

The full toxicological significance of protein phosphatase inhibition is not completely understood as these essential enzymes are highly conserved throughout the eukaryotes and are essentially involved in many cellular processes. However, a number of compounds are known to inhibit protein phosphatases, including MCs and NODs from cyanobacteria. Although MCs are some of the most studied protein phosphatase inhibitors, other naturally occurring, largely marine polyketide compounds also have the potential to inhibit protein phosphatases. Calyculin A, and the related acanthifolicin (9, 10-epithio-okadaic acid; [4]) are both known to be produced by marine sponges, with okadaic acid being responsible for diarrhetic shellfish poisoning. Other diarrhetic shellfish toxins, the dinophysis toxins (DTX 1 and 2 [5]) are also potent inhibitors of protein phosphatases [6]. A further inhibitor, tautomycin, is produced by the bacterium *Streptomyces spiroverticilatus* and is also often found in shellfish.

26.3 Effects of Microcystins on Cyanobacterial Protein Phosphatases

As protein phosphatases are highly conserved among the eukaryotes and as these enzymes are found in cyanobacteria, the possibility exists that MCs, produced by the cyanobacteria themselves may affect their own enzymatic activity. The protein phosphatase gene sequence from cyanobacteria was cloned, expressed in *Escherichia coli* and the enzyme purified. Assaying this enzyme showed that cyanobacterial protein phosphatases were not affected by MCs or okadaic acid [7].

26.4 The Basis of the PPIA Assay for Microcystins and Its Evolution

The protein phosphatase inhibition assay (PPIA) is an *in vitro* assay used to measure the effect of inhibitors on the activity of protein phosphatases. The original assay used the release of ^{32}P from labelled substrates, with inhibition of the enzyme resulting in reduced release of radioactive phosphate [1]. Although this worked well with good sensitivity, the use of radioactivity did not always lend itself to routine analysis of environmental samples. An and Carmichael [8] used a colorimetric version of the assay, in which the protein phosphatase enzyme could hydrolyse *para*-nitrophenyl phosphate, releasing phosphate and producing *para*-nitrophenol, a colorimetric product, whose production could be used to measure the effect of MCs on protein phosphatases.

26.5 Comparison of PPIA with Other Analytical Methods for Microcystins

The colorimetric PPIA assay can be useful for quantitative screening of MCs and NODs in environmental samples. Although the assay is offered commercially in microplate format, the PPIA is different from other commercial tests such as ELISA, since it provides information about the bioactivity of toxins in the sample, rather than structural recognition of compounds. On the other hand, PPIA cannot differentiate between different variants of MCs and as already discussed PPIA is not absolutely specific to MCs. For this reason, additional confirmatory analysis with LC-MS/MS may be needed for PPIA-positive samples. However, PPIA has been shown to have good correlation with HPLC-PDA for MCs [9]. The main advantage of PPIA as a screening tool is ease of use. Commercial PPIA kits using a 96-well plate format include all necessary reagents and only require a single 30-minute incubation step. A large number of samples can be analysed in less than

1 to 2 hours, especially when sample preparation is not needed (e.g. drinking water), using facile equipment (microplate reader) and less-qualified personnel than required for HPLC or LC-MS analysis. The reported limits of detection of PPIA are around 0.25 $\mu g\,L^{-1}$ in water, comparable to or even better than LC-MS/MS with direct injection. This means that PPIA can be used to demonstrate compliance with the WHO provisional Guideline Value of 1 $\mu g\,L^{-1}$ MC-LR in drinking water. However, depending on the sample matrix, a clean-up procedure may be needed to avoid interferences that generally lead to false positive results. Such results when comparing PPIA to LC-MS/MS in environmental samples may be due to sample matrix interferences, to differences in inhibitory activity between MC variants, or to the presence of other compounds in the sample that are inhibitors of protein phosphatases [9, 10].

26.6 Commercially Available Kits for Microcystins

At least two PPIA kits are commercially available in 96-well microplate and tube formats (Zeu Immunotec and Abraxis). Both are based on the inhibition of protein phosphatase 2A (PP2A), which is more sensitive than PP1 in its inhibition by MCs [11] and use *para*-nitrophenyl phosphate (p-NPP) as the substrate [12]. The assay protocols include a single incubation step at 37°C for 30 minutes followed by measurement of absorbance at 405 nm. Calibration is carried out with solutions of MC-LR included in the kits and results are calculated from calibration curves which are plotted as percentage inhibition of PP2A, expressed as %B/Bo, versus MC-LR concentration on a semi-log scale, where B and Bo are the absorbance values of the sample and the blank respectively. The kits are suitable for drinking water and raw-surface waters with a range of measurement from 0.25 to 2.5 $\mu g\,L^{-1}$. Data regarding verification of the assay according to the Environmental Technology Verification (ETV) Program established by the EPA have been published for one PPIA kit [13]. As performance of PPIA is widely affected by sample matrices and MC concentrations, it is recommended that laboratories carry out their own in-house validation study that fits their intended use of the assay.

26.7 Improvements to the PPIA Assay to Make It More Specific to Microcystins

Although the protein phosphatase inhibition assay can measure the activity of inhibitors such as MCs, due to the number of other compounds that can affect this enzyme, modifications which improve the detection of MCs and NODs by this inhibition assay are required. Rivasseau et al. [14] used a combination of solid-phase extraction with C_{18} SPE cartridges, followed by the protein phosphatase inhibition assay, allowing clean-up and purified extracts to be assessed by the enzyme inhibition assay. An alternative approach was taken by Metcalf et al. [15], where antibodies that were reactive against MCs and NODs were able to discriminate among cyanobacterial protein phosphatase inhibitors and okadaic acid, calyculin A, and tautomycin.

26.8 Conclusions about the Effectiveness of the PPIA Assay for Microcystins and Nodularins in Different Matrices

As with any analytical system, often the better the extracts used for analysis, the better the data that are produced. Although, with the majority of cyanobacterial extracts, the protein phosphatase inhibition assay works well, and gives good agreement with HPLC-PDA for MCs [9]. If animal extracts are used, then there may be issues with using such an assay. This is largely due to the potential presence of endogenous phosphatases, which may affect the release of phosphate-containing substrates in the assay. Furthermore, using SPE-cleaned extracts, or even those from immunoaffinity columns (e.g. [16]), would be expected to improve the reliability of the results from the PPIA assay. A further issue is when comparing toxin concentrations to toxicity equivalents. The protein phosphatase inhibition assay will give an indication of the toxicity of samples, whereas other analytical methods such as HPLC or ELISA will give toxin concentration equivalents, often in comparison with a single MC standard. Both toxin concentration and toxicity equivalents are useful, and can give invaluable information, depending on the situation. With the analysis of marine samples, an understanding of the potential for other non-cyanobacterial protein phosphatase inhibitors should also be considered, with confirmatory techniques used to identify these other inhibitors or the presence of NOD.

References

[1] MacKintosh, C., Beattie, K.A., Klumpp, S. *et al.* (1990) Cyanobacterial microcystin-LR is a potent and specific inhibitor of protein phosphatases 1 and 2A from both mammals and higher plants. *FEBS Letters* **264**, 187–192.

[2] Hastie, C.J., Borthwick, E.B., Morrison, L.F. *et al.* (2005) Inhibition of several protein phosphatases by a non-covalently interacting microcystin and a novel cyanobacterial peptide, nostocyclin. *Biochimica et Biophysica Acta* **1726**, 187–193.

[3] MacKintosh, R.W., Dalby, K.N., Campbell, D.G. *et al.* (1995) The cyanobacterial toxin microcystin binds covalently to cysteine-273 on protein phosphatase 1. *FEBS Letters* **371**, 236–240.

[4] Holmes, C.F.B., Luu, H.A., Carrier, F. and Schmitz, F.J. (1990) Inhibition of protein phosphatases-1 and -2A with acanthifolicin: Comparisons with diarrhetic shellfish toxins and identification of a region on okadaic acid important for phosphatase inhibition. *FEBS Letters* **270**, 216–218.

[5] Garibo, D., de la Iglesia, P., Diogène, J. and Campàs M. (2013) Inhibition equivalency factors for dinophysistoxin-1 and dinophysistoxin-2 in protein phosphatase assays: applicability to the analysis of shellfish samples and comparisons with LC-MS/MS. *Journal of Agricultural and Food Chemistry* **61**, 2572–2579.

[6] Smienk, H., Dominguez, E., Rodriguez-Velasco, M.L. *et al.* (2013) Quantitative determination of the okadaic acid toxins group by a colorimetric phosphatase inhibition assay: interlaboratory study. *Journal of the AOAC International* **96**, 77–85.

[7] Shi, L., Carmichael, W.W. and Kennelly, P.J. (1999) Cyanobacterial PPP family protein phosphatases possess multifunctional capabilities and are resistant to microcystin-LR. *The Journal of Biological Chemistry* **274**, 10039–10046.

[8] An, J. and Carmichael, W.W. (1994) Use of a colorimetric protein phosphatase inhibition assay and enzyme linked immunosorbent assay for the study of microcystins and nodularins. *Toxicon* **32**, 1495–1507.

[9] Ward, C.J., Beattie, K.A., Lee, E.Y.C. *et al.* (1997) Colorimetric protein phosphatase inhibition assay of laboratory strains and environmental blooms of cyanobacteria: comparisons with high performance liquid chromatographic analysis for microcystins. *FEMS Microbiology Letters* **153**, 465–473.

[10] Kaloudis, T., Zervou, S., Tsimeli, K. *et al.* (2013) Determination of microcystins and nodularin (cyanobacterial toxins) in water by LC-MS/MS. Monitoring of Lake Marathonas, a water reservoir of Athens, Greece. *Journal of Hazardous Materials* **263**(1), 105–115.

[11] Honkanen, R.E., Codispoti, B.A., Tse, K. and Boynton, A.L. (1994) Characterization of natural toxins with inhibitory activity against serine/threonine protein phosphatases. *Toxicon* **32**(4), 339–350.

[12] Heresztyn, T. and Nicholson, B.C. (2001) Determination of cyanobacterial hepatotoxins directly in water using a protein phosphatase inhibition assay. *Water Research* **35**(13), 3049–3056.

[13] James, R., Gregg, A., Dindal, A. and McKernan, J. (2011) ETV Report (Environmental Technology Verification Report) Microcystest kit.

[14] Rivasseau, C., Racaud, P., Deguin, A. *et al.* (1999) Development of a bioanalytical phosphatase inhibition test for the monitoring of microcystins in environmental water samples. *Analytica Chimica Acta* **394**, 243–257.

[15] Metcalf, J.S., Bell, S.G. and Codd, G.A. (2001) Colorimetric Immuno-protein phosphatase inhibition assay for specific detection of microcystins and nodularins of cyanobacteria. *Applied and Environmental Microbiology* **67**, 904–909.

[16] Aranda-Rodriguez, R., Kubwabo, C. and Benoit, F.M. (2003) Extraction of 15 microcystins and nodularin using immunoaffinity columns. *Toxicon* **42**, 587–599.

27

Bioassay Use in the Field of Toxic Cyanobacteria

Luděk Bláha[1], Ana Maria Cameán[2], Valérie Fessard[3], Daniel Gutiérrez-Praena[2], Ángeles Jos[2], Benjamin Marie[4], James S. Metcalf[5], Silvia Pichardo[2], María Puerto[2], Andrea Törökné[6], Gábor Vasas[7], and Bojana Žegura[8]

[1]Faculty of Science, RECETOX, Masaryk University, Brno, Czech Republic
[2]Area of Toxicology, Faculty of Pharmacy, University of Sevilla, Spain
[3]Unit Toxicology of Contaminants, French Agency for Food, Environmental and Occupational Health and Safety, Fougères laboratory, France
[4]UMR 7245 MCAM CNRS-MNHN, Muséum National d'Histoire Naturelle, Paris, France
[5]Institute for Ethnomedicine, Jackson, USA
[6]Knot Ltd., Budapest, Hungary
[7]Department of Botany, University of Debrecen, Hungary
[8]Department of Genetic Toxicology and Cancer Biology, National Institute of Biology, Ljubljana, Slovenia

27.1 Introduction

Physicochemical analytical methods and immunochemical assays discussed elsewhere in this handbook allow determination of individual molecules (e.g. cyanotoxins) based on their chemical and structural properties (e.g. size and hydrophobicity in reverse-phased high-performance liquid chromatography [HPLC] or structural domains in the ELISA). On the other hand, tools based on biological units (bioassays) do not identify individual toxins but provide important information on the biological effects, including toxicity of the investigated material (isolated toxin, water sample, cyanobacterial biomass, etc.). In the present chapter, we focus on bioassays that are based on cells, tissues and whole organisms. Macromolecular tools, such as enzyme "mechanism-based" assays like protein phosphatase inhibition are discussed elsewhere.

Handbook of Cyanobacterial Monitoring and Cyanotoxin Analysis, First Edition.
Edited by Jussi Meriluoto, Lisa Spoof and Geoffrey A. Codd.

By definition, toxicity is the degree to which a substance (or sample, mixture, extract, etc.) damages the biological unit (i.e. cell or organism). As postulated by Paracelsus in the early sixteenth century, it is only the dose or concentration which determines if the substance is poisonous or not. Each bioassay should therefore be able to provide dose-response data, and knowledge of the effective concentrations (e.g. IC_{10}, IC_{50}: the concentration causing 10% or 50% inhibition, the LOEL: the lowest observable effect level, etc.) is of premier importance. The results should always be clearly described and critically evaluated in terms of the relevance of the conditions of the exposure: the exposure route, exposure medium, concentrations of test material used, duration of exposure, and the kinetics data obtained. Both exposure investigation and toxic effects (hazard) assessment are required to estimate the risk of intoxication and some specific bioassays can be used for these two purposes.

With respect to the effective dose, two types of toxicity can be distinguished. For most of the biological effects at different levels (cytotoxicity, lethality, reproduction toxicity etc.), a threshold, no-effect, or safe concentration can be derived. On the other hand, chemical-induced genotoxicity or mutagenicity (the initiation steps in carcinogenesis) bioassays are in most cases considered inducing "no-threshold" stochastic events, when even a single molecule can cause damage with some probability. However in some cases, as for aneugenic compounds, mutagenicity is induced only when a threshold level is reached.

Numerous bioassays have been used in the investigation of cyanobacterial samples including pure or purified cyanotoxins, filtered water samples or crude extracts prepared from cyanobacterial cultures or from bloom biomass, etc. With the exception of pure cyanotoxins, all the other samples represent very complex mixtures of diverse chemicals, which may be specific to cyanobacteria and/or may co-occur in the sample (e.g. toxic metals or anthropogenic pollutants can accumulate in cyanobacterial cells or organisms; toxic ammonia can be formed during decay of organic material in the biomass). Both matrix effects and co-occurrence with anthropogenic pollutants, should be carefully considered when interpreting the bioassay response because several studies showed high "toxic effects" for cyanobacterial blooms that did not contain any known toxins as microcystins (MCs) or cylindrospermopsin (CYN) [1, 2].

27.2 Drivers and Objectives for Bioassay Use

With respect to the application of bioassays, one should differentiate between the bioassays devoted to detailed effect evaluation studies and those set up for monitoring. The former are intended to give indication of the response of cyanotoxins or extracts for various toxicity endpoints. The majority of the standardized assays dealing with hazard assessment belong to the OECD guidelines for chemicals and are expected to follow the described protocols. However, other bioassays can also be relevant and bring key information for understanding the overall toxicity or mechanism of action of cyanotoxins. The bioassays used for monitoring aim to be easily and rapidly applied to larger numbers of samples for detecting a specific bioactivity or toxicity and to thereby serve as early warning tools for potential exposure. To summarise, two classes of bioassays can be distinguished:

- Bioassays for evaluating *effects* of bioactive molecules (including cyanotoxins) and investigating hazardous or beneficial properties (specific in vitro and in vivo tests).
- Bioassays for *monitoring the presence of toxic compounds* in environmental samples of cyanobacteria. The responses should be specific to cyanobacterial metabolites and compared with a reference "positive" sample (e.g. pure cyanotoxin).

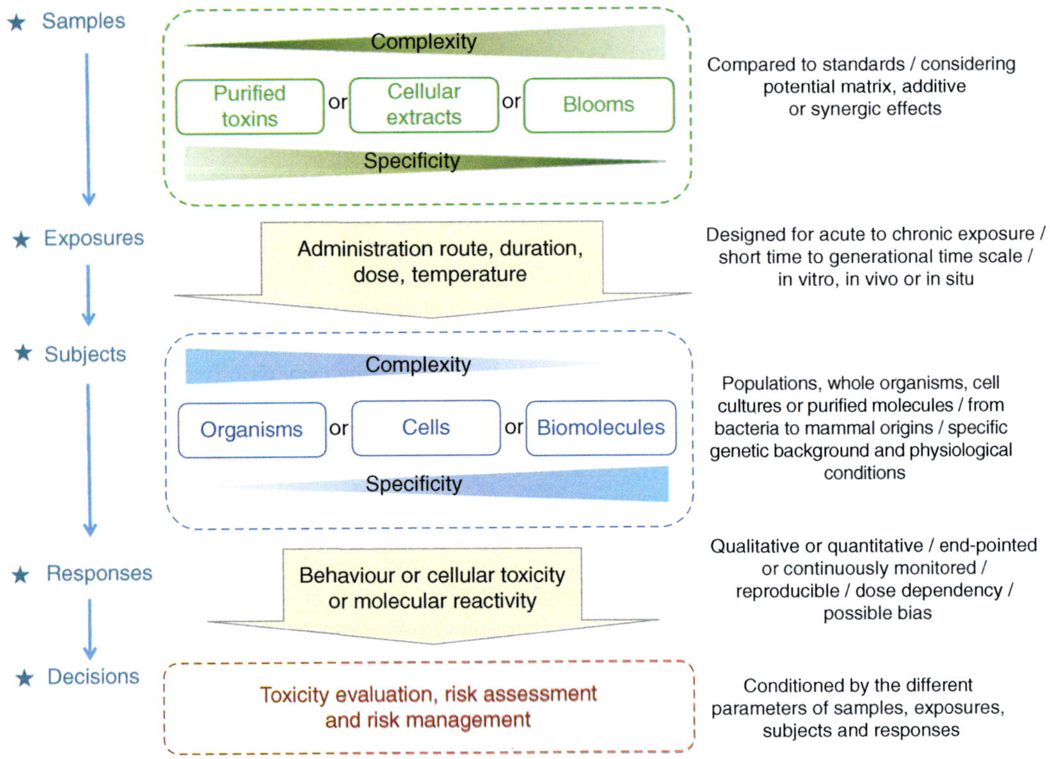

Figure 27.1 *The drivers for bioassay design and application*

More specifically, the design of the bioassay strategy integrates different levels (sample, exposure, subject, response, and decision-making) described by several parameters (Fig. 27.1). All of these parameters need to be carefully considered according to the objectives of each assessment as they can directly affect the interpretation and the significance of the bioassay.

27.3 Classification and Terminology

Various classifications of bioassays can be found in the literature, considering the exposure, mechanism of toxicity, target organism or target tissue, level of biological organization, etc. Key terminology is briefly explained in the following paragraphs.

Toxicology and ecotoxicology. Bioassays are used both for toxicological and ecotoxicological purposes. Toxicology (or specifically toxinology focused on natural toxic compounds) aims to protect human health, and correspondingly uses mammalian toxicological in vivo animal bioassays (e.g. rat, mouse, rabbit) or in vitro tools (e.g. tissues or cell cultures derived from humans or other mammals). On the other hand, ecotoxicology does not primarily aim to protect individuals but rather the populations and functioning of diverse species in ecosystems. Therefore, one should examine responses in multiple ecotoxicological models that represent different trophic levels including primary producers (including algae and plants), invertebrate consumers (aquatic cladocerans like *Daphnia*, soil earthworms, and many others), vertebrate consumers (e.g. fish, amphibians, birds), and decomposers (bacterial bioassays). For ecotoxicology, in vivo studies are certainly more common than in vitro assays.

Bioassay types. For both toxicology and ecotoxicology, acute or chronic bioassays are designed reflecting short- or long-term exposures (or high- and low-dose situations), respectively. They inform either on the integrative response of the whole organism or provide specific information such as on effects at cellular level and on the molecular mechanism(s) of action. In addition, higher tier testing systems have been developed especially in ecotoxicology and include, for example, whole life cycle or multigenerational tests, multispecies microcosms, and mesocosms. For the most common in vivo single species bioassays (in both toxicological and ecotoxicological studies), two major groups of endpoints can be distinguished. First parameters inform about the fitness of the individuals (lethality, growth parameters, or observations of diverse physiological and organ-specific effects such as hepatotoxicity, neurotoxicity, and immunosuppression), while the second group are populational parameters (most often reproduction and fecundity). Another classification scheme reflects the practical use of the tests discriminating formally validated and standardized bioassays (only few bioassays are listed, for example, by OECD or ISO and accepted in legislation especially for ecotoxicology) and large numbers of experimental, research, or alternative bioassays, which provide scientifically unique information, although their acceptance by official authorities is limited.

Mechanisms of toxicity. With respect to toxicity mechanisms, two broad groups of bioassay responses are known. First, the response will largely depend on the affinity of the toxicant for a specific biological target (e.g. protein phosphatase inhibition by MCs leading to liver damage; activation of nicotinic receptors by anatoxin-a, which results in lethal neurotoxic symptoms). Second, the reactivity of the toxic component with biological macromolecules (proteins, nucleic acids, phospholipids) can affect key pathways inducing the toxic effects. Advanced molecular and "-omics" techniques (genomic, transcriptomic, proteomic, metabolomics, etc.) are currently being used to investigate in depth effects of cyanobacterial metabolites or to search for new biomarkers.

This overview on diverse types and classes of bioassays has, however, multiple overlaps with no clear boarders. An enormous diversity of biota is used for bioassays testing, which may produce different responses. Commonly, there are clear correlations in the effects across different organisms (e.g. general potency of MCs to inhibit protein phosphatases in eukaryotes), but there are also examples of chemicals that exhibit highly species-specific effects in certain sensitive species (e.g. diclofenac toxicity in vultures, toxicity of insecticides to certain arthropods).

The following sections provide some selected examples of bioassays and their specific applications for the two different purposes described here, that is, detailed effect evaluation and monitoring of the presence of toxic compounds.

27.4 Bioassays for the Effect Evaluation

Diverse bioassays can be used for detailed characterization of toxicity and kinetic parameters of purified compounds including toxins (e.g. the minimum effect concentration and LD_{50}), or for in-depth research of modes of action (MoA). Validated bioassays, for which a panel of guidelines is established from international organizations such as OECD, ISO, or ASTM, should preferably be recommended for standard toxicological and ecotoxicological studies. However, detailed research into MoA often requires less standardized experimental approaches using specific species and diverse cellular and molecular targets (e.g. antitumor cytotoxicity, chemoprotective effects, teratogenicity, carcinogenicity, or endocrine disruptor activity).

Toxicology. Mammalian bioassays (especially involving rodents) are standard models in human toxicology described e.g. in OECD guidelines for testing of chemicals (Section 4: Health effects). Other tools such as cell cultures are irreplaceable in the identification of toxicity mechanisms. For example, several standardized OECD methods exist for the detection of acute irritation effects (guidelines No. 430, 431, or 493), mutagenicity (guideline No. 471), or genotoxicity (guidelines No. 473, 474, 476, 479, 482, 487, and 489; all guidelines are available for downloand from the OECD.org website). Other exploratory studies of cyanotoxins are used for investigating MoA, for example research on oxidative stress (by measuring diverse biomarker responses both in vivo and in vitro), mechanisms of carcinogenicity and endocrine disrupting effects (using, for example, reporter gene bioassays for estrogenicity detection with mammalian cell lines).

Ecotoxicology. Also for the assessment of cyanotoxin ecotoxicological effects, standardized OECD guidelines assays exist for different trophic levels including algae or plants (guidelines No. 201, 221, and others), invertebrates such as *Daphnia magna* (No. 202 and 211) and fish (No. 203, 204, 210, 212, 215, 229, and 230). However, most of the published research on cyanotoxin ecotoxicity has used other experimental biotests and endpoints. With respect to autotrophs [3], evaluated endpoints included alterations of growth, development, anatomy, cell structure, physiological, biochemical, and enzymatic processes in both unicellular algae and higher plants.

Oxidative stress has been shown to be a general phenomenon in autotrophic organisms exposed to MCs, cyanobacteria, β-.-methylamino-L-alanine (BMAA), and anatoxin-a. Aquatic invertebrates are probably the most widely used organisms in general ecotoxicology research as well as in specific investigations of cyanotoxins [4]. Especially bioassays with cladocerans such as *Daphnia* sp. or *Ceriodaphnia* sp. have been widely used due to their small size, sensitivity, fast parthenogenetic reproduction, ease of culturing, large clutch size, high population density etc. Toxic cyanobacteria were shown to induce diverse adverse effects in invertebrate models including mortality, physiological changes, and the suppression of reproduction as well as transgenerational manifestation of adaptive responses. Experimental studies with vertebrates have used fish (including trout, common carp, tilapia, zebrafish, and many others) as well as amphibians (e.g. *Xenopus laevis*) showing genotoxicity, modulations of oxidative stress biomarkers, alterations of gene expression patterns, as well as disruption of physiological processes, development, or reproduction [5].

27.5 Bioassays for Monitoring

The bioassays used for monitoring purposes should be able to rapidly assess larger numbers of samples and detect bioactivity or toxicity with some added value if specific for particular toxins. A long list of diverse bioassays has been used for the screening of cyanobacterial samples.

In vivo tests with laboratory rodents. The mouse bioassay using Swiss Albino males has historically been used to detect toxicity of blooms [6, 7]. The mouse bioassay relies on the acute toxicity, within 24 hours, providing an integrative response. On the other hand, it is not very specific and co-occurring compounds in bloom samples may induce interferences. The method uses intraperitoneal injection (i.p.) of 0.1–1.0 mL of a lysate of cyanobacteria sterilized by membrane ultrafiltration, and toxicity is expressed as LD_{50} or approximate/minimal lethal dose (A/MLD) (mg cell dry weight per kg mouse body weight). The presence of neurotoxic cyanotoxins is often manifested by strong spasms followed by fast death. MCs and CYN induce characteristic changes in liver (swollen, blood-engorged, disruption of liver structure, and haemorrhage) and in other organs such as kidney and lungs [8]. The mouse bioassay has become increasingly disfavoured for screening purposes due to ethical issues, and broader adoption of the 3R principles of replacement, reduction and refinement, substantially limited use of such animal bioassays.

Cell lines and in vitro bioassays. A wide range of established cell lines and primary cultures from human and animal organs has been used to evaluate pure cyanotoxins and cyanobacterial extracts [9]. They provide an alternative for reducing the use of vertebrate models (3R principle) and several in vitro toxicity methods have been validated by OECD or ECVAM. However, none of the available methods would nowadays be directly recommended for the monitoring of cyanotoxins. Primary cell cultures better reflect physiological in vivo situations than permanent immortal cell lines due to the maintenance of the uptake and metabolic machinery underlying the toxic effects. Cytotoxicity can be evaluated with a variety of assays such as neutral red uptake, reduction of tetrazolium salts (MTT, MTS, XTT tests), protein, or ATP content and lactate dehydrogenase (LDH) leakage. Neurotoxicity caused by saxitoxins can be detected for example with the Neuro2A assay or the $DiBAC_4$ membrane potential assay [10]. Diverse cellular tests have also been used to monitor genotoxicity (e.g. by micronucleus assay or comet assay [11]).

Plants and algae. Different plants have been used in cyanobacterial toxicity studies such as *Brassica* sp., *Ceratophyllum* sp., *Lemna* sp., *Myriophyllum* sp., and many others. With respect to green algae, which can be direct competitors of cyanobacteria for resources, a number or organisms have been used such as *Scenedesmus* sp., *Chlorella* sp., and *Dunaliella* sp. However, the results seem to be controversial showing rather low sensitivity to known cyanotoxins [3]. Taking into account the requirements and expectations of the plant test for monitoring purposes (fast, efficient, size), the test with white mustard *Sinapis alba*, strain Budakalászi Sárga, has been shown to have good sensitivity for MCs (IC_{50} 5 μg mL^{-1}) and CYN (IC_{50} 18 μg mL^{-1}) [12].

Invertebrates. A number of studies have investigated the use of diverse crustaceans also for monitoring purposes. From a range of approaches, a commercially available biotest employing larvae of the fairy shrimp *Thamnocephalus platyurus* has successfully been used in the screening of cyanobacterial toxicity [13]. It is a ready-to-use 24-hour exposure method whose reliability was assessed in an interlaboratory trial resulting in a specific scoring system [14]. Another model is the brine shrimp, *Artemia salina* [15, 16], which has successfully been used to test MCs, anatoxin-a, CYN, and BMAA. The cysts of this invertebrate can be stored at 4°C and hatched prior to toxicity testing using mortality over periods of up to 96 hours.

Fish and amphibians. While current European legislation aims to minimize the use of adult vertebrates in toxicity testing, embryos are considered an alternative. Several fish embryo toxicity (FET) tests have been suggested, including fathead minnow *Pimephales promelas*, Japanese medaka *Oryzias latipes*, or zebrafish *Danio rerio* [17]. In Germany, for example, the zebrafish FET is being routinely employed for monitoring the toxicity of waste water effluents. However, their use for cyanobacterial monitoring would need further research. Regarding amphibians, toxic and embryonal damage by cyanotoxins has been studied with, for example, the Frog Embryo Teratogenesis Assay Xenopus (FETAX) model with *Xenopus laevis* embryos, which allows determination of mortality, malformations, and growth inhibition [18].

27.6 Conclusions and Future Perspectives

Bioassays are important tools for investigating complex laboratory and environmental cyanobacterial biomass samples as well as in the detailed characterization of individual cyanotoxins. They provide unique information on the biological effects (toxicity), which clearly adds additional and important value to chemical analytical methods. Both approaches, physicochemical and bioassays, should ideally be complemented and interpreted together. This would allow establishment of a clear causal link between the toxic effects and presence of a certain toxin.

At the present time, no single method (or a set of specific bioassays) is available to detect all known cyanotoxins and there is no clear agreement among scientists on the list of recommended bioassays. Further validation and comparison of methods are therefore needed before general recommendations on their application can be given. Future efforts should also focus on better integration and interpretation of the results obtained from advanced biological methods including microarrays, next-generation sequencing, and various "-omics" techniques but these are still far from giving a reliable indication of the toxicity level of samples.

Nevertheless, bioassays can be used for primary screening of toxicity under some of these conditions:

- When there is limited access to adequate equipment, expertise, or possibilities for chemical methods or subcontracting analysis.
- When the possibility exists of toxic metabolites other than (or in addition to) the known cyanotoxins.
- When cyanobacterial blooms are dominated by less-studied species, which may contain unknown metabolites with high toxicity.

References

[1] Falconer, I.R. (2007) Cyanobacterial toxins present in *Microcystis aeruginosa* extracts - more than microcystins!, *Toxicon* **50**, 585–588.

[2] Bláha, L., Babica, P., Hilscherová, K. *et al.* (2010) Inhibition of gap-junctional intercellular communication and activation of mitogen-activated protein kinases by cyanobacterial extracts - indications of novel tumor promoting cyanotoxins ? *Toxicon* **55** (1), 126–134.

[3] Babica, P., Bláha, L. and Maršálek, B. (2006) Exploring the natural role of microcystins – a review of effects on photoautotrophic organisms. *Journal of Phycology* **42**, 9–20.

[4] Marsalek, B. and Blaha, L. (2004) Comparison of 17 biotests for detection of cyanobacterial toxicity. *Environmental Toxicology* **19** (4), 310–317.

[5] Ziková, A., Lorenz, C., Lutz, I. *et al.* (2013) Physiological responses of *Xenopus laevis* tadpoles exposed to cyanobacterial biomass containing microcystin-LR. *Aquatic Toxicology* **128–129**, 25–33.

[6] Falconer, I.R., Smith, J.V., Jackson A.R.B. *et al.* (1988) Oral toxicity of a bloom of the cyanobacterium *Microcystis aeruginosa* administered to mice over periods up to 1 year. *Journal of Toxicology and Environmental Health* **24**, 291–305.

[7] Falconer, I.R. (1993) Measurement of toxins from blue-green algae in water and foodstuffs. In: I.R. Falconer [Ed.] Algal Toxins in Seafood and Drinking Water. Academic Press, London, 165–175.

[8] Runnegar, M.T.C. and Falconer, I.R. (1982) The in vivo and in vitro biological effects of the peptide hepatotoxin from the blue-green alga *Microcystis-aeruginosa*. *South African Journal of Science* **78**, 363–366.

[9] Štraser, A., Filipic, M. and Zegura, B. (2013) Cylindrospermopsin induced transcriptional responses in human hepatoma HepG2 cells. *Toxicology In Vitro* **27**, 1809–1819.

[10] Humpage, A.R., Ledreux, A., Fanok, S. *et al.* (2007) Application of the neuroblastoma assay for paralytic shellfish poisons to neurotoxic freshwater cyanobacteria: interlaboratory calibration and comparison with other methods of analysis. *Environmental Toxicology and Chemistry* **26**, 1512–1519.

[11] Žegura, B., Štraser, A., Filipič, M. (2011) Genotoxicity and potential carcinogenicity of cyanobacterial toxins – a review. *Mutation Research/Reviews in Mutation Research* **727**, 16–41.
[12] Vasas, G., Gáspár, A. Surányi, G. *et al.* (2002) Capillary Electrophoretic assay and purification of cylindrospermopsin, a cyanobacterial toxin from *Aphanizomenon ovalisporum* by plant test (Blue-Green Sinapis Test). *Analytical Biochemistry* **302**, 95–103.
[13] Törökné, A.K., Vasdinyei, R. and Asztalos, B.M. (2007) A rapid microbiotest for the detection of cyanobacterial toxins. *Environmental Toxicology* **22** (1), 64–68.
[14] Törökné, K.A. (2000) The potential of the Thamnotoxkit microbiotest for routine detection of cyanobacterial toxins. In: New Microbiotests for Routine Toxicity Screening and Biomonitoring, eds. G. Persoone, C. Janssen W. De Coen, Published Kluwer Academic/Plenum Publishers, 533–539.
[15] Metcalf, J.S., Lindsay, J., Beattie, K.A. *et al.* (2002). Toxicity of cylindrospermopsin to the brine shrimp *Artemia salina*: comparisons with protein synthesis inhibitors and microcystins. *Toxicon* **40** (8), 1115–1120.
[16] Purdie, E.L., Metcalf, J.S., Kashmiri, S. *et al.* (2009) Toxicity of the cyanobacterial neurotoxin beta-*N*-methylamino-L-alanine to three aquatic animal species. *Amyotrophic Lateral Sclerosis* **10** (S2), 67–70.
[17] Ács, A., Kovács, A.W., Csepregi, J.Z. *et al.* (2013) The ecotoxicological evaluation of *Cylindrospermopsis raciborskii* from Lake Balaton (Hungary) employing a battery of bioassays and chemical screening. *Toxicon* **70**, 98–106.
[18] Burýsková, B., Hilscherová, K., Babica, P. *et al.* (2006) Toxicity of complex cyanobacterial samples and their fractions in *Xenopus laevis* embryos and the role of microcystins. *Aquatic Toxicology* **80**, 346–354.

28

Molecular Tools for the Detection of Toxigenic Cyanobacteria in Natural Ecosystems

Jean-François Humbert

Institute of Ecology and Environmental Sciences, UPMC, Paris, France

28.1 Introduction

In freshwater ecosystems, several cyanobacterial genera are able to produce hepatotoxins, in particular the small peptide microcystins and the alkaloid cylindrospermopsin, and neurotoxins including anatoxins (anatoxin-a and homoanatoxin-a), anatoxin-a(*S*), and saxitoxins [1]. The survey of the potential toxicity of cyanobacteria during blooms is presented with different kinds of problems, one of them being that toxin concentrations are generally not positively correlated with cyanobacterial biomass (e.g. [2]). This lack of correlation is partly explained by the existence, within cyanobacterial populations of a given species, of toxin-producing and non–toxin-producing cells. Another challenge in environmental monitoring concerns the early detection of potentially toxigenic cyanobacteria and cyanotoxins at the beginning of bloom events. To help to overcome these problems, molecular approaches based on a PCR (polymerase chain reaction) reaction targeting genes involved in the biosynthesis of cyanotoxins have been developed in the past 15 years. This was achieved through the increasing number of complete microbial genome sequences available in databanks that have permitted the identification of the genes involved in the biosynthetic pathways for cyanotoxins. In this chapter, a brief overview is presented on molecular tools, which can be used for the detection of potential cyanotoxins. More detailed accounts are given in an accompanying handbook [3].

Handbook of Cyanobacterial Monitoring and Cyanotoxin Analysis, First Edition.
Edited by Jussi Meriluoto, Lisa Spoof and Geoffrey A. Codd.

28.2 Molecular Methods for the Monitoring of Potentially Toxic Cyanobacteria

Three different PCR-based methods have been used for the detection of freshwater cyanobacteria potentially producing cyanotoxins in natural ecosystems [4–6]. The first approach is based on the use of a conventional PCR targeting only one gene. Concerning the choice of the targeted genes, two main strategies have been implemented. In some methods, the targeted genes are not directly involved in the biosynthesis of cyanotoxins (but for example, on the intergenic spacer region in the phycocyanin operon, the *rpoC1* gene or on rRNA 16S-23S or ITS). These can permit the identification of cyanobacteria, including potentially toxic species in environmental samples. However, in most studies, the targeted genes belong to the gene clusters involved in the biosynthesis of cyanotoxins, in particular microcystins (*mcyA*, *mcyB*, *mcyC*, *mcyD*, or *mcyE*). These conventional PCR-based methods provide qualitative results (presence/absence of the genes in a sample) that can be useful for the early detection of potentially toxic cells just at the beginning of the blooms when the toxin concentrations are too low to be detected. On the other hand, these methods do not permit to quantify the abundance of potentially toxic cells per unit of a water volume or the proportions of these potentially toxic cells in the wider population of cyanobacteria.

In order to get a quantitative estimation of cyanotoxin gene abundance, the second approach that has been developed is based on the use of qPCR (quantitative PCR or real-time PCR) targeting a gene involved in the biosynthesis of cyanotoxins. In qPCR, the amount of DNA is measured after each cycle by the use of fluorescent markers that are incorporated into the PCR product, generating a fluorescence amplification curve (Fig. 28.1). As the progression of the fluorescence curve is directly correlated to the amount of PCR product amplified in each cycle, an analysis of this curve enables calculation of the original starting quantity of target DNA. Consequently, qPCR can

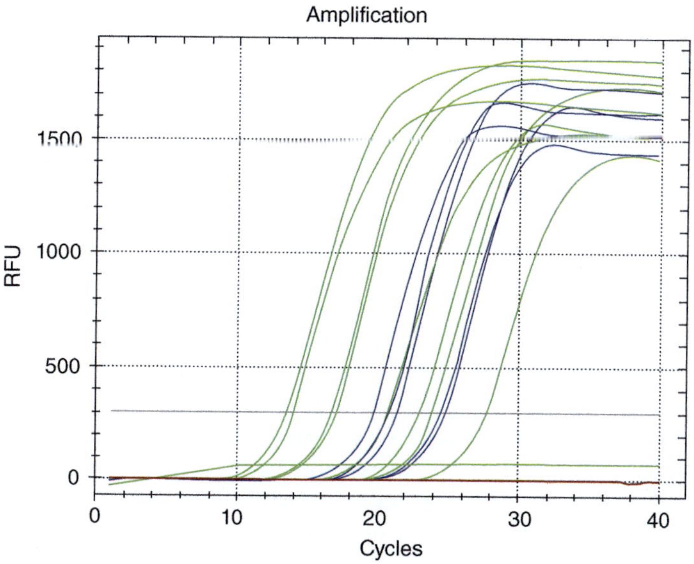

Figure 28.1 *Relative fluorescence (RFU) vs. cycle number in a qPCR reaction (green: standards; blue: environmental samples; red: control)*

provide data on the gene copy number, on the proportion of cells containing toxin genes in the total population of cyanobacteria, or on the levels of transcription of genes involved in cyanotoxin biosynthesis. qPCR can also permit the estimation of the proportions of potential microcystin-producing cells among the total population of cyanobacteria by a multiplex approach targeting one gene involved in the biosynthesis of microcystins and one gene present in all cyanobacterial cells (e.g. 16S rRNA or phycocyanin; see, for example, [7]). Whilst numerous reports have been published in recent years on the use of qPCR for the quantification of genes involved in microcystin biosynthesis, only a few applications have been directed to genes for the biosynthesis of nodularin, cylindrospermopsin, and saxitoxin [4].

Finally, the combined use of PCR and DNA microarray has been investigated for the detection of potentially toxic cyanobacterial species or for the specific detection of two genes involved in the biosynthesis of microcystins and nodularins [8]. This approach enables to reduce the detection limit to be lowered and the specificity of the detection to be increased.

28.3 Strengths and Limitation of These Molecular Approaches

The two main strengths of these molecular approaches for the monitoring of toxigenic cyanobacteria are: (i) their sensitivity that permits early detection and (ii) the speed of data acquisition since all of the stages from DNA extraction to obtaining PCR results can take less than 4 hours. In addition, the unit cost per sample is already relatively low and can be expected to decrease further in future years. It can also be expected that, in the short term, fully automatic processes will be implemented to make these analyses more widely accessible.

One the other hand, there are still many limitations on their routine use. The first constraint is a possible overestimation of the potential toxicity of the sample. Indeed, despite a positive detection by PCR of the toxic genotypes, mutations, insertions, or deletions that are not detected by the PCR can inactivate genes involved in biosynthesis of toxins and hamper toxin production [9]. The second limitation in the use of these approaches for the monitoring of toxigenic cyanobacteria is that they provide data on the potential toxicity of the microorganisms present, but not on the actual toxicity of the samples. Consequently, these molecular data can be used for a rapid assessment of the potential toxicity during the course of the blooms but the identification and quantification of the toxins using physicochemical methods (e.g. HPLC, LC-MS) will always be needed in the framework of managerial actions. The third limitation of these molecular approaches is that they are still reserved for research laboratories (in particular qPCR) due, for example, to the relatively high levels of technical skill required for the optimization of the assays or to their current costs.

28.4 Conclusions

Despite the high potential of molecular approaches for the monitoring of toxigenic cyanobacteria, including potential toxin-producers, in natural ecosystems, it remains true that these tools are still mainly limited to research work. However, these approaches can be already used in the support of classical methods of monitoring toxigenic cyanobacteria, for example if an early detection of toxigenic cells is needed. The development of automated tools will make these methods more accessible in the near future.

References

[1] Dittmann, E., Fewer, D.P. and Neilan, B.A. (2013) Cyanobacterial toxins: biosynthetic routes and evolutionary roots. *FEMS Microbiology Reviews* **37**, 23–43.

[2] Sinang, S.C., Reichwaldt, E.S. and Ghadouani, A. (2013) Spatial and temporal variability in the relationship between cyanobacterial biomass and microcystins. *Environmental Monitoring and Assessment* **185**, 6379–6395.

[3] Sivonen, K., Wilmotte, A. and Kurmayer, R. (Eds.) (2016) A handbook on molecular tools for the detection and quantification of toxigenic cyanobacteria. Handbook in preparation in the framework of CYANOCOST – COST Action ES1105.

[4] Humbert, J.F., Quiblier, C. and Gugger, M. (2010) Molecular approaches for monitoring potentially toxic marine and freshwater phytoplankton species. *Analytical and Bioanalytical Chemestry* **397**, 1723–1732.

[5] Pearson, L.A. and Neilan, B.A. (2008) The molecular genetics of cyanobacterial toxicity as a basis for monitoring water quality and public health risk. *Current Opinion in Biotechnology* **19**, 281–288.

[6] Srivastava, A., Singh, S., Ahn, C.Y. *et al.* (2013) Monitoring approaches for a toxic cyanobacterial bloom. *Environmental Science and Technology* **47**, 8999–9013.

[7] Briand, E., Escoffier, N., Straub, C. *et al.* (2009) Spatiotemporal changes in the genetic diversity of a bloom-forming *Microcystis aeruginosa* (cyanobacteria) population. *The ISME Journal* **3**, 419–429.

[8] Rantala-Ylinen, A., Sipari, H. and Sivonen, K. (2011) Molecular methods: chip assay and quantitative real-time PCR: In detecting hepatotoxic cyanobacteria. *Methods in Molecular Biology* **739**, 73–86.

[9] Christiansen, G., Molitor, C., Philmus, B. *et al.* (2008) Nontoxic strains of cyanobacteria are the result of major gene deletion events induced by a transposable elements. *Molecular Biology and Evolution* **25**, 1695–1704.

Section VI

Methodological Considerations

29

Method Validation Guidelines for the Analysis of Cyanotoxins

Theodoros M. Triantis[1], Triantafyllos Kaloudis[2], and Anastasia Hiskia[1]

[1]Institute of Nanoscience & Nanotechnology, National Center for Scientific Research "DEMOKRITOS," Athens, Greece
[2]Water Quality Control, Athens Water Supply and Sewerage Company – EYDAP SA, Athens, Greece

29.1 Introduction: Method Validation as a Requirement for Laboratory Accreditation

In the International Vocabulary of Metrology (VIM), *validation* is defined as "verification, where the specified requirements are adequate for an intended use" [1]. Referring to analytical laboratories, validation is the verification or check that a method of analysis (a measurement procedure) fulfills the specified performance criteria and requirements [2]. Validation of analytical methods is an essential requirement for laboratory accreditation according to ISO 17025, where *validation* is defined as "confirmation by examination and provision of objective evidence that the particular requirements for a specific intended use are fulfilled" [3]. Method validation is also a requirement of analytical methods used for the purposes of chemical monitoring programs carried out under Directive 2000/60/EC (Water Framework Directive, WFD) [4] where it is stated that methods should be validated and documented in accordance with ISO 17025 [5].

All methods (standard and nonstandard methods, laboratory in-house methods, modified standard methods) should be validated/verified in-house; however, standard methods require a validation to a lesser extent, sometimes referred to as "verification." Validation studies follow method development and optimization and all measurements needed for validation are carried out according to the method's documented SOP. Method performance parameters that are commonly studied during validation are measurement range (measuring interval), selectivity, trueness,

Handbook of Cyanobacterial Monitoring and Cyanotoxin Analysis, First Edition.
Edited by Jussi Meriluoto, Lisa Spoof and Geoffrey A. Codd.

precision, limit of detection/quantification (LOD/LOQ), and robustness (ruggedness). An estimation of uncertainty of measurement is also required, which can be largely based on validation data.

29.2 Performance Criteria and Validation Protocols for the Analysis of Cyanotoxins in Environmental Studies

Performance criteria of methods for analysis of cyanotoxins in environmental samples should be established by the laboratory according to the intended purpose of analysis. For example, a research laboratory carrying out a qualitative study of the presence of cyanotoxins in cyanobacterial strains with results given as "positive/negative," could focus mainly on selectivity and LOD/LOQ. On the other hand, if the method is applied to demonstrate compliance with regulatory limits, usually a more detailed study is needed to assess method performance. If the method is used to assess the presence of cyanotoxins in drinking water (e.g. with relevance to the 1 μg L^{-1} MC-LR limit of the World Health Organization), relevant performance criteria of Directive 98/83/EC (Drinking Water Directive, DWD) [6] should be used, although cyanotoxins are not directly included in this directive. However, DWD includes criteria for analysis of organic pollutants, that is, for trueness (25%), precision (25%) and LOD (10–25%) as percent of the parametric values (regulation limits). If the method is applied within the context of WFD, the estimated uncertainty should be generally below 50%, whereas in the case of cyanotoxins where no environmental quality standards are yet established, monitoring should be carried out "using best available techniques not entailing excessive costs" [5]. Further performance criteria can be adopted from guidelines on validation procedures for organic pollutant analysis, such as for pesticides in food [7].

A validation must be as extensive as needed to meet the needs of a given application; however, in practice this is not always possible due to limited resources or other technical constraints. The laboratory has to decide which performance parameters will be studied and to what extent, so that validation meets the requirements of regulations and customers. The laboratory should prepare a validation protocol (plan) that includes detailed information about the samples that will be analysed, the series of experiments and measurements, the performance parameters that will be assessed, the calculations, and acceptance criteria. SOP 29 describes an example of a basic validation protocol for the analysis of cyanotoxins in environmental samples. If the method will be applied to different sample matrices or groups of matrices, then a separate validation study is needed for every different matrix or matrix group (e.g. matrix groups could be water, fish tissue, cyanobacterial biomass). The laboratory should keep detailed records of validation data and results that should be traceable to primary observations (e.g. chromatograms).

Performance parameters that are usually assessed in validation studies of methods for analysis of cyanotoxins are as follow:

- *Trueness* (bias, accuracy) is usually expressed as percent recovery (%R), that is, the percent fraction of the measured concentration to the known concentration of a sample. Trueness is assessed by analysis of samples having known concentrations of cyanotoxins. Certified reference materials (CRMs) or in-house prepared samples spiked with known amounts of cyanotoxins can be used to assess trueness. Mean recovery should generally be within the range of 70–120% [7] or within 25% of the parametric value [6]. Further data regarding trueness can be obtained through interlaboratory comparisons (proficiency testing schemes [PT]).
- *Precision* is a measure of the spread of results, usually estimated under repeatability and reproducibility conditions and expressed as %RSDr and %RSDR, respectively. Precision is assessed by replicate measurements of the same sample by the same analyst/equipment in a short time

(repeatability) or by different analysts, equipment, at different days (reproducibility). Replicate measurements should be independent and should reflect the whole analysis procedure, including sample preparation. Samples should be representative of real test samples, in terms of matrix and cyanotoxin concentrations. Usually at least 6 replicates per sample matrix are required to assess precision.

- *LOD and LOQ*. Definitions of LOD and LOQ vary widely between sectors and guidance documents. LOD and LOQ can be statistically estimated by determination of the standard deviation of measurements at low concentrations (SD) and multiplication of "SD" by a suitable factor. However, a more reliable procedure is proposed, where the laboratory defines LOQ as the lowest concentration level that meets performance criteria for trueness and precision, that is, the lowest validated level [7]. LODs/LOQs should reflect the whole measurement procedure, while clear distinction should be made from "instrument detection limit." LODs and LOQs should always be experimentally verified, especially when their estimation is based on "SD."
- *Measurement range* (working range, measuring interval) is the concentration interval in which the method provides results with acceptable performance. The lower end of measurement range is the LOQ, but other levels should be included in the validation study (e.g. at least one level per order of magnitude of concentration). In order to determine measurement range the laboratory should take into account the method's intended use, linearity range and calibration procedures. The laboratory should assess working range by performing a validation study for concentration levels that are relevant to this range (LOQ is one of those concentration levels). When additional sample dilutions or preconcentrations are needed so that the final measurement falls within working range, those should be documented and taken into consideration in validation.
- *Robustness (ruggedness)* is a measure of the method's capacity to remain unaffected when experimental parameter values vary from those defined in the standardized procedure. Robustness studies are carried out to assess the effect of not-easily controlled experimental parameters (factors), or parameters that could vary unpredictably in long-term routine application of the method. Robustness studies are carried out by suitable experimental designs, where the effect of variation of a number of parameters on results is evaluated. If effects are statistically not significant, the method is assumed "robust" to the factors studied. When effects are significant, measures should be taken to control the experimental parameters in order to avoid bias in results.
- *Selectivity (specificity)* is the ability of the method to determine cyanotoxins in mixtures or matrices of other components of similar behavior. The highest selectivity for cyanotoxins analysis is achieved with mass-spectrometric techniques (e.g. LC-MS/MS), where chromatographic separation in addition to multiple reaction monitoring and ion-ratio controls largely eliminate interferences. On the other hand, poor selectivity can be observed with ELISA methods, where it is recommended that positive results should be verified with use of MS techniques. Selectivity is assessed by analysing samples with mixtures of targeted cyanotoxins and samples of different matrices and origins. Further data regarding selectivity can be obtained through interlaboratory comparisons and proficiency testing schemes.

Apart from method performance parameters assessed in the validation study, laboratories should also estimate uncertainty of measurement [3]. *Uncertainty* can be defined as a "parameter associated with the result of a measurement that characterizes the dispersion of the values that could reasonably be attributed to the measurand" [8]. The general principles of measurement uncertainty estimation are described in *Guide to Expression of Uncertainty in Measurement* [9]. The laboratory must identify the uncertainty sources and make a quantitative estimation of the overall uncertainty of measurement. There are basically two strategies to estimate uncertainty: (a) estimate uncertainties arising from different sources and then combine them and (b) estimate directly the combined uncertainty using method performance data. The latter approach, known as "Nordtest" approach, provides a practical, understandable and simple procedure for estimation of uncertainty that is largely based on validation and quality control data. The "Nordtest" procedure is described in a handbook that was developed for environmental laboratories and translated in several languages [10]. The Finnish Environment Institute (SYKE) has developed "MUkit" (measurement uncertainty kit), which is a practical, user-friendly, open-source, freeware software package based on the Nordtest measurement uncertainty estimation procedure [11, 12].

29.3 Validation Issues Concerning the Analysis of Cyanotoxins

Standardized methods or methods of cyanotoxins analysis that provide adequate validation data are rare in literature, although it has been shown that incomplete documentation or lack of validation data can be a source of major uncertainties and debates in cyanotoxins research (e.g. [13]). International standard methods are limited to ISO 21079:2005, which describes a procedure for the determination of microcystins in water using solid-phase extraction (SPE) and high-performance liquid chromatography (HPLC) with UV detection [14]. The method is validated for MC-LR, MC-RR, and MC-YR, while some basic information for LC-MS detection is also included in an annex. ELISA methods are also widely used, especially following incidents such as the Toledo water crisis in the United States, where the Environmental Protection Agency (EPA) proposed the use of ELISA for quantitative screening of microcystins [15]. EPA requires that commercially available ELISA kits for cyanotoxins are approved by the Environmental Technology Verification (ETV) Program [16]. Currently, performance data are available from ETV for some ELISA and PPIA test kits. Already existing validation data can be used by laboratories to assess the fit-for-purpose of methods according to their needs; however, it is strongly recommended that laboratories validate in-house all methods to prove their capability in achieving the required performance.

Commercial availability of cyanotoxins (calibration standards or reference materials) is rather limited, but during the last few years, the number of available compounds has increased, for example, there were 15 microcystin variants commercially available by the end of 2014. As cyanotoxins standards come from several different vendors and accompanying purity data are not always sufficient, there are issues regarding purity of standards. When purchasing standard solutions, the laboratory is recommended to request certification for purity and measurement of concentration. In many cases where pure compounds are purchased the laboratory should determine concentrations of in-house prepared standard solutions using generally accepted procedures (e.g. by spectrophotometry). In this case, all details of determinations should be recorded to ensure traceability and facilitate comparisons between laboratories. This is important, since, for example, published extinction coefficient values can vary widely for the same cyanotoxin (e.g. [17]). A good source of certified reference materials for cyanotoxin analyses is the Certified Reference Materials (CRM) program of the National Research Council of Canada (NRC) [18].

PT schemes (interlaboratory comparisons) have only recently been available for microcystins. Currently, there are at least three schemes available for microcystins, while there are not any PT schemes for other cyanotoxins. Participation in PT schemes is important because it provides further data for assessment of trueness and compares the performance of a laboratory to other laboratories. In selecting a PT scheme, laboratories should take into consideration European guidelines [19] and prefer schemes that include cyanotoxins in their scope of accreditation. Updated information on available PT schemes for analysis of cyanotoxins can be found on the EPTIS database [20].

References

[1] JCGM 200:2012 (2012) *International vocabulary of metrology – Basic and general concepts and associated terms (VIM)*, 3rd edn, Joint Committee for Guides in Metrology (JCGM).

[2] Barwick, V. and Prichard, E. (eds) (2011) *Terminology in Analytical Measurement – Introduction to VIM3*, Eurachem.

[3] ISO/IEC 17025:2005. (2005) *General requirements for the competence of testing and calibration laboratories*. International Standards Organization, ISO/CASCO.

[4] Directive 2000/60/EC of the European parliament and of the council of 23 October 2000 establishing a framework for Community action in the field of water policy. Official Journal of the European Communities, L 327/1, 22.12.2000.

[5] Commission Directive 2009/90/EC of 31 July 2009 laying down, pursuant to Directive 2000/60/EC of the European Parliament and of the Council, technical specifications for chemical analysis and monitoring of water status. Official Journal of the European Union, L201/36, 1.08.2009, pp. 36–38.

[6] Council Directive 98/83/EC of 3 November 1998 on the quality of water intended for human consumption. Official Journal of the European Union, L330/32, 5.12.98. pp. 32–54.

[7] SANCO/12571/2013 (2013) *Guidance document on analytical quality control and validation procedures for pesticide residues analysis in food and feed*. European Commission.

[8] Ellison S.L.R. and Williams A. (eds) (2012) *CITAC Guide CG 4 - Quantifying Uncertainty in Analytical Measurement*. 3rd edn, EURACHEM.

[9] JCGM 100:2008 (2008) *Evaluation of measurement data — Guide to the expression of uncertainty in measurement*. Joint Committee for Guides in Metrology (JCGM) (available from http://www.bipm.org) (accessed 26 April 2015).

[10] Magnusson, B., Näykk, T., Hovind, H. and Krysell, M. (2012) *Handbook for Calculation of Measurement Uncertainty in Environmental Laboratories (NTR537-Edition 3.1)*. Nordtest (available from http://www.nordtest.info) (accessed 26 April 2015).

[11] SYKE Finnish Environment Institute (2012) *MUkit – Measurement Uncertainty Kit – User's Manual, Version 0.5*, (available from http://www.syke.fi) (accessed 26 April 2015).

[12] Naykki, T., Virtanen, A. and Leito, I. (2012) Software support for the Nordtest method of measurement uncertainty evaluation. *Accreditation and Quality Assurance*, **17** (6), 603–612.

[13] Faasen, E.J. (2014) Presence of the neurotoxin BMAA in aquatic ecosystems: what do we really know? *Toxins*, **6** (3), 1109–1138.

[14] ISO 20179:2005. (2005) *Water quality - Determination of microcystins- Method using solid phase extraction (SPE) and high performance liquid chromatography (HPLC) with ultraviolet (UV) detection*. International Standards Organization, Technical Committee ISO TC147/SC2.

[15] Kasich, J.R., Taylor, M. and Butler, C.W. (2014) *Public Water System Harmful Algal Bloom Response Strategy*. Ohio Environmental Protection Agency (US-EPA).

[16] United States Environmental Protection Agency (US-EPA) *Environmental Technology Verification Program*, http://www.epa.gov/etv/ (accessed 26 April 2015).

[17] Sano, T., Kikuchi, S., Kubo, T. *et al.* (2008) New values of molecular extinction coefficient and specific rotation for cyanobacterial toxin Cylindrospermopsin. *Toxicon*, **51** (4), 717–719.

[18] National Research Council Canada (NRC) http://www.nrc-cnrc.gc.ca (accessed 26 April 2015).

[19] Mann, I. and Brookman, B. (eds.) (2011) *Selection, Use and Interpretation of Proficiency Testing (PT) Schemes*. 2nd edn, Eurachem.

[20] EPTIS Database www.eptis.bam.de (accessed 26 April 2015).

30

Interpretation, Significance, and Reporting of Results

Geoffrey A. Codd[1,2], *Jutta Fastner*[3], *Tore Lindholm*[4], *Jussi Meriluoto*[5,6], *and James S. Metcalf*[7]

[1]*Biological and Environmental Sciences, University of Stirling, Scotland, United Kingdom*
[2]*School of the Environment, Flinders University, Adelaide, Australia*
[3]*Section Drinking Water Resources and Water Treatment, German Environment Agency, Berlin, Germany*
[4]*Environmental and Marine Biology, Faculty of Science and Engineering, Åbo Akademi University, Turku, Finland*
[5]*Biochemistry, Faculty of Science and Engineering, Åbo Akademi University, Turku, Finland*
[6]*Laboratory for Paleoenvironmental Reconstruction, Faculty of Sciences, University of Novi Sad, Serbia*
[7]*Institute for Ethnomedicine, Jackson, USA*

30.1 Introduction

The monitoring and analysis of cyanobacteria and cyanotoxins in natural and controlled waterbodies are carried out to (i) advance scientific knowledge through primary research; (ii) contribute to policies for environmental characterisation, restoration, and protection; (iii) provide data for the quality, health, and safety assessment of water resources for human and animal health protection; (iv) contribute to investigations into human and animal health incidents in which exposure to cyanobacteria or cyanotoxins has occurred, or been suspected; (v) guide the use of waterbodies required for activities potentially at risk of adverse health effects due to exposure to cyanobacterial mass populations, such as recreation, aquaculture, and irrigation for agriculture; and (vi) inform the design, operation, and effectiveness of abstraction and treatment processes for the supply of treated drinking water from sources containing, or potentially containing cyanobacteria.

Handbook of Cyanobacterial Monitoring and Cyanotoxin Analysis, First Edition.
Edited by Jussi Meriluoto, Lisa Spoof and Geoffrey A. Codd.

In addition to the need for high-quality technical performance at the waterside and in the laboratory, the value of the data obtained depends upon their interpretation, evaluation, and communication.

30.2 Interpretation and Significance of Results

The interpretation of the results of monitoring or analysis activities and estimation of their significance should be carried out with due reference to a wide range of relevant, influential factors. The detailed circumstances of the investigation(s) including the reasons for the investigation (whether reactive or incident driven or proactive as part of a proscribed schedule with trigger points for action) should be taken into full consideration.In addition to the contemporary and recent circumstances relating to the investigation, knowledge of the history of the study site can be useful if long-term datasets (years, decades) are available. The data resulting from the particular investigation(s) should also be evaluated with reference to the broader principles of epidemiological and ecotoxicological investigation [1] and to the wider body of knowledge of the ecology and toxicology of cyanobacteria [2]. The interpretation and significance of the results will also be influenced by the experimental design, competence of performance, and whether any known technical or material problems were encountered, as well as by the recognised uncertainties and gaps in knowledge.

An appropriate sampling scheme is essential to ensure the value of the data gathered. It requires consideration of the adequacy of the selected time intervals and sampling sites in relation to health risks from potential cyanotoxin exposure in a specific setting. Knowledge of the possible spatial and temporal heterogeneity of certain cyanobacteria and consequently toxin occurrence as well as the toxins` characteristics (mainly cell bound and/or dissolved, stability, etc.) is thus necessary. For example, data on cyanobacteria/cyanotoxin occurrence at the surface of a reservoir are not significant for assessing the risk of cyanotoxin intake when raw water is abstracted from 10-m depth from a layer of toxic *Planktothrix rubescens*. Also, surveillance based on cyanobacterial biomass alone is not adequate for assessing the health risk due to cylindrospermopsin as this toxin may still be present in waterbodies several weeks after the decline of the producer-organism.

Methods for cyanotoxin analysis differ in their sensitivity, specificity, and accuracy. Scientific questions such as the production of a toxin in cyanobacteria usually require sensitive, specific, and accurate methods such as HPLC-PDA or LC-MS, which are often time-consuming and expensive and need a high level of technical expertise. Methods being less accurate or specific would make the results less significant. On the other hand, for the surveillance of cyanobacteria and cyanotoxins more widely available and quick methods, such as fluoroprobes for biomass or immunoassay for microcystins, may be preferable. Here, a lower accuracy is negligible when, for example, biomass of scum-forming *Microcystis* can change over three orders of magnitude within hours at a sampling- or water-intake site due to changes of wind direction and speed. Also, the lower selectivity, but high ease-of-use, of an immunoassay may be advantageous in surveillance as it can potentially respond to all microcystin congeners in a sample (though to varying extents), while the highly specific LC-MS analysis may underestimate total microcystin concentration due to the shortage of certified reference standards for many of the microcystin congeners.

Numerous gaps exist in the knowledge of the health effects of cyanobacteria and cyanotoxins. One of the most significant points is that cyanotoxin data are being interpreted on the basis of the provisional WHO guideline value (GV) of 1 $\mu g\ L^{-1}$ for microcystin-LR in drinking water [3]. However, no GVs exist for the other microcystin congeners and current practice is to apply the microcystin-LR GV to all the other congeners, thus often leading to a toxicological "worst-case" estimation as not all microcystins are as toxic to mammals as microcystin-LR.

For cylindrospermopsin also, a GV of 1 μg L^{-1} has been proposed [4]. Since other GVs do not exist, it is prudent to apply the GV for microcystin-LR to all other cyanotoxins until sufficient data for the latter are available.

30.3 Reporting of Results and Maximization of Benefits

In addition to the need for the results to be evaluated by the investigators or laboratory heads, according to knowledge of cyanobacteriology and cyanotoxins, and of the circumstances involved, the value of such results for decision-makers and the wide range of potential end-users depends on effective communication. This principle applies to the data from both proactive (scheduled, routine, structured) investigations and to reactive (post-event) sampling and analyses. In some cases, the reporting of data alone may be sufficient. However the value of information is usually greatly enhanced if accompanied by its interpretation and with recommendations arising, if the authors are experienced in cyanobacterial biology and cyanotoxin ecotoxicology. In cases where such personal experience is not available locally, this Handbook should assist. Contacting recognised experts in the field for the interpretation of data and recommendations is also now quicker and easier via the Internet.

Response times for the reporting of data can also greatly affect their value. In the case of proactive investigations, especially during periods of minimal cyanobacterial development in temperate latitudes, a lag of some days in reporting analytical results may not be critical. However, the reduction of response times by the laboratory to a minimum (e.g. on day of sample arrival or within 24 hours) is encouraged. This can be highly beneficial due to the dynamic nature of cyanobacterial mass distribution in waterbodies, where changes in the vertical and horizontal concentrations of cells and cyanotoxins (when largely intracellular), can occur within a few hours. Such changes may influence whether cyanobacterial cell and cyanotoxin concentrations result in their values exceeding, or complying with management guidelines, or legislation. Such rapid feedback of results is also useful: (i) in the days before major scheduled events, such as water sports or fishing competitions, where cyanobacterial bloom and cyanotoxin concentrations need to be taken into account in finally deciding whether to proceed; (ii) in reactive investigations into adverse health episodes where cyanobacterial cells/cyanotoxins are suspected; (iii) as an operational aid in assessing or upgrading drinking water treatment trains. Since not all analytical equipment (e.g. LC-MS) is widely available or designated for the rapid supply of results, interest is increasing in in situ and point-of-use methods (e.g. phycocyanin, immuno-, and biosensor methods) for the initial supply of results, with the necessary confirmation (e.g. LC-MS) to follow.

The effective communication of monitoring and analytical data can be further increased by: (i) increasing the awareness and basic knowledge of nonspecialist decision-makers (e.g. water treatment plant and recreation park managers, public health officers) via workshops and jargon-free literature (e.g. [5–9]); (ii) use of previously prepared standard letters to end-users for rapid uploading of data [8]); and (iii) relating the communicated data and recommendations to contingency plans for risk management. These should be previously drawn up by the end-users in collaboration with cyanobacterial/cyanotoxin experts [7–9].

30.4 Examples, Debriefing

One of the earliest intensively studied areas with toxic cyanobacterial blooms is Åland Islands, SW Finland. This region has around 130 lakes, many of which are used for drinking water abstraction, recreation, and irrigation purposes. Studies at Åbo Akademi University have shown that 10–15% of the lakes in the region have detectable concentrations of microcystins [10]. The following examples illustrate how cyanobacterial blooms can affect society and the natural environment and how environmental monitoring can help to recognize and manage the problems. Here, we briefly present some cases with a focus on lessons learnt. We have found that open discussion of cyanobacterial problems with stakeholders, solid research combined with reactive/proactive alleviative measures, and ample information to the general public on cyanobacterial questions, is a fruitful approach.

30.4.1 Problems with Drinking Water Production

In the 1980s and 1990s, Lake Östra Kyrksundet suffered from extensive blooms of toxic *Planktothrix agardhii*, often with maximum density and high microcystin concentrations in the metalimnion [11, 12]. The rudimentary drinking water treatment for the local community consisted of sand filtration and chlorination only, and *Planktothrix* filaments passed through the process and were present in the drinking water. In 1987, the lake was closed as a drinking water source.

Lessons learned:

i. Not all cyanobacteria produce visible surface scums. Particularly, *Planktothrix agardhii* may be present in deeper layers in stratified lakes. The water intake to a waterworks is often located in deeper water and thus is vulnerable to hidden cyanobacterial subsurface maxima.
ii. It is important to produce toxin data per volume unit (not only per plankton biomass) in order to evaluate the risk for consumers.
iii. There must be courage and preparedness to close a lake when necessary in order to protect consumers. Moreover, the decision-making system must be available throughout the holiday season and beyond.
iv. *Planktothrix agardhii* is often abundant throughout the year and may also form marked maxima under the ice.

In the 1970s and 1980s, Markusbölefjärden and Långsjön suffered from extensive cyanobacterial blooms and associated animal mortalities. Despite this, the lakes were used for drinking water abstraction for the town of Mariehamn and central Åland. Complaints of water taste and odour were commonplace.

Lessons learned:

i. It cannot be assumed that simple water treatment technologies can remove noxious or toxic cyanobacterial metabolites to an adequate extent. The recurrent cyanobacterial blooms triggered and justified the construction of a new waterworks. This decision improved the safety and quality of life of the citizens and thousands of tourists.
ii. A rethinking concerning basin management resulted in a system where the water company now relies on three different lakes, Markusbölefjärden, Långsjön, and Dalkarby träsk, for drinking water abstraction. As the limnological profiles of the lakes differ from each other, selected water abstraction provides a strategy to avoid acute cyanobacterial problems. Moreover, the use of water from all three basins simultaneously means a more stable raw water quality, which has resulted in decreased use of chemicals.

30.4.2 Animal Deaths and Multiple Toxic Species/Multiple Toxins

Vargsundet is a brackish-water lake that in July 1997 suffered from simultaneous mass occurrences of the ichthyotoxic microalga *Prymnesium* sp. in the epilimnion and the hepatotoxic cyanobacterium *Planktothrix agardhii* in the metalimnion. Fish died in the relatively clear epilimnion but survived in the metalimnion [13].

Lessons learned:

i. Toxic episodes can involve several harmful plankton species and several kinds of toxins. Such complex interplay and possible involvement of novel compounds may obscure the identification of the main causative agent(s).
ii. Different organism groups react differently to toxic agents. The responses of aquatic organisms cannot easily be extrapolated to those of mammals and vice versa.
iii. *Prymnesium* sp. is a small-celled warm water species that easily escapes detection (it is hardly present in plankton net samples).
iv. As global climate change continues and novel harmful algal/cyanobacterial species spread to new geographical areas, there should be a preparedness to identify and manage new challenges.

30.4.3 Irrigation and Water Quality Problems

Untreated water from tens of lakes in Åland, and many thousands worldwide, is used for irrigation purposes. In dry seasons, when the need for water is high, lake water levels tend to be naturally low, making small and shallow lakes highly vulnerable. Low water levels may lead to fast ecological changes, such as an increase of certain macrophytes, increased water turbidity, or cyanobacterial blooms. Any nutrient loading in the growth season is also most efficient in bloom promotion when it is mixed into the smallest possible water volume. It is thus common that the demands concerning water quality, water quantity and lake ecosystem health are hard or impossible to meet simultaneously in dry seasons and arid areas.

Water from Vargata träsk has long been used for the irrigation of crop plants. This lake has suffered from extensive *Anabaena* blooms and occasional fish kills of unknown etiology. The problems were not only aesthetic. Moreover, researchers were not fully trusted by farmers and rational solutions were difficult to find.

Lessons learned:

i. Although the possible transfer of cyanotoxins from irrigation water into crop plants is rather poorly understood, there are reasons to avoid the use of cyanobacterial blooms for crop irrigation. Plants with aerial edible parts that are directly subjected to cyanobacterial material in the irrigation water are prone to contamination by cyanobacterial off-flavours and cyanotoxins.
ii. Lakes adjacent to crop fields are easily exposed to nutrient-rich runoff from the fields. This process promotes eutrophication, which in turn may have negative effects for agriculture. Sustainable agriculture requires an understanding of the nutrient cycle and its coupling to the possible production of noxious and toxic cyanobacterial compounds.

References

[1] Bradford-Hill, A (1965) The environment and disease: association or causation. *Proceedings of the Royal Society of Medicine*, **58** (5), 295–300.
[2] Whitton, B.A. (Ed.) (2012) *Ecology of Cyanobacteria II: Their Diversity in Space and Time*, Springer, Dordrecht, Heidelberg, New York, London.
[3] World Health Organization. http://www.who.int/water_sanitation_health/dwq/chemicals/microcystin/en/.
[4] Humpage, A.R. and Falconer, I.R. (2003) Oral toxicity of the cyanobacterial toxin cylindrospermopsin in male Swiss albino mice: determination of no observed adverse effect level for deriving a drinking water guideline value. *Environmental Toxicology* **18**, 94–103.
[5] Australian Water Quality Centre: Cyanobacterial toxins. Water quality solutions. Australian Water Quality Centre, Adelaide. www.awqc.au

[6] Metcalf, J.S. and Codd, G.A. (2014) *Cyanobacterial Toxins (Cyanotoxins) in Water*. FR/R0009. Foundation for Water Research, Marlow, Bucks., UK. www.fwr.org.
[7] Newcombe, G., House, J., Ho, L. *et al.* (2010) *Management Strategies for Cyanobacteria (Blue-Green Algae): A Guide for Water Utilities*. South Australian Water, Adelaide. www.wqra.com.au.
[8] Scottish Government (2012) *Cyanobacteria (Blue-Green Algae) in Inland and Inshore Waters: Assessment and Minimisation of Risks to Human Health*. Revised Guidance. Scottish Government, Edinburgh, UK.
[9] Federal Environment Agency, Germany: Decision Support Tool for the Development of a Setting-specific Strategy from Catchment to Consumer against the Occurrence of Cyanotoxins in Drinking-Water. www.toxische-cyanobakterien.de/en/.
[10] Lindholm, T., Vesterkvist, P., Spoof, L., Lundberg-Niinistö, C. and Meriluoto, J. (2003) Microcystin occurrence in lakes in Åland, SW Finland. *Hydrobiologia* **505** (1–3), 129–138.
[11] Lindholm, T., Eriksson, J. E. and Meriluoto, J. A. O. (1989) Toxic cyanobacteria and water quality problems – Examples from a eutrophic lake on Åland, SW Finland. *Water Research* **23** (4), 481–486.
[12] Lindholm, T. and Meriluoto, J. A. O. (1991) Recurrent depth maxima of the hepatotoxic cyanobacterium *Oscillatoria agardhii*. *Canadian Journal of Fisheries and Aquatic Sciences* **48** (9), 1629–1634.
[13] Lindholm, T., Öhman, P., Kurki-Helasmo, K., Kincaid, B. and Meriluoto, J. (1999) Toxic algae and fish mortality in a brackish-water lake in Åland, SW Finland. *Hydrobiologia* **397** 109–120.

31

Lessons from the Užice Case: How to Complement Analytical Data

Zorica Svirčev [1,2], Damjana Drobac [1], Nada Tokodi [1], Dunja Đenić [3], Jelica Simeunović [1], Anastasia Hiskia [4], Triantafyllos Kaloudis [5], Biljana Mijović [6,7], Stamenko Šušak [8], Mlađan Protić [8,9], Milka Vidović [10], Antonije Onjia [11], Sonja Nybom [2], Tamara Važić [1], Tamara Palanački Malešević [1], Tamara Dulić [1], Dijana Pantelić [1], Marina Vukašinović [12], and Jussi Meriluoto [1,2]

[1] Laboratory for Paleoenvironmental Reconstruction, Faculty of Sciences, University of Novi Sad, Serbia
[2] Biochemistry, Faculty of Science and Engineering, Åbo Akademi University, Turku, Finland
[3] Council for Environmental Protection and Energy Efficiency, Užice, Serbia
[4] Institute of Nanoscience & Nanotechnology, National Center for Scientific Research "DEMOKRITOS", Athens, Greece
[5] Water Quality Control, Athens Water Supply and Sewerage Company - EYDAP SA, Athens, Greece
[6] Institute of Public Health, Užice, Serbia
[7] Medical Faculty Foca, University of East Sarajevo, Foca, Republic of Srpska
[8] Medical Faculty, University of Novi Sad, Serbia
[9] Oncology Institute of Vojvodina, Sremska Kamenica, Serbia
[10] Institute of Chemistry, Technology and Metallurgy, Department of Ecology and Technoeconomics, University of Belgrade, Serbia
[11] Anahem Laboratory, Belgrade, Serbia
[12] Waterworks Užice, Water treatment plant "Petar Antonijević", Public Utility Company Užice, Serbia

Handbook of Cyanobacterial Monitoring and Cyanotoxin Analysis, First Edition.
Edited by Jussi Meriluoto, Lisa Spoof and Geoffrey A. Codd.

31.1 Introduction

Highly toxic metabolites produced by cyanobacteria, cyanotoxins, have been connected to several well-known health incidents such as the Caruaru case (Brazil), where lethal microcystin/cylindrospermopsin exposure occurred via hemodialysis [1–3]; the Palm Island disease, later revealed to be caused by cylindrospermopsin contamination of drinking water supply [4–6]; an extensive outbreak of gastroenteritis after drinking water contamination by cyanobacteria in Sweden [7]; liver damage in Armidale, Australia, after a hepatotoxic cyanobacterial bloom in a reservoir used for water supply [8]; and others. Hepatotoxins in drinking water could be a risk factor for primary liver cancer, but only a limited number of epidemiological studies have confronted this issue, although results from China [9] and Serbia [10–12] indicate that cyanobacterial hepatotoxins can lead to a higher incidence of primary liver cancer. Along with the more serious cyanotoxin poisonings, cyanobacterial toxins have been detected in raw and final tap water throughout the world including countries such as Argentina, Australia, Bangladesh, Canada, Czech Republic, China, Finland, France, Germany, Latvia, Poland, Thailand, Turkey, Spain, Switzerland, and the United States [13–16]. High risk to human health is linked to the acute ingestion of large cyanotoxin quantities or the intake of small doses during an extended exposure [17]. Furthermore, cyanotoxins can reach humans in several ways [18], and affect other organisms (e.g. [19]) and ecosystems in general.

Not all countries have appropriate means to fight problems caused by cyanobacteria [20]. In particular, sensitive and precise analytical methods for cyanotoxin detection (such as LC-MS) are expensive and still mostly inaccessible for countries like Serbia. Even in cases where sufficient analytical resources are available but regular monitoring is absent, some additional information is required for the proper interpretation and reconstruction of cyanobacterial bloom events. Therefore, some kind of alternative, supplementary approaches are needed.

Cyanobacterial blooms are common and widespread in Serbia [21–23]. Recently, cyanobacterial blooms have occurred in Vrutci reservoir used for the water supply of the city of Užice, Serbia. In the Užice case, 70,000 inhabitants were potentially exposed to cyanotoxins, possibly even before the observed bloom. As it will be described in this chapter, the Užice case shows how important it is to complement analytical results with supplementary data, such as bioassays, epidemiological data, and questionnaires, in order to gain insight into the overall situation and the possible health consequences of cyanobacterial blooms.

31.2 Vrutci Reservoir and the Cyanobacterial Bloom Detected in December 2013

Vrutci reservoir provided water for a population of more than 70,000 people in Užice, the largest city in western Serbia. The Vrutci dam was built in 1986 at the entrance to the gorge of the Đetinja River, 12 km upstream from Užice. According to the initial plan reservoir Vrutci had multiple purposes: flood control, sediment retention, and water supply of Užice including its surroundings. According to its nominal morphometric values, a volume of 54 million m^3, an area of about 1.92 km^2, and a mean depth of 20.8 m, Vrutci should be relatively oligotrophic. However, reservoir Vrutci is also used for recreational purposes such as fishing and swimming in the summer months, and these activities have contributed to the eutrophication of the reservoir. Restocking of Vrutci has been done repeatedly in order to manage the fish stocks in the reservoir. In addition, a small hydropower plant was built, which was not foreseen by the project. It is assumed that a drastic lowering of the water level twice a year, in order to accept the torrential influx, enables eroded material and suspended particles to reach the dam, with negative consequences for the overall quality of water in the reservoir. Land use including weekend houses in the catchment area and the absence of basic measures to protect water quality favor the process of eutrophication and the general deterioration of water quality in Vrutci [24].

On 14 December 2013, a large, thick "red carpet" was spotted on the surface of reservoir Vrutci (Fig. 31.1A). The blooming species was *Planktothrix rubescens*. During the following days there was an increase of the floating matter, and moving blue and red clusters were detected. Dark purple-red floating formations were transported to the coastal area by wind. The water color was noticeably changed into dark purplish-red, not only on the surface but also along the water column [24, 25].

Upon the order from the Republic Water Management Inspection, water in reservoir Vrutci was sampled by the Agency for Environmental Protection on 31 December 2013. The bloom-forming cyanobacterial species *P. rubescens* was determined by microscopic examination. The number of cells was highest (107,900 cells mL^{-1}) at the entrance of the reservoir at 0.5 m depth and near the dam at 10 m depth (98,936 cells mL^{-1}) [26].

The first complete analysis of the water in the piped-water network during the bloom (sample from 23 December 2013) was carried out at the Institute of Public Health of Serbia "Dr Milan Jovanović Batut," the reference institution for controlling the safety of water. The results showed the presence of the cyanobacterium *P. rubescens*, with about 10,000 cells L^{-1} in the treated water,

Figure 31.1 (A) Planktothrix rubescens bloom in Vrutci reservoir December 2013. (B) Redness in gills of fish from the reservoir (photographs: by Tijana Jevtić, used with permission)

and approximately 1000 cells L^{-1} in a sample from the piped-water network. The presence of microcystin-LR was noted in a piped-water network sample from 27 December 2013, but below the guideline value issued by World Health Organization, that is, less than 1 μg L^{-1} [27].

From 26 December 2013, the Sanitary Inspection of Republic of Serbia prohibited the use of tap water for drinking and cooking purposes. The tap water should be used for sanitary and technical purposes only. This action resulted in panic and crowd-forming among the people, and all the bottled water in town was instantly sold out, since residents were advised to drink water from water tanks and bottled water. However, the situation was manageable and the provision of drinking water was established for a temporary period of time. In the absence of a satisfactory long-term solution for controlling the bloom in reservoir Vrutci, possible alternative sources of water supply for Užice were considered. Since 7 February 2014, people drink water from this alternative source, Sušičko vrelo, while reservoir Vrutci is still blooming. Having in mind the importance of such a radical solution for the best of citizens and for the protection of human health, we propose to call the action of switching the abstraction from a blooming reservoir to an alternative source the "Užice solution".

31.3 Analytical Work: Toxin Analyses of Water, Cyanobacterial Biomass, and Fish from Reservoir Vrutci

The presence of microcystins in biomass, water, and fish from reservoir Vrutci was confirmed by LC-MS/MS analyses performed at Anahem laboratories/Belgrade, Demokritos/Athens and Åbo Akademi University/Turku (Table 31.1) [28]. The samples were taken at different time after and before the cyanobacterial bloom in December 2013 (Table 31.1).

Based on the number of cells per mL and concentration of microcystins, according to World Health Organization [29], water from reservoir Vrutci is classified as a high-risk water for recreation and drinking water abstraction purposes. The associated risks include short-term health effects (skin irritation, intestinal disorders), chronic effects, and possibility for acute poisonings with lethal effects [29].

Table 31.1 Detected (LC-MS/MS) microcystins in cyanobacterial biomass, water, and fish samples from Lake Vrutci

Sample	dmMC-RR	MC-YR	dmMC-LR	MC-LR
Biomass (shoreline) January 2014	11700 μg L^{-1}	20 μg L^{-1}	3700 μg L^{-1}	53 μg L^{-1}
Tap water January 2014 (during the prohibition)	5.7 μg L^{-1}	—	—	—
Reservoir water (shoreline) April 2014	2.1 μg L^{-1}	—	—	—
Fish muscle tissue October 2013	detected	—	—	—
Fish April 2014	detected	—	detected	—

Adapted from [28].

Only modest fish mortality was observed during the cyanobacterial bloom, but fish were observed swimming disoriented in circles. Dead fish had intensively red-coloured gills (Fig. 31.1B).

Microcystin was also detected in fish, including muscle of frozen fish from October 2013, which could indicate the presence of cyanobacteria even before the confirmed bloom incident in December 2013.

31.4 Complementary Data on Toxicity and Observed Health Problems

31.4.1 Bioassays

Artemia salina bioassays were performed at the University of Novi Sad, Department of Biology and Ecology, on biomass extracts from the samples taken at 23 December 2013. The results showed significant intracellular and extracellular toxicity of the investigated cyanobacterial biomass after 24 hours of exposure. Based on the results of the *A. salina* bioassay, the samples from Vrutci reservoir were subjected to analyses other than those of microcystin-LR (since that one was negative according to the official reports as it was shown in the previous text). Other analyses were performed, such as an epidemiological study and a questionnaire survey.

31.4.2 Questionnaire

Questionnaires represent a simple and understandable method for the production of complementary data supporting analytical results. They constitute a good base for evaluating the possible effects of cyanobacterial blooms on human and animal health.

There are some examples of questionnaires concerning the health effects of cyanobacteria and cyanotoxins [30, 31], and they were used during the preparation of a questionnaire intended to survey the health of the inhabitants of Užice. The questionnaire was designed shortly after the bloom event, and distributed during March–June 2014 among citizens of Užice either by home delivery or electronically via email or as an online survey. The intent was to collect data for testing the hypothesis whether cyanobacteria and their toxins were associated with some health issues in the region. The information gathered from the respondents (N = 640) gave much insight into problems potentially caused by cyanobacteria in Vrutci reservoir (Table 31.2).

Based on the questionnaire, it was concluded that people continued to use contaminated water, primarily for showering and personal hygiene, and they consumed fish even after the announced ban of water use. Some respondents and their animals also came into contact with the reservoir water during recreational activities. In accordance with the main hypothesis concerning a possible connection between the bloom and health problems, some citizens expressed several health issues. However, the health problems were similar before and after December 2013. This raised the following questions: Is it possible that the cyanobacterial bloom did not significantly affect the health of the respondents? Or, did cyanobacterial blooms occur even before December 2013?

31.4.3 Epidemiological Survey

In order to investigate the possible influence of cyanobacterial blooms in Vrutci on the health status of citizens of Užice, statistical data on the number of patients of certain diseases was obtained from the Institute of Public Health Užice and analyzed. The investigation was carried out for two periods of time: 2008–2011 and after the suspected/detected bloom, 2012–2015. The reasons for this division related to the statistical analysis were the following: (1) during the monitoring of Vrutci in 2012, a high oxygen saturation (62%) in the hypolimnion pointed to a possibility of

Table 31.2 Results of the questionnaire

Questions	In the past 3 years	Since December 2013
1. Tap water was used for?		
a) Drinking	–	daily **8.7%** occasionally **22.0%** never **69.3%**
b) Food preparation	–	daily **19.0%** occasionally **27.5%** never **53.5%**
c) Personal hygiene (washing of teeth, hands, face)	–	daily **76.7%** occasionally **15.4%** never **7.9%**
d) Showering	–	daily **80.4%** occasionally **16.0%** never **3.6%**
2. Did the tap water change?	–	colour **18.4%** taste **30.5%** smell **65.2%** remained the same **45.0%**
3. Fish from reservoir Vrutci is consumed	daily **5.9%** few times a week **1.6%** few times a month **7.5%** few times a year **17.3%** never **67.8%**	daily **1.9%** few times a week **1.9%** few times a month **2.6%** few times a year **-** never **-**
4. Recreational activities in/on/near reservoir Vrutci*		
a) How often do you visit reservoir Vrutci?	few times a week **3.7%** few times a month **12.9%** few times a year **41.7%** never **41.7%**	
b) Which activities do you do in/on/near reservoir Vrutci?	swimming **48.4%** fishing **18.8%** staying near lake water **46.9%** other (name which) **21.1% (rowing, cycling, hiking, walking)**	
c) Did you notice changes in the water of reservoir Vrutci?	unpleasant smell **15.1%** colour change **50.7%** foaming, scum and mats **63.0%**	
d) Did you get in contact with the water during these changes?	no **77.0%** yes **23.0%**	
e) Did you see dead animals in the water or near the water of reservoir Vrutci?	no **87.0%** yes (name which animals) **13.0% (fish 55.2%, ducks, rooster, hawk, ram, sheep, goat, cow, horse, badger)**	

(*Continued*)

Table 31.2 (Continued)

Questions	In the past 3 years	Since December 2013
f) Did your animals get in contact with water from reservoir Vrutci?	no **92.0%** yes (name which animals) **8.0% (dogs)**	
g) Did your animals have any health problems after contact with water from reservoir Vrutci?	no **92.0%** yes (name which health problems) **8.0% (skin lesions, allergies, stomach pain)**	
5. Did you have any of the following health problems?		
a) Skin irritations	**24.0%**	**25.5%**
b) Eye irritations	**14.5%**	**15.1%**
c) Digestive disorders	**26.0%**	**27.3%**
d) Fever	**2.3%**	**1.5%**
e) Chronic headaches	**15.5%**	**13.7%**
f) Kidney problems	**6.6%**	**7.4%**
g) Liver problems	**3.0%**	**1.8%**
h) Other (write)	**8.2%** **dizziness, problems with blood pressure and bladder, uterus cancer, bloating, tonsils allergies, dry cough, increased number of moles**	**7.7%** **problems with blood pressure and bladder, dizziness, long-lasting cough**
6. Did other members of the family have any of the following health problems?		
a) Skin irritations	**18.6%**	**23.1%**
b) Eye irritations	**14.0%**	**14.0%**
c) Digestive disorders	**22.2%**	**32.1%**
d) Fever	**0.5%**	**1.8%**
e) Chronic headaches	**7.2%**	**10.9%**
f) Kidney problems	**11.3%**	**3.2%**
g) Liver problems	**1.8%**	**1.4%**
h) Other (write)	**problems with blood pressure and other circulatory problems, prostate, digestive system, dizziness**	

*N.B. The answers to Question 4 are not connected to the time periods (three years before or after the incident).

P. rubescens biomass development in the deeper layers [32]; (2) microcystin was detected in the fish sample caught already in October 2013; and (3) there were questionnaire results showing that similar health problems occurred before and after the observed bloom in December 2013.

According to statistical analyses (*t*-test) of epidemiological data, the number of people suffering from diseases of digestive system, skin and subcutaneous tissue was significantly increased ($p<0.05$) in the period 2012–2015 compared to the previous period of four years (Figs. 31.2 and 31.3).

The infections of skin and subcutaneous tissue were not statistically significantly increased during the investigated period, suggesting that the bloom did not cause an increase of patients with these infections. However, other diseases of skin and subcutaneous tissue were significantly increased during the later 4 years.

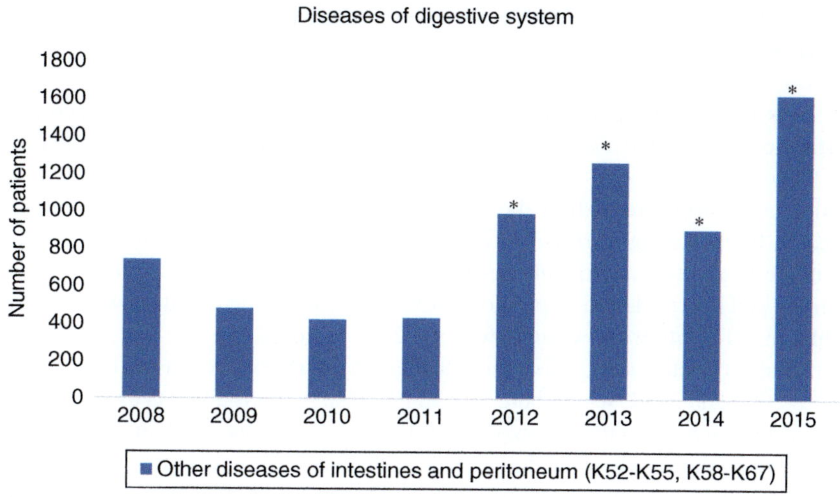

Figure 31.2 Number of patients with diseases of digestive system from 2008 to 2015. *Denotes significantly higher number of patients according to t-test ($p<0.05$)

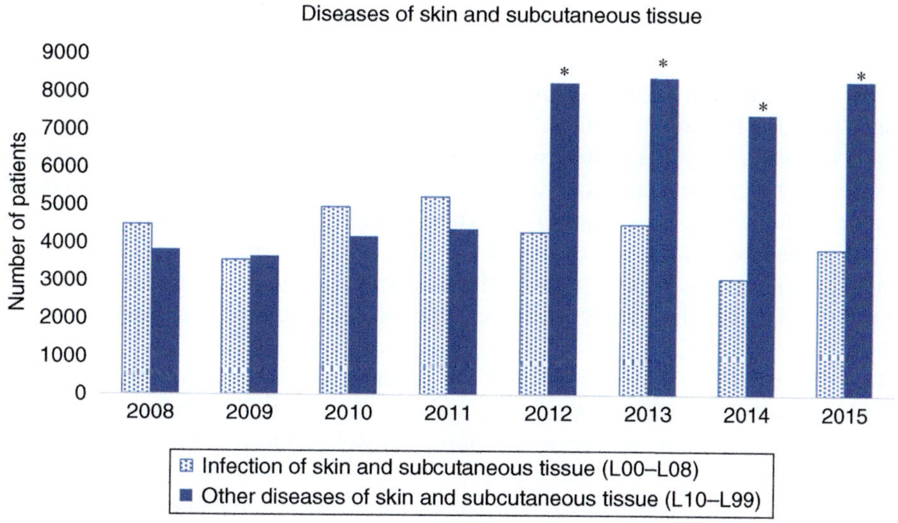

Figure 31.3 Number of patients with diseases of skin and subcutaneous tissue from 2008 to 2015. *Denotes significantly higher number of patients according to t-test ($p<0.05$)

In the light of the epidemiological investigation, it is possible to assume that blooms occurred before December 2013 and that the health issues of the citizens of Užice could be the result of exposure to cyanobacteria and cyanotoxins.

Our epidemiological study was not performed using the data presenting the incidence of examined diseases but the total number of registered patients per year (both new cases and follow-up of therapy-related cases). For this reason, our data should not be used without further detailed examination to prove/reject the relation between blooms and disease but rather as an example how to beneficially use epidemiology to complement analytical results.

31.5 Analytical and Supplementary Results Combined: A Plausible Reconstruction of Events in Vrutci Reservoir and the City of Užice

Combined results from analytical and supplementary methods can enable the reconstruction of some important events related to cyanobacteria, cyanotoxins, and their potential influence on human life.

The questionnaire results regarding the latest three years and obtained epidemiological data initiated a search for fish caught from Vrutci before December 2013 and stored frozen. Detected microcystin in the frozen fish and the above-mentioned questionnaire/epidemiological data, led to a detailed mining and interpretation of previous monitoring data. The intention was to find evidence of possible blooms of *P. rubescens* even before the 2013 incident.

According to all collected data, it is assumed that *P. rubescens* began to develop in reservoir Vrutci between 2009 and 2011, when there was an increase in internal loads as a result of a regime change in water discharge from the reservoir [32]. In the late August 2012, an increased content of oxygen was found at the depth of 26.5 m, which indicates the possibility of *P. rubescens* mass occurrence in the deeper layers of the reservoir [32]. The observed oxygen maximum, the detected microcystin in fish muscle and the occurrence of some health issues in the citizens of Užice before the confirmed cyanobacterial bloom during December 2013 suggest that cyanobacterial blooms might have occurred even before December 2013. A bloom in the deeper water could easily be overseen by visual inspection as it is not causing major changes in the waterbody.

The World Health Organization has provided a guideline value solely for microcystin-LR. However, over 200 analogues of microcystins have been identified (as reported in Appendix 3), but only some of them are monitored on an individual basis. The problem of lacking guideline values for microcystins other than microcystin-LR was clearly demonstrated in the Užice case, where the reservoir and tap water as well as fish tissues contained mostly other microcystins than microcystin-LR (Table 31.1). Therefore, any conclusions based only on the presence of microcystin-LR can be misleading in the Užice case and generally. There is a clear need for establishing and introducing guideline values for other forms of microcystins as well. A further complication is that there are no limits for cyanotoxins in the Serbian national legislation. This should be changed in the near future since many cities in Serbia face the problem of cyanobacteria and their toxins in waterbodies.

The Užice case describes the complexity of the situation and why analytical data from a short period of time alone are not enough. Analytical methods showed the presence of microcystins, but the questionnaire raised many important questions that were addressed by epidemiological data and some additional analyses (e.g. analyses of frozen fish). The additional observations indicated that the population of Užice might have been exposed to cyanobacteria and cyanotoxins even two years before the December 2013 bloom.

31.6 Conclusions from the Užice Case

Analytical data are not always sufficient to describe, interpret, explain, and reconstruct cases of cyanobacterial blooms, especially when the blooms occur in drinking water reservoirs and when regular monitoring has not been established. This is particularly evident in cases where the analytical and toxicological interest concerns one microcystin only (microcystin-LR), like it was the case with the first water analyses of Vrutci where microcystin-LR was not observed at concentrations exceeding the World Health Organization guideline. However, bioassays conducted as supportive measurements were positive for toxins, which triggered the analyses of further cyanotoxins. These further analyses were positive for the period of blooming starting in December 2013. Other complementary data including the questionnaire and epidemiological results showed that health problems possibly related to cyanotoxins occurred already at least 2 years prior to the bloom detected at the end of 2013. Microcystin analysis of frozen fish meat, together with earlier chemical analyses conducted as a part of reservoir monitoring, supported the hypothesis of earlier occurrence of cyanotoxins.

The results of the analytical work, the complementary work (the questionnaire, epidemiological data and bioassays), as well as other observations and analyses by the involved scientists corroborated each other and gave additional value to the public health study of the Užice case.

References

[1] Jochimsen, E.M., Carmichael, W.W., An, J.S. *et al.* (1998) Liver failure and death following exposure to microcystin toxins at a hemodialysis center in Brazil. *The New England Journal of Medicine* **338** (13), 873–879.

[2] Pouria, S., de Andrade, A., Barbosa, J. *et al.* (1998) Fatal microcystin intoxication in haemodialysis unit in Caruaru, Brazil. *The Lancet* **352** (9121), 21–26.

[3] Azevedo, S.M., Carmichael, W.W., Jochimsen, E.M. *et al.* (2002) Human intoxication by microcystins during renal dialysis treatment in Caruaru-Brazil. *Toxicology* **181–182**, 441–446.

[4] Hawkins, P.R., Runnegar, M.T.C., Jackson, A.R.B. and Falconer, I.R. (1985) Severe hepatotoxicity caused by the tropical cyanobacterium (blue-green alga) *Cylindrospermopsis raciborskii* (Wolozynska) Seenaya and Subba Raju isolated from a domestic water supply reservoir. *Applied and Environmental Microbiology* **50** (5), 1292–1295.

[5] Ohtani, I., Moore, R.E. and Runnegar, M.T.C. (1992) Cylindrospermopsin-a potent hepatotoxin from the blue-green algae *Cylindrospermopsis raciborskii*. *Journal of American Chemical Society*, **114** (20), 7941–7942.

[6] Griffiths, D.J. and Saker, M.L. (2003) The Palm Island mystery disease 20 years on: a review of research on the cyanotoxin cylindrospermopsin. *Environmental Toxicology* **18** (2), 78–93.

[7] Annadotter, H., Cronberg, G., Lawton, L.A. *et al.* (2001) An extensive outbreak of gastroenteritis associated with the toxic cyanobacterium *Planktothrix agardhii* (Oscillatoriales, Cyanophyceae) in Scania, South Sweden, in *Cyanotoxins, occurrence, causes, consequences* (ed. I. Chorus), Springer, Berlin, pp. 200–208.

[8] Falconer, I.R., Beresford, A.M. and Runnegar, M.T.C. (1983) Evidence of liver damage by toxin from a bloom of the blue-green alga, *Microcystis aeruginosa*. *Medical Journal of Australia* **1** (11), 511–514.

[9] Yu, S.Z. (1995) Primary prevention of hepatocellular carcinoma. *Journal of Gastroenterology and Hepatology* **10** (6), 674–682.

[10] Svirčev, Z., Krstić, S., Miladinov-Mikov, M., *et al.* (2009) Freshwater cyanobacterial blooms and primary liver cancer epidemiological studies in Serbia. *Journal of Environmental Science and Health Part C* **27** (1), 36–55.

[11] Svirčev, Z., Drobac, D., Tokodi, N. *et al.* (2013) Epidemiology of primary liver cancer in Serbia and possible connection with cyanobacterial blooms. *Journal of Environmental Science and Health Part C* **31** (3), 181–200.

[12] Svirčev, Z., Drobac, D., Tokodi, N. *et al.* (2014) Epidemiology of cancers in Serbia and possible connection with cyanobacterial blooms. *Journal of Environmental Science and Health Part C* **32** (4), 319–337.

[13] Westrick, J.A. (2003) Everything a manager should know about algal toxins but was afraid to ask. *Journal American Water Works Association* **95**, 26–34.

[14] Hoeger, S.J., Hitzfeld, B.C. and Dietrich, D.R. (2005) Occurrence and elimination of cyanobacterial toxins in drinking water treatment plants. *Toxicology and Applied Pharmacology* **203**, 231–242.

[15] Hoeger, S.J., Shaw, G., Hitzfeld, B.C. and Dietrich, D.R. (2004) Occurrence and elimination of cyanobacterial toxins in two Australian drinking water treatment plants. *Toxicon* **43**, 639–649.

[16] Dietrich, D. and Hoeger, S. (2005) Guidance values for microcystins in water and cyanobacterial supplement products (blue-green algal supplements): a reasonable or misguided approach? *Toxicology and Applied Pharmacology* **203**, 273–289.
[17] Svirčev, Z., Baltić, V., Gantar, M. *et al.* (2010) Molecular aspects of microcystin-induced hepatotoxicity and hepatocarcinogenesis. *Journal of Environmental Science and Health Part C* **28**, 39–59.
[18] Drobac, D., Tokodi, N., Simeunović, J. *et al.* (2013) Human exposure to cyanotoxins and their health effects. *Archives of Industrial Hygiene and Toxicology* **64** (2), 305–316.
[19] Svirčev, Z., Lujić, J., Marinović, Z. *et al.* (2015) Toxicopathology induced by microcystins and nodularin: a histopathological review. *Journal of Environmental Science and Health Part C* **33** (2), 125–167.
[20] Pantelić, D., Svirčev, Z., Simeunović, J. *et al.* (2013) Cyanotoxins: characteristics, production and degradation routes in drinking water treatment. *Chemosphere Journal* **91** (4), 421–441.
[21] Simeunović, J., Svirčev, Z., Karaman, M. *et al.* (2010) Cyanobacterial blooms and first observation of microcystin occurrences in freshwater ecosystems in Vojvodina region (Serbia). *Fresenius Environmental Bulletin* **19** (2), 198–207.
[22] Svirčev, Z., Simeunović, J., Subakov-Simić, G. *et al.* (2013) Cyanobacterial blooms and their toxicity in Vojvodina lakes, Serbia. *International Journal of Environmental Research* **7** (3), 745–758.
[23] Svirčev Z., Tokodi N., Drobac D. and Geoffrey A. Codd, GA. (2014) Cyanobacteria in aquatic ecosystems in Serbia: effects on water quality, human health and biodiversity. *Systematics and Biodiversity* **12** (3), 261–270.
[24] Đenić, D (2014) *Cyanobacterial blooms in Vrutci reservoir (Cvetanje cijanobakterija u hidroakumulaciji Vrutci)*. Master's thesis, University of Novi Sad, Serbia.
[25] The Agency for Environmental Protection (Agencija za zaštitu životne sredine) (2013) *Ispitivanje kvaliteta vode akumulacije Vrutci, na teritoriji opštine Užice*, http://www.sepa.gov.rs/download/akcidenti/vode/Havarija_Vrutci_final.pdf (accessed 21 January 2016).
[26] The Agency for Environmental Protection (Agencija za zaštitu životne sredine) (2013) http://www.sepa.gov.rs/index.php?menu=504000000&id=20013&akcija=showArhiva (accessed 21 January 2016).
[27] Institute of Public Health Serbia "Dr Milan Jovanović Batut" (2014) Report on the request for access to information of public importance No.8301/1, Belgrade. http://uimenaroda.rs/predmeti/li%C4%8Dni-stav/190-u%C5%BEice-vodosnabdevanje-slika-srbije.html (accessed 21 January 2016).
[28] Drobac, D. (2015) *Human exposure to cyanotoxins and their health effects (Putevi izloženosti čoveka cijanotoksinima i njihov uticaj na zdravlje)*. Doctoral dissertation. University of Novi Sad.
[29] Chorus, I. and Bartram, J. (eds) (1999) *Toxic cyanobacteria in water. A guide to their public health consequences, monitoring and management*, E&FN Spon, WHO, London, England.
[30] http://toxische-cyanobakterien.de/en/system-assessment/quick-hazard-analysis-via-online-questionnaire/.
[31] Stewart, I. (2004) Recreational exposure to freshwater cyanobacteria: epidemiology, dermal toxicity and biological activity of cyanobacterial lipopolysaccharides. Doctoral dissertation. University of Queensland.
[32] Jaroslav Černi Institute for the Development of Water Resources (Institut za vodoprivredu, Jaroslav Černi") (2014) http://www.graduzice.org/userfiles/files/Izvestaj%20instituta%20J%20Cerni.pdf (accessed 21 January 2016).

32

Selection of Analytical Methodology for Cyanotoxin Analysis

Jussi Meriluoto[1,2], James S. Metcalf[3], and Geoffrey A. Codd[4,5]

[1]*Biochemistry, Faculty of Science and Engineering, Åbo Akademi University, Turku, Finland*
[2]*Laboratory for Paleoenvironmental Reconstruction, Faculty of Sciences, University of Novi Sad, Serbia*
[3]*Institute for Ethnomedicine, Jackson, USA*
[4]*Biological and Environmental Sciences, University of Stirling, Scotland, United Kingdom*
[5]*School of the Environment, Flinders University, Adelaide, Australia*

32.1 Introduction

A wide range of analytical methods is available for the recovery, identification, and quantification of cyanotoxins. No single procedure is optimal for all of the cyanotoxin classes, the most suitable method being influenced by the cyanotoxin(s) under investigation, their abundance, and the matrices from which they are to be extracted. Method selection is also necessarily influenced by the availability of particular analytical equipment, specialist reagents, and, not least, the levels of technical and scientific skill available for method performance and data interpretation.

32.2 General Comparison of Physicochemical Analyses, Biochemical Methods, and Bioassays

Analytical methodology has to balance between the desire to obtain a maximum amount of valid information from the sample and the aim to minimize resource needs in doing so. Response time in the provision of the data may also be an important factor. The best methodological choice

Handbook of Cyanobacterial Monitoring and Cyanotoxin Analysis, First Edition.
Edited by Jussi Meriluoto, Lisa Spoof and Geoffrey A. Codd.

clearly depends on the sample, the target compounds, the nature of analytical answers required, and the level of uncertainty allowed. In a regulatory setting, the methodology may be enforced and fixed to national or international standard methods. In real life, some compromises often have to be made and the methodology depends on the available resources. There are great differences between the proposed detection methods for cyanotoxins in terms of sensitivity, speed, ability to identify chemical species or toxic principles, handling of multiple samples, need for reference materials, costs including capital investments, operator training required, and running costs.

Screening: This is the process of testing a large sample set for a specific group of compounds or a specific property, and it calls for quick and simple methods. Screening methods are often based on biochemical principles. For instance, there are convenient enzyme-linked immunosorbent assays (ELISAs) or protein phosphatase inhibition assays for microcystins and nodularins. Screening for unspecified toxicity has traditionally been based on whole-organism bioassays usually using mice. While such bioassays have been unavoidable for some toxin groups, at least until sufficient knowledge of the relevant compounds has been attained, there are ethical and humanitarian concerns and usually statutory requirements regarding the use of whole-animal vertebrate bioassays. Bioassays (and biochemical methods) also tend to produce somewhat less consistent results than physicochemical methods, which are easier to standardize and validate.

Trace analysis: Regulatory needs often necessitate the use of physicochemical analytical methods as the concentration of a specific compound may be regulated in drinking water or foodstuffs. A physicochemical method that targets and identifies single analytes (e.g. microcystin-LR or anatoxin-a) is then the right methodological choice. Most physicochemical methods for cyanotoxins are based on high-performance liquid chromatography (HPLC) coupled to different detectors such as a photodiode array detector, fluorescence detector, or mass spectrometer. It is technically possible to include several analytes or analyte groups in the same HPLC analysis as the chromatographic and detection parameters can be varied during a single run. Sometimes, it is also possible to detect common chemical moieties in a family of analytes either through detection of specific product ions in mass spectrometry or through derivatization and subsequent analysis of the derivatization product. One example of the latter is the analysis of MMPB, an oxidative cleavage product of ADDA present in microcystins and nodularins.

32.3 Guidance for Selecting and Using Standard Operating Procedures Found in this Handbook

This Handbook contains 29 Standard Operating Procedures (SOPs) for cyanobacterial monitoring and analyses of chemical compounds produced by cyanobacteria. The SOPs come from internationally recognized scientists and laboratories with a long track record, and they represent current state-of-the-art. The SOPs are thus excellent starting points for introducing the methods in a laboratory planning to be engaged in this type of work. In any case, **the SOPs must be adapted to local requirements and validated in local conditions by the end-user** (see Chapter 29 and SOP 29). No responsibility for the fitness of the methods for a particular purpose can be assumed by the authors, the editors, or the publisher of this Handbook. Once again, it is also necessary to remind the reader of the potent toxicity of characterised and emerging cyanotoxins and the general risks associated with any laboratory work. **The editors urge the laboratory using the SOPs to make a risk assessment of the planned work with toxic cyanobacteria and cyanotoxins according to local regulations, and to take the necessary precautions to guarantee the safety of the personnel involved**. Several of the SOPs are likely to be used by, for example, doctoral students who do not necessarily have an extensive knowledge of the risks involved. It is especially important to protect the health of young people because some of the cyanobacterial compounds are suspected carcinogens and novel compounds of unknown character are likely to exist.

The first five SOPs (1–5) concern fundamental work with environmental samples containing cyanobacteria and with pigment analysis. These SOPs should be read together with text Chapters 1–10 and Appendix 1. Laboratory work with cyanotoxins and other cyanobacterial compounds is described in SOPs 6–29. These in turn should be read together with Chapters 11–32 and Appendices 2 and 3.

As cyanotoxins are being discovered in an increasing number of cyanobacterial species (Appendix 2), it is not safe to assume that certain cyanotoxins are associated with certain species only. In a critical setting, such as when drinking water is being abstracted and purified from a source with a high cyanobacterial biomass, a wide range of assays and analyses should be performed to protect public health.

32.4 Methodology versus Required Response Time

All analytical methods and the interpretation of their results require time, training, and an understanding of the risks involved. Even microscopy to determine the identity and abundance of cyanobacteria present can require many hours of practice, especially when performing cyanobacterial identification to species level. Increasingly, there is also a development of analytical methods, which are being moved from the laboratory to the field. These include the use of lateral-flow immunoassays, which can allow a nonspecialist person to perform toxin analysis. Further, with outreach to members of the public, a greater understanding of cyanobacteria and the inherent risks which they can present has occurred, with the possibility that the public can assist in the reporting of blooms, potentially shortening the response time.

In general, the greater the amount of information that is required, the longer it takes to perform the analysis. Response times can certainly be increased if legislation is in place to regulate compounds that have the potential to cause deleterious effects.

For general monitoring, this is often routine, the methods for sample collection and analysis are already in place, and there is often a schedule that can be followed to analyse the samples and obtain results. The issue of time becomes more important when results are required concerning, for example, after possible poisoning incidents, to aid decision-making before major water-based recreational events and issues with water quality and agricultural produce that may be contaminated. Time-sensitive situations can range from the reporting of incidents, to the analysis of known toxins, through to an understanding of the significance of the data. With water treatment, if drinking water does contain cyanotoxins, then the faster that accurate information can be obtained can save time and money and minimise potential exposure that may lead to adverse health effects. With poisonings, rapid collection of samples is required, so that the site of the incident and the likely cause of the poisoning can be identified and preventative measures carried out, such as possible medical intervention or the closure of an area of a waterbody to avoid further incidents.

Indications, such as from bioassays, that further new bioactive compounds continue to be discovered and that understanding of known cyanotoxin toxicity and exposure routes continues to develop present gaps in knowledge and continuing uncertainties in cyanobacterial risk management. However, continuing research into new modes of action, and into relevant exposure media and routes (e.g. via inhalation) should contribute to the further formulation of appropriate risk management measures.

32.5 Influence of Waterbody History on the Choice of Methods

Although the methods used for cyanotoxin analysis and cyanobacterial identification may be precisely defined, a number of possibilities may exist that could change the nature of the methods and the sampling regimen used. One of the major considerations concerns the history of the waterbody. If this has a history of cyanobacterial blooms, then the potential for cyanotoxin production may be increased, possibly leading to the need to increase the frequency of monitoring/sampling or to the provision of rapid analytical techniques that could be used in the event of a bloom. In this connection, should a waterbody have a history of poisoning incidents, then greater monitoring and analysis may be advantageous, especially in light of the potential for adverse reporting and public concern. The use of the waterbody is also a potential issue, especially if this has changed recently, as different uses may affect the potential for blooms, but also whether contact with the water by people or animals has changed. Waterbody use with particular relevance to cyanotoxins encompasses drinking water, recreational waters, fisheries, livestock watering and spray irrigation as examples. Depending on how the water is used, and what is done to it, such as with commercial fish farms that feed fish and thereby increase nutrient input, may affect the potential for blooms and cyanotoxin production and abundance.

Sometimes, waterbodies may require special attention in the form of observing whether blooms have formed or are likely to do so, and concerning the production of cyanotoxins. Often this is required when special water contact events occur at the lake, such as training exercises and major sporting events, like triathlons and aquatic Olympic events. During such periods it may be necessary to monitor for and have intensive analysis, in addition to contingencies, so that if cyanotoxin thresholds are exceeded, then alternative venues can be used or in-lake remediation measures can be taken.

If a waterbody has been modified, this may lead to the potential for increased or decreased blooms. Measures such as dredging or sediment sealing can remove or render inaccessible nutrients present within the bed of the lake, so that in subsequent years, the potential for blooms can be diminished. Other less drastic measures also include the installation of destratification units, or the addition of barley straw, possibly preventing bloom formation for short periods of time.

Although the analysis and choice of methods may alter with the history of the waterbody, the need to analyse for cyanobacteria and cyanotoxins is likely to be reduced when the catchment or basin of a waterbody can be altered or controlled, so that nutrient inputs, which are a major primer for cyanobacterial blooms can be reduced.

32.6 Integration of the Results Obtained: Making Sense

Where a single method or, more ideally, multiple methods have been applied to the same analyte, comparability should be sought between the data obtained by each method, taking into account:

1. Each method's strengths and limitations;
2. Any performance problems encountered;
3. Consistency of the data with prior knowledge of the environment or materials under investigation;
4. Consistency with wider physicochemical, biological, and toxicological knowledge of the occurrence, identity, and potential abundance of cyanotoxins.

However, in seeking consistency and logicality, a preparedness is also necessary to account for the continuing discovery of novel cyanotoxins.

Section VII

Standard Operating Procedures (SOPs)

SOP 1

Cyanobacterial Samples: Preservation, Enumeration, and Biovolume Measurements

Arnaud Catherine[1], *Selma Maloufi*[1], *Roberta Congestri*[2], *Emanuela Viaggiu*[2,3], *and Renata Pilkaityte*[4]

[1] *UMR 7245 MCAM CNRS-MNHN, Muséum National d'Histoire Naturelle, Paris, France*
[2] *Department of Biology, University of Rome 'Tor Vergata', Italy*
[3] *AlgaRes srl, Spin-off of University of Rome 'Tor Vergata', Italy*
[4] *Open Access Centre for Marine Research, Klaipeda University, Klaipeda, Lithuania*

SOP 1.1 Introduction

Estimating the occurrence, cell concentration, and biomass of cyanobacteria in natural samples is a fundamental requirement to assess potential risks linked to cyanobacterial bloom events. Special attention should be given to ensure that a reliable sampling technique is used (see Chapter 6). Aspects covered in this SOP (Fig. SOP 1.1) include sample preservation, step-by-step procedures for cyanobacterial enumeration by common methods, and available methods for cyanobacteria biovolume measurements.

SOP 1.2 Sample Preservation

SOP 1.2.1 Introduction

Cyanobacterial identification should preferably be performed with live samples. However, if samples cannot be processed immediately, they should be preserved on site, as their composition

Handbook of Cyanobacterial Monitoring and Cyanotoxin Analysis, First Edition.
Edited by Jussi Meriluoto, Lisa Spoof and Geoffrey A. Codd.

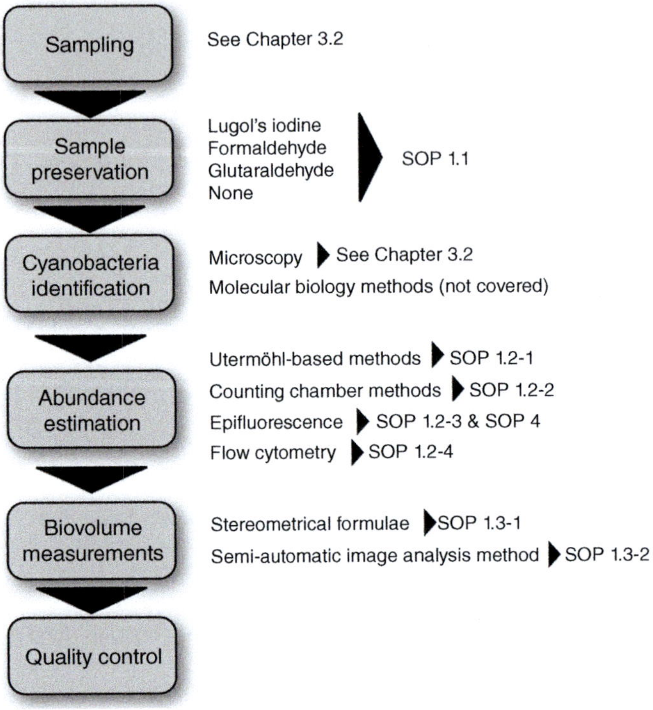

Figure SOP 1.1 Workflow diagram for cyanobacterial sample preservation, identification, enumeration, and biovolume measurements

may change within hours (grazing, cell death etc.). In case the on-site addition of preservative is not possible, samples should be kept refrigerated (4°C).

No single preservative is ideal for all purposes or for all cyanobacterial taxa. Depending on the method used for the quantitative analysis of samples, and on available safety equipment, different preservation reagents may be used (Table SOP 1.1).

Among preservative reagents, acid Lugol's iodine solution is the most commonly used. However, Lugol's iodine is known to induce cell shrinkage [1] and should be avoided when cell volume measurements are required. If cell volume measurements are to be obtained from Lugol's iodine–preserved samples, modifications to the form of the cells due to shrinkage should be taken into account. In any case, the most reliable volume estimates are obtained from fresh samples. Buffered formaldehyde allows for preserving cell structural details and is adequate for long-term storage of samples. However, special care must be taken when using formaldehyde, because the addition of this preservative can destroy small delicate species, such as flagellates (small cryptomonads and others). In addition, like Lugol's solution, it can significantly distort shapes, causing problems in the estimation of biovolumes. Despite modifying cell fluorescence, formaldehyde may be used for epifluorescence analysis of samples. Glutaraldehyde is usually preferred for flow cytometry. Preserved samples should be stored in the dark, at temperatures around 4°C.

SOP 1.2.2 Preservation in Acid Lugol's Iodine Solution

The volume of Lugol's iodine solution to be added depends on the organic matter content and phytoplankton biomass of the samples. Suitable preservation requires the sample to reach a dark

Table SOP 1.1 Advantages and limits of common water sample preservatives

Preservative	Advantages	Limitations	Common usage
Acid Lugol's iodine	• Disintegrates colonies • Decreases cell buoyancy • Low cell loss	• Disintegrates colonies • Induces cell shrinkage • Not adapted for long-term (>1 year storage) unless fresh Lugol's iodine is regularly added • Masking of some structural details	• Cyanobacterial identification (see Chapter 6) • Cyanobacterial enumeration by microscopy (see SOP 1.1.2-1 and -2) • Biovolume measurements (see SOP 1.1.3-1 and -2)
Buffered formaldehyde	• Structural details well preserved • Moderate cell loss with most species	• Modifies cell shape • Affects cell fluorescence to some extent • pH decreases with time	• Cyanobacterial identification (see Chapter 6) • Cyanobacterial enumeration by microscopy (see SOP 1.1.2-1 and -2) • Epifluorescence (see SOP 1.1.2-3 and SOP 1.4) • Biovolume measurements (see SOP 1.2.1.3-1 and -2)
Glutaraldehyde	• Cell shape well preserved • Cell fluorescence preserved • Long-term storage	• Higher cell loss for large size species may occur • Requires low temperature storage	• Epifluorescence (see SOP 1.1.2-3) • Flow cytometry (see SOP 1.1.2-4)

tea colour. Samples stored for several months should be checked regularly and extra Lugol's iodine added if required. If kept refrigerated (4°C) in the dark, samples may be preserved for up to 1 year before counting. The total volume of Lugol's iodine added to the samples should be determined and taken into account in cell concentration calculation.

SOP 1.2.2.1 Materials

a. Potassium iodide (CAS number: 7681-11-0).
b. Iodine crystals (CAS number: 7553-56-2).
c. Anhydrous acetic acid (CAS number: 64-19-7).
d. Purified water (distilled or demineralized water).
e. Small glass or plastic containers with caps or lids (glass containers are preferred since plastic will take up iodine from the solution).

SOP 1.2.2.2 Stock Solutions

a. For 1 litre of acid Lugol's Iodine solution, add 100 g potassium iodide, 50 g iodine crystals, and 100 mL of anhydrous acetic acid to 900 mL of purified water.
b. Mix in written order and ensure that each added reagent is completely dissolved before adding the next.
c. Store the solution up to 1 year in the dark at room temperature.

SOP 1.2.2.3 Sample Fixation

a. Add 200 mL of a water sample in a glass or plastic container equipped with a water-tight lid (e.g. screw cap universal bottle).
b. Add 2 mL of acid Lugol's iodine solution and mix.

SOP 1.2.3 Preservation in Buffered Formaldehyde

Formaldehyde is a category 2 carcinogen: use suitable protective equipment at all steps. If samples are stored for several months, check pH on a regular basis as formaldehyde readily oxidizes to formic acid.

SOP 1.2.3.1 Materials

a. Formaldehyde solution 37% w/v in H_2O (CAS number: 50-00-0).
b. Hexamethylenetetramine (CAS number: 100-97-0).
c. Purified water (distilled or demineralized water).
d. Small glass or plastic containers with caps or lids.

SOP 1.2.3.2 Stock Solutions

a. Prepare a 1:5 buffered dilution (hereafter called working solution) by adding 200 mL of formaldehyde solution and 100 g of hexamethylenetetramine in 800 mL of purified water (final pH 7.3–7.9).
b. Store the solution up to 1 year at room temperature.

SOP 1.2.3.3 Sample Fixation

a. Add 100 mL of a water sample in a glass or plastic container equipped with a water-tight lid (e.g. screw cap universal bottle).
b. Add 100 mL of the working solution (to reach a final formaldehyde concentration of 3.7% w/v) and mix.
c. For long-term storage, check pH every few months as pH may decrease and adjust pH by alkali addition if necessary.

SOP 1.2.4 Preservation in Glutaraldehyde

Glutaraldehyde is toxic: use suitable protective equipment at all steps.

SOP 1.2.4.1 Materials

a. Glutaraldehyde solution 50% w/v in H_2O (CAS number: 111-30-8).
b. Purified water (distilled or demineralized water).
c. Plastic tubes suitable for low temperature storage.
d. Liquid nitrogen container.

SOP 1.2.4.2 Sample Fixation

a. Add 9.5 mL of a water sample in a plastic tube suitable for low temperature storage (Cryotube).
b. Add 0.5 mL of 50% glutaraldehyde solution.
c. Store in liquid nitrogen. Samples may be refrozen with limited damage.

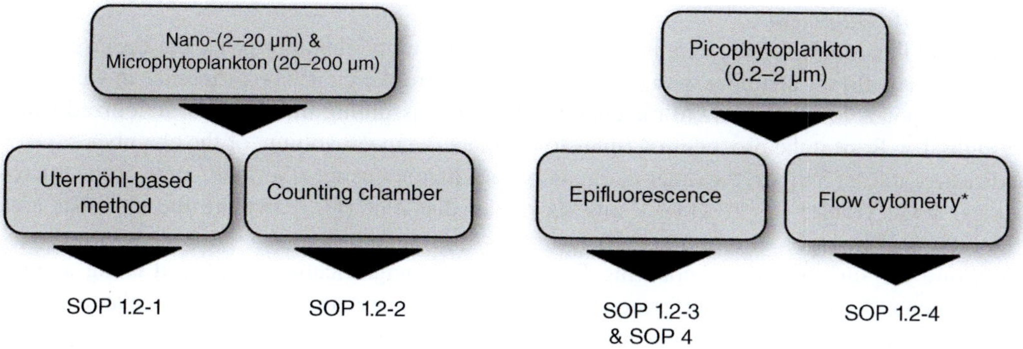

Figure SOP 1.2 *Workflow diagram for cyanobacteria sample enumeration. *Flow cytometry may also be used for larger organisms*

SOP 1.3 Methods for Cyanobacterial Enumeration

SOP 1.3.1 Introduction

Depending on the kind of cyanobacteria to be enumerated, different methods may be used (Fig. SOP 1.2). The Utermöhl method requires the sedimentation of a known volume of sample in a counting chamber and its subsequent enumeration using an inverted microscope. Cyanobacteria may also be enumerated using counting chamber slides using an upright microscope. Small cyanobacteria (picocyanobacteria) are preferably counted using epifluorescence or flow cytometry.

SOP 1.3.2 Utermöhl-based Method

SOP 1.3.2.1 Introduction

The Utermöhl method [2] is considered as one of the most reliable approaches for cyanobacterial cell enumeration. It allows cyanobacterial enumeration with limited sample handling and relatively standard equipment. Additional information may be found in [3]. Prior to cyanobacterial enumeration, a qualitative analysis of the sample should be performed using an upright microscope. If identification to species level cannot be performed with sufficient certainty, identification to genus (or higher) level is preferred to prevent misidentification.

SOP 1.3.2.2 Experimental

Materials

a. Inverted microscope equipped with a movable stage and magnification objectives (10×, 20×, and 40×).
b. Measuring cylinders.
c. Sonication bath or sonication probe.
d. Potter-Elvehjem (PTFE pestle and glass tube).
e. Sedimentation chambers (Utermöhl chamber; e.g. Hydro-Bios No. 435 025 or equivalent).
f. Calibrated stage micrometer.
g. Calibrated ocular micrometer.
h. Reticle with a single 10-mm square.
i. Cell counters.
j. Taxonomic keys (see list in Chapter 6).

Calibration of Equipment

a. Calibrate the volume of the chamber by weighing an empty chamber (with the cover glass in place). Then, fill the chamber with deionized water and re-weigh it. The difference in weight (in g) corresponds to the volume of the chamber (in mL), assuming that 1 mL of deionized water weighs 1 g. Repeat the procedure 4 times to estimate the mean volume of the chamber.
b. Measure, at least 5 times, the inner diameter of the chamber using a Vernier gauge or the microscope stage Vernier (if present). Using the mean diameter (D), calculate the chamber area $[A=\pi*(D/2)^2]$.
c. Measure the surface area of counting fields for each magnification employed using a stage micrometer. If transects are performed in the counting process, the transect surface area at each magnification should also be measured.

Sample Preparation

a. Allow the sample to equilibrate to room temperature to prevent air bubble formation and convection currents, which may prevent homogeneous sedimentation. If large colonies of cyanobacteria are present, a pre-treatment of the sample may be required using either a sonication bath or a sonication probe (preferred) for disrupting *Microcystis* spp. colonies or a Potter-Elvehjem for separating colonial filamentous cyanobacteria.
b. Check that the chamber cylinder and the bottom glass plate are free from contamination from previous samples.
c. Select an appropriate cylinder based on the density of the sample to be analysed (2-, 5-, 10-, 25-, 50-, and 100-mL chambers are commercially available).
d. Gently invert the sample bottle 30–50 times to mix contents.
e. Fill the chamber with the sample and seal using a cover glass (a small amount of grease may be used to achieve a tight seal). Check that there is no air bubble under the cover glass.
f. Allow a settling time of approx. 4h per cm height of the chamber. If the sample contains high numbers of buoyant cyanobacteria a drop of glacial acetic acid (CAS number: 64-19-7) may be added before sealing the chamber.
g. After sedimentation, if a combined plate chamber is used, then slide the column aside to remove the supernatant. Place a cover slip on the chamber. Check that there is no air bubble under the cover glass.
h. Check that no significant numbers of buoyant cyanobacteria remain in the supernatant. If large numbers of cyanobacteria are observed, sonicate the sample for 10 s and prepare a new column.
i. Move the chamber gently to the microscope stage. Scan the chamber to visually check for random distribution of phytoplankton using a low magnification objective. A random (Poisson) distribution of the particles should be observed (a detailed method to check for random cell distribution can be found in the Appendix of Chapter 6 and in Annex F of [3]).
j. On completion of analysis discard the sample and soak the chamber in detergent overnight.

Sample Enumeration Depending on the abundance of the organisms to be counted, different magnification and analysis procedures may be used. To ensure data reliability between samples, always count a specific taxon at the same magnification. Filaments or colonies should be recorded as objects. For unicellular colonial cyanobacteria (e.g. *Microcystis* sp.), the number of individual cells may be roughly estimated based on colony size [4], colony volume [5], or by counting small fractions of colonies [6]. More accurate cell counts may be obtained by disrupting colonies into single cells (or small cell clumps). A number of methods have been proposed including alkaline hydrolysis [4], high speed blending [7], sonication (preferably using a sonication probe) [4], and heating at 80°C [5, 8]. For filamentous cyanobacteria, cell concentrations can be obtained either by estimating the mean number of cells per filament (on at least 30 filaments) or by dividing the total length of filaments (evaluated during counting) by the average length of the cells.

Depending on the density of cyanobacterial objects, the following steps should be followed:

a. Count abundant taxa at 400× magnification in random selected optical fields or along four to six transects (diameters of the counting chamber).
b. Count rare or large taxa at 100–200× magnification in larger volumes (i.e. more random selected fields or transects).

Use the 10-mm-square reticle for transect delimitation. Any objects that cross the top line are counted. Objects crossing the bottom line are not counted and will be counted on the next transect. Randomly rotate the chamber for transect selection.

For random selected fields, fields are delimited by the 10-mm-square reticle. The objects crossing the lines of the square reticle should be counted according to the following figure:

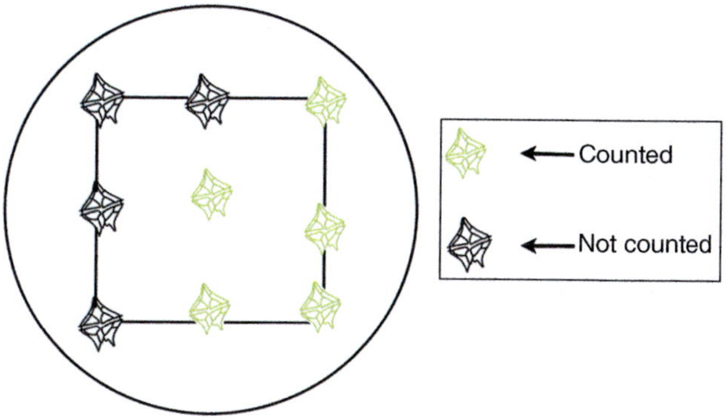

Count precision depends on the number of cyanobacterial units counted (unicells/colonies/filaments). The estimation of the number of cells to be enumerated in order to obtain a certain precision generally assumes a nonaggregated, random distribution of individuals, which can be described by a Poisson distribution (see Chapter 6 for details). Under this assumption, the precision, defined as ±95% confidence limits, is approximated by $2/\sqrt{N} \times 100$, where N is the number of units counted.

Cell Population Density Calculation The following formula should be used to calculate the number of cells per unit of volume:

$$\text{Cells. L}^{-1} = N * \left(\frac{A_t}{A_c}\right) * \frac{d}{V} * 1000$$

where:

N: total number of cells counted for a given species,
A_t: total area of the counting chamber,
A_c: counted area of the counting chamber (number of fields/transects counted * area of a field/transect),
d: dilution factor (if required),
V: volume of the counting chamber in mL.

SOP 1.3.3 Abundance Estimation Using Counting Chambers

SOP 1.3.3.1 Introduction

Counting chamber methods are common methods for phytoplankton enumeration. They are well adapted for samples of high cell population density; however, low-density samples may be analysed following a concentration procedure (e.g. sedimentation and removal of supernatant). Various counting chambers may be used including the Sedgewick-Rafter counting slide, the Lund counting slide, the Palmer-Maloney counting slide, and the haemocytometer counting slide. This section proposes a standard operating procedure to be used with a Sedgewick-Rafter counting slide and was proposed according to the EN 15204:2006 standard [3]. Other method descriptions can be found in [9]. Prior to cyanobacteria enumeration, a qualitative analysis of the sample should be performed using an upright microscope. If identification to species level cannot be performed with sufficient certainty, identification to genus (or higher) level is preferred to prevent misidentification.

SOP 1.3.3.2 Experimental

Materials

a. Upright microscope equipped with a movable stage and magnification objectives (10×, 20×, and 40×).
b. Sedgewick-Rafter chamber with a volume of 1 mL. The base of the chamber is divided in 1000 squares (50 × 20 squares), each representing 1 μL.
c. Cell counters.
d. Taxonomic keys (see list in Chapter 6).

Calibration of Equipment

a. Weigh an empty chamber with its coverslip.
b. Measure the area of the chamber (nominally 50 × 20 mm = 1000 mm^2) and its depth (nominally 1 mm).
c. Fill the chamber with distilled water at one corner with the coverslip at an angle across the chamber.
d. Move the coverslip to cover the entire chamber and reweigh the chamber. The difference in weight (in g) corresponds to the volume of the chamber (in mL), assuming that 1 mL of deionized water weighs 1 g. Repeat the procedure 4 times to estimate the mean volume of the chamber.
e. If the mean volume of the chamber differs from 1 mL by 5% or more, apply the corresponding correcting factor. Write the correcting factor on the chamber for future use.

Sample Preparation

a. Allow the sample to equilibrate to room temperature to prevent air bubble formation and convection currents, which may prevent homogeneous sedimentation. If large colonies of cyanobacteria are present, a pretreatment of the sample may be required using either a sonication bath or a sonication probe (preferred) for disrupting *Microcystis* spp. colonies or a Potter-Elvehjem for separating colonial filamentous cyanobacteria.
b. Check that the chamber is free from contamination from previous samples.
c. Gently invert the sample bottle 30–50 times to homogenise its contents.
d. Use a pipette to subsample 1 mL of the sample to be analysed (the sample may require to be concentrated or diluted to achieve an appropriate cell density in the chamber).

e. Fill the chamber with the sample at one corner with the coverslip at an angle across the chamber.
f. Move the coverslip to cover the entire chamber.
g. Allow the sample to settle for 15 minutes. To prevent any evaporation of the sample, place the chamber in a closed Petri dish containing a piece of moistened paper.

Sample Enumeration and Cell Population Density Calculation

a. The chamber may display a nonrandom distribution of phytoplankton units. It is thus advised to count the contents of the entire chamber if possible. If not, at least four rows of 50 squares should be counted.
b. Filaments or colonies should be recorded as objects. The average number of cells per filament or colonies should be estimated by counting the number of cells in at least 30 individuals at appropriate magnification.
c. The following formula should be used to calculate the number of cells per unit of volume:

$$\text{Cells. L}^{-1} = \frac{N * V}{A_f * N_f * d}$$

where:
N: total number of cells counted,
V: volume of the chamber in mm^3 (nominally 1000 mm^3),
A_f: area of a field in mm^2 (nominally 1 mm^2),
N_f: number of fields counted,
d: depth of the chamber in mm (nominally 1 mm).

SOP 1.3.4 Epifluorescence Microscopy Analysis

SOP 1.3.4.1 Introduction

Quantitative epifluorescence techniques, involving green or blue excitation (546 or 450 nm, respectively) after sample filtration on polycarbonate membranes, are commonly used to enumerate the picocyanobacterial fraction of phytoplankton communities (see also SOP 1.4). The method can be performed by direct counts or by image analysis. When pigment excitation with blue light is used, phycoerythrin-rich cells emit a yellow to orange light (Λ_{max} = 560 nm), while phycocyanin-rich cells appear dull red (A_{max} = 625 nm) or are not visible at all. When excited by green light, both phycoerythrin-rich and phycocyanin-rich picocyanobacteria emit an intense orange to red light. Both groups of picocyanobacteria can be easily distinguished from eukaryotic picoplankton and prochlorophytes, which fluoresce a very faint red or do not fluoresce at all [10, 11]. Computer-mediated procedures, based on digital recording of optical fields and automated image post-processing, allow the assessment of individual cell morphometry and, ultimately, to evaluate picocyanobacterial biovolume. The method also provides valuable information on *in situ* cell size distribution and taxonomic descriptions [12–14]. Epifluorescence microscopy combined with image analysis can also be applied to quantitative investigation of filamentous cyanobacteria and their morphometric analysis [15–19].

SOP 1.3.4.2 Experimental

Materials

a. Polycarbonate membrane filters (pore size 0.2 μm).
b. Filtration unit.

c. Vacuum pump.
d. Glass microscope slides and cover slips.
e. Immersion oil (type FF).
f. Epifluorescence microscope equipped with a mercury lamp, green filter set (excitation 546 nm, cut-off 580 nm) and magnification objectives (25×, 40×, and 100×).
g. For computerized image analysis, a CCD camera mounted on the camera column of the microscope.

Sample Fixation and Storage Use glutaraldehyde fixation or formaldehyde as previously described.

Slide Preparation

a. Filter a known volume of sample through a polycarbonate membrane filter under vacuum below 10 kPa.
b. Place a drop of immersion oil on a glass slide. Mount the filter on top of the oil drop.
c. Place a drop of immersion oil on the filter and place the coverslip.
d. Phytoplankton density should be of 10–20 units per counting field or grid under 1000× magnification. If a nonrandom distribution is observed or if cell density is inappropriate, a new sample should be prepared.

Epifluorescence Microscopy Analysis

a. Direct cell counts are performed using a 100× objective and a green filter set (546 nm excitation, 580–590 nm emission). Digital images of optical fields may be recorded using a high-definition CCD camera and stored to be processed by an automatic image analyser (see earlier for details).
b. When performing direct cell counts, *at least* 30 randomly selected fields or grids at high (1000×) magnification should be enumerated (i.e. corresponding to at least 400 units with the recommended sample concentration).
c. If appropriate, dividing cells in each grid may be counted to estimate the ratio of dividing cells to the total cell abundance. Cells with any visible sign of wall invagination, but not yet separated into two visibly distinct cells, should be considered to be dividing. Cell aggregates may also be present in the samples and are more easily estimated by image analysis of digitalized fields [13].

SOP 1.3.5 Flow Cytometry

SOP 1.3.5.1 Introduction

During the last two decades, flow cytometry microscopy (FCM) has been recognized as a powerful tool for the study of phytoplankton ecology, especially for the identification and characterisation of spatial and seasonal trends. Due to variations in light scattering and fluorescence emission (originating from differences in chlorophyll and phycobilin contents) of phytoplankton cells, the composition of mixed populations can be discriminated. FCM is especially suited for the study of small organisms including picophytoplankton, which are difficult to observe and count using conventional microscopy-based techniques [20].

The majority of commercially available instruments use an air-cooled argon gas laser, emitting a monochromatic beam of light at 488 nm with 15 mW of power. As particles or cells cross the light beam, light is scattered in various directions. Particle fluorescence originating from (i) cell autofluorescence, (ii) fluorescent *in situ* hybridization (FISH) techniques, or (iii) fluorochrome-labelled monoclonal antibodies may also be measured. The resulting signals are processed to gather information about the relative size of cells in the mixture (forward light scatter or FSC), their shape or internal complexity (side light scatter or SSC), and their fluorescence profile.

Most instruments are capable of measuring six parameters: FSC, SSC, and three fluorescent emission wavelengths (green, yellow-orange, and red) utilizing the first laser. A smaller diode-laser emitting red light at 635 nm is used to excite compounds that fluoresce above 650 nm, which are detected with a fourth fluorescent detector [21]. Flow cytometry can be coupled with digital image analysis (e.g. FlowCAM from Fluid Imaging Technologies Inc.), which enables various cell parameters to be measured, such as length, width, equivalent spherical diameter, and fluorescence. Additional information on flow cytometry may be found in [22–26].

SOP 1.3.5.2 Experimental

Materials

a. Flow cytometer equipped with air-cooling and emitting a monochromatic beam of light at 488 nm with 15 mW of power using an argon laser (three-colour FACSCalibur equipped with an autoloader and a FACSFlow Supply System or equivalent).
b. Calibration beads (yellow-green fluorescent beads; diameter: 0.92 μm).
c. Sheath fluid as recommended by the flow cytometer manufacturer.

Sample Preparation and Analysis

a. Samples are fixed with paraformaldehyde (0.2% final concentration), kept in the dark at 4°C for <48 h before being processed.
b. Set photomultipliers to quantify: red (>650 nm), orange (564–606 nm), and green fluorescence (515–545 nm).
c. Set the data acquisition software to record FSC, SSC, and fluorescence parameters. Peak heights should be recorded on four-decade logarithmic scales for green, orange, and red fluorescence. Peak area of orange fluorescence is recorded on a linear scale.
d. Prepare subsamples into plastic tubes containing 3 mL of samples and 100 μL of calibration beads stock suspension (final concentration of 3000 beads mL^{-1}). Vortex gently and load on the autoloader carousel.
e. Using the manufacturer's software, delay data acquisition for 15 s prior to the analysis of each sample to allow the autoloader to gently shake the sample (5 s), to flush the sample interrogation cuvette with new sample and to allow the flow rate to stabilize (10 s).
f. Calibrate the flow rate using an adaptation of the gravimetric protocol of [27]. Specifically, the flow rate corresponds to the regression slope between sheath fluid weight change and the operating time (for periods of between 0.5 and 12 min). The procedure should be repeated 6 times to estimate flow rate. The procedure should be repeated daily (or more often if fluctuations in sample voltage are observed). Sheath fluid quality and potential bias due to tube position should be assessed using duplicate control tubes added to every sample set.
g. Run the analysis at high flow rate for 3 min to analyse about 200 μL of sample.
h. Phytoplankton abundance is calculated from flow rates and the duration of analysis as this provides more reliable results than values calculated from bead concentrations.

SOP 1.4 Methods for Biovolume Measurement

SOP 1.4.1 Introduction

Phytoplankton cell size varies greatly among cyanobacterial genera and even between different individuals of the same species. For some species, mean cell biovolume measured in different sampling sites may show a 10-fold variation [28]. Biovolume measurements can be obtained from measurements of cell dimensions and corresponding stereometrical formulae. If epifluorescence is used, semiautomated image analysis may also be used.

SOP 1.4.2 Stereometrical Formulae

SOP 1.4.2.1 Introduction

Microscopic observation is a direct and convenient approach to perform cell biovolume calculation. A geometric shape is assigned to the species under study [28–31] and the dimensions necessary to calculate cell biovolume are obtained using a calibrated ocular micrometer or, if available, image analysis software.

SOP 1.4.2.2 Experimental

Materials

a. Upright microscope equipped with a movable stage and high-magnification objectives (40× and 100×).
b. Calibrated stage micrometer.
c. Calibrated ocular micrometer.
d. Image analysis software (if available).

Sample Preparation and Biovolume Measurements

a. Add a drop of the sample onto a microscope slide, cover with a coverslip and place the slide on the microscope stage.
b. Select an appropriate stereometrical formula for the species under study. A few examples frequently used for cyanobacteria are shown in Table SOP 1.2.
c. Measure the corresponding stereometrical dimensions (e.g. diameter, length) for individual cells (or trichomes when appropriate) on at least 30 individuals.
d. Calculate the mean cell/trichome biovolume using the corresponding stereometrical formula.

SOP 1.4.3 Semiautomated Image Analysis

SOP 1.4.3.1 Introduction

Epifluorescence microscopy, based on phycobiliprotein autofluorescence, can be combined with image processing to measure linear dimensions of either coccoid or filamentous cyanobacteria, and to ultimately estimate their biovolume in water samples. Several semi-automated or automated methods have been proposed to calculate biovolume of picocyanobacteria [13] and filamentous taxa [15–17]. These procedures involve a series of commands and computer operations common to most image analysis software. The method proposed in this section was developed for Baltic Sea cyanobacteria [16, 17].

Table SOP 1.2 *Some stereometric formulae used to estimate the biovolume of cyanobacterial cells*

Shape	Formula	Example
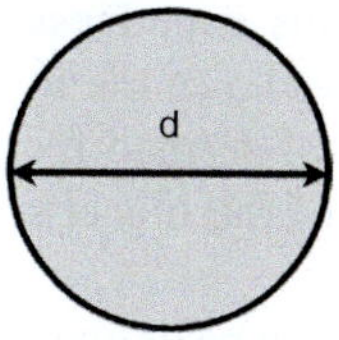 Sphere	$V = \frac{\pi \cdot d^2 \cdot h}{4}$	*Chroococcus, Gomphosphaeria, Merismopedia, Microcystis*
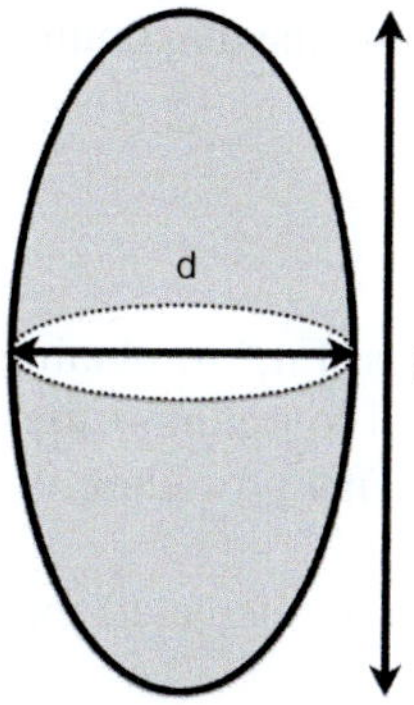 Prolate spheroid	$V = \frac{\pi \cdot d^2 \cdot h}{6}$	*Cyanodictyon, Gomphosphaeria, Lemmermanniella, Snowella, Woronichinia*
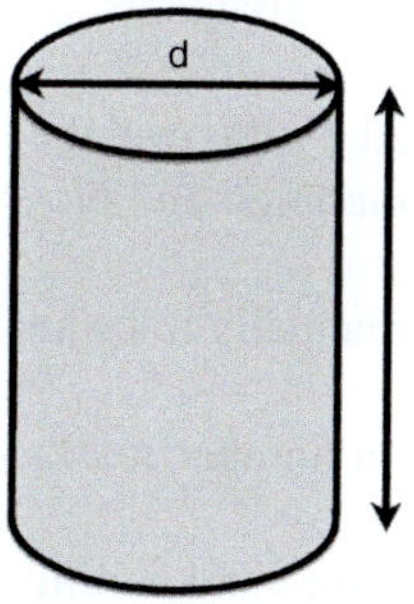 Cylinder	$V = \frac{\pi \cdot d^2 \cdot h}{4}$	*Achroonema, Aphanizomenon, Lyngbya, Nodularia, Nostoc, Oscillatoria, Planktolyngbya, Phormidium, Pseudoanabaena, Planktothrix*

SOP 1.4.3.2 Experimental

Materials

The procedure reported here involves a series of commands and computer operations, which are common to most image analysis software. In the present method, the Image ProPlus 3.0 software for Windows was used.

Data Acquisition

a. Acquire epifluorescence images from at least 30 random non-overlapping optical fields.
b. Keep the brightness setting constant throughout image acquisition using the gain level to standardise the procedure.
c. Calibrate the recorded images to ensure that subsequent measurements are stored in the same dimensional units. Save the calibration values in a calibration file and use these constants for image analysis.

Software Settings

a. Prior to image processing, define a set of morphometric parameters of detectable objects in the images (i.e. the dimensions to be measured).
b. Area, minimum and maximum diameter, perimeter, minimum and maximum radius, roundness index, size-width, and size-length are commonly selected to this purpose and saved as a file to be loaded at the beginning of the process. Minimum and maximum diameter are defined for most software as the length of the shortest and longest line joining two points of the object's outline and passing through the centroid.
c. If picocyanobacteria are to be measured, a fixed range for the minimum diameter can be defined between 0 and 2.3 μm in order to exclude from the analysis any object exceeding the maximum cell diameter of phototrophic picoplankton.

Image Analysis

a. Once an image is opened, recall the "calibration" and "setting" files.
b. The contrast of the image is automatically enhanced by using "Equalize/Best Fit," to obtain good detection of the cells/filaments/objects. Manual adjustment can be used to optimise the computer detection and highlight the objects below the machine threshold. This procedure is crucial for the accuracy of the following morphometric analysis.
c. Using the tool "Measure, Count/Size, Count," the set of selected parameters is calculated for detected objects.
d. All numerical data, figures, and morphometric measurements of individual cells, filaments, or filament segments are eventually saved in an Excel spread sheet.
e. If aggregates of picocyanobacteria are present, the aggregate's area is divided by the average area of the individual cells present in the same optical field. This allows enumeration and biovolume calculation of the aggregate-forming cells.
f. If different taxa of filamentous cyanobacteria are present in the same image, manually trace an AOI (area of interest) of the region of the image containing only one taxon.
g. Each trichome present in the AOI is automatically selected and highlighted, a number marks each object, and then the command [step (f)] can be operated.
h. Further manual interactions can be required to correct problems linked to curvature and coiling (e.g. *Nodularia*). Such coiled filaments can be manually split using the facility in order to obtain sections as straight as possible, and steps (f) and (g) can then be performed.
i. The presence of overlapping areas may require further manual adjustments. Overlapping regions of filaments will appear brighter due to autofluorescence enhancement. This can be corrected by drawing additional lines to delimit filament outlines. These regions, marked by additional numbers, have to be added to the final calculation of the biovolume. Diluting the samples before filtration can decrease the occurrence of crossing and overlapping filaments.

Biovolume Calculation

a. For coccoid cyanobacteria, biovolume is obtained using the formula for calculating the volume of a sphere, a spheroid, or a cylinder with hemispherical ends.
b. For filamentous cyanobacteria with straight trichomes, biovolume is obtained by calculating the volume of a cylinder whose base diameter (trichome width) and length (trichome length) are measured by the software for every trichome or section of trichome in the image.
c. To calculate the biovolume of a coiled filament, individual filaments are split into straight segments. The volume of each segment is then measured as described in (b). The trichome volume is then obtained by summing the volume of each individual segment.

References

[1] Hawkins, P.R., Holliday, J., Kathuria, A. and Bowling, L. (2005) Change in cyanobacterial biovolume due to preservation by Lugol's Iodine. *Harmful Algae* **4** (6), 1033–1043.

[2] Utermöhl, H. (1958) Zur vervollkommnung der quantitativen phytoplankton-methodik. *Mitteilung Internationale Vereinigung fuer Theoretische unde Amgewandte Limnologie* **9**, 1–38.

[3] EN 15204:2006 (2006) *Water Quality – Guidance standard on the enumeration of phytoplankton using inverted microscope (Utermöhl technique)*, European Committee for Standardization, Brussels.

[4] Reynolds, C.S. and Jaworski, G.H.M. (1978) Enumeration of natural *Microcystis* populations. *British Phycological Journal* **13** (3), 269–277.

[5] Joung, S.-H., Kim, C.-J., Ahn, C.-Y. *et al.* (2006) Simple method for a cell count of the colonial cyanobacterium, *Microcystis* sp. *Journal of Microbiology* **44** (5), 562–565.

[6] Hötzel, G. and Croome, R. (1999) A phytoplankton Methods Manual for Australian Freshwaters. Land and Water Resources Research and Development Corporation, Canberra. http://lwa.gov.au/products/pr990300

[7] Zohary, T. and Pais Madeira, A.M. (2001) Counting natural populations of *Microcystis aeruginosa*: a simple method for colony disruption into single cells and its effect on cell counts of other species. *Southern African Journal of Child and Adolescent Mental Health* **13** (2), 75–77.

[8] Humphries, S.E. and Widjaja, F. (1979) A simple method for separating cells of *Microcystis aeruginosa* for counting. *British Phycological Journal* **14** (4), 313–316.

[9] Edler, L. and Elbrächter, M. (2010) The Utermöhl method for quantitative phytoplankton analysis, in *Microscopic and Molecular Methods for Quantitative Phytoplankton Analysis* (eds. B. Karlson, C. Cusack C and Bresnan E.), UNESCO, Paris, pp. 13–20.

[10] Callieri, C. and Piscia, R. (2002) Photosynthetic efficiency and seasonality of autotrophic picoplankton in Lago Maggiore after its recovery. *Freshwater Biology* **47** (5), 941–956.

[11] Stomp, M., Huisman, J., Vörös, L. *et al.* (2007) Colourful coexistence of red and green picocyanobacteria in lakes and seas. *Ecology Letters* **10** (4), 290–298.

[12] Viles, C.L. and Sieracki, M.E. (1992) Measurement of marine picoplankton cell size by using a cooled, charge-coupled device camera with image-analyzed fluorescence microscopy. *Applied and Environmental Microbiology* **58** (2), 584–592.

[13] Albertano, P., Somma, D.D. and Capucci, E. (1997) Cyanobacterial picoplankton from the Central Baltic Sea: cell size classification by image-analyzed fluorescence microscopy. *Journal of Plankton Research* **19** (10), 1405–1416.

[14] Sherry, N.D. and Wood, A.M. (1995) Phycoerythrin-containing picocyanobacteria in the Arabian Sea in February 1995: diel patterns, spatial variability, and growth rates. *Deep Sea Research Part II Tropical Studies in Oceanography* **48** (6–7), 1263–1283.

[15] Walsby, A.E. and Avery, A. (1996) Measurement of filamentous cyanobacteria by image analysis. *Journal of Microbiological Methods* **26** (1–2), 11–20.

[16] Congestri, R., Federici, R. and Albertano, P. (2000) Evaluating biomass of Baltic filamentous cyanobacteria by image analysis. *Aquatic Microbial Ecology* **22** (3), 283–290.

[17] Congestri, R., Capucci, E. and Albertano, P. (2003) Morphometric variability of the genus *Nodularia* (Cyanobacteria) in Baltic natural communities. *Aquatic Microbial Ecology* **32** (3), 251–259.

[18] Almesjö, L. and Rolff, C. (2007) Automated measurements of filamentous cyanobacteria by digital image analysis. *Limnology and Oceanography Methods* **5**, 217–224.

[19] Zeder, M., Van den Wyngaert, S., Koster, O. *et al.* (2010) Automated quantification and sizing of unbranched filamentous cyanobacteria by model-based object-oriented image analysis. *Applied and Environmental Microbiology* **76** (5), 1615–1622.

[20] Cellamare, M., Rolland, A. and Jacquet, S. (2010) Flow cytometry sorting of freshwater phytoplankton. *Journal of Applied Phycology* **22** (1), 87–100.

[21] Nunez, R. (2001) Flow cytometry: principles and instrumentation. *Current Issues in Molecular Biology* **3** (2), 39–45.

[22] Li, W.K.W. and Wood, A.M. (1988) Vertical distribution of North Atlantic ultraphytoplankton: analysis by flow cytometry and epifluorescence microscopy. *Deep Sea Research Part A. Oceanographic Research Papers* **35** (9), 1615–1638.

[23] Moreira-Turcq, P.F., Cauwet, G. and Martin, J.M. (2011) Contribution of flow cytometry to estimate picoplankton biomass in estuarine systems. *Hydrobiologia* **462** (1–3), 157–168.

[24] Crosbie, N., Teubner, K. and Weisse, T. (2003) Flow-cytometric mapping provides novel insights into the seasonal and vertical distributions of freshwater autotrophic picoplankton. *Aquatic Microbial Ecology* **33** (1), 53–66.

[25] Wang, B., Liu, C.-Q., Wang, F. *et al.* (2009) Flow cytometric observation of picophytoplankton community structure in the cascade reservoirs along the Wujiang River, SW China. *Journal of Limnology* **68** (1), 53–63.

[26] Poulton, N.J. and Martin, J.L. (2010) Imaging flow cytometry for quantitative phytoplankton analysis – FlowCAM, in *Microscopic and Molecular Methods for Quantitative Phytoplankton Analysis* (eds. B. Karlson, C. Cusack C and Bresnan E.), UNESCO, Paris, pp. 47–54.

[27] Olson, R.J., Zettler, E.R. and DuRand, M.D. (1993) Phytoplankton analysis using flow cytometry, in *Handbook of Methods in Aquatic Microbial Ecology* (eds. P.F. Kemp, B.F. Sherr, E.B. Sherr and J.J. Cole JJ). Lewis Publishers, Boca Raton, pp. 175–186.

[28] Olenina, I., Hajdu, S., Edler, L., Andersson A. *et al.* (2006) Biovolumes and Size-Classes of Phytoplankton in the Baltic Sea. HELCOM, Helsinki. http://helcom.fi/Lists/Publications/BSEP106.pdf

[29] Rott, E. (1981) Some results from phytoplankton counting intercalibrations. *Schweizerische Zeitschrift für Hydrologie* **43** (1), 34–62.

[30] Hillebrand, H., Dürselen, C.-D., Kirschtel, D. *et al.* (1999) Biovolume calculation for pelagic and benthic microalgae. *Journal of Phycology* **35** (2), 403–424.

[31] Sun, J. and Liu, D. (2003) Geometric models for calculating cell biovolume and surface area for phytoplankton. *Journal of Plankton Research* **25** (11), 1331–1346.

SOP 2

Chlorophyll a Extraction and Determination

Claude Yéprémian[1], Arnaud Catherine[1], Cécile Bernard[1], Roberta Congestri[2], Tina Elersek[3], and Renata Pilkaityte[4]

[1] *UMR 7245 MCAM CNRS-MNHN, Muséum National d'Histoire Naturelle, Paris, France*
[2] *Department of Biology, University of Rome 'Tor Vergata', Italy*
[3] *Department of Genetic Toxicology and Cancer Biology, National Institute of Biology, Ljubljana, Slovenia*
[4] *Open Access Centre for Marine Research, Klaipeda University, Klaipeda, Lithuania*

SOP 2.1 Introduction

Chlorophyll a (Chl a) is a widely used and accepted indirect indicator of phytoplankton biomass including cyanobacteria. The method is less time consuming than microscopic biomass determination, and Chl a concentration is commonly used as an indicator of biomass during cyanobacterial blooms. Although less accurate than HPLC methods and less sensitive than fluorimetric ones, spectrophotometric techniques are easy to perform. Up-to-date sets of equations are provided, and a simple method using 100% ethanol for Chl a extraction is detailed.

SOP 2.2 Experimental

SOP 2.2.1 Materials

- Analytical grade 100% ethanol, saturated with $MgCO_3$ (around $5\,mg\,L^{-1}$)
- Hydrochloric acid (HCl), analytical grade, $1\,mol\,L^{-1}$ in MilliQ water
- Sodium hydroxide (NaOH), analytical grade, $1\,mol\,L^{-1}$ in MilliQ water
- (GF/C) glass fibre filters, 47-mm diameter
- Graduated 15-mL polypropylene centrifuge tubes

Handbook of Cyanobacterial Monitoring and Cyanotoxin Analysis, First Edition.
Edited by Jussi Meriluoto, Lisa Spoof and Geoffrey A. Codd.

- Thermo-regulated water bath or block heater
- Vortex Mixer
- Centrifuge (4000 rpm = 3220 × *g*)
- Ultrasonic bath
- Calibrated spectrophotometer suitable for readings at 632, 649, 665, 696, and 750 nm (for calibration see [1])
- Glass semi-microcuvettes or disposable cuvettes

SOP 2.2.2 Spectrophotometer Calibration

Absorbance measurements of the spectrophotometer can be checked by using a fresh solution of copper sulphate pentahydrate ($CuSO_4$, $5H_2O$) or more easily by using a neutral glass filter. Both methods are described in [1].

SOP 2.2.3 Sample Preparation

The following steps have to be performed under dimmed light to prevent Chl a degradation.

a. Filter 0.1 to 0.5 L of water sample through a GF/C filter. To avoid cell lysis, the maximum vacuum should not exceed 310 millibar (residual pressure: 700 millibar).
b. Dry the filter by pressing on absorbent paper and place in a 15-mL polypropylene centrifuge tube. Cover the tube with aluminum foil to avoid photochemical degradation.
c. Freeze-thaw the sample to induce cell lysis (filters can be stored for up to 7 days at –20°C or up to 3 months at –80°C if necessary).
d. Add 10 mL of 100% ethanol and vortex alternately the top and the bottom of the tube to break down the filter paper.
e. Place the tube for 10 min at 75°C for chlorophyll extraction (water bath or other heating device).
f. Sonicate for 10 min in an ultrasonic bath with tap water and ice (this step reduces the turbidity of the extracts).
g. For recalcitrant samples, extraction can be prolonged for 16 h at 4°C.
h. Centrifuge the sample for 10 min, at 10°C, 4000 rpm. Make sure that no filter particles remain in the upper part of the tube.

SOP 2.2.4 Spectrophotometric Readings and Calculation

a. Blank spectrophotometer with the extraction solvent.
b. Place the sample in the cuvette and record absorbance at 632, 649, 665, 696, and 750 nm.
c. Absorbance values at 632, 649, 665, and 696 nm should be corrected by subtracting the value recorded at 750 nm. The absorbance of samples at 750 nm is indicative of the residual nonspecific turbidity of the sample and should in all cases be below 0.010 a.u. Following correction, measured absorbance at 632, 649, 665, and 696 nm should range between 0.03 and 1.00.
d. *For Chl a analysis:* Use corrected absorbance values at 665 nm and apply Equation SOP 2.1 to obtain concentrations.
e. *For Chl a quantification including Chls b, c, and d:* use corrected absorbance values at 632, 649, 665, and 696 nm and apply Equation SOP 2.2 to obtain concentrations.

f. *For pheophytin a quantification:*

- For 1 mL of 100% ethanol extract, add 10 µL of 1 mol L^{-1} HCl in the cuvette, agitate gently, wait 2 min,
- Neutralize with 10 µL of 1 mol L^{-1} NaOH, agitate gently, wait 2 min,
- Record absorbance at 665 nm and 750 nm before acidification and after neutralization,
- Apply Equation SOP 2.3 to obtain pheophytin-corrected Chl a concentrations,
- Apply Equation SOP 2.4 to obtain pheophytin concentrations.

SOP 2.2.5 Equations

Equation SOP 2.1 [3]:

$$\text{Chl a } (\mu g\, L^{-1}) = \frac{11.90 \times \text{Abs}665 \times \text{Ve}}{\text{Vs} \times \text{I}}$$

where

Ve = volume of ethanol extract (mL)
Vs = volume of water sample (litre)
I = path length of cuvette (cm)

Equation SOP 2.2 [4]:

$$\text{Chl a } (\mu g\, L^{-1}) = \frac{\left[(0.06 \times \text{Abs}632) - (4.52 \times \text{Abs}649) + (13.29 \times \text{Abs}665) - (1.74 \times \text{Abs}696)\right] \times \text{Ve}}{\text{Vs} \times \text{I}}$$

Equation SOP 2.3 [3, 5, 6]:

$$\text{Chl a}_c\ (\mu g\, L^{-1}) = \frac{28.44 \times (\text{Abs } 665\text{b} - \text{Abs } 665\text{a}) \times \text{Ve}}{\text{Vs} \times \text{I}}$$

where

Abs 665b: absorbance at 665 before acidification
Abs 665a: absorbance at 665 after neutralization
Chl $a_{c:}$ pheophytin-corrected Chl a concentration

Equation SOP 2.4 [3, 5, 6]

$$\text{Pheophytin a } (\mu g\, L^{-1}) = \frac{20.47 \times \text{Abs } 665\text{a} \times \text{Ve}}{\text{Vs} \times \text{I}} - \text{Chl a}_c$$

Specific equations to quantify Chls b, c, and d can be found in [4].

SOP 2.2.6 Performance Indicators

This analytical method was validated using typical validation characteristics (Table SOP 2.1).

Table SOP 2.1 *Validation assay*

Method	Chl a extraction and determination	
SOP	2	
Sample matrix	Abiogenic and biogenic particulate matter	
Performance Parameter and Results		
Repeatability[a]: Intraday variability	% RSD_r	Equation SOP 2.1: 1.4 Equation SOP 2.2: 1.5 Equation SOP 2.3: 1.7 Equation SOP 2.4: 12.2
Reproductibility[b]: Interlaboratory variability	% RSD_R	Equation SOP 2.1 : 4.5 Equation SOP 2.2 : 2.3 Equation SOP 2.3 : 4.1 Equation SOP 2.4 : 16.4
LOD[c] Limit of detection Equation SOP 2.1	µg Chl a mL^{-1} ethanol	0.12 µg mL^{-1}
LOQ[c] Limit of quantification Equation SOP 2.1	µg Chl a mL^{-1}ethanol	0.33 µg mL^{-1}
Range of linearity[d] Equation SOP 2.1	µg Chl a mL^{-1} ethanol	0.33–12 µg mL^{-1}
Selectivity 1[e] Without spiking pure Chl a	Eq. dry wt cyanobacteria detected	0.43% eq. DW cyanobacteria, < LOD
Selectivity 2[f] By spiking pure Chl a	Chl a% recovery	105.2% (1.5 µg Chl a mL^{-1}) 99.3% (3 µg Chl a mL^{-1})
Efficiency of extraction Equation SOP 2.1	[first extraction]/[first extraction + second extraction])%	>90%

[a, b] Four different laboratories were involved and analyzed, respectively, five, four, four, and four samples of *Microcystis aeruginosa* (PMC 570-08). Each sample corresponded to 5 mL of live culture filtered on 24/10/2013 and freeze-dried on 25/10/2013.
[c] Determined after pooling three independent calibration curves using pure Chl a standard within a concentration range of 0–30 mg L^{-1} in ethanol 100%; 11 calibration points were defined and concentrations were calculated according to [7].
[d] Determined using pure Chl a calibration curves, *Spirulina* powder capsules and live cultures of *Euglena gracilis* (72,000 µg Chl a L^{-1}).
[e] Determined by analyzing a live sample of *Aspergillus versicolor* (which does not contain Chl a); 87.5 mg (dry matter) was extracted in 100% ethanol, with the assumption that cyanobacteria contain 1% dry weight of Chl a.
[f] As in [e], but the sample was spiked with two different concentrations of pure Chl a (µg Chl a mL^{-1} ethanol).

References

[1] International Accreditation New Zealand (IANZ). (2005) Technical Guide – UV/VIS. Spectrophotometer Calibration Procedures, second edition.

[2] Algalweb. (2012) http://algalweb.net/Methods/Chlorophyll-ethanol-extraction.pdf. Accessed 6 March 2015.

[3] Ritchie, R.J. (2006) Consistent sets of spectrophotometric chlorophyll equations for acetone, methanol and ethanol solvents. *Photosynthesis Research* **89**, 27–41.

[4] Ritchie, R.J. (2008) Universal chlorophyll equations for estimating chlorophylls a, b, c, and and total chlorophylls in natural assemblages of photosynthetic organisms using acetone, methanol, or ethanol solvents. *Photosynthetica* **46** (1), 115–126.

[5] ISO 10260. (1992) *Water Quality: Measurement of biochemical parameter: Spectrometric* determination of the chlorophyll-a concentrations, International Standards Organisation, Geneva

[6] Sartory, D.P. and Grobbelaar,U.(1984) Extraction of chlorophyll a from freshwater phytoplankton for spectrophotometric analysis. *Hydrobiologia* **114**, 177–187.

[7] Miller, J. and Miller J.C. (2005) Statistics and Chemometrics for Analytical Chemistry, 5th edition, Pearson Education Limited, Harlow, UK.

SOP 3

Phycocyanin Extraction and Determination

Claude Yéprémian[1], Arnaud Catherine[1], Cécile Bernard[1], Roberta Congestri[2], Tina Elersek[3], and Renata Pilkaityte[4]

[1] UMR 7245 MCAM CNRS-MNHN, Muséum National d'Histoire Naturelle, Paris, France
[2] Department of Biology, University of Rome 'Tor Vergata', Italy
[3] Department of Genetic Toxicology and Cancer Biology, National Institute of Biology, Ljubljana, Slovenia
[4] Open Access Centre for Marine Research, Klaipeda University, Klaipeda, Lithuania

SOP 3.1 Introduction

Cyanobacteria contain specific pigments that differ from most eukaryotic phytoplanktonic cells. These light-harvesting pigments, the phycobiliproteins, are hydrophilic and can be estimated using spectrophotometry [1, 3]. Among the phycobiliproteins, phycocyanin provides a convenient marker for cyanobacterial presence in water samples and may be used as an indicator of cyanobacterial biomass. However, phycocyanin may also be found in Cryptophyta, Rhodophyta, Glaucophyta, and Dinophyta. The method described in this SOP provides a simple and rapid procedure to estimate phycocyanin concentration in water samples.

SOP 3.2 Experimental

SOP 3.2.1 Materials

- Sodium chloride (NaCl)
- Potassium chloride (KCl)
- Sodium phosphate anhydrous dibasic (Na_2HPO_4)

Handbook of Cyanobacterial Monitoring and Cyanotoxin Analysis, First Edition.
Edited by Jussi Meriluoto, Lisa Spoof and Geoffrey A. Codd.

- Potassium phosphate anhydrous monobasic (KH_2PO_4)
- EDTA disodium salt
- 47-mm GF/C glass fibre filters
- Graduated 15-mL polypropylene centrifuge tubes
- Centrifuge (4000 rpm = 3220 × *g*)
- Ultrasonic bath
- Calibrated spectrophotometer suitable for readings at 620, 650, and 750 nm
- Glass semi-microcuvettes or disposable cuvettes

The absorbance measurements of the spectrophotometer can be checked by using a fresh solution of copper sulphate pentahydrate ($CuSO_4$, $5H_2O$) or more easily by using a neutral glass filter. Both methods are described in [2].

SOP 3.2.2 Saline Buffer Preparation

a. To a final volume of 1 L of MilliQ water, successively add and dissolve 8.77 g of NaCl (final concentration 0.15 mol L^{-1}), 2.01 g of KCl (final concentration 27 mmol L^{-1}), 11.36 g of Na_2HPO_4 (final concentration 80 mmol L^{-1}), 2.72 g of KH_2PO_4 (final concentration 20 mmol L^{-1}) and 3.72 g of Na_2EDTA (final concentration 10 mmol L^{-1}). Increasing the pH to 7.45–7.50 will help the solubilisation of the latter.
b. Adjust the pH to 7.45–7.50 using 3–4 mL of 5 mol L^{-1} NaOH.
c. Store the solution at 4 °C for up to 6 months.

SOP 3.2.3 Sample Preparation

The following steps should be performed under dim light to avoid the photochemical degradation of phycocyanin.

a. Filter 0.1–0.5 L of water sample through a GF/C filter. To avoid cells lysis, the maximum vacuum should not exceed 310 millibar (residual pressure: 700 millibar).
b. Dry the filter by pressing on absorbent paper and place in a 15-mL polypropylene centrifuge tube. Cover the tube with aluminum foil to avoid photochemical degradation.
c. Freeze-thaw the sample to induce cell lysis (filters can be stored for up to 7 days at –20°).
d. Add 10 mL extraction saline buffer and vortex alternately to the top and the bottom of the tube to break down the filter paper.
e. Place the tube at 4°C for 16 hours.
f. Sonicate for 10 min in an ultrasonic bath with tap water and ice (this step reduces the turbidity of the extracts).
g. Centrifuge the sample, 20 min, at 4°C, 4000 rpm (3220 × *g*). Make sure that no filter particles remain in the upper part of the tube.

SOP 3.2.4 Spectrophotometric Readings

a. Blank spectrophotometer with the extraction saline buffer.
b. Absorbance values at 620 and 650 nm should be corrected by subtracting the value recorded at 750 nm. The absorbance of samples at 750 nm is indicative of the residual nonspecific turbidity of the sample and should in all cases be below 0.010 a.u. Readings should range between 0.03 and 1.5 for the corrected absorbance units.

SOP 3.3 Calculation

The following equation was proposed by [3]:

$$\text{Phycocyanin}\left(\text{mg L}^{-1}\right)=\frac{\left(\text{Abs }620-0.7\times\text{Abs }650\right)\times\text{Ve}}{7.38\times\text{Vs}\times\text{I}}$$

where
Ve = volume of buffer extract (mL)
Vs = volume of water sample (litre)
I = path length of cuvette (cm)

As Cyanobacteria may contain other phycobiliproteins in addition to phycocyanin, equations allowing the quantification of allophycocyanin (APC) and phycoerythrin (PE) can be found in [3].

SOP 3.4 Performance Indicators

This analytical method was validated using typical validation characteristics (Table SOP 3.1).

Table SOP 3.1 *Validation assay*

Method	Phycocyanin (PC) extraction and determination	
SOP	3	
Sample matrix	Abiogenic and biogenic particulate matter	
Performance parameter and results		
Repeatability[a]: Intraday variability	m, mg L^{-1} culture % RSD_r	21.4 2.8
Reproductibility[b]: Interlaboratory variability	m, mg L^{-1} culture % RSD_R	19.7 16.6
LOD[c] (limit of detection)	PC μg mL^{-1} saline buffer	1.2
LOQ[c] (limit of quantification)	PC μg mL^{-1} saline buffer	4.0
Range of linearity[d]	PC μg mL^{-1} saline buffer	4–180
Selectivity 1[e] without spiking pure phycocyanin	PC μg L^{-1} culture detected	51.5 μg L^{-1} culture *Euglena gracilis* = 24× lower than LOD
Selectivity 2[f] by spiking pure phycocyanin	PC % recovery	83% (5 μg mL^{-1}) 86% (10 μg mL^{-1})
Efficiency of extraction	[first extraction]/[first extraction + second extraction]) %	>90 %

[a,b] Two laboratories were involved and analyzed, respectively, five and six samples of *Microcystis aeruginosa* (PMC 570-08). Each sample corresponded to 5 mL of live culture filtered on 24/10/2013 and freeze-dried on 25/10/2013.
[c] Determined after pooling three independent calibration curves using a pure PC standard within a concentration range of 0–40 mg L^{-1} in saline buffer. Nine calibration points were defined and concentrations were calculated according to [4].
[d] Determined using pure PC calibration curves and *Spirulina* powder capsules.
[e] Determined by analyzing a live sample of *Euglena gracilis* (72,000 μg Chl a L^{-1}, which does not contain PC).
[f] As in [e], but samples were spiked with two different concentrations of PC standard (5 and 10 μg mL^{-1} buffer).

References

[1] Bennett, A. and Bogorad, L. (1973) Complementary chromatic adaptation in a filamentous blue-green alga. *The Journal of Cell Biology* **58**, 419–435.
[2] International Accreditation New Zealand (IANZ). (2005) *Technical Guide – UV/VIS. Spectrophotometer Calibration Procedures*, second edition.
[3] Tandeau de Marsac, N. and Houmard J. (1988) Complementary chromatic adaptation: physiological conditions and action spectra, in *Methods in Enzymology* (eds. L. Packer, A.N. Glazer), Academic Press, London/NewYork, **167**, pp 318–328.
[4] Miller, J. and Miller J.C. (2005) *Statistics and Chemometrics for Analytical Chemistry*, 5th edition, Pearson Education Limited, Harlow, UK.

SOP 4

Analysis of Picocyanobacteria Abundance in Epifluorescence Microscopy

Iwona Jasser[1] and Cristiana Callieri[2]

[1] *Faculty of Biology, University of Warsaw, Poland*
[2] *CNR – Institute of Ecosystem Study, Verbania, Italy*

SOP 4.1 Introduction

Picocyanobacteria, the smallest cyanobacteria in the size range 0.2–2.0 μm, are both numerous and ubiquitous in freshwater, brackish and marine environments. Their contribution to primary production decreases with the increase of trophic level; however, they were shown to form persisting blooms in some eutrophicated coastal lagoons [1]. Picocyanobacteria have been shown to produce hepatotoxins of the microcystin group, neurotoxins (β-*N*-methylamino-L-alanine), lipopolysaccharides, and taste and odour compounds, as well as cytotoxic and antimicrobial compounds. These findings suggest that picocyanobacteria can be a new potential, and until now overlooked, source of cyanotoxins and other bioactive compounds in freshwaters and marine environments and, further, that microcystin occurrence in marine environments may be more common than recognized so far [see Chapter 3].

Because of their minute size, picocyanobacteria are difficult to be counted using an inverted microscope. Instead, they are enumerated by epifluorescence microscopy, which distinguishes phototrophic cells from aplastidic ones by using the natural autofluorescence of phycobiliprotein pigments and chlorophyll a, through their autofluorescence [2, 3]. Depending on the prevailing pigment and the light used for excitation, the cells fluoresce in different colours (Table SOP 4.1). Picocyanobacteria can be also identified and enumerated by flow cytometry. This aspect is discussed in the COST handbook [4] concerning molecular detection of toxic cyanobacteria and cyanotoxins.

Handbook of Cyanobacterial Monitoring and Cyanotoxin Analysis, First Edition.
Edited by Jussi Meriluoto, Lisa Spoof and Geoffrey A. Codd.

Table SOP 4.1 Filter sets and fluorescence of light-harvesting pigments in epifluorescence microscopy

Prevalent pigment	Filter set	Fluorescence colour
Phycocyanin	Blue	Dim red or invisible
	Green	Bright red
	CY3	Bright red
Phycoerythrin	Blue	Yellow
	Green	Orange/red
	CY3	Yellow/orange
Chlorophyll *a*	Blue	Red
	Green	Red
	CY3	Faded red

SOP 4.2 Experimental

SOP 4.2.1 Materials

a. Picocyanobacteria samples
b. 20% formaldehyde v/v (0.2 μm filtered) buffered with sodium cacodylate 0.1 M (final concentration 2% v/v), stored in darkness at 4°C
c. DAPI (4′,6-diamidino-2-phenylindole (final concentration 2% v/v) [5]
d. Polycarbonate membrane filters (0.2-μm pore size)
e. Filtration units for 25-mm-diameter polycarbonate membranes with 25 (Millipore, Sartorius) connected to vacuum pump
f. Cargille immersion oil for epifluorescence microscopy
g. 50% v/v glycerol
h. Microscope slides
i. Cover slips
j. 5-mL pipette
k. Disposable tips
l. Distilled water

SOP 4.2.2 Equipment

a. Epifluorescence microscope with 100- or 50-W halogen lamp or with LED (e.g. NIKON, Olympus, Zeiss, Leica)
b. 100× magnifying objective, plan Fluor and if possible additional lens (e.g. OPTOVAR) for 1.25×
c. Filter sets:
   Blue (Ex450–490 nm, DM510 nm, BA520 nm)
   Green (Ex510–560 nm, DM580 nm, BA590 nm)
   Narrow banded green (Ex530–560 nm, DM570 nm, BA573–648 nm)
   UV – (Ex365 nm, DM395 nm, BA420 nm).
d. Image analysis system: colour digital camera (8 megapixel)
e. Computer and software for image analysis (e.g. NIKON – Lucia or Nis Elements, Image-PRO plus, Media Cybernetics)

SOP 4.2.3 Procedure

a. In order to count picocyanobacteria a raw, unfiltered water sample is taken in a vertical profile, including to depths with 1% surface PAR or less.
b. Samples are stored in dark glass bottles in a cool place. In the laboratory the samples should be counted immediately or fixed with 20% formaldehyde buffered with sodium cacodylate 0.1 mol L^{-1} [6] to a final concentration of 1% v/v.
c. From 1 to 10 mL of samples (depending on the expected number of picocyanobacterial cells) are filtered under low vacuum (~100 mm Hg) onto membranes of 0.2-μm pore size.
d. The membrane is taken by forceps, dried for about 30 s, and mounted on the microscope slide on a drop of Cargille oil or 50% glycerol–water solution. Then an additional drop of oil or glycerol is added on top of the membrane and this is covered by a coverslip. The slide with the membrane has to be turned upside down and pressed slightly onto absorbent paper to remove excess oil/glycerol.
e. A drop of immersion oil is put onto the coverslip and the preparation observed under an epifluorescence microscope at 1000–1250×.
f. Manual or automated counting may be used for picocyanobacteria. For manual counting at least 400 cells or 20 fields, should be analysed. For automated counting snapshots of at least 10 fields, with densely, but evenly distributed cells should be taken by colour digital camera. Dividing cells should be separated in the editing option and then the cells counted automatically. If PC and PE fluorescing cells are in the same sample, they should be counted separately. PE cells are enumerated in blue light, while PC cells are better detected in green light. Both PE- and PC-containing cells can be identified and counted using a green narrow banded CY3 filter (Fig. SOP 4.1).
g. To calculate the biovolume of picocyanobacteria the snapshots of cells are analysed in an image analysis system according to the manufacturer's instructions.
h. Equations to calculate number of cells and biomass of picocyanobacteria are given below.

SOP 4.2.4 Further Notes and Biomass Calculation

a. Usually formaldehyde is used but other possible fixatives, including glutaraldehyde, have been suggested by various authors [7].
b. The samples should be counted immediately after they are brought to the laboratory or after fixation within 1 week, otherwise the cells lose fluorescence. Alternatively they can be fixed, filtered onto 0.2 μm membranes, mounted on a slide and kept frozen (–20°C) until they are analysed.
c. The best filters for picocyanobacterial enumeration are white polycarbonate membranes (e.g. Nuclepore, Unipore, etc.). Aluminium oxide filters (e.g. Anopore) and black polycarbonate

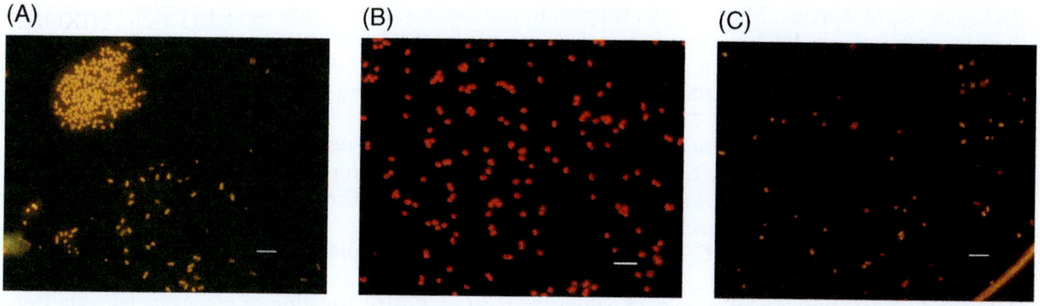

Figure SOP 4.1 *PE cells in blue filer (A), PC cells in green filter (B), and PE and PC cells in narrow banded green filter CY3 (C) (photographs by I. Jasser and C. Callieri)*

membranes are also used in case of subsequent analysis of carbon or counting with DAPI respectively [8].

d. For identification and enumeration, blue, green, or narrow banded green filter sets are used [7, 9]. The narrow banded green filter set (CY3) allows the enumeration of PE- and PC-containing cells under the same excitation.
e. Old samples can be investigated after DAPI staining in UV light (final concentration, 0.1 μg mL^{-1}) [7].
f. Equation 1 is used to calculate abundance of picocyanobacteria:

$$1: A = F \times N / V \times 0.95$$

Where:
A = cell abundance (cells mL^{-1})
F = conversion factor: filtration area of the apparatus/microscopic field area
N = average cell number per field
V = filtered volume (mL)
0.95 = dilution correction factor for fixative addition

g. Cell volumes of picocyanobacteria are calculated as ellipsoids from cell length and width (obtained by image analysis). Cell volumes are converted to carbon using a conversion factor of 200 fg C μm^{-3} [10] according to Equation 2:

$$2: B = \left(N \times V\right) \times 0.0002$$

Where:
B = biomass (μg C L^{-1})
N = cell abundance (cells mL^{-1})
V = mean cell biovolume (μm^3 $cell^{-1}$)
0.0002 = conversion factor

References

[1] Callieri, C., Cronberg, G. and Stockner, J. (2012) Freshwater Picocyanobacteria: Single Cells, Microcolonies and Colonial Forms. In: *Ecology of Cyanobacteria: Their Diversity in Time and Space*. 2nd edn. (ed B.A. Whitton), Springer Publishers, pp. 229–269.
[2] Wood, A.M., Horan, P.K., Muirhead, K. et al. (1985) Discrimination between types of pigments in marine *Synechococcus* spp. by scanning spectroscopy, epifluorescence microscopy and flow cytometry. *Limnology and Oceanography* **30**, 1303–1315.
[3] Ernst, A. (1991) Cyanobacterial picoplankton from Lake Constance isolation by fluorescence characteristics. *Journal of Plankton Research* **13** 1307–1312.
[4] Haande, S., M. Gugger, M., Jasser, I., Rippka, R. and Svircev, Z. Isolation and cultivation of strains & isolation techniques. In: *A handbook on molecular tools for the detection and quantification of toxigenic cyanobacteria* (eds. K. Sivonen, A. Wilmotte and R. Kurmayer).
[5] Porter, K.G. and Feig, Y.S. (1980) The use of DAPI for identifying and counting aquatic microflora. *Limnology and Oceanography* **25**, 943–948.
[6] Hayat, M.A. (1981) *Fixation for Electron Microscopy*. Academic Press, New York, 500 pp.
[7] MacIsaac, E.A. and Stockner, J.G. (1993) Enumeration of phototrophic picoplankton by autofluorescence microscopy In *Handbook of Methods in Aquatic Microbial Ecology* (eds. P.F. Kemp, B.F. Sherr, E.B. Sherr and J.J. Cole). Lewis Publishing, pp. 187–197.
[8] Callieri, C. and Stockner, J.G. (2002) Freshwater autotrophic picoplankton: a review. *Journal of Limnology* **61**, 1–14.
[9] Jasser, I., Karnkowska-Ishikawa, A., Kozłowska, E. *et al.* (2010) Isolation of picocyanobacteria from Great Mazurian Lake System – comparison of two methods. *Polish Journal of Microbiology* **59**, 21--31.
[10] Weisse, T.: Dynamics of autotrophic picoplankton in marine and freshwater ecosystems. In *Advances in Microbial Ecology*, (ed. J.G. Jones), Plenum Press, New York, Vol. **13**. pp. 327–370.

SOP 5

Estimation of Cyanobacteria Biomass by Marker Pigment Analysis

Jean-Pierre Descy

Research Unit in Environmental and Evolutionary Biology (URBE), University of Namur, Belgium

SOP 5.1 Introduction

HPLC analysis of marker pigments is a powerful tool for characterization and quantification of freshwater and marine phytoplankton [1–3]. The method is based on the chromatographic separation of pigment markers that are specific at the phylum or class level, from extracts obtained by a procedure similar to that of SOP 2. Owing to the solvent used for pigment extraction, and to the nature of most phytoplankton pigments, the markers generally used are chlorophylls and carotenoids, chiefly xanthophylls, which are often specific to one class or few classes [2]. Once the identification and quantification of pigments have been achieved, estimates of abundance of phytoplankton groups is obtained by computing techniques that assign a given pigment to one class or several phytoplankton classes and calculate a contribution of those groups to total chlorophyll a, based on marker:chlorophyll a ratios [1, 3, 4].

Quantification of biomass among phytoplankton taxa is not straightforward, however. Indeed, if some carotenoid markers are specific to a given class, most markers are distributed among several classes, and some variation in pigment composition in a given class may occur. Moreover the amount of pigment per cell may vary depending on cell response to environmental factors (e.g. light and nutrients), affecting the marker:chlorophyll a ratio. These problems can be circumvented by careful data processing using algorithms which take into account these variations. Also, intercalibration with other methods for assessment of phytoplankton biomass (see SOP 1) may be helpful.

Handbook of Cyanobacterial Monitoring and Cyanotoxin Analysis, First Edition.
Edited by Jussi Meriluoto, Lisa Spoof and Geoffrey A. Codd.

Despite these potential shortcomings, pigment analysis has been used in countless marine and freshwater studies, not only for the automated analysis of large numbers of samples in surveys of phytoplankton assemblages – which would not be feasible with microscopy – but also in studies of zooplankton grazing, of fate of the organic matter in aquatic systems, and in palaeolimnology, thanks to the analysis of pigments or their degradation products in lake sediment. Still, the taxonomic resolution remains low, and the pigment approach is best combined with microscopy [3, 5], flow cytometry, or DNA sequencing.

Here, we present a method for quantification of cyanobacteria biomass based on HPLC analysis of phytoplankton extracts. Two "pigment-types" of cyanobacteria have been described: the "*Synechococcus* type," or type 2 (T2) [6], which contains β,β-carotene and zeaxanthin, and type 1 (T1), composed mainly of freshwater colonial and filamentous taxa, which contains in addition variable amount of carotenoids found only in cyanobacteria such as echinenone, canthaxanthin, myxoxanthophyll, and others.

SOP 5.2 Experimental

SOP 5.2.1 Sample Processing and Pigment Extraction

Several procedures similar to that described in SOP 2 can be used for pigment extraction from water samples, as detailed in [7]. However, if extraction can be done with ethanol or methanol, 90% acetone is preferred as it allows best pigment preservation after extraction. It is recommended to proceed with filtration of water samples and pigment extraction on the sampling day and to protect the samples from light during processing. The extracts should then be kept in dark vials at –25°C until analysis. Alternatively, the filters can be frozen in liquid nitrogen and kept frozen until extraction. To optimise pigment extraction, it is recommended to submit the vials containing the filters and the extraction solvent to sonication in a sonication bath.

SOP 5.2.2 Materials

a. Methanol, HPLC grade
b. Acetonitrile, HPLC grade
c. MilliQ water
d. Ethyl acetate, HPLC grade
e. Ammonium acetate $0.5\,mol\,L^{-1}$
f. C_{18} HPLC column, such as Waters NovaPak C18, 4-μm particle size, 250×4.6mm; a compatible guard column is also required.
g. Borosilicate glass chromatographic vials: e.g. 1.5mL dark glass with writing surface. For small sample volumes, borosilicate glass inserts can be used.
h. Pigment standards: most standards of marker pigments can be provided by DHI, Denmark (c14@dhigroup.com).

SOP 5.2.3 Special Equipment

a. High-performance liquid chromatograph equipped with high-pressure or low-pressure gradient pump, autosampler, column oven, and photodiode-array (PDA) detector
b. Chromatography analysis software

SOP 5.2.4 HPLC Mobile Phase [7]

a. HPLC mobile phase component A: 80:20 methanol:0.5 mol L^{-1} ammonium acetate
b. HPLC mobile phase component B: 90:10 acetonitrile:water
c. HPLC mobile phase component C: ethyl acetate
d. Disposal of spent solvents should conform to local regulations.

SOP 5.2.5 Chromatography

SOP 5.2.5.1 General Procedure

a. The HPLC system should be set up as described in the manufacturer's instructions including degassing, priming, and changing columns.
b. Always use a guard column. Change the guard column if the back-pressure rises or peak forms deteriorate.
c. Set column oven at 30°C.
d. Change the HPLC gradually up to starting conditions and allow to condition.
e. Chromatograph the samples and standards according to the recommended HPLC gradients (see below); use 10- to 50-μL injections.
f. Analyse the chromatogram. Compare retention times and spectra to standards.

Calculate the pigment concentration using the pigment standards used for the calibration.

SOP 5.2.5.2 HPLC Gradient

The gradient provided below (Table SOP 5.1) allows separation of critical pigments such as lutein (green algae) and zeaxanthin (abudant in cyanobacteria), and is convenient for a Waters HPLC system, but other methods are available. For a comparison of methods, see [6].

Table SOP 5.1 *Example of analytical gradient protocol [7]*

Time (min)	Flow rate (mL min^{-1})	%A	%B	%C	Conditions
0	1.0	100	0	0	Injection
4	1.0	0	100	0	Linear gradient
18	1.0	0	20	80	Linear gradient
21	1.0	0	100	0	Linear gradient
24	1.0	100	0	0	Linear gradient
29	1.0	100	0	0	Equilibration

SOP 5.2.5.3 Example of Chromatogram

An illustrative chromatogram of phytoplankton pigments can be found in Figure SOP 5.1.

SOP 5.2.5.4 Cyanobacteria Biomass Determination with CHEMTAX

Several methods can be used to infer biomass of phytoplankton classes from marker pigment concentrations [7]. CHEMTAX [4] is a matrix factorisation program, which estimates the contribution of each specified phytoplankton class to the total chlorophyll a concentration in water samples. CHEMTAX allocates specific pigment markers to phytoplankton classes and

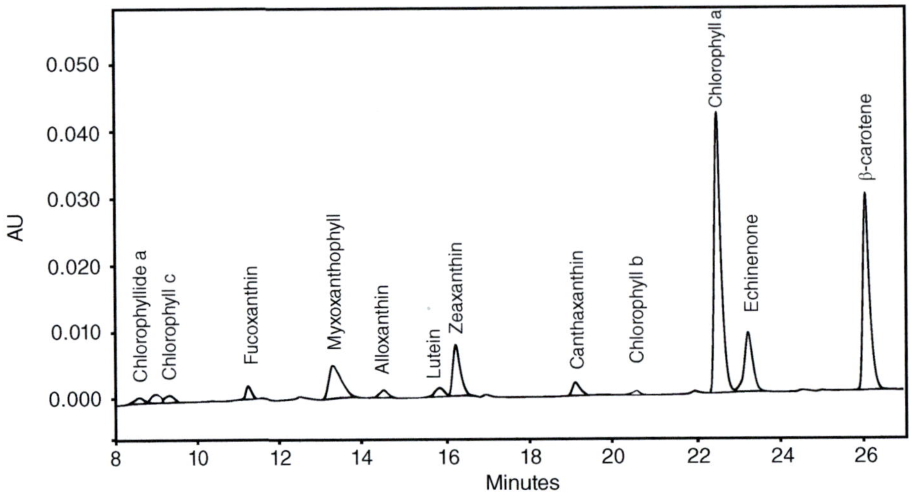

Figure SOP 5.1 *Chromatogram (a.u. [absorbance units]) of an extract of a phytoplankton sample from a eutrophic lake, dominated by cyanobacteria. Cyanobacterial pigments are myxoxanthophyll, zeaxanthin, canthaxanthin, and echinenone. Also note the large peak of β-carotene, abundant in cyanobacteria. Only myxoxanthophyll and echinenone are specific to cyanobacteria, so that further processing of pigment concentrations is needed for correct biomass assessment*

can solve the problem of pigments that are distributed among several phytoplankton classes. For instance, zeaxanthin, a key marker pigment of cyanobacteria, is shared with green algae, whereas echinenone is specific to cyanobacteria T1. Therefore, the data processing must be carried out for all phytoplankton classes to achieve correct estimates of cyanobacteria biomass.

CHEMTAX requires several input matrices. The input data matrix (Table SOP 5.2a) contains the pigment concentrations ($\mu g L^{-1}$) in a set of samples. If data originate from sampling a stratified water column, it is recommended to process the data in separate depth bins, in order to take into account a possible variation of markers:chlorophyll a ratios with the light climate and/or the nutrient concentrations. Samples containing more than 25% chlorophyll a degradation products (phaeophorbides a and phaeophytins a) should be discarded.

A second matrix (Table SOP5.2b) contains initial values of the markers:chlorophyll a ratios for the expected phytoplankton classes. CHEMTAX will start from these initial values and adjust the ratio values that fit the input data.

A third matrix, the ratio limit matrix, contains limits within which CHEMTAX is allowed to adjust the ratios [4].

A typical CHEMTAX run performs several successive iterations, until the residues are found stable, and provides an output ratio matrix containing the final ratio values (and the deviations from the initial values) and the output data matrix, with the contribution of each phytoplankton class to total chlorophyll a (Table SOP 5.3). The results are expressed as a relative proportion of total chlorophyll a or as an absolute biomass values, in chlorophyll a equivalents (μg Chl a L^{-1}). If the deviations of the final ratios from the initial ratios are too high (>20%) for some pigments, it is recommended to make new CHEMTAX runs using the final ratio matrix as initial ratio matrix, until deviations are minimised.

Table SOP 5.2 *Examples of input matrices in a CHEMTAX run*

(a) Input data matrix: pigment concentration (μg L^{-1})

Date	peri	fuco	neo	myxo	viol	ddx	allo	lut	zea	crox	echi	acar	chl_c	chl_b	chl_a
17-Apr-08	0.000	5.065	0.048	0.110	0.306	1.726	2.999	0.256	0.072	0.140	0.000	0.509	0.000	0.350	24.905
30-Apr-08	0.000	4.996	0.085	0.000	0.546	1.164	1.986	0.159	0.031	0.139	0.000	0.352	7.383	0.153	19.764
7-May-08	0.000	2.980	0.000	0.000	0.057	1.109	1.122	0.042	0.000	0.000	0.000	0.163	4.813	0.000	11.231
21-May-08	0.000	0.386	0.355	0.094	0.764	0.222	11.195	1.163	0.000	0.000	0.121	2.596	11.695	2.542	53.403
29-May-08	0.000	0.918	0.056	0.900	0.171	0.288	3.235	0.154	0.000	0.000	0.273	0.478	3.185	0.186	16.144
4-Jun-08	0.000	0.190	0.055	2.231	0.175	0.000	0.711	0.128	0.000	0.000	0.653	0.120	1.001	0.275	13.632
11-Jun-08	0.000	2.761	1.256	1.387	2.329	1.208	1.642	2.623	0.000	0.000	0.480	0.305	4.161	12.156	54.368
18-Jun-08	0.000	1.443	0.161	0.591	0.316	0.732	0.462	0.477	0.557	0.000	0.225	0.051	1.477	0.476	10.142
25-Jun-08	0.000	8.246	0.348	0.000	4.070	0.640	1.225	1.145	1.152	0.000	0.079	0.179	9.300	0.937	48.345
3-Jul-08	0.283	3.390	0.329	0.000	1.177	0.941	0.747	1.152	0.895	0.000	0.074	0.089	4.412	1.282	20.314
10-Jul-08	0.000	1.864	0.414	0.000	1.092	0.678	0.884	1.752	0.866	0.000	0.092	0.231	2.166	1.431	17.182
17-Jul-08	0.156	1.299	0.542	0.159	0.970	0.432	1.600	1.625	0.647	0.000	0.243	0.315	2.867	2.466	20.729
24-Jul-08	0.562	1.208	1.281	0.759	3.580	1.252	5.218	3.840	1.076	0.000	1.041	0.956	7.009	9.201	56.313
30-Jul-08	1.144	0.412	0.287	0.000	0.812	0.893	4.748	1.234	1.422	0.000	0.160	0.669	5.729	1.986	22.835
7-Aug-08	0.453	8.120	0.670	0.000	1.272	1.032	12.084	1.527	1.005	2.004	0.331	2.716	20.545	2.457	71.761
14-Aug-08	0.000	15.918	0.978	0.000	2.569	3.983	1.740	2.265	0.422	0.000	0.285	0.591	12.217	2.908	41.671
20-Aug-08	0.000	9.380	0.289	0.000	0.564	1.480	1.172	0.539	0.236	0.000	0.156	0.372	8.185	1.033	29.678
28-Aug-08	0.000	9.566	0.000	0.000	0.340	1.923	3.321	0.276	0.000	0.000	0.000	0.516	9.726	0.586	39.589
4-Sep-08	0.520	7.940	0.130	0.000	0.188	1.863	4.672	0.404	0.000	0.000	0.000	0.970	11.903	0.866	35.813
11-Sep-08	0.970	2.871	0.000	0.000	0.103	1.377	3.070	0.183	0.000	0.000	0.067	0.484	7.051	0.376	22.010
17-Apr-08	0.000	5.233	0.093	0.000	0.689	1.529	3.230	0.281	0.070	0.000	0.000	0.460	0.000	0.289	25.647

(b) Initial ratio matrix

Class / Pigment	peri	fuco	neo	myxo	viol	ddx	allo	lut	zea	crox	echi	acar	chl_c	chl_b	chl_a
Chlorophytes	0.000	0.000	0.018	0 000	0.023	0.000	0.000	0.117	0.018	0.000	0.000	0.000	0.000	0.300	1.000
Chrysophytes	0.000	0.220	0.000	0.000	0.060	0.000	0.000	0.000	0.000	0.000	0.000	0.000	0.030	0.000	1.000
Cryptophytes	0.000	0.000	0.000	0.000	0.000	0.000	0.224	0.000	0.000	0.043	0.000	0.024	0.234	0.000	1.000
Cyanobacteria_T2	0.000	0.000	0.000	0.000	0.000	0.000	0.000	0.000	0.109	0.000	0.000	0.000	0.000	0.000	1.000
Cyanobacteria_T1	0.000	0.000	0.000	0.018	0.000	0.000	0.000	0.000	0.043	0.000	0.097	0.000	0.000	0.000	1.000
Diatoms	0.000	0.563	0.000	0.000	0.000	0.129	0.000	0.000	0.000	0.000	0.000	0.000	0.100	0.000	1.000
Dinoflagellates	0.629	0.000	0.000	0.000	0.000	0.250	0.000	0.000	0.000	0.000	0.000	0.000	0.150	0.000	1.000
Euglenophytes	0.000	0.000	0.030	0.000	0.000	0.250	0.000	0.000	0.000	0.000	0.000	0.000	0.000	0.200	1.000

(a) Input data matrix containing the pigment concentrations in μg L^{-1}. peri = peridinin; fuco = fucoxanthin; neo = neoxanthin; myxo = myxoxanthophylls; viol = violaxanthin; ddx = diatoxanthin + diadinoxanthin; allo = alloxanthin; lut = lutein; zea = zeaxanthin; crox = crocoxanthin; echi = echinenone; acar = α-carotene; chl_c = chlorophyll c; chl_b = chlorophyll b; chl_a = chlorophyll a.

Table SOP 5.3 *Example of output data matrix resulting from the CHEMTAX run on the data of Table SOP 5.2a*

Date	Chloro	Chryso	Crypto	Cyanobacteria_T2	Cyanobacteria_T1	Diatoms	Dino	Eugleno
17-Apr-08	0.315	3.463	9.913	0.302	0.061	8.158	0.014	2.679
30-Apr-08	0.232	4.872	8.326	0.000	0.000	5.670	0.036	0.628
7-May-08	0.000	0.000	5.220	0.000	0.000	5.177	0.056	0.779
21-May-08	8.195	2.772	41.798	0.000	0.000	0.000	0.000	0.638
29-May-08	0.424	0.755	10.378	0.000	3.272	0.932	0.000	0.383
4-Jun-08	0.721	1.294	2.396	0.000	9.184	0.000	0.000	0.037
11-Jun-08	29.156	13.442	3.026	0.000	2.102	0.000	0.000	6.643
18-Jun-08	1.455	0.743	0.499	2.146	2.472	1.862	0.033	0.932
25-Jun-08	3.813	37.807	1.492	4.519	0.000	0.000	0.078	0.635
3-Jul-08	4.356	7.845	0.999	4.097	0.000	1.782	0.419	0.817
10-Jul-08	6.502	4.981	0.485	3.674	0.000	1.219	0.102	0.219
17-Jul-08	8.402	5.042	3.566	1.710	1.205	0.000	0.243	0.560
24-Jul-08	25.736	13.541	8.933	0.000	4.572	0.000	0.615	2.915
30-Jul-08	5.446	0.000	9.394	6.094	0.000	0.000	1.380	0.521
7-Aug-08	7.916	7.591	47.027	2.627	0.000	6.139	0.462	0.000
14-Aug-08	7.786	14.712	2.002	0.000	0.000	14.413	0.072	2.686
20-Aug-08	3.352	7.166	6.896	0.442	0.898	10.923	0.000	0.000
28-Aug-08	2.156	5.947	16.493	0.000	0.386	14.575	0.033	0.000
4-Sep-08	2.671	1.122	19.480	0.000	0.000	11.822	0.717	0.000
11-Sep-08	0.854	0.471	12.996	0.000	0.349	5.034	1.450	0.856

In units of μg chlorophyll a L^{-1}. Chloro = Chlorophytes; Chryso = Chrysophytes; Crypto = Cryptophytes; Dino = Dinophytes; Eugleno = Euglenophytes.

Great care must be taken when choosing the initial pigment ratios since the system is underdetermined and the program is likely to find local minima if incorrect pigment ratios are chosen. Examples of initial ratio matrices for freshwater phytoplankton, including cyanobacteria, can be found in various publications (e.g. [1, 3, 5]).

References

[1] Descy, J.-P., Higgins, H.W., Mackey, D.J. *et al.* (2000) Pigment ratios and phytoplankton assessment in northern Wisconsin lakes. *Journal of Phycology* **36**, 274–286.

[2] Jeffrey, S.W., Mantoura, R.F.C. and Wright, S.W. (1997) Phytoplankton pigments in oceanography: guidelines to modern methods, SCOR–UNESCO, Paris, 661 p.

[3] Schlüter, L., Lauridsen T.L., Krogh, G. *et al.* (2006) Identification and quantification of phytoplankton groups in lakes using new pigment ratios–a comparison between pigment analysis by HPLC and microscopy. *Freshwater Biology* **51**, 1474–1485.

[4] Mackey, M.D., Mackey, D.J., Higgins, H.W. *et al.* (1996) CHEMTAX–a program for estimating class abundances from chemical markers: application to HPLC measurements of phytoplankton. *Marine Ecology Progress Series* **144**, 265–283.

[5] Sarmento, H. and Descy, J.-P. (2009) Use of marker pigments and functional groups for assessing the status of phytoplankton assemblages in lakes. *Journal of Applied Phycology* **20**, 1001–1011.

[6] Wright, S.W. and Jeffrey, S.W. (2006) Pigment markers for phytoplankton production. In: Volhman, J.H. (ed.) *Marine organic matter: Biomarkers, isotopes and DNA*. Springer Verlag, Berlin, Heidelberg, pp. 71–104.

[7] Wright, S.W., Jeffrey, S.W., Mantoura, R.F.C. *et al.* (1991) Improved HPLC method for the analysis of chlorophylls and carotenoids from marine phytoplankton. *Marine Ecology Progress Series* **77**, 183–196.

SOP 6

Extraction of Cyanotoxins from Cyanobacterial Biomass

Leonardo Cerasino[1], Jussi Meriluoto[2,3], Luděk Bláha[4], Shmuel Carmeli[5], Triantafyllos Kaloudis[6], and Hanna Mazur-Marzec[7]

[1]*IASMA Research and Innovation Centre, Fondazione Edmund Mach-Istituto Agrario di S. Michele all'Adige, S. Michele all'Adige, Trento, Italy*
[2]*Biochemistry, Faculty of Science and Engineering, Åbo Akademi University, Turku, Finland*
[3]*Laboratory for Paleoenvironmental Reconstruction, Faculty of Sciences, University of Novi Sad, Serbia*
[4]*Faculty of Science, RECETOX, Masaryk University, Brno, Czech Republic*
[5]*Raymond and Beverly Sackler School of Chemistry, Tel Aviv University, Tel Aviv, Israel*
[6]*Water Quality Control, Athens Water Supply and Sewerage Company – EYDAP SA, Athens, Greece*
[7]*Institute of Oceanography, University of Gdansk, Gdynia, Poland*

SOP 6.1 Introduction

This SOP describes recommended approaches for the extraction of the most prominent cyanobacterial toxins from lyophilized material. Lyophilization is highly recommended for comparative purposes, as it allows calculating the toxin concentrations in terms of toxin amount per biomass dry weight unit.

The SOP will consider the following toxins: microcystins (MCs), nodularins (NODs), cylindrospermopsin (CYN), anatoxin-a (ATX-a), and saxitoxins (STXs). The scope of this SOP is to provide guidance on extraction procedures for analytical purposes. The recommended method of analysis after the extraction is liquid chromatography (LC) with UV or MS detection.

The different toxins can be extracted following a common extraction scheme, although each of them has specific requirements (solvents, solid-phase extraction [SPE] cartridges, etc.). Toxins are

Handbook of Cyanobacterial Monitoring and Cyanotoxin Analysis, First Edition.
Edited by Jussi Meriluoto, Lisa Spoof and Geoffrey A. Codd.

extracted from the biomass with an appropriate solvent enforced by sonication. The extract is then separated from the solid matter by centrifugation, and the solvent is evaporated by nitrogen stream or rotatory evaporator. Dry material is then redissolved in the appropriate solvent for SPE purification. After SPE, samples can be analyzed by LC methods. When aqueous solvents are used for extraction (e.g. 5% acetic acid), evaporation is not required; therefore, after centrifugation, the sample can be directly used for SPE.

LC analysis (in particular, LC-MS) can be also carried out without SPE purification, by directly analysing the sample after centrifugation or filtration. However, preliminary SPE is always recommended as it allows to get an enriched sample without many possible interfering compounds.

SOP 6.2 Experimental

SOP 6.2.1 Materials

- Water HPLC grade, 18 MΩ cm
- Acetic acid
- Methanol HPLC grade
- Nitrogen or argon gas, >99.99%
- Borosilicate glass vials/tubes
- Syringe filters (before analysis by LC-UV or LC-MS) equivalent to GHP Acrodisc syringe filters 13 mm in diameter and with 0.2 µm pore size (Pall Life Sciences, Ann Arbor, MI)
- 2-mL all-plastic (no rubber piston) single-use syringes
- Nitrogen evaporation unit (or rotary evaporator, centrifugal evaporator)
- Ultrasonic bath (or probe sonicator with appropriate tip)

Specific reagents and materials for SPE and LC analysis, as well as standard laboratory equipment, are not mentioned (e.g. analytical balance, microcentrifuge, pipettes, graduated cylinders).

SOP 6.2.2 Solvents

Solvent mixtures mentioned in the procedure are prepared on a volumetric basis (e.g. 75% methanol is obtained mixing 75 mL of methanol with 25 mL of water).

SOP 6.2.3 General Procedure

The procedure is basically the same for all considered toxins (Table SOP 6.1). The dry material is extracted with an appropriate solvent mixture, whose composition is specific for each toxin. After extraction and evaporation of the solvent, the sample has to be redissolved in appropriate solvent for LC analysis or SPE. No indication about the solvent to be used in this step is given in Table SOP 6.1, as different LC and SPE protocols can be adopted, with different sample requirements (always specified in the protocols). Some suitable LC and SPE procedures are available in several SOPs of this handbook as suggested in Table SOP 6.1.

SOP 6.3 Quality Control

The use of an internal standard is highly recommended during the analysis for verifying the analytical performance (cfr. Chapter 23 of this handbook). They can be added at different steps of the process to control the extraction efficiency and/or the analytical accuracy. In case of LC-MS

Table SOP 6.1 *Comparative scheme for the extraction of different toxins from cyanobacterial biomass*

Step	Description	MC/Nod	CYN	ATX-a	STX
1	Weighing of dry material into a vial or tube[a]	10 mg			
2	Addition of extraction solvent[b]	1 mL 75% methanol	1 mL 75% methanol	1 mL 0.3% acetic acid in 50% methanol	1 mL 0.2% acetic acid in 30% methanol
3	Extraction in ultrasonic bath[c]	15 min			
4	Centrifugation and collection of the supernatant	10,000 × *g*/10 min			
5	Repetition of steps 2–4; combination of the two extracts	Yes			
6	Evaporation of the combined extracts using the nitrogen evaporation unit at 40°C[d]	Yes			
7	Dissolution in adequate solvent suitable for either: • direct analysis • preliminary SPE purification (recommended)	Refer to the specific protocols for either LC or SPE			
8	Clarification of the sample before LC-UV or LC-MS analysis by centrifugation or filtration[f]	10,000 × *g*/10 min or use syringe filters			
	Suggested protocols for purification and analysis				
SPE purification		SOP 7	SOP 15	SOP 17	SOP 20[e]
LC analysis		SOPs 10, 11	SOPs 16, 19	SOPs 18, 19	SOP 20
	References				
		[1–5]	[5–7]	[5,8,9]	[5,8,10,11]

[a] Glass vials are recommended to prevent sorption from aqueous extracts but using glass may be a problem in other steps (e.g. centrifugation).
[b] The most common extraction solvent for each toxin is reported. Some authors, however, have proved that all toxins can be simultaneously extracted in an efficient way by using a single solvent (e.g. 50% methanol [5]).
[c] Probe sonicator can be used as an alternative (30 W, ~20 kHz, pulsed mode).
[d] Other evaporation equipments (such as rotatory evaporator, centrifugal evaporator) can be used as alternatives.
[e] SOP20 describes an online-SPE purification.
[f] The filters should not show any adsorption of toxins from the methanolic solvents.

analysis, isotopically labeled toxins represent the best choice but have a very limited commercial availability. More common is to use surrogates, such as nodularin, which is often used as internal standard in the analysis of microcystins with LC-MS methods.

In addition, reference material for quality control purposes can be prepared in house, such as a large volume of well-homogenized bloom sample with known toxin concentration. The material should be stored in adequate conditions for assuring stability for a long period of time, and toxin concentration periodically verified. The reference material should be processed together with each batch of samples. The quality of the analysis will be verified by the use of a quality control chart.

SOP 6.4 Validation Data

Readers can consult references in this and in cited SOPs to find data about validation of the entire analytical protocol from extraction to LC analysis.

SOP 6.5 Special Issues

- Cyanobacterial bloom material is hazardous – use personal protection devices (gloves, goggles, face mask).
- Lyophilized material should be handled with extreme care. Avoid dispersion into the air; preferentially work in a fume hood. Gloves should be worn through the entire process, and face mask usage is recommended.
- Some cyanobacterial samples can contain large amount of mucilage matrix. which will interfere with the extraction (agar-like properties). These samples are difficult to handle. and no generally valid recommendation can be made. In specific cases. diluting the sample with a larger volume of the extraction solvent may provide a solution.

References

[1] Spoof, L., Vesterkvist, P., Lindholm, T. and Meriluoto, J. (2003) Screening for cyanobacterial hepatotoxins microcystins and nodularin in environmental water samples by reversed-phase liquid chromatography–electrospray ionisation mass spectrometry. *Journal of Chromatography* **1020**, 105–119.

[2] ISO 20179 (2005) Water Quality – Determination of Microcystins – Method Using Solid Phase Extraction (SPE) and High Performance Liquid Chromatography (HPLC) with Ultraviolet (UV) Detection. Geneva: International Organization for Standardization.

[3] Kaloudis, T., Zervou, S. K., Tsimeli, K. *et al.* (2013) Determination of microcystins and nodularin (cyanobacterial toxins) in water by LC–MS/MS. Monitoring of Lake Marathonas, a water reservoir of Athens, Greece. *Journal of Hazardous Materials* **263** (1), 105–115.

[4] Shams, S., Capelli, C., Cerasino, L. *et al.* (2015) Anatoxin-a producing *Tychonema* (Cyanobacteria) in European water bodies. *Water Research* **69**, 68–79.

[5] Dahlmann, J., Budakowsky, W.R. and Luckas, B. (2003) Liquid chromatography–electrospray ionisation-mass spectrometry based method for the simultaneous determination of algal and cyanobacterial toxins in phytoplankton from marine waters and lakes followed by tentative structural elucidation of microcystins. *Journal of Chromatography A* **994**, 45–57.

[6] Törökné, A., Asztalos, M., Bánkiné, M. *et al.* (2004) Interlaboratory comparison trial on cylindrospermopsin measurement. *Analytical Biochemistry* **332** (2), 280–284.

[7] Wörmer, L., Carrasco, D., Cirés, S. and Quesada, A. (2009) Advances in solid phase extraction of the cyanobacterial toxin cylindrospermopsin. *Limnology and Oceanography: Methods* **7**, 568–575.

[8] Hiller, S., Krock, B., Cembella, A. and Luckas, B. (2007) Rapid detection of cyanobacterial toxins in precursor ion mode by liquid chromatography tandem mass spectrometry. *Journal of Mass Spectrometry* **42**, 1238–1250.

[9] Dimitrakopoulos, I.K., Triantafyllos, K., Hiskia, A., *et al.* (2010) Development of a fast and selective method for the sensitive determination of anatoxin-a in lake waters using liquid chromatography–tandem mass spectrometry and phenylalanine-d5 as internal standard. *Analytical and Bioanalytical Chemistry* **397** (6), 2245–2252.

[10] Ballot, A., Fastner, J. and Wiedner C. (2010) Paralytic shellfish poisoning toxin-producing cyanobacterium *Aphazinomenon gracile* in Northeast Germany. *Applied and Environmental Microbiology* 1173–1180.

[11] Halme, M., Rapinoja, M.L., Karjalainen, M. and Vanninen, P. (2012) Verification and quantification of saxitoxin from algal samples using fast and validated hydrophilic interaction liquid chromatography-tandem mass spectrometry method. *Journal of Chromatography B* **880**, 50–57.

SOP 7

Solid-Phase Extraction of Microcystins and Nodularin from Drinking Water

Theodoros M. Triantis[1], Triantafyllos Kaloudis[2], Sevasti-Kiriaki Zervou[1], and Anastasia Hiskia[1]

[1] Institute of Nanoscience & Nanotechnology, National Center for Scientific Research "DEMOKRITOS," Athens, Greece

[2] Water Quality Control, Athens Water Supply and Sewerage Company – EYDAP SA, Athens, Greece

SOP 7.1 Introduction

This SOP specifies a method for the solid phase extraction (SPE) of microcystins (MCs) and nodularins (NODs) from drinking water or filtered water samples. The extracts produced by this method are suitable for subsequent analysis by LC-MS/MS (SOP 10) or HPLC-PDA. This extraction method is suitable for the determination of MCs and NOD that are in solution (exotoxins) and not inside the cells (endotoxins) or bound to insoluble material.

The method has been tested for the target compounds shown in Table SOP 7.1. When other MCs are targeted, such as MC-LW or other nonpolar MC variants, modifications in the extraction procedure may be needed (e.g. in the elution solvent).

The target compounds are extracted from water with use of solid phase extraction cartridges. The compounds are eluted from cartridges with methanol and redissolved in a water–methanol solution. This is an in-house method based on methods described in [1–3].

Handbook of Cyanobacterial Monitoring and Cyanotoxin Analysis, First Edition.
Edited by Jussi Meriluoto, Lisa Spoof and Geoffrey A. Codd.

Table SOP 7.1 Registry information of targeted compounds

Name	Formula	M.W.	CAS Number
Microcystin-RR (MC-RR)	$C_{49}H_{75}N_{13}O_{12}$	1038.20	111755-37-4
Nodularin (NOD)	$C_{41}H_{60}N_{8}O_{10}$	824.96	118399-22-7
Microcystin-YR (MC-YR)	$C_{52}H_{72}N_{10}O_{13}$	1045.19	101064-48-6
Microcystin-LR (MC-LR)	$C_{49}H_{74}N_{10}O_{12}$	995.17	101043-37-2
Microcystin-LA (MC-LA)	$C_{46}H_{67}N_{7}O_{12}$	910.06	96180-79-9

SOP 7.2 Experimental

SOP 7.2.1 Materials

- Ultrapure water, 18 MΩ cm
- Sodium thiosulphate ($Na_2S_2O_3$)
- Formic acid (HCOOH), analytical grade
- Ammonium hydroxide (NH_4OH), analytical grade
- Methanol HPLC grade
- Nitrogen gas (>99.999%)
- Glass microfibre filters, 47-mm diameter
- Volumetric cylinders, 10 mL and 500 mL
- Pipette 1–10 mL
- Stopwatch
- Glass sample bottles, 0.5 or 1 L
- SPE Cartridges: Oasis HLB, size 200 mg in 6-mL reservoir (Waters) or equivalent performance cartridges
- Teflon tubing and adaptors for connection of sample bottles with SPE cartridges
- Glass test tubes, 10 mL
- Vacuum collection bottle (waste aspirator bottle), e.g. 10 L, plastic recommended

SOP 7.2.2 Special Equipment

- Glass filtration apparatus (vacuum) with slit-sieve disc for mounting glass-fiber filters (e.g. 47 mm) and 1 L filtering flask
- Solid phase extraction (SPE) vacuum manifold e.g. 12 port
- Vacuum diaphragm pump (e.g. 5 L min^{-1} pumping speed, oil-free and Teflon diaphragm recommended)
- Nitrogen evaporation unit with thermostated bath or heat block (e.g. 12 port) for use with 10-mL glass test tubes
- pH-meter
- Ultrasonic bath

The configuration of the SPE manifold with cartridges connected to glass sample vials is shown in Figure SOP 7.1.

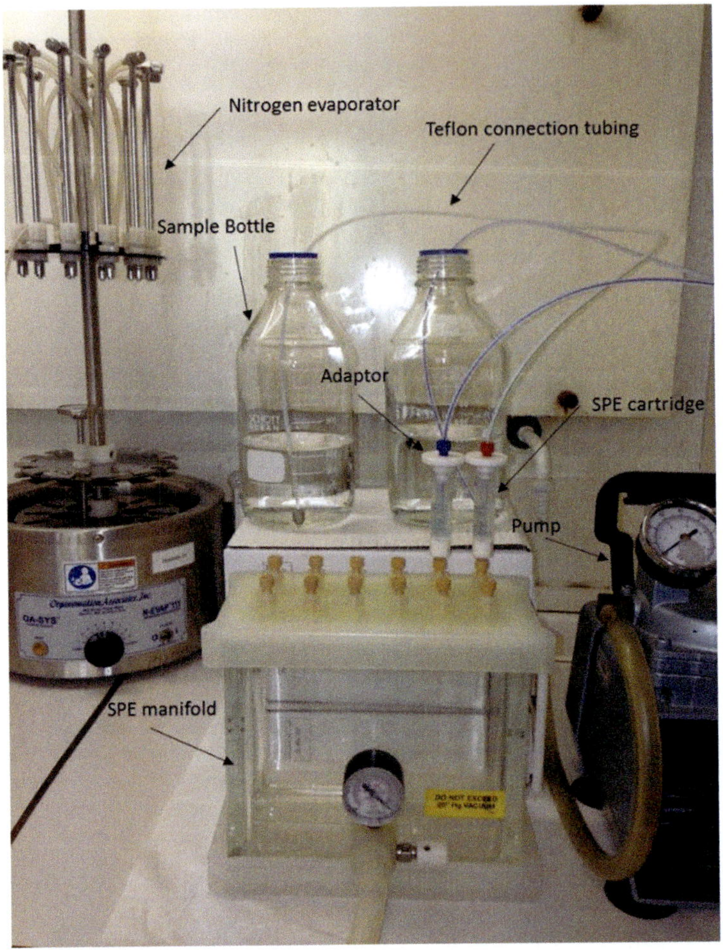

Figure SOP 7.1 Arrangement of the SPE manifold

SOP 7.2.3 Solutions

- 1% w/v sodium thiosulphate ($Na_2S_2O_3$) solution. Dissolve 1 g sodium thiosulphate in 100 mL of water.
- 1 mol L^{-1} formic acid in water
- Ammonium hydroxide solution (e.g. 25%)
- Methanol/water 20:80 v/v

SOP 7.2.4 General Procedure

a. Clean thoroughly the glass sample bottles, the SPE manifold and connecting tubing with suitable solvents (e.g. acetone) followed by rinsing with ultrapure water.
b. Filter surface water samples and turbid samples through glass fiber filters.
c. In case of samples containing residual chlorine (e.g. treated water samples), add 500 µL of sodium thiosulphate solution per 500-mL sample. Shake the samples vigorously and let stand for 5 min.

d. Measure the pH of samples which should be 5–8. If it is outside this range, adjust with HCOOH or NH_4OH solutions.
e. Transfer the sample (500 mL) in a glass bottle, add 5 mL of methanol, and mix the contents by shaking.
f. Condition the SPE cartridges by passing 4 mL of methanol followed by 4 mL of water at a flow of about 5 mL min^{-1}. Note. If you are working with cartridges other than HLB e.g. C_{18}, do not let the cartridge go dry during conditioning, sample application, and wash.
g. Connect the sample bottle with Teflon tubing and adaptor to the SPE cartridge. Pass the sample through the cartridge at a flow rate of about 10 mL min^{-1} (visible drops) with help of the vacuum generated by the pump. The flow rate can be measured with a graduated cylinder and a stopwatch, and it is adjusted with use of regulating valves in the pump or manifold.
h. When the entire sample has passed through the cartridge, remove the connecting tubing and adaptor. Wash the cartridge with 4 mL of 20% methanol.
i. Elute the dried cartridge with 4 mL methanol at a flow rate of about 5 mL min^{-1} and the eluate is collected in a glass test tube.
j. Evaporate the eluates to dryness under a gentle nitrogen stream at 40°C in a nitrogen evaporation unit.
k. Redissolve the extracts in 500 µL of 20% methanol and sonicate for 5 min. Centrifuge 10 min at 10,000 × *g*, or, alternatively filter and transfer the final solutions to autosampler vials (e.g. 2 mL volume) for subsequent analysis. It is recommended that analysis is carried out immediately after SPE. In cases where this is not possible, solutions are kept at –20°C.

SOP 7.3 Quality Control

With every batch of extractions a spiked sample is prepared by spiking ultrapure water or one of the samples (duplicate) with a mixture of the target compounds at suitable concentrations (e.g. 0, 100, or 1.0 µg L^{-1}). The results of this quality control sample are evaluated against quality control chart limits.

SOP 7.4 Validation Data

This SOP describes only the extraction process. It has been shown that this extraction process, when coupled to LC-MS/MS measurements, meets the criteria of 70–120% mean recovery and <20% RSD. Validation data of this extraction process followed by LC-MS/MS measurements are presented in SOP 10.

References

[1] ISO 20179:2005. (2005) Water quality–Determination of microcystins–Method using solid phase extraction (SPE) and high performance liquid chromatography (HPLC) with ultraviolet (UV) detection. *International Standards Organization, Technical Committee ISO TC147/SC2.*
[2] Meriluoto, J. and Spoof, L. (2005) Solid phase extraction of microcystins in water samples, in *TOXIC–Cyanobacterial Monitoring and Cyanotoxin Analysis* (eds. J. Meriluoto and G.A. Codd), Åbo Akademi University Press, Åbo, pp. 73–76.
[3] Kaloudis, T., Zervou, S.-K., Tsimeli, K. *et al.* (2013) Determination of microcystins and nodularin (cyanobacterial toxins) in water by LC–MS/MS. Monitoring of Lake Marathonas, a water reservoir of Athens, Greece, *Journal of Hazardous Materials* **263** (1), 105–115.

SOP 8

Extraction of Microcystins from Animal Tissues

Ondřej Adamovský and Luděk Bláha

Faculty of Science, RECETOX, Masaryk University, Brno, Czech Republic

SOP 8.1 Introduction

Microcystins (MCs) have been shown to accumulate in various aquatic organisms, including mussels and fish [1, 2]. MCs accumulate preferably in liver, although high concentrations have also been found in other animal tissues such as muscles [3]. MCs are routinely extracted by methanol or mixture of methanol–water enhanced by ultrasonication in a bath or by a probe-type sonicator [4, 5]. MCs are amphipathic molecules with hydrophobic groups in combination with polar groups of hydrophilic character. Combination water–methanol is suitable mixture for extraction of a large spectrum of MC congeners. The extraction recovery may be affected by several aspects such as metabolisation of the MCs or covalent binding of MCs to macromolecules (proteins and peptides) [4]. High MC-binding capacity in mice liver was demonstrated in several studies [6]. Tissue extracts can be analysed for MCs by HPLC-MS method as described, for example, in SOP 12.

SOP 8.2 Experimental

SOP 8.2.1 Materials

- Distilled water (or preferentially water purified to 18.2 MΩ cm)
- Hexane HPLC grade
- Methanol HPLC grade

Handbook of Cyanobacterial Monitoring and Cyanotoxin Analysis, First Edition.
Edited by Jussi Meriluoto, Lisa Spoof and Geoffrey A. Codd.

- Glass tubes (5–10 mL volume) suitable for homogenisation and centrifugation
- Glass flasks compatible with rotary evaporator
- Variable volume pipettes

SOP 8.2.2 Special Equipment

- Tissue homogeniser for volumes up to 20 mL able to disperse 5- to 25-µm particles
- Rotary vacuum evaporator
- Centrifuge with swing-bucket rotor (2900 × *g* force required)
- Ultrasonic bath with ultrasonic peak output minimum of 320 W (cooled or ice should be added)
- Vortex mixer

SOP 8.2.3 General Procedure

a. A schematic presentation of the different steps of the procedure is shown in Figure SOP 8.1.
b. Take approximately 0.5 g of the frozen tissue (fresh weight) and record the exact weight at three-digit accuracy.
c. Add 3 mL methanol to the frozen and weighted tissue (in glass tube). The tubes used should be compatible with the homogenizer.
d. Homogenize the sample for approximately 30 s by moving the tube containing the sample up and down along the blades of the homogenizer. Use glass tubes to obtain suspension without visible solid particles. Time of homogenization vary depending on the tissue and the power of homogenizer. Keep the tube with the sample on ice during all homogenization.

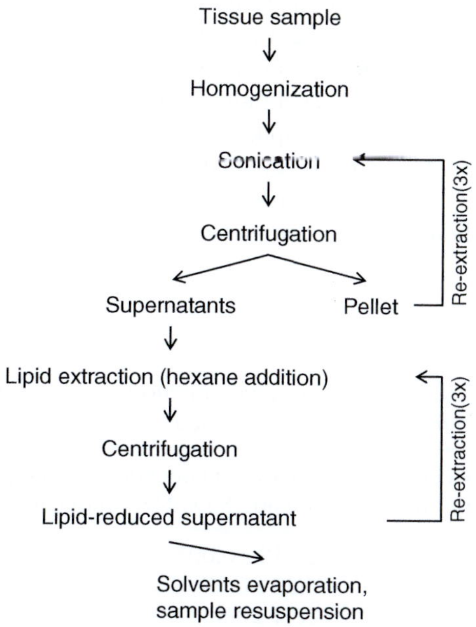

Figure SOP 8.1 Scheme of the extraction procedure

e. Sonicate the homogenate for 30 min in cooled water (10°C) in ultrasonic bath with ultrasonic peak output minimum of 320 W. Time of the sonication may be adjusted depending on the energy of the ultrasonic output.
f. Centrifuge the homogenate at 2900 × *g* for 10 min in the centrifuge with swing-bucket rotor. Slow start/end is recommended to avoid stirring up the pellet.
g. Collect the supernatant and re-suspend the pellet with another 3 mL methanol.
h. Reextract the pellet by repeating the procedure described (d–g) a total of two times.
i. Pool all the three supernatants (step g; three times ~2.5–3 mL) having thus total extract of approximate volume 7–9 mL.
j. Add hexane (1 mL) and extract to remove lipids. After 1 minute intensive vortex mixing (shaking) of sample with hexane, centrifuge the sample at 2900 × *g* for 10 min.
k. Remove gently the upper lipid (hexane) layer with pipette.
l. Repeat the lipid removal steps (j–k) additionally two times.
m. Evaporate the pre-cleaned supernatant (at 50°C in vacuum rotary evaporator) and dissolve the residue in 300 μL of 50% aqueous methanol (v/v). The desired solvent can be modified according to the following analytical method (e.g. LC-MS/MS).

SOP 8.3 Validation

Recovery of extraction for each type of the tissue should be checked.

a. Prepare the tissue homogenate as described above.
b. Incubate the tissue homogenate with the spiked MC-LR in variable concentrations (e.g. 0.1, 1, and 10 μg MC g tissue^{-1}) in continuously rotating glass vials at room temperature for 24 h in order to achieve a representative amount of covalently bound MC [4].
c. Prepare each validation sample in three independent replicates.
d. Extract the homogenate according to the extraction procedure described earlier.
e. Analyse the concentration of MCs in the samples (e.g. using HPLC-MS/MS method described in SOP 12) and calculate the recovery as a ratio of extracted and originally spiked amount of MC. For example, 0.5 g of the tissue was spiked with 5 μg of MC-LR to achieve concentration 10 μg MC g tissue^{-1}. LC-MS/MS analysis revealed 1.14 μg of MC-LR in 300-μL sample, which represents 0.5 g of tissue. The recovery is thus (1.14/5)100 = 22.8%.
f. Results of the example validation using common carp liver spiked with MC-LR is presented in Table SOP 8.1.

Table SOP 8.1 *Results of validation of the extraction method using common carp liver; analysis of MCs using HPLC-MSMS described in SOP 12 was used*

Amount spiked per 0.5 g f.w. tissue^{-1}	Concentration (μg g f.w.$^{-1}$)	Measured concentration μg g f.w.$^{-1}$ (mean ± S.D., N = 3 replicates)	Mean recovery (%)
0.05	0.1	0.019 ± 0.008	19
0.5	1	0.23 ± 0.07	23
5	10	2.41 ± 0.53	24
		Average recovery	***22***

SOP 8.4 Special Issues

- Homogenize the tissue with the homogenizer as long as necessary to obtain fine homogenate without any visible solid particles (puree appearance).
- In case of sample with high content of fat, repeat removal of lipids [SOP 8.2.3 (j–k)] more than three times to obtain clear samples.
- It is recommended to centrifuge or filter (small syringe filters) the final extract prior to LC-MS/MS analysis to remove any residual particulate matter in the sample.
- Any changes in extraction should be reflected also in the validation procedures.

References

[1] Adamovsky, O., Kopp, R., Hilscherova, K. *et al.* (2007) Microcystin kinetics (bioaccumulation, elimination) and biochemical responses in common carp (*Cyprinus carpio*) and silver carp (*Hypophthalmichthys molitrix*) exposed to toxic cyanobacterial blooms. *Environmental Toxicology and Chemistry* **26** (12), 2687–2693.

[2] Vareli, K., E. Zarali, G.S.A. Zacharioudakis *et al.* (2012) Microcystin producing cyanobacterial communities in Amvrakikos Gulf (Mediterranean Sea, NW Greece) and toxin accumulation in mussels (*Mytilus galloprovincialis*). *Harmful Algae* **15**, 109–118.

[3] Poste, A.E., R.E. Hecky and Guildford SJ. (2011) Evaluating microcystin exposure risk through fish consumption. *Environmental Science and Technology* **45** (13), 5806–5811.

[4] Ernst, B., Dietz, L., Hoeger, S.J. *et al.* (2005) Recovery of MC-LR in fish liver tissue. *Environmental Toxicology* **20** (4), 449–458.

[5] Schmidt, J., M. Shaskus, J.F. Estenik *et al.* (2013) Variations in the microcystin content of different fish species collected from a eutrophic lake. *Toxins* **5** (5), 992–1009.

[6] Imanishi, S. and K. Harada (2004). Proteomics approach on microcystin binding proteins in mouse liver for investigation of microcystin toxicity. *Toxicon* **43** (6), 651–659.

SOP 9

Analysis of Microcystins by Online Solid Phase Extraction–Liquid Chromatography Tandem Mass Spectrometry

Cintia Flores and Josep Caixach

Mass Spectrometry Laboratory/Organic Pollutants, IDAEA-CSIC, Barcelona, Spain

SOP 9.1 Introduction

High-performance liquid chromatography tandem mass spectrometry (HPLC-MS/MS) has been widely used to analyse microcystins for its robust quantitative analysis and good sensitivity and specificity [1]. The routine analysis of aqueous samples for extracellular microcystins requires a sample extraction, concentration, and clean-up. A variety of techniques have been reported for this purpose and among them, solid-phase extraction (SPE) is the most commonly used [1]. Online SPE-HPLC allows a fast and reliable approach to the monitoring of trace pollutants in water and has been applied to the matrix removal and pre-concentration of microcystins from water samples prior to HPLC [2–5]. In summary, online SPE offers practical advantages over offline methods: low amounts of sample required and solvent consumption; reduction of analysis times and sample manipulation (no contamination or loss of sample occurs during the treatment steps); improvement of the sample throughput; excellent precision and accuracy; and automation of the method. Additionally, HPLC method for online trace enrichment carried out by means of column switching technique is very simple, easily configured and allows the reduction in the analysis expenses. On the other hand, tandem mass spectrometry provides high sensitivity, specificity, and selectivity, large versatility of MS/MS experiments, and robust quantitative analysis. In particular, the selected reaction monitoring (SRM) mode by triple quadrupole instruments offers high signal-to-noise ratio and an improved accuracy and reproducibility of quantitation.

Handbook of Cyanobacterial Monitoring and Cyanotoxin Analysis, First Edition.
Edited by Jussi Meriluoto, Lisa Spoof and Geoffrey A. Codd.

The aim of this SOP is to describe the general systematic followed for the analysis of microcystins in aqueous samples by online SPE HPLC-ESI-MS/MS in SRM quantitation mode. This SOP applies to aqueous samples—surface, treated, and drinking waters—for analysis of microcystin (especially microcystin-LR, -RR, and -YR but also demethylated-microcystin-RR, microcystin-WR, -LA, -LY, -LW, and -LF). The range of concentrations for all analytes is 0.02 to 10 μg L^{-1}.

SOP 9.2 Experimental

SOP 9.2.1 Materials

- High purity water produced with a MilliQ system
- Acetonitrile gradient HPLC grade
- Formic acid (HCOOH), analytical grade
- L(+)-Ascorbic acid, analytical grade
- Argon or nitrogen, 99.999% purity
- GF/F filters (0.7 μm)
- C_{18} end-capped HPLC column [Luna C_{18} (2) 150×2.0 mm, 5-μm particle size, from Phenomenex, Torrance, CA, is mentioned in this SOP as possible alternative]. A compatible guard column is also required
- C_{18} load column (like Hypersil Gold C_{18} 20×2.1 mm, 12-μm particle size, from Thermo Scientific, Bellefonte, PA)
- Pyrex borosilicate glass bottles
- Borosilicate glass chromatographic vials: e.g. 10 mL clear glass
- Microcystin (MC) and nodularin (NOD) standards

SOP 9.2.2 Special Equipment

- Online SPE HPLC-ESI-MS/MS system consisting of an autosampler, two high-pressure or low-pressure pump systems, and one switching device unit (six-port valve). All connected to triple quadrupole mass spectrometer equipped with an electrospray ionisation (ESI) source as shown in Figure SOP 9.1. Two columns are used, load and HPLC columns, the former for pre-concentration of the sample and the latter for the analysis.
- HPLC-MS analysis software
- Filtering system

SOP 9.2.3 HPLC Mobile Phase

- HPLC mobile phase component A: HPLC or Milli Q water + 0.08% HCOOH.
  a. Add 1600 μL of HCOOH in 2 L of water.
  b. Replace every week.
- HPLC mobile phase component B: acetonitrile + 0.08% HCOOH.
  a. Take 850 μL of HCOOH in a borosilicate glass vial. Add 800 μL of HCOOH in 1 L of ACN. Discard the rest of the HCOOH.
  b. Replace every week.

Please observe the following:

- Work in fume hood with HCOOH or acetonitrile.
- Dispense the HCOOH under the surface of acetonitrile.

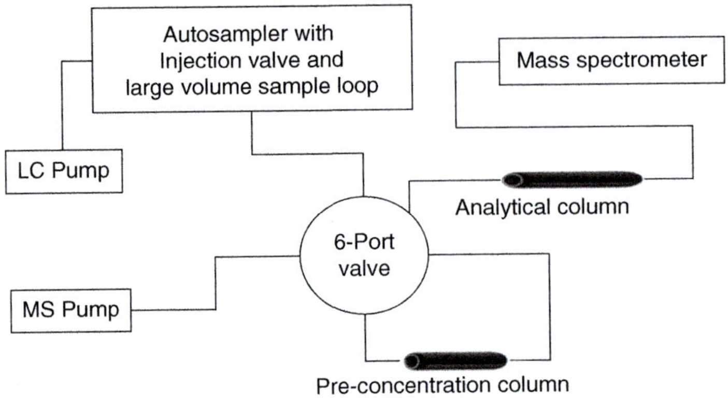

Figure SOP 9.1 Scheme of online SPE HPLC-ESI-MS/MS system device connections

- Do not contaminate the original HCOOH bottle with acetonitrile.
- Disposal of acetonitrile and HCOOH should conform to local regulations.

SOP 9.2.4 Online SPE HPLC-ESI-MS/MS Analysis

SOP 9.2.4.1 General Procedure

Sample Preparation.

a. Filter the water samples through the GF/F filters if it contains particulates or cyanobacterial cells. Collect the water samples in Pyrex borosilicate glass bottles and immediately place in darkness at 4°C until analyses (typically within 48 h of collection). Stored aqueous and sestonic (defined as the particles retained on 0.7-μm filters) fractions are analysed separately. The analysis of sestonic fraction is not subject of this SOP.
b. Remove residual chlorine: Chlorinated waters are usually treated with ascorbic acid or sodium thiosulphate before pre-concentration in order to prevent further degradation of microcystins (2 mL ascorbic acid L^{-1} per 2 ppm Cl_2).
c. Add an internal standard: Spike 10 mL of filtered water samples with nodularin as internal standard to obtain a final concentration of 1 μg L^{-1} (10 μL at 1 μg mL^{-1}) and then process directly using the online SPE HPLC-ESI-MS/MS system.

HPLC setup. The HPLC system should be set up according to good HPLC practice: e.g., unnecessary dead volumes should be avoided and the mobile phases should be degassed. All components and parameters should be monitored and documented. Always use a guard column. Change the guard column if the back-pressure rises or peak forms deteriorate. Change the HPLC gradually up to starting conditions and allow to condition. The flow is directed to waste while conditioning.

MS setup. The MS instrument with the ESI source should be set up and tuned as described in the manufacturer instructions. The following ionisation parameters are given just as an example for one triple quadupole instrument (Thermo Fisher Scientific TSQ Quantum Discovery, San Jose, CA, USA) in the analysis of microcystins. REMEMBER: EVERY INSTRUMENT MODEL MUST BE TUNED INDIVIDUALLY because the resulting optimal parameters may be slightly different on distinct systems, even of the same model. The analyses are performed operating in the positive ESI (+) mode. Examples of ionisation parameters: spray voltage 4.5 kV, optimum tube lens voltages (TL) for each *m/z* (Table SOP 9.1), ion transfer tube temperature 250°C, sheath gas (N_2), ion sweep gas (N_2) and auxiliary gas (N_2) flow rates 30 psi, 0 and 5 a.u. (arbitrary units), respectively. Scan, SIR, or SRM

Table SOP 9.1 *MS/MS conditions of some typical microcystins*

Compound	Precursor ion (m/z_1)	State of molecular ion	Tube Lens (V)	Product ion (m/z_2)	Collision energy (eV)
Microcystin-RR	519.7	$[M+2H]^{2+}$	137	127.0	41
				135.1	28
Nodularin	825.3	$[M+H]^{+}$	225	135.1	50
				227.1	50
Microcystin-LR	995.5	$[M+H]^{+}$	236	135.1	55
				213.1	45
Microcystin-YR	1045.5	$[M+H]^{+}$	224	135.1	57
				213.1	40

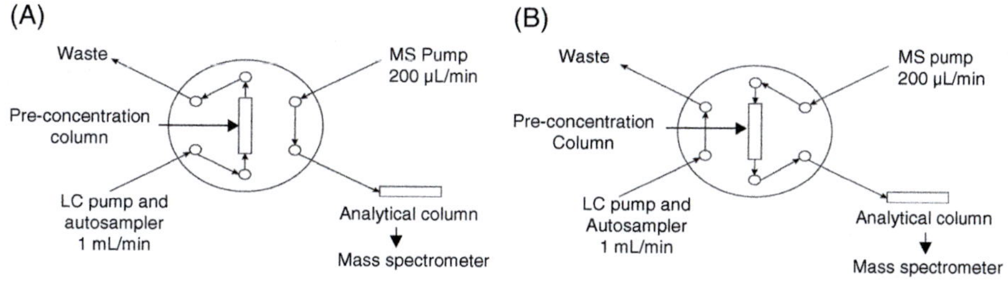

Figure SOP 9.2 *Six-port valve positions during online SPE analysis: (A) load position for pre-concentration and (B) elution position for analysis*

acquisition modes can be applied depending on the purpose of analyses. High purity Ar is used as collision-induced gas with a pressure of 1.5 mtorr and the optimum collision energy (CE) for each transition is selected. For a reliable quantitation by tandem mass spectrometry, SRM acquisition mode must be chosen, and two transitions, one for quantitation and a second one for confirmation, for each compound should be monitored to avoid false positives due to matrix interferences in complex samples. A relation between transitions is calculated (Commission Decision 2002/657/CE) [6]. Table SOP 9.1 shows the MS/MS conditions of some typical microcystins.

Automatic online SPE HPLC-ESI-MS/MS analysis. Briefly, 1 mL of water sample is injected directly onto load column (pre-concentration) in a high aqueous mobile phase (95%) at a flow rate of 1 mL min^{-1} (Figure SOP 9.2A). Afterwards, the six-port valve on the mass spectrometer is switched by instrument control software. This enabled the load column to be back flushed onto the analytical column, where the compounds are separated prior the introduction into the mass spectrometer (Figure SOP 9.2B). After all compounds are eluted from the analytical column, the six-port valve is switched back to the starting position, and the loading and analytical columns are re-equilibrated to their starting conditions. After separation, detection is carried out using a triple quadrupole mass spectrometer equipped with an ESI source.

SOP 9.2.4.2 Analysis of Data

Identification. For identification purposes comply with the following conditions:

a. Analyte retention time in the sample compared to analyte retention time in the calibration curve should be in agreement (the ratio of the chromatographic retention time of the analyte to that of the internal standard, shall correspond to that of the calibration solution at a tolerance of ± 2.5 according Commission Decision 2002/657/EC) [6].

b. Two *m/z* transitions are confirmed for every analyte (Table SOP 9.1).
c. The ratio between the area of two transitions in the sample compared to ratio in the calibration curve should be in agreement to [calibration curve average ± SD (calibration curve)] taking into account the maximum tolerances included in the Commission Decision 2002/657/EC [6].

Quantitation. For quantitation purposes calculate the microcystin concentration according to internal calibration based on isotope dilution using nodularin as internal standard (IS). As there are no isotopically labelled microcystins commercially available, nodularin is currently used as internal standard (SOP 9.2.4.1c).

a. Prepare a set of standard solutions ($n \geq 5$) for microcystin mixture (analytes) as calibration curve method, and add a constant amount of nodularin (IS) to each solution.
b. Prepare a plot of (area quantitation transition of the analyte/area quantitation transition of the IS) versus the analyte concentration, for each microcystin. The quantitation transition is chosen as the most specific, selective and intense of the two transitions monitored.

SOP 9.2.4.3 Quality Control

Blank samples to test the specificity of the method and blank spikes should be used as quality controls.

a. A blank sample should always be extracted with every batch of samples ensuring good repeatability of the method.
b. Instrumental blanks (MilliQ water) should be injected between samples and three additional blanks should be injected after the most concentrated calibration standard solution in order to avoid cross-talking between injections.
c. To check the recoveries in all samples, the recovery of nodularin, added before extraction, may be calculated.
d. To evaluate possible changes in sensitivity during the sample sequence and to ensure a proper quantitation, a fortified sample of medium concentration (1 $\mu g L^{-1}$) is injected and quantified like a sample. If the calculated concentration using the calibration curve is not consistent with the theoretical concentration with a tolerance of 25%, reinject the calibration curve.
e. External quality assurance may be evaluated by participation in intercomparison exercises.

SOP 9.2.5 Online SPE HPLC-ESI-MS/MS Chromatograms of Microcystins on Isocratic Mode: Suitable for Samples of Medium Complexity

Do not use unnecessarily long HPLC gradients with easy samples. The use of an isocratic mode saves time and solvents. See suggested gradient programs for load and analytical columns in Tables SOP 9.2 and SOP 9.3, respectively. Figure SOP 9.3 shows a chromatogram of a MC standard mixture as an example.

SOP 9.2.6 Online SPE HPLC-ESI-MS/MS Chromatograms of Microcystins on Elution Gradient: Suitable for Complex Samples

Complex field or strain samples may require the use of a long elution gradient and thus be able elucidate the toxin profile [7]. See suggested gradient programs for load and analytical columns in Tables SOP 9.4 and 9.5, respectively.

Table SOP 9.2 *Suggested gradient programme and operation mode of switching valve for C_{18} load column like Thermo Hypersil Gold C_{18} 20×2.1 mm, 12-µm particle size on Thermo Electron's EQuan environmental quantitation online SPE HPLC-ESI-MS/MS system. Linear gradient at a flow rate of 1 mL min^{-1}; injection cycle about 22 min*

Time (min)	Switching valve mode	Flow (mL min^{-1})	%A	%B
0	Loading	1	95	5
1.2	Eluting	1	95	5
2.0		1	95	5
2.2		0	95	5
13.0	Loading	0	95	5
14.9		0	95	5
15.0		1	5	95
16.0		1	5	95
16.1		1	95	5
22.0		1	95	5

Table SOP 9.3 *Suggested isocratic programme at flow rate of 200 µL min^{-1}, for an analytical column like Phenomenex Luna C_{18} (2) 150×2.0 mm, 5-µm particle size; injection cycle about 22 min*

Time (min)	%A	%B
0	70	30
22	70	30

Figures SOP 9.4 and 9.5 show chromatograms of commercial standard mixture and a natural bloom sample as examples.

SOP 9.3 Validation

Linearity: The calibration solutions of microcystins should be prepared daily in an appropriate concentration range ($n \geq 5$) covering the WHO maximum guideline level of 1 µg L^{-1}, with nodularin as the internal standard, and should be subjected to the overall online procedure. When microcystin standards with concentrations between 0.02 and 10 µg L^{-1} ($n = 7$) and with 1 µg L^{-1} nodularin were analysed, satisfactory linearities for all tested microcystins were achieved with coefficient of determination values (r^2) higher than 0.98 and residuals lower than 25%.

Matrix effect: Calibration standard solutions in three different matrices: drinking, surface and bottled water was prepared, obtaining equivalent calibration curves for the three matrices. As no matrix effect was observed the validation in this SOP was conducted only in bottled water.

Method accuracy and precision: Recovery experiments ($n \geq 3$) using several matrix samples (for example surface and drinking water) spiked at two or three concentration levels (low, medium and high concentrations) should be performed to evaluate method accuracy and precision. The calculated recoveries should be in the 50–130% range, and precision can be considered acceptable if relative standard deviations (RSDs) are lower than 25%.

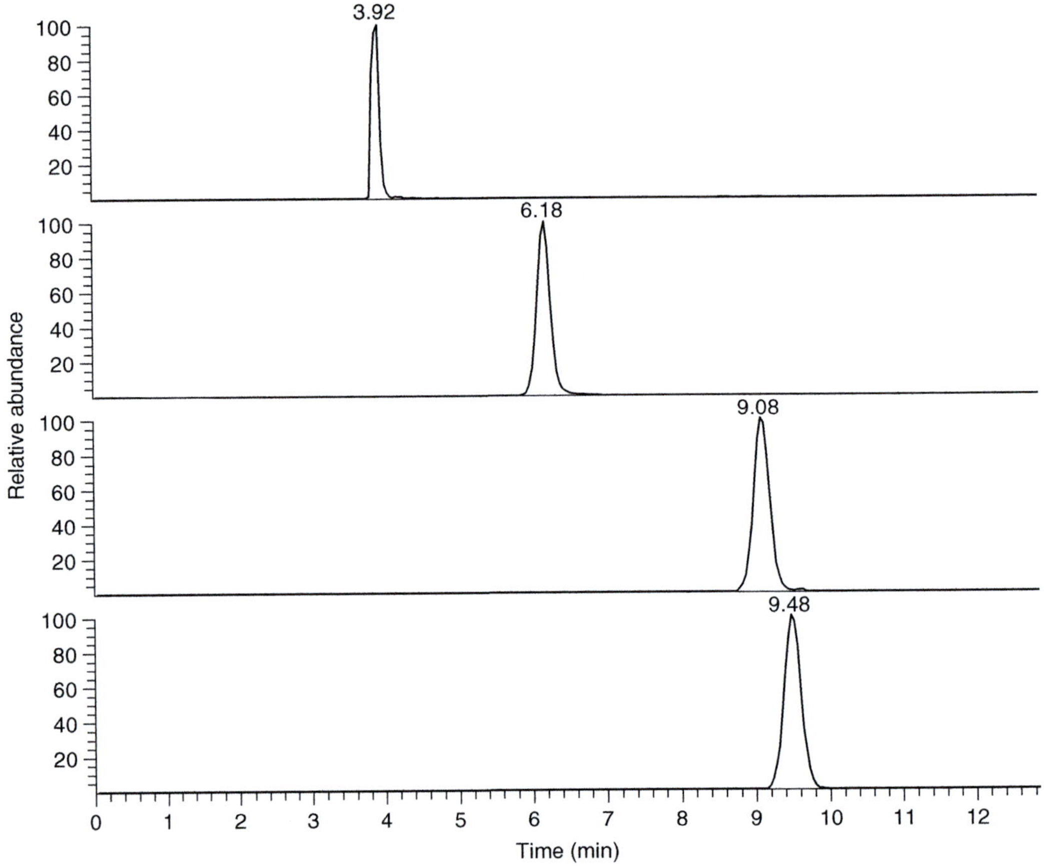

Figure SOP 9.3 Online SPE HPLC-ESI-MS/MS chromatograms of microcystin extracted SRM traces from a standard mixture of 100 ng L^{-1} (-RR, -YR, and -LR and nodularin) on an isocratic mode. Load column: Thermo Hypersil Gold C_{18} 20 × 2.1 mm, 12-μm particle size and analytical column: Phenomenex Luna C_{18} (2) 150 × 2.0 mm, 5-μm particle size. Retention times: MC-RR (3.9 min); NOD (6.2 min); MC-YR (9.1 min) and MC-LR (9.5 min). The total time is 11 min for the analysis of MC-LR, -RR and -YR and 22 min including pre-equilibration process

Table SOP 9.4 Suggested gradient programme and operation mode of switching valve for C_{18} load column like Thermo Hypersil Gold C_{18} 20 × 2.1 mm, 12-μm particle size on Thermo Electron's EQuan environmental quantitation online SPE HPLC-ESI-MS/MS system. Linear gradient at a flow rate of 1 mL min^{-1}; injection cycle about 70 min

Time (min)	Switching valve mode	Flow (mL min^{-1})	%A	%B
0	Loading	1	95	5
1.2	Eluting	1	95	5
2.0		1	95	5
2.2		0	95	5
63.4	Loading	0	95	5
64.9		0	95	5
65.0		1	5	95
66.0		1	5	95
66.1		1	95	5
70.0		1	95	5

Table SOP 9.5 *One possible gradient programme for an analytical column like Phenomenex Luna C_{18} (2) 150 × 2.0 mm, 5-μm particle size. Linear gradient at a flow rate of 200 μL min^{-1}; injection cycle about 70 min*

Time (min)	%A	%B
0	90	10
2.0	90	10
12.0	70	30
32.0	65	35
47.0	45	55
52.0	45	55
54.0	10	90
57.0	10	90
57.1	90	10
70.0	90	10

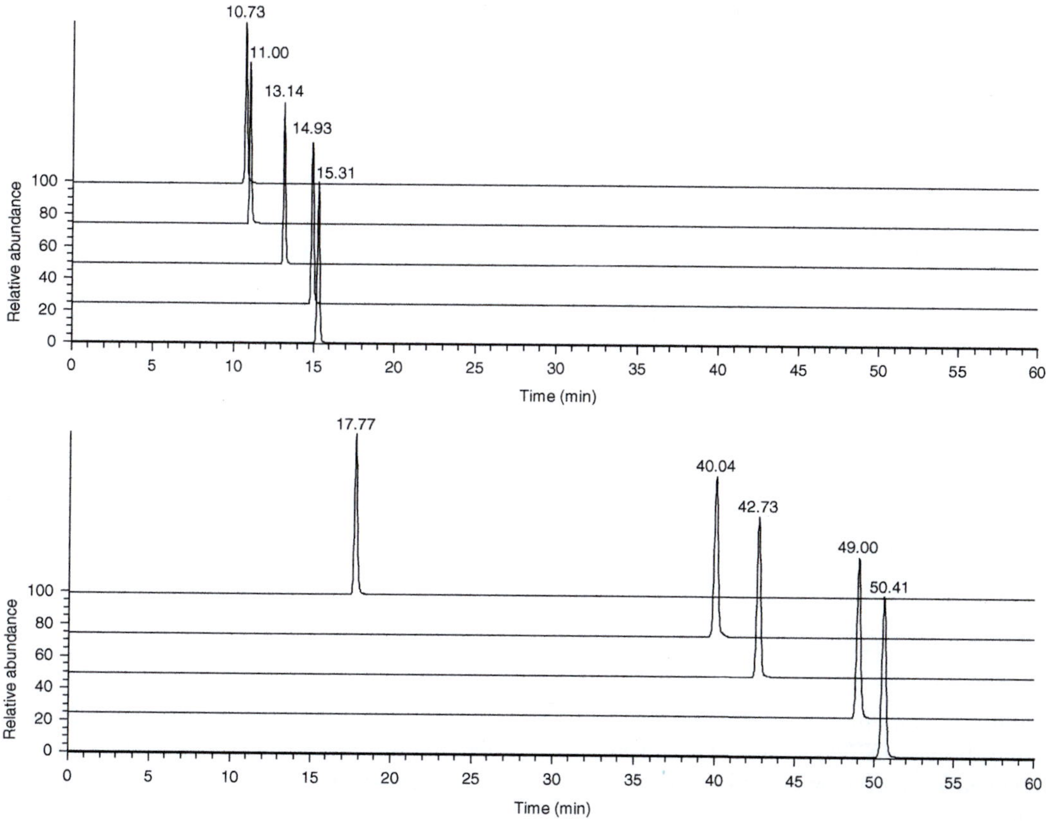

Figure SOP 9.4 *Overlaid extracted SRM traces of microcystins from a mixture of commercial standards at 100 ng L^{-1} by online SPE HPLC-ESI-MS/MS analysis using an elution gradient. Load column: Thermo Hypersil Gold C_{18} 20 × 2.1 mm, 12-μm particle size; and analytical column: Phenomenex Luna C_{18} (2) 150 × 2.0 mm, 5-μm particle size. Retention times: dm-MC-RR (10.7 min); MC-RR (11.0 min); nodularin (13.1 min); MC-YR (14.9 min); MC-LR (15.3 min); MC-WR (17.8 min); MC-LA (40.0 min); MC-LY (42.7 min); MC-LW (49.0 min); and MC-LF (50.4 min). The total analysis time is 70 min including pre-equilibration process*

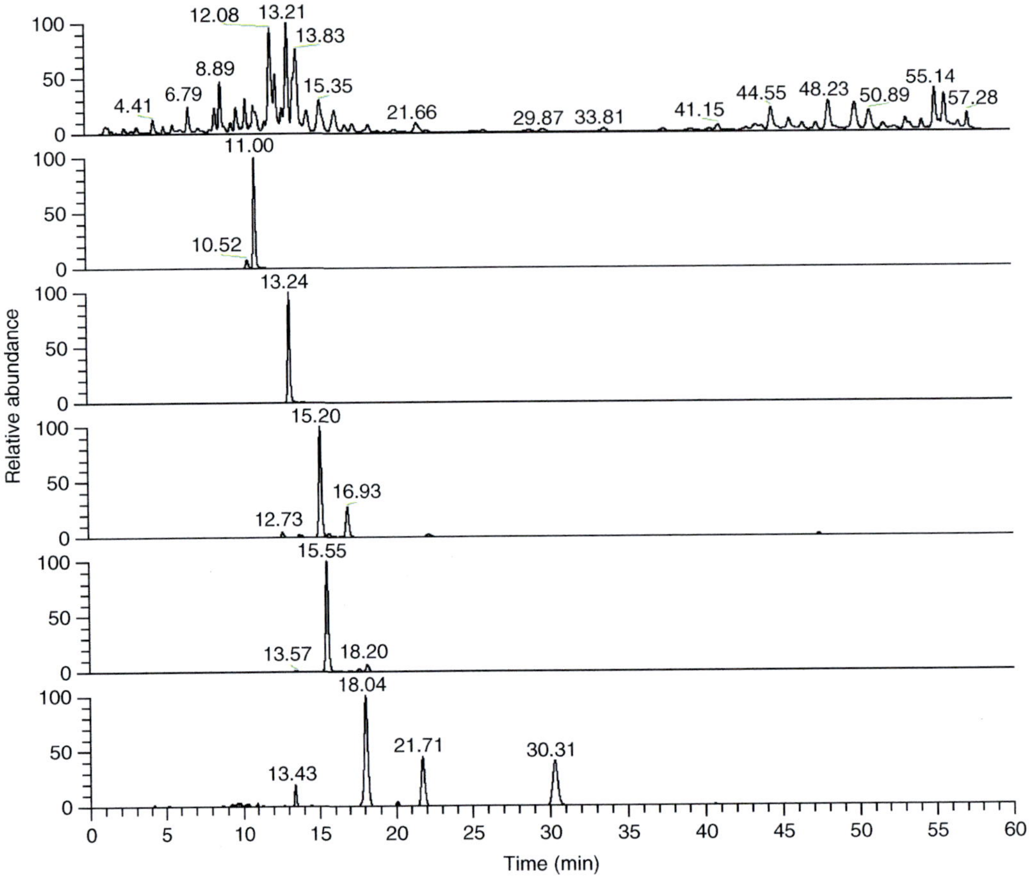

Figure SOP 9.5 Online SPE HPLC-ESI-MS/MS total ion and SRM extracted chromatograms of microcystins detected in a natural bloom sample using an elution gradient. The different microcystin concentrations ranging from 0.1 to 15 μg L^{-1}. Load column: Thermo Hypersil Gold C_{18} 20×2.1 mm, 12-μm particle size; and analytical column: Phenomenex Luna C_{18} (2) 150×2.0 mm, 5-μm particle size. Retention times: MC-RR (11.0 min); nodularin (internal standard) (13.2 min); MC-YR (15.2 min); MC-LR (15.5 min); and MC-WR (18.0 min). The total analysis time is 70 min including pre-equilibration process

Limit of quantitation (LOQ) and limit of detection (LOD): The LOD may be estimated for a signal-to-noise (S/N) ratio equal to 3 from the chromatograms of the samples spiked at the lowest validated level. The lowest validated level (0.02 μg L^{-1}) of the linear range may be considered as the LOQ.

Table SOP 9.6 presents a summary of the main analytical parameters validated for some typical microcystins.

N.B. The reader is also advised to follow up the development of the ISO 20179:2005 standard method for microcystin analysis [8].

Table SOP 9.6 Validation parameters

Compound	LOD (ng L^{-1})	LOQ (ng L^{-1})	% Recovery			Precision (% RSD) 1 μg L^{-1}	
			0.02 μg L^{-1}	1 μg L^{-1}	10 μg L^{-1}	Interday (n=5)	Intraday (n=5)
Microcystin-LR	6.0	20	101	102	94	17	9
Microcystin-RR	3.0	20	99	97	102	7	26
Microcystin-YR	14	20	90	121	89	16	22

References

[1] Meriluoto, J. and Codd, G.A. (eds.) (2005) TOXIC–Cyanobacterial Monitoring and Cyanotoxin Analysis, Abo Akademi University Press.

[2] Beltrán, E., Ibáñez, M., Sancho, J.V. *et al.* (2012) Determination of six microcystins and nodularin in surface and drinking waters by online solid-phase extraction-ultra high pressure liquid chromatography tandem mass spectrometry. *Journal of Chromatography A* **1266**, 61–68.

[3] Lee, H.S., Jeong, C.K., Lee, H.M. *et al.* (1999) Online trace enrichment for the simultaneous determination of microcystins in aqueous samples using high-performance liquid chromatography with diode-array detection. *Journal of Chromatography A* **848**, 179–184.

[4] Rivasseau, C., Vanhoenacker, G., Sandra, P., *et al.* (2000) Online solid-phase extraction in microcolumn-liquid chromatography coupled to UV or MS detection: Application to the analysis of cyanobacterial toxins. *Journal of Microcolumn Separations* **12**, 323–332.

[5] Shan, Y., Shi, X., Dou, A., *et al.* (2011) A fully automated system with online micro solid-phase extraction combined with capillary liquid chromatography-tandem mass spectrometry for high throughput analysis of microcystins and nodularin-R in tap water and lake water. *Journal of Chromatography A* **1218**, 1743–1748.

[6] Commission Decision 2002/657/EC (2002) implementing Council Directive 96/23/EC concerning the performance of analytical methods and the interpretation of results.

[7] Flores, C. and Caixach, J. (2015) An integrated strategy for rapid and accurate determination of free and cell-bound microcystins and related peptides in natural blooms by liquid chromatography-electrospray-high resolution mass spectrometry and matrix-assisted laser desorption/ionization time-of-flight/time-of-flight mass spectrometry using both positive and negative ionization modes. *Journal of Chromatography A* **1407**, 76–89.

[8] ISO 20179: 2005 Water quality–Determination of microcystins–method using solid-phase extraction (SPE) and high performance liquid chromatography (HPLC) with ultraviolet (UV) detection.

SOP 10

Determination of Microcystins and Nodularin in Filtered and Drinking Water by LC-MS/MS

Theodoros M. Triantis[1], Triantafyllos Kaloudis[2], Sevasti-Kiriaki Zervou[1], and Anastasia Hiskia[1]

[1] *Institute of Nanoscience & Nanotechnology, National Center for Scientific Research "DEMOKRITOS," Athens, Greece*

[2] *Water Quality Control, Athens Water Supply and Sewerage Company – EYDAP SA, Athens, Greece*

SOP 10.1 Introduction

This SOP specifies a method for the determination of cyclic peptide cyanotoxins, microcystins (MCs), and nodularin (NODs) by LC-MS/MS in filtered water and drinking water samples. The method can be applied after solid-phase extraction of the target compounds using SOP 7.

The method has been tested for the target compounds shown in Table SOP 10.1. When additional MCs are targeted, it is recommended that sufficient chromatographic separation is achieved, as well as specific MRM transitions selected and recorded for every target compound.

Extracts produced by SOP 7 are analyzed using reversed phase HPLC and tandem triple-quadrupole mass spectrometry (MS/MS). This is an in-house method based on methods described in [1–3].

SOP 10.2 Experimental

SOP 10.2.1 Materials

- Ultrapure water, 18.2 MΩ cm
- Acetonitrile HPLC grade

Handbook of Cyanobacterial Monitoring and Cyanotoxin Analysis, First Edition.
Edited by Jussi Meriluoto, Lisa Spoof and Geoffrey A. Codd.

Table SOP 10.1 *Registry information of target compounds*

Name	Formula	M.W.	CAS number
Microcystin-RR (MC-RR)	$C_{49}H_{75}N_{13}O_{12}$	1038.20	111755-37-4
Nodularin (NOD)	$C_{41}H_{60}N_{8}O_{10}$	824.96	118399-22-7
Microcystin-YR (MC-YR)	$C_{52}H_{72}N_{10}O_{13}$	1045.19	101064-48-6
Microcystin-LR (MC-LR)	$C_{49}H_{74}N_{10}O_{12}$	995.17	101043-37-2
Microcystin-LA (MC-LA)	$C_{46}H_{67}N_{7}O_{12}$	910.06	96180-79-9

- Formic acid (HCOOH), analytical grade
- Methanol analytical grade
- Volumetric flasks 10 mL, A Class
- Pipettes 10–100 μL, 100–1000 μL, 1–10 mL, with confirmed calibration status
- Chromatographic column: reverse phase (e.g. C_{18} 30 mm × 2.1 mm, 3-μm particle size)

SOP 10.2.2 Special Equipment

- Spectrophotometer UV/Vis with quartz cells, suitable for measurements of absorbance at 220–250 nm, with confirmed calibration status
- Ultrasonic bath
- LC-MS/MS system (triple quadrupole), consisting of:
  a. HPLC pump and mixer suitable for gradient elution (mixing of two solvents)
  b. Tandem mass spectrometric detector, consisting of an Electrospray ionization source (ESI) and a triple quadrupole mass analyzer
  c. Autosampler, for liquid samples (e.g. 2-mL vials) is recommended.
  d. Software for analyzing the results

SOP 10.2.3 Solutions

- Dilution solvent: methanol/water 20:80 v/v

SOP 10.2.4 Calibration Standards and Quality Control Standard

Calibration standards can be prepared by dilutions of a standard mixture of target compounds. The standard mixture is prepared by mixing individual standard stock solutions of MCs and NODs. The individual stock solutions are prepared by dissolving suitable amounts of the pure compounds in methanol and measuring the exact concentrations with photometry. Commercially available standard solutions for calibration (e.g. 10 μg mL^{-1} in methanol) that are accompanied by appropriate certificates of analysis can be used as an alternative.

In the case where pure compounds are used to prepare calibration standards, individual stock solutions of the target compounds (e.g. 25–100 μg mL^{-1}) are prepared by dilution in methanol, sonication for 15 min and measurement of the absorbance at 220–250 nm in a spectrophotometer using a quartz cell (e.g. with 0.2-cm optical path).

The concentration of target compounds (mg L^{-1}) in the individual stock solutions is determined according to the formula:

$$C = A_{max} M_i x 1000 / \varepsilon_i x d \qquad \text{(SOP 10.1)}$$

where:

A_{max} = the absorbance at the maximum point (corrected for blank)

M_i = the molar mass of the target compound, g mol^{-1} (Table SOP 10.1)

ε_i = the molar extinction coefficient of the compounds in methanol, e.g. equal to 39,800 L mol^{-1} cm^{-1} for MC-LR. Since different ε_i values are reported in the literature or in certificates of analysis, those used by the laboratory should be documented so that comparison of results between laboratories is possible.

d = the optical path of the cell, in cm

Individual stock solutions are kept at –20°C.

Depending on the range of calibration, a standard mixture solution is prepared (e.g. 200 µg L^{-1}) which is further diluted to produce the calibration standard solutions (e.g. 5–100 µg L^{-1}). Dilutions are carried out in such a way that the final calibration solutions are in methanol/water 20:80 v/v.

A quality control standard (e.g. 200 µg L^{-1}) is prepared using the above procedure. The quality control standard is used for spiking blank or real samples (e.g. with 100 ng L^{-1}) in order to routinely check the performance of the analysis (see also SOP 7). It is recommended that the quality control standard is prepared by independent standards from those used for calibration.

Whenever significant matrix effects are observed, it is recommended that calibration and quality control is carried out with matrix-matched samples.

SOP 10.2.5 General Procedure

SOP 10.2.5.1 Liquid Chromatography

Reversed-phase chromatographic separation of the target compounds can be achieved under various experimental conditions, e.g.:

Column: C_{18} 30 mm × 2.1 mm, 3-µm particle size

Column temperature: 30°C

LC mobile phase: Solvent A: acetonitrile - 0.5% HCOOH, Solvent B: water - 0.5% HCOOH

Gradient: 25% Solvent A (0–1 min) to 70% A (10.5 min) to 90% A (12.5–13.5 min) to initial conditions

Flow rate: 0.2 mL min^{-1}

Injection volume: 20 µL

SOP 10.2.5.2 Mass Spectrometry

Mass spectrometry operation parameters for the detection and quantitative measurement of the target compounds in positive ionization mode, ESI+ (e.g. spray voltage, ion transfer capillary temperature, sheath gas, auxiliary gas, probe position, collision gas pressure, peak width) should be optimized according to manufacturers' guides and laboratory practices.

Identification and quantitative determination of target compounds is based on the precursor to product ion transitions (SRMs) and the chromatographic retention times. Indicative retention times (obtained through experimental conditions) and typical transitions of the target analytes are shown in Table SOP 10.2.

Typical MRM chromatograms obtained under the above experimental conditions are shown in Figure SOP 10.1.

Table SOP 10.2 Retention times and selected transitions for the determination of target compounds

Compound	Retention time (min)	Precursor ion *m/z*	Product ions *m/z*
MC-RR	4.7	519.8 $[M+2H]^{2+}$	135*
			213
Nodularin	8.3	825.3 $[M+H]^{+}$	135*
			227
MC-YR	9.3	1045.4 $[M+H]^{+}$	135*
			213*
MC-LR	9.7	995.5 $[M+H]^{+}$	135*
			213
MC-LA	11.7	932.2 $[M+Na]^{+}$	508*
			347*
		910.4 $[M+H]^{+}$	776*
			375

* Quantifier ions (indicative; the laboratory should use the most abundant/reproducible transitions achieved in the laboratory's analytical system).

SOP 10.3 Calibration

Calibration is carried out prior to every batch of sample analysis. The following calibration alternatives can be used:

- Single-point calibration: A calibration standard (e.g. 100 µg L^{-1}) can be used. It is recommended that the sample response should be within ±50% of the calibration response unless further extrapolation is supported by evidence of acceptable linearity of response.
- Two-point "bracketing" calibration. The responses of samples should fall within the bracketing range.
- Multiple point calibration: A calibration function is calculated and used between the lowest and highest calibration levels. In this method the calibration curve is linear and should not be forced through the origin.

SOP 10.4 Sample Analysis – Calculations

A typical sequence of analysis is shown in Table SOP 10.3. Calibration and quality control are repeated every 10 samples. Detection and identification of target compounds is based on the retention times, characteristic ions, and ion ratios of the compounds. Quantification of target compounds is based on the areas of the quantifier ions (see Table SOP 10.2).

SOP 10.5 Quality Control

A recommended quality control scheme to be carried out with every batch of analysis is shown in Table SOP 10.4.

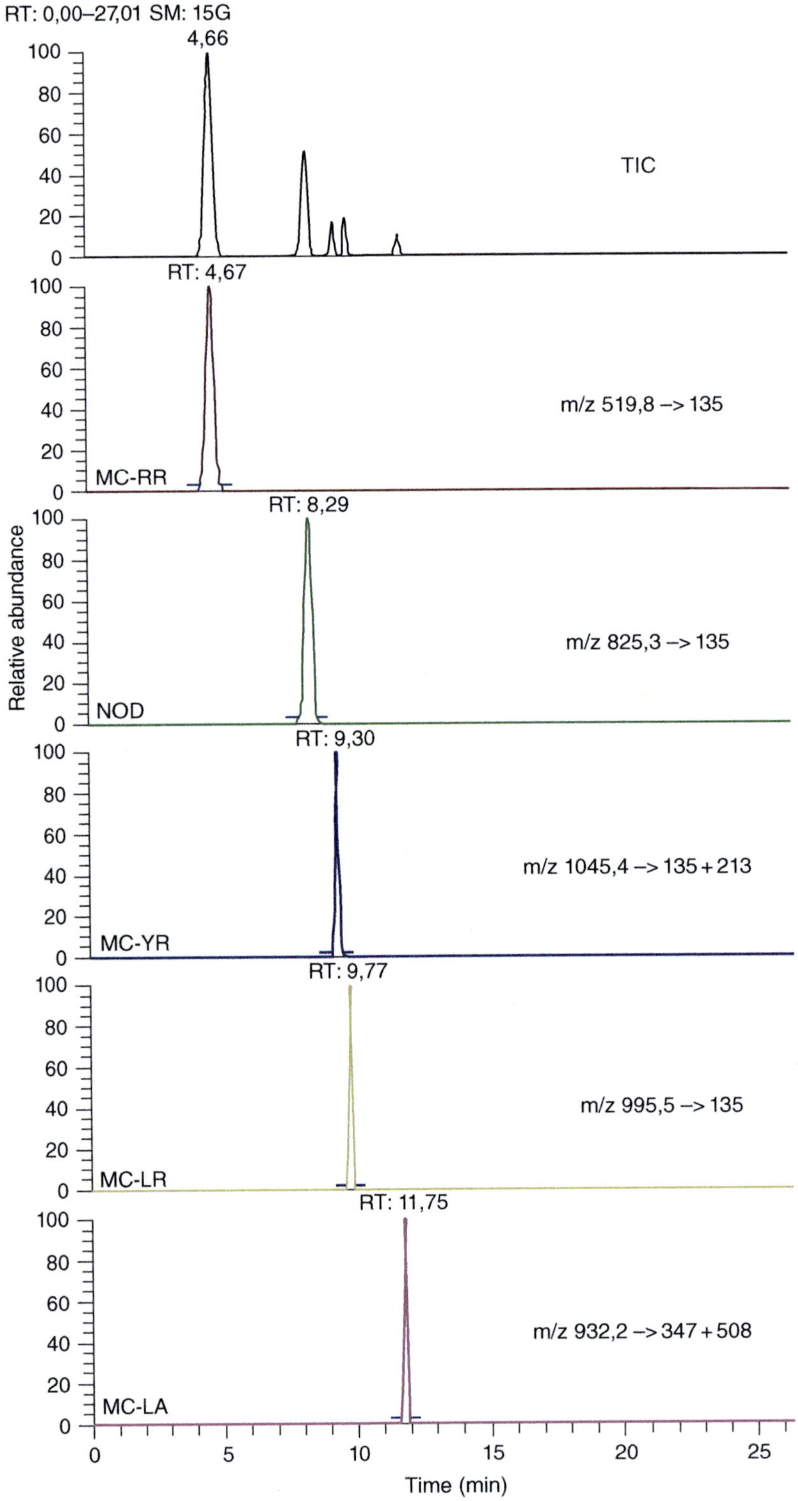

Figure SOP 10.1 Total ion chromatogram and typical MRM chromatograms of the five target compounds (100 μg L^{-1})

Table SOP 10.3 Recommended sequence of analysis

No.	Sample	Comments
1	Blank (solvent methanol/water 20:80 (v/v)	Check for carryover
2	Calibration Standard(s)	Calibration – System suitability checks
3	Quality control sample	Check for recovery
4	Blank sample	Check for carryover and contamination
5	Sample 1	Detection, Identification, Quantitation
6	Sample 2	Detection, identification, quantitation
...	....	....
13	Sample 10	Detection, identification, quantitation
Go back to No. 1 (recalibration)		

Table SOP 10.4 Quality control scheme (recommended)

No.	Check	Measurements	Criteria
1	MS/MS performance	MS/MS calibration	Manufacturer's criteria[a]
2	Carryover and contamination	Blanks (rows 1, 4 Table SOP 10.3)	Target compounds not detected.
3	Calibration	Calibration standard(s), (row 2, Table SOP 10.3)	r^2 and slope of calibration curve
4	System suitability	One calibration standard (row 2, Table SOP 10.3)	Resolution (e.g. MC-YR/MC-LR) or theoretical plates
5	Recoveries	QC sample (row 3, Table SOP 10.3)	Control chart limits
6	Positive samples	Identification parameters	Retention times, ion ratios

[a] Frequency of calibration according to manufacturer and laboratory experience.

SOP 10.6 Expression of Results

Results are expressed in μg L^{-1} with three decimal digits. Results below limits of detection (LOD) are expressed as "not detected." Results above LOD and below limits of quantification (LOQ) are expressed as "< LOQ."

SOP 10.7 Validation Data

This method was validated in-house using filtered water and drinking water as matrices. Summarized validation results can be found in [3], obtained from the combination of SOP 7 (SPE extraction) and the present SOP 10 (determination by LC-MS/MS). The combination of these SOPs is proved to be suitable for determinations at the low ng L^{-1} levels of the target compounds. LOQs and LODs for drinking water are well below (<<10%) the provisional limit of WHO for MC-LR (1 μg L^{-1}).

References

[1] ISO 20179:2005. (2005) Water quality–Determination of microcystins–Method using solid phase extraction (SPE) and high performance liquid chromatography (HPLC) with ultraviolet (UV) detection. International Standards Organization, Technical Committee ISO TC147/SC2.

[2] Sjövall, O. and Meriluoto, J. (2005) Analysis of microcystins by liquid chromatography-mass spectrometry, in *TOXIC–Cyanobacterial Monitoring and Cyanotoxin Analysis*, (eds. J. Meriluoto and G. A. Codd) Åbo Akademi University Press, Åbo, pp. 105–110.

[3] Kaloudis, T., Zervou, S.-K., Tsimeli, K. *et al.* (2013) Determination of microcystins and nodularin (cyanobacterial toxins) in water by LC–MS/MS. Monitoring of Lake Marathonas, a water reservoir of Athens, Greece, *Journal of Hazardous Materials* **263** (1), 105–115.

SOP 11

Analysis of Microcystins and Nodularin by Ultra High-Performance Liquid Chromatography Tandem Mass Spectrometry

Leonardo Cerasino

IASMA Research and Innovation Centre, Fondazione Edmund Mach-Istituto Agrario di S. Michele all'Adige, S. Michele all'Adige, Trento, Italy

SOP 11.1 Introduction

Ultra high-performance LC (UHPLC) is a recent introduction in the LC technology and offers some advantages (reduction of experiment times, reduction of solvent consumption) compared with conventional HPLC, maintaining the same chromatographic efficiency. In this SOP, the application of UHPLC-MS/MS methodology to the analysis of microcystins (MCs) and nodularins (NODs) in water samples will be described. The proposed method allows the determination of nine MCs and NOD-R in 7-min run time [1–4]. The SOP reports details about instrumentation and LC column that have been used for the method development and optimization; the method can be, however, adapted to any equivalent instrumental configuration.

SOP 11.2 Experimental

SOP 11.2.1 Materials

- Acetonitrile (ACN) LC-MS grade
- Methanol LC-MS grade
- Ultrapure water, 18.2 MΩ cm

Handbook of Cyanobacterial Monitoring and Cyanotoxin Analysis, First Edition.
Edited by Jussi Meriluoto, Lisa Spoof and Geoffrey A. Codd.

- Formic acid LC-MS grade
- Microcystin standards (-[D-Asp^3]RR, -RR, -YR, -LR, -WR, -LA, -LY, -LW, -LF) and NOD-R (25 μg each)

SOP 11.2.2 Special Equipment

- The UHPLC used for the development of the present SOP was a Waters Acquity UPLC system equipped with an autosampler carrying a 10-μL loop, a binary pump, a column manager equipped with a Phenomenex Kinetex XB-C18 column (50×2.1 mm, 1.7-μm particle size) or equivalent. The LC was connected without splitting to the mass detector.
- The mass spectrometer was an AB Sciex 4000 QTRAP triple quadrupole, equipped with a TurboIonspray (ESI) source. High purity nitrogen was used as curtain, heater, nebulizer, and collision gas.

SOP 11.2.3 Solutions

a. 30% Methanol in water. Mix 30 volumes of methanol with 70 volumes of water.
b. 50% Methanol in water. Mix 50 volumes of methanol with 50 volumes of water.

SOP 11.2.4 Calibration Standards

a. Bulk solutions of individual standards (5 mL, concentration 5 μg mL^{-1}). Dissolve 25 μg of MCs in 5 mL of 30% methanol. Nod-R is dissolved preferably in 50% methanol. The solutions can be stored at −20°C for several months.
b. Multistandard bulk solution (5 mL, concentration 100 ng mL^{-1}). Mix together 100 μL of each of the 10 standard solutions (a) and add 4000 μL of 30% methanol. Prepare from single standard solutions (a) at every analytical session.
c. Calibration solutions (1 mL, concentrations 100, 75, 50, 25, 10, 5, 1 ng mL^{-1}). Dilute the bulk solution (b) with the appropriate amount of 30% methanol.

SOP 11.2.5 HPLC Mobile Phases

a. Component A: 0.1% formic acid in 5% ACN. Add 50 mL of ACN to 950 mL of water; then, add 1 mL of formic acid.
b. Component B: 0.1% formic acid in ACN. Add 1 mL of formic acid to 1000 mL ACN.

Please observe the following:

- Eluents must be renewed weekly.
- Formic acid and ACN must be handled in a fume-hood.

SOP 11.2.6 UHPLC-MS/MS

SOP 11.2.6.1 Example of LC Settings

Injection volume: 2 μL
Injection mode: partial loop with needle overfill
Column compartment temperature: 40°C
Flow rate: 0.250 mL min^{-1}

Table SOP 11.1 Gradient description

Time (min)	% A	% B	Curve
0.0	80	20	
4.5	30	70	7
6.5	80	20	6
7.0	80	20	6

Mobile phase: Gradient of two components, A (SOP 11.2.5.a) and B (SOP 11.2.5.b), according to the scheme reported in Table SOP 11.1.
Run time: 7 min

SOP 11.2.6.2 Example of Mass Detector Settings

Note: Please observe that every instrument model must be tuned individually.

Source settings:

Ionization mode: positive electro spray (ESI+)
Ion spray voltage: 5000 V
Heater gas temperature: 300°C
Curtain gas (CUR): 20
Nebulizer gas (GS1): 20
Heater gas (GS2): 20

Triple quadrupole settings:

Scan type: scheduled MRM
MRM detection window: 60 sec
Target scan time: 60 sec
Settling time: 0 msec
Entrance potential (EP): 10 V
Collision cell exit potential (CXP): 10 V

Compound-specific settings are reported in Table SOP 11.2.

SOP 11.3 Identification of Analytes

The identification of the analytes is based on perfect matching with the characteristics of the corresponding standard: presence of the two MRM transitions (Q and q), same Q/q ratio, and same retention time. See Figure SOP 11.1.

SOP 11.4 Quantification of Analytes

Quantification of analytes is performed by the external standard procedure, and for each analyte the area under the curve of the Q transition is considered. A standard curve is prepared using the multistandard solutions (SOP 11.2.4.b) in the range 1–100 ng mL^{-1}. A regression

Table SOP 11.2 *Compound-specific mass detector settings*

Toxin variant	Precursor ion (m/z)	RT (min)	Product ion (m/z)	DP (V)	CE (V)	Q/q ratio
[D-Asp^3]RR	512.8 $[M+2H]^{2+}$	1.20	135 Q	85	44	17
			213 q	85	50	
RR	520.1 $[M+2H]^{2+}$	1.43	135 Q	85	44	15
			213 q	85	50	
NOD-R	825.6 $[M+H]^{+}$	2.20	135 Q	140	83	1
			70 q	90	100	
YR	523.6 $[M+2H]^{2+}$	2.63	135 Q	45	20	6
			213 q	45	20	
LR	498.6 $[M+2H]^{2+}$	2.78	135 Q	40	19	17
			213 q	40	43	
WR	535.0 $[M+2H]^{2+}$	3.06	135 Q	40	18	18
			213 q	40	42	
LA	911.6 $[M+H]^{+}$	4.30	135 Q	85	90	2
			213 q	85	65	
LY	1002.6 $[M+H]^{+}$	4.40	135 Q	106	96	2
			213 q	106	75	
LW	1025.6 $[M+H]^{+}$	4.80	135 Q	111	100	2
			213 q	111	80	
LF	986.6 $[M+H]^{+}$	4.93	135 Q	96	95	2.5
			213 q	96	73	

Two transitions have been monitored per each analyte; the most intense is used for quantification (Q), the less intense for confirmation (q). RT = chromatographic retention time. DP = declustering potential. CE = collision energy.

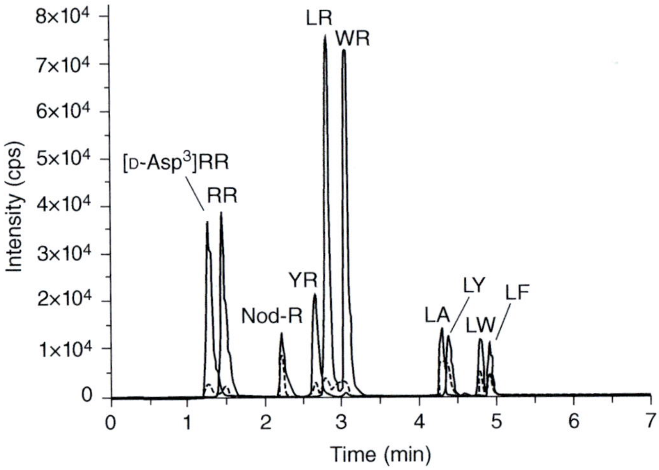

Figure SOP 11.1 *Representative LC-MS/MS chromatogram showing the MRM transitions in a multistandard sample (100 ng mL^{-1}). Two MRM transitions are monitored for each analyte (solid and dashed lines)*

Table SOP 11.3 Data for quantification of toxins

Toxin	Regression eq.	R^2	LOD (ng mL^{-1})	LOQ (ng mL^{-1})
[D-Asp^3]RR	$y = 993.59x + 468.92$	0.9999	0.40	1.26
RR	$y = 1766.2x + 6157.2$	0.9932	0.64	2.20
NOD-R	$y = 20.702x - 3.7008$	0.9990	0.80	2.60
YR	$y = 252.73x + 254.99$	0.9994	0.90	2.86
LR	$y = 943.28x + 507.16$	0.9997	0.24	0.54
WR	$y = 454.4x - 197.57$	0.9987	0.18	0.62
LA	$y = 149.69x + 49.649$	0.9986	1.36	4.60
LY	$y = 76.804x + 79.295$	0.9997	1.52	5.00
LW	$y = 42.536x - 82.656$	0.9985	2.50	8.40
LF	$y = 55.326x + 33.074$	0.9980	2.64	9.00

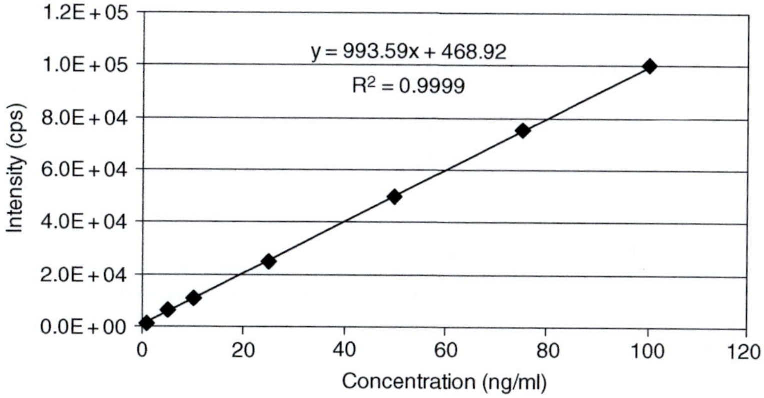

Figure SOP 11.2 Representative standard curve for quantification. The case of MC-[D-Asp^3]RR is reported

equation is then derived for each toxin, which is used for calculating the concentrations of unknown samples. In Table SOP 11.3, regression equations for the toxins considered in this SOP are reported. Limits of detections (LOD) and limits of quantifications (LOQ) have been calculated for each toxin as the concentrations for which a 1:3 and a 1:10 signal-to-noise ratio has been observed, respectively (Table SOP 11.3). In Figure SOP 11.2, an example of calibration curve is reported.

Whether higher sensitivity is required, a preliminary preconcentration by SPE step can be performed, according to the procedures described in other SOPs in the present handbook.

SOP 11.5 Validation

In Table SOP 11.4, the validation data obtained for the 10 MCs at two concentrations are reported.

Table SOP 11.4 *Validation data for 12 replicates recorded in two working sessions in different days*

Toxin variant	Spiked concentration ($ng\,mL^{-1}$)	Recovery (%)	RSD (%)
[D-Asp³]RR	2	117	10
	20	126	1
RR	2	105	10
	20	104	3
NOD-R	2	85	20
	20	89	7
YR	2	122	15
	20	127	5
LR	1	127	12
	10	128	7
WR	2	99	9
	20	117	5
LA	4	94	9
	40	91	3
LY	9	96	14
	90	113	6
LW	9	103	18
	90	94	4
LF	9	86	8
	90	126	4

Matrix: surface water. Recovery and relative standard deviation (RSD) are reported.

References

[1] Shams, S., Capelli, C., Cerasino, L. *et al.* (2015) Anatoxin-a producing *Tychonema* (Cyanobacteria) in European water bodies. *Water Research* **69**, 68–79.

[2] Cerasino, L. and Salmaso, N. (2012) Diversity and distribution of cyanobacterial toxins in the Italian sub-alpine lacustrine district. *Oceanological and Hydrobiological Studies* **41** (3), 54–63.

[3] Oehrle, S.A., Southwell, B. and Westrick, J. (2010) Detection of various freshwater cyanobacterial toxins using ultra-performance liquid chromatography tandem mass spectrometry. *Toxicon* **55** (5), 965–972.

[4] Spoof, L., Neffling, M.R. and Meriluoto, J. (2009) Separation of microcystins and nodularins by ultra performance liquid chromatography. *Journal of Chromatography B* **877** (30), 3822–3830.

SOP 12

Analysis of Microcystins in Animal Tissues Using LC-MS/MS

Jiří Kohoutek and Luděk Bláha

Faculty of Science, RECETOX, Masaryk University, Brno, Czech Republic

SOP 12.1 Introduction

Microcystins have been shown to accumulate in biota and their transfer in the food webs (including biomagnification) has been discussed [1]. Some authors reported fairly high concentrations of microcystins accumulated in fish tissues, which might be of toxicological concern for humans [2]. Analyses of microcystins in the food and food supplements are therefore important for critical assessment of health risks.

A number of methods for microcystin analyses in biotic matrices are described of which two major groups of analytical approaches are routinely used. The first group consist of biological based methods such as protein phosphatase inhibition assay (PPIA) and immunochemical ELISA [3]. The second group consists of instrumental analytical methods, such as liquid chromatography (LC) with UV detection. Microcystins have characteristic absorbance spectrum with the peak at 238 nm, which can be easily detected with optical detectors operating in the UV range. However, both biological and traditional instrumental methods suffer from interferences in complex matrices, and their application for microcystin analyses in biota is limited.

More recent approaches are based on liquid chromatography with tandem mass spectrometric detection, which has a number of advantages including selectivity, sensitivity, and robustness and seem to be a suitable tool for minimizing the above mentioned complications [4–6]. Assessment of microcystins by HPLC-MS/MS uses a high-efficiency C_{18} columns, which are able to separate

Handbook of Cyanobacterial Monitoring and Cyanotoxin Analysis, First Edition.
Edited by Jussi Meriluoto, Lisa Spoof and Geoffrey A. Codd.

interfering compounds and the target analytes and which are also relatively resistant to high protein content in tissue extracts. Due to the lack of isotopically labelled microcystin standards, quantification by mass spectrometry is based on the use of an external calibration.

SOP 12.2 Experimental

SOP 12.2.1 Materials

- Ultrapure water (conductivity lower than 0.1 μS cm^{-1})
- Methanol LC/MS grade
- Acetonitrile LC/MS grade
- Ammonium acetate p.a. grade
- Reverse-phase HPLC column. Suitable option is Supelcosil ABZ+Plus RP-18 endcapped (5 μm) 150×4.6 mm i.d. (Supelco, Bellefonte, PA) or similar. A compatible guard column is also recommended (e.g. SecureGuard C18 - Phenomenex, Torrance, CA).
- Borosilicate glass chromatographic vials: e.g. 1.8-mL amber glass vials. For small sample volumes, 0.3-mL borosilicate glass inserts can be used.
- Microcystin standards

SOP 12.2.2 Special Equipment

a. High-performance liquid chromatograph equipped with gradient pump, autosampler, and column oven
b. Tandem mass spectrometer equipped with electrospray type ion source

SOP 12.2.3 HPLC Mobile Phase

a. HPLC mobile phase component A: 5 mmol L^{-1} ammonium acetate in water, pH 4.
   i. Add 0.38 g of ammonium acetate into 1.0 L of water.
   ii. Keep in dark, cool place.
b. HPLC mobile phase component B: 5 mol L^{-1} ammonium acetate in acetonitrile
   i. Add 0.38 g of ammonium acetate in 1.0 L of acetonitrile.
   ii. Keep in dark, cool place.

Please observe the following:

- Work in fume-hood with acetonitrile and methanol.
- Disposal of acetonitrile and methanol should conform to local regulations.

SOP 12.2.4 HPLC-MS/MS

SOP 12.2.4.1 Chromatographic Separation

The parameters of the recommended linear gradient of mobile phase are given in Table SOP 12.1. The flow of a mobile phase is 0.4 mL min^{-1}. The temperature of column oven is set to 30°C; 10 to 20 μL of individual sample dissolved in 50% methanol in water is injected for the analysis.

Table SOP 12.1 *Suggested gradient programme (using the column Supelcosil ABZ+Plus RP-18 endcapped (5 μm) 150 × 4.6 mm i.d. linear gradient at a flow rate of 0.4 mL min^{-1})*

Time (min)	%A	%B
0.00	68	32
12.00	60	40
20.00	58	42
20.01	10	90
30.00	10	90
30.01	68	32
45.00	68	32

Table SOP 12.2 *Ion source parameters (typical values for Agilent 6410 Triple Quadrupole)*

Ion mode	ESI
Polarity	Positive
Capillary voltage (V)	5500
Drying gas temperature (°C)	350
Drying gas flow (L min^{-1})	11

SOP 12.2.4.2 Tandem Mass Spectrometric Detection

Ion source parameters are given in Table SOP 12.2. Observe that every instrument model must be tuned individually.

In multiple reaction monitoring (MRM) mode, the following *m/z* are monitored: MC-RR $[M+2H]^{2+}$ 519.8, MC-YR $[M+H]^{+}$ 1045.5, and MC-LR $[M+H]^{+}$ 995.5.

The transitions from the protonated molecular ion to a fragment of amino acid Adda at *m/z* 135.2 and fragment at *m/z* 127.1 are used for quantitation and qualification, respectively.

Collision energies (CE, V) used for fragmentation: MC-RR CE = 50 V; MC-YR and MC-LR CE = 40 V.

Quantification of analytes is based on external standards of MC-RR, MC-YR, and MC-LR in matrix (final extract of microcystin-free fish tissue). The mixture of microcystins is added into extract prior to analysis. The typical chromatogram is shown in Figure SOP 12.1.

SOP 12.3 Calibration and Method Validation

Calibration standard solutions are prepared in 1.8-mL amber vials—prepare eight calibration solutions (Cal1 to Cal8) containing microcystins at concentrations given in Table SOP 12.3. The Cal4 calibration standard solution is used as a daily verification standard.

Validation of the analytical method has been performed by repeated extraction and analyses of the spiked fish liver tissue ($N = 5$) using the extraction protocol described in SOP 8 (Extraction of Microcystins From Animal Tissues). The results of the validation are in Table SOP 12.4.

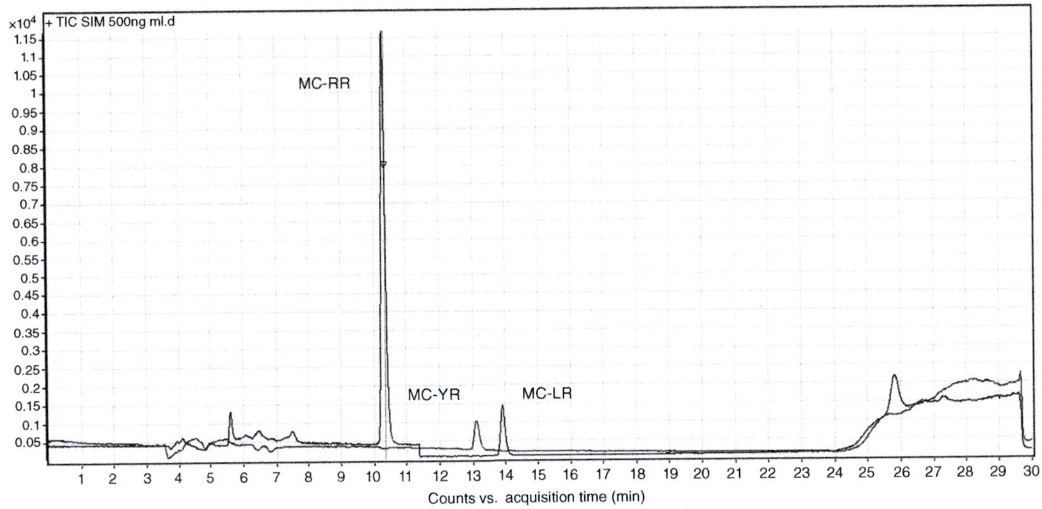

Figure SOP 12.1 Extracted ion chromatogram of MC-RR, -YR, and -LR in the extract of fish hepatopancreas (2.5 ng of each analyte per column was injected). Column Supelcosil ABZ+Plus RP-18 endcapped (5 μm) 150 × 4.6 mm i.d. Retention times: MC-RR 10.35 min, MC-YR 13.17 min, and MC-LR 13.92 min

Table SOP 12.3 Concentrations of microcystins in calibration solutions for determination by HPLC-MS/MS

Analyte	Concentration of calibration solution (ng mL^{-1}, in 50% methanol)							
	Cal1	Cal2	Cal3	Cal4	Cal5	Cal6	Cal7	Cal8
MC-RR	1.0	5.0	10.0	25.0	50.0	100.0	250.0	500.0
MC-YR	1.0	5.0	10.0	25.0	50.0	100.0	250.0	500.0
MC-LR	1.0	5.0	10.0	25.0	50.0	100.0	250.0	500.0

Table SOP 12.4 Method detection limit (MDL; per gram of tissue, fresh weight), recovery of analytes and respective RSD

Analyte	MDL (ng g f.w.$^{-1}$)	Recovery (%)	RSD (%)
MC-RR	1.2	69.5	22.5
MC-YR	5.4	107.0	35.0
MC-LR	5.4	89.8	22.6

References

[1] Mohamed, Z.A., Carmichael, W.W. and Hussein, A.A. (2003). Estimation of microcystins in the freshwater fish *Oreochromis niloticus* in an Egyptian fish farm containing a *Microcystis* bloom. *Environmental Toxicology* **18** (2), 137–141.

[2] Xie, L.Q., Xie, P., Guo, L.G. *et al.* (2005). Organ distribution and bioaccumulation of microcystins in freshwater fish at different trophic levels from the eutrophic Lake Chaohu, China. *Environmental Toxicology* **20** (3), 293–300.

[3] Carmichael, W.W. and An, J.S. (1999). Using an enzyme linked immunosorbent assay (ELISA) and a protein phosphatase inhibition assay (PPIA) for the detection of microcystins and nodularins. *Natural Toxins* **7** (6), 377–385.
[4] Vasconcelos, J.F., Barbosa, J.E.L., Lira, W. *et al.* (2013) Microcystin bioaccumulation can cause potential mutagenic effects in farm fish. *Egyptian Journal of Aquatic Research* **39** (3), 185–192.
[5] Kim, I.S., Nguyen, G.H., Kim, S.-J. *et al.* (2015) Qualitative analysis of the most toxic and abundant microcystin variants (LR, RR, and YR) by using LCMS–IT–TOF. *Journal of Industrial and Engineering Chemistry* **29** (1), 375–381.
[6] Kohoutek, J., Adamovský, O., Oravec M. *et al.* (2010) LC-MS analyses of microcystins in fish tissues overestimate toxin levels–critical comparison with LC-MS/MS. *Analytical and Bioanalytical Chemistry* **398** (3), 1231–1237.

SOP 13

Quantitative Screening of Microcystins and Nodularin in Water Samples with Commercially Available ELISA Kits

Triantafyllos Kaloudis[1], Theodoros M. Triantis[2], and Anastasia Hiskia[2]

[1] *Water Quality Control, Athens Water Supply and Sewerage Company – EYDAP SA, Athens, Greece*
[2] *Institute of Nanoscience & Nanotechnology, National Center for Scientific Research "DEMOKRITOS," Athens, Greece*

SOP 13.1 Introduction

This SOP includes a procedure for the quantitative screening of microcystins (MCs) and nodularins (NODs) in water samples (surface and drinking waters) with use of commercially available ELISA kits. Commercially available ELISA kits are developed and tested in order to be used according to the analytical protocols that accompany the kits; therefore, it is recommended that users follow the manufacturer's instructions. This SOP gives additional guidance and recommendations regarding the analysis of MCs and NODs by ELISA kits, including quality control and validation procedures. Since ELISA methods for MCs and NODs should be considered as "screening methods," positive samples should be analysed with LC-MS/MS, especially when there is no other information about the sample or the history of the tested water body regarding the presence of MCs and NOD.

Commercially available ELISA kits are competitive assays in which the target toxins in the sample compete with enzyme-labelled toxins or microcystin-protein analogues immobilized on the plate for the binding sites of antibodies. Anti–MC-LR or anti-ADDA antibodies are used depending on the kit. The final step is an enzyme (HRP)–substrate (TMB) reaction where a colour is developed. The intensity of the colour is reversely proportional to the concentration of target toxins in the sample.

Handbook of Cyanobacterial Monitoring and Cyanotoxin Analysis, First Edition.
Edited by Jussi Meriluoto, Lisa Spoof and Geoffrey A. Codd.

SOP 13.2 Experimental

SOP 13.2.1 Materials

- Ultra-pure water, 18 MΩ cm
- ELISA kits in microplate format (e.g. Envirologix Quantiplate kit for Microcystins, Abraxis Microcystins-ADDA ELISA). These kits contain all reagents, calibrators, and quality controls for the assay. Care should be taken not to exceed the shelf-life of the kits. ELISA kits are usually stored in refrigeration.
- Pipette, 10–100 µL, with confirmed calibration status
- Multichannel pipette (eight channels), 50–250 µL, optional
- Stopwatch

SOP 13.2.2 Special Equipment

- Microplate reader (also called plate reader, microtiter plate reader, or ELISA reader) for 96-well microplates, equipped with monochromator or optical filters, capable of measuring absorbance (optical density [OD]) at 450 nm. Temperature control and shaking options are recommended. The microplate reader must have a confirmed calibration or performance status.
- Microplate washer for 96-well plates, optional.

SOP 13.2.3 Standard Solutions

- Standard solution of MC-LR (or other MC, or NOD variant) for independent calibration–quality control checks (recommended). The preparation of standard solutions of MCs and NOD is described in SOP 10. Alternatively, certified standard solutions of MC-LR can be used.

SOP 13.2.4 General Procedure

- Read carefully and follow the instructions of the kit regarding the handling of microplate, reagents, calibrators, antibody solutions, enzyme, and substrate solutions.
- Perform calibration according to the kit's instructions in every analytical batch. Alternatively, calibration can be carried out with laboratory's own standard solutions. Plot %B/Bo for each calibration sample, where B is the optical density (OD) of the sample and Bo is the OD of the blank sample (negative control). Construct a calibration curve by plotting %B/Bo (*y* axis) versus the concentration of MC-LR in logarithmic *x* axis. Calibration curves can be plotted automatically by software.
- Carry out the assay according to the manufacturer's instructions. Samples should be analysed in duplicates or triplicates.

SOP 13.3 Quality Control

- Carry out quality control checks that are included in the kit, or preferably analyse samples of known concentration of MC-LR or NODs, such as certified reference solutions or in-house prepared samples.
- Evaluate the results of quality control samples using control charts, manufacturer's specifications or information from validation data.

SOP 13.4 Validation

Validation data can be available by the manufacturers. However, laboratories are advised to carry out their own in-house validation studies in order to evaluate the performance of ELISA kits according to their intended use.

A typical validation study of an ELISA kit for MCs and NODs can include:

- Performance of the kit in different water matrices (e.g. Ultrapure, tap, surface water)
- Evaluation of trueness (% recovery), repeatability and reproducibility
- Estimation of LOD and LOQ
- Estimation of % false-positives and % false-negatives at certain "decision levels"
- Evaluation of cross-reactivity, i.e. response to various MC variants or other compounds

Results of an in-house validation of two different ELISA kits for MCs and NODs in water are presented in [1] (Method A: Envirologix Quantiplate kit, Method B: Abraxis Anti-ADDA ELISA kit). It is shown that while both kits perform well in high-purity (HP) water (mean recoveries 78–110%), performance of one kit (Method B) changes significantly when used with tap and lake water samples, where recoveries >130% are observed.

Reference

[1] Triantis, T.M., Tsimeli, K., Kaloudis, T., *et al* (2010) Development of an integrated laboratory system for the monitoring of cyanotoxins in surface and drinking waters. *Toxicon* **55** (5), 979–989.

SOP 14

Quantitative Screening of Microcystins and Nodularin in Water Samples with Commercially Available PPIA Kits

Triantafyllos Kaloudis[1], Theodoros M. Triantis[2], and Anastasia Hiskia[2]

[1] *Water Quality Control, Athens Water Supply and Sewerage Company – EYDAP SA, Athens, Greece*
[2] *Institute of Nanoscience & Nanotechnology, National Center for Scientific Research "DEMOKRITOS," Athens, Greece*

SOP 14.1 Introduction

This SOP includes a procedure for the quantitative screening of microcystins (MCs) and nodularin (NOD) in water samples (surface and drinking waters) with use of commercially available Protein Phosphatase Inhibition Assay (PPIA) kits in microplate format. PPIA responds to compounds that inhibit protein phosphatases (PPs); these compounds are not limited to MCs and NOD. For example, okadaic acid is also a potent PP inhibitor. PPIA assays can be developed with use of PP1 or PP2A enzymes. When a PPIA kit is not available, the procedures described in [1] and [2] can be followed. Commercially available PPIA kits are developed and tested in order to be used according to the analytical protocols that accompany the kits; therefore it is recommended that users follow the manufacturer's instructions. This SOP gives additional guidance and recommendations regarding the analysis of MCs and NOD by PPIA kits, including quality control and validation data. Since PPIA assay for MCs and NOD should be considered as "screening method," positive samples should be analysed with LC-MS/MS, especially when there is no other information about the sample or the history of the tested water body regarding the presence of MCs and NOD.

PP1A and PP2A are capable of hydrolysing a chromogenic substrate, *p*-nitrophenyl phosphate to *p*-nitrophenol which can be detected at 405 nm. Samples containing MCs or NOD inhibit PPs

Handbook of Cyanobacterial Monitoring and Cyanotoxin Analysis, First Edition.
Edited by Jussi Meriluoto, Lisa Spoof and Geoffrey A. Codd.

proportionally to the concentration of toxins in the sample. The assay validated in this SOP is based on the intensity of color (absorbance) after incubation of water samples with PP2A and subsequent addition of the substrate. Absorbance is reversely proportional to the concentration of target toxins in the sample.

SOP 14.2 Experimental

SOP 14.2.1 Materials

- Ultra-pure water, 18 MΩ cm
- PPIA kits in microplate format (e.g. Zeu Immunotec MicroCystest kit, Abraxis Microcystins/Nodularins PP2A or others). These kits contain all reagents, calibrators and quality controls for the assay. Care should be taken not to exceed the shelf-life of the kits. PPIA kits are usually stored in refrigeration.
- Pipette, 10–100 µL, with confirmed calibration status
- Multichannel pipette (eight channels), 50–250 µL, optional, recommended
- Stopwatch

SOP 14.2.2 Special Equipment

- Microplate reader (also called plate reader, microtiter plate reader, or ELISA reader) for 96-well microplates, equipped with monochromator or optical filters, capable of measuring absorbance (optical density [OD]) at 405 nm. Temperature control (incubation) and shaking options are recommended. The microplate reader must have a confirmed calibration or performance status.
- Microplate incubator (37°C), when incubation option is not included in the microplate reader.

SOP 14.2.3 Standard Solutions

- Standard solution of MC-LR (or other MC or NOD variant) for independent calibration – quality control checks (recommended). The preparation of standard solutions of MCs and NOD is described in SOP 10. Alternatively, certified standard solutions of MC-LR can be used.

SOP 14.2.4 General Procedure

Read carefully and follow the instructions of the kit regarding the handling of microplate, reagents, calibrators, enzyme, and substrate solutions.

Perform calibration according to the kit's instructions in every analytical batch. Alternatively, calibration can be carried out with laboratory's own standard solutions. A calibration curve is constructed by plotting absorbance at 405 nm in the y axis versus concentration of MC-LR (or other target compound) in a logarithmic *x* axis. The linear part of the curve is usually used for measurements. Calibration curves can be plotted automatically by software provided by the manufacturer or other source.

Carry out the assay according to the manufacturer's instructions. Samples should be analyzed in duplicates.

SOP 14.3 Quality Control

Carry out quality control checks that are included in the kit, or preferably analyze samples of known concentration of MC-LR or NOD (e.g. certified reference solutions or in-house prepared samples).

Evaluate the results of quality control samples using control charts, manufacturer's specifications or information from validation data.

SOP 14.4 Validation

Validation data can be available by the manufacturers. However, laboratories are advised to carry out their own in-house validation studies in order to evaluate the performance of PPIA kits according to their intended use.

A typical validation study of a PPIA kit for MCs and NOD can include:

- Performance of the kit in different water matrices (e.g. ultrapure, tap, surface water).
- Evaluation of trueness (% Recovery), repeatability and reproducibility.
- Estimation of LOD and LOQ.
- Estimation of % False Positives and % False Negatives at certain "decision levels".
- Evaluation of cross-reactivity, i.e. response to various MC variants or other compounds.
- Cross-validation studies (e.g. comparison with LC-MS/MS results).

Examples of summarized validation results for the Zeu-Immunotec MicroCystest kit are reported in detail in [3]. It is recommended that laboratories carry out their own in-house validation study to prove the applicability of the kits for their intended use.

References

[1] An, J., Carmichael, W.W., (1994) Use of a colorimetric protein phosphatase inhibition assay and enzyme linked immunosorbent assay for the study of microcystins and nodularins. *Toxicon* **32** (12), 1495–1507.

[2] Heresztyn, T., Nicholson, B.C., (200) Determination of cyanobacterial hepatotoxins directly in water using a protein phosphatase inhibition assay. *Water Research* **35** (13), 3049–3056.

[3] James, R., Gregg, A., Dindal, A. and McKernan, J., (2011) Performance Summary for the ZEU-INMUNOTEC MicroCystest in *Environmental Technology Verification Report ZEU INMUNOTEC Microcystin Test Kit: MicroCystest*, Battelle & U.S. Environmental Protection Agency (EPA), pp. 40.

SOP 15

Solid-Phase Extraction of Cylindrospermopsin from Filtered and Drinking Water

Theodoros M. Triantis[1], *Triantafyllos Kaloudis*[2], *and Anastasia Hiskia*[1]

[1] *Institute of Nanoscience & Nanotechnology, National Center for Scientific Research "DEMOKRITOS," Athens, Greece*

[2] *Water Quality Control, Athens Water Supply and Sewerage Company – EYDAP SA, Athens, Greece*

SOP 15.1 Introduction

This SOP specifies a method for the solid-phase extraction (SPE) of cylindrospermopsin (CYN) from drinking water or filtered water (e.g. filtered surface water) samples. The extracts produced by this method are suitable for subsequent analysis by LC-MS/MS (SOP 16). This extraction method is suitable for the determination of cylindrospermopsin that is dissolved in water (not inside cells or bound to other material). Table SOP 15.1 gives the basic registry information of the compound cylindrospermopsin.

CYN is extracted from water samples with use of graphitic carbon SPE cartridges and then eluted with dichloromethane–methanol and redissolved in water. The final solution is suitable for LC-MS/MS analysis (SOP 16). This should be considered an in-house method based on the procedure described in [1] and [2].

Handbook of Cyanobacterial Monitoring and Cyanotoxin Analysis, First Edition.
Edited by Jussi Meriluoto, Lisa Spoof and Geoffrey A. Codd.

Table SOP 15.1 *Registry information of cylindrospermopsin*

Name	Formula	M.W.	CAS Number
Cylindrospermopsin	$C_{15}H_{21}N_5O_7S$	415.43	143545-90-8

SOP 15.2 Experimental

SOP 15.2.1 Materials

- Ultrapure water, 18 MΩ cm
- Methanol HPLC grade
- Dichloromethane HPLC grade
- Formic acid (HCOOH), analytical grade
- Sodium chloride (NaCl), analytical grade
- Nitrogen gas (99.999%)
- Glass microfibre filters, 47 mm diameter
- Vacuum collection bottle (waste aspirator bottle), e.g. 10 L, plastic recommended
- Volumetric cylinders, 10 mL and 500 mL
- Pipette, 1–10 mL
- Stopwatch
- Glass sample bottles, 0.5–1 L
- Teflon tubing and adaptors for connection of sample bottles with SPE cartridges
- Glass test tubes, 10 mL
- SPE cartridges: graphite carbon (GC) cartridges, 250 mg, 3 or 6 mL (e.g., Interchrom or equivalent)

SOP 15.2.2 Special Equipment

- Glass filtration apparatus (vacuum) with slit-sieve disc for mounting glass-fibre filters (e.g. 47 mm) and 1 L filtering flask
- SPE vacuum manifold e.g. 12 port. An example setup of the SPE manifold is shown in SOP 7, Figure SOP 7.1
- Vacuum diaphragm pump (e.g. 5 L min^{-1} pumping speed, oil-free and Teflon diaphragm recommended)
- Nitrogen evaporation unit with thermostated bath or heat block (e.g. 12 port) for use with 10-mL glass test tubes
- Ultrasonic bath

SOP 15.2.3 Solutions

- Methanol/dichloromethane 20:80% (v/v) containing 5% HCOOH

SOP 15.2.4 General Procedure

a. Clean glass sample bottles, the SPE manifold and connecting tubing thoroughly with suitable solvents (e.g. acetone) followed by rinsing with ultrapure water.
b. Filter surface water samples and turbid samples through glass-fibre filters.

c. Add 500 mL of sample in a glass bottle with 0.5 g NaCl and 5 mL HCOOH. Condition the SPE by passing 10 mL of methanol followed by 10 mL of water at a flow of about 5 mL min^{-1}.
d. Connect the sample bottle with Teflon tubing and adaptor to the SPE cartridge. Pass the sample through the cartridge at a flow rate of about 10 mL min^{-1} (visible drops) with help of the vacuum generated by the pump. The flow rate can be measured with a graduated cylinder and a stopwatch and adjusted with use of regulating valves in the pump or manifold.
e. Remove the connecting tubing and adaptor after the entire sample has passed through the cartridge. Wash the cartridge with 10 mL of water and let it dry by passing air under vacuum for 5 min.
f. Elute the dried cartridge with 10 mL of methanol/dichloromethane 20:80% (v/v) solution containing 5% HCOOH at a flow rate of about 5 mL min^{-1} and collect the eluate in a glass test tube.
g. Evaporate the eluate to dryness under a gentle nitrogen stream at 50°C in a nitrogen evaporation unit.
h. Reconstitute the extract with 500 µL of water and sonicate for 10 min in an ultrasonic bath.
i. Centrifuge 10 min 10,000 × g, or, alternatively filter and transfer the final solutions to autosampler vials (e.g. 2 mL volume) for subsequent LC-MS/MS analysis (SOP 16). It is recommended that analysis is carried out immediately after SPE. In cases where this is not possible, solutions are kept at –20°C.

SOP 15.3 Quality Data

With every batch of extractions a spiked sample is prepared by spiking ultrapure water or one of the samples (duplicate) with a solution of CYN at a relevant concentration (e.g. 0.100 µg L^{-1}). The results of this quality control sample are evaluated against quality control chart limits.

SOP 15.4 Validation

This SOP describes only the extraction process. It has been shown that this extraction process, when coupled to LC-MS/MS measurements, meets the criteria of 70–120% mean recovery and <20% RSD. Validation data of this extraction process followed by LC-MS/MS measurements are presented in SOP 16.

References

[1] Wörmer, L., Carrasco, D, Cirés, S., Quesada, A., (2009) Advances in solid phase extraction of the cyanobacterial toxin cylindrospermopsin. *Limnology and Oceanography: Methods* **7** (7), 568–575.
[2] Triantis, T.M., Zervou, S.-K. and Hiskia, A., (2012) Determination of CYN in water by SPE-LC-MS/MS, SOP MD5, Laboratory of Environmental Analysis, NCSR "Demokritos", Athens, Greece.

SOP 16

Determination of Cylindrospermopsin in Filtered and Drinking Water by LC-MS/MS

Theodoros M. Triantis[1], Triantafyllos Kaloudis[2], and Anastasia Hiskia[1]

[1]Institute of Nanoscience & Nanotechnology, National Center for Scientific Research "DEMOKRITOS," Athens, Greece

[2]Water Quality Control, Athens Water Supply and Sewerage Company – EYDAP SA, Athens, Greece

SOP 16.1 Introduction

This SOP specifies a method for the determination of cylindrospermopsin (CYN) in filtered water (e.g. filtered surface water) and drinking water samples by LC-MS/MS. The method can be applied after solid-phase extraction (SPE) of CYN according to SOP 15.

Extracts produced by SOP 15 are analysed using reversed-phase HPLC and tandem triple quadrupole mass spectrometry (MS/MS). This is an in-house method based on the procedure described in [1].

SOP 16.2 Experimental

SOP 16.2.1 Materials

- Ultrapure water 18.2 MΩ cm
- Acetonitrile HPLC grade

Handbook of Cyanobacterial Monitoring and Cyanotoxin Analysis, First Edition.
Edited by Jussi Meriluoto, Lisa Spoof and Geoffrey A. Codd.

- Formic acid (HCOOH), analytical grade
- Volumetric flasks 10 mL, A class
- Pipettes 10–100 μL, 100–1000 μL, 1–10 mL, with confirmed calibration status
- Chromatographic column: C_{18}, 100 mm × 2.1 mm, 3 μm particle size (e.g. Atlantis T3, Waters)

SOP 16.2.2 Special Equipment

- Spectrophotometer UV/Vis with quartz cells, suitable for measurements of absorbance at 262 nm, with confirmed calibration status
- LC-MS/MS system (triple quadrupole) consisting of:
  i. HPLC pump and mixer suitable for gradient elution (mixing of two solvents)
  ii. Tandem mass spectrometric detector, consisting of an electrospray ionisation (ESI) source and a triple quadrupole mass analyser
  iii. Recommended autosampler for liquid samples (e.g. 2 mL-vials)
  iv. Software for analysing the results

SOP 16.2.3 Calibration Standards and Quality Control Standard

CYN is available commercially as pure compound (e.g. from Abraxis, Sigma-Aldrich and others). It is recommended that the specified purity of the standard compound is >95% and a measurement of the actual concentration of a stock standard solution is carried out in the laboratory, when there are no sufficient data regarding the purity and quantity of the purchased compound. A procedure to carry out this measurement is the following:

Dissolve pure CYN in its own vial (as purchased, usually 0.1–0.5 mg) with water (e.g. 1 mL). Sonicate the vial for 5 min in a sonication bath. Measure the absorbance (optical density) of this solution at 262 nm using a calibrated UV spectrophotometer with a quartz cell (e.g. 1 cm length). Calculate the exact concentration of the solution using the formula

$$C = \frac{A_{max} \times M_{CYN} \times 1000}{\varepsilon \times 1}$$

where

A_{max} is the absorbance of CYN solution at 262 nm (corrected for blank).
C is the concentration of CYN in mg L^{-1}
M_{CYN} is the molar mass of CYN, that is, 415.43 g mol^{-1}
ε is the molar extinction coefficient of CYN at 262 nm in water, that is, 9800 L mol^{-1} cm^{-1} according to [2]
l is the optical path of the cell

Using the prementioned procedure, stock standard solutions of CYN (e.g. 0.1–0.5 mg mL^{-1}) can be prepared. Stock solutions of CYN are stored at –20°C.

A quality control standard (e.g. 100 μg L^{-1}) is prepared using the prementioned procedure. The quality control standard is used for spiking blank or real samples (e.g. with 100 ng L^{-1}) in order to routinely check the performance of the analysis. It is recommended that the quality control standard be prepared by independent stock standard solution from those used for calibration.

Table SOP 16.1 Retention times and selected transitions for the determination of CYN

Compound	Retention time, min	Precursor ion, *m/z*	Product ions, *m/z*
CYN	2.59	416	194[1], 176, 336, 274

[1] Quantifier ion.

SOP 16.2.4 General Procedure

SOP 16.2.4.1 Liquid Chromatography

Reversed-phase chromatography separation of CYN can be achieved under the following conditions:

Column: C_{18}, 100 mm × 2.1 mm, 3 µm particle size
Column temp.: 35°C
Solvent A: water – 0.5% HCOOH
Solvent B: acetonitrile – 0.5% HCOOH
Isocratic elution A:B (95:5% v/v)
Flow: 0.2 mL min^{-1}
Injection volume: 20 µL

SOP 16.2.4.2 Mass Spectrometry

Mass spectrometry operation parameters for the detection and quantitative measurement of CYN in ESI positive ionisation mode (e.g. spray voltage, ion transfer capillary temperature, sheath gas, auxiliary gas, probe position, collision gas pressure, peak width) should be optimised according to manufacturer's guides and laboratory practices.

Identification and quantitative determination of CYN (that co-elutes with Phe) is based on the precursor-to-product ion transitions (SRMs) and the chromatographic retention times. Indicative retention times obtained through experimental conditions and typical transitions of CYN are shown in Table SOP 16.1.

Typical chromatograms obtained under the prementioned experimental conditions are shown in Figure SOP 16.1.

SOP 16.3 Calibration

Calibration is carried out prior to every batch of sample analysis. The following calibration alternatives can be used:

a. *Single-point calibration*: A calibration standard (e.g. 100 µg L^{-1}) can be used. It is recommended that the sample response should be within ±50% of the calibration response unless further extrapolation is supported by evidence of acceptable linearity of response.
b. *Two-point 'bracketing' calibration*: The responses of samples should fall within the bracketing range.
c. *Multiple-point calibration*: A calibration function is calculated and used between the lowest and highest calibration levels. In this method the calibration curve is linear and should not be forced through the origin.

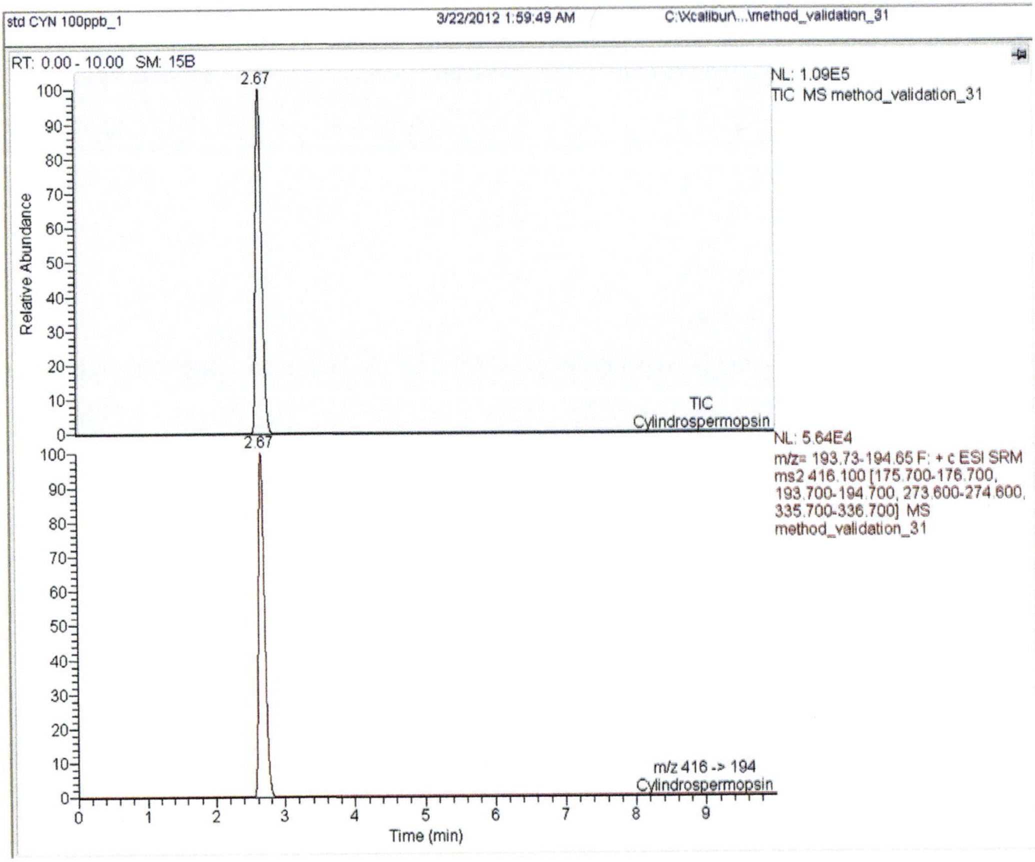

Figure SOP 16.1 *Typical total ion chromatogram and SRM chromatograms of CYN (100 μg L^{-1})*

SOP 16.3.1 Multiple-Point Calibration

A stock standard solution (S) of CYN (e.g. 500 mg L^{-1}) is prepared as described in Section 16.2.3.

An intermediate standard solution (S1) of CYN (e.g. 5 mg L^{-1}) can be prepared by diluting a stock solution (S) volume (e.g. 100 μL) with water (e.g. 10 mL final volume). Solution S1 can be further diluted with water (e.g. 1–10 mL) to solution (S2) (e.g. 0.5 mg L^{-1}).

A series of calibration standard solutions to cover the desired range (e.g. 0.5–250 μg L^{-1}) is prepared by further dilution of S2. These solutions are analysed by LC-MS/MS (see Section 16.2.4.2), and a calibration curve is constructed by plotting the peak area of the quantifier ion of CYN versus CYN concentration.

SOP 16.4 Sample Analysis: Calculations

A typical sequence of analysis is shown in Table SOP 16.2. Calibration checks and quality control are repeated every 10 samples. Detection and identification of CYN are based on the retention time, characteristic ions and ion ratios. Quantification of CYN is based on the quantifier ion peak area of CYN (Table SOP 16.1), and the concentration of CYN is calculated from calibration.

Table SOP 16.2 *Recommended sequence of analysis*

No	Sample	Comments
1	Blank (solvent water)	Check for carry-over
2	Calibration or calibration check standards	Calibration or calibration check System suitability checks
3	Quality control sample	Check for recovery
4	Blank sample	Check for carry-over and contamination
5	Sample 1	Detection, identification, quantitation
6	Sample 2	Detection, identification, quantitation
…	…………	……………..
13	Sample 10	Detection, identification, quantitation
Go back to no. 1 (recalibration)		

Table SOP 16.3 *Quality control scheme (recommended)*

No	Check	Measurements	Criteria
1	MS/MS performance	MS/MS calibration	Manufacturer's criteria
2	Carry-over and contamination	Blanks (rows 1, 4, Table SOP 16.2)	Target compounds not detected
3	Calibration	Calibration standard(s), (row 2, Table SOP 16.2)	r^2 and slope of calibration curve, predefined criteria for calibration check
4	Recoveries	QC sample, for example, 0.100 $\mu g\,L^{-1}$ (row 3, Table SOP 16.2)	Control chart limits
5	Positive samples	Identification parameters	Retention times, ion ratios

SOP 16.5 Quality Control

A recommended quality control scheme to be carried out with every batch of analysis is shown in Table SOP 16.3.

SOP 16.6 Expression of Results

Results are expressed in $\mu g\,L^{-1}$ with three decimal digits. Results below limits of detection (LOD) are expressed as 'not detected'. Results above LOD and below limits of quantification (LOQ) are expressed as '<LOQ'.

SOP 16.7 Validation Data

This method was validated in-house using potable bottled water as a matrix. Summarised validation results are presented in Table SOP 16.4. Results in this table are obtained from the combination of

Table SOP 16.4 *Summarised validation results of CYN in potable bottled water at two concentration levels*

Compound	Cylindrospermopsin	
	$0.010\,\mu g\,L^{-1}$	$0.100\,\mu g\,L^{-1}$
R%[1]	67	85
%RSR$_r$[2]	18	9
%RSD$_R$[3]	20	13

Linearity range of measurement: 1–200 $\mu g\,L^{-1}$ (concentration of injected sample), $r^2 = 0.996$ (seven concentration levels, two replicates each)

LOD: 0.001 $\mu g\,L^{-1}$ (based on S/N > 3)

LOQ: 0.002 $\mu g\,L^{-1}$ (based on S/N > 10) or 0.010 $\mu g\,L^{-1}$ (lowest validated level)

[1] Expressed as % mean recovery, N = 12.
[2] Repeatability (from pooled SD), N = 12.
[3] Reproducibility, N = 12.

SOP 15 (SPE extraction of CYN) and the present SOP 16 (determination of CYN by LC-MS/MS). The combination of these SOPs is proved to be suitable for determinations at the low ng L^{-1} levels of CYN.

References

[1] Triantis, T.M., Zervou, S.-K. and Hiskia, A. (2012) Determination of CYN in water by SPE–LC–MS/MS, SOP MD5, Laboratory of Environmental Analysis, NCSR "Demokritos," Athens, Greece.

[2] Sano, T., Kikuchi, S., Kubo, T. *et al.* (2008) New values of molecular extinction coefficient and specific rotation for cyanobacterial toxin cylindrospermopsin. *Toxicon* **51** (4), 717–719.

SOP 17

Solid-Phase Extraction of Anatoxin-a from Filtered and Drinking Water

Theodoros M. Triantis[1], Triantafyllos Kaloudis[2], and Anastasia Hiskia[1]

[1] *Institute of Nanoscience & Nanotechnology, National Center for Scientific Research "DEMOKRITOS," Athens, Greece*

[2] *Water Quality Control, Athens Water Supply and Sewerage Company – EYDAP SA, Athens, Greece*

SOP 17.1 Introduction

This SOP specifies a method for the solid-phase extraction (SPE) of anatoxin-a from drinking water or filtered water samples. The extracts produced by this method are suitable for subsequent analysis by LC-MS/MS (see SOP 18). This extraction method is suitable for the determination of anatoxin-a that is dissolved in water (not inside cells or bound to other material). Table SOP 17.1 gives the basic registry information of the compound anatoxin-a.

Anatoxin-a is extracted from water samples after addition of phenylalanine-*d5* (internal–surrogate standard), with use of graphitic carbon solid phase extraction cartridges and then eluted with methanol and redissolved in water. The final solution is suitable for LC-MS/MS analysis (SOP 18). This should be considered an in-house method based on the procedure described in [1].

SOP 17.2 Experimental

SOP 17.2.1 Materials

- Phenylalanine-*d5* (Phe-*d5*) isotope-labeled surrogate standard (Ring-*d5* 98% atom D, Cambridge Isotope Laboratories Andower, USA)

Handbook of Cyanobacterial Monitoring and Cyanotoxin Analysis, First Edition.
Edited by Jussi Meriluoto, Lisa Spoof and Geoffrey A. Codd.

Table SOP 17.1 Registry information of anatoxin-a

Name	Formula	M.W.	CAS Number
Anatoxin-a	$C_{10}H_{15}NO$	165.232	64285-06-9

- Anatoxin-a fumarate (98%) or purified anatoxin-a (e.g. from Tocris Bioscience, Sigma-Aldrich and others)
- Ultrapure water, 18 MΩ cm
- Methanol HPLC grade
- Sodium hydroxide (NaOH), analytical grade
- Formic acid (HCOOH), analytical grade
- Nitrogen gas (99.999%)
- Glass Microfiber filters, 47-mm diameter
- Volumetric cylinders, 10 mL and 500 mL
- Pipette 1–10 mL
- Stopwatch
- Vacuum collection bottle (waste aspirator bottle), e.g. 10 L, plastic recommended.
- Glass sample bottles, 0.25–0.5 L
- Teflon tubing and adaptors for connection of sample bottles with SPE cartridges
- Glass test tubes, 10 mL
- SPE Cartridges: Graphitized Carbon cartridges, size 200 mg in 3 mL-reservoir e.g. Hypersep PGC, Thermo Fisher Scientific, (Waltham, USA) or equivalent.

SOP 17.2.2 Special Equipment

- Glass filtration apparatus (vacuum) with slit-sieve disc for mounting glass-fiber filters (e.g. 47 mm) and 1-L filtering flask
- SPE vacuum manifold e.g. 12 port. An example setup of the SPE manifold is shown in SOP 7, Figure SOP 7.1
- Vacuum diaphragm pump (e.g. 5 L min^{-1} pumping speed, oil-free and Teflon diaphragm recommended)
- Nitrogen evaporation unit with thermostated bath or heat block (e.g. 12 port) for use with 10-mL glass test tubes
- pH-meter
- Ultrasonic bath
- Microcentrifuge

SOP 17.2.3 Solutions

- Phe-*d5* in water, 0.60 mg L^{-1}
- NaOH in water, 2 mol L^{-1}
- Methanol containing 0.1% HCOOH

SOP 17.2.4 Standard Anatoxin-a Solution

Anatoxin-a is available commercially as anatoxin-a fumarate or anatoxin-a hydrochloride. It is recommended that the specified purity of the standard compound is >95% and a measurement of the actual concentration of a standard solution is carried out by the laboratory, when there are no sufficient data regarding the purity and quantity of the purchased compound. A procedure to carry out this measurement in the case of anatoxin-a fumarate is described in [1].

Stock solution (e.g. 100 mg L^{-1}) of anatoxin-a prepared in water is used for further dilutions. This solution is stable for at least one month when refrigerated.

SOP 17.2.5 General Procedure

a. Clean thoroughly glass sample bottles, the SPE manifold and connecting tubing should be with suitable solvents (e.g. acetone) followed by rinsing with ultrapure water.
b. Filter surface water samples and turbid samples through glass-fibre filters.
c. Add 200 µL of the Phe-*d5* solution to 200 mL of sample in a glass bottle.
d. Adjust pH to 10.5 using the 2 mol L^{-1} NaOH solution.
e. Condition the SPE cartridges by passing 6 mL of methanol followed by 6 ml of water at a flow of about 5 mL min^{-1}.
   Note: Do not let the cartridges dry under conditioning and sample application.
f. Connect the sample bottle with Teflon tubing and adaptor to the SPE cartridge. Pass the sample through the cartridge at a flow rate of about 10 mL min^{-1} (visible drops) with help of the vacuum generated by the pump. The flow rate can be measured with a graduated cylinder and a stopwatch, and is adjusted by using the regulating valves in the pump or manifold.
g. Remove the connecting tubing and adaptor after the entire sample has passed through the cartridge and dry the cartridge by passing air under vacuum for 15 min.
h. Elute the dried cartridge with 3 mL of methanol containing 0.1% HCOOH at a flow rate of about 5 mL min^{-1}. Collect the eluate in a glass test tube.
i. The eluate is evaporated to dryness under a gentle nitrogen stream at 40°C in a nitrogen evaporation unit.
j. Reconstitute the extract with 200 µL of water.
k. Centrifugate 10,000 × g for 10 min or alternatively, filtrate the sample.
l. Transfer the sample to autosampler vials for subsequent LC-MS/MS analysis (SOP 18). It is recommended that analysis is carried out immediately after SPE. In cases where this is not possible, solutions are kept at –20°C.

SOP 17.3 Quality Control

With every batch of extractions, a spiked sample is prepared by spiking ultrapure water or one of the samples (duplicate) with a solution of anatoxin-a at a relevant concentration (e.g. 0.100 µg L^{-1}). The results of this quality control sample are evaluated against quality control chart limits.

SOP 17.4 Validation

This SOP describes only the extraction process. It has been shown that this extraction process, when coupled to LC-MS/MS measurements, meets the criteria of 70–120% mean recovery and <20% RSD. Validation data of this extraction process followed by LC-MS/MS measurements are presented in SOP 18.

Reference

[1] Dimitrakopoulos, I., Kaloudis, T., Hiskia, A., *et al.* (2010) Development of a fast and selective method for the sensitive determination of anatoxin-a in lake waters using liquid chromatography–tandem mass spectrometry and phenylalanine-d5 as internal standard. *Analytical and Bioanalytical Chemistry* **397** (6), 2245–2252.

SOP 18

Determination of Anatoxin-a in Filtered and Drinking Water by LC-MS/MS

Theodoros M. Triantis[1], Triantafyllos Kaloudis[2], and Anastasia Hiskia[1]

[1] *Institute of Nanoscience & Nanotechnology, National Center for Scientific Research "DEMOKRITOS," Athens, Greece*

[2] *Water Quality Control, Athens Water Supply and Sewerage Company – EYDAP SA, Athens, Greece*

SOP 18.1 Introduction

This SOP specifies a method for the determination of anatoxin-a in filtered water (e.g. filtered surface water) and drinking water samples by LC-MS/MS. This method avoids interferences from the presence of phenylalanine that has the same mass fragmentation pattern with anatoxin-a. This is achieved through chromatographic separation of anatoxin-a and phenylalanine. The method can be applied after solid-phase extraction (SPE) of anatoxin-a according to SOP17.

Extracts produced by SOP17 are analyzed using reversed phase HPLC and tandem triple-quadrupole mass spectrometry (MS/MS). Phenylalanine-*d5* is used as a surrogate standard. This is an in-house method based on the procedure described in [1].

SOP 18.2 Experimental

SOP 18.2.1 Materials

- Ultrapure water, 18.2 MΩ cm
- Acetonitrile HPLC grade
- Formic acid (HCOOH), analytical grade
- Methanol analytical grade

Handbook of Cyanobacterial Monitoring and Cyanotoxin Analysis, First Edition.
Edited by Jussi Meriluoto, Lisa Spoof and Geoffrey A. Codd.

- Volumetric flasks 10 mL, Class A
- Pipettes 10–100 µL, 100–1000 µL, 1–10 mL, with confirmed calibration status
- Chromatographic column: C_{18}, 50 mm × 2.1 mm, 1.8-µm particles (e.g. Zorbax Rapid Resolution, Agilent)

SOP 18.2.2 Special Equipment

- LC-MS/MS system (triple quadrupole), consisting of:
  a. HPLC pump and mixer suitable for gradient elution (mixing of two solvents)
  b. Tandem mass spectrometric detector, consisting of an electrospray ionization source (ESI) and a triple quadrupole mass analyzer
  c. Autosampler, for liquid samples (e.g 2-mL vials) is recommended
  d. Software for analysing the results

SOP 18.2.3 LC Solutions

- Solvent A: methanol containing 0.1% HCOOH
- Solvent B: water containing 0.1% HCOOH

SOP 18.2.4 Calibration Standards and Quality Control Standard

Anatoxin-a is available commercially as anatoxin-a fumarate or anatoxin-a hydrochloride (e.g. from Tocris Bioscience, Sigma-Aldrich and others). It is recommended that the specified purity of the standard compound is >95% and a measurement of the actual concentration of a standard solution is carried out by the laboratory, when there are no sufficient data regarding the purity and quantity of the purchased compound. A procedure to carry out this measurement in the case of anatoxin-a fumarate is described in [1].

Phenylalanine-*d5* (Phe-*d5*) isotope-labeled surrogate/internal standard (Ring-*d5* 98% atom D, e.g. Cambridge Isotope Laboratories or equivalent) is used.

Stock solutions (e.g. 100 mg L^{-1}) of anatoxin-a and Phe-*d5* are prepared in water and are used for further dilutions. These solutions are stable for at least 1 month when refrigerated.

A series of calibration standard solutions containing different concentrations of anatoxin-a (e.g. from 0.5 to 2000 µg L^{-1}) and a constant concentration of Phe-*d5* (e.g. 600 µg L^{-1}) in water are prepared. These solutions are analyzed by LC-MS/MS (see later) and a calibration curve is constructed by plotting the ratio of anatoxin-a peak area to Phe-*d5* peak area versus anatoxin-a concentration.

A quality control standard (e.g. 200 µg L^{-1}) is prepared using the above procedure. The quality control standard is used for spiking blank or real samples (e.g. with 100 ng L^{-1}) in order to routinely check the performance of the analysis. It is recommended that the quality control standard is prepared by independent standards from those used for calibration.

SOP 18.2.5 General Procedure

SOP 18.2.5.1 Liquid Chromatography

Reversed-phase chromatographic separation of anatoxin-a and phenylalanine-*d5* can be achieved under the following conditions:

Column: 50 mm × 2.1 mm C_{18} column, 1.8-µm particles
Column temperature: 35°C

Isocratic elution Solvent A: Solvent B (7:93% v/v)
Flow rate: 0.2 mL min^{-1}
Injection volume: 10 μL

SOP 18.2.5.2 Mass Spectrometry

Mass spectrometry operation parameters for the detection and quantitative measurement of anatoxin-a and Phe-*d5* in ESI positive ionization mode (e.g. spray voltage, ion transfer capillary temperature, sheath gas, auxiliary gas, probe position, collision gas pressure, peak width) should be optimized according to manufacturer's guides and laboratory practices.

Identification and quantitative determination of anatoxin-a and Phe-*d5* (that co-elutes with Phe) are based on the precursor to product ion transitions (SRMs) and the chromatographic retention times. Indicative retention times (obtained through experimental conditions) and typical transitions of the target analytes are shown in Table SOP 18.1.

A typical chromatogram obtained under these experimental conditions is shown in Figure SOP 18.1.

SOP 18.3 Calibration

A calibration curve with use of Phe-*d5* as internal standard (IS) is constructed as follows:

A series of calibration standard solutions containing different concentrations of anatoxin-a (e.g. from 0.5 to 2000 μg L^{-1}) and a constant concentration of Phe-*d5* (e.g. 600 μg L^{-1}) in water. These solutions are analyzed by LC-MS/MS (see later) and a calibration curve is constructed by plotting the ratio of anatoxin-a peak area to Phe-*d5* peak area versus anatoxin-a concentration.

A check of the calibration status is carried out prior to every batch of sample analysis. This can be done with measurement of one of the calibration standards (containing both anatoxin-a and Phe-*d5*). The measured concentration should be within specified limits (e.g. ±20%) of the actual concentration of the standard solution.

Table SOP 18.1 *Retention times and selected transitions for the determination of anatoxin-a using Phe-d5 as internal standard*

Compound	Retention time (min)	Precursor ion *m/z*	Product ions *m/z*
Anatoxin-a	1.39	166 $[M+H]^+$	149* $[M+H\text{-}NH_3]^+$ 131 $[M+H\text{-}NH_3\text{-}H_2O]^+$ 107 $[M+H\text{-}NH_3\text{-}COCHCH_3]^+$ 91 $[C_7H_7]^+$
Phe-d5 (IS)	1.89	171	125*

*Quantifier ion.

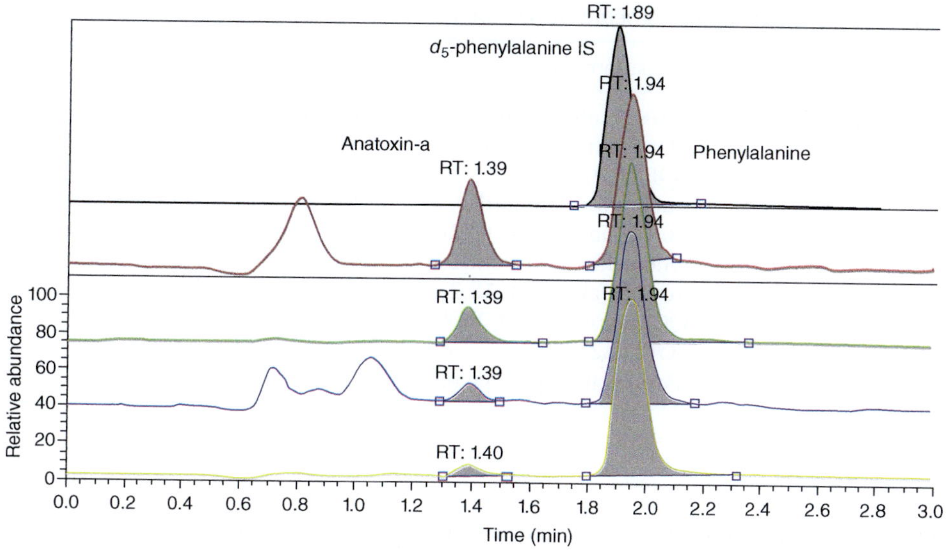

Figure SOP 18.1 *SRM chromatograms of anatoxin-a (tR = 1.39 min) at a level of 10 ng L⁻¹ from a spiked lake water sample containing 0.60 µg L⁻¹ of IS (phenylalanine-d5; tR = 1.89 min). The peak of phenylalanine from the lake water sample is also shown (tR = 1.94 min). Reproduced from [1]. With kind permission from Springer Science and Business Media*

Table SOP 18.2 *Recommended sequence of analysis*

No.	Sample	Comments
1	Blank (solvent water)	Check for carryover
2	Calibration or calibration check standards	Calibration or calibration check – System suitability checks
3	Quality control sample	Check for recovery
4	Blank sample	Check for carryover and contamination
5	Sample 1	Detection, identification, quantitation
6	Sample 2	Detection, identification, quantitation
…	…………	………………
13	Sample 10	Detection, identification, quantitation
Go back to No. 1 (recalibration)		

SOP 18.4 Sample Analysis—Calculations

A typical sequence of analysis is shown in Table SOP 18.2. Calibration and quality control are repeated every 10 samples. Detection and identification of anatoxin-a is based on the retention time, characteristic ions and ion ratios [1]. Quantification of anatoxin-a is based on the ratio of the anatoxin-a quantifier ion peak area to Phe-*d5* peak area (see Table SOP 18.1) and the concentration of anatoxin-a is derived from the calibration curve.

Table SOP 18. 3 *Quality control scheme (recommended)*

No.	Check	Measurements	Criteria
1	MS/MS performance	MS/MS calibration	Manufacturer's criteria
2	Carryover and contamination	Blanks (rows 1, 4 Table SOP 18.2)	Target compounds not detected
3	Calibration	Calibration standard(s), (row 2, Table SOP 18.2)	r^2 and slope of calibration curve, predefined criteria for calibration check
4	System suitability	One calibration standard (row 2, Table SOP 18.2)	Resolution (e.g. anatoxin-a from Phe-*d5*) or theoretical plates
5	Recoveries	QC sample (row 3, Table SOP 18.2)	Control chart limits
6	Positive samples	Identification parameters	Retention times, ion ratios

SOP 18.5 Quality Control

A recommended quality control scheme to be carried out with every batch of analysis is shown in Table SOP 18.3.

SOP 18.6 Expression of Results

Results are expressed in μg L^{-1} with three decimal digits. Results below limits of detection (LOD) are expressed as "not detected." Results above LOD and below limits of quantification (LOQ) are expressed as "<LOQ."

SOP 18.7 Validation

This method was validated in-house using filtered surface water as a matrix. Summarized validation results are presented in [1], as obtained from the combination of SOP 17 (SPE) and the present SOP18 (determination by LC-MS/MS). The method LOD is 1 ng L^{-1} with mean recoveries 73–136% and %RSDs <10 in the range 0.01–1.0 μg L^{-1} and is suitable for determinations of anatoxin-a at low ng L^{-1} levels.

Reference

[1] Dimitrakopoulos, I., Kaloudis, T., Hiskia, A., *et al.* (2010) Development of a fast and selective method for the sensitive determination of anatoxin-a in lake waters using liquid chromatography–tandem mass spectrometry and phenylalanine-d5 as internal standard. *Analytical and Bioanalytical Chemistry* **397** (6), 2245–2252.

SOP 19

Analysis of Anatoxin-a and Cylindrospermopsin by Ultra High-Performance Liquid Chromatography Tandem Mass Spectrometry

Leonardo Cerasino

IASMA Research and Innovation Centre, Fondazione Edmund Mach-Istituto Agrario di S. Michele all'Adige, S. Michele all'Adige, Trento, Italy

SOP 19.1 Introduction

UHPLC is a recent introduction in the LC technology and offers some advantages (reduction of experiment times, reduction of solvent consumption) compared with conventional HPLC, maintaining the same chromatographic efficiency. This SOP describes the procedure for analysing anatoxin-a (ATX-a) and cylindrospermopsin (CYN) in water samples based on UHPLC-MS/MS methodology. The proposed method allows the determination of ATX-a and CYN in 10 min run time [1–4]. The SOP reports details about instrumentation and LC column that have been used for the method development and optimization; the method can be however adapted to any equivalent instrumental configuration.

SOP 19.2 Experimental

SOP 19.2.1 Materials

- Acetonitrile (ACN) LC-MS grade
- Ultrapure water, 18.2 MΩ cm

Handbook of Cyanobacterial Monitoring and Cyanotoxin Analysis, First Edition.
Edited by Jussi Meriluoto, Lisa Spoof and Geoffrey A. Codd.

- Ammonium acetate, LC-MS grade
- Acetic acid, LC-MS grade
- Pure commercial (±) ATX-a fumarate, CYN, phenylalanine

SOP 19.2.2 Special Equipment

a. The UHPLC used for the development of the present SOP was a Waters Acquity UPLC system equipped with an autosampler carrying a 10 μL loop, a binary pump and a column manager equipped with a Phenomenex Kinetex HILIC column (50 × 2.1 mm, 1.7 μm particle size). The LC was connected without splitting to the mass detector.
b. The mass detector was an AB Sciex 4000 QTRAP triple quadrupole, equipped with a Turbo spray (ESI) source. High-purity nitrogen was employed as curtain, heater, nebulizer and collision gas.

SOP 19.2.3 Calibration Standards

a. Bulk solutions of individual standards (5 mL, concentration 5 μg mL^{-1}). Dissolve 25 μg of standard in 5 mL of ACN.
b. Multistandard bulk solution (5 mL, concentration 100 ng mL^{-1} CYN and 200 ng mL^{-1} ATX-a). Add CYN and phenylalanine bulk solutions (a), 100 μL of each, to 2000 μL of water; then add 200 μL of ATX-a bulk solution (a); and finally adjust the volume to 5000 μL with water.
c. Calibration solutions (1 mL, concentrations 100, 75, 50, 25, 10, 5 and 1 ng mL^{-1} for CYN and phenylalanine; 200, 150, 100, 50, 20, 10 and 2 ng mL^{-1} for ATX-a). Dilute the multistandard bulk solution (b) with the appropriate amount of water.

SOP 19.2.4 HPLC Mobile Phases

a. Component A: 10 mM ammonium acetate/10 mM acetic acid in 1% ACN. Dissolve 770 mg of ammonium acetate in 500 mL of water, add 570 μL of acetic acid, then add 10 mL ACN and adjust the volume to 1000 mL with water. Finally, filter (0.2 μm pore size) the solution.
b. Component B: ACN.

Please observe the following:

- Eluents must be renewed weekly.
- Acetic acid and ACN must be handled under a fume hood.

SOP 19.2.5 UHPLC-MS/MS

SOP 19.2.5.1 Example of LC Settings

Injection volume: 2 μL
Injection mode: partial loop with needle overfill
Column compartment temperature: 30°C
Flow rate: 0.250 mL min^{-1}
Mobile phase: gradient of two components, A (SOP 19.2.4.a) and B (SOP 19.2.4.b), according to the scheme reported in Table SOP 19.1.
Run time: 10 min

Table SOP 19.1 *Gradient description*

Time (min)	% A	% B	Curve
0.0	10	90	
0.5	10	90	6
1.0	25	75	6
2.0	25	75	6
3.5	60	40	6
5.5	60	40	6
8.0	10	90	6
10.0	10	90	6

Table SOP 19.2 *Compound-specific mass detector settings*

Toxin variant	Precursor ion (m/z)	RT (min)	Product ion (m/z)	DP (V)	CE (V)	Q/q ratio
CYN	416.3 $[M+H]^+$	1.84	194.2 Q	80	53	1.5
			336.2 q			
Phenylalanine[1]	166.1 $[M+H]^+$	2.20	103.0	70	53	
ATX-a	166.1 $[M+H]^+$	4.30	149.0 Q	60	21	1.6
			131.0 q	60	24	

Two transitions have been monitored per analyte; the most intense is used for quantification (Q), the less intense for confirmation (q). RT = chromatographic retention time. DP = declustering potential. CE = collision energy.

[1] Phenylalanine is a nontoxic compound which is normally present in all living organisms. It is isobaric with anatoxin-a, has a similar fragmentation pattern and therefore can be confused with the toxin.

SOP 19.2.5.2 Example of Mass Detector Settings (AB Sciex 4000 QTRAP Triple Quadrupole)

Note: Please observe that every instrument model must be tuned individually.
Source settings:

Ionization mode: positive electro spray (ESI+)
Ion spray voltage: 5000 V
Heater gas temperature: 300°C
Curtain gas (CUR): 20
Nebulizer gas (GS1): 20
Heater gas (GS2): 20

Triple quadrupole settings:

Scan type: scheduled MRM
MRM detection window: 60 s
Target scan time: 60 s
Settling time: 0 ms
Entrance potential (EP): 10 V
Collision cell exit potential (CXP): 10 V

Compound-specific settings are reported for clarity in Table SOP 19.2.

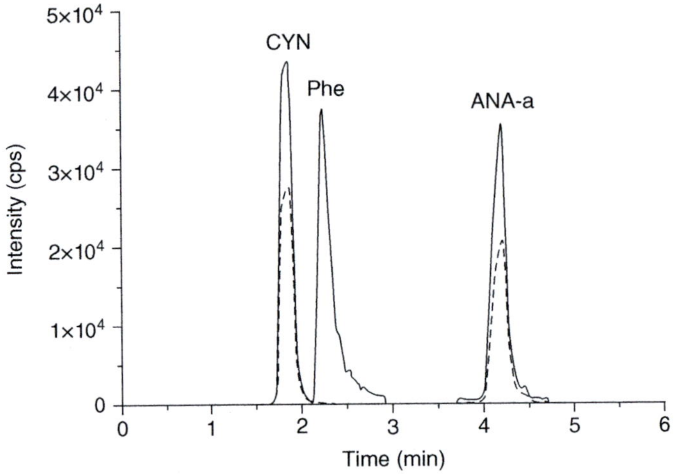

Figure SOP 19.1 Representative LC-MS/MS chromatogram showing the MRM transitions of CYN ($100\,ng\,mL^{-1}$) and ANA-a ($330\,ng\,mL^{-1}$). Two MRM transitions are monitored for each analyte (solid and dashed lines). Phenylalanine (Phe) is also present

Table SOP 19.3 Data for quantification of toxins

Toxin	Regression equation	R^2	LOD ($ng\,mL^{-1}$)	LOQ ($ng\,mL^{-1}$)
CYN	y = 5419.7x + 7042.7	0.9997	0.1	0.5
ATX-a	y = 1396.1x + 1597.7	0.9993	1.0	4.0

SOP 19.3 Identification of Analytes

Analytes are identified when there is a perfect matching with the characteristics of the corresponding standard: presence of the two MRM transitions (Q and q), same Q/q ratio and same retention time (Fig. SOP 19.1).

SOP 19.4 Quantification of Analytes

Quantification of analytes is performed by the external standard procedure, and for each analyte the area under the curve of the Q transition is considered. A standard curve is prepared using the multistandard solutions (SOP 19.2.3.b) in the range 1–100 $ng\,mL^{-1}$ for CYN and 2–200 $ng\,mL^{-1}$ for ATX-a. A regression equation is then derived for each toxin, which is used for calculating the concentrations of unknown samples. In Table SOP 19.3 regression equations for CYN and ATX-a are reported. Limits of detections (LOD) and limits of quantifications (LOQ) have been calculated for each toxin as the concentrations for which a 1:3 and a 1:10 signal to noise ratio has been observed, respectively. In Figure SOP 19.2 calibration curves are reported.

If higher sensitivity is required, a preliminary preconcentration by SPE step can be performed, according to other SOPs in the present handbook.

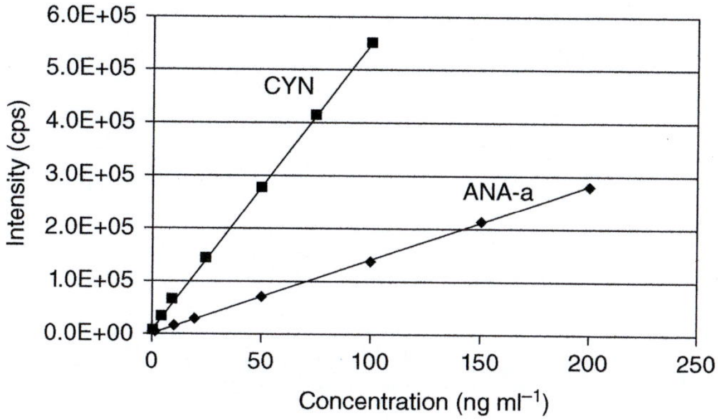

Figure SOP 19.2 Standard curves for quantification of CYN and ATX-a

Table SOP 19.4 Validation data for 12 replicates recorded in 2 working sessions in different days

Toxin variant	Spiked conc. (ng mL^{-1})	Recovery (%)	RSD (%)
CYN	1	106	1.4
	10	105	2.6
ATX-a	6	102	3.5
	60	104	2.7

Matrix: Surface water. Recovery and relative standard deviation (RSD) are reported.

SOP 19.5 Validation

In Table SOP 19.4 the validation data obtained for the two toxins at two concentrations are reported.

References

[1] Shams, S., Capelli, C., Cerasino, L. *et al.* (2015) Anatoxin-a producing *Tychonema* (Cyanobacteria) in European waterbodies. *Water Research* **69**, 68–79.
[2] De la Cruz, A.A., Hiskia, A., Kaloudis, T. *et al.* (2013) A review on cylindrospermopsin: the global occurrence, detection, toxicity and degradation of a potent cyanotoxin. *Environmental Science: Processes & Impacts* **15** (11), 1979–2003.
[3] Dell'Aversano, C., Eaglesham, G.K. and Quilliam, M.A. (2004) Analysis of cyanobacterial toxins by hydrophilic interaction liquid chromatography–mass spectrometry. *Journal of Chromatography A* **1028** (1), 155–164.
[4] Furey, A., Crowley, J., Hamilton, B. *et al.* (2005) Strategies to avoid the mis-identification of anatoxin-a using mass spectrometry in the forensic investigation of acute neurotoxic poisoning. *Journal of Chromatography A* **1082** (1), 91–97.

SOP 20

Extraction and Chemical Analysis of Saxitoxin and Analogues in Water

Lutz Imhof and Wido Schmidt

DVGW Technologiezentrum Wasser TZW, Branch Office Dresden, Germany

The approach for trace analysis of saxitoxins is subdivided into the following SOPs:

SOP 20.1 Preparation of Aqueous Standard Solutions and Calibration Conditions for Dissolved and Cell-Bound Saxitoxins
SOP 20.2 Online Solid-Phase Extraction of Dissolved Saxitoxins
SOP 20.3 Extraction of Cell-Bound Saxitoxins From Biomass Filtered on Glass-Fibre Filters
SOP 20.4 Analysis of Saxitoxins by High-Performance Liquid Chromatography Coupled to Mass Spectrometry

SOP 20.1 Preparation of Aqueous Standard Solutions and Calibration Conditions for Dissolved and Cell-Bound Saxitoxins

SOP 20.1.1 Introduction

The molecular weight of saxitoxins and its analoques (STXs) range between 300 and 450 a.u. The structures are characterized by amino and sulfonic acid groups responsible for the high polarity of these compounds making their extraction from water quite difficult.

For preparation of stock solutions certified reference standards are used. The concentration of certified toxin standard solutions can differ between manufacturing batches. In general they cover the range from 50 to 300 μmol L^{-1}. The preparation of a stock solution containing all congeners is

Handbook of Cyanobacterial Monitoring and Cyanotoxin Analysis, First Edition.
Edited by Jussi Meriluoto, Lisa Spoof and Geoffrey A. Codd.

favorable. This stock solution can be used for calibration of cell-bound and dissolved toxins as well as for a daily quality control.

SOP 20.1.2 Experimental

SOP 20.1.2.1 Materials

- Saxitoxin (STX) from National Research Council (NRC) Institute for Marine Biosciences Halifax, Canada
- Neosaxitoxin (NEO), NRC
- Gonyautoxin-1 & -4 (GTX 1/4), NRC
- Gonyautoxin-2 & -3 (GTX 2/3), NRC
- Benzoyl-L-arginyl-4-amino-benzoic acid (BAAB), e. g. from Fluka (Buchs, Schwitzerland)
- Water purified (MilliQ Plus quality)
- Acetonitrile HPLC grade
- Methanol HPLC grade
- Sodium chloride puriss.
- Hydrochloric acid
- Acetic acid puriss.
- Glass-fibre filters, equivalent to Whatman (Maidstone, UK) GF/C, diameter 47 mm
- 20-mL vials with screw cap
- Volumetric flask, 100 mL
- Measuring cylinder, 250 mL
- Glass flasks, 250 mL, 500 mL
- Borosilicate glass vials (V-Vials) with green/red top cap, 3 mL
- Glass autosampler vials with 0.2-mL borosilicate glass insert and screw caps with Si/PTFE septa
- Microliter syringes 5 µL, 10 µL, 25 µL, 100 µL

SOP 20.1.2.2 Special Equipment

- Electronic laboratory scale with precision of 0.0001 g

SOP 20.1.2.3 Solutions

- 3 mmol L^{-1} hydrochloric acid (HCl)
- Acetonitrile/MilliQ-water (50:50, v/v)
- 30% methanol/70% MilliQ water (v/v), containing 0.2% acetic acid

SOP 20.1.2.4 Internal Standard Stock Solution (BAAB), Concentration 50 ng μL^{-1}

a. 5 mg BAAB is dissolved in a 100 mL volumetric flask in a mixture of 50% (acetonitrile/50% MilliQ water (v/v). Store in the freezer (50%/50% remains as liquid at −18°C).
b. For daily use, transfer an aliquot in a 3-mL V-Vial with green/red top cap and store in the fridge.

SOP 20.1.2.5 STXs Standard Mix Stock Solution

The following standard STXs mix stock solution (5 ng μL^{-1} per single substance) is used for a control standard (see Table 20.1.1). The total amount of the solution should be 0.5 mL. This solution is stable for about 4 weeks at 4°C.

Table SOP 20.1.1 *Preparation of calibration standard and daily quality control standard*

Standard	Conc. CRM standard, μmol L^{-1}	Conc. CRM standard, μg mL^{-1}	Add of each component, μL
HCl 3 mmol L^{-1}			166.7
STX	65	19.5	128
NEO	65	20.5	122
GTX 1 and 4	~141	58	43.1
GTX 2 and 3	ca. 157	62.1	40.2
Total			500

SOP 20.1.2.6 Calibration Conditions for the Analysis of Dissolved Toxins

For calibration of the whole procedure nonchlorinated tap water is preferably used. The use of matrix free ultrapure water, e.g. MilliQ is not optimal, because it does not reflect conditions of practice.

In the case of calibration in sea water, sodium chloride is added to the tap water with a concentration covering the level between 35 and 40 g L^{-1}.

The concentration of STXs for calibration should vary between 0.05 and 1.25 μg L^{-1} with at least 6 dilutions and 10 in maximum.

The volume of the final concentrated sample depends on the equipment of online enrichment and the sensitivity of the mass spectrometer. The proposed range covers a level between 1 and 10 mL, which requires a suitable sample loop.

Preparation of Standard Dilutions:

a. For each concentration fill one glass flask with 100 mL of water.
b. Add standard mix to the flasks using microliter syringe according to Table SOP 20.1.2.
c. Add internal standard (BAAB, 20.1.2.4.) into the flasks. Concentration of the IS in the dilutions is always 2.5 μg L^{-1}.

For online SPE, see SOP 20.4 and for LC-MS/MS see SOP 20.4.

The calculated calibration curve should show a linear form with a correlation coefficient, r^2>0.99. Examples of calculated calibration parameters for dissolved STXs by LC-MS/MS are presented in Table 20.1.3.

Note: Check the HPLC-MS system regularly by injection of known samples and repeat the calibration as necessary, typically after 1 or 2 months.

SOP 20.1.2.7 Calibration Conditions for the Analysis of Cell-Bound (Filter extracted) Saxitoxins

Preparation of standard dilutions:

a. Wash glass-fibre filters (GF/C Whatman) carefully with 10–20 mL MilliQ-water in vacuum as with real samples and place them in 20-mL vials, in order to simulate the whole procedure.
b. Add 3 mL of 30% methanol containing 0.2% acetic acid in each vial.
c. Add standard mix stock solution (20.1.2.5) according to Table SOP 20.1.4 into the vials using a microliter syringe, the range of calibration covers 0.06–2.5 μg L^{-1} with 6 points in minimum and 10 in maximum.
d. Add internal standard (BAAB, 20.1.2.4.) in each vial.

Table SOP 20.1.2 *Dilutions of the mix stock solution for calibration*

Flask No.	Concentration, µg L^{-1}	Add stock solution, µL	Add IS, µL
1	—	—	5
2	0.05	1	5
3	0.075	1.5	5
4	0.1	2	5
5	0.2	4	5
6	0.3	6	5
7	0.5	10	5
8	0.75	15	5
9	1.0	20	5
10	1.25	25	5

Table SOP 20.1.3 *Example of calibration parameters for analysis of dissolved toxins and IS by LC-MS/MS*[a]

Compound	Q1 ion>Q3 ion *m/z*>*m/z*	Calibration equation, 8 data points (slope)	Limit of detection (LOD), µg L^{-1}	Limit of determination, µg L^{-1}	Recovery, %	Correlation coefficient
Saxitoxin	300>204	$y=0.211\,x-0.002$	0.04	0.12	70	0.993
Neosaxitoxin	316>238	$y=0.072\,x-0.002$	0.07	0.25	60	0.991
GTX 2/3	396>316;298	$y=0.514\,x-0.04$	0.05	0.16	86	0.996
GTX 1/4	412>332;314	$y=0.216\,x-0.03$	0.09	0.27	72	0.989

Q1 = $[M+H]^+$: protonated molecule ion; Q3 = product ion.
[a]The values of Table SOP 20.1.3 are produced by online enrichment of 10-mL sample and LC-MS/MS System API 2000.

Table SOP 20.1.4 *Preparation of dilutions of the mix stock solution (20.1.2.5) for the analysis of cell-bound saxitoxins*

Flask No.	Concentration, µg L^{-1}	Absolute amount of each toxin in 250-mL sample, ng	Add mix stock solution, /µL	Add IS BAAB, µL
1	—	—	—	10
2	0.06	15	3	10
3	0.1	25	5	10
4	0.3	75	15	10
5	0.5	125	25	10
6	0.75	187.5	37.5	10
7	1	250	50	10
8	1.5	375	75	10
9	2.0	500	100	10
10	2.5	625	125	10

Table SOP 20.1.5 Example of calibration parameters for analysis of cell-bound saxitoxins and IS by LC-MS/MS[a]

Compound	Q1 ion > Q3 ion $m/z > m/z$	calibration equation, 8 data points (slope)	Limit of detection (LOD), µg L^{-1}	Limit of determination, µg L^{-1}	Recovery, %	Correlation coefficient
Saxitoxin	300 > 204	$y = 0.161x$	0.05	0.18	83	0.993
Neosaxitoxin	316 > 238	$y = 0.049x$	0.09	0.31	79	0.991
GTX 2/3	396 > 316; 298	$y = 0.217x$	0.06	0.22	87	0.987
GTX 1/4	412 > 332; 314	$y = 0.089x$	0.09	0.31	85	0.993

Q1 = $[M+H]^+$: protonated molecule ion; Q3 = product ion.
[a]The values of Table SOP 20.1.5 are produced by filter extraction of 250 mL water sample and LC-MS/MS System Api 2000.

The concentration of cell-bound toxins are calculated in µg L^{-1} water. The values in Table SOP 20.1.4 are based on filtration of 250 mL water sample. The sample volume filtered depends on the amount of cyanobacteria in the water.

For further extraction of cell-bound toxins, see SOP 20.3.2.4.c and for LC-MS/MS, see SOP 20.4.

The calculated calibration curve should show a linear form with a correlation coefficient, $R^2 > 0.99$. Examples of calculated calibration parameters for cell-bound STXs by LC-MS/MS are presented in Table 20.1.5.

SOP 20.2 Online Solid-Phase Extraction of Dissolved Saxitoxins

SOP 20.2.1 Introduction

STXs are produced by marine dinoflagellates and by freshwater cyanobacteria [1]. The high polarity of structures makes the preconcentration of STXs from aqueous phase very difficult [2]. A combination of different SPE-column materials is recommended [3,4]. In this SOP the online extraction is done by combination of two solid phases with commercial online extraction equipment. The use of online concentrators guarantees the constancy of pressure needed, and allows to use lower sample volumes for pre-concentration.

SOP 20.2.2 Experimental

SOP 20.2.2.1 Materials

- Water (MilliQ Plus quality)
- Methanol HPLC grade
- Acetonitrile HPLC grade
- Sodium hydroxide
- Hydrochloric acid
- Acetic acid, puriss.
- Ammonia solution (25%)
- Heptafluorobutyric acid (HFBA), puriss.
- Glass-fibre filters, equivalent to Whatman (Maidstone, UK) GF/C, diameter 47 mm
- Measuring cylinder, 250 mL
- Glass flasks, 250 mL, 500 mL, 1000 mL
- 20-mL vial with screw cap

- Glass autosampler vials with 0.2-mL borosilicate glass insert and screw caps with Si/PTFE septa
- Pipettes 1–2 mL
- Microliter syringes 5 µL, 100 µL, 500 µL
- Borosilicate glass vials (V-vials), 3 mL and 5 mL capacities
- Online cartridges equivalent to 20 mm × 2 mm Strata-X (Phenomenex, USA), part No. 00M-S033-B0-C8 with cartridge holder
- Online cartridges equivalent to 20 mm × 2 mm, Idex Corporation (Upchurch Scientific, Oak Harbor, USA) self filled with PGC Hypercarb Thermo Scientific (Waltham, MA, USA) or HPLC cartridge 20 mm × 2.1 mm Hypercarb Dash-HTS column (Thermo Scientific)

SOP 20.2.2.2 Special Equipment

- Vacuum filtration unit, e.g. from Sartorius (Germany)
- Magnetic stirrer
- pH-meter
- Online SPE equipment e.g. Symbiosis™ (Spark Holland) or online SPE System from Axel Semrau (Germany) or GERSTEL SPE^{XOS} from Gerstel GmbH (Germany) or equivivalent combined with a large volume autosampler, e.g. from Gilson (Canada) or alternatively large volume autosampler with additional 2-port/6-position switch valve and HPLC-pump

SOP 20.2.2.3 Solutions

- 0.2 mol L^{-1} Sodium hydroxide solution
- 0.2 mol L^{-1} Hydrochloric acid
- Solvent A for online SPE: 95% methanol

SOP 20.2.2.4 General Procedure

Sample Preparation

a. Water samples in 0.5–1.0 L glass flasks, can be stored 1 day in maximum at 4°C.
b. Filter 250 mL of water sample by vacuum using a glass fibre filter.
   Note: The reproducibility of the whole procedure deteriorate if too much biomass is on the filter. If the biovolume is too high or in the case of dense blooms reduce the filtrated water volume.
c. Wash the filter carefully with 2 × 1 mL of MilliQ water.
d. The filter is stored in a 20-mL glass vial at –18°C for further analysis of cell-bound toxins. Store also the filtrates that are not analyzed immediately at –18°C.
e. Take 100 mL from the filtered (frozen samples should be thawed at room temperature) sample and place it in a 250-mL flask.
f. Add 10 µL of heptafluorobutyric acid, 20 µL of acetic acid and 25 µL of ammonia solution (25%).
g. Adjust the pH-value with 0.2 mol L^{-1} HCl or 0.2 mol L^{-1} NaOH to the range between 5.2 and 5.4.
h. Add 5 µL of the internal standard (BAAB, SOP 20.2.2.4.1) to the sample.

The sample is ready for online SPE.

Online SPE of Water Samples

This SOP is based on the use of the first generation of Spark Holland online concentrators. A further possibility is the use of a large-volume autosampler with sample loop and additional HPLC-pump.

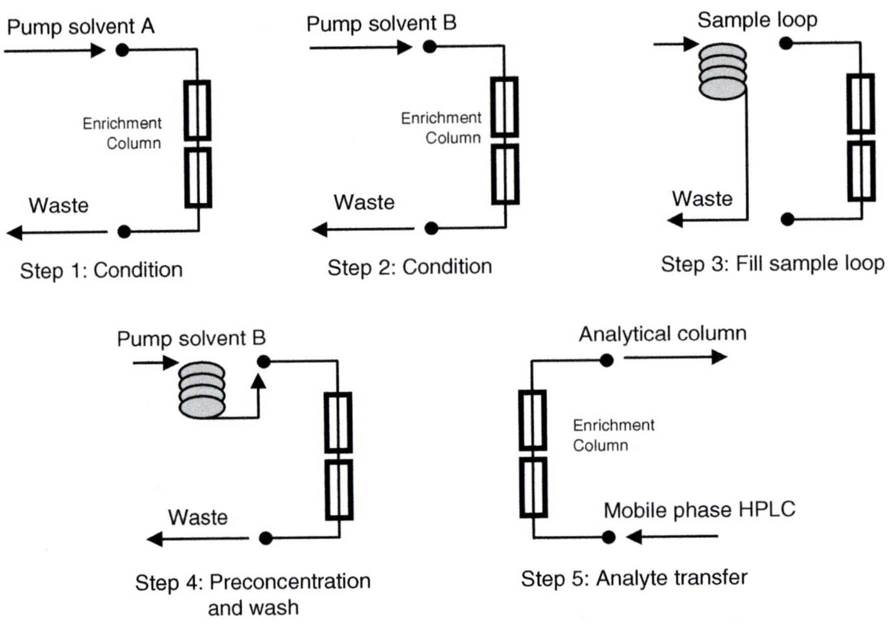

Figure SOP 20.2.1 *Scheme of online-concentration, elution, and chromatography*

The solid-phase system consists of a combination of an online extraction cartridge 20 mm × 2 mm Strata-X and an online 20 mm × 2 mm guard cartridge filled with PGC Hypercarb.

Step 1: Conditioning of online SPE-cartridges with ~5 mL of 95% methanol with maximum flow rate 2 mL min^{-1} (Solvent A).
Step 2: Conditioning of online SPE-cartridges with ~ 5 mL of water (MilliQ quality) with maximum flow rate 1.2 mL min^{-1} (Solvent B).
Step 3: Filling of sample loop.
Step 4: Solid phase extraction and washing with ~ 2 mL of water with maximum flow rate 1.2 mL min^{-1} (Solvent B).
Step 5: Analyte transfer, start of the chromatographic cyclus (see SOP 20.4). The volume of sample loop can differ between 1 and 10 mL in maximum.

Special Issues
The life time of the cartridge combination depends on:

- the sample matrix
- the number of samples
- the sample volume needed for extraction. In the case of 5-mL sample extraction up to 100 runs are possible
- the strong pressure changes between concentration and elution

The quality of SPE is controlled by estimation of the recovery of STX spiked standard sample. The sequence of control samples depends on the sample throughput.

The carryover of analytes can be avoided by flushing with solvent which is checked by control measurements. Control measurements are defined as blank runs, e.g. the enrichment of analyte-free natural sample in form of tap water.

SOP 20.3 Extraction of Cell-Bound Saxitoxins From Biomass Filtered on Glass-Fibre Filters

SOP 20.3.1 Introduction

The determination of cell-bound STX-toxins requires a mild exploration of the algal cells avoiding toxin destruction. An optimal version with a combination of freezing, ultrasonication and solvent extraction is proposed.

SOP 20.3.2 Experimental

SOP 20.3.2.1 Materials

- Water purified (MilliQ Plus quality)
- Nitrogen gas (>99.9%)
- Methanol HPLC-grade
- Acetic acid puriss.
- 20-mL vials with screw cap
- Pipette 1–2 mL
- Single use polypropylene syringes, 5 mL
- Syringe filters 17 mm or 25 mm in diameter (combined glass fibre/PVDF 0.45 μm), e.g. Macherey-Nagel (Düren, Germany)
- Microliter syringes 5 μL, 100 μL, 500 μL
- Borosilicate glass vials (V-Vials), 3-mL capacity
- Glass autosampler vials with 0.2-mL borosilicate glass insert and screw caps with Si/PTFE septa

SOP 20.3.2.2 Special Equipment

- –20°C freezer
- Bath ultrasonicator
- Evaporator (hot block and gas apparatus)
- Rotary shaker

SOP 20.3.2.3 Solutions

- 30% methanol, 70% MilliQ (v/v), containing 0.2% acetic acid
- Acetonitrile/MilliQ (5:95, v/v) containing 0.1% acetic acid.

SOP 20.3.2.4 General Procedure

The filters preserved frozen at –18°C (see SOP 20.2.2.4.1) will be used.

a. Add 3 mL of 30% methanol containing 0.2% acetic acid in the 20-mL vial containing the filter.
b. Add 10 μL of the internal standard (BAAB, SOP 20.2.2.4).

c. Extract the sample by ultrasonication ~15 min.
d. Continue the extraction by shaking ~12 h.
e. Transfer the liquid phase with a 5-mL single-use polypropylene syringe via a 25-mm syringe filter (glass fibre/PVDF 0.45 µm) in a borosilicate glass vial.
f. Dry the sample completely in an evaporator at 50–55°C by nitrogen stream.
g. Dissolve the dry residue in 0.2 mL solution of 5% acetonitrile containing 0.1% acetic acid.
h. After sedimentation by gravity (20 to 30 minutes) of the solid residuals transfer the clear phase (0.1 mL at minimum) by a Pasteur pipette in a glass autosampler vial with 0.2-mL borosilicate glass insert.

The sample is ready for analysis.

SOP 20.4 Analysis of Saxitoxins by High-Performance Liquid Chromatography Coupled to Mass Spectrometry

SOP 20.4.1 Introduction

The complete chromatographic separation of STX analoques, which are structurally very similar, is challenging and requires methods such as hydrophilic interaction chromatography [5] or ion pair chromatography [6] coupled to fluorescence detection or mass spectrometry [7, 8].

Because of the online enrichment step used in this SOP, the optimal chromatographic separation of the compounds GTX 1 and GTX 4 as well as GTX 2 and GTX 3 are not possible.Therefore, the identification and quantification of these compounds by MS/MS detection are given as the sum of GTX 1/4 and GTX 2/3, respectively.

The described chromatographic conditions can be used for both cell-bound and dissolved toxins. The injection volume depends on the LC-MS/MS system used and should be adapted. For cell-bound toxins, 10–20 µL sample is injected using the gradient program in Table SOP 20.4.7.

SOP 20.4.2 Experimental

SOP 20.4.2.1 Materials

- Water (Millipore MilliQ Plus quality)
- Acetonitrile HPLC quality grade
- Heptafluorobutyric acid (HFBA), protein sequence analysis grade
- Acetic acid HPLC quality grade
- Ammonia solution 25% HPLC quality grade
- Column: Phenomenex Luna C_{18}(2) 3 µm 150 mm × 3 mm I.D. Cat. No. 00F-4251-Y0 or equivalent
- Guard-column: C_{18} 4 mm × 2 mm I.D. Cat. No. AJ0-4286, Phenomenex, USA or equivalent
- Glass autosampler vials with 0.2-mL borosilicate glass insert and screw caps with Si/PTFE septa

SOP 20.4.2.2 Special Equipment

The Online-SPE-LC-MS/MS system is composed of:

- High-performance liquid chromatograph with gradient pump system, a degasser, a autosampler and a column oven, e.g. Agilent.

- A triple-quadrupole mass analyser with electrospray ion source, e.g. API2000, API4500 from AB Sciex.
- Online SPE-System, e.g. Spark Holland.

SOP 20.4.2.3 HPLC Mobile Phase

Mobile phase component A: 95% MilliQ+5% acetonitrile (v/v), containing 0.1% acetic acid and 0.025% HFBA

a. Add 1 mL of acetic acid in 950 mL MilliQ
b. Add 250 µL HFBA
c. Add 50 mL acetonitrile
d. Adjust the pH value to 3.8 with ammonia solution
e. Replace every week

Mobile phase component B: acetonitrile + 0.1% acetic acid

a. Add 500 µL acetic acid in 500 mL acetonitrile
b. Replace every week

Note: Work in the fume-hood and do not contaminate HFBA with acetonitrile.

SOP 20.4.2.4 LC-MS/MS Procedure for Standards and Real Samples

- The HPLC system should be set up as described in the manufacturer's instructions including degassing, priming, and changing columns.
- Always use a guard column (4 mm × 2 mm, C_{18}). Change guard column if the back-pressure rises or peak forms deteriorate.
- Set column oven at room temperature.
- Examples of LC-gradients for dissolved and cell-bound STXs are described in Table SOP 20.4.6 and SOP 20.4.7, respectively.

The identification and quantification of the saxitoxins are done by MS/MS system in MRM (multireaction monitoring) mode.

Examples of MS parameters are for the system API 2000 (see Tables SOP 20.4.8 and SOP 20.4.9). If other MS systems are used, all parameters have to be adjusted.

Table SOP 20.4.6 *Gradient program for analysis of dissolved toxins; Phenomenex Luna C_{18}(2) 3 µm 150 × 3 mm ID, flow rate 0.2 mL min^{-1}*

Time (min)	%A (aqueous phase)	%B (organic phase)
START	100	0
0.01	95	5
7	75	25
12	70	30
22	5	95
24	5	95
26	95	5
26.01	100	0
40	100	0

Table SOP 20.4.7 *Gradient program for analysis of cell-bound saxitoxins; Phenomenex Luna C_{18} (2) 3 μm 150 mm × 3 mm ID*

Time (min)	%A (aqueous phase)	%B (organic phase)
START	100	0
0.01	95	5
10	25	75
15	70	30
20	25	75
21	5	95
24	5	95
26	95	5
26.01	100	0
40	100	0

Table SOP 20.4.8 *Source-dependent parameters for API 2000 with electron spray interface*

Curtain gas (CUR)	25
Nebulizer gas (GS1)	30
Heater gas (GS2)	50
Collision gas (CAD)	3
Temperature heater gas (TEM)	350°C
Ion spray voltage (IS)	+5200V
Focusing potential (FP)	350
Entrance potential (EP)	10

Table SOP 20.4.9 *Compound-dependent LC-MS/MS parameters for API 2000*

Analyte	Q1 mass *m/z*	Q3 mass *m/z*	Declustering potential (DP), V	Collision energy (CE), V
STX	300.2	204.1	25	35
STX	300.2	138.1	25	40
STX	300.2	282	25	20
NEO	316.3	220.4	20	30
NEO	316.3	238.3	20	25
NEO	316.3	298	20	20
GTX 1/4	412	332	15	20
GTX 1/4	412	314	15	30
GTX 2/3	396	316	15	20
GTX 2/3	396	298	15	25
IS BAAB	398	244.1	50	28
IS BAAB	398	216.1	50	33

Q1 = $[M+H]^+$: protonated molecule ion; Q3 = product ion.

Table SOP 20.4.10 *Mass transitions*

Compound	Q1 ion > Q3 ion *m/z* > *m/z*	Integration as	Used for
Saxitoxin	300 > 204 300 > 138		Quantification Qualification
Neosaxitoxin	316 > 238 316 > 220		Quantification Qualification
GTX 2/3	396 > 316; 298	Sum of two ions	Quantification
GTX 1/4	412 > 332; 314	Sum of two ions	Quantification
IS BAAB	398 > 216 398 > 244		Quantification Qualification

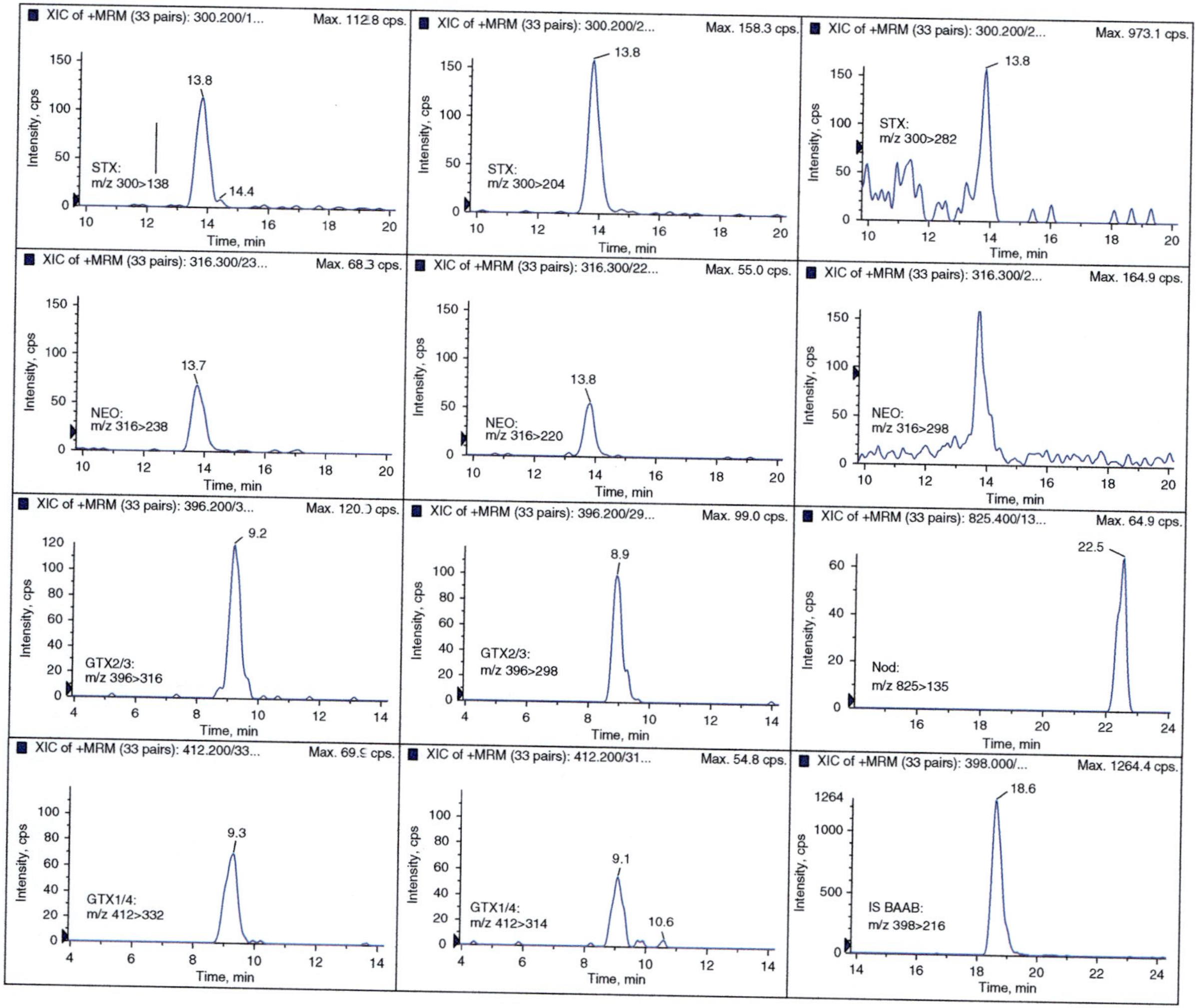

Figure SOP 20.4.2 *Standard chromatograms of toxins corresponding to 0.1 µg L^{-1} per substance by online SPE of spiked real surface water. (The occurrence of nodularin in the model waters is caused by project specific reasons.)*

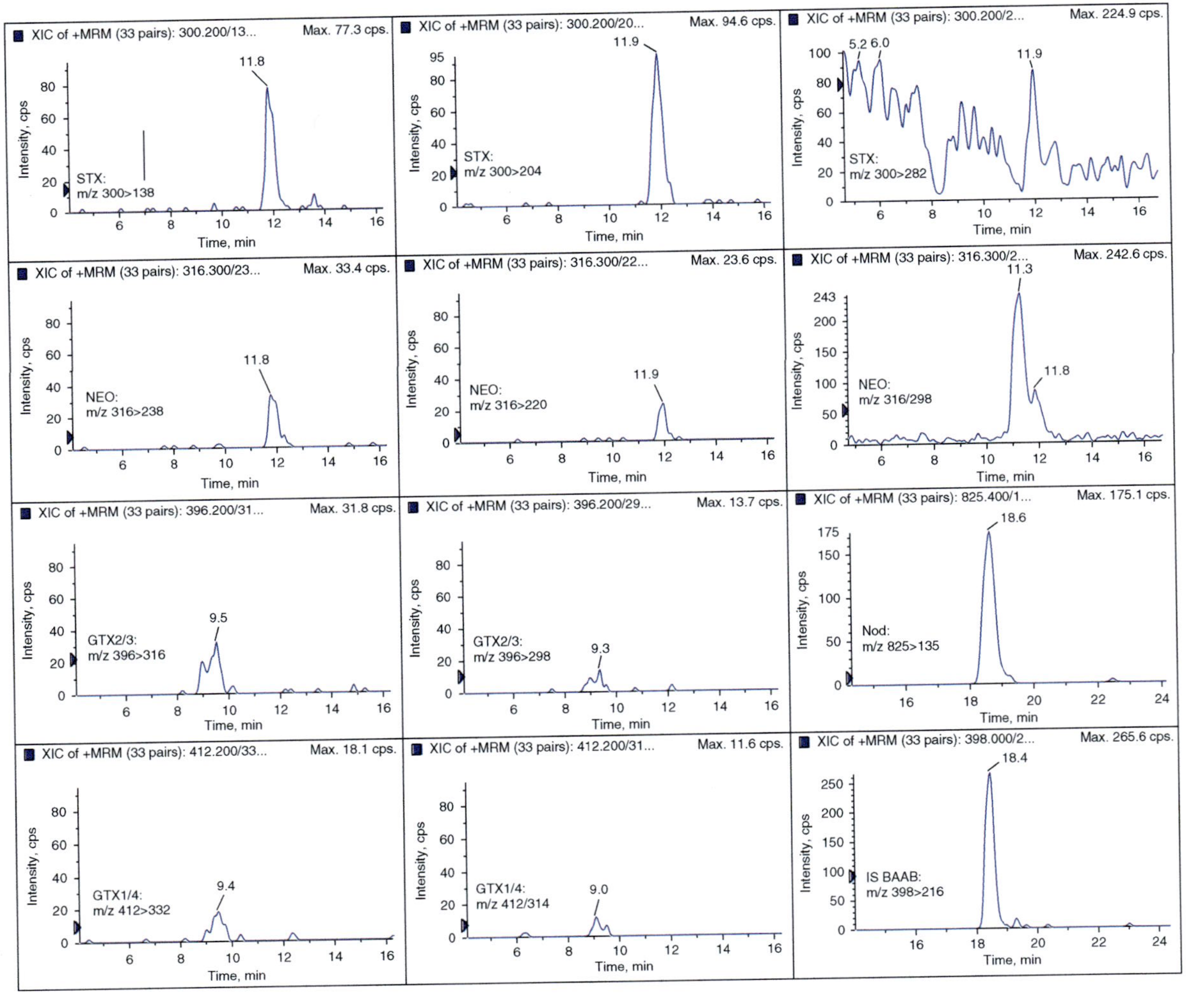

Figure SOP 20.4.3 *Standard chromatogram of toxins corresponding 0.3 μg L^{-1} per substance, comparable to cell-bound toxins (filter extraction). (The occurrence of nodularin in the model waters is caused by project specific reasons.)*

SOP 20.4.3 Quantification for Standards and Real Samples

The chromatographic separation of GTX 1/4 and GTX 2/3 is not possible with the method described in this SOP. As alternative the peak areas of single mass transitions will be summarized (see Table SOP 20.4.10).

In Figure SOP 20.4.2, an example of online extraction standard chromatograms corresponding to 0.1 $\mu g\,L^{-1}$ for each analyte is shown.

In Figure SOP 20.4.3, an example of filter extraction standard chromatograms corresponding to 0.3 $\mu g\,L^{-1}$ for each analyte is shown.

References

[1] Ballot A., Fastner J., and Wiedner C., (2010) Paralytic shellfish poisoning toxin-producing cyanobacterium *Aphazinomenon gracile* in Northeast Germany. *Applied and Environmental Microbiology* **76** (4), 1173–1180.

[2] Chan I., Lam P., Cheung R., Lam M. *et al.* (2005) Application of solid phase microextraction in the determination of paralytic shellfish poisoning toxins. *Analyst* **130**, 1524–1529.

[3] Thermo Scientific Hypercarb columns, (2009) Applications Notebook Issue 1.

[4] Thermo Scientific, (2007) Method development guide for hybercarb columns.

[5] Diener M., Erler K.,Christian B. *et al.* (2007) Application of a new zwitterionic hydrophilic interaction chromatography column for determination of paralytic shellfish poisoning toxins. *Journal of Separation Sciences* **30**, 1821–1826.

[6] Jaime E., Hummert C., Hess P. *et al.* (2001) Determination of paralytic shellfish poisoning toxins by high-performance ion-exchange chromatography. *Journal of Chromatography* **929**, 43–49.

[7] Diener M., Erler K., Hiller S., *et al.* (2006) Determination of paralytic shellfish poisoning (PSP) toxins in dietary supplements by application of a new HPLC/FD method. *European Food Research and Technology* **224**, 147–151.

[8] Hiller S., Krock B., Cembella A. *et al.* (2007) Rapid detection of cyanobacterial toxins in precursor ion mode by liquid chromatography tandem mass spectrometry. *Journal of Massspectrometry* **42**, 1238–1250.

SOP 21

Extraction of BMAA from Cyanobacteria

James S. Metcalf, Sandra A. Banack, and Paul A. Cox

Institute for Ethnomedicine, Jackson, USA

SOP 21.1 Introduction

β-*N*-Methylamino-L-alanine (BMAA) is a neurotoxic nonprotein amino acid associated with human neurodegenerative disease, produced by cyanobacteria [1, 2]. BMAA exists as a free amino acid and associated with proteins. Consequently, methods are required to extract both BMAA fractions for separate analysis by a variety of analytical methods.

SOP 21.2 Experimental

SOP 21.2.1 Materials

- Deionized water (18.2 MΩ cm)
- HCl
- Trichloroacetic acid (TCA)
- Microcentrifuge tubes
- Ultrafree MC centrifuge filters
- Pipettes
- Pipette tips
- Cyanobacteria
- Screw-capped glass vials

Handbook of Cyanobacterial Monitoring and Cyanotoxin Analysis, First Edition.
Edited by Jussi Meriluoto, Lisa Spoof and Geoffrey A. Codd.

SOP 21.2.2 Special Equipment

- Probe sonicator
- Freeze-drier
- SpeedVac
- Microcentrifuge
- Balance
- Vortex
- Hot block

SOP 21.2.3 Solutions

- 20% TCA in water. Prepare 20% (w/v) TCA by dissolving 20 g trichloroacetic acid in 100 mL of deionized water
- 6 mol L^{-1} HCl in water. Prepare 6 mol L^{-1} HCl by adding an equal volume of deionized water to a solution of 12 mol L^{-1} HCl

SOP 21.2.4 General Procedure

a. Remove cyanobacterial material from cultures or environmental samples and freeze-dry.
b. Weigh out approximately 50 mg freeze-dried cyanobacterial material into a microcentrifuge tube.
c. Add 400 µL 20% (w/v) TCA to the cyanobacterial material and vortex to resuspend the material.
d. Sonicate the material with a probe sonicator for approximately 20 to 30 s on a medium to low setting, sufficient to rupture the cells.
e. Place the microcentrifuge tube containing the sample at 4°C overnight.
f. Remove the microcentrifuge tube and centrifuge at full speed for 5 min at room temperature.
g. Remove the supernatant and transfer to a 0.2-µm spin-filter microcentrifuge tube.
h. Add 250 µL 20% (w/v) TCA to the pelleted cyanobacterial material and vortex to resuspend the material.
i. Sonicate the suspension for 20 s and leave at room temperature for 2 h.
j. Centrifuge the suspension and transfer the supernatant to the spin-filter microcentrifuge tube containing the first extraction, pooling both extractions.
k. Centrifuge filter the supernatant and store the 20% (w/v) TCA extract at 4°C for analysis. If analysis will not occur until days to weeks later, the extract should be stored at –20°C.
l. Resuspend the pelleted cyanobacterial material in 400 µL 6 mol L^{-1} HCl.
m. Vortex and transfer the suspension to a screw-capped glass vial.
n. Place in the hot block and heat at 110°C for 16 h.
o. Remove from the hot block and allow to cool to room temperature.
p. Remove an aliquot and place in the 0.2-µm microcentrifuge filter. Centrifuge 5 min at 12,000 rpm.
q. Once all of the liquid has been filtered, remove the filter and dry the filtrate in the speedvac, within the filter microcentrifuge tube. Alternatively, the TCA extract can be diluted 1:50 with deionized water for analysis.
r. Once dry, store at –20°C until analysed.

SOP 21.3 Validation

BMAA has been detected in a wide range of cyanobacteria, from various environments. As of yet, no systematic study concerning the extraction of BMAA from cyanobacteria has been carried out.

Reported concentrations of BMAA as a free amino acid range from "Not Detected" to 6478 $\mu g\,g^{-1}$, with a median of 32 $\mu g\,g^{-1}$ and a mode of not detected, based on the analysis of 30 strains [1]. Concentrations for protein-bound BMAA had a median of 54 $\mu g\,g^{-1}$ with a mode of not detected [1]. Further studies will examine the extractability of BMAA using various methods of extraction. Using HPLC, BMAA can be observed, often through the spiking of samples.

SOP 21.4 Special Issues

a. Occasionally, when acids are added to cyanobacterial material, especially with 6 $mol\,L^{-1}$ HCl, gas production (e.g. CO_2) can result, neutralizing the acid. Consequently, incomplete hydrolysis may occur. In order to ensure complete hydrolysis, further additions of 6 $mol\,L^{-1}M$ HCl and heating may be required so that complete liberation of amino acids is achieved.
b. A further issue can occur with screw-capped glass vials, which require that the cap is secure when placed on the vial, so that drying of the hydrolysate does not occur during hydrolysis at 110°C.
c. A final issue concerns the centrifuge filtering of extracts, which may become blocked and it therefore may be necessary to use multiple filter inserts to filter all of the material. For drying solutions in the microcentrifuge tubes after filtration, a hole can be poked into the lid using a hot needle, so that the solutions can be dried with the lid closed, with care taken to ensure damage does not occur to the drying device through drying concentrated acid samples. Once dried, a small piece of laboratory film can be used to make a seal under the cap until analysis.

References

[1] Cox, P.A., Banack, S.A., Murch, S.J. *et al.* (2005). Diverse taxa of cyanobacteria produce β-N-methylamino-l-alanine, a neurotoxic amino acid. *Proceedings of the National Academy of Sciences USA* **102** (14), 5074–5078.

[2] Banack, S.A., Johnson, H.E., Cheng, R. and Cox, P.A. (2007). Production of the neurotoxin BMAA by a marine cyanbacterium. *Marine Drugs* **5** (2), 180–196.

SOP 22

Analysis of β-*N*-Methylamino-L-Alanine by UHPLC-MS/MS

James S. Metcalf, William B. Glover, Sandra A. Banack, and Paul A. Cox

Institute for Ethnomedicine, Jackson, USA

SOP 22.1 Introduction

Many methods have been used for the analysis of β-*N*-methylamino-L-alanine (BMAA), but LC-MS/MS has become the method of choice, largely through its ability to specifically detect BMAA. This is achieved using retention time and the correct ratio of product ions to lead to the accurate, sensitive detection of this neurotoxic amino acid. This SOP describes a method for analysis of BMAA based on precolumn AccQ-TAG derivatisation with 6-aminoquinolyl-N-hydroxysuccinimidyl carbamate (AQC) followed by ultra high-performance chromatography (UHPLC) coupled to ESI-MS/MS detection.

SOP 22.2 Experimental

SOP 22.2.1 Materials

- Deionized water (18.2 MΩ cm^{-1})
- Methanol LC-MS grade
- Ammonium acetate
- HCl
- Glacial acetic acid
- Amino acid standard, e.g. lysine

Handbook of Cyanobacterial Monitoring and Cyanotoxin Analysis, First Edition.
Edited by Jussi Meriluoto, Lisa Spoof and Geoffrey A. Codd.

- L-2,4-Diaminobutyric acid (DAB)
- *N*-(2-aminoethyl)glycine (AEG)
- Microcentrifuge tubes
- Autosampler vials
- Waters AccQtag reagent kit (PN WAT052880),
- Column, e.g. Phenomenex Kinetex 1.7 µm, C18, 100Å, 100×2.1 mm (part 00D-4475-AN, although check isomer separation on every new column).

SOP 22.2.2 Special Equipment

- Hot block
- UHPLC-MS/MS system, such as a Waters Acquity UPLC system, TSQ Quantiva (Thermo Scientific), HESI-II probe

SOP 22.2.3 Solutions

- 20 mmol L^{-1} HCl
- 20 mmol L^{-1} ammonium acetate (pH 5, adjusted with glacial acetic acid)

SOP 22.2.3.1 LC Mobile Phase

- 20 mmol L^{-1} ammonium acetate (pH 5):methanol (62:38)

SOP 22.2.4 General Procedure

SOP 22.2.4.1 Sample Preparation

a. Resuspend and dilute extracts with 20 mmol L^{-1} HCl
b. Following the instructions supplied with the Waters AccQtag kit, derivatise aliquots of the extract for LC-MS/MS analysis

SOP 22.2.4.2 UPLC-MS/MS

a. Equilibrate the UPLC system with the mobile phase, use isocratic flow of 20 mmol L^{-1} ammonium acetate:methanol at a flow rate of 0.35 mL min^{-1} and the column at 65°C (run time 5 min)
b. Supply nitrogen gas to the heated electrospray ionization (H-ESI-II) probe using the following parameters: sheath gas = 40 arb; aux gas = 2 arb; sweep gas = 1 arb; vaporizer temperature of 400°C
c. Operate the mass spectrometer under the following conditions: capillary temperature set at 350°C and spray voltage 3500 V
d. Use a divert valve, except during the selected reaction monitoring (SRM) scans. The second quadrupole was pressurized to 1.0 mTorr with argon
e. Isolate ion *m/z* 459 in the first quadrupole filter as the precursor ion and subject to collision induced dissociation (CID). Perform second step mass filtering using selective reaction monitoring (SRM) of BMAA after CID in the collision cell. Monitor the following *m/z* transitions: 459 > 119, CE 25 eV; 459 > 289 CE 23 eV; 459 > 171 CE 45 eV. Detect the resultant three product ions originating from derivatized BMAA (*m/z* 119, 289, 171), after entering the third quadrupole and quantify their relative abundances. Compound specific *m/z* transitions can be added; 459 > 258 CE 36 eV (BMAA), 459 > 188 CE 38 eV (DAB), and 459 > 214 CE 35 eV (AEG) to increase selectivity. Furthermore, doubly protonated BMAA isomers can be investigated using compound transitions of *m/z* 230 > 171 CE 27 eV

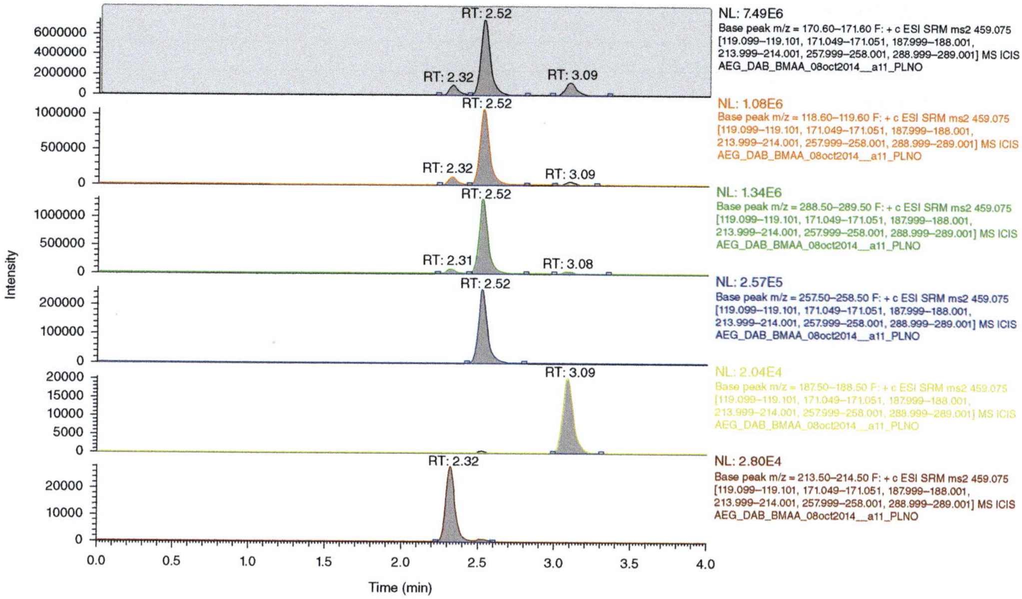

Figure SOP 22.1 *Example of LC-MS/MS analysis of synthetic AEG, BMAA, and DAB (each 10 ng mL^{-1}), isolating the derivatised precursor molecule at m/z 459, and identification of the product ions*

f. Small injection volumes, e.g. 5 µL should be performed on the column
g. For assessment of the reaction in complex samples, additional amino acids can be included such as lysine (single derivatised, *m/z* 317; double derivatised, *m/z* 487)

SOP 22.3 Validation

BMAA is generally observed by triple quadrupole LC-MS/MS, where the product ions can be visualized and the ratios observed, in comparison with synthetic standards. LC separation should be achieved so that the isomers DAB and AEG can be separated (Figure SOP 22.1). Synthetic BMAA standards can also be used to calibrate the machine by creation of standard curves (Figure SOP 22.2).

The EPA Method Detection Level (MDL) can be used, which defines the LOD as the minimum concentration of substance that can be measured and reported with 99% confidence that the analyte concentration is greater than 0. The MDL (48 fmol) was calculated using the standard deviation of replicates multiplied by the t statistic with $\alpha = 0.01$ and $n = 1$ degrees of freedom. The LOQ (0.48 pmol) was calculated by multiplying the MDL by 10.

SOP 22.4 Special Issues

a. As all UHPLC systems and columns are different, testing of the gradient and modification, based on the results, should be carried out. Testing is usually attained through the use of the BMAA isomers, DAB and AEG.
b. With AQC and all derivatisation methods, dilution of the extract may be necessary, as a balance between the amino acids present and the concentration of tag applied is required. Using LC-MS/

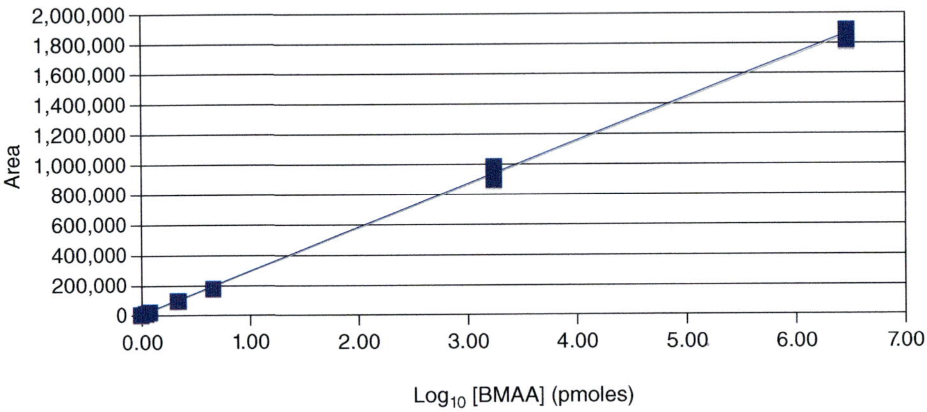

Figure SOP 22.2 *BMAA standard curve as analysed by LC-MS/MS monitoring the area of the transition of* m/z *459 to 171* ($y = 288555x - 2387.45$, $R^2 = 1$; $n = 7$)

MS, the correct AQC balance can be monitored with lysine, which also double derivatizes. Upon analysis, most of the compound should be found in the double derivatized form if sufficient AQC is present relative to other nitrogen compounds. Single dervatized lysine (*m/z* 317) has two peaks at earlier retention times relative to BMAA. Double derivatised lysine (*m/z* 487) has one peak at a later retention time than BMAA.

c. Spiking biological matrices with small amounts of synthetic BMAA can also be helpful in order to identify BMAA in the case of small retention time shifts. Note that the timing of a BMAA spike can affect the recovery of the spike as spikes that are left in the matrix for long periods can react with other compounds that are present and form adducts and complexes [2]. Specific detection can be achieved through monitoring of the product ions, present at the correct ratios.
d. Even though the major percursor ion analysed is *m/z* 459, double charged BMAA (*m/z* 230), can be abundant and the presence of metal adducts and other complexes should be considered [2].
e. Modification of the cone voltage may also be necessary in order to maximize the detection of BMAA. Such testing may prevent BMAA underestimation in cyanobacterial samples, especially from marine environments.

References

[1] Cox, P. A., Banack, S.A., Murch, S.J. *et al.* (2005). Diverse taxa of cyanobacteria produce β-N-methylamino-l-alanine, a neurotoxic amino acid. *Proceedings of National Academy of Science USA* **102** (5), 5074–5078.

[2] Glover, W.B., Liberto, C.M., McNeil, W.S. *et al.* (2012). Reactivity of β-methylamino-L-alanine in complex sample matrixes complicating detection and quantification by mass spectrometry. *Analytical Chemistry* **84**, 7946–7953.

SOP 23

Extraction and LC-MS/MS Analysis of Underivatised BMAA

Elisabeth J. Faassen

Aquatic Ecology and Water Quality Management Group, Wageningen University, The Netherlands

SOP 23.1 Introduction

Liquid chromatography coupled to tandem mass-spectrometry detection (LC-MS/MS) is a suitable, commonly used technique for the quantification of the neurotoxin β-*N*-methylamino-L-alanine (BMAA) in cyanobacteria [1]. BMAA can either be derivatised before analysis, for instance with 6-aminoquinolyl-*N*-hydroxysuccinimidyl carbamate (AQC) [2, 3], propyl chloroformate [4], and dansyl chloride [5], or it can be analysed without pre-column derivatisation [6, 7].

In LC-MS/MS analysis, BMAA should be properly separated from isobaric compounds such as α,γ-diaminobutyric acid (DAB), *N*-(2-aminoethyl) glycine (AEG), and β-amino-*N*-methylalanine (BAMA) [3,8]. Furthermore, attention should be paid to the form in which BMAA can be present in natural samples: either free, bound in a fraction that precipitates with proteins, or bound in a (0.1 mol L^{-1} trichloroacetic acid [TCA]) soluble fraction.

The extraction methods used in this SOP (0.1 mol L^{-1} TCA extraction and liquid 6 mol L^{-1} HCl hydrolysis) are commonly used in BMAA research. However, it is possible that method details like incubation times and conditions will still be optimized. Furthermore, the precursors of the bound BMAA fractions are still unknown and therefore, more effective extraction methods for bound BMAA could be developed [9]. It is recommended to check recent literature before setting up an analytical method for BMAA.

Handbook of Cyanobacterial Monitoring and Cyanotoxin Analysis, First Edition.
Edited by Jussi Meriluoto, Lisa Spoof and Geoffrey A. Codd.

SOP 23.2 Experimental

SOP 23.2.1 Materials

- Ultrapure water, e.g. with a Qpod (Millipore)
- Acetonitrile at least of HPLC grade
- HCl, TCA, and formic acid (FA) of analytical grade
- Teflon forceps
- Glass Pasteur pipettes
- 2-mL plastic tubes (e.g. from Eppendorf)
- 2-mL spin filter tubes with 0.2-μm cellulose acetate filter
- Small glass tubes for hydrolysis (6×50 mm, when hydrolysis workstation is used)
- HPLC/UHPLC glass vials
- Calibration standards of BMAA, DAB, AEG, and deuterium-labelled BMAA as an internal standard (D_3BMAA), all standards are commercially available

SOP 23.2.2 Special Equipment

- Freeze-drier
- SpeedVac
- Microcentrifuge
- Hydrolysis workstation (e.g. Eldex); alternatively, samples can be hydrolysed in a stove, in glass vials with screw caps (in which case samples are analysed under atmospheric conditions, the steps describing nitrogen flushing and application of vacuum can be skipped)
- HPLC or UPLC coupled to a triple quadrupole mass spectrometer
- Integration software

SOP 23.2.3 Solutions

- 1 mol L^{-1} TCA
- 6 mol L^{-1} HCl
- 20 mmol L^{-1} HCl
- 67% acetonitrile:33% water:0.1% FA solution (v/v/v)
- D_3BMAA spike solution in 20 mmol L^{-1} HCl. The concentration of the spike solution and the volume of spiking solution added during sample processing depend on the method LOD. In this SOP, a concentration of 1 μg mL^{-1} is used. When the spiking volumes described in section SOP 23.2.5 are used, this results in a concentration of 40 ng mL^{-1} in the solution to be analysed (if 100% recovery would be achieved).

SOP 23.2.4 LC Mobile Phase

- Mobile phase A: acetonitrile with 0.1% FA (v/v)
- Mobile phase B: water with 0.1% FA (v/v)

SOP 23.2.5 General Procedure

In sections SOP 23.2.5.1 and SOP 23.2.5.2, the extraction of free BMAA, total 0.1 mol L^{-1} TCA-dissolved BMAA (including the soluble bound fraction), and total BMAA in cyanobacterial samples is described step by step. The amount of bound BMAA present in the precipitated fraction can be determined by subtracting the amounts of free and dissolved bound BMAA

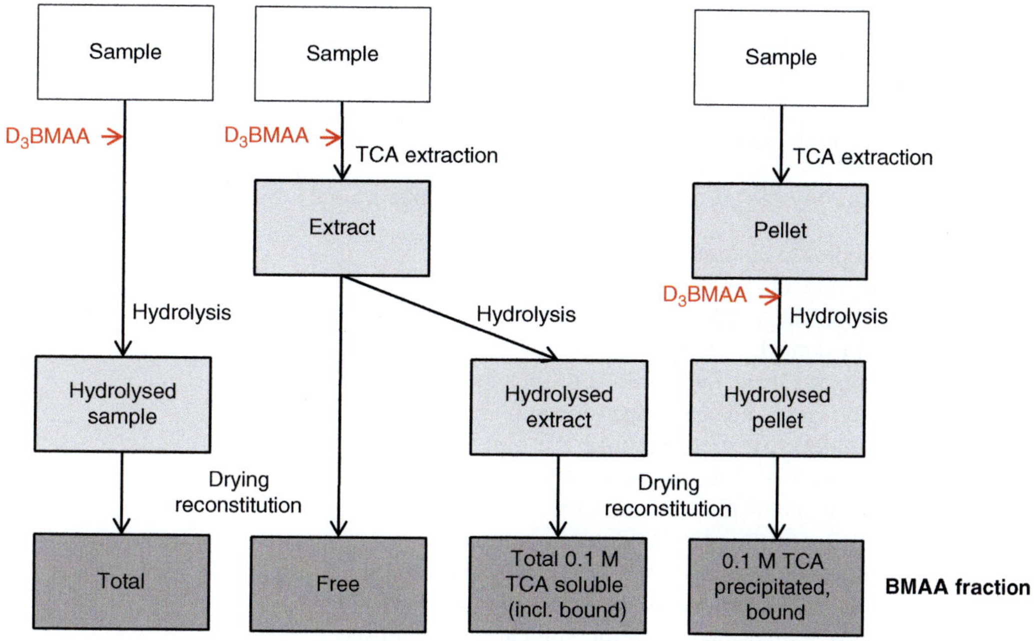

Figure SOP 23.1 Sample preparation scheme for the extraction of different BMAA fractions.

from the total amount in the sample. But preferably, this fraction is directly determined, as described in section SOP 23.2.5.3 (see Figure SOP 23.1).

For all extraction methods, it is recommended to include blank samples (no matrix) and positive controls (e.g. cycad seeds, *Cycas micronesica* [Hill]). Hydrolysis of sample material collected on GF/C filters with the methods described below results in very low recovery, cyanobacterial samples are therefore better concentrated in other ways, e.g. by centrifugation. Furthermore, all extractions should be performed under acidic conditions to ensure BMAA stability.

SOP 23.2.5.1 Total BMAA

a. Freeze dry samples to disrupt the cell walls
b. Weigh 1 mg into a small glass tube
c. Add D_3BMAA spike (40 µL)
d. Dry the sample under vacuum in the hydrolysis workstation
e. Add 30 µL 6 mol L^{-1} HCl to each glass tube. After HCl addition, use Teflon forceps to handle the tubes
f. Flush sample three times with nitrogen in the hydrolysis workstation, apply vacuum between flushing steps
g. Apply a vacuum of approximately 0.7 mbar, without drying down the acid
h. Hydrolyse the sample for 20 h at 105°C
i. After hydrolysis, dry the samples under vacuum in the hydrolysis workstation
j. Add 500 µL 67% acetonitrile:33% water:0.1% FA to each sample
k. Vortex until sample has detached from the glass (use pipette tip if needed) and transfer suspension to spin filter tube with glass Pasteur pipette
l. Centrifuge sample for 5 min at maximum speed in a micro centrifuge

m. Transfer filtrate to a glass vial
n. Add again 500 µL 67% acetonitrile:33% water:0.1% FA to each sample tube
o. Vortex and transfer suspension to same spin filter tube
p. Centrifuge again for 5 min at maximum speed
q. Transfer filtrate to the same vial, mix and close the vial

SOP 23.2.5.2 Free BMAA and Total 0.1 mol L^{-1} TCA-soluble BMAA

a. Freeze dry samples to disrupt the cell walls
b. Weigh 12.5 mg into a plastic 2-mL tube
c. Add 50 µL of D_3BMAA spike solution
d. Add 600 µL of 0.1 mol L^{-1} TCA
e. Vortex for 1 min and leave for 10 min at room temperature
f. Vortex again for 1 min and centrifuge for 5 min at maximum speed
g. Transfer supernatant to a spin filter tube
h. Centrifuge the spin filter tube for 5 min at maximum speed
i. Transfer filtrate to a new plastic 2-mL tube
j. Add again 600 µL of 0.1 mol L^{-1} TCA to the sample
k. Vortex for 1 min and centrifuge for 5 min at maximum speed
l. Transfer the second supernatant to the same spin filter tube
m. Centrifuge the spin filter tube again for 5 min at maximum speed
n. Transfer the second filtrate to the same tube as the first filtrate (total volume 1250 µL)
o. Vortex and split the extract in two fractions:

Free BMAA:

p. Transfer 500 µL of the extract to a new plastic 2-mL tube
q. Dry these samples down in SpeedVac
r. Add 500 µL of a 67% acetonitrile:33% water:0.1% FA solution to each tube
s. Vortex for 2 min and transfer samples to HPLC/UHPLC vials for analysis

Total soluble BMAA:

t. Transfer 500 µL of the extract to a small glass tube for hydrolysis
u. Put the samples in the freezer until they are completely frozen and subsequently freeze dry them (overnight)
v. Perform hydrolysis and reconstitute as described for total BMAA (steps SOP 23.2.5.1.e to SOP 23.2.5.1.q) but use 2×250 µL of 67% acetonitrile/33% water/0.1% FA (final volume in HPLC/UHPLC vial is 500 µL)

SOP 23.2.5.3 Bound BMAA in the Precipitate

Bound BMAA in the pellet created during 0.1 mol L^{-1} TCA extraction can best be determined by extracting another 1 mg of lyophilized sample as described in the following steps.

a. Freeze dry samples to disrupt the cell walls
b. Weigh 1 mg into a plastic 2-mL tube
c. Add 150 µL of 0.1 mol L^{-1} TCA
d. Vortex for 1 min and leave for 10 min at room temperature
e. Vortex again for 1 min and centrifuge for 5 min at maximum speed
f. Discard the supernatant

g. Add again 150 µL of 0.1 mol L^{-1} TCA to the sample
h. Vortex for 1 min and centrifuge for 5 min at maximum speed
i. Discard the supernatant
j. Transfer the pellet to small glass tube for hydrolysis, use 20 m mol L^{-1} HCl to make sure all material is transferred
k. Add 40 µL D_3BMAA spike
l. Dry the sample under vacuum in the hydrolysis workstation
m. Perform hydrolysis and reconstitute as described for total BMAA (steps SOP 23.2.5.1.e to SOP 23.2.5.1.q)

Alternatively, the pellet created during the extraction for free and total 0.1 mol L^{-1} TCA soluble BMAA (SOP 23.2.5.2) can be transferred to small glass tubes, dried and hydrolysed as described in steps SOP 23.2.5.1.d to SOP 23.2.5.1.q. In this case, a portion of the pellet equivalent to 1 mg of dry weight should be used and D_3BMAA should be added to the pellet after it was transferred to the tube for hydrolysis.

SOP 23.2.6 Chromatography

Underivatised BMAA can effectively be separated from DAB and AEG by hydrophilic liquid interaction chromatography (HILIC). For HPLC equipment, chromatography can be performed on a 2.1 × 150 mm, 5-µm ZIC-HILIC column (SeQuant, Sweden) at 40°C. No guard column was used due to the centrifuge filtering of the samples (e.g. SOP 23.2.5.1.o–q). Use an injection volume of 5 µL and a flow rate of 0.4 mL min^{-1}. Mobile phases are acetonitrile with 0.1% FA (eluent A) and water with 0.1% FA (eluent B). Use the following elution program [10]:

- 0–2 min: 95% A
- 4 min: 65% A
- 8–17 min 55% A
- 17–23 min 95% A (equilibration)

with linear decreases between the time steps and directing the flow to waste during the first 4 and last 6 min. Chromatograms of a calibration standard and a cycad seed sarcotesta extracted for free BMAA are shown in Figure SOP 23.2. Separation from isobaric compounds other than DAB and AEG is not described in this SOP, but see [8] for more information. For LC systems operated at higher pressure, similar columns, but with smaller particle diameter can be used, and the elution program should be adjusted accordingly.

Tips for HILIC chromatography:

a. The solution used for dissolving the samples and calibration standards should have a low water content to ensure good retention.
b. Elute under acidic conditions to ensure BMAA is positively charged and will be well retained on the column.
c. Using a relatively high percentage of water at the end of each run and a sufficient equilibration time before the next injection ensures retention time stability.

SOP 23.2.7 Detection by MS/MS

Settings of the tandem mass spectrometer should be optimized for the specific equipment used; commercially available standards can be used for this. For detection and quantification of BMAA, electrospray ionisation (ESI) in positive mode is usually employed. The transitions that should be monitored in multiple reaction mode are displayed in Table SOP 23.1. Preferably, two qualifiers

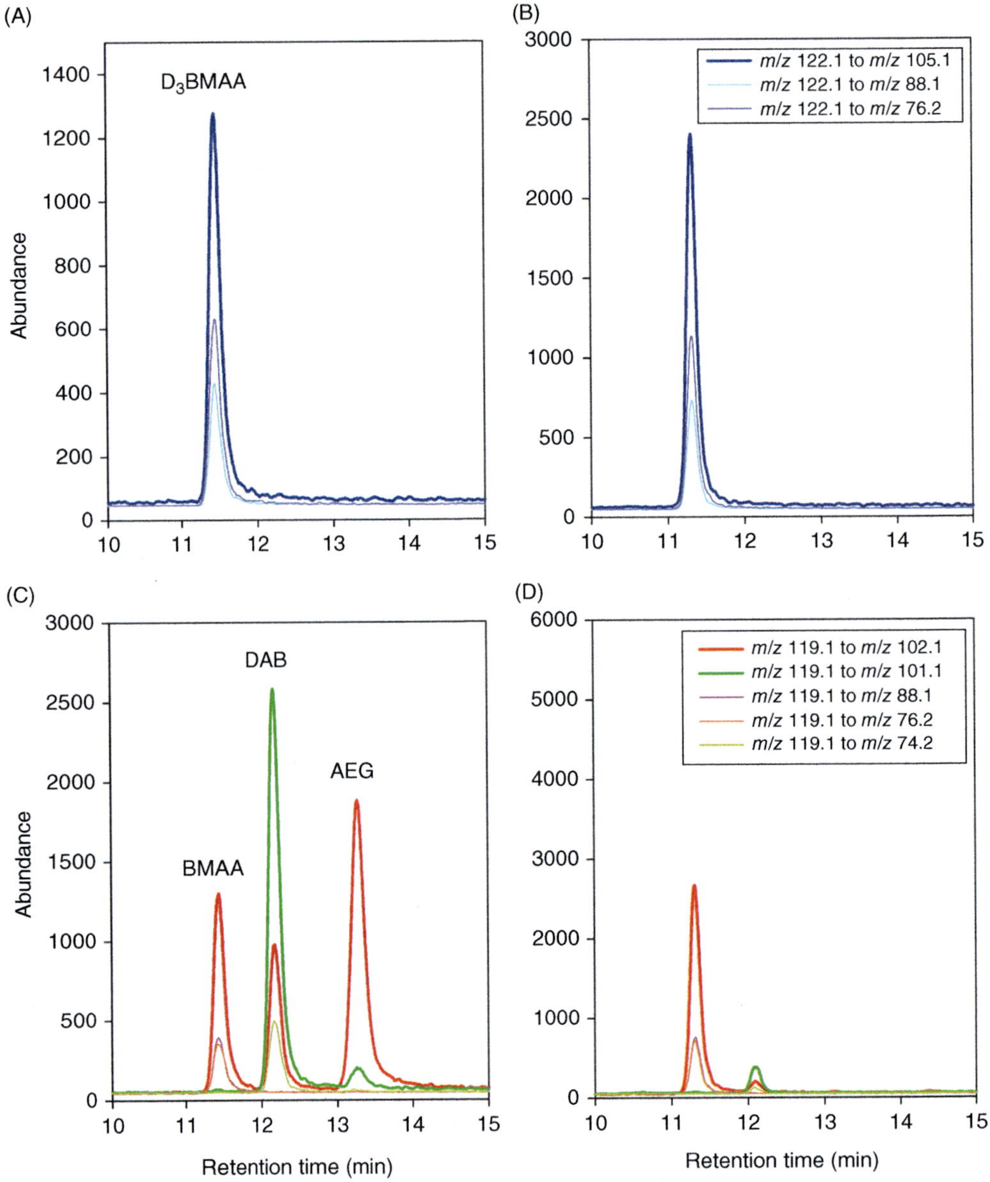

Figure SOP 23.2 *Chromatograms of a 50 μg L⁻¹ BMAA, D_3BMAA, DAB, and AEG standard (panels A and C) and a cycad seed sarcotesta* (Cycas micronesica), *extracted for free BMAA as positive control (8.1 mg used, concentration determined at 11.4 μg g⁻¹) and spiked with D_3BMAA at 80 μg L⁻¹ (panels B and D). Quantifiers are indicated by bold lines, qualifiers are indicated by normal lines. Chromatography was performed as described in SOP 23.2.6 on an Agilent 1260 LC. Compounds were detected on an Agilent 6460 triple quadrupole mass spectrometer with the following settings: drying gas temperature 230°C, sheath gas temperature 250°C, drying gas flow and sheath gas flow 12 L min⁻¹, nebulizer pressure 40 psi, capillary voltage 2500 V, nozzle voltage 500 V. Both quadrupoles were operated in unit mode. Fragmentor voltage was 50 V, collision energy was 4 V for the transitions* m/z *122.1 to 105.1 and* m/z *119.1 to 102.1, 101.1, and 74.2 and was 8 V for all other transitions.*

Table SOP 23.1 *MS/MS transitions for the analysis of underivatised D_3BMAA, BMAA, DAB, and AEG*

Compound	Precursor *m/z*[a]	Quantifier *m/z*	Qualifier m/z	Qualifier *m/z*
D_3BMAA	122.1	105.1	88.1	76.2
BMAA	119.1	102.1	88.1	76.2
DAB	119.1	101.1	74.2	–
AEG	119.1	102.1	–	–

[a] Mass-to-charge ratio.

are used for BMAA identification. For reliable identification, it is recommended to set the resolution of both mass filters as narrow as possible. For BMAA identification, the following criteria should be met for each sample:

a. BMAA retention time should be similar to the D_3BMAA retention time in the same sample
b. The signal-to-noise ratio of all product ions should be at least 3
c. The ratio between *m/z* 102.1 and *m/z* 88.1 should not deviate more than 20% from the ratio between the same ions in the BMAA calibration standards
d. The ratio between *m/z* 102.1 and *m/z* 76.2 should not deviate more than 20% from the ratio between the same ions in the BMAA calibration standards

SOP 23.3 Quantification

For BMAA quantification, a calibration curve containing BMAA and D_3BMAA should be constructed. The solvent should be the same as the one used for the samples (e.g. a 67% acetonitrile:33% water:0.1% FA solution). Also add DAB and AEG standards to the calibration curve, either for semiquantification of these compounds or to make sure that no coelution with BMAA has taken place. As an example, a dilution series of 0, 0.5, 1, 5, 10, 50, and 100 $\mu g\,L^{-1}$ could be used, in which the reported concentration of the standard corresponds to the concentration of each individual component of the mixture. For example, standard 1 $\mu g\,L^{-1}$ contains BMAA, D_3BMAA, DAB, and AEG, 1 $\mu g\,L^{-1}$ of each.

The LC-MS/MS worklist should start and end with calibration standards, as well as after a fixed number of injected samples (e.g. 10), a set of calibration standards should be analysed. In this way, it is possible to correct for changes in signal intensity that might occur during the run. Include vials with only solvent to check for signal carry over.

Quantify BMAA by determining the BMAA and D_3BMAA concentration of each sample against the external calibration curves (signal-to-noise ratio should be at least 10 for the quantifier ions). Next, calculate D_3BMAA recovery for each sample, and correct the determined BMAA concentration for this value. Finally, correct for the exact amount of sample used, the final volume in which the extract was dissolved before analysis, and in case of free and total 0.1 $mol\,L^{-1}$ TCA-soluble BMAA, the fraction of the extract used; 500/1250 μL (steps SOP 23.2.5.2.p and SOP 23.2.5.2.t) (Equation SOP 23.1).

$$[BMAA]_{sample} = [BMAA]_{analysed} \times \frac{[D_3BMAA]_{expected}}{[D_3BMAA]_{analysed}} \times \frac{v_{in\ vial}}{w_{sample}} \times \frac{v_{total\ extract}}{v_{used\ extract}} \qquad \text{(Eq. SOP 23.1)}$$

In which $[BMAA]_{sample}$ is expressed as $\mu g\,g^{-1}$, $[BMAA]_{analysed}$, $[D_3BMAA]_{expected}$ and $[D_3BMAA]_{analysed}$ as ng mL^{-1}, $V_{in\ vial}$ as mL, W_{sample} as mg, and $V_{total\ extract}$ and $V_{used\ extract}$ as μL.

References

[1] Faassen, E.J. (2014) Presence of the neurotoxin BMAA in aquatic ecosystems: what do we really know? *Toxins* **6** (3), 1109–38.

[2] Banack, S.A., Johnson, H.E., Cheng, R. and Cox, P.A. (2007) Production of the neurotoxin BMAA by a marine cyanobacterium. *Marine Drugs* **5** (4), 180–96.

[3] Jiang, L., Aigret, B., De Borggraeve, W.M., *et al.* (2012) Selective LC-MS/MS method for the identification of BMAA from its isomers in biological samples. *Analytical and Bioanalytical Chemistry* **403** (6),1719–30.

[4] Downing, S., Banack, S.A., Metcalf, J.S. *et al.* (2011) Nitrogen starvation of cyanobacteria results in the production of β-*N*-methylamino-L-alanine. *Toxicon* **58** (2), 187–94.

[5] Lampinen Salomonsson, M., Hansson, A. and Bondesson, U. (2013) Development and in-house validation of a method for quantification of BMAA in mussels using dansyl chloride derivatization and ultra performance liquid chromatography tandem mass spectrometry. *Analytical Methods* **5** (18), 4865–74.

[6] Rosén, J. and Hellenäs, K.E. (2008) Determination of the neurotoxin BMAA (β-*N*-methylamino-L-alanine) in cycad seed and cyanobacteria by LC-MS/MS (liquid chromatography tandem mass spectrometry). *Analyst*, **133** (12), 1785–9.

[7] Faassen, E.J., Gillissen, F., Zweers, H.A.J. and Lürling, M. (2009) Determination of the neurotoxins BMAA (β-*N*-methylamino-L-alanine) and DAB (α-,γ-diaminobutyric acid) by LC-MSMS in Dutch urban waters with cyanobacterial blooms. *Amyotrophic Lateral Sclerosis* **10** (SUPPL. 2), 79–84.

[8] Beach, D.G., Kerrin, E.S. and Quilliam, M.A. (2015). Selective quantitiation of the neurotoxin BMAA by use of hydrophilic-interaction liquid chromatography-differential mobility spectrometry-tandem mass spectrometry (HILIC-DMS-MS/MS). *Analytical and Bioanalytical Chemistry* **407**, 8379–8409.

[9] Rosén, J, Westerberg, E., Schmiedt, S. and Hellenäs, K.E. (2016). BMAA deteted as neither free nor protein bound amino acid in blue mussels. *Toxicon* **109**, 45–50.

[10] Faassen, E.J., Gillissen, F. and Lürling, M. (2012) A comparative study on three analytical methods for the determination of the neurotoxin BMAA in cyanobacteria. *PLoS ONE* **7** (5).

SOP 24

Extraction, Purification, and Testing of LPS from Cyanobacterial Samples

Lucie Bláhová and Luděk Bláha

Faculty of Science, RECETOX, Masaryk University, Brno, Czech Republic

SOP 24.1 Introduction

Lipopolysaccharides (LPS) are major components of cyanobacterial cell walls. LPS are known to induce pyrogenicity (fewer) and diverse biological responses in vivo [1]. They have been shown to potentiate toxicity of heavy metals [2] and modulate biotransformation and toxicity of microcystins [3].

The most often discussed method of LPS extraction and purification from cyanobacteria is the hot phenol–water (HPW) extraction [4–6]. Other organic solvents have also been used such as less hazardous 1-butanol (HBW extraction) or *N'*-dimethylformamide (DMF extraction) [5]. Organic solvent in combination with appropriate temperature and pH separates hydrophobic compounds from cyanobacterial extract (organic layer) while LPS is concentrated in the water phase. Hydrophilic interferences in the water (such as RNA, glycans, or residual proteins) can be removed by enzymatic activities of RNAase A or β-glucosidase (cellulase) [7].

Various LPS share similar mode of biological action (toxicity) but have diverse chemical structures among different cyanobacterial strains. Therefore their analysis by traditional approaches such as chromatography is not common. Isolated LPS can be characterized and analyzed by various other approaches such as in vivo toxicity test, standardized pyrogenicity test with Limulus amoebocyte lysate (LAL), advanced pyrogenicity test using the PyroGene rFC endotoxin system (PyroGene), or testing the potency to activate human leukocytes [8].

Handbook of Cyanobacterial Monitoring and Cyanotoxin Analysis, First Edition.
Edited by Jussi Meriluoto, Lisa Spoof and Geoffrey A. Codd.

SOP 24.2 Experimental

SOP 24.2.1 Materials

- Phenol
- Toluene
- Distilled water
- NaOH
- Cellulose membranes for dialysis (33 × 21 mm, available, e.g., from Sigma-Aldrich)
- Tris(2-hydroxyethyl)amine hydrochloride (Tris-HCl, CAS 637-39-8, pH 7.4)
- Ribonuclease A (R4642 ribonuclease A, CAS 9001-99-4, available, e.g., from Sigma-Aldrich)
- LAL reagent water (LRW, endotoxin free, supplied, e.g., by Cape Cod Inc.)
- Pipettes
- Apyrogenic tips and sterile glass (supplied, e.g., by Cape Cod Inc.)
- 96 microplates for testing of LPS activity (LAL test—supplied, e.g., by Associates of Cape Cod, Inc., CA961)
- Glass beakers—200 and 1000 mL volumes
- Polypropylene 50 mL centrifuge tubes with cap
- Polypropylene 2 mL centrifuge microtubes

SOP 24.2.2 Special Equipment

- Freeze dryer
- Vortex
- Magnetic stirrer with heating
- Centrifuge with cooling function for larger tubes (50 mL)
- Fume hood
- Ice
- Shaker
- Incubator (37°C)
- Thermometer (to check 68°C temperature)
- Ultrasonic bath
- Analytical scales

SOP 24.2.3 Solutions

- 0.1 M NaOH
- 0.1 M Tris-HCl (pH 7.4)
- 90% Phenol in water (pH 7.8 adjusted by 0.1 M NaOH; v/v)

Please observe the following:

- Phenol is toxic and hazardous material and it is important to work with it in fume hood; disposal of phenol should conform to appropriate regulations.
- Cyanobacteria are hazardous material—use personal protection devices (gloves, goggles).

SOP 24.2.4 Extraction, Purification, and Determination of LPS

SOP 24.2.4.1 Extraction Procedure

a. Collect biomass of natural water blooms or laboratory cultures (using glass fiber filter or plankton net, 25 μm) and store frozen at −18°C. Freeze-dry (lyophilize) biomass prior to extraction.

b. Resuspend lyophilized biomass—approximately 200 mg—in 50 mL of distilled water in a glass beaker using the ice-cooled sonication bath.
c. Heat the suspension to 68°C and mix with 50 mL of 90% phenol (preheated to approximately 68°C). Mixing of two preheated materials is recommended for practical reasons (immediate start of the extraction at 68°C). Stir the mixture at 68°C for 20 min in a fume hood.
d. Cool down the suspension to 4°C (using the ice bath) and centrifuge for 30 min at 4000× g at 4°C (polypropylene 50 mL centrifuge tubes).
e. Collect the supernatant (aqueous phase with LPS) which has separated from the phenol during centrifugation.
f. Reextract the phenol layer once more with distilled water (50 mL) by repeating steps (c) to (e). Supernatants from both extraction steps are pooled.
g. Purify the pooled supernatants (aqueous phase) from residual phenol by dialysis. Prior to dialysis make the cellulose membrane wet with distilled water, and close one of the ends of the dialysis tube (e.g., by tying a knot). Apply the LPS extract to the dialysis bag using pipette, close the membrane bag (e.g., by tying a knot also on the other side of the membrane tube), and immerse it into distilled water (1000 mL in large glass beaker) at the magnetic stirrer. Stir for 48 h at laboratory temperature. Add 10 µL of toluene to 1000 mL of water to avoid bacterial contamination. Change water after 24 h.
h. Collect the semipurified LPS from the cellulose membrane and centrifuge (polypropylene 50 mL centrifuge tubes) at 4000× g (30 min, 4°C).
i. Freeze and lyophilize the supernatant and keep the dry residue at −18°C until further processing.

SOP 24.2.4.2 Purification Procedure

a. Resuspend the semipurified lyophilized residue of LPS (SOP 24.2.4.1.i) in 750 µL of 0.1 M Tris-HCl buffer containing 25 µg mL^{-1} ribonuclease A (use, e.g., 2 mL polypropylene tube). The solution is incubated with slow shaking for 16 h at 37°C.
b. Stop the reaction of enzyme ribonuclease A by adding 750 µL of 90% phenol in water (pH 7.8) in to the Tris-HCl reaction solution. The mixture is vortexed and incubated for 4 min at room temperature. The solution is then centrifuged (18 407× g, 15 min, room temperature).
c. Purify the upper aqueous phase with LPS from residual phenol by dialysis and lyophilize—see steps SOP 24.2.4.1.g to SOP 24.2.4.1.i.
d. The freeze-dried powder constituting of purified LPS is weighed by analytical scales for the assessment of the LPS content (amount) in the biomass (µg LPS/mg dry biomass) and kept at −18°C.
e. Prior to testing of LPS activity, the lyophilized LPS is resuspended in LRW water.

SOP 24.2.4.3 Pyrogenicity Tests

The natural variability and complexity of LPS brings major challenges to its detection and quantification by conventional chemical methods. Properties of the isolated and purified LPS can be tested by different approaches [8], which all have their advantages and limitations.

i. Standard and broadly used test of the endotoxin activity with the Limulus amoebocyte lysate test (LAL). The method is based on the precipitation of the hemolymph of *Limulus polyphemus* after the exposure to endotoxin. Precipitation, which is proportional to the concentration and pyrogenicity of the tested sample, is determined by color reaction or by measurement of turbidity using microwell plate reader. The results are expressed in endotoxin units (EU mL^{-1}). The LAL test is sensitive and fast but often nonspecific because

Limulus hemolymph precipitation can be induced also by other compounds than LPS. LAL tests are commercially available, and users are advised to use protocols provided by the manufacturer.

ii. An advanced alternative to LAL test is using isolated recombinant Factor C (rFC), which tests the first step of the precipitating cascade of the LAL test. Activation of rFC by the binding of endotoxin is determined by colorimetric detection using conventional microwell plate reader. The rFC assay is more specific than LAL test, and it is commercially available with detailed protocols provided by the manufacturer—for example, PyroGene test.

iii. Pyrogenicity can also be tested in rabbits or other laboratory animals in vivo, but this approach is becoming less popular with respect to ethical issues and implementation of 3R principles—that is, replacement, reduction, and refinement of the animal testing.

iv. Alternatively to in vivo assays, in vitro tests have been introduced employing various permanent cell lines, isolated primary cells, or whole blood. In these systems, the activation of cells by the tested pyrogenic samples is evaluated as the release of various cytokines, production of nitric oxide, or reactive oxygen species. In comparison to the LAL and rFC tests, in vitro cellular tests better reflect the reaction of the biological system to a pyrogenic agent. However, this might also be a limitation, because the response of cells in vitro can be induced also by a number of other compounds than LPS (such as cell wall components muramyl dipeptide and peptidoglycan, DNA, and others).

SOP 24.3 Validation

Reproducibility is checked by measuring the amount and LPS activity in the pooled extracts after repeated extractions of the same cyanobacterial biomass.

a. Extract the lyophilized biomass (200 mg) according to the extraction (section SOP 24.2.4.1) and purification (section SOP 24.2.4.2) steps in three independent replicates.
b. Analyze the replicates for amount of isolated LPS by using precise laboratory analytical scales (LPS mass isolated from 200 mg of the original biomass).
c. Analyze the activity (pyrogenicity) of the isolate by a standardized LAL assay or other advanced techniques such as PyroGene rFC endotoxin system (PyroGene).
d. Results of the example validation are shown in Table SOP 24.1.

Table SOP 24.1 *Results of validation of reproducibility of the extraction and purification method for isolation of LPS from cyanobacterial laboratory culture Microcystis aeruginosa PCC 7806*

Sample	LPS amount mg g^{-1} biomass d.w. (mean ± S.D., N = 3)	LPS %RSD	LAL endotoxin activity × 10^4 EU mg^{-1} LPS (mean ± S.D., N = 3)	LAL %RSD	PyroGene endotoxin activity × 10^4 EU mg^{-1} LPS (mean ± S.D., N = 3)	PyroGene %RSD
M. aeruginosa PCC 7806	3.75 ± 1.06	28%	4.61 ± 0.59	13%	0.22 ± 0.12	55%

EU, endotoxin units.

References

[1] Weckesser, J., Drews, G. and Mayer, H. (1979) Lipopolysaccharides of photosynthetic prokaryotes. *Annual Review of Microbiology* **33**, 215–239.

[2] Notch, E.G., Miniutti, D.M., Berry, J.P. *et al.* (2011) Cyanobacterial LPS potentiates cadmium toxicity in zebrafish (*Danio rerio*) embryos. *Environmental Toxicology* **26** (5), 498–505.

[3] Best, J.H., Pflugmacher, S., Wiegand, C. *et al.* (2002) Effects of enteric bacterial and cyanobacterial lipopolysaccharides, and of microcystin-LR, on glutathione S-transferase activities in zebrafish (*Danio rerio*). *Aquatic Toxicology* **60**, 223–231.

[4] Westphal, O. and Jann, K. (1965) Bacterial lipopolysaccharides: extraction with phenol-water and further applications of the procedure. *Methods in Carbohydrate Chemistry* **5**, 83–91.

[5] Papageorgiou, J., Linke, T.A., Kapralos, C. *et al.* (2004) Extraction of cyanobacterial endotoxin. *Environmental Toxicology* **19**, 82–87.

[6] Bernardová, K., Babica, P., Maršálek, B. *et al.* (2008) Isolation and endotoxin activities of lipopolysaccharides from cyanobacterial cultures and complex water blooms and comparison with effects of heterotrophic bacteria and green alga. *Journal of Applied Toxicology* **28**, 72–77.

[7] Wu, L., Tsai, C.M. and Frasch, C.E. (1987) A method for purification of bacterial R-type lipopolysaccharides (lipooligosaccharides). *Analytical Biochemistry* **160**, 281–289.

[8] Bláhová, L., Adamovský, O., Kubala, L. *et al.* (2013) The isolation and characterization of lipopolysaccharides from *Microcystis aeruginosa*, a prominent toxic water bloom forming cyanobacteria. *Toxicon* **76**, 187–196.

SOP 25

Extraction and Chemical Analysis of Planktopeptin and Anabaenopeptins

Hanna Mazur-Marzec[1], Tina Elersek[2], and Agata Błaszczyk[1]

[1] Institute of Oceanography, University of Gdansk, Gdynia, Poland
[2] Department of Genetic Toxicology and Cancer Biology, National Institute of Biology, Ljubljana, Slovenia

SOP 25.1 Solid-Phase Extraction of Anabaenopeptins and Planktopeptin

SOP 25.1.1 Introduction

Extraction and analysis of peptides from cyanobacterial material is a multistage process (Fig. SOP 25.1) and requires high level of experience and application of advanced chromatographic methods [1–3]. In the first steps, the content of the matrix is reduced by extraction of the peptides to the proper solvent enhanced by sonication and followed by solid-phase extraction (SPE). SPE is also recommended before any type of chromatographic separation, especially when mass detector is used and online SPE is not available. The preliminary purification of cyanobacterial extract can significantly enhance compound identification and improve the accuracy of quantitative analysis. This SOP describes the method of SPE of anabaenopeptins and planktopeptins from a complex cyanobacterial sample.

SOP 25.1.2 Experimental

SOP 25.1.2.1 Materials

- Ultrapure water, 18.2 MΩ cm
- Methanol HPLC grade
- Acetic acid, analytical grade

Handbook of Cyanobacterial Monitoring and Cyanotoxin Analysis, First Edition.
Edited by Jussi Meriluoto, Lisa Spoof and Geoffrey A. Codd.

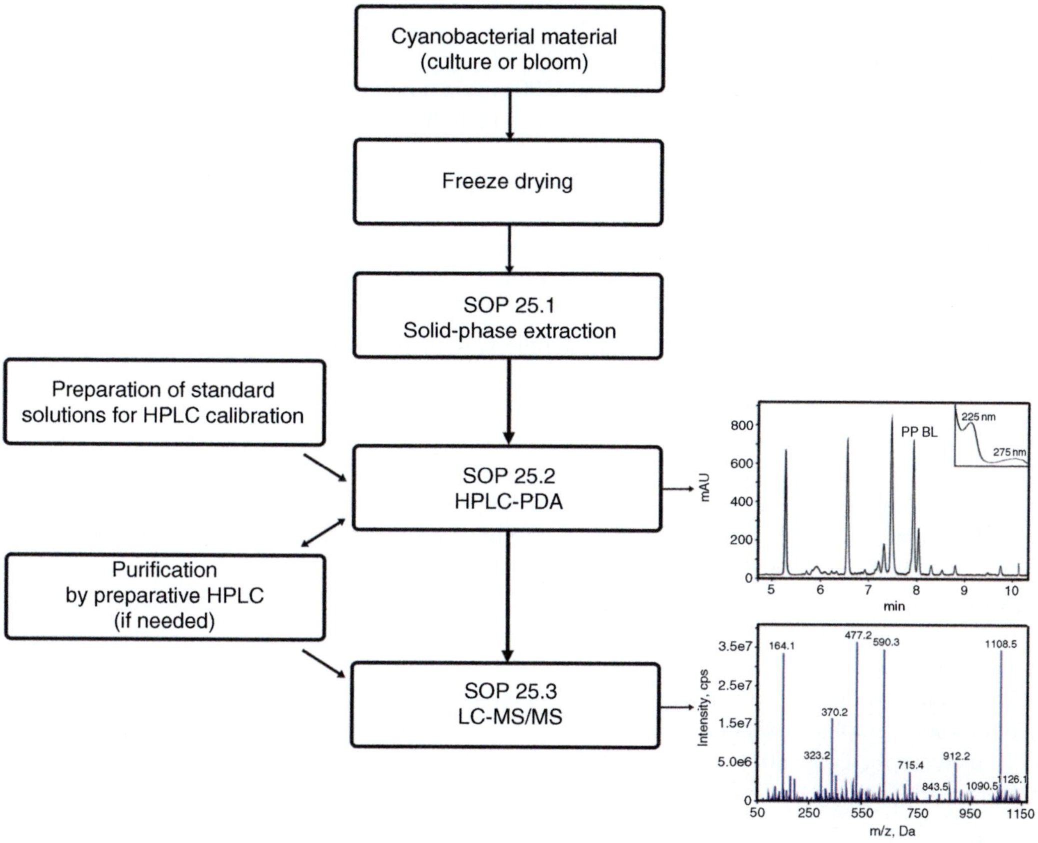

Figure SOP 25.1 Workflow diagram of extraction and analysis of planktopeptin and anabaenopeptins

- Centrifuge tubes (polypropylene) approx. 10 mL
- Borosilicate glass tube or beakers, approx. 15–25 mL
- Adjustable pipette capable of accurately dispensing of 500–5000 μL
- Solid-phase extraction cartridges with polymeric materials equivalent to Oasis HLB (Waters, Milford, MA, USA), with 500 mg sorbent mass

SOP 25.1.2.2 Special Equipment

- Centrifuge (alternative to filtration)
- Ultrasonic bath
- Solid-phase extraction vacuum manifold equipped with stopcocks and vacuum source
- Rotary evaporator or vacuum concentrator
- Vortex mixer

SOP 25.1.2.3 Solutions

- 20% aqueous methanol solution
- 5% aqueous acetic acid solution

SOP 25.1.2.4 General Procedure

Sample Preparation

a. Place freeze-dried cyanobacterial material (200 mg) in a polypropylene tube.
b. Add 3 mL of 5% aqueous acetic acid, close the tube tightly with the screw cap, and vortex it for 10 min.
c. Extract the sample for further 15 min in ultrasonic bath. Keep the bath temperature low with ice.
d. Centrifuge the extract at 10 000× g for 10 min.
e. Transfer the supernatant to a borosilicate glass tube or beaker.
f. Reconstitute the remaining sediment in 3 mL of 5% aqueous acetic acid and repeat the extraction procedure (b–e).
g. Combine the supernatants and adjust the sample volume to 10 mL with water.

Solid-Phase Extraction

a. Condition the solid-phase extraction cartridge (500 mg sorbent mass) with 5 mL of methanol followed by 5 mL of water.
b. Apply the sample to the cartridge. Wash the vessel (where the sample was kept) three times with 1 mL of water and apply to the cartridge. During the whole extraction procedure, use a flow rate at which the drops are visible. Regulate the flow with the stopcocks.
c. Wash the cartridge with 10 mL of water and subsequently with 10 mL of 20% aqueous methanol.
d. Elute the peptides with 10 mL of 100% methanol.
e. Before chromatographic analysis, dilute a portion of the sample with water or aqueous methanol. The content of organic solvent in the sample should be similar as in the mobile phase at the start of chromatographic analysis.
f. To increase the concentration of peptides in the extract or to preserve the sample before longer storage, the solvent can be removed in rotary evaporator or vacuum concentrator.

SOP 25.1.3 Validation

The method of solid-phase extraction was validated using six replicates of cyanobacterial extracts containing anabaenopeptin AP F, planktopeptin PP BL1125, and microcystins MC-LR and MC-RR. For quantitative analysis, LC-MS/MS system operating in MRM mode was used. When standards of the peptides are not available, their recoveries can be determined on the basis of the peak areas measured before and after SPE.

The calculated mean recoveries for all peptides and samples (R) were higher than 90%; the repeatabilities (RSD_r) and the reproducibilities (RSD_R) were lower than 10 and 20%, respectively (Table SOP 25.1).

SOP 25.1.4 Special Issues

Instead of 5% aqueous acetic acid, 75% aqueous methanol can be used for the extraction of peptides from cyanobacterial samples. In this case, the extract should be diluted in water so that methanol content in the sample subjected to SPE does not exceed 20%.

The recovery and reproducibility of the procedure can be lowered when the cartridge is overloaded. Therefore, it is recommended to adjust the sorbent mass to the concentration and volume of the sample. Each time the method is modified, after passing the sample through the cartridge, check the eluate for breakthrough. Also the volume of the eluting solvent should be optimized.

Table SOP 25.1 Recoveries of peptides from bloom sample of cyanobacteria extracted with Oasis HLB cartridge

Peptide	PP BL1125	AP F	MC-RR	MC-LR
MRM transition	1126→477	851→201	519→135	995→213
Average peak area × 10^5	0.49	385.0	0.12	0.03
S.d. (n = 6)	0.02	9.32	0.01	0.001
Mean recovery R (%)	93.8	99.4	94.2	97.6
RSD_r (%)	4.1	2.4	8.3	3.3
RSD_R(%)	8.2	5.9	15.5	15.8

SOP 25.2 Analysis of Anabaenopeptins and Planktopeptin by HPLC with PDA Detector

SOP 25.2.1 Introduction

Reversed-phase chromatography with RP C_{18} column and acetonitrile as a mobile-phase component is a common choice for separating smaller peptides. When HPLC combined with photodiode array detection (PDA) is used, the recorded specific retention times and absorption spectra enable to perform a preliminary characterization of cyanobacterial peptides such as microcystins, anabaenopeptins, and planktopeptins [4]. For qualitative and quantitative analyses of the compounds, the application of standards is required. As different cyanobacterial communities are characterized by different peptide profiles, this method has broad application in environmental studies and in monitoring of cyanobacterial blooms. At the same time, when used for preparative purposes, it provides a fast and simple isolation tool.

The method described is recommended for analytical qualitative or semiquantitative analysis of cyanobacterial peptides. The procedure for preparative HPLC method is very similar but uses a preparative column with higher flow rates (e.g., 10 mL min^{-1}).

SOP 25.2.2 Experimental

SOP 25.2.2.1 Materials

- HPLC grade water purified to 18.2 MΩ cm
- Methanol HPLC grade
- Acetonitrile (ACN) HPLC grade
- Trifluoroacetic acid (TFA), analytical grade
- C_{18} endcapped HPLC column and a compatible guard column
- Borosilicate glass chromatographic vials: for example, 1.5 mL clear glass with writing surface
- A positive displacement pipette with a Teflon-coated piston working in a glass capillary
- Standards (purchased or previously isolated cyclic peptides in pure form, determined by MS) dissolved in methanol and diluted in 20% aqueous methanol before analysis

SOP 25.2.2.2 LC Mobile Phase

A: 0.05% TFA in 5% aqueous ACN
B: 0.05% TFA in 100% ACN
Note: Replace the mobile phases every week.

SOP 25.2.2.3 Special Equipment

- High-performance liquid chromatograph equipped with high-pressure or low-pressure gradient pump, auto sampler, column oven, and photodiode array detector
- Chromatography analysis software

SOP 25.2.2.4 Chromatography

General Procedure

a. Separate the extracts on an analytical column using HPLC–PDA to visualize cyclic peptides.
b. Use appropriate flow rate (e.g., 0.5 mL min^{-1} for the column Phenomenex Aeris PEPTIDE XB-C18 (4.6 mm × 150 mm, 3.5 μm) or equivalent).
c. Use gradient elution (example in Table SOP 25.2).

Identification and Visualization of the Cyclic Peptides, Anabaenopeptins, Planktopeptin BL1125, and Microcystin, with a Photodiode Array Detector (PDA)

i. The column eluate should be monitored at several different wavelengths: at 275, 238, 225, 220, and 215 nm in order to locate and distinguish microcystins from other bioactive cyclic peptides of interest.
ii. The wavelengths at which the compounds are showing maximum absorptivity, λ_{max}, are characteristic for individual cyclic peptide groups; microcystins have an absorption maximum at 238 nm, while representatives of the other two groups have absorption maxima at lower wavelengths. The depsipeptide planktopeptin BL1125 is detected at 225 nm and anabaenopeptins B and F at 220 nm and 215 nm, respectively (Figure SOP 25.2).
iii. Planktopeptins and anabaenopeptins have an additional characteristic absorption maximum at 274–279 nm. The presence of this absorption maximum confirms the preliminary identification.
iv. The amounts of the cyclic peptides can be calculated by comparison of the integrated peak areas at λ_{max} with the values from the calibration curves that were standardized by previously isolated cyclic peptides in pure form.

Table SOP 25.2 *Suggested gradient program for Phenomenex Aeris PEPTIDE XB-C18 (4.6 mm × 150 mm, 3.5 μm), at a flow rate of 0.5 mL min^{-1}*

Time (min)	%A	%B
0	85	15
10.0	50	50
15.0	1	99
20.0	1	99
22.0	85	15
30.0	85	15

A = 0.05% TFA in 5% ACN; B = 0.05% TFA in 100% ACN.

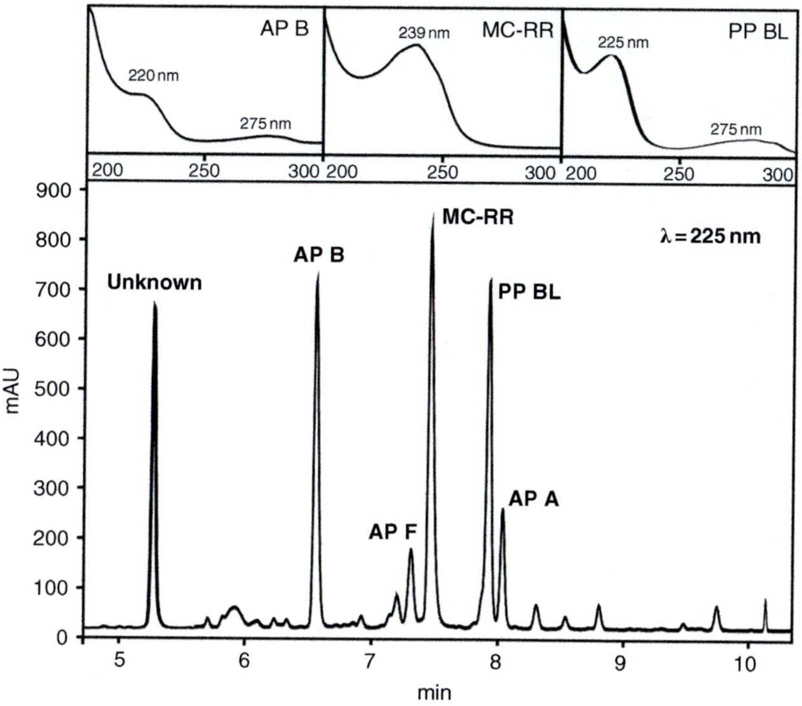

Figure SOP 25.2 Absorption spectra (upper part) and HPLC chromatogram at 225 nm (lower part) with planktopeptin BL1125 (PP BL), three anabaenopeptins (AP A, B, and F), one microcystin (MC-RR), and unknown compound

Table SOP 25.3 Recoveries (R) and reproducibility (RSD_R) of HPLC analyses with selected cyanobacterial peptides from bloom sample

Peptide	PP BL	AP F	AP B
Average peak area × 10^3	17.4	18.0	15.9
RSD_R (%)	15.7	22.8	20.0
R (%)	96.5	99.9	88.4

SOP 25.2.3 Validation

This analytical method was validated using only limited basic validation data. Cyanobacterial extract containing planktopeptin PP BL1125 (PP BL) and anabaenopeptins B (AP B) and F (AP F) was analyzed at five different days, with two different analysts. The calculated recoveries (R) were 88–100% and the reproducibility (RSD_R) were 15–23% (Table SOP 25.3).

SOP 25.2.4 Special Issues

To obtain a better resolution of the peaks of interest, modify the gradient. The recovery and reproducibility of the procedure can drop when the HPLC column is overloaded.

SOP 25.3 Analysis of Anabaenopeptins and Planktopeptin by LC-MC/MC Using Information-Dependent Acquisition (IDA) Mode

SOP 25.3.1 Introduction

Many structures of cyanobacterial peptides have already been identified, but new variants are still being discovered (see Chapter 19). LC-MS/MS is a method of choice in the analysis of novel peptides [5–7]. In this subchapter, data- or information-dependent acquisition (DDA or IDA) is described as the type of MS experiment useful in the analysis of unknown peptides in complex samples. In IDA experiments, product ion mass spectra for numerous precursor ions are collected in one chromatographic run. First, a survey scan (full MS scan) is used to register ions that give peaks in the chromatogram. Then, according to the user-defined criteria, selected ions are fragmented, and the obtained fragment ions are scanned (dependent scan).

The MS and IDA criteria were optimized for the detection and characterization of anabaenopeptins and planktopeptin PP BL1125.

SOP 25.3.2 Experimental

SOP 25.3.2.1 Materials

- Ultrapure water 18.2 MΩ cm
- Acetonitrile (ACN) LC-MS grade
- Formic acid (FA) for mass spectrometry, approx. 98%
- Centrifuge tubes (polypropylene) approx. 10 mL
- Borosilicate glass chromatographic vials (2 mL) with glass conical inserts (100 µm) and screw caps with polyethylene/silicone septa
- Adjustable pipette capable of accurately dispensing of formic acid (100–1000 µL)
- Adjustable pipette for dispensing 100 µL of cyanobacterial extract (20–200 µL)
- Neutral pipette tips (100–1000 µL and 20–200 µL)
- Chromatographic column: C_{18} HPLC column, for example, 4.6 mm × 150 mm, 5 µm, or any C_{18} column suitable for the chromatographic system and flow rates that are used. Some chromatographic columns are specially dedicated to separation of peptides.

SOP 25.3.2.2 Special Equipment

- Centrifuge (alternative to filtration)
- Bath ultrasonicator for solvent degassing
- LC-MS/MS system: Chromatographic system equipped with binary pump, autosampler, and column oven. Mass spectrometer with electrospray ion source and triple quadrupole, ion trap, or hybrid Qtof or Qtrap mass analyzer

SOP 25.3.2.3 LC Mobile Phase

A: 0.1% FA in 5% aqueous ACN
B: 0.1% FA in 100% ACN

Optimize the gradient program for the best resolution of the peaks (Table SOP 25.4).
Replace the mobile phase every week.

Table SOP 25.4 Suggested gradient program for Agilent ZORBAX Eclipse XDB-C18 (4.6 mm × 150 mm, 5 µm), at a flow rate of 0.6 mL min⁻¹

Time (min)	%A	%B
0	85	15
5.0	50	50
6.0	1	99
13.0	1	99
14.0	85	15
20.0	85	15

A = 0.1% FA in 5% ACN; B = 0.1% FA in 100% ACN.

SOP 25.3.2.4 LC-ESI-MS/MS in IDA Mode

Sample Preparation

a. Use cyanobacterial extract prepared as described in SOP 25.1.
b. Centrifuge the extract at 10 000× g for 10 min, if stored for longer time or if the extract is reconstituted after solvent evaporation. Note: Dilution of the extract may also cause precipitation of some particles.
c. Transfer the supernatant to glass chromatographic vials (2 mL) with glass conical inserts.

LC-MS/MS Settings The LC-ESI-MS/MS system should be set up according to the instructions of the manufacturers of both instruments: LC and MS. Cyanobacterial peptides are most frequently analyzed in positive ionization mode (ESI+). The ionization parameters should be optimized for each type of MS instrument individually. Here, the parameters used for QTRAP5500 with Analyst software are given as an example: IonSpray voltage (IS) 5500 V; ion source temperature (TEM) 550 °C; ion source gas 1 (GS1) and 2 (GS2), 60 and 65 L h^{-1}, respectively; curtain gas (CUR) 20 L h^{-1}; declustering potential (DP) set at 80 V; and entrance potential (EP) at 10 V.

a. Select a suitable survey scan to detect ions within the range from 50 to 1200 Da. In QTRAP5500, Q3 linear ion trap (LIT) can be used for scanning (Figure SOP 25.3).
b. For IDA experiment select singly and doubly protonated ions, ions greater than 500 Da and smaller than 1200 Da. This range should be selected according to the expected molecular mass of the analyzed peptides. Define the minimal intensity of ion to be fragmented. In the case of QTRAP5500, the intensity exceeding 500 000 counts and LIT fill time of 50 ms can be selected. The collision-induced dissociation energy (CE) was 60 eV with collision energy spread (CES) 20 eV and collision exit potential (CEX) 12 V; collision gas (CAD) 9.
c. Inject 5 µL sample.
d. The results of the analysis can be viewed in IDA window (Figure SOP 25.4).

SOP 25.3.3 Special Issues

The MS and IDA criteria should be selected depending on the molecular weight and properties of the analyzed peptide. For further structural analyses, other MS/MS experiments are suggested, for example, the product ion mode with optimize CE (Figure SOP 25.5) or multistage mass spectrometry (MS^n).

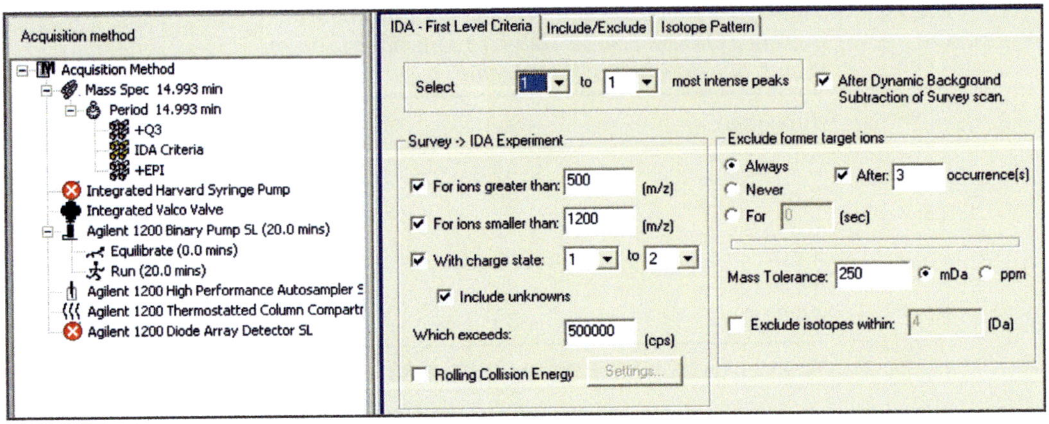

Figure SOP 25.3 Selection of criteria for information-dependent acquisition

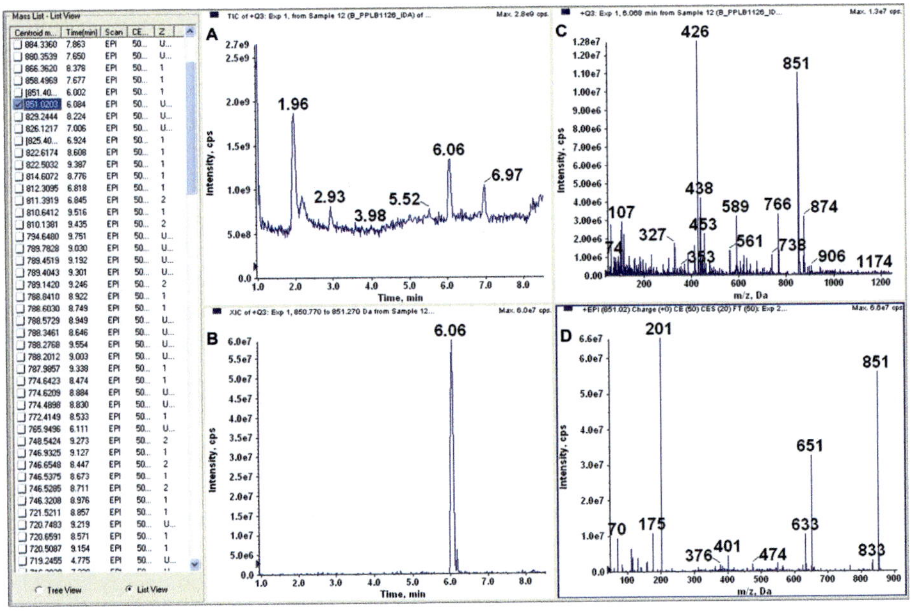

Figure SOP 25.4 The information-dependent acquisition window with the list of ions (m/z) sent to the dependent scan (left part) and the graphical information: total ion chromatogram (TIC) (A), extracted ion chromatogram (XIC) (B), Q3 scan (C), and enhanced product ion spectrum of anabaenopeptin F (D). The mass signals were assigned to the following fragments: 851 $[M+H]^+$, 833 $[M+H-H_2O]^+$, 651 $[Phe+MeAla+Hty+Ile+Lys+H]^+$, 474 $[Ile+Lys+Phe+MeAla+H]^+$, 376 $[MeAla+Hty+Ile+H]^+$, 201 $[Arg+CO]^+$, and 175 $[Arg+H]^+$

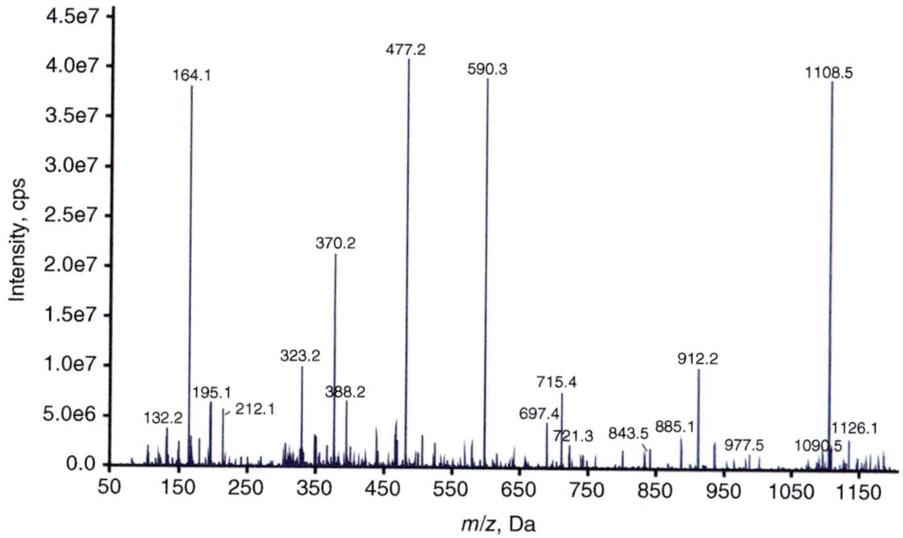

Figure SOP 25.5 Product ion mass spectra of planktopeptin PP BL1125 with $[M+H]^+$ ion at m/z 1126 (Qtrap enhanced product ion mass range limited to 1000 Da). The mass signals were assigned to the following fragments: 843 $[M+H-(H_2O)-(GA+HTyr)]^+$, 715 $[M+H-(H_2O)-(GA+HTyr+Gln)]^+$, 590 $[GA+HTyr+Gln+Thr+Leu+H-(H_2O)]^+$, 477 $[GA+HTyr+Gln+Thr+H-(H_2O)]^+$, and glyceric acid (GA)

References

[1] Harada, K.-I. (2004) Production of secondary metabolites by freshwater cyanobacteria. *Chemical Pharmaceutical Bulletin* **52** (8), 889–899.

[2]. Lin, Y., Schiavo, S., Orjala, J. *et al.* (2008). Microscale LC-MS-NMR platform applied to the identification of active cyanobacterial metabolites. *Analytical Chemistry* **80** (21), 8045–8054.

[3] Morrison, L.F., Parkin, G., and Codd, G.A. (2006). Optimization of anabaenopeptin extraction from cyanobacteria and the effect of methanol on laboratory manipulation. *Peptides* **27** (1), 10–17.

[4] Sedmak, B., Eleršek, T., Grach-Pogrebinsky, O. *et al.* (2008) Ecotoxicologically relevant cyclic peptides from cyanobacterial bloom (*Planktothrix rubescens*)—a threat to human and environmental health. *Radiology and Oncology* **42** (2), 102–113.

[5] Mayumi, T., Kato, H., Kawasaki, Y. and Harada, K.-I. (2007) Formation of diagnostic product ions from cyanobacterial cyclic peptides by the two-bond fission mechanism using ion trap liquid chromatography/multi-stage mass spectrometry. *Rapid Communication in Mass Spectrometry* **21** (6), 1025–1033.

[6] Mazur-Marzec, H., Kaczkowska, M.J., Błaszczyk, A. *et al.* (2013) Diversity of peptides produced by *Nodularia spumigena* from various geographical regions. *Marine Drugs* **11** (1), 1–19.

[7] Spoof, L., Błaszczyk, A., Meriluoto, J. *et al.* (2016) Structures and activity of new anabaenopeptins produced by the Baltic Sea cyanobacteria. *Marine Drugs* **14** (8), 1–14.

SOP 26

Thamnocephalus Test

Andrea Töröknė

Knot Ltd., Budapest, Hungary

SOP 26.1 Introduction

The purpose of this test is to make an initial screening to determine whether cyanobacterial bloom material is toxic or not to the fairy shrimp *Thamnocephalus platyurus* (commercially available as the Thamnotoxkit F™). At the end of twentieth century, there was a great need to have a simple, routinely used method to replace the mouse bioassay which was expensive, and the legal background restricted the use of mammals as test organisms. Numerous trials were carried out to identify useful alternative procedure solutions which can be used for testing the toxicity of cyanobacterial bloom materials. Some researchers tried out the commercially available Thamnotoxkit F™ for sensitivity against cyanotoxins. Firstly Törökne [1,2] proved the suitability of *T. platyurus* larvae as test organisms as used in the Thamnotoxkit. Later, other authors [3,4] compared the sensitivity of a number of invertebrates. The *Thamnocephalus* larvae were the most sensitive of the invertebrates examined for a preliminary screening method for cyanobacterial scums. The principle feature of the Thamnotoxkit is the freshly hatched larvae of *Thamnocephalus* which are exposed to a dilution series of cyanobacterial biomass in standard freshwater for 24 h. The end point of the test is the evaluation of the mortality of the test organisms. The response is compared to the survival of the test larvae in the control dilution water not containing cyanobacterial material. The concentration of cyanobacterial material which kills 50% of the larvae (LC_{50}) is calculated.

Note: Annex 2 helps to make decisions as to what steps should be taken if a cyanobacterial bloom occurs in bathing water.

Handbook of Cyanobacterial Monitoring and Cyanotoxin Analysis, First Edition.
Edited by Jussi Meriluoto, Lisa Spoof and Geoffrey A. Codd.

SOP 26.2 Test Material

Cyanobacterial biomass freeze-dried or dried at room temperature [1].

SOP 26.3 Validity of the Test

1. If the mortality in the controls exceeds 10%, the bioassay is considered invalid, and the test must be repeated.
2. Coefficients of variation (CV) should be under 20% (see Annex 1) [2].

SOP 26.4 Description of Method [5]

1. Equipment and materials:
   a. Test tubes with cysts (dormant eggs) of *T. platyurus*
   b. Polystyrene plates (9 × 13 cm) with 24 wells (3 mL) serving as test containers
   c. Petri dishes (5 cm) for the hatching of *Thamnocephalus* cysts
   d. Polyethylene micropipettes for transferring the larvae
   e. Parafilm strips for sealing multiwell plates to minimize evaporation during the incubation period
   f. Dissection microscope with magnification 10–12×
   g. Incubator (25°C)
   h. Test animals: *Thamnocephalus* larvae, less than 24 h-old, hatched from cysts (dormant eggs)
2. Preparation of dilution water (standard freshwater):
   a. Put 750 mL distilled (or deionized) water in a 1000 mL calibrated flask
   b. Add and dissolve 96 mg $NaHCO_3$, 60 mg $CaSO_4$, 123 mg $MgSO_4$, and 4 mg KCl to the water in the flask and fill up to 1000 mL with distilled (or deionized) water.
   c. The standard freshwater must be aerated for at least 15 min prior to use for hatching of the cysts and the preparation of the toxicant dilutions.
3. Test organisms:
   a. Cyst hatching should be initiated 16–18 h before the start of toxicity testing.
   b. Prehydrate the cysts by adding 1 mL standard freshwater to the tube containing the cysts, shaking regularly for 30 min.
   c. Transfer the prehydrated cysts to a Petri dish containing 10 mL standard freshwater.
   d. Incubate the covered Petri dish at 25°C for 16–18 h under continuous illumination (40–75 $\mu mol\, m^{-2}\, s^{-1}$).
4. Preparation of cyanobacterial sample for testing.
   a. Filter the cyanobacterial biomass through a membrane filter (0.45 μm pore size) and subsequently dry at room temperature or freeze-dry.
   b. Suspend 30 mg of the dry cyanobacterial biomass in 6 mL standard freshwater (stock solution). Freeze and thaw the suspension twice to release the toxic material from the cyanobacteria.
   c. Centrifuge for 10 min at 1000 rpm and recover the supernatant.

Note: High concentrations of cell-free extract can give false positives, probably due to oxygen depletion caused by the activity of bacteria. Sample cleanup is therefore necessary to detect low concentrations of toxin successfully. Sample cleanup, for example, the use of solid-phase extraction, can also increase the specificity of the assay because it selectively concentrates microcystins and nodularins, reducing interference by other compounds [4].

SOP 26.5 Procedure

SOP 26.5.1 Concentrations of Test Substances

Prepare a dilution series from the stock solution (it is advised to make a dilution series of 0.1, 0.3, 0.5, 1 and 3 mg mL^{-1}).

SOP 26.5.2 Replicates and Controls

a. Take the 24-well plate and add 1 mL dilution water to each of the 4 wells in the first column as controls.
b. Add 1 mL cyanobacterial sample of the lowest test concentration to the four wells of the second column.
c. Proceed in the same way for the further columns, with addition of 1 mL of the higher toxin concentrations to each well.

SOP 26.5.3 Transfer of the Larvae

a. Separate the hatched larvae from the unhatched and empty cysts by using the phototactic behavior of the test organisms to concentrate them on one side of the Petri dish by unilateral illumination with a desk lamp.
b. Transfer the larvae with the aid of a micropipette into another Petri dish containing standard freshwater.
c. Transfer approximately 50 larvae from the Petri dish into each of the six wells of the bottom row of the plate (rinsing wells).
d. Transfer 10 larvae from each rinsing well into the three wells of the same column (three parallels).

Note: 30 larvae are hence tested for each concentration.
Cover the plate with a sheet of parafilm and put the cover on tightly.

SOP 26.5.4 Incubation Conditions

Incubate the plate in darkness at 25–27°C for 24 h.

SOP 26.6 Scoring of Results

1. Take the multiwell plate out of the incubator and put it under the dissection microscope.
2. Check all the wells (except the rinsing wells) and record the number of dead and living larvae in each well.
3. Calculate the LC_{50} (e.g., by probit analysis).

SOP 26.7 Evaluation of Results [2]

$LC_{50} < 0.1\,mg\,mL^{-1} \rightarrow$ extremely toxic sample
$0.1\,mg\,mL^{-1} < LC_{50} < 1\,mg\,mL^{-1} \rightarrow$ very toxic sample
$1\,mg\,mL^{-1} < LC_{50} < 2\,mg\,mL^{-1} \rightarrow$ toxic sample

References

[1] Törökné, A.K. (1999) A new culture-free microbiotest for routine detection of cyanobacterial toxins. *Environmental Toxicology* **14** (5), 466–472.
[2] Törökné, A.K. (2000) The potential of the Thamnotoxkit microbiotest for routine detection of cyanobacterial toxins, in *New Microbiotests for Routine Toxicity Screening and Biomonitoring* (eds G. Persoone, C. Janssen and W. De Coen), Kluwer Academic/Plenum Publishers, New York, pp. 533–539.
[3] Tarczynska, M., Nalecz-Jawecki, G., Romanowska-Duda, Z. *et al.* (2001) Tests for the toxicity assessment of cyanobacterial bloom samples. *Environmental Toxicology* **16**, 383–390.
[4] Marsalek, B. and Blaha, L. (2004) Comparison of 17 biotests for detection of cyanobacterial toxicity. *Environmental Toxicology* **19**, 310–317.
[5] Standard Operational Procedure provided in the commercial kit of Thamnotoxkit F™ by MicroBioTest Inc.

Annex 1

Interlaboratory Trial [2]

An interlaboratory trial with eight participating laboratories was organized in 1997 for evaluating the potential of the Thamnotoxkit F™ microbiotest for toxicity detection of five toxin-producing cyanobacteria, *Dolichospermum flos-aquae* (Brébisson ex Bornet & Flahault) P. Wacklin, L. Hoffmann & J. Komárek 2009 (Anabaena flos-aquae Brébisson ex Bornet & Flauhault 1888), *Microcystis aeruginosa* (Kützing) Kützing 1846, *Cylindrospermopsis raciborskii* (Woloszynska) Seenayya & Subba Raju 1972, *Aphanizomenon flos-aquae* Ralfs ex Bornet & Flahault 1886, *Tychonema bourrellyi* (J.W.G. Lund) K. Anagnostidis & J. Komárek 1988, and the green alga *Raphidocelis subcapitata* (Korsikov) nov.comb (*Selenastrum capricornutum*) as a control. The Thamnotoxkit test is a 24-h bioassay using as test organisms larvae of the freshwater anostracan crustacean *Thamnocephalus platyurus* hatched from cysts. Dilution series were prepared from the freeze-dried samples of microalgae. Repeated freezing and melting (twice) were applied to the material before extraction of the water soluble cyanotoxins. The concentration series used for each sample of microalgae was 0.3, 0.5, 1.0, 3.0, and 5.0 mg mL^{-1}. Each sample of the microalgae was tested twice at different time by each examiner. Every sample eventually yielded 16 independent results obtained from three parallels (48 data for each sample). An exception was the samples of *C. raciborskii* which were only studied by four examiners, hence giving eight results (24 data). The water extracts of *D. flos-aquae* and *M. aeruginosa* caused the most toxic effect on *T. platyurus*. Extracts of *Anabaena flos-aquae* were almost totally toxic in all the concentrations used. Extracts of *M. aeruginosa* caused 73.5% mortality in the 0.3 mg mL^{-1} dilution but 100% in all other concentrations. Water extracts of *C. raciborskii* gave 100% mortality in the 1.0, 3.0 and 5.0 mg mL^{-1} dilutions, and *A. flos-aquae* were also 100% effective at the 3.0 mg mL^{-1} and the 5.0 mg mL^{-1} dilutions. Water extracts of *T. bourrellyi* caused 97% mortality at the 5.0 mg mL^{-1} dilution. The water extract of *R. subcapitata*, used as control, gave a minimal effect only in the highest concentration. Standard deviations (SD) were calculated for all concentrations and for each cyanobacterial and microalgal sample; the highest SD was found with *Selenastrum* in the most concentrated water extract. Relatively high SD values were also found for *Tychonema* in the two highest concentrations, for *Cylindrospermopsis* and *Aphanizomenon* in the median concentrations and for *Microcystis* at the lowest concentration. Coefficients of variation (CV) ranged between 0 and 17% at biologically important toxicity levels (over 75% and under 10% mortality, i.e. the absolute toxic level and the negative level) except for *Tychonema*, where the CV was 31% at 81% mortality level (Fig. SOP 26.1).

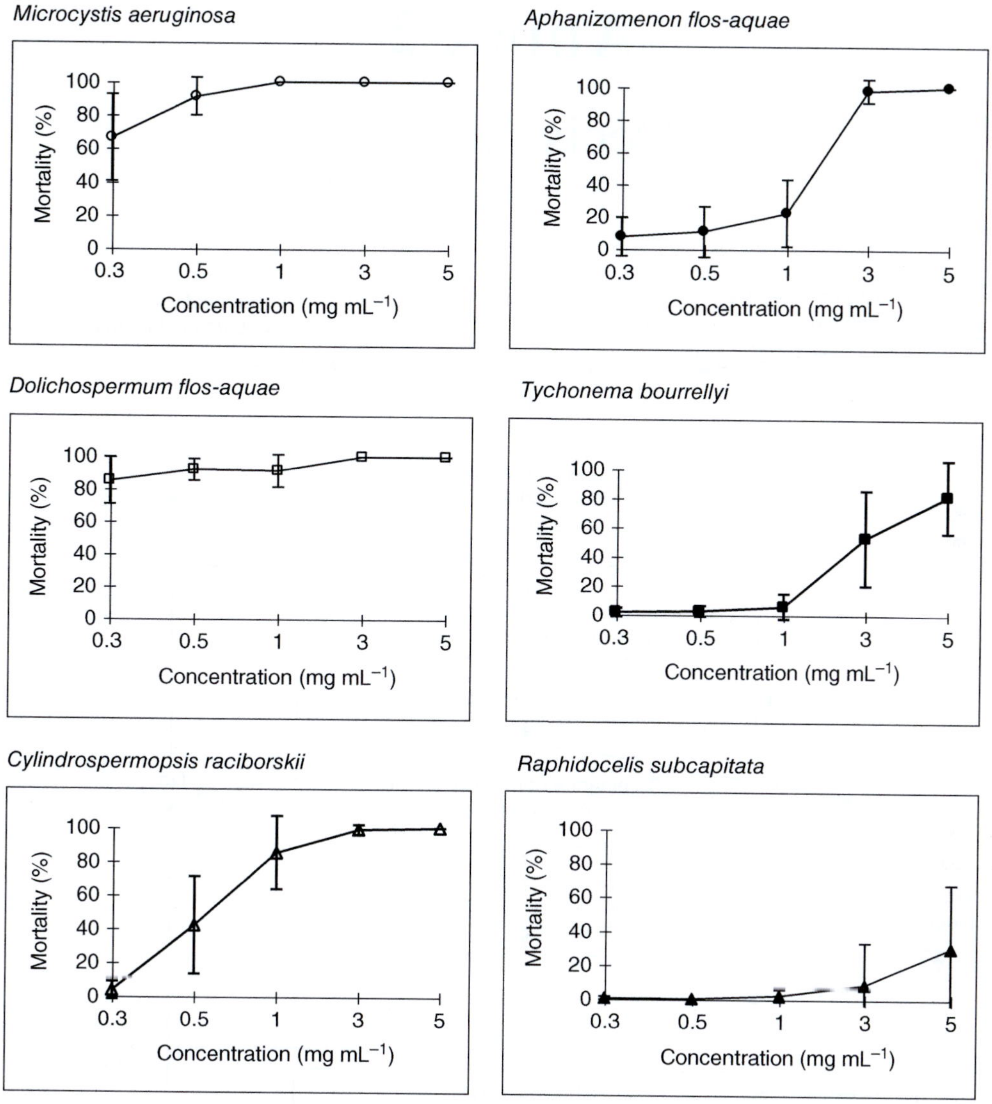

Figure SOP 26.1 *Mean mortality (with standard deviations) obtained with the Thamnotoxkit F™ for the different dilutions of water extracts of* Microcystis aeruginosa, Dolichospermum flos-aquae, Aphanizomenon flos-aquae, Cylindrospermopsis raciborskii, Tychonema bourrellyi, *and* Raphidocelis subcapitata *(control)*

Annex 2

Decision Tree

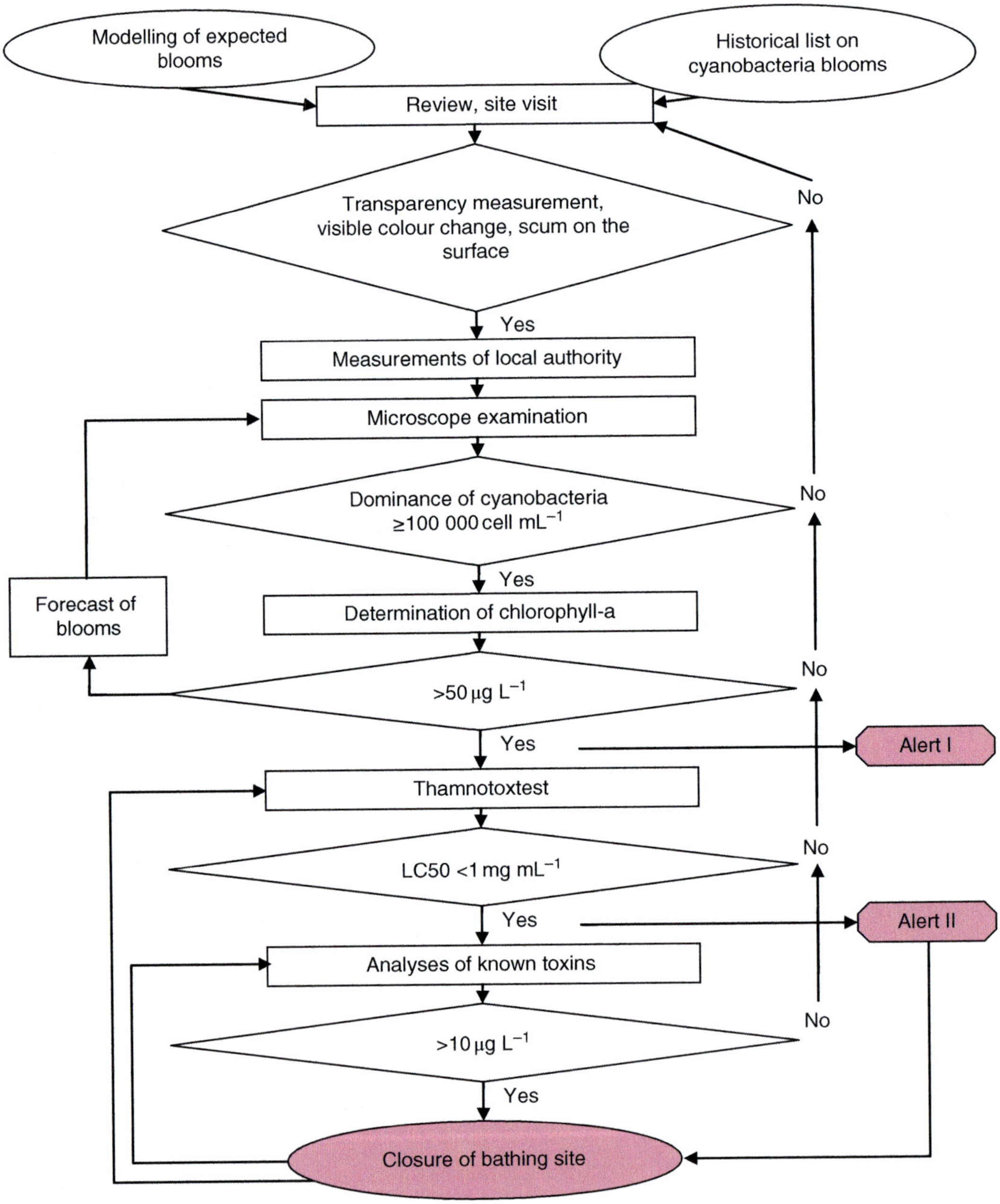

Figure SOP 26.2 *Decision tree helps to decide what steps should be taken if cyanobacterial bloom occurs at bathing site*

SOP 27

Determination of Geosmin and 2-Methylisoborneol in Water by HS-SPME-GC/MS

Triantafyllos Kaloudis[1], *Theodoros M. Triantis*[2], *and Anastasia Hiskia*[2]

[1] *Water Quality Control, Athens Water Supply and Sewerage Company – EYDAP SA, Athens, Greece*
[2] *Institute of Nanoscience & Nanotechnology, National Center for Scientific Research "DEMOKRITOS," Athens, Greece*

SOP 27.1 Introduction

This SOP specifies a method for the determination (detection, identification and quantification) of geosmin (GSM) and 2-methylisoborneol (2-MIB) in water (Table SOP 27.1). Geosmin and 2-MIB are produced by cyanobacteria and actinomycetes and give unfavorable earthy (GSM) and musty (2-MIB) odours in water. The odour thresholds of the two compounds are very low, i.e. $4\,\text{ng}\,\text{L}^{-1}$ and $9\,\text{ng}\,\text{L}^{-1}$, respectively. The method can be applied to surface water and drinking water samples. This is an in-house developed method based on methods [1] and [2].

The target compounds are extracted from the headspace (HS) of a water sample by solid-phase microextraction (SPME) with use of a suitable SPME fiber. The fiber is then inserted in the injector of a gas chromatography (GC) system coupled to a mass spectrometric detector (MS), the compounds are desorbed from the fiber and chromatographic separation, detection, identification and quantification is achieved.

Note: This SOP describes the manual SPME procedure, however, the method can be automated with use of an SPME autosampler.

Handbook of Cyanobacterial Monitoring and Cyanotoxin Analysis, First Edition.
Edited by Jussi Meriluoto, Lisa Spoof and Geoffrey A. Codd.

Table SOP 27.1 Registry information of geosmin and 2-MIB

Name	Formula	M.W.	CAS Number
Geosmin	$C_{12}H_{22}O$	182.30	16423-19-1
2-Methylisoborneol	$C_{11}H_{20}O$	168.28	2371-42-8

SOP 27.2 Experimental

SOP 27.2.1 Materials

- Ultrapure water, 18.2 MΩ cm
- Standard mixture solution of GSM and 2-MIB 100 µg mL^{-1} each in methanol (Stock, “S” solution). Alternatively, a standard mixture solution can be prepared by mixing individual standard solutions.
- Sodium chloride (NaCl) 99.5%
- SPME fiber: This method has been validated with use of divinylbenzene/carboxen/polydimethylsiloxane (DVB/CAR/PDMS), 50/30-µm-thick Stableflex fiber for manual holder which presents optimal performance [1]. Other types of fibers can be used (e.g. PDMS) bearing in mind that the performance may not be equal.
- SPME vials, 4 mL, glass, screw top, with PTFE/Silicone septa
- SPME magnetic stirrers, teflon coated 10 mm × 3 mm
- GC column e.g. HP-5 MS, 30 m × 0.25 mm × 0.25 µm (Agilent) or equivalent. Other columns may be used if they present satisfactory chromatographic performance.
- Pipettes 100 µL, 1 mL
- Volumetric flasks, 10 mL, Class A

SOP 27.2.2 Special Equipment

- SPME fiber holder for manual sampling
- SPME sampling stand for 4-mL vials (optional, recommended)
- Thermostatic water bath with thermometer, 0–100°C. Alternatively, a heating block can be used
- Hot plate with magnetic stirrer
- Gas chromatography–mass spectrometry integrated system (GC-MS), e.g. with single quadrupole mass spectrometer

SOP 27.2.3 General Procedure

SOP 27.2.3.1 Preparation of Standard Solutions

The following standard solutions (mixtures) are prepared from stock solutions of GSM and 2-MIB in ultrapure water (Table SOP 27.2).

SOP 27.2.3.2 SPME Fiber Conditioning and Extraction of Samples by SPME

A new fiber is conditioned according to manufacturer’s instructions, e.g. in a GC injector set at 270°C with He flow (split) 20 mL min^{-1} for 2 h. Before each use, the fibre is conditioned in a GC injector set at 250°C with He flow (split) 20 mL min^{-1} for 15 min.

Table SOP 27.2 *Preparation of standard solutions*

Solution	Preparation	Concentration
Stock, S	See 2.1	100 μg mL^{-1}
S1	100 μL S + 9.9 mL H_2O	1 μg mL^{-1}
S2	100 μL S1 + 9.9 mL H_2O	10 ng mL^{-1}
S3	1 mL S2 + 9 mL H_2O	1 ng mL^{-1}
S4	3 mL S3 + 3 mL H_2O	0.500 ng mL^{-1}
S5	1 mL S3 + 9 mL H_2O	0.100 ng mL^{-1}

Table SOP 27.3 *GC conditions (indicative)*

Injector temperature	250°C
Injection mode	Splitless – At 2-min split 20 mL min^{-1}
Duration of injection	2 min
Column	HP-5 MS, 30 m × 0.25 mm i.d. × 0.25 μm film thickness
He flow rate	1 mL min^{-1} constant flow
Oven temperature program	50°C for 1 min, 12°C min^{-1} to 250°C, hold for 6 min

Table SOP 27.4 *MS conditions (indicative)*

Tuning mode	Autotune, standard spectra
Acquisition mode	Single ion monitoring (SIM)
Solvent delay	5 min
Monitored ions (m/z)	GSM: 112 (base peak – quantifier ion), 126 (qualifier ions) 2-MIB: 95 (base peak-quantifier ion), 108, 135 (qualifier ions)

In a 4-mL vial, 0.75 g NaCl and a magnetic stirrer are added. Using a pipette, 2 mL of sample or standard is added. The vial is closed and shaken gently and then placed in the sampling stand in such a way that the vial is immersed in the water bath. The SPME fiber is fixed on the holder and inserted in the vial with its lower end being 2–3 mm above the surface of the water sample. The temperature of the bath is set at 70°C and stirring is set at 1000 rpm. The sample is extracted for 30 min. At the end of the extraction, the fiber is withdrawn from the vial and then inserted in the GC injector.

SOP 27.2.3.3 GC–MS

Examples of GC-MS conditions are presented in Tables SOP 27.3 and SOP 27.4.

Under the above conditions, complete chromatographic separation of geosmin and 2-MIB is achieved as shown in Figure SOP 27.1. Mass spectra of GSM and 2-MIB are presented in Figures SOP 27.2 and SOP 27.3, respectively.

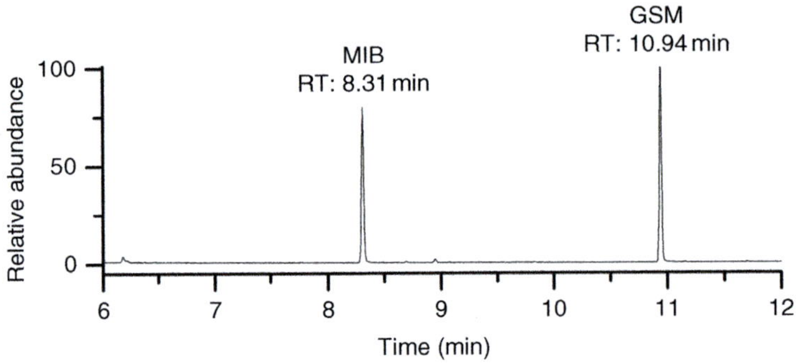

Figure SOP 27.1 Total ion chromatogram of a standard solution containing GSM and 2-MIB

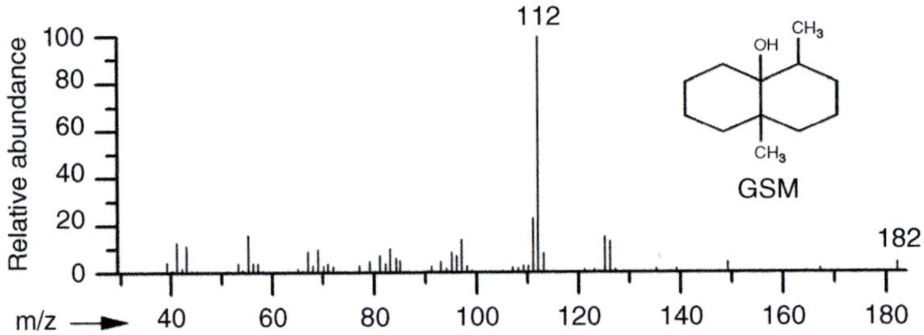

Figure SOP 27.2 Mass spectrum (SIM) of GSM determined by HS-SPME and GC-MS (m/z 112)

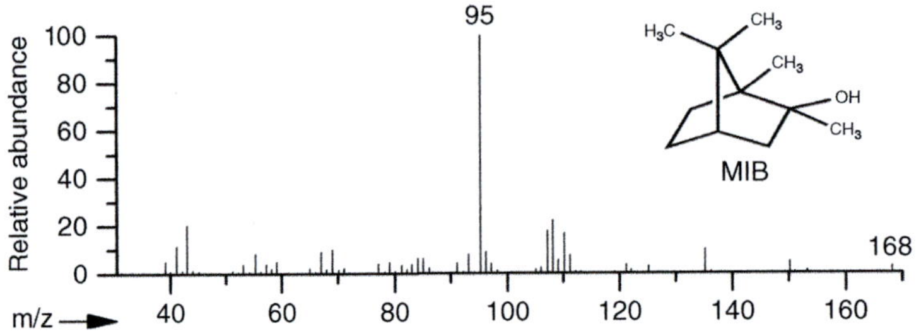

Figure SOP 27.3 Mass spectrum (SIM) of 2-MIB determined by HS-SPME and GC-MS (m/z 95)

SOP 27.3 Calibration

The following calibration alternatives can be used:

a. Single-level calibration: A calibration standard (e.g. S5, 100 ng L^{-1} can be used. It is recommended that the sample response should be within ±50% of the calibration response unless further extrapolation is supported by evidence of acceptable linearity of response.
b. Multiple-level calibration: A calibration function is calculated and used between the lowest and highest calibration levels (e.g. 10–100 ng L^{-1}). In this method the calibration curve is linear and should not be forced through the origin.

c. Standard addition technique: Known amounts of standards are added to the sample and the peak areas are plotted vs concentrations (x-axis). The point of interception of the resulting line with the x-axis is the concentration of the sample.

SOP 27.4 Sample Analysis – Calculations

Detection and identification of GSM and 2-MIB are based on the retention times, characteristic ions, and ion ratios of the compounds. Quantification of GSM and 2-MIB is based on the areas of the base-quantifier ions (112 and 95, respectively). A sequence of analysis as shown in Table SOP 27.5 is recommended in the case of a single-level calibration (100 ng L^{-1}). Calibration and quality control are repeated every 6–10 samples.

SOP 27.5 Quality Control

A recommended quality control scheme to be carried out with every batch of analysis is shown in Table SOP 27.6.

Table SOP 27.5 *Recommended sequence of analysis*

No.	Sample	Comments
1	Fiber after conditioning, (no extraction)	Check for carryover
2	Standard S5	Calibration – quality control
3	Blank sample (ultrapure water)	Check for carryover and contamination
3	Sample 1	
4	Sample 2	
…	…………	
8	Sample 6	
	Go back to No. 1 (recalibration)	

Table SOP 27.6 *Quality control scheme*

No.	Check	Measurements	Criteria
1	MS performance	Evaluation of tuning	Manufacturer's criteria
2	Fiber carryover	Sample No. 1, Table SOP 27.5	GSM and 2-MIB not detected
3	Calibration check	Sample No. 2, Table SOP 27.5 (Standard S5)	Difference from linearity plot, e.g. < ±20% or within ±3 σ (or ±3 SD) in quality control charts
4	System suitability	Sample No. 2, Table SOP 27.5 (Standard S5)	Resolution (GSM–2 MIB) or theoretical plates.
5	Positive samples	Identification parameters	Retention times, ion ratios.

Table SOP 27.7 Summarized validation results in drinking water

Compound	Linearity, r^2	Trueness[a] %		RSD$_r$[b] %		LOQ[c] $ng\,L^{-1}$	LOD[d] $ng\,L^{-1}$
Concentration level:	1–5 $ng\,L^{-1}$	5 $ng\,L^{-1}$	100 $ng\,L^{-1}$	5 $ng\,L^{-1}$	100 $ng\,L^{-1}$		
GSM	0.998	99.6	101.0	6.0	4.8	5	1
2-MIB	0.998	116.9	96.6	9.0	1.9	5	1
Criteria:	0.98	70–120	70–120	<20	<20	<10	<5

[a] Expressed as % mean recovery, $N=6$.
[b] Under repeatability conditions, $N=6$.
[c] Defined as the lowest validated level that meets validation criteria.
[d] Lowest concentration achieving S/N >3.

SOP 27.6 Expression of Results

Results are expressed as $ng\,L^{-1}$ or as $\mu g\,L^{-1}$ with three decimal digits. Results below limits of detection (LOD) are expressed as "Not Detected." Results above LOD and below limits of quantification (LOQ) are expressed as "<LOQ."

SOP 27.7 Validation

This method was validated in-house using drinking water as a matrix and the single level calibration process described in SOP 27.3. Summarized validation results are presented in Table SOP 27.7. The method is proved to be suitable for determinations at the low $ng\,L^{-1}$ levels below the odour thresholds of the target compounds.

References

[1] Saito, K., Okamura, K., Kataoka, H., (2008) Determination of musty odorants, 2-methylisoborneol and geosmin, in environmental water by headspace solid-phase microextraction and gas chromatography-mass spectrometry. *Journal of Chromatography A* **1186** (1–2), 434–437.
[2] Greenberg, A., *et al.* (2012) Method 6040 D, in *Standard Methods for the Examination of Water and Wastewater*, 22nd ed., American Public Health Association (APHA), American Water Works Association (AWWA) & Water Environment Federation (WEF), USA.

SOP 28

Rapid Analysis of Geosmin and 2-Methylisoborneol from Aqueous Samples Using Solid-Phase Extraction and GC-MS

Christine Edwards, Craig McKenzie, Carlos Joao Pestana, Kyari Yates, and Linda A. Lawton

School of Pharmacy and Life Sciences, Robert Gordon University, Aberdeen, United Kingdom

SOP 28.1 Introduction

Taste and odour compounds, geosmin, 2β,6α-dimethylbicyclo[4.4.0]decan-1β-ol (GSM) and (1*R*-exo)-1,2,7,7-tetramethylbicyclo[2.2.1]heptan-2-ol (2-MIB), produced by Actinobacteria and cyanobacteria, frequently cause problems in drinking water production and aquaculture. Although not toxic, the musty-earthy smell is associated with poor water quality resulting in customer complaints and financial impact [1]. Consequently it is important to monitor these compounds which are typically detected by humans at 2–10 $ng\,L^{-1}$. The majority of methods rely on preconcentration using solid-phase microextraction (SPME) followed by gas chromatography–mass spectrometry (GC-MS), achieving limits of detection <0.5 $ng\,L^{-1}$ [2]. However, SPME is expensive and requires expertise, so SPE extraction offers a simple alternative [3].

Handbook of Cyanobacterial Monitoring and Cyanotoxin Analysis, First Edition.
Edited by Jussi Meriluoto, Lisa Spoof and Geoffrey A. Codd.

SOP 28.2 Experimental

SOP 28.2.1 Materials

- Geosmin, 2β,6α-dimethylbicyclo[4.4.0]decan-1β-ol (>97% by GC)
- 2-MIB, (1*R*-exo)-1,2,7,7-tetramethylbicyclo[2.2.1]heptan-2-ol (>97% by GC)
- Tetrachloroanisole (TCA), instrument internal standard
- MilliQ water (or equivalent purified to 18.2 mΩ cm^{-1})
- HPLC-grade solvents (methanol, hexane, dichloromethane and acetone)
- Ethyl acetate puriss., absolute, over molecular sieve (H_2O ≤0.005%), ≥99.5% (by GC) (e.g. Sigma-Aldrich)
- High purity helium (99.99%)
- ISOLUTE C8 SPE cartridges, end capped, 100 mg, 1-mL cartridges (Biotage, Uppsala, Sweden)
- GF/C filters, diameter 5.5 cm, equivalent to Whatman (Maidstone, UK)
- Clean 250-mL Pyrex bottles
- Calibrated pipettes
- 1.5-mL GC-MS vials
- Volumetric flasks

SOP 28.2.2 Special Equipment

- Fused silica capillary column DB-5MS, 5% phenyl polysiloxane as nonpolar stationary phase (30 m × 0.25 mm i.d. and 0.25-μm film thickness) or equivalent
- GC-MS with single quadrupole (data presented is from an Agilent 6890 GC coupled with a 5975 MS
- Vacuum manifold and PFTE tubes
- Calibrated four-place analytical balance

SOP 28.2.3 General Procedure

SOP 28.2.3.1 Sample Preparation

a. Collect water samples in rinsed, acid washed bottles. Store for a maximum of 24 h at 4 °C prior to processing
b. Remove particulates by GF/C filtration before SPE
c. Prepare replicate water samples (200 mL)
d. Weigh GC-MS vials (without lid)

SOP 28.2.3.2 SPE

a. Condition the SPE cartridges with a sequence of solvents: 2 mL of hexane, 2 mL acetone, 2 mL methanol
b. Wash with 10 mL of MilliQ water
c. Load prefiltered water samples (200 mL) in clean Pyrex bottles under vacuum to the SPE cartridges via PTFE tubing at a flow rate of 5 mL min^{-1}
d. Dry under vacuum for 5 min
e. Elute with anhydrous ethyl acetate (0.5 mL) directly into a preweighed GC-MS vial
f. Weigh the vial
g. Add TCA (50 μL) as internal standard to the GC-MS vial
h. Reweigh the vial

Note: It is important that the exposure of the ethyl acetate to the atmosphere is minimised at all times as it is a hygroscopic solvent and the presence of water in the extract can affect the quality of the instrumental analysis (it deactivates the GC injection liner and causes peak tailing, particularly of the more polar MIB).

SOP 28.2.3.3 Calculation of Sample Concentration Factor

a. Calculate the volume of the extract
   Extract volume = Weight of extract (g)/density of ethyl acetate (0.902 g/mL)
b. Calculate the sample concentration factor
   Sample starting volume (200 mL)/Extract volume (e.g. 0.55 mL) e.g. 200/0.55 = 364

SOP 28.2.3.4 GC-MS Analysis

GC-MS parameters are listed in Tables SOP 28.1 and SOP 28.2.

Three ions were recorded for each compound, one quantifier (*) and two qualifier ions. The presence of analytes in samples was confirmed by their retention time and the identification of target and qualifier ions. Retention times must be within ±0.2 min of the expected time and qualifier-to-target ratios within a 20% range for positive confirmation.

Table SOP 28.1 *GC-MS settings for Agilent 6890 GC coupled with a 5975 MS and fitted with a 4-mm i.d. single tapered splitless injection liner*

Parameter	Setting
Injector temperature	250 °C
Injector mode	Pulsed splitless
Injection volume	3 µL
Inlet pressure	10 psi
Injection parameters	25 psi for 0.5 min preinjection and 1 min postinjection, purge flow to split vent 50 mL min^{-1}. Wash solvent: Dichloromethane
Column	DB-5MS (30 m × 0.25 mm i.d.)
Gas	Helium (1 mL min^{-1})
Gradient	Initial 60 °C for 1 min, 20 °C min^{-1} to 100 °C, 7.5 °C min^{-1} to 152 °C (held for 2 min).

Table SOP 28.2 *Mass spectrometry conditions (electron impact (EI) ionization mode at 70 eV)*

Parameter	Setting
Solvent delay	5 min
Acquisition mode	Single ion monitoring (SIM)
Ions monitored	2-MIB (95*, 107, 135) TCA (167*, 195, 210) GSM (112*, 125, 182)

SOP 28.3 Internal and External Standard Calibration

a. Prepare standards in ethyl acetate (1 to 100 μg L^{-1})
b. Add internal standard (tetrachloroanisole, TCA) at a concentration of 35 μg L^{-1}
c. The GC-MS was calibrated using either a linear or nonlinear curve as appropriate (MS detector linearity and sensitivity is dependent on the individual instrument)

The instruments limit of detection (LOD) and limit of quantification (LOQ) was calculated from the mean signal–to–root mean squared (RMS) noise signal ratio calculated from ten injections of a calibration standard at 2 μg L^{-1}. These were determined by calculating the concentration of MIB and GSM that gave a detector response 3 times and 10 times the root mean–noise signal ratio respectively (Table SOP 28.3). The method LOD and LOQ were determined by dividing the instrument LOD and LOQ by the sample concentration factor.

SOP 28.4 Calculating the Concentration of MIB and GSM in the Water Sample

The concentration of MIB and GSM in the sample extract (μg L^{-1}) is calculated from the calibration curve using the instrument software. The concentration in the water sample is then calculated using the following equation:

Concentration in the water sample (ng L^{-1}) = (concentration in sample extract (μg L^{-1}) × sample concentration factor) × 1000.

SOP 28.5 Quality Assurance Requirements

To ensure the batch quality control analysis:

- A method blank (a distilled water sample free from MIB and GSM, negative control),
- A recovery check sample (a water sample spiked to give 25 ng L^{-1} MIB and GSM, positive control)
- A replicate test sample

To ensure the instrument calibration accuracy analysis:

- Instrumental blank (ethyl acetate, negative control).
- Two calibration check standards (7.5 and 75 ng L^{-1} MIB and geosmin)

All positive control samples should meet minimum quality standards e.g. maximum allowable bias and precision). Negative control samples should not contain MIB and GSM at concentrations

Table SOP 28.3 Instrument and method limits of detection (LOD) and quantification (LOQ) for 2-MIB and GSM

	Instrument LOD (μg L^{-1})	Instrument LOQ (μg L^{-1})	Method LOD (ng L^{-1})	Method LOQ (ng L^{-1})
2-MIB	0.35	1.16	0.96	3.19
GSM	0.36	1.17	0.99	3.21

Table SOP 28.4 Examples of mean values (± SD; n = 3) of MIB and GSM in natural water samples

Source Water	MIB (ng L^{-1})	GSM (ng L^{-1})
Rescobie Loch	5.3 ± 0.04	1.68 ± 0.05*
Outdoor aquaculture pond	6.18 ± 1.47	6.80 ± 0.73
Aquaculture pond A[a]	3.19 ± 0.50	4.42 ± 0.46
Aquaculture pond B[a]	10.67 ± 0.28	8.36 ± 0.78
Aquaculture pond A[b]	3.21 ± 0.49	7.57 ± 0.78
Aquaculture pond B[b]	10.44 ± 0.21	7.80 ± 0.34

* **Note**: Concentration lower than method LOQ but above method LOD would be reported as <3.21 ng L^{-1}.
[a] Sampled 1/7/11.
[b] Sampled 14/7/11.

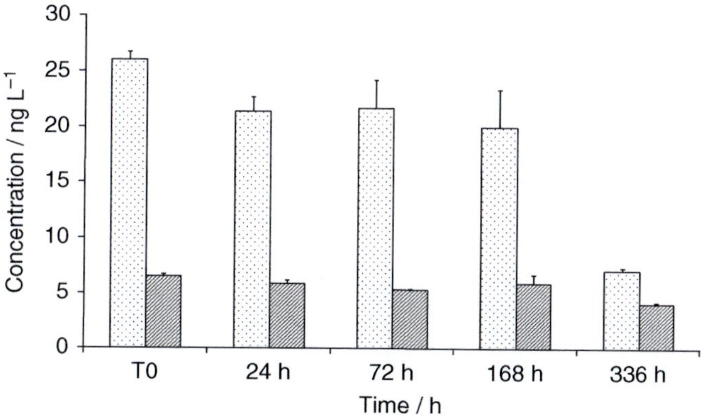

Figure SOP 28.1 Stability of 2-MIB (spots) and GSM (stripes) from environmental sample (Rescobie Loch) on C8 (EC) 100 mg, 1 mL Isolute cartridges stored at 4 °C (refrigerator) for different amounts of time; T_0 immediately after SPE, 24 h, 72 h, 168 h (1 week), 336 h (2 weeks). Error bars = 1 SD; n = 3

above the instrumental limit of detection. Reproducibility of SPE method in selected water samples illustrated method robustness as shown in Table SOP 28.4.

SOP 28.6 Storage and Shipping

In addition to the simplicity of this method, a key advantage is that samples can be prepared on site using simple vacuum apparatus and shipped for analysis as the compounds are stable on dry cartridges for up to 7 days (Figure SOP 28.1).

SOP 28.7 Special Issues

This method provides consistently high recoveries (>90%) of MIB and GSM from water samples. Analyte losses can be corrected through the use of deuterated MIB and GSM surrogate standards, which may be added to the water samples prior to SPE extraction.

References

[1] Hanson, T. (2003) Economic Impact of Off-Flavor to the U.S. Catfish Industry, in Off Flavours in Aquaculture, ACS Symposium Series 848, 13–29. Chapter DOI: 10.1021/bk-2003-0848.ch002, ISBN13: 9780841238213.

[2] Pestana, C.J. (2012) Monitoring and regulating cyanobacterial metabolites (microcystins and geosmin) in aquatic systems. PhD Thesis, Robert Gordon University, Aberdeen, Scotland.

[3] Pestana, C.J., Robertson, P.K. J. Edwards, C. *et al.* (2014) A continuous flow packed bed photocatalytic reactor for the destruction of 2-methylisoborneol and geosmin utilising pelletised TiO2. *Chemical Engineering Journal* **235**, 393–398.

SOP 29

Basic Validation Protocol for the Analysis of Cyanotoxins in Environmental Samples

Triantafyllos Kaloudis[1], Theodoros M. Triantis[2], and Anastasia Hiskia[2]

[1] Water Quality Control, Athens Water Supply and Sewerage Company – EYDAP SA, Athens, Greece
[2] Institute of Nanoscience & Nanotechnology, National Center for Scientific Research "DEMOKRITOS," Athens, Greece

SOP 29.1 Scope

This SOP outlines a basic protocol for in-house validation of methods for cyanotoxins analysis in environmental samples. This protocol can be used for laboratory accreditation according to ISO 17025. The protocol assesses repeatability, reproducibility, trueness, limit of quantitation, limit of detection, range of measurement and can be used with any analytical technique (chromatographic, spectrophotometric, etc.). However, expansion of the protocol may be needed depending on the method of analysis or other specific requirements (e.g. ELISA methods may need further assessment of cross-reactivity). Further study of method robustness may be carried out if needed by using suitable experimental designs that are not covered in this SOP. It is emphasised that the protocol presented in this SOP includes a minimum assessment of method performance in order to meet accreditation criteria. As validation is always a balance among costs, risks, and technical possibilities, more detailed studies may be needed depending on the situation or specific requirements of authorities or national accreditation bodies. This in-house validation should also be complemented with results from inter-laboratory tests where available. Verification of validation or complete revalidation should be carried out at fixed time intervals, usually every 3 to 4 years. This SOP complements Chapter 29, "Method Validation Guidelines for the Analysis of Cyanotoxins."

Handbook of Cyanobacterial Monitoring and Cyanotoxin Analysis, First Edition.
Edited by Jussi Meriluoto, Lisa Spoof and Geoffrey A. Codd.

WARNING: For practical reasons, the protocol is presented taking the analysis of microcystins in drinking water as an example. It is stressed that this **serves only as example** and laboratories should adjust the protocol according to their intended purpose.

SOP 29.2 Principle

The laboratory defines the intended use of the method to be validated and sets the performance requirements for this intended use. Performance parameters to be studied are determined and validation criteria are set. Validation experiments (measurements) are designed so that they represent the true performance of the method in real samples and in such a way that performance parameters can be estimated. Validation experiments are carried out and performance parameters are compared to validation criteria. If validation criteria are met, then a statement about the validity of the method for its intended use can be made.

SOP 29.3 General Considerations

- The validation study is carried out with measurements of laboratory reference samples, that is, real samples spiked with known quantities of the target cyanotoxins.
- When the scope of the method includes different sample matrices (e.g. drinking water, fish tissue, etc.), a complete validation for every different matrix type should be carried out.
- Validation should be carried out at minimum two concentration levels: the one (low level) should be the desired limit of quantitation (LOQ) and the other (high level) should be the regulation limit (if any) or other relevant high concentration level.
- Validation measurements are carried out by following strictly the method (SOP) that is to be validated. This includes sample preparation, calibration procedures, materials, equipment, and authorized personnel.
- The laboratory should keep records of all validation data and results. Results must be traceable to primary data (e.g. chromatograms).

SOP 29.4 Performance Parameters and Validation Criteria

This protocol assesses the following method performance parameters: repeatability, reproducibility, trueness, LOQ, limit of detection (LOD), and range of measurement. Repeatability and reproducibility (precision) may be expressed as %RSD, while trueness (referred to also as accuracy) may be expressed as %Recovery. The laboratory should set validation criteria for these parameters. Validation criteria are requirements for method performance that are derived by regulations, authorities, customer requirements or the laboratory's quality system.

As an example, Table SOP 29.1 presents the performance parameters and validation criteria for the analysis of microcystins in drinking water by LC-MS/MS (SOP 7 followed by SOP 10).

SOP 29.5 Validation Experiments

Validation experiments are carried out in samples spiked with the target cyanotoxins at two concentration levels. Samples should be as close as possible to real samples and preferably they should not have detectable amounts of the target compounds. Repeated measurements of spiked

Table SOP 29.1 *Example of performance parameters and criteria for the validation study*

Method	Determination of microcystins in drinking water by SPE-LC-MS/MS	
SOPs	7 and 10	
Sample matrix	Drinking water	
Validation levels	0.010 μg L^{-1} (low) and 0.100 μg L^{-1} (high)	
	Performance Parameters and Criteria	
Parameter	Expression	Validation Criteria
Repeatability	% RSD_r	<20[1]
Reproducibility	% RSD_R	<20[1]
Trueness	% Mean Recovery	70–120[1]
LOQ	Concentration, μg L^{-1}	0.010[2]
LOD	Concentration, μg L^{-1}	<0.010 (based on S/N)
Meas. range	Concentration, μg L^{-1}	0.010–1.0[3]

[1] Since there are no specific requirements for the analysis of microcystins, this is a general requirement for analysis of organic residues/pollutants *[1]*. It also covers the requirement of the Drinking Water Directive for pesticides/PAHs *[2]*.
[2] This LOQ was set with the objective to quantify MCs at levels well below the WHO limit (1 μg L^{-1}). LOQ could be higher (e.g. 0.100 μg L^{-1}) when the aim is only to prove conformance to WHO limit. LOQ is a validated level of concentration in this protocol.
[3] Range of measurement in this example is set from desired LOQ to the WHO limit for MC-LR in drinking water.

Table SOP 29.2 *Example setup of validation experiments*

Method	Determination of microcystins in drinking water by SPE-LC-MS/MS	
SOPs	7 and 10	
Sample matrix	Drinking water	
	Spiking concentration level (μg L^{-1})	
	0.010	**1.0**
Day 1	Blank	Blank
	Sample 1a	Sample 1b
	Sample 2a	Sample 2b
	Sample 3a	Sample 3b
	Sample 4a	Sample 4b
	Sample 5a	Sample 5b
	Sample 6a	Sample 6b
Day 2	As in day 1	As in day 1
Day 3	As in day 1	As in day 1
	Samples spiked at desired LOD level (two or three replicates).	

samples are carried out in repeatability conditions (same day, some analyst) and reproducibility conditions (different days; second analyst upon availability). All determinations are carried out by following the corresponding SOP without any modifications. This includes calibration procedures.

Depending on the type of analysis, if the laboratory can carry out 10–12 determinations in one day, then 3 days will be needed to complete this basic validation. An example of the setup of validation experiments for the determination of microcystins in drinking water by SPE-LC-MS/MS is shown in Table SOP 29.2.

SOP 29.6 Assessment of Method Performance Parameters

Assessment of validation results is done for every separate sample matrix, individual target compound and concentration level.

Repeatability: Repeatability should reflect the "within-day" variability. Because experiments are carried out in 3 different days, $\%RSD_r$ should preferably be estimated from the pooled data, using the pooled standard deviation formula:

$$s_{pooled} = \left(\frac{(n_1 - 1)s_1^2 + (n_2 - 1)s_2^2 + (n_3 - 1)s_3^2}{n_1 + n_2 + n_3 - 3} \right)^{1/2}$$

where s_1, s_2, s_3 are the individually calculated standard deviations within each day (under repeatability conditions) and n_1, n_2, n_3 are the number of measurements in each day.

$$\%RSD_r = \left(\frac{s_{pooled}}{Mean\ conc.} \right) \times 100$$

Reproducibility: Calculate $\%RSD_R$ by using all measurement data that are derived from different days (under reproducibility conditions).

$$\%RSD_R = \left(\frac{s_{all\ data}}{Mean\ conc.} \right) \times 100$$

Trueness: Calculate % mean recovery for each target compound/sample matrix using all data that are derived from different days.

$$\%\ Mean\ Recovery = \left(\sum_{1}^{n} \frac{x_i}{\mu} \times 100 \right) / n$$

where n is the number of measurements, x_i the measurement results and μ the known concentration of the sample.

LOQ: Check if the lowest validated level conforms to the validation criteria. If yes, then LOQ is defined as the lowest validated level.

LOD: Check that target compounds are detected with an S/N >3 in samples spiked at LOD level.

Range of measurement: Check if both low and high concentration levels meet validation criteria. Calibration procedures that cover this range should be documented in the SOP.

Uncertainty of measurement: Estimation of uncertainty of measurement is a requirement of ISO 17025 and of regulations for analysis of drinking water [3] and surface water [4]. There are different approaches in estimating uncertainty of measurement. A practical approach that largely makes use of validation data is the "Nordtest" method [5] that is accompanied by an easy to use freeware software, the MUkit [6].

An example of calculations of performance parameters is shown in Table SOP 29.3. It is stressed that the values in this table do not represent real validation results and this table should be used only as an example of calculations. In this example, all performance parameters meet validation criteria (Table SOP 29.1) therefore the method is valid for microcystin-LR in drinking water with an LOQ of 0.010 μg L^{-1}.

Table SOP 29.3 Example of calculations

Method	Determination of microcystins in drinking water by SPE-LC-MS/MS		
SOPs	7 and 10		
Sample matrix	Drinking water		
Target compound	Microcystin-LR		
		Validation Results ($\mu g\,L^{-1}$)	
	Sample	Concentration level: 0.010 $\mu g\,L^{-1}$	Concentration level: 1.0 $\mu g\,L^{-1}$
Day 1	1	0.010	0.977
Analyst 1	2	0.009	0.862
	3	0.010	0.996
	4	0.009	0.701
	5	0.009	0.711
	6	0.009	1.117
Day 2	7	0.007	0.777
Analyst 2	8	0.009	0.741
	9	0.008	0.798
	10	0.008	1.011
	11	0.008	1.003
	12	0.008	0.979
Day 3	13	0.009	0.989
Analyst 3	14	0.009	0.838
	15	0.009	0.961
	16	0.009	1.127
	17	0.010	0.847
	18	0.007	1.003
		Calculated Performance Parameters	
Mean concentration (all days)		0.009	0.913
Standard Deviation (all days)		0.000895	0.132014
Standard Deviation, day 1 (s1)		0.000516	0.166604
Standard Deviation, day 2 (s2)		0.000632	0.125383
Standard Deviation, day 3 (s3)		0.000983	0.107916
Pooled Std. Deviation (Spooled)		0.000738	0.135553
% Mean Recovery (all days)		89.44	91.32
$\%RSD_r$		8.20	14.85
$\%RSD_R$		10.26	14.46

References

[1] SANTE/11945/2015 (2015) *Guidance document on analytical quality control and method validation procedures for pesticides residues analysis in food and feed.* European Commission.

[2] Council Directive 98/83/EC of 3 November 1998 on the quality of water intended for human consumption. *Official Journal of the European Union* L330/32-54, 5.12.98.

[3] Commission Directive (EU) 2015/1787 of 6 October 2015, amending Annexes II and III to Council Directive 98/83/EC on the quality of water intended for human consumption. *Official Journal of the European Union* L260/6-17, 7.10.2015.

[4] Commission Directive 2009/90/EC of 31 July 2009, laying down, pursuant to Directive 2000/60/EC of the European Parliament and of the Council, technical specifications for chemical analysis and monitoring of water status. *Official Journal of the European Union* L201/36-38, 1.8.2009.

[5] Nordtest, 2012, "Handbook for Calculation of Measurement Uncertainty in Environmental Laboratories", NT Techn. Report 537, Ed. 3.1., 11/2012, available from http://www.nordtest.info

[6] SYKE 2012, "MUkit – Measurement Uncertainty Kit – User's Manual", v.0.5, Finnish Environment Institute, available from http://www.syke.fi

Section VIII

Appendices

Appendix 1

Cyanobacterial Species and Recent Synonyms

Nico Salmaso[1], Reyhan Akçaalan[2], Cécile Bernard[3], Tina Elersek[4], Svetislav Krstić[5], Renata Pilkaityte[6], Antonio Quesada[7], Ricardo Santos[8], Andrea Törökné[9], Gábor Vasas[10], and Geoffrey A. Codd[11,12]

[1] *IASMA Research and Innovation Centre, Fondazione Edmund Mach-Istituto Agrario di S. Michele all'Adige, S. Michele all'Adige, Trento, Italy*
[2] *Faculty of Fisheries, Istanbul University, Laleli-Istanbul, Turkey*
[3] *UMR 7245 MCAM CNRS-MNHN, Muséum National d'Histoire Naturelle, Paris, France*
[4] *Department of Genetic Toxicology and Cancer Biology, National Institute of Biology, Ljubljana, Slovenia*
[5] *Faculty of Natural Sciences and Mathematics, "St.Cyril and Methodius" University, Skopje, Macedonia*
[6] *Open Access Centre for Marine Research, Klaipeda University, Klaipeda, Lithuania*
[7] *Departamento de Biología, Universidad Autónoma de Madrid, Spain*
[8] *Instituto Superior Técnico, Laboratorio de Análises, Lisboa, Portugal*
[9] *Knot Ltd. Budapest, Hungary*
[10] *Department of Botany, University of Debrecen, Debrecen, Hungary*
[11] *Biological and Environmental Sciences, University of Stirling, Scotland, United Kingdom*
[12] *School of the Environment, Flinders University, Adelaide, Australia*

The nomenclature and taxonomic classification of cyanobacteria have always been in a dynamic state. However, with the increasing application of combined, polyphasic approaches that include microscopy, genetic and phylogenetic analyses, they are now in a phase of increasingly rapid and wide-scale change (e.g. Komárek *et al.*, 2014; Lee *et al.*, 2014; Soo *et al.*, 2014). The new names apply to many of the most familiar toxigenic cyanobacteria in freshwaters and marine waters and are beginning to be used as the primary descriptors in ecotoxicological publications (e.g. Brutemark *et al.*, 2015). It should be anticipated that this trend will continue. The present list includes the

Handbook of Cyanobacterial Monitoring and Cyanotoxin Analysis, First Edition.
Edited by Jussi Meriluoto, Lisa Spoof and Geoffrey A. Codd.

Table A1.1 *Cyanobacterial species and recent synonyms*

Traditional groups	Present orders	Present name (January 2016)	Recent or common synonyms	Prevalent habitat	Main recent reference
Chroococcales	Synechococcales	*Anathece clathrata* (W. West & G.S. West) J. Komárek, Kaštovský & Jezberová 2011	*Aphanothece clathrata* W. West & G.S. West 1906	F	Komárek *et al.* (2011)
	Synechococcales	*Aphanocapsa cumulus* J. Komárek & G. Cronberg 2001		F	Komárek & Cronberg (2001)
	Synechococcales	*Cyanobium rubescens* (T.P. Chang) J. Komárek, J. Kopecký & V. Cepák 1999	*Synechococcus rubescens* T.P. Chang 1981	F	Komárek *et al.* (1999)
	Synechococcales	*Cyanodictyon tropicale* P.A.C. Senna, A. Delazari, & C.L. Sant'Anna 1999		F	Senna *et al.* (1999)
	Synechococcales	*Merismopedia tenuissima* ("*tenuissimum*") Lemmermann 1898		F/B	(I)
	Chroococcales	*Microcystis aeruginosa* (Kützing) Kützing 1846		F/B	(I)
	Chroococcales	*Microcystis bengalensis* Banerji 1936		F	(I)
	Chroococcales	*Microcystis botrys* Teiling 1942		F/(B)	(I)
	Chroococcales	*Microcystis flos-aquae* (Wittrock) Kirchner 1898		F	(I)
	Chroococcales	*Microcystis ichthyoblabe* (Kunze) Kützing 1843		F	(I); Index Nominum Algarum (2015)
	Chroococcales	*Microcystis novacekii* (Komárek) Compère 1974		F	(I)
	Chroococcales	*Microcystis panniformis* J. Komárek, J. Komárková-Legnerová, C.L. Sant'Anna, M.T.P. Azevedo, & P.A.C. Senna 2002		F	Komárek *et al.* (2002)
	Chroococcales	*Microcystis pseudofilamentosa* Crow 1923		F	(I)
	Chroococcales	*Microcystis ramosa* Bharadwaya 1935 (?). (Probably synonymous with *M. pseudofilamentosa* Crow 1923)		F	(I)

	Chroococcales	*Microcystis smithii* Komárek & Anagnostidis 1995		F	(I)
	Chroococcales	*Microcystis viridis* (A. Braun) Lemmermann 1903		F	(I)
	Chroococcales	*Microcystis wesenbergii* (Komárek) Komárek ex Komárek 2006	*Microcystis wesenbergii* (Komárek) Komárek in Kondrateva 1968	F	(I); Joosten (2006); Guiry & Guiry (2015)
	Chroococcales	*Radiocystis fernandoi* Komárek & Komárková-Legnerová 1993		F	Komárek & Komárková-Legnerová (1993)
	Synechococcales	*Synechococcus capitatus* A.E. Bailey-Watts & J. Komárek 1991		F	(I)
	Synechococcales	*Synechococcus lividus* J.J. Copeland 1936		F (thermophilic)	(I)
	Synechococcales	*Synechococcus nidulans* (Pringsheim) Komárek in Bourrelly 1970		F	(I)
	Synechococcales	*Woronichinia compacta* (Lemmermann) Komárek et Hindák 1988	*Gomphosphaeria lacustris* var. *compacta* Lemmermann 1899; *Gomphosphaeria compacta* (Lammermann) Ström 1923; *Gomphosphaeria lacustris* f. *compacta* (Lemmermann) Elenkin 1938	F(B)	(I)
	Synechococcales	*Woronichinia naegeliana* (Unger) Elenkin 1933	*Coelosphaerium naegelianum* Unger 1854; *Gomphosphaeria naegeliana* (Unger) Lemmermann 1907	F	(I)
Oscillatoriales	Oscillatoriales	*Annamia toxica* L.T.T. Nguyen 2013		F	Nguyen *et al.* (2013)
	Oscillatoriales	*Arthrospira fusiformis* (Voronikhin) J. Komárek & J.W.G. Lund 1990	*Spirulina fusiformis* Woronichin 1934	F	(II)
	Oscillatoriales	*Blennothrix lyngbyacea* (Kützing ex Gomont) K. Anagnostidis & J. Komárek 1988	*Hydrocoleum lyngbyaceum* Kützing ex Gomont 1892	M	(II)

(*Continued*)

Table A1.1 *(Continued)*

Traditional groups	Present orders	Present name (January 2016)	Recent or common synonyms	Prevalent habitat	Main recent reference
	Oscillatoriales	*Geitlerinema amphibium* (C.Agardh ex Gomont) Anagnostidis 1989	*Geitlerinema unigranulatum* (R.N. Singh) J. Komárek & M.T.P. Azevedo 2000	F	Bittencourt-Oliveira & Piccin-Santos (2012)
	Oscillatoriales	*Kamptonema formosum* (Bory ex Gomont) O. Strunecký, J. Komárek, & J. Šmarda 2014	*Phormidium formosum* (Bory ex Gomont) K. Anagnostidis & J. Komárek 1988. *Oscillatoria formosa* Bory ex Gomont 1892	F(B)	(II); Strunecký *et al.* (2014)
	Synechococcales	*Leibleinia gracilis* (Rabenhorst ex Gomont) K. Anagnostidis & J. Komárek 1988	*Lyngbya gracilis* Rabenhorst ex Gomont 1892; *Phormidium gracile* (Rabenhorst ex Gomont) Anagnostidis 2001	M	(II); Guiry & Guiry (2015)
	Synechococcales	*Leptolyngbya boryana* (Gomont) K. Anagnostidis & J. Komárek 1988	*Plectonema boryanum* Gomont 1899	F	(II)
	Synechococcales	*Leptolyngbya tenuis* (Gomont) K. Anagnostidis & J. Komárek 1988	*Phormidium tenue* Gomont 1892	F/B	(II)
	Synechococcales	*Limnothrix redekei* (van Goor) M.-E. Meffert 1988		F	(II)
	Oscillatoriales	*Lyngbya majuscula* Harvey ex Gomont 1892		M	(II)
	Oscillatoriales	*Lyngbya robusta* (Parakutty) J. Komárek & J. Komárková-Legnerová 2007		F	Komárek & Komárková-Legnerová (2007)
	Oscillatoriales	*Lyngbya wollei* (Farlow ex Gomont) B.J. Speziale & L.A. Dyck 1992	*Plectonema wollei* Farlow ex Gomont 1892; (but see *P. wollei* Farlow ex Gomont 1892)	F	Speziale & Dyck (1992)

Oscillatoriales	*Microcoleus autumnalis* (Gomont) Strunecký, Komárek & J.R. Johansen 2013	*Phormidium autumnale* Gomont 1892	F	(II); Strunecký *et al.* (2013); Guiry & Guiry (2015).
Oscillatoriales	*Moorea producens* Engene, Rottacker, Kastovsky, Byrum, Choi, Ellisman, Komárek & Gerwick 2012		M	Engene *et al.* (2012)
Oscillatoriales	*Oscillatoria limosa* C. Agardh ex Gomont, 1892	*Phormidium limosum* (Dillwyn) P.C. Silva (?)	F	(II); Guiry & Guiry (2015)
Oscillatoriales	*Oscillatoria tenuis* C. Agardh ex Gomont 1892		F	(II)
Oscillatoriales	*Phormidium aerugineo-caeruleum* (Gomont) K. Anagnostidis & J. Komárek 1988	*Lyngbya aerugineo-caerulea* Gomont 1892;	F	(II)
Oscillatoriales	*Phormidium corium* Gomont 1892	*Oscillatoria corium* C. Agardh 1812; *Lyngbya corium* Hansgirg 1892	F	(II)
Oscillatoriales	*Phormidium favosum* Gomont 1892		F	(II)
Oscillatoriales	*Phormidium nigroviride* (Thwaites ex Gomont) Anagnostidis & Komárek 1988	*Oscillatoria nigroviridis* Thwaites ex Gomont 1892	M	(II)
Oscillatoriales	*Phormidium uncinatum* Gomont ex Gomont 1892	*Lyngbya uncinata* (Gomont ex Gomont) Compère 1980	F	(II)
Synechococcales	*Planktolyngbya circumcreta* (G.S. West) K. Anagnostidis & J. Komárek 1988	*Lyngbya circumcreta* G.S.West 1907	F	(II)
Synechococcales	*Planktolyngbya limnetica* (Lemmermann) J. Komárková-Legnerová & G. Cronberg 1992	*Lyngbya limnetica* Lemmermann 1898; *Planktolyngbya subtilis* (W. West) K. Anagnostidis & J. Komárek 1988	F	(II)
Synechococcales	*Planktolyngbya microspira* J. Komárek & G. Cronberg 2001		F	(II)
Oscillatoriales	*Planktothrix agardhii* (Gomont) K. Anagnostidis & J. Komárek 1988	*Oscillatoria agardhii* Gomont 1892	F	(II)
Oscillatoriales	*Planktothrix rubescens* (de Candolle ex Gomont) K. Anagnostidis & J. Komárek 1988	*Oscillatoria rubescens* de Candolle ex Gomont 1892	F	(II)

(*Continued*)

Table A1.1 (*Continued*)

Traditional groups	Present orders	Present name (January 2016)	Recent or common synonyms	Prevalent habitat	Main recent reference
	Oscillatoriales	*Plectonema wollei* Farlow ex Gomont 1892	*Lyngbya wollei* Farlow 1880; (but see *Lyn. wollei* (Farlow ex Gomont) B.J. Speziale & L.A. Dyck 1992)	F	(II)
	Synechococcales	*Pseudanabaena limnetica* (Lemmermann) Komárek 1974	*Oscillatoria limnetica* Lemmermann 1900	F	(II)
	Synechococcales	*Romeria leopoliensis* (Raciborski) Koczwara 1932	*Synechococcus leopoliensis* (Raciborski) Komárek 1970	F	(I, II); Komárek & Hauer (2013)
	Synechococcales	*Schizothrix calcicola* Gomont 1892		M	(II)
	Oscillatoriales	*Trichodesmium erythraeum* Ehrenberg ex Gomont 1892		M	(II)
	Oscillatoriales	*Trichodesmium thiebautii* Gomont ex Gomont 1892		M	(II)
	Oscillatoriales	*Tychonema bourrellyi* (J.W.G. Lund) K. Anagnostidis & J. Komárek 1988	*Oscillatoria bourrellyi* J.W.G. Lund 1955	F	(II)
	Oscillatoriales	*Tychonema tenue* (Skuja) K. Anagnostidis & J. Komárek 1988		F	(II)
Heterocytous genera	Nostocales	*Anabaena cylindrica* Lemmermann 1896	*Anabaena subcylindrica* Borge 1921	F	Whitton (2011); Guiry and Guiry (2015)
	Nostocales	*Anabaena lapponica* Borge 1913		F	(III)
	Nostocales	*Anabaena oscillarioides* Bory ex Bornet & Flahault 1888		F	(III)
	Nostocales	*Anabaenopsis arnoldii* Aptekar' 1926		F	(III)
	Nostocales	*Anabaenopsis milleri* Woronichin 1929		F	(III)
	Nostocales	*Aphanizomenon flos-aquae* Ralfs ex Bornet & Flahault 1886		F	(III)
	Nostocales	*Aphanizomenon gracile* (Lemmermann) Lemmermann 1907		F	(III)
	Nostocales	*Aphanizomenon klebahnii* Elenkin ex Pechar 2008	*Aphanizomenon flos-aquae* var. *klebahnii* Elenkin 1909	F	(III)

Nostocales	*Aphanizomenon skujae* J. Komárková-Legnerová & G. Cronberg 1992		F	(III)
Nostocales	*Calothrix parietina* Thuret ex Bornet & Flahault 1886		F	(III)
Nostocales	*Chrysosporum bergii* (Ostenfeld) E. Zapomelová, O. Skácelová, P. Pumann, R. Kopp & E. Janecek 2012	*Aphanizomenon bergii* (Ostenfeld) Komárek 1983; *Anabaena bergii* Ostenfeld 1908	B/F-high salinity	
Nostocales	*Chrysosporum ovalisporum* (Forti) E. Zapomelová, O. Skácelová, P. Pumann, R. Kopp & E. Janecek 2012	*Aphanizomenon ovalisporum* Forti 1911	F	(III)
Nostocales	*Cuspidothrix issatschenkoi* (Usachev) P. Rajaniemi, Komárek, R. Willame, P. Hrouzek, K. Kastovská, L. Hoffmann & K. Sivonen 2005	*Aphanizomenon issatschenkoi* (Usačev) Proshkina-Lavrenko 1968	F	(III)
Nostocales	*Cylindrospermopsis raciborskii* (Woloszynska) Seenayya & Subba Raju 1972	*Anabaena raciborskii* Woloszynska 1912	F	(III)
Nostocales	*Cylindrospermum stagnale* Bornet & Flahault 1888		F	(III)
Nostocales	*Dichothrix utahensis* Tilden 1898		B	(III)
Nostocales	*Dolichospermum affine* (Lemmermann) P. Wacklin, L. Hoffmann & J. Komárek 2009	*Anabaena affinis* Lemmermann 1897		
Nostocales	*Dolichospermum circinale* (Rabenhorst ex Bornet & Flahault) P. Wacklin, L. Hoffmann & J. Komárek 2009	*Anabaena circinalis* Rabenhorst ex Bornet & Flahault 1888	F	(III)
Nostocales	*Dolichospermum compactum* (Nygaard) P. Wacklin, L. Hoffmann & J. Komárek 2009	*Anabaena compacta* (Nygaard) Hickel 1985	F	(III)
Nostocales	*Dolichospermum crassum* (Lemmermann) P. Wacklin, L. Hoffmann & J. Komárek 2009	*Anabaena crassa* (Lemmermann) J. Komárková-Legnerová & G. Cronberg 1992	F	(III)

(Continued)

Table A1.1 (*Continued*)

Traditional groups	Present orders	Present name (January 2016)	Recent or common synonyms	Prevalent habitat	Main recent reference
	Nostocales	*Dolichospermum flos-aquae* (Brébisson ex Bornet & Flahault) P. Wacklin, L. Hoffmann & J. Komárek 2009	*Anabaena flos-aquae* Brébisson ex Bornet & Flauhault 1888	F	(III)
	Nostocales	*Dolichospermum lemmermannii* (P.G. Richter) P. Wacklin, L. Hoffmann & J. Komárek 2009	*Anabaena lemmermannii* P.G. Richter 1903; *Anabaena flos-aquae* f. *lemmermannii* (P.G. Richter) Canabaeus 1929; *Anabaena utermoehlii* Geitler 1925	F	(III)
	Nostocales	*Dolichospermum macrosporum* (Klebahn)) P. Wacklin, L. Hoffmann & J. Komárek 2009	*Anabaena macrospora* Klebahn 1895	F	(III)
	Nostocales	*Dolichospermum mendotae* (W. Trelease) P. Wacklin, L. Hoffmann & J. Komárek 2009	*Anabaena mendotae* W. Trelease 1889	F	(III)
	Nostocales	*Dolichospermum planctonicum* (Brunnthaler) P. Wacklin, L. Hoffmann & J. Komárek 2009	*Anabaena planctonica* Brunnthaler 1903; *Anabaena limnetica* G.M. Smith 1916	F	(III)
	Nostocales	*Dolichospermum smithii* (Komárek) P. Wacklin, L. Hoffmann & J. Komárek 2009	*Anabaena smithii* (Komárek) M. Watanabe 1992; *Anabaena solitaria* f. *smithii* Komarek 1958	F	(III)
	Nostocales	*Dolichospermum solitarium* (Klebahn) P. Wacklin, L. Hoffmann & J. Komárek 2009	*Anabaena solitaria* Klebahn 1895	F	(III)
	Nostocales	*Dolichospermum spiroides* (Klebhan) P. Wacklin, L. Hoffmann & J. Komárek 2009	*Anabaena spiroides* Klebahn 1895	F	(III)

Nostocales	*Gloeotrichia echinulata* P.G. Richter 1894	*Conferva echinulata* Smith 1804	F	
Nostocales	*Hapalosiphon hibernicus* W. West & G.S. West 1896	?	F	(III)
Nostocales	*Nodularia baltica* Komárek, M. Hübel, H. Hübel & Šmarda 1993		M/B/F-high salinity	
Nostocales	*Nodularia harveyana* Thuret ex Bornet & Flahault 1888		M/F-high salinity	(III)
Nostocales	*Nodularia sphaerocarpa* Bornet & Flahault 1888		F	(III)
Nostocales	*Nodularia spumigena* Mertens ex Bornet & Flahault 1888		M/B/F-high salinity	(III)
Nostocales	*Nostoc carneum* C. Agardh ex Bornet & Flahault 1888	*Nostoc spongiaeforme* C. Agardh ex Bornet & Flahault 1888; *Anabaena rufescens* (C. Agardh) Kirchner 1878	F	(III)
Nostocales	*Raphidiopsis brookii* P.J. Hill 1972		F	(III)
	Raphidiopsis curvata F.E. Fritsch & M.F. Rich 1929	*Raphidiopsis spiralis* F.E. Fritsch & M.F. Rich in Skuja 1949	F	(III)
Nostocales	*Raphidiopsis mediterranea* Skuja 1937[1]	*Raphidiopsis subrecta* Frémy ex Skuja 1949	F	(III)
Nostocales	*Scytonema crispum* (C. Agardh) Bornet 1889		F	(III)
Nostocales	*Sphaerospermopsis aphanizomenoides* (Forti) Zapomelová, Jezberová, Hrouzek, Hisem, Reháková & Komárková 2010	*Anabaena aphanizomenoides* Forti 1912; *Aphanizomenon aphanizomenoides* (Forti) Horecká & Komárek 1979; *Sphaerospermum aphanizomenoides* (Forti) Zapomelová, Jezberová, Hrouzek, Hisem, Reháková & Komárková 2009	F	(III)

(*Continued*)

Table A1.1 (*Continued*)

Traditional groups	Present orders	Present name (January 2016)	Recent or common synonyms	Prevalent habitat	Main recent reference
	Nostocales	*Sphaerospermopsis reniformis* (Lemmermann) Zapomelová, Jezberová, Hrouzek, Hisem, Reháková & Komárková 2010	*Anabaena reniformis* Lemmermann 1898; *Sphaerospermum reniforme* (Lemmermann) Zapomelová, Jezberová, Hrouzek, Hisem, Reháková & Komárková 2009	F/F-high salinity	(III)
	Nostocales	*Trichormus variabilis* (Kützing ex Bornet & Flahault) Komárek & Anagnostidis 1989	*Anabaena variabilis* Kützing ex Bornet & Flahault 1886	F	(III)
	Nostocales	*Umezakia natans* M. Watanabe 1987		F	(III)
	Nostocales	*Westiellopsis prolifica* Janet 1941		F	(III)

(1) *Raphidiopsis mediterranea* var. *grandis* H. Hill 1970 was suggested to coincide with non-heterocytous *Cuspidothrix issatschenkoi* (Usachev) P. Rajaniemi, Komárek, R. Willame, P. Hrouzek, K. Kastovská, L. Hoffmann & K. Sivonen 2005 (III; Hodoki *et al.*, 2013).

species cited in this handbook and a few other relevant names. The list is based and ordered (first column) on the basis of the book series Süßwasserflora von Mitteleuropa, integrated with further and most recent literature: (I), "Chroococcales," Komárek & Anagnostidis (1999); (II), "Oscillatoriales," Komárek & Anagnostidis (2005); (III) "Heterocytous genera," Komárek (2013). For a complete list of historical synonyms, see the main references, and Komárek & Hauer (2013) and Guiry and Guiry (2015). The "Present orders" are based on the work by Komárek *et al.* (2014). Author names have been further checked also consulting the Index Nominum Algarum (2015). Prevalent habitat: F, freshwater (including moist substrata); M, marine; B, brackish waters.

References

Bittencourt-Oliveira, M. do C. and Piccin-Santos, V. (2012) Genetic diversity of Brazilian cyanobacteria revealed by phylogenetic analysis. In *Genetic Diversity in Microorganisms* (ed. Mahmut Caliskan), InTech Europe University Campus STeP Ri, Rijeka, Croatia, pp. 275–290.

Brutemark, A., Engstrom-Öst, J., Vehmaa, A. and Gorokhova, E. (2015) Growth, toxicity and oxidative stress of a cultured cyanobacterium (*Dolichospermum* sp.) under different CO_2/pH and temperature conditions. *Phycological Research* **63**, 56–63.

Engene, N., Rottacker, E.C., Kastovsky, J. *et al.* (2012) *Moorea producens* gen. nov., sp. nov. and *Moorea bouillonii* comb. nov., tropical marine cyanobacteria rich in bioactive secondary metabolites. *International Journal of Systematic and Evolutionary Microbiology* **62**: 1171–1178.

Guiry, M.D., and Guiry, G.M. (2015) "AlgaeBase. World-wide electronic publication - National University of Ireland, Galway." http://www.algaebase.org.

Hodoki, Y., Ohbayashi, K., Kobayashi, Y., *et al.* (2013) Anatoxin-a-producing *Raphidiopsis mediterranea* Skuja var. *grandis* Hill is one ecotype of non-heterocytous *Cuspidothrix issatschenkoi* (Usačev) Rajaniemi *et al.* in Japanese lakes. *Harmful Algae* **21-22**, 44–53.

Index Nominum Algarum (2015) *University Herbarium*, University of California, Berkeley. Compiled by Paul Silva. Available online at http://ucjeps.berkeley.edu/CPD/.

Joosten, A.M.T. (2006) *Flora of the blue-green algae of the Netherlands I The non-filamentous species of inland waters*. pp. [1-]5-237. Utrecht: KNNV Publishing.

Komárek, J. (2013) Cyanoprokariota. part 3: Heterocytous Genera, in Süßwasserflora von Mitteleuropa, Band 19/3, Springer Spektrum.

Komárek, J., and Anagnostidis, K. (1999) Cyanoprokaryota Teil 1 (Chroococcales), in Süßwasserflora von Mitteleuropa, Band 19, Gustav Fischer Verlag, Stuttgart.

Komárek, J., and Anagnostidis, K. (2005) Cyanoprokariota. part 2: Oscillatoriales, in Süßwasserflora von Mitteleuropa, Band 19/2., Springer Spektrum.

Komárek, J. and Hauer, T. (2013) *CyanoDB.cz - On-line database of cyanobacterial genera*. - Word-wide electronic publication, Univ. of South Bohemia & Inst. of Botany AS CR, http://www.cyanodb.cz.

Komárek, J., Kaštovský, J., and Jezberová, J. (2011) Phylogenetic and taxonomic delimitation of the cyanobacterial genus *Aphanothece* and description of *Anathece* gen. nov . *European Journal of Phycology* **46**(3), 315–326.

Komárek, J., Kaštovský, J., Mareš, J., and Johansen, J.R. (2014) Taxonomic classification of cyanoprokaryotes (cyanobacterial genera) 2014, using a polyphasic approach. *Preslia* **86**, 295–335.

Komárek, J., Kopeky, J., and Cepák, V. (1999) Generic characters of the simplest cyanoprokaryotes *Cyanobium, Cyanobacterium* and *Synechococcus*. *Cryptogamie Algologie* **20**(3), 209–222.

Komárek, J. and Cronberg, G. (2001) Some Chroococcalean and Oscillatorialean Cyanoprokaryotes from Southern African lakes, ponds and pools. *Nova Hedwigia* **73**, 129–160.

Komárek, J., and Komárková-Legnerová, J. (1993) *Radiocystis fernandoi,* a new planktic cyanoprokaryotic species from tropical freshwater reservoirs. *Preslia* **65**, 355–357.

Komárek, J., and Komárková-Legnerová, J. (2007) Several rare freshwater planktic Cyanobacteria (Cyanoprokaryotes) from reservoirs in South America. *Hoehnea* **34**(1), 49–58.

Komárek, J., Komárková-Legnerová, J., Sant'Anna, C.L., *et al.* (2002) Two common *Microcystis* species (Chroococcales, Cyanobacteria) from tropical America, including *M. panniformis* sp. nov.. *Cryptogamie Algologie* **23**(2), 159–177.

Lee, E., Ryan, U.M., Monis, P., *et al.* (2014) Polyphasic identification of cyanobacterial isolates from Australia. *Water Research* **59**, 248–61.

Nguyen, L.T.T., Cronberg, G., Moestrup, O., and Daugbjerg, N. (2013) *Annamia toxica* gen. et sp. nov. (Cyanobacteria), a freshwater cyanobacterium from Vietnam that produces microcystins: ultrastructure, toxicity and molecular phylogenetics. *Phycologia* **52**(1), 25–36.

Senna, P.A.C., Delazari, A., and Sant'Anna, C.L. (1999). A new planktic cyanoprokaryotic species, *Cyanodictyon tropicalis* (Cyanophyceae) from Espírito Santo State (Southeastern), Brazil. *Algological Studies/Archiv für Hydrobiologie,* Supplement Volumes **94**, 293–303.

Soo, R.M., Skennerton, C.T., Sekiguchi, Y., *et al.* (2014) An expanded genomic representation of the phylum cyanobacteria. *Genome Biology and Evolution* **6** (5), 1031–45.

Speziale, B.J. and Dyck, L.A. (1992) *Lyngbya* infestations: comparative taxonomy of *Lyngbya wollei* comb. nov. (Cyanobacteria). *Journal of Phycology* **28**, 693–706.

Strunecký, O., Komárek, J., and Smarda, J. (2014) *Kamptonema* (Microcoleaceae, Cyanobacteria), a new genus derived from the polyphyletic *Phormidium* on the basis of combined molecular and cytomorphological markers. *Preslia)* **86**, 193–207.

Strunecký, O., Komárek, J., Johansen, J., *et al.* (2013) Molecular and morphological criteria for revision of the genus *Microcoleus* (Oscillatoriales, Cyanobacteria). *Journal of Phycology* **49**(6), 1167–1180.

Whitton, B.A. (2011) Cyanobacteria (Cyanophyta). In: *The Freshwater Algal Flora of the British Isles.* An Identification Guide to Freshwater and Terrestrial Algae. 2nd edn (eds. D.M. John, B.A. Whitton and A.J. Brook), Cambridge University Press, Cambridge, pp. 31–158.

Appendix 2

Cyanobacteria Associated With the Production of Cyanotoxins

Cécile Bernard[1], Andreas Ballot[2], Solène Thomazeau[1], Selma Maloufi[1], Ambrose Furey[3], Joanna Mankiewicz-Boczek[4,5], Barbara Pawlik-Skowrońska[6], Camilla Capelli[7], and Nico Salmaso[7]

[1] UMR 7245 MCAM CNRS-MNHN, Muséum National d'Histoire Naturelle, Paris, France
[2] Norwegian Institute for Water Research (NIVA), Oslo, Norway
[3] Mass Spectrometry Research Centre, Department of Physical Sciences, Cork Institute of Technology (CIT), Cork, Ireland
[4] European Regional Centre for Ecohydrology of the Polish Academy of Sciences, Łódź, Poland
[5] Department of Applied Ecology, Faculty of Biology and Environmental Protection, University of Łódź, Poland
[6] Department of Hydrobiology, University of Life Sciences in Lublin, Poland
[7] IASMA Research and Innovation Centre, Fondazione Edmund Mach-Istituto Agrario di S. Michele all'Adige, S. Michele all'Adige, Trento, Italy

Cyanobacteria associated with the production of cyanotoxins have been exhaustively listed by Sivonen and Jones (1999) and Carmichael (2008). In Table Appendix 2.1, we report an updated list of cyanobacterial species or genera (i) producing toxin from isolated strains or (ii) suspected to produce toxins, based on data obtained from environmental samples. The original species names assigned (or successively modified) by the indicated authors have been updated and ordered following the more recent changes in the taxonomic literature (Appendix 1).

On isolated strains, the toxin presence (highlighted **in bold**) has been verified by laboratory studies. On environmental samples, the toxin synthesis or toxicity of the species has not been verified by strain isolation, but has been only determined in the environmental samples. In these cases, the cyanobacterial species have been suggested by the authors as the probable toxin-producers, but these results should be treated with great care, until pure strains are isolated and studied.

Handbook of Cyanobacterial Monitoring and Cyanotoxin Analysis, First Edition.
Edited by Jussi Meriluoto, Lisa Spoof and Geoffrey A. Codd.

Table A2.1 *Cyanobacterial genera producing or suspected to produce cyanotoxins.* ***In bold:*** *toxin presence verified on isolated strains by laboratory studies*

Present name	Toxic species (names originally assigned by authors)	Toxin or toxic effect	Location	Identification based on isolated strains	Identification based on collection of environmental samples (suspected toxin-producers)
Anabaena cylindrica	*Anabaena sub-cylindrica* (benthic)	**MCs**	Egypt	Mohamed *et al.* 2006	
Anabaena lapponica	*Anabaena lapponica*	**CYN**	Finland	Spoof *et al.* 2006	
Anabaena sp.	*Anabaena* sp.	**Antx-a**	Japan	Park *et al.* 1993	
	Anabaena sp.	**Antx-a**	Portugal	Osswald *et al.* 2009	
	Anabaena sp.	Antx-a	Ireland		James *et al.* 1997
	Anabaena sp.	**MCs**	Scandinavia	Sivonen *et al.* 1992a (cf. *A. flos-aquae, A. circinalis* and *A. lemmermannii*)	Sivonen and Jones 1999
Anabaenopsis arnoldii	*Anabaenopsis arnoldi*	**MCs**	Saudi Arabia	Mohamed & Al Shehri 2009	
Anabaenopsis milleri	*Anabaenopsis milleri*	MCs	Greece		Lanaras & Cook 1994
Annamia toxica	*Annamia toxica*	**MCs**	Vietnam	Nguyen *et al.* 2013	
Aphanizomenon flos-aquae	*Aphanizomenon flos-aquae*	**Antx-a**	Finland	Rapala *et al.* 1993	
	Aphanizomenon flos-aquae	**Antx-a**	Portugal	Osswald *et al.* 2009	
	Aphanizomenon flos-aquae	CYN	France		Brient *et al.* 2009
	Aphanizomenon flos-aquae	MCs	France		Maatouk *et al.* 2002
Aphanizomenon gracile	*Aphanizomenon gracile*	**CYN**	Poland	Kokociński *et al.* 2013	
	Aphanizomenon gracile	**STXs**	Portugal	Pereira *et al.* 2004	
	Aphanizomenon gracile	**STXs**	France	Ledreux *et al.* 2010	
	Aphanizomenon gracile	**STXs**	Germany	Ballot *et al.* 2010a	
	Aphanizomenon gracile	**STXs (STX)**	Spain	Casero *et al.* 2014	
Aphanizomenon mendotae	*Aphanizomenon flos-aquae*	**STXs**	Portugal	Ferreira *et al.* 2001, Li R. *et al.* 2000, 2003	
Aphanizomenon sp.	*Aphanizomenon* sp.	**Antx-a**	Finland	Sivonen *et al.* 1989a	
	Aphanizomenon sp.	**Antx-a**	Portugal	Osswald *et al.* 2009	
	Aphanizomenon spp. (either *Aphan. flos-aquae* or *Aphan. gracile*)	**CYN**	Germany	Preußel *et al.* 2006; Stüken *et al.* 2009.	

	Aphanizomenon sp. (strains NH-1 and NH5) (originally classified as *Aphan. flos-aquae*)	**STXs**	USA	Ikawa *et al.* 1982; Mahmood & Carmichael 1986a; Li R. *et al.* 2000;	
	Aphanizomenon sp. (strain DC-1)	**STXs**	China	Liu *et al.* 2006	
Aphanocapsa cumulus	*Aphanocapsa cumulus*	**MCs**	Brazil	Domingos *et al.* 1999	
Arthrospira fusiformis	*Arthrospira fusiformis*	**MCs & Antx-a**	Kenya	Ballot *et al.* 2004, 2005	Ballot *et al.* 2004, 2005
	Arthrospira fusiformis	Toxic	Tanzania		Lugomela *et al.* 2006
Blennothrix lyngbyacea	*Hydrocoleum lyngbyaceum* (marine strain)	Homo-Antx-a	France (New Caledonia)		Méjean *et al.* 2010
Calothrix parietina	*Calothrix parietina*	**MCs**	Saudi Arabia	Mohamed and Al Shehri 2007;	
	Calothrix parietina	**Neurotoxic**	Egypt	Mohamed *et al.* 2006	
Chrysosporum ovalisporum	*Aphanizomenon ovalisporum*	**CYN**	USA	Yilmaz *et al.* 2008	
	Aphanizomenon ovalisporum	**CYN**	Spain	Stüken *et al.* 2009	Quesada *et al.* 2006
	Aphanizomenon ovalisporum previous identification as *Anabaena bergii* (strains ANA283A and ANA366B)	**CYN**	Australia	Stüken *et al.* 2009	
	Aphanizomenon ovalisporum	**CYN**	Australia	Stuken *et al.* 2009	Shaw *et al.* 1999
	Aphanizomenon ovalisporum	**CYN**	Israel	Banker *et al.* 1997; Stüken *et al.* 2009	
	Chrysosporum ovalisporum	CYN	Turkey		Akcaalan *et al.* 2014b
Cuspidothrix issatschenkoi	*Aphanizomenon issatschenkoi*	**Antx-a**	New Zealand	Wood *et al.* 2007a; Selwood *et al.* 2007	
	Aphanizomenon issatschenkoi	**Antx-a**	Germany	Ballot *et al.* 2010b	

(*Continued*)

Table A2.1 (*Continued*)

Present name	Toxic species (names originally assigned by authors)	Toxin or toxic effect	Location	Identification based on isolated strains	Identification based on collection of environmental samples (suspected toxin-producers)
	Aphanizomenon issatschenkoi (previous identification as *Aphan. flos-aquae* and *Aphanizomenon* sp.)	**STXs**	Portugal	Pereira *et al.* 2000; Dias *et al.* 2002 and Li R. *et al.* 2003	
	Aphanizomenon issatschenkoi	**STXs**	Portugal	Nogueira *et al.* 2004	
Cuspidothrix issatschenkoi with no heterocytes	*Raphidiopsis mediterranea* var. *grandis* (several strains)	**Antx-a**	Japan	Hodoki *et al.* 2013	
	Raphidiopsis mediterranea (strain LBRI48)	**Antx-a, Homo-Antx-a**	Japan	Namikoshi *et al.* 2003	
	Raphidiopsis mediterranea var. *grandis* (strain LBRI48)	**Homo-Antx-a**	Japan	Watanabe *et al.* 2003	
Cylindrospermopsis raciborskii	*Cylindrospermopsis raciborskii*	**CYN**	Australia	Hawkins *et al.* 1985, 1997, 2001; Ohtani *et al.* 1992; Saker *et al.* 1999; Schembri *et al.* 2001	
	Cylindrospermopsis raciborskii	**CYN & deoxy-CYN**	Thailand	Li R. *et al.* 2001a	
	Cylindrospermopsis raciborskii	**CYN *like***	France	Bernard *et al.* 2003	
	Cylindrospermopsis raciborskii	**deoxy-CYN**	Japan	Zarenezhad *et al.* 2012	
	Cylindrospermopsis raciborskii	**Hepatotoxic**	Portugal	Saker *et al.* 2003	
	Cylindrospermopsis raciborskii	**Hepatotoxic**	Egypt	Mohamed 2007	

	Cylindrospermopsis raciborskii	**STXs**	Brazil	Lagos *et al*. 1999; Molica *et al*. 2002; Castro *et al*. 2004; Pomati *et al*. 2004	
Cylindrospermum sp.	*Cylindrospermum* sp.	**Hepatotoxins, and Antx-a (small amount)**	Finland	Sivonen *et al*. 1989a	
	Cylindrospermum stagnale	**STXs**	Brazil	Borges *et al*. 2015	
Dolichospermum circinale	*Anabaena circinalis*	**Antx-a**	Finland	Sivonen *et al*. 1989a,	
	Anabaena circinalis	**MCs**	France	Vezie *et al*. 1998	
	Anabaena circinalis	**STXs**	Australia	Humpage *et al*. 1994; Negri *et al*. 1997; Testé *et al*. 2002; Pomati *et al*. 2006	Negri *et al*. 1995; Jones and Negri, 1997
Dolichospermum flos-aquae	*Anabaena flos-aquae*	**Antx-a**	Finland	Sivonen *et al*. 1989a; Rapala *et al*. 1993;	
	Anabaena flos-aquae	**Antx-a**	Japan	Park *et al*. 1993;	
	Anabaena flos-aquae	**Antx-a**	Canada	Devlin *et al*. 1977; Carmichael *et al*. 1979; Lakshmana Rao *et al*. 2002;	
	Anabaena flos-aquae	**Antx-a(S)**	Canada	Carmichael and Gorham 1978; Mahmood and Carmichael 1986b, 1987; Matsunaga et al, 1989.	
	Anabaena flos-aquae	Antx-a(S)	USA		Mahmood *et al*. 1988
	Anabaena flos-aquae	**MCs**	Canada	Harada *et al*. 1991	
Dolichospermum lemmermannii	*Anabaena lemmermannii*	**Antx-a(S)**	Denmark	Henriksen *et al*. 1997; Onodera *et al*. 1997a	
	Anabaena flos-aquae f. lemmermannii (NIVA-CYA 83/1)	**MCs**	Norway	Sivonen *et al*. 1992b	
	Anabaena lemmermannii	STXs	Finland		Rapala *et al*. 2005
Dolichospermum macrosporum	*Anabaena macrospora*	**Antx-a**	Japan	Park *et al*. 1993	

(*Continued*)

Table A2.1 (*Continued*)

Present name	Toxic species (names originally assigned by authors)	Toxin or toxic effect	Location	Identification based on isolated strains	Identification based on collection of environmental samples (suspected toxin-producers)
Dolichospermum mendotae	*Anabaena mendotae*	**Antx-a**	Finland	Rapala *et al.* 1993	
	Dolichospermum mendotae	CYN	Turkey		Akcaalan *et al.* 2014b
Dolichospermum planctonicum	*Anabaena planctonica*	**Antx-a**	Japan	Park *et al.* 1993	Brient *et al.* 2009
	Anabaena planctonica	CYN	France		
Dolichospermum spiroides	*Anabaena spiroides*	**Antx-a**	Japan	Park *et al.* 1993	
	Anabaena spiroides	**Antx-a(s)**	Brazil	Molica *et al.* 2005	Montserrat *et al.* 2001
Fischerella sp.	*Fischerella* sp.	**MCs**	Brazil	Fiore *et al.* 2009	
	Fischerella sp.	**MCs**	Australia	Cirés *et al.* 2014	
Geitlerinema sp.	*Geitlerinema* sp. (corals marine strains)	**MCs**	USA	Myers *et al.* 2007; Richardson *et al.* 2007; Gantar *et al.* 2009	
	Geitlerinema amphibium	**STXs**	Brazil	Borges *et al.* 2015	
	Geitlerinema lemmermannii	**STXs**	Brazil	Borges *et al.* 2015	
Hapalosiphon hibernicus	*Hapalosiphon hibernicus* (soil extract strain)	**MCs**	USA	Prinsep *et al.* 1992	
Hydrocoleum sp.	*Hydrocoleum* sp.	STXs, neurotoxins and CTX-like compound	New Caledonia		Laurent *et al.* 2008
Leibleinia gracilis	*Lyngbya gracilis*	Debromoaplysiatoxin			Mynderse *et al.* 1977
Leptolyngbya boryana	*Plectonema boryanum* (benthic)	**MCs**	Egypt	Mohamed *et al.* 2006	
Leptolyngbya sp.	*Leptolyngbya sp.* (corals marine strains)	**MCs**	USA	Myers *et al.* 2007; Richardson *et al.* 2007; Gantar *et al.* 2009	

	Leptolyngbya sp.	**MCs**	Brazil	Furtado *et al.* 2009	
Leptolyngbya tenuis	*Phormidium tenue*	**Neurotoxicity on mouse**	Egypt	Mohamed *et al.* 2006	
Limnothrix redekei (?)	*Limnothrix redekei* (?) (molecular identification; originally identified as *Planktothrix* sp. FP1)	**STXs**	Italy	Pomati *et al.* 2000; Gkelis *et al.* 2005	
Limnothrix sp.	*Limnothrix* sp. (morphological identification as *Geitlerinema unigranulatum*)	**Limnothrixin (?)**	Australia	Bernard *et al.* 2011; Humpage *et al.* 2012	
	Limnothrix sp.	**MCs**	Brazil	Furtado *et al.* 2009	
Lyngbya majuscula	*Lyngbya majuscula*	Aplysiatoxin & debromoaplysiatoxin	Japan		Fujiki *et al.* 1985
	Lyngbya majuscula	Lyngbyatoxin A	USA		Cardellina *et al.* 1979; Aimi *et al.* 1990
	Lyngbya majuscula	Lyngbyatoxin-A & debromoaplysiatoxin	Australia		Osborne *et al.* 2001
	Lyngbya majuscula	Lyngbyatoxins A,B,C	USA		Aimi *et al.* 1990
Lyngbya robusta	*Lyngbya robusta*	STXs, CYN (traces)	Guatemala		Rejmánková *et al.* 2011
Lyngbya wollei	*Lyngbya wollei*	**CYN & deoxy-CYN**	Australia	Seifert *et al.* 2007	
	Lyngbya wollei	**STXs**	USA	Carmichael *et al.* 1997	Onodera *et al.* 1997b; Foss *et al.* 2012
	Lyngbya wollei	STXs	Canada		Lajeunesse *et al.* 2012
Merismopedia sp.	*Merismopedia* sp.	**MCs**	Brazil	Furtado *et al.* 2009	
Microcoleus autumnalis	*Phormidium autumnale* (benthic)	**Antxa**	New Zeland	Harland *et al.* 2013	
	Phormidium autumnale (benthic)	Antxa, Homo-Antx-a	New Zeland		Wood *et al.* 2007b; Heath *et al.* 2011
Microcystis aeruginosa	*Microcystis aeruginosa*	**Antx-a**	Japan	Park *et al.* 1993	
	Microcystis aeruginosa	**MCs**	Worldwide (see Chapter 11)	Several, e.g. Park *et al.* 1993; Vezie *et al.* 1998; Sivonen & Jones 1999; Via-Ordorika *et al.* 2004; Chorus 2012; Nguyen *et al.* 2012	Several, e.g. Sivonen & Jones 1999; Messineo *et al.* 2009
Microcystis bengalensis	*Microcystis bengalensis*	**MCs**	Vietnam	Nguyen *et al.* 2012	

(*Continued*)

Table A2.1 (*Continued*)

Present name	Toxic species (names originally assigned by authors)	Toxin or toxic effect	Location	Identification based on isolated strains	Identification based on collection of environmental samples (suspected toxin-producers)
Microcystis botrys	*Microcystis botrys*	**MCs**	Europe	Sivonen and Jones 1999; Via-Ordorika *et al.* 2004	
Microcystis flos-aquae	*Microcystis flos-aquae*	**MCs**	Worldwide	Sivonen and Jones 1999; Via-Ordorika *et al.* 2004	
Microcystis ichthyoblabe	*Microcystis ichthyoblabe*	**MCs**	Europe	Via-Ordorika *et al.* 2004	
	Microcystis ichthyoblabe	**MCs**	Vietnam	Nguyen *et al.* 2012	
Microcystis novacekii	*Microcystis novacekii*	**MCs**	Strain UAM250	Li H. *et al.* 2009	
Microcystis panniformis	*Microcystis panniformis*	**MCs**	Italy	Via-Ordorika *et al.* 2004	
Microcystis pseudofilamentosa	*Microcystis pseudofilamentosa*	**MCs**	Vietnam	Nguyen *et al.* 2012	
Microcystis ramosa	*Microcystis ramosa*	**MCs**	Vietnam	Nguyen *et al.* 2012	
Microcystis sp.	*Microcystis* sp.	**Antx-a**	Portugal	Osswald *et al.* 2009	
Microcystis viridis	*Microcystis viridis*	**MCs**	Europe	Via-Ordorika *et al.* 2004	
	Microcystis viridis	**MCs**	Japan	Park *et al.* 1993	
Moorea producens	*Moorea producens* (often incorrectly cited as *Lyngbya majuscula* or *L. sordida*) (marine benthic)	**Lyngbyatoxins**	widely distributed between the Tropic of Cancer and the Tropic of Capricorn	Engene *et al.* 2012	
Nodularia sphaerocarpa	*Nodularia sphaerocarpa* (PCC7804) (freshwater thermal source)	**Nodularin, [L-Har2] nodularin**	France	Beattie *et al.* 2000;	

Nodularia spumigena	*Nodularia spumigena*	**Nodularin**	UK	Twist and Codd 1997	
	Nodularia spumigena	**Nodularin**	Baltic Sea	Sivonen *et al.* 1989b	
	Nodularia spumigena	**Nodularin**	Australia	McGregor *et al.* 2012	Baker and Humpage 1994; Van Buynder *et al.* 2001
	Nodularia spumigena	**Nodularin**	New Zealand	Carmichael *et al.* 1988	
	Nodularia spumigena	**Nodularin**	Turkey	Akcaalan *et al.* 2009	
	Nodularia spumigena	Nodularin	Gulf of Gdańsk		Mazur and Plinski 2003
Nostoc carneum	*Nostoc spongiaeforme* (benthic)	**MC**	Egypt	Mohamed *et al.* 2006	
Nostoc sp.	*Nostoc* sp. (freshwater)	**MC**	Finland	Sivonen *et al.* 1990b	
	Nostoc sp. (lichen associate)	**MC**	Finland	Oksanen *et al.* 2004	
	Nostoc sp. (brackish water)	**MC**	UK	Beattie *et al.* 1998	
	Nostoc sp. (benthic)	MC	Antarctica		Wood *et al.* 2008
	Nostoc sp. (symbiont strains)	**Nodularin, [L-Har2] nodularin**	Australia	Gehringer *et al.* 2012	
Oscillatoria limosa	*Oscillatoria limosa* (benthic)	**MC**	Saudi Arabia	Mohamed 2008	
Oscillatoria sp.	*Oscillatoria* sp.	**Antx-a**	Finland	Sivonen *et al.* 1989a	
	Oscillatoria sp.	**Antx-a**	Portugal	Osswald *et al.* 2009	
	Oscillatoria sp. (benthic)	**Antx-a**	Scotland	Edwards *et al.* 1992	
	Oscillatoria sp.	**Antx-a**	Several strains from Finland, Netherland, USA	Araóz *et al.* 2005	
	« *Oscillatoria* sp. » (benthic mats)	Antx-a	Ireland		James *et al.* 1997
	Oscillatoria-like (benthic)	Antx-a	New Zealand		Hamill *et al.* 2001
	« *Oscillatoria* sp. » (strains PCC-6407, -6506, -7926, -10702)	**CYN & 7-epi-CYN**	Europe, USA	Mazmouz *et al.* 2010	
	Oscillatoria sp. (benthic)	**Homo-Antx-a, Antx-a**	France	Cadel-Six *et al.* 2007	

(*Continued*)

Table A2.1 (*Continued*)

Present name	Toxic species (names originally assigned by authors)	Toxin or toxic effect	Location	Identification based on isolated strains	Identification based on collection of environmental samples (suspected toxin-producers)
	Oscillatoria sp. (benthic)	**Homo-Antx-a, Antx-a, CYN & 7-deoxy-CYN & 7-epi-CYN**	France	Méjan *et al.* 2009; Mazmouz *et al.* 2010	
Oscillatoria tenuis	*Oscillatoria tenuis*	**MCs**	Egypt	Mohamed *et al.* 1999; Brittain *et al.* 2000	
Oscillatoriaceae	Oscillatoriaceae (benthic)	Antx-a	USA		Puschner *et al.* 2008
Phormidium aerugineo-caeruleum	*Lyngbya aerugineo-coerulea*	**Neuro- & hepatotoxicity**	Strain PACC8601	Teneva *et al.* 2003	
Phormidium cf. *uncinatum*	*Phormidium* cf. *uncinatum*	**Antx-a, Homo-Antx-a**	New Zeland	Harland *et al.* 2014	
Phormidium corium	*Phormidium corium* (benthic)	**MCs**	Egypt	Mohamed *et al.* 2006	
Phormidium favosum	*Phormidium favosum* (benthic)	**Antx-a**	France	Gugger *et al.* 2005	
Phormidium formosum (*Kamptonema formosum*)	*Phormidium formosum*	**Homo-Antx-a**	Norway	Skulberg *et al.* 1992	
	Oscillatoria formosa	**Homo-Antx-a**	Sweden	Araóz *et al.* 2005	
Phormidium sp.	*Phormidium* sp. (benthic)	**Antx-a, Homo-Antx-a**	France	Cadel-Six *et al.* 2007	
	Phormidium sp. (benthic)	Antx-a, Homo-Antx-a	Netherland		Faassen *et al.* 2012
	Phormidium sp. (marine strain)	**MCs**	USA	Gantar *et al.* 2009	
	Phormidium sp. (?)	**MCs**	USA	Izaguirre *et al.* 2007	
Phormidium uncinatum		**STXs**	Brazil	Borges *et al.* 2015	
Planktothrix agardhii	*Oscillatoria agardhii*	**Hepatotoxins, and Antx-a** (traces)	Finland	Sivonen *et al.* 1989a	
	Planktothrix agardhii	**MCs**	France	Yéprémian *et al.* 2007	Briand *et al.* 2002

	Oscillatoria agardhii	**MCs**	Finland	Sivonen 1990a; Luukkainen *et al.* 1993	
	Oscillatoria agardhii	**MCs**	Norway (strain CYA-29)	Meriluoto *et al.* 1989	
	Planktothrix agardhii	**MCs**	Strain NIES 595	Akçaalan *et al.* 2006	
	Planktothrix agardhii	MCs	Denmark		Henriksen *et al.* 1996
	Planktothrix agardhii	MCs	Germany		Fastner *et al.* 1999; Dolman *et al.* 2012
	Planktothrix agardhii	MCs	Poland		Pawlik-Skowrońska *et al.* 2008, Kokociński *et al.* 2011
	Planktothrix agardhii	MCs	United Kingdom		Akçaalan *et al.* 2006
Planktothrix rubescens	*Planktothrix rubescens*	Antx-a	Italy		Viaggiu *et al.* 2004
	Planktothrix rubescens	**MCs**	France	Briand *et al.* 2005	
	Planktothrix rubescens	**MCs**	Germany	Dadheech *et al.* 2014	Fastner *et al.* 1999
	Planktothrix rubescens	**MCs**	Hungary	Vasas *et al.* 2013	
	Planktothrix rubescens	**MCs**	Italy	Shams *et al.* 2014	Naselli-Flores *et al.* 2007; Messineo *et al.* 2009; Manganelli *et al.* 2010; Salmaso *et al.* 2013, 2014
	Planktothrix rubescens	**MCs**	Turkey	Akcaalan *et al.* 2006	Akcaalan *et al.* 2014a
	Planktothrix rubescens	MCs	Spain		Barco *et al.* 2004
Planktothrix sp.	*Planktothrix* sp.	MCs	Germany		Ernst *et al.* 2001
Planktothrix sp. (?)	Oscillatoriales (cf. *Planktothrix* (?) (benthic)	**MCs**	New Zealand	Wood *et al.* 2010	Wood *et al.* 2010
Pseudanabaena limnetica	*Oscillatoria limnetica*	**Antx-a**	Portugal	Osswald *et al.* 2009	
Pseudanabaena sp.	*Pseudanabaena* sp. (marine strain)	**MCs**	USA	Gantar *et al.* 2009	
Radiocystis fernandoi	*Radiocystis fernandoi*	**MCs**	Brazil	Vieira *et al.* 2003; Pereira *et al.* 2012	

(*Continued*)

Table A2.1 (*Continued*)

Present name	Toxic species (names originally assigned by authors)	Toxin or toxic effect	Location	Identification based on isolated strains	Identification based on collection of environmental samples (suspected toxin-producers)
Raphidiopsis brookii	*Raphidiopsis brookii*	**STXs**	Brazil	Yunes *et al.* 2009; Soto-Liebe *et al.* 2012	
Raphidiopsis curvata	*Raphidiopsis curvata*	**CYN & deoxy-CYN**	China	Li R. *et al.* 2001b	
Raphidiopsis mediterranea	*Raphidiopsis mediterranea*	**CYN & deoxy-CYN**	Australia	McGregor *et al.* 2011	
	Raphidiopsis mediterranea	**Neurotoxic**	Egypt	Mohamed 2007	
Scytonema cf. *crispum*	*Scytonema* cf. *crispum*	**STX**	New-Zealand	Smith *et al.* 2011, 2012	
Spirulina sp.	*Spirulina* sp. (marine strain)	**MCs**	USA	Gantar *et al.* 2009	
Synechococcus lividus	*Synechococcus lividus*	**MCs**	Saudi Arabia (hot springs)	Mohamed 2008	
Synechococcus sp.	*Synechococcus* sp. (marine strain)	**MCs**	USA	Gantar *et al.* 2009	
	Synechococcus sp.	**MCs**	Brazil	Furtado *et al.* 2009	
Trichodesmium erythraeum	*Trichodesmium erythraeum*	CTX-like	Australia		Endean *et al.* 1993
	Trichodesmium erythraeum	MCs	Canary Islands (Spain)		Ramos *et al.* 2005
	Trichodesmium erythraeum	PLTX & 42-OH-PLTX	New Caledonia		Kerbrat *et al.* 2011
	Trichodesmium erythraeum	STX and MC (low concentrations)	Brazil		Proença *et al.* 2009
Trichodesmium spp.	*Trichodesmium thiebautii* and *T. erythraeum*	Neurotoxicity	Virgin Islands		Hawser *et al.* 1991
Trichodesmium thiebautii	*Trichodesmium thiebautii*	Trichotoxin	Gulf of Mexico		Schock *et al.* 2011

Trichormus variabilis	*Anabaena variabilis* (benthic)	**MCs**	Egypt	Mohamed *et al*. 2006	
Tychonema bourrellyi	*Tychonema bourrellyi*	**Antx-a**	Italy, Norway	Shams *et al*. 2015	
Umezakia natans	*Umezakia natans*	**CYN**	Japan	Harada *et al*. 1994; Terao *et al*. 1994	
Woronichinia naegeliana	*Woronichinia naegeliana*	MCs	Norway		Willame *et al*. 2005
	Woronichinia naegeliana	Toxic (mouse bioassay)			Skulberg *et al*. 1994

Table A2.2 *Cyanotoxins identified in different cyanobacterial assemblages*

Toxin	Location	Environmental samples (with cyanobacterial assemblages)
Antx-a	Argentina	Ruiz *et al*. 2013
Antx-a	China	Li *et al*. 2010
Antx-a	Germany	Bumke-Vogt *et al*. 1999
Antx-a	Italy	Bruno *et al*. 1994; Messineo *et al*. 2009
Antx-a	Kenya	Krienitz *et al*. 2003
Antx-a	Poland	Pawlik-Skowronska *et al*. 2004, 2012, 2013
Antx-a	South Korea	Park *et al*. 1998
Antx-a	Spain	Carrasco *et al*. 2007
Antx-a	USA	Bellinger and Hagerthey 2010
CYN	Australia	Everson *et al*. 2011
CYN	Czech Republic	Bláhová *et al*. 2008, 2009; Jančula et al. 2014
CYN	France	Brient *et al*. 2009
CYN	Germany	Fastner *et al*. 2007; Wiedner *et al*. 2008
CYN	Greece	Gkelis *et al*. 2014
CYN	Italy	Messineo *et al*. 2009
CYN	Mexico	Berry and Lind 2010
CYN	Poland	Kokociński *et al*. 2009
CYN	USA	Bellinger *et al*. 2010
Debromoaplysiatoxin, oscillatoxin A	Atoll of Marakei (Pacific Ocean)	Mynderse and Moore 1978
MCs	Antarctica	Jungblut *et al*. 2006
MCs	Argentina	Ruiz *et al*. 2013
MCs	Australia	Dasey *et al*. 2005
MCs	Bangladesh	Welker *et al*. 2005
MCs	Belgium, Luxembourg	Willame *et al*. 2005
MCs	Brazil	Costa *et al*. 2006; Sotero-Santos *et al*., 2008
MCs	Bulgaria	Pavlova *et al*. 2014
MCs	China	Li Z. *et al*. 2010; Li H. *et al*. 2011; Ueno *et al*. 1996
MCs	Czech Republic	Jancula et al. 2014; Blaha and Marsalek 2003; Bláhová *et al*. 2008
MCs	Denmark	Kaas and Henriksen, 2000
MCs	Greece	Gkelis *et al*. 2005; Gkelis and Zaoutsos 2014
MCs	Ireland	Mooney *et al*. 2011
MCs	Italy	Messineo *et al*. 2009
MCs	Kenya	Krienitz *et al*. 2003
MCs	Morocco	Oudra *et al*. 2009
MCs	New Zealand	Wood *et al*. 2006
MCs	Poland	Pawlik-Skowronska *et al*. 2004, 2012, 2013; Mankiewicz *et al*. 2005
MCs	Qatar (desert crusts)	Metcalf *et al*. 2012
MCs	Russia	Belykh *et al*. 2011
MCs	South Korea	Park *et al*. 1998
MCs	Spain	Aboal and Puig 2005

Table A2.2 (*Continued*)

Toxin	Location	Environmental samples (with cyanobacterial assemblages)
MCs	Sweden	Cronberg *et al.* 1999
MCs	Swiss	Mez *et al.* 1997
MCs	Tunisia	El Herry *et al.* 2008
MCs	USA	Bellinger *et al.* 2010
MCs/NODs	Antarctica	Hitzfeld *et al.* 2000
NODs	South Africa	Harding *et al.* 1995
STX	Greece	Gkelis and Zaoutsos 2014
STX	USA	Bellinger and Hagerthey 2010
STXs	Czech Republic	Jancula *et al.* 2014
STXs	Denmarkrk	Kaas and Henriksen, 2000
STXs	Brazil	Costa *et al.* 2006
STXs	Mexico	Berry and Lind 2010

Finally, when water bodies are dominated by different cyanobacterial species, i.e. when it is not possible to identify a clear dominant taxon, the identification of cyanotoxins-producers can be very difficult, if not impractical. A selected list of papers reporting the presence of different cyanotoxins measured in samples containing cyanobacterial assemblages is reported in Table Appendix 2.2.

The geographic distribution of toxins can be considered only coarsely indicative, being based on the selected papers included in this review and on a biased selection of environments where the investigations were carried out.

References concerning BMAA production are not listed. For other bioactive molecules produced by various cyanobacteria see the review published by Dixit and Suseela (2013). For more data concerning environmental benthic cyanobacteria mats and the associated toxins, see the review published by Quiblier *et al.* (2013).

MCs: microcystins. Antx-a: anatoxin-a. STXs: STX (saxitoxin) and analogues ("Paralytic Shellfish Toxins", e.g. gonyautoxins and derivatives). Homo-Antx-a: homo-anatoxin-a. CYN: cylindrospermopsin. CTX: ciguatoxins. PLTX: palytoxin and derivates.

References

Aboal, M., and Puig, M.-A. (2005) Intracellular and dissolved microcystin in reservoirs of the River Segura basin, Murcia, SE Spain. *Toxicon* **45** (4), 509–18.

Aimi, N., Odaka, H., Sakai, S., *et al.* (1990) Lyngbyatoxins B and C, Two new Irritants from *Lyngbya majuscula*. *Journal of Natural Products* **53** (6), 1593–1596.

Akcaalan, R., Köker, L., Gürevin, C. and Albay, M. (2014a). *Planktothrix rubescens*: a perennial presence and toxicity in Lake Sapanca. *Turkish Journal of Botany* **38** (4), 782–9.

Akcaalan, R., Köker, L., Oğuz, A., *et al.* (2014b) First report of cylindrospermopsin production by two cyanobacteria (*Dolichospermum mendotae* and *Chrysosporum ovalisporum*) in Lake Iznik, Turkey. *Toxins (Basel)* **6** (11), 3173–86.

Akcaalan, R., Mazur-Marzec, H., Zalewska, A. and Albay, M. (2009) Phenotypic and toxicological characterization of toxic *Nodularia spumigena* from a freshwater lake in Turkey. *Harmful Algae* **8** (2), 273–278.

Akcaalan, R., Young, F.M., Metcalf, J.S., *et al.* (2006) Microcystin analysis in single filaments of *Planktothrix* spp. in laboratory cultures and environmental blooms. *Water Research* **40** (8), 1583–1590.

Araoz, R. (2005) Neurotoxins in axenic oscillatorian cyanobacteria: coexistence of anatoxin-a and homoanatoxin-a determined by ligand-binding assay and GC/MS. *Microbiology* **151** (4), 1263–1273.

Baker, P. and Humpage, A. (1994) Toxicity associated with commonly occurring cyanobacteria in surface waters of the Murray-Darling Basin, Australia. *Marine and Freshwater Research* **45** (5), 773.

Ballot, A. (2004) Cyanobacteria and cyanobacterial toxins in three alkaline Rift Valley lakes of Kenya-Lakes Bogoria, Nakuru and Elmenteita. *Journal of Plankton Research* **26** (8), 925–935.

Ballot, A., Fastner, J., and Wiedner, C. (2010a) Paralytic shellfish poisoning toxin-producing cyanobacterium *Aphanizomenon gracile* in northeast Germany. *Applied and Environmental Microbiology* **76** (4), 1173–80.

Ballot, A., Fastner, J., Lentz, M., and Wiedner, C. (2010b) First report of anatoxin-a-producing cyanobacterium *Aphanizomenon issatschenkoi* in northeastern Germany. *Toxicon* **56** (6), 964–71.

Ballot, A., Krienitz, L., Kotut, K., *et al.* (2005) Cyanobacteria and cyanobacterial toxins in the alkaline crater lakes Sonachi and Simbi, Kenya. *Harmful Algae* **4** (1), 139–150.

Banker, R., Carmeli, S., Hadas, O. *et al.* (1997) Identification of cylindrospermopsin in *Aphanizomenon ovalisporum* (cyanophyceae) isolated from Lake Kinneret, Israel. *Journal of Phycology* **33** (4), 613–616.

Barco, M., Flores, C., Rivera, J., and Caixach, J. (2004) Determination of microcystin variants and related peptides present in a water bloom of *Planktothrix (Oscillatoria) rubescens* in a Spanish drinking water reservoir by LC/ESI-MS. *Toxicon* **44** (8), 881–6.

Beattie, K.A., Kaya, K., and Codd, G.A. (2000) The cyanobacterium *Nodularia* PCC 7804, of freshwater origin, produces [L-Har2]nodularin. *Phytochemistry* **54** (1), 57–61.

Beattie, K.A., Kaya, K., Sano, T., and Codd, G.A. (1998) Three dehydrobutyrine-containing microcystins from *Nostoc*. *Phytochemistry* **47** (7), 1289–1292.

Bellinger, B.J., and Hagerthey, S.E. (2010) Presence and diversity of algal toxins in subtropical peatland periphyton: the Florida Everglades, USA. *Journal of Phycology* **46** (4), 674–678.

Belykh, O.I., Sorokovikova, E.G., Fedorova, G.A., *et al.* (2011) Presence and genetic diversity of microcystin-producing cyanobacteria (*Anabaena* and *Microcystis*) in Lake Kotokel (Russia, Lake Baikal Region). *Hydrobiologia* **671** (1), 241–252.

Bernard, C., Froscio, S., Campbell, R., *et al.* (2011) Novel toxic effects associated with a tropical *Limnothrix/Geitlerinema*-like cyanobacterium. *Environmental Toxicology* **26** (3), 260–70.

Bernard, C., Harvey, M., Briand, J.F., *et al.* (2003) Toxicological comparison of diverse *Cylindrospermopsis raciborskii* strains: Evidence of liver damage caused by a French *C. raciborskii* strain. *Environmental Toxicology* **18** (3), 176–186.

Berry, J.P., and Lind, O. (2010) First evidence of "paralytic shellfish toxins" and cylindrospermopsin in a Mexican freshwater system, Lago Catemaco, and apparent bioaccumulation of the toxins in "tegogolo" snails (Pomacea patula catemacensis). *Toxicon* **55** (5), 930–8.

Bláha, L., and Maršálek, B. (2003) Contamination of drinking water in the Czech Republic by microcystins. *Archiv für Hydrobiologie* **158** (3), 421–429.

Bláhová, L., Babica, P., Adamovský, O., *et al.* (2008) Analyses of cyanobacterial toxins (microcystins, cylindrospermopsin) in the reservoirs of the Czech Republic and evaluation of health risks. *Environmental Chemistry Letters* **6** (4), 223–227.

Bláhová, L., Oravec, M., Maršálek, B., *et al.* (2009) The first occurrence of the cyanobacterial alkaloid toxin cylindrospermopsin in the Czech Republic as determined by immunochemical and LC/MS methods. *Toxicon* **53** (5), 519–524.

Borges, H.L.F., Branco, L.H.Z., Martins, M.D. *et al.* (2015) Cyanotoxin production and phylogeny of benthic cyanobacterial strains isolated from the northeast of Brazil. *Harmful Algae* **43**, 46–57.

Briand, J.-F., Jacquet, S., Flinois, C., *et al.* (2005) Variations in the microcystin production of *Planktothrix rubescens* (cyanobacteria) assessed from a four-year survey of Lac du Bourget (France) and from laboratory experiments. *Microbial Ecology* **50** (3), 418–28.

Briand, J.F., Robillot, C., Quiblier-Lloberas, C., and Bernard, C. (2002) A perennial bloom of *Planktothrix agardhii* (Cyanobacteria) in a shallow eutrophic French lake: limnological and microcystin production studies. *Archiv für Hydrobiologie* **153** (4), 605–622.

Brient, L., Lengronne, M., Bormans, M. and Fastner, J. (2009) First occurrence of cylindrospermopsin in freshwater in France. *Environmental Toxicology* **24** (4), 415–20.

Brittain, S., Mohamed, Z.A., Wang, J., *et al.* (2000) Isolation and characterization of microcystins from a River Nile strain of *Oscillatoria tenuis* Agardh ex Gomont. *Toxicon* **38** (12), 1759–1771.

Bruno, M., Barbini, D.A., Pierdominici, E., *et al.* (1994) Anatoxin-a and a previously unknown toxin in *Anabaena planctonica* from blooms found in Lake Mulargia (Italy). *Toxicon* **32** (3), 369–373.

Bumke-Vogt, C., Mailahn, W. and Chorus, I. (1999) Anatoxin-a and neurotoxic cyanobacteria in German lakes and reservoirs. *Environmental Toxicology* **14** (1), 117–125.

Cadel-Six, S., Peyraud-Thomas, C., Brient, L., *et al.* (2007) Different genotypes of anatoxin-producing cyanobacteria coexist in the Tarn River, France. *Applied and Environmental Microbiology* **73** (23), 7605–14.

Cardellina, J., Marner, F., and Moore, R. (1979) Seaweed dermatitis: structure of lyngbyatoxin A. *Science* **204** (4389), 193–195.

Carmichael W.W. and Gorham P.R. (1978) Anatoxins from clones of *Anabaena flos-aquae* isolated from lakes of western Canada. *Verhandlungen des Internationalen Verein Limnologie* **21**, 285–295.

Carmichael, W., Evans, W., Yin, Q., *et al.* (1997) Evidence for paralytic shellfish poisons in the freshwater cyanobacterium *Lyngbya wollei* (Farlow ex Gomont) comb. nov. *Applied and Environmental Microbiology* **63** (8), 3104–3110.

Carmichael, W.W., 2008. A world overview — One-hundred-twenty-seven years of research on toxic cyanobacteria — Where do we go from here?, In: Hudnell, K.H. (Ed.), Proceedings of the Interagency, International Symposium on Cyanobacterial Harmful Algal Blooms, pp. 99–114.

Carmichael, W.W., Biggs, D.F., and Peterson, M.A. (1979) Pharmacology of Anatoxin-a, produced by the freshwater cyanophyte *Anabaena flos-aquae* NRC-44-1. *Toxicon* **17** (3), 229–236.

Carmichael, W.W., Eschedor, J.T., Patterson, G.M., and Moore, R.E. (1988) Toxicity and partial structure of a hepatotoxic peptide produced by the cyanobacterium *Nodularia spumigena* Mertens emend. L575 from New Zealand. *Applied and Environmental Microbiology* **54** (9), 2257–2263.

Carrasco, D., Moreno, E., Paniagua, T., *et al.* (2007) Anatoxin-a occurrence and potential cyanobacterial anatoxin-a producers in Spanish reservoirs 1. *Journal of Phycology* **43** (6), 1120–1125.

Casero, M.C., Ballot, A., Agha, R., *et al.* (2014) Characterization of saxitoxin production and release and phylogeny of sxt genes in paralytic shellfish poisoning toxin-producing *Aphanizomenon gracile*. *Harmful Algae* **37**, 28–37.

Castro, D., Vera, D., Lagos, N., *et al.* (2004) The effect of temperature on growth and production of paralytic shellfish poisoning toxins by the cyanobacterium *Cylindrospermopsis raciborskii* C10. *Toxicon* **44** (5), 483–9.

Chorus I. (2012) Current approaches to Cyanotoxin risk assessment, risk management and regulations in different countries. Federal Environmental Agency, Germany http://www.uba.de/uba-info-medien-e/4390.html.

Cirés, S., Alvarez-Roa, C., Wood, S.A., *et al.* (2014) First report of microcystin-producing *Fischerella* sp. (Stigonematales, Cyanobacteria) in tropical Australia. *Toxicon* **88**, 62–66.

Costa, I.A.S., Azevedo, S.M.F.O., Senna, P.A.C., *et al.* (2006) Occurrence of toxin-producing cyanobacteria blooms in a Brazilian semiarid reservoir. *Brazilian Journal of Biology* **66** (1b), 211–219.

Cronberg, G., Annadotter, H. and Lawton, L.A. (1999) The occurrence of toxic blue-green algae in Lake Ringsjön, southern Sweden, despite nutrient reduction and fish biomanipulation. *Hydrobiologia* **404**, 123–129.

Dadheech, P. K., Selmeczy, G. B., Vasas, G., *et al.* (2014) Presence of potential toxin-producing cyanobacteria in an oligo-mesotrophic lake in Baltic lake district, Germany: an ecological, genetic and toxicological survey. *Toxins* **6** (10), 2912–2931.

Dasey, M., Ryan, N., Wilson, J., *et al.* (2005) Investigations into the taxonomy, toxicity and ecology of benthic cyanobacterial accumulations in Myall Lake, Australia. *Marine and Freshwater Research* **56** (1), 45.

Devlin, J.P., Edwards, O.E., Gorham, P.R., *et al.* (1977) Anatoxin- a, a toxic alkaloid from *Anabaena flos-aquae* NRC-44h. *Canadian Journal of Chemistry* **55** (8), 1367–1371.

Dias, E., Pereira, P., and Franca, S. (2002) Production of paralytic shellfish toxins by *Aphanizomenon* sp. Lmecya 31 (cyanobacteria). *Journal of Phycology* **38** (4), 705–712.

Dixit, R.B., and Suseela, M.R. (2013) Cyanobacteria: potential candidates for drug discovery. *Antonie van Leeuwenhoek* **103** (5), 947–61.

Dolman, A.M., Rücker, J., Pick, F.R. *et al.* (2012) Cyanobacteria and cyanotoxins: the influence of nitrogen versus phosphorus. *PLoS ONE* **7** (6), e38757.

Domingos, P., Rubim, T.K., Molica, R.J.R. *et al.* (1999) First report of microcystin production by picoplanktonic cyanobacteria isolated from a northeast Brazilian drinking water supply. *Environmental Toxicology* **14**, 31–35.

Edwards, C., Beattie, K.A., Scrimgeour, C.M., and Codd, G.A. (1992) Identification of anatoxin-A in benthic cyanobacteria (blue-green algae) and in associated dog poisonings at Loch Insh, Scotland. *Toxicon* **30** (10), 1165–1175.

El Herry, S., Fathalli, A., Rejeb, A.J.B. and Bouaïcha, N. (2008) Seasonal occurrence and toxicity of *Microcystis* spp. and *Oscillatoria tenuis* in the Lebna Dam, Tunisia. *Water Research* **42** (4-5), 1263–73.

Endean, R., Monks, S.A., Griffith, J.K., and Llewellyn, L.E. (1993) Apparent relationships between toxins elaborated by the cyanobacterium *Trichodesmium erythraeum* and those present in the flesh of the narrow-barred Spanish mackerel *Scomberomorus commersoni*. *Toxicon* **31** (9), 1155–65.

Engene, N., Rottacker, E.C., Kaštovský, J., *et al.* (2012) *Moorea producens* gen. nov., sp. nov. and *Moorea bouillonii* comb. nov., tropical marine cyanobacteria rich in bioactive secondary metabolites. *International Journal of Systematic and Evolutionary Microbiology* **62** (Pt 5), 1171–8.

Ernst, B., Hitzfeld, B., and Dietrich, D. (2001) Presence of *Planktothrix* sp. and cyanobacterial toxins in Lake Ammersee, Germany and their impact on whitefish (*Coregonus lavaretus* L.). *Environmental Toxicology* **16** (6), 483–8.

Everson, S., Fabbro, L., Kinnear, S. and Wright, P. (2011) Extreme differences in akinete, heterocyte and cylindrospermopsin concentrations with depth in a successive bloom involving *Aphanizomenon ovalisporum* (Forti) and *Cylindrospermopsis raciborskii* (Woloszynska) Seenaya and Subba Raju. *Harmful Algae* **10** (3), 265–276.

Faassen, E.J., Harkema, L., Begeman, L., and Lurling, M. (2012) First report of (homo)anatoxin-a and dog neurotoxicosis after ingestion of benthic cyanobacteria in The Netherlands. *Toxicon* **60** (3), 378–84.

Fastner, J., Erhard, M., Carmichael, W.W., *et al.* (1999) Characterization and diversity of microcystins in natural blooms and strains of the genera *Microcystis* and *Planktothrix* from German freshwaters. *Archiv für Hydrobiologie* **145** (2), 147–163.

Fastner, J., Rücker, J., Stüken, A., *et al.* (2007) Occurrence of the cyanobacterial toxin cylindrospermopsin in northeast Germany. *Environmental Toxicology* **22** (1), 26–32.

Ferreira, F.M.B., Soler, J.M.F., Fidalgo, M.L., and Fernández-Vila, P. (2001) PSP toxins from *Aphanizomenon flos-aquae* (cyanobacteria) collected in the Crestuma-Lever reservoir (Douro river, northern Portugal). *Toxicon* **39** (6), 757–761.

Fiore, M.F., Genuário, D.B., da Silva, C.S.P., *et al.* (2009) Microcystin production by a freshwater spring cyanobacterium of the genus *Fischerella*. *Toxicon* **53** (7-8), 754–61.

Foss, A.J., Phlips, E.J., Yilmaz, M., and Chapman, A. (2012) Characterization of paralytic shellfish toxins from *Lyngbya wollei* dominated mats collected from two Florida springs. *Harmful Algae* **16**, 98–107.

Fujiki, H., Ikegami, K., Hakii, H. *et al.* (1985) A blue-green-alga from Okinawa contains aplysiatoxins, the 3rd class of tumor promoters. *Japanese Journal of Cancer Research* **76** (4), 257–259.

Furtado, A., Calijuri, M.D., Lorenzi, A.S. *et al.* (2009) Morphological and molecular characterization of cyanobacteria from a Brazilian facultative wastewater stabilization pond and evaluation of microcystin production. *Hydrobiologia* **627** (1), 195–209.

Gantar, M., Sekar, R., and Richardson, L.L. (2009) Cyanotoxins from black band disease of corals and from other coral reef environments. *Microbial Ecology* **58** (4), 856–64.

Gehringer, M.M., Adler, L., Roberts, A.A., *et al.* (2012) Nodularin, a cyanobacterial toxin, is synthesized in planta by symbiotic *Nostoc* sp. *The International Society for Microbial Ecology Journal Journal* **6** (10), 1834–1847.

Gkelis, S. and Zaoutsos, N. (2014) Cyanotoxin occurrence and potentially toxin producing cyanobacteria in freshwaters of Greece: a multi-disciplinary approach. *Toxicon* **78**, 1–9.

Gkelis, S., Rajaniemi, P., Vardaka, E. *et al.* (2005) *Limnothrix redekei* (Van Goor) Meffert (Cyanobacteria) strains from Lake Kastoria, Greece form a separate phylogenetic group. *Microbial Ecology* **49** (1), 176–82.

Gugger, M., Lenoir, S., Berger, C. *et al.* (2005) First report in a river in France of the benthic cyanobacterium *Phormidium favosum* producing anatoxin-a associated with dog neurotoxicosis. *Toxicon* **45** (7), 919–28.

Hamill, K.D. (2001) Toxicity in benthic freshwater cyanobacteria (blue-green algae): First observations in New Zealand. *New Zeal. Journal of Marine and Freshwater Research* **35** (5), 1057–1059.

Harada, K., Ogawa K., Kimura Y. *et al.* (1991). Microcystins from *Anabaena flos-aquae* NRC 525-17. *Chemical Research in Toxicology* **4** (5), 535–540.

Harada, K.I., Othani, I., Iwamoto, K. *et al.* (1994) Isolation of cylindrospermopsin from a cyanobacterium *Umezakia natans* and its screening method. *Toxicon* **32**, 73–84.

Harding, W.R., Rowe, N., Wessels, J.C., *et al.* (1995) Death of a dog attributed to the cyanobacterial (blue-green algal) hepatotoxin nodularin in South Africa. *Journal of the South African Veterinary Association* **66** (4), 256–9.

Harland, F., Wood, S., Broady, P. *et al.* (2014) Polyphasic studies of cyanobacterial strains isolated from benthic freshwater mats in Canterbury, New Zealand. *New Zealand Journal of Botany* **52** (1), 116–135.

Harland, F.M.J., Wood, S.A., Moltchanova, E. *et al.* (2013) *Phormidium autumnale* growth and anatoxin-a production under iron and copper stress. *Toxins (Basel)* **5** (12), 2504–21.

Hawkins, P.R., Chandrasena, N.R., Jones, G.J. *et al.* (1997) Isolation and toxicity of *Cylindrospermopsis raciborskii* from an ornamental lake. *Toxicon* **35** (3), 341–346.

Hawkins, P.R., Putt, E., Falconer, I.R. and Humpage, A. (2001). Phenotypical variation in a toxic strain of the phytoplankter, *Cylindrospermopsis raciborskii* (Nostocales, Cyanophyceae) during batch culture. *Environmental Toxicology* **16**, 460–467.

Hawkins, P.R., Runnegar, M.T., Jackson, A.R., and Falconer, I.R. (1985) Severe hepatotoxicity caused by the tropical cyanobacterium (blue-green alga) *Cylindrospermopsis raciborskii* (Woloszynska) Seenaya and Subba Raju isolated from a domestic water supply reservoir. *Applied and Environmental Microbiology* **50** (5), 1292–1295.

Hawser, S.P., Codd, G.A., Capone, D.G., and Carpenter, E.J. (1991) A neurotoxic factor associated with the bloom-forming cyanobacterium *Trichodesmium*. *Toxicon* **29** (3), 277–8.

Heath, M.W., Wood, S.A., and Ryan, K.G. (2011) Spatial and temporal variability in *Phormidium* mats and associated anatoxin-a and homoanatoxin-a in two New Zealand rivers. *Aquatic Microbial Ecology* **64** (1), 69–79.

Henriksen, P. (1996) Microcystin profiles and contents in Danish populations of cyanobacteria/blue-green algae as determined by HPLC. *Phycologia* **35** (6S), 102–110.

Henriksen, P., Carmichael, W.W., An, J., and Moestrup, O. (1997) Detection of an anatoxin-a(s)-like anticholinesterase in natural blooms and cultures of cyanobacteria/blue-green algae from danish lakes and in the stomach contents of poisoned birds. *Toxicon* **35** (6), 901–913.

Hitzfeld, B.C., Lampert, C.S., Spaeth, N., *et al.* (2000) Toxin production in cyanobacterial mats from ponds on the McMurdo Ice Shelf, Antarctica. *Toxicon* **38** (12), 1731–1748.

Hodoki, Y., Ohbayashi, K., Kobayashi, Y., *et al.* (2013) Anatoxin-a-producing *Raphidiopsis mediterranea* Skuja var. grandis Hill is one ecotype of non-heterocytous *Cuspidothrix issatschenkoi* (Usačev) Rajaniemi *et al.* in Japanese lakes. *Harmful Algae* **21-22**, 44–53.

Humpage, A., Falconer, I., Bernard, C. *et al.* (2012) Toxicity of the cyanobacterium *Limnothrix* AC0243 to male Balb/c mice. *Water Research* **46** (5), 1576–83.

Humpage, A., Rositano, J., Bretag, A. *et al.* (1994) Paralytic shellfish poisons from Australian cyanobacterial blooms. *Marine and Freshwater Research* **45** (5), 761.

Ikawa, M., Wegener, K., Foxall, T.L., and Sasner, J.J. (1982) Comparison of the toxins of the blue-green alga *Aphanizomenon flos-aquae* with the *Gonyaulax* toxins. *Toxicon* **20** (4), 747–752.

Izaguirre, G., Jungblut, A.-D., and Neilan, B.A. (2007) Benthic cyanobacteria (Oscillatoriaceae) that produce microcystin-LR, isolated from four reservoirs in southern California. *Water Research* **41** (2), 492–8.

James, K.J., Sherlock, I.R., and Stack, M.A. (1997) Anatoxin-a in Irish freshwater and cyanobacteria, determined using a new fluorimetric liquid chromatographic method. *Toxicon* **35** (6), 963–971.

Jančula, D., Straková, L., Sadílek, J., *et al.* (2014) Survey of cyanobacterial toxins in Czech water reservoirs–the first observation of neurotoxic saxitoxins. *Environmental Science and Pollution Research* **21** (13), 8006–15.

Jones, G.J., and Negri, A.P. (1997) Persistence and degradation of cyanobacterial paralytic shellfish poisons (PSPs) in freshwaters. *Water Research* **31** (3), 525–533.

Jungblut, A.D., Hoeger, S.J., Mountfort, D., *et al.* (2006) Characterization of microcystin production in an Antarctic cyanobacterial mat community. *Toxicon* **47** (3), 271–8.

Kaas, H. and Henriksen, P. (2000) Saxitoxins (PSP toxins) in Danish lakes. *Water Research* **34** (7), 2089–2097.

Kerbrat, A.S., Amzil, Z., Pawlowiez, R. *et al.* (2011) First evidence of palytoxin and 42-hydroxy-palytoxin in the marine cyanobacterium *Trichodesmium*. *Marine Drugs* **9** (12), 543–560.

Kokociński, M., Dziga, D., Spoof, L., *et al.* (2009) First report of the cyanobacterial toxin cylindrospermopsin in the shallow, eutrophic lakes of western Poland. *Chemosphere* **74** (5), 669–675.

Kokociński, M., Mankiewicz-Boczek, J., Jurczak, T. *et al.* (2013) *Aphanizomenon gracile* (Nostocales), a cylindrospermopsin-producing cyanobacterium in Polish lakes. *Environmental Science and Pollution Research* **20** (8), 5243–64.

Kokociński, M., Stefaniak, K., Izydorczyk, K. *et al.* (2011). Temporal variation in microcystin production by *Planktothrix agardhii* (Gomont) Anagnostidis and Komárek (Cyanobacteria, Oscillatoriales) in a temperate lake. *Annales de Limnologie* **47**, 363–371.

Krienitz, L., Ballot, A., Kotut, K., *et al.* (2003) Contribution of hot spring cyanobacteria to the mysterious deaths of Lesser Flamingos at Lake Bogoria, Kenya. *FEMS Microbiology Ecology* **43** (2), 141–8.

Lagos, N., Onodera, H., Zagatto, P.A. *et al.* (1999) The first evidence of paralytic shellfish toxins in the freshwater cyanobacterium *Cylindrospermopsis raciborskii*, isolated from Brazil. *Toxicon* **37** (10), 1359–1373.

Lajeunesse, A., Segura, P.A., Gélinas, M. *et al.* (2012) Detection and confirmation of saxitoxin analogues in freshwater benthic *Lyngbya wollei* algae collected in the St. Lawrence River (Canada) by liquid chromatography-tandem mass spectrometry. *Journal of Chromatography A* **1219**, 93–103.

Lakshmana Rao, P.V., Bhattacharya, R., Gupta, N. *et al.* (2002) Involvement of caspase and reactive oxygen species in cyanobacterial toxin anatoxin-a-induced cytotoxicity and apoptosis in rat thymocytes and Vero cells. *Archives of Toxicology* **76** (4), 227–35.

Lanaras, T., and Cook, C.M. (1994) Toxin extraction from an *Anabaenopsis milleri* — dominated bloom. *Science of the Total Environment* **142** (3), 163–169.

Laurent, D., Kerbrat, A.-S., Darius, H.T. *et al.* (2008) Are cyanobacteria involved in Ciguatera Fish Poisoning-like outbreaks in New Caledonia? *Harmful Algae* **7** (6), 827–838.

Ledreux, A., Thomazeau, S., Catherine, A. *et al.* (2010) Evidence for saxitoxins production by the cyanobacterium *Aphanizomenon gracile* in a French recreational water body. *Harmful Algae* **10** (1), 88–97.

Li, H., Du, G., Wu, Y., *et al.* (2011) Correlation of microcystins and water environment factors in Miyun reservoir of Beijing, China. *Environmental Engineering and Management Journal*, **10** (6), 849–853.

Li, H., Murphy, T., Guo, J. *et al.* (2009) Iron-stimulated growth and microcystin production of *Microcystis novacekii* UAM 250. *Limnologica-Ecology and Management of Inland Waters* **39** (3), 255–259.

Li, R., Carmichael, W.W. and Peirera P. (2003) Morphological and 16S rRNA gene evidence for reclassification of the paralytic shellfish toxin producing *Aphanizomenon flos-aquae* LMECYA 31 as *Aphanizomenon issatschenkoi* (Cyanophyceae). *Journal of Phycology* **39**, 814–818.

Li, R., Carmichael, W.W., Brittain, S. *et al.* (2001a) Isolation and identification of the cyanotoxin cylindrospermopsin and deoxy-cylindrospermopsin from a Thailand strain of *Cylindrospermopsis raciborskii* (Cyanobacteria). *Toxicon* **39** (7), 973–980.

Li, R., Carmichael, W.W., Brittain, S. *et al.* (2001b) First report of the cyanotoxins cylindrospermopsin and deoxycylindrospermopsin from *Raphidiopsis curvata* (cyanobacteria). *Journal of Phycology* **37** (6), 1121–1126.

Li, R., Carmichael, W.W., Liu, Y. and Watanabe, M.M. (2000) Taxonomic re-evaluation of *Aphanizomenon flos-aquae* NH-5 based on morphology and 16S rRNA gene sequences. *Hydrobiologia* **438**, 99–105.

Li, Z., Yu, J., Yang, M., *et al.* (2010). Cyanobacterial population and harmful metabolites dynamics during a bloom in Yanghe Reservoir, North China. *Harmful Algae* **9**, 481–488.

Liu, Y., Chen, W., Li, D. *et al.* (2006) Analysis of paralytic shellfish toxins in *Aphanizomenon* DC-1 from Lake Dianchi, China. *Environmental Toxicology* **21** (3), 289–95.

Lugomela, C., Pratap, H.B., and Mgaya, Y.D. (2006) Cyanobacteria blooms—A possible cause of mass mortality of Lesser Flamingos in Lake Manyara and Lake Big Momela, Tanzania. *Harmful Algae* **5** (5), 534–541.

Luukkainen, R., Sivonen, K., Namikoshi, M. *et al.* (1993) Isolation and identification of eight microcystins from thirteen *Oscillatoria agardhii* strains and structure of a new microcystin. *Applied and Environmental Microbiology.* **59** (7), 2204–2209.

Maatouk, I. (2002) Seasonal variation of microcystin concentrations in the Saint-Caprais reservoir (France) and their removal in a small full-scale treatment plant. *Water Research* **36** (11), 2891–2897.

Mahmood, N.A., and Carmichael, W.W. (1986a) Paralytic shellfish poisons produced by the freshwater cyanobacterium *Aphanizomenon flos-aquae* NH-5. *Toxicon* **24** (2), 175–186.

Mahmood, N.A., and Carmichael, W.W. (1986b) The pharmacology of anatoxin-a(s), a neurotoxin produced by the freshwater cyanobacterium *Anabaena flos-aquae* NRC 525-17. *Toxicon* **24** (5), 425–34.

Mahmood, N.A., and Carmichael, W.W. (1987) Anatoxin-a(s), an anticholinesterase from the cyanobacterium *Anabaena flos-aquae* NRC-525-17. *Toxicon* **25** (11), 1221–7.

Mahmood, N.A., Carmichael, W.W., and Pfahler, D. (1988) Anticholinesterase poisonings in dogs from a cyanobacterial (blue-green algae) bloom dominated by *Anabaena flos-aquae. American Journal of Veterinary Research* **49** (4), 500–3.

Manganelli, M., Scardala S., Stefanelli M. *et al.* (2010) Health risk evaluation associated to *Planktothrix rubescens*: An integrated approach to design tailored monitoring programs for human exposure to cyanotoxins. *Water Research* **44** (5), 1297–1306.

Mankiewicz, J., Komárková, J., Izydorczyk, K., *et al.* (2005) Hepatotoxic cyanobacterial blooms in the lakes of northern Poland. *Environmental Toxicology* **20** (5), 499–506.

Matsunaga, S., Moore, R.E., Niemczura, W.P., and Carmichael, W.W. (1989) Anatoxin-a(s), a potent anticholinesterase from *Anabaena flos-aquae. Journal of the American Chemical Society* **111** (20), 8021–8023.

Mazmouz, R., Chapuis-Hugon, F., Mann, S., *et al.* (2010) Biosynthesis of cylindrospermopsin and 7-epicylindrospermopsin in *Oscillatoria* sp. strain PCC 6506: identification of the *cyr* gene cluster and toxin analysis. *Applied and Environmental Microbiology* **76** (15), 4943–9.

Mazur, H. and Plinski, M. (2003) *Nodularia spumigena* blooms and the occurrence of hepatotoxin in the Gulf of Gdańsk. *Oceanologia* **45**, 305–316.

McGregor, G.B., Sendall, B.C., Hunt, L.T. and Eaglesham, G.K. (2011) Report of the cyanotoxins cylindrospermopsin and deoxy-cylindrospermopsin from *Raphidiopsis mediterranea* Skuja (Cyanobacteria/ Nostocales). *Harmful Algae* **10**, 402–410.

McGregor, G.B., Stewart, I., Sendall, B.C., *et al.* (2012) First report of a toxic *Nodularia spumigena* (Nostocales/ Cyanobacteria) bloom in sub-tropical Australia. I. Phycological and public health investigations. *International Journal of Environmental Research and Public Health* **9** (7), 2396–411.

Méjean, A., Mann, S., Maldiney, T., *et al.* (2009) Evidence that biosynthesis of the neurotoxic alkaloids anatoxin-a and homoanatoxin-a in the cyanobacterium *Oscillatoria* PCC 6506 occurs on a modular polyketide synthase initiated by L-proline. *Journal of the American Chemical Society* **131** (22), 7512–3.

Méjean, A., Peyraud-Thomas, C., Kerbrat, A.S., *et al.* (2010) First identification of the neurotoxin homoanatoxin-a from mats of *Hydrocoleum lyngbyaceum* (marine cyanobacterium) possibly linked to giant clam poisoning in New Caledonia. *Toxicon* **56** (5), 829–35.

Meriluoto, J.A.O., Sandström, A., Eriksson, J.E., *et al.* (1989) Structure and toxicity of a peptide hepatotoxin from the cyanobacterium *Oscillatoria agardhii. Toxicon* **27** (9), 1021–1034.

Messineo, V., Bogialli, S., Melchiorre, S., *et al.* (2009) Cyanobacterial toxins in Italian freshwaters. *Limnologica - Ecology and Management of Inland Waters* **39** (2), 95–106.

Metcalf, J.S., Richer, R., Cox P.A. and Codd G.A. (2012) Cyanotoxins in desert environments may present a risk to human health. *The Science of the Total Environment* **421**, 118–123.

Mez, K., Beattie, K.A., Codd, G.A., *et al.* (1997) Identification of a microcystin in benthic cyanobacteria linked to cattle deaths on alpine pastures in Switzerland. *European Journal of Phycology* **32** (02), 111–117.

Mohamed, Z.A. (2007) First report of toxic *Cylindrospermopsis raciborskii* and *Raphidiopsis mediterranea* (Cyanoprokaryota) in Egyptian fresh waters. *FEMS Microbiology Ecology* **59** (3), 749–761.

Mohamed, Z.A. (2008) Toxic cyanobacteria and cyanotoxins in public hot springs in Saudi Arabia. *Toxicon* **51** (1), 17–27.

Mohamed, Z.A., and Al Shehri, A.M. (2007) Cyanobacteria and their toxins in treated-water storage reservoirs in Abha city, Saudi Arabia. *Toxicon* **50** (1), 75–84.

Mohamed, Z.A., and Al Shehri, A.M. (2009) Microcystin-producing blooms of *Anabaenopsis arnoldi* in a potable mountain lake in Saudi Arabia. *Federation of European Microbiological Societies* **69** (1), 98–105.

Mohamed, Z.A., Carmichael, W.W., An, J., and El-Sharouny, H.M. (1999) Activated carbon removal efficiency of microcystins in an aqueous cell extract of *Microcystis aeruginosa* and *Oscillatoria tenuis* strains isolated from Egyptian freshwaters. *Environmental Toxicology* **14** (1), 197–201.

Mohamed, Z.A., el-Sharouny, H.M., and Ali, W.S.M. (2006) Microcystin production in benthic mats of cyanobacteria in the Nile River and irrigation canals, Egypt. *Toxicon* **47** (5), 584–90.

Molica, R., Onodera, H., García, C., *et al.* (2002) Toxins in the freshwater cyanobacterium *Cylindrospermopsis raciborskii* (Cyanophyceae) isolated from Tabocas reservoir in Caruaru, Brazil, including demonstration of a new saxitoxin analogue. *Phycologia* **41** (6), 606–611.

Molica, R.J.R., Oliveira, E.J.A., Carvalho, P.V.V.C., *et al.* (2005) Occurrence of saxitoxins and an anatoxin-a(s)-like anticholinesterase in a Brazilian drinking water supply. *Harmful Algae* **4** (4), 743–753.

Monserrat, J.M., Yunes, J.S., and Bianchini, A. (2001) Effects of *Anabaena spiroides* (cyanobacteria) aqueous extracts on the acetylcholinesterase activity of aquatic species. *Environmental Toxicology and Chemistry* **20** (6), 1228–1235.

Mooney, K.M., Hamilton, J.T.G., Floyd, S.D., *et al.* (2011) Initial studies on the occurrence of cyanobacteria and microcystins in Irish lakes. *Environmental Toxicology* **26** (5), 566–70.

Myers, J.L., Sekar, R., and Richardson, L.L. (2007) Molecular detection and ecological significance of the cyanobacterial genera *Geitlerinema* and *Leptolyngbya* in black band disease of corals. *Applied and Environmental Microbiology* **73** (16), 5173–82.

Mynderse J.S. and Moore R.E. (1978) Toxins from blue-green algae: structures of oscillatoxin A and three related bromine-containing toxins. *The Journal of Organic Chemistry* **43**, 2301.

Mynderse, J., Moore, R., Kashiwagi, M., and Norton, T. (1977) Antileukemia activity in the Osillatoriaceae: isolation of Debromoaplysiatoxin from *Lyngbya. Science* **196** (4289), 538–540.

Namikoshi, M., Murakami, T., Watanabe, M.F., *et al.* (2003) Simultaneous production of homoanatoxin-a, anatoxin-a, and a new non-toxic 4-hydroxyhomoanatoxin-a by the cyanobacterium *Raphidiopsis mediterranea* Skuja. *Toxicon* **42** (5), 533–538.

Naselli-Flores, L., Barone, R., Chorus, I., and Kurmayer, R. (2007) Toxic cyanobacterial blooms in reservoirs under a semiarid mediterranean climate: the magnification of a problem. *Environmental Toxicology* **22** (4), 399–404.

Negri, A.P., Jones, G.J., and Hindmarsh, M. (1995) Sheep mortality associated with paralytic shellfish poisons from the cyanobacterium *Anabaena circinalis*. *Toxicon* **33** (10), 1321–1329.

Negri, A.P., Jones, G.J., Blackburn, S.I., *et al.* (1997) Effect of culture and bloom development and of sample storage on paralytic shellfish poisons in the cyanobacterium *Anabaena circinalis*. *Journal of Phycology* **33** (1), 26–35.

Nguyen V. L. A., Tanabe, Y., Matsuura, H. *et al.* (2012) Morphological, biochemical and phylogenetic assessments of water-bloom-forming tropical morphospecies of *Microcystis* (Chroococcales, Cyanobacteria). *Phycological Research* **60**, 208–222.

Nguyen, L.T.T., Cronberg, G., Moestrup, Ø., and Daugbjerg, N. (2013) *Annamia toxica* gen. et sp. nov. (Cyanobacteria), a freshwater cyanobacterium from Vietnam that produces microcystins: ultrastructure, toxicity and molecular phylogenetics. *Phycologia* **52** (1), 25–36.

Nogueira, I.C.G., Pereira, P., Dias, E., *et al.* (2004) Accumulation of paralytic shellfish toxins (PST) from the cyanobacterium *Aphanizomenon issatschenkoi* by the cladoceran *Daphnia magna*. *Toxicon* **44** (7), 773–80.

Ohtani, I., Moore, R.E., and Runnegar, M.T.C. (1992) Cylindrospermopsin: a potent hepatotoxin from the blue-green alga *Cylindrospermopsis raciborskii*. *Journal of the American Chemical Society* **114** (20), 7941–7942.

Oksanen, I., Jokela, J., Fewer, D.P., *et al.* (2004) Discovery of rare and highly toxic microcystins from lichen-associated cyanobacterium *Nostoc* sp. strain IO-102-I. *Applied and Environmental Microbiology* **70** (10), 5756–63.

Onodera, H., Oshima, Y., Henriksen, P., and Yasumoto, T. (1997a) Confirmation of anatoxin-a(s), in the cyanobacterium *Anabaena lemmermannii*, as the cause of bird kills in Danish Lakes. *Toxicon* **35** (11), 1645–1648.

Onodera, H., Satake, M., Oshima, Y., *et al.* (1997b) New saxitoxin analogues from the freshwater filamentous cyanobacterium *Lyngbya wollei*. *Natural Toxins* **5** (4), 146–151.

Osborne, N.J.T., Webb, P.M., and Shaw, G.R. (2001) The toxins of *Lyngbya majuscula* and their human and ecological health effects. *Environment International* **27** (5), 381–392.

Osswald, J., Rellán, S., Gago-Martinez, A., and Vasconcelos, V. (2009) Production of anatoxin-a by cyanobacterial strains isolated from Portuguese fresh water systems. *Ecotoxicology* **18** (8), 1110–5.

Oudra, B., Dadi-El Andaloussi, M. and Vasconcelos, V.M. (2009) Identification and quantification of microcystins from a *Nostoc muscorum* bloom occurring in Oukaïmeden River (High-Atlas mountains of Marrakech, Morocco). *Environmental Monitoring and Assessment* **149** (1–4), 437–44.

Park, H.D., Kim, B., Kim, E. and Okino, T. (1998) Hepatotoxic microcystins and neurotoxic anatoxin-a in cyanobacterial blooms from Korean lakes. *Environmental ToxicologyWater Quality* **13** (3), 225–234.

Park, H.D., Watanabe, M.F., Harada, K.-I., *et al.* (1993) Hepatotoxin (microcystin) and neurotoxin (anatoxin-a) contained in natural blooms and strains of cyanobacteria from Japanese freshwaters. *Natural Toxins* **1** (6), 353–360.

Pavlova, V., Stoyneva M., Georgieva V. *et al.* (2014) New records of microcystins in some bulgarian water bodies of health and conservational importance. *Journal of Water Resource and Protection* **06** (05), 446–453.

Pawlik-Skowrońska B., Pirszel J., and Kornijów R. (2008) Spatial and temporal variation in microcystin concentrations during perennial bloom of *Planktothrix agardhii* in a hypertrophic lake. *Annales de Limnologie – International Journal of Limnology* **44**, 63–68.

Pawlik-Skowrońska, B., Kalinowska, R. and Skowroński, T. (2013) Cyanotoxin diversity and food web bioaccumulation in a reservoir with decreasing phosphorus concentrations and perennial cyanobacterial blooms. *Harmful Algae* **28**, 118–125.

Pawlik-Skowrońska, B., Skowronski, T., Pirszel, J. and Adamczyk, A. (2004). Relationship between cyanobacterial bloom composition and anatoxin-a and microcystin occurrence in the eutrophic dam reservoir (SE Poland). *Polish Journal of Ecology* **52**, 479–490.

Pawlik-Skowrońska, B., Toporowska, M., and Rechulicz, J. (2012) Simultaneous accumulation of anatoxin-a and microcystins in three fish species indigenous to lakes affected by cyanobacterial blooms. *Oceanological and Hydrobiological Studies* **41** (4), 53–65.

Pereira, D.A., Pimenta, A. M.C. and Giani A. (2012) Profiles of toxic and non-toxic oligopeptides of *Radiocystis fernandoii* (Cyanobacteria) exposed to three different light intensities. *Microbiological Research* **167**, 413–421.

Pereira, P., Li, R., Carmichael, W., *et al.* (2004) Taxonomy and production of paralytic shellfish toxins by the freshwater cyanobacterium *Aphanizomenon gracile* LMECYA40. *European Journal of Phycology* **39** (4), 361–368.

Pereira, P., Onodera, H., Andrinolo, D., *et al.* (2000) Paralytic shellfish toxins in the freshwater cyanobacterium *Aphanizomenon flos-aquae*, isolated from Montargil reservoir, Portugal. *Toxicon* **38** (12), 1689–1702.

Pomati, F., Kellmann, R., Cavalieri, R., *et al.* (2006) Comparative gene expression of PSP-toxin producing and non-toxic *Anabaena circinalis* strains. *Environment International* **32** (6), 743–8.

Pomati, F., Moffitt, M.C., Cavaliere, R., and Neilan, B.A. (2004) Evidence for differences in the metabolism of saxitoxin and C1+2 toxins in the freshwater cyanobacterium *Cylindrospermopsis raciborskii* T3. *Biochimica et Biophysica Acta - General Subjects* **1674** (1), 60–67.

Pomati, F., Sacchi, S., Rossetti, C., *et al.* (2000) The freshwater cyanobacterium *Planktothrix* sp. Fp1: molecular identification and detection of paralytic shellfish poisoning toxins. *Journal of Phycology* **36** (3), 553–562.

Preussel, K., Stüken, A., Wiedner, C., *et al.* (2006) First report on cylindrospermopsin producing *Aphanizomenon flos-aquae* (Cyanobacteria) isolated from two German lakes. *Toxicon* **47** (2), 156–62.

Prinsep, M.R., Caplan, F.R., Moore, R.E. *et al.* (1992) Microcystin-LR from a blue-green alga belonging to the Stigonematales. *Phytochemestry* **31**, 1247–1248.

Proença, L., Tamanaha, M., and Fonseca, R. (2009) Screening the toxicity and toxin content of blooms of the cyanobacterium *Trichodesmium erythraeum* (Ehrenberg) in northeast Brasil. *Journal of Venomous Animals and Toxins including Tropical Diseases* **15** (2), 204–215.

Puschner, B., Hoff, B., and Tor, E.R. (2008) Diagnosis of anatoxin-a poisoning in dogs from North America. *Journal of Veterinary Diagnostic Investigation* **20** (1), 89–92.

Quesada, A., Moreno, E., Carrasco, D., *et al.* (2006) Toxicity of *Aphanizomenon ovalisporum* (Cyanobacteria) in a Spanish water reservoir. *Eur. Journal of Phycology* **41** (1), 39–45.

Quiblier, C., Wood, S., Echenique-Subiabre, I., *et al.* (2013) A review of current knowledge on toxic benthic freshwater cyanobacteria - ecology, toxin production and risk management. *Water Research* **47** (15), 5464–79.

Ramos, A., Martel, A., Codd, G., *et al.* (2005) Bloom of the marine diazotrophic cyanobacterium *Trichodesmium erythraeum* in the Northwest African Upwelling. *Marine Ecology Progress Series* **301**, 303–305.

Rapala, J., Robertson, A., Negri, A.P., *et al.* (2005) First report of saxitoxin in Finnish lakes and possible associated effects on human health. *Environmental Toxicology* **20** (3), 331–40.

Rapala, J., Sivonen, K., Luukkainen, R., and Niemelä, S.I. (1993) Anatoxin-a concentration in *Anabaena* and *Aphanizomenon* under different environmental conditions and comparison of growth by toxic and non-toxic *Anabaena*-strains — a laboratory study. *Journal of Applied Phycology* **5** (6), 581–591.

Rejmánková, E., Komárek, J., Dix, M., *et al.* (2011) Cyanobacterial blooms in Lake Atitlan, Guatemala. *Limnologica - Ecology and Management of Inland Waters* **41** (4), 296–302.

Richardson, L.L., Sekar, R., Myers, J.L., *et al.* (2007) The presence of the cyanobacterial toxin microcystin in black band disease of corals. *FEMS Microbiology Letters* **272** (2), 182–7.

Ruiz, M., Galanti, L., Ruibal, A.L., *et al.* (2013) First Report of Microcystins and Anatoxin-a Co-occurrence in San Roque Reservoir (Córdoba, Argentina). *Water, Air, & Soil Pollution* **224** (6), 1593.

Saker, M.L., Thomas, A.D. and Norton, J.H. (1999) Cattle mortality attributed to the toxic cyanobacterium *Cylindrospermopsis raciborskii* in an outback region of North Queensland. *Environmental Toxicology* **14**, 179–182.

Saker, M.L., Nogueira, I.C.G., Vasconcelos, V.M., *et al.* (2003) First report and toxicological assessment of the cyanobacterium *Cylindrospermopsis raciborskii* from Portuguese freshwaters. *Ecotoxicology and Environmental Safety* **55** (2), 243–250.

Salmaso N., Copetti D., Cerasino L. *et al.* (2014) Variability of microcystin cell quota in metapopulations of *Planktothrix rubescens*: Causes and implications for water management. *Toxicon* **90,** 82–96.

Salmaso, N., Boscaini, A., Shams, S., and Cerasino, L. (2013) Strict coupling between the development of *Planktothrix rubescens* and microcystin content in two nearby lakes south of the Alps (lakes Garda and Ledro). *Annales de Limnologie - International Journal of Limnology* **49** (4), 309–318.

Schembri, M.A., Neilan, B.A., and Saint, C.P. (2001) Identification of genes implicated in toxin production in the cyanobacterium *Cylindrospermopsis raciborskii. Environmental Toxicology* **16** (5), 413–21.

Schock, T.B., Huncik, K., Beauchesne, K.R., *et al.* (2011) Identification of trichotoxin, a novel chlorinated compound associated with the bloom forming Cyanobacterium, *Trichodesmium thiebautii. Environmental Science & Technology* **45** (17), 7503–9.

Seifert, M., McGregor, G., Eaglesham, G. *et al.* (2007) First evidence for the production of cylindrospermopsin and deoxy-cylindrospermop sin by the freshwater benthic cyanobacterium, *Lyngbya wollei* (Farlow ex Gornont) Speziale and Dyck. *Harmful Algae* **6** (1), 73–80.

Selwood, A.I., Holland, P.T., Wood, S.A., *et al.* (2007) Production of anatoxin-a and a novel biosynthetic precursor by the cyanobacterium *Aphanizomenon issatschenkoi. Environmental Science & Technology* **41** (2), 506–510.

Shams S., Cerasino L., Salmaso N. and Dietrich, D.R. (2014) Experimental models of microcystin accumulation in *Daphnia magna* grazing on *Planktothrix rubescens*: Implications for water management. *Aquatic Toxicology* **148**, 9–15.

Shams, S., Capelli, C. Cerasino, L., *et al.* (2015). Anatoxin-a producing *Tychonema* (cyanobacteria) in European waterbodies. *Water Research* **69,** 68–79.

Shaw, G.R., Sukenik, A., Livne, A., *et al.* (1999) Blooms of the cylindrospermopsin containing cyanobacterium, *Aphanizomenon ovalisporum* (Forti), in newly constructed lakes, Queensland, Australia. *Environmental Toxicology* **14** (1), 167–177.

Sivonen, K. (1990a) Effects of light, temperature, nitrate, orthophosphate, and bacteria on growth of and hepatotoxin production by *Oscillatoria agardhii* strains. *Applied and Environmental Microbiology.* **56** (9), 2658–2666.

Sivonen, K. and Jones, G. (1999) Cyanobacterial toxins, In: Chorus, I., Bartram, J. (Eds.), Toxic cyanobacteria in water. FN Spon, pp. 41–111.

Sivonen, K., Carmichael, W.W., Namikoshi, M., *et al.* (1990b) Isolation and characterization of hepatotoxic microcystin homologs from the filamentous freshwater cyanobacterium *Nostoc* sp. strain 152. *Applied and Environmental Microbiology* **56** (9), 2650–2657.

Sivonen, K., Himberg, K., Luukkainen, R., *et al.* (1989a) Preliminary characterization of neurotoxic cyanobacteria blooms and strains from Finland. *Toxicity Assessment* **4** (3), 339–352.

Sivonen, K., Kononen, K., Carmichael, W.W., *et al.* (1989b) Occurrence of the hepatotoxic cyanobacterium *Nodularia spumigena* in the Baltic Sea and structure of the toxin. *Applied and Environmental Microbiology* **55** (8), 1990–1995.

Sivonen, K., Namikoshi, M., Evans, W.R., *et al.* (1992a) Isolation and characterization of a variety of microcystins from seven strains of the cyanobacterial genus *Anabaena. Applied and Environmental Microbiology* **58** (8), 2495–2500.

Sivonen, K., Skulberg, O.M., Namikoshi, M., *et al.* (1992b) Two methyl ester derivatives of microcystins, cyclic heptapeptide hepatotoxins, isolated from *Anabaena flos-aquae strain* CYA 83/1. *Toxicon* **30** (11), 1465–1471.

Skulberg, O.M., Skulberg, R., Carmichael, W.W., *et al.* (1992) Investigations of a neurotoxic oscillatorialean strain (Cyanophyceae) and its toxin. Isolation and characterization of homoanatoxin-a. *Environmental Toxicology and Chemistry* **11** (3), 321–329.

Skulberg, O.M., Underdal, B. and Utkilen, H. (1994) Toxic waterblooms with cyanophytes in Norway - current knowledge. *Archives für Hydrobiology / Algological Studies* **75**, 279–289.

Smith, F. M.J., Wood, S.A., van Ginkel, R. *et al.* (2011) First report of saxitoxin production by a species of the freshwater benthic cyanobacterium, *Scytonema* Agardh. *Toxicon* **57** (4), 566–573.

Smith, F., Wood, S.A., Wilks, T., Kelly, D., Broady, P., Gaw, S. (2012) Survey of *Scytonema* (Cyanobacteria) and associated saxitoxins in the littoral zone of recreational lakes in Canterbury (New Zealand). *Phycologia* **51**, 542–551.

Sotero-Santos, R.B., Carvalho, E.G., Dellamano-Oliveira, M.J. and Rocha, O. (2008) Occurrence and toxicity of an *Anabaena bloom* in a tropical reservoir (Southeast Brazil). *Harmful Algae* **7** (5), 590–598.

Soto-Liebe, K., Méndez, M.A., Fuenzalida, L., *et al.* (2012) PSP toxin release from the cyanobacterium *Raphidiopsis brookii* D9 (Nostocales) can be induced by sodium and potassium ions. *Toxicon* **60** (7), 1324–34.

Spoof, L., Berg, K.A., Rapala, J., *et al.* (2006) First observation of cylindrospermopsin in *Anabaena lapponica* isolated from the boreal environment (Finland). *Environmental Toxicology* **21** (6), 552–60.

Stuken, A., Campbell, R.J., Quesada, A., *et al.* (2009) Genetic and morphologic characterization of four putative cylindrospermopsin producing species of the cyanobacterial genera *Anabaena* and *Aphanizomenon. Journal of Plankton Research* **31** (5), 465–480.

Teneva, I., Asparuhova, D., Dzhambazov, B., *et al.* (2003) The freshwater cyanobacterium *Lyngbya aerugineo-coerulea* produces compounds toxic to mice and to mammalian and fish cells. *Environmental Toxicology* **18** (1), 9–20.

Terao, K., Ohmori, S., Igarashi, K., *et al.* (1994) Electron microscopic studies on experimental poisoning in mice induced by cylindrospermopsin isolated from blue-green alga *Umezakia natans*. *Toxicon* **32** (7), 833–843.

Testé, V., Briand, J.F., Nicholson, B.C., and Puiseux-Dao, S. (2002) Comparison of changes in toxicity during growth of *Anabaena circinalis* (cyanobacteria) determined by mouse neuroblastoma bioassay and HPLC. *Journal of Applied Phycology* **14** (5), 399–407.

Twist, H., and Codd, G.A. (1997) Degradation of the cyanobacterial hepatotoxin, nodularin, under light and dark conditions. *FEMS Microbiology Letters* **151** (1), 83–88.

Ueno, Y., Nagata, S., Tsutsumi, T., *et al.* (1996) Detection of microcystins, a blue-green algal hepatotoxin, in drinking water sampled in Haimen and Fusui, endemic areas of primary liver cancer in China, by highly sensitive immunoassay. *Carcinogenesis* **17** (6), 1317–1321.

Van Buynder, P.G., Oughtred, T., Kirkby, B., *et al.* (2001) Nodularin uptake by seafood during a cyanobacterial bloom. *Environmental Toxicology* **16** (6), 468–71.

Vasas, G., Farkas, O., Borics, G. *et al.* (2013) Appearance of *Planktothrix rubescens* Bloom with [D-Asp3, Mdha7] MC–RR in Gravel Pit Pond of a Shallow Lake-Dominated Area. *Toxins* **5** (12), 2434–2455.

Vezie, C., Brient, L., Sivonen, K., *et al.* (1998) Variation of microcystin content of cyanobacterial blooms and isolated strains in lake Grand-Lieu (France). *Microbial Ecology* **35** (2), 126–135.

Viaggiu, E., Melchiorre, S., Volpi, F., *et al.* (2004) Anatoxin-a toxin in the cyanobacterium *Planktothrix rubescens* from a fishing pond in northern Italy. *Environmental Toxicology* **19** (3), 191–197.

Via-Ordorika, L., Fastner, J., Kurmayer, R., *et al.* (2004) Distribution of microcystin-producing and non-microcystin-producing *Microcystis* sp. in European freshwater bodies: detection of microcystins and microcystin genes in individual colonies. *Systematic and Applied Microbiology* **27** (5), 592–602.

Vieira, J.M. dos S., Azevedo, M.T. de P., Azevedo, S.M.F. de O., *et al.* (2003) Microcystin production by *Radiocystis fernandoi* (Chroococcales, Cyanobacteria) isolated from a drinking water reservoir in the city of Belém, PA, Brazilian Amazonia region. *Toxicon* **42** (7), 709–713.

Watanabe, M.F., Tsujimura S., Oishi S., *et al.* (2003) Isolation and identification of homoanatoxin-a from a toxic strain of the cyanobacterium *Raphidiopsis mediterranea* Skuja isolated from Lake Biwa, Japan. *Phycologia* **42** (4), 364–369.

Welker, M., Chorus, I., Fastner, J., *et al.* (2005) Microcystins (cyanobacterial toxins) in surface waters of rural Bangladesh: pilot study. *Journal of Water and Health* **3**, 325–337.

Wiedner, C., Rücker, J., Fastner, J., *et al.* (2008) Seasonal dynamics of cylindrospermopsin and cyanobacteria in two German lakes. *Toxicon* **52** (6), 677–86.

Willame, R., Jurczak, T., Iffly, J.-F., *et al.* (2005) Distribution of hepatotoxic cyanobacterial blooms in Belgium and Luxembourg. *Hydrobiologia* **551** (1), 99–117.

Wood, S.A., Holland, P.T., Stirling, D.J., *et al.* (2006). Survey of cyanotoxins in New Zealand waterbodies between 2001 and 2004. *New Zealand Journal of Marine and Freshwater Research* **40**, 585–597.

Wood, S.A., Mountfort, D., Selwood, A.I., *et al.* (2008) Widespread distribution and identification of eight novel microcystins in antarctic cyanobacterial mats. *Applied and Environmental Microbiology* **74** (23), 7243–51.

Wood, S.A., Rasmussen, J.P., Holland, P.T., *et al.* (2007a) First report of the cyanotoxin anatoxin-a from *Aphanizomenon issatschenkoi* (cyanobacteria). *Journal of Phycology* **43** (2), 356–365.

Wood, S.A., Selwood, A.I., Rueckert, A., *et al.* (2007b) First report of homoanatoxin-a and associated dog neurotoxicosis in New Zealand. *Toxicon* **50** (2), 292–301.

Yéprémian, C., Gugger, M.F., Briand, E., *et al.* (2007) Microcystin ecotypes in a perennial *Planktothrix agardhii* bloom. *Water Research* **41** (19), 4446–56.

Yilmaz, M., Phlips, E.J., Szabo, N.J., and Badylak, S. (2008) A comparative study of Florida strains of *Cylindrospermopsis* and *Aphanizomenon* for cylindrospermopsin production. *Toxicon* **51** (1), 130–9.

Yunes, J.S., De La Rocha, S., Giroldo, D., *et al.* (2009) Release of carbohydrates and proteins by a subtropical strain of *Raphidiopsis brookii* (cyanobacteria) able to produce saxitoxin at three nitrate concentrations. *Journal of Phycology* **45** (3), 585–591.

Zarenezhad, S., Sano, T., Watanabe, M.M., and Kawachi, M. (2012) Evidence of the existence of a toxic form of *Cylindrospermopsis raciborskii* (Nostocales, Cyanobacteria) in Japan. *Phycological Research* **60** (2), 98–104.

Appendix 3

Tables of Microcystins and Nodularins

Lisa Spoof[1] *and Arnaud Catherine*[2]

[1]*Biochemistry, Faculty of Science and Engineering, Åbo Akademi University, Turku, Finland*
[2]*UMR 7245 MCAM CNRS-MNHN, Muséum National d'Histoire Naturelle, Paris, France*

Tables 1 and 2 in Appendix 3 list microcystins (MCs) and nodularins (NODs) found in the literature.

The number of known MC congeners has increased rapidly in recent years. More sensitive mass spectrometric techniques have enabled the detection of MCs and NODs present in trace amounts in cyanobacterial samples. However, structural determinations by tandem mass spectrometry alone are tentative. Mass spectrometric techniques cannot, for example, differentiate between isobaric amino acids or elucidate the stereochemistry.

As MC and NOD identification are usually performed by mass spectrometry, the tables report the protonated *m/z* (mass-to-charge) of MCs and NODs, and the corresponding chemical formula. In some cases, especially with MC-RR and related congeners, the dominant ion observed in mass spectrometry is the doubly protonated ion $[M+2H]^{2+}$.

Handbook of Cyanobacterial Monitoring and Cyanotoxin Analysis, First Edition.
Edited by Jussi Meriluoto, Lisa Spoof and Geoffrey A. Codd.

Table A3.1 *Microcystins*

Number	Microcystin	Calculated* $[M+H]^+$	Corresponding composition*	References
1	[D-Asp³, DMAdda⁵]MC-LA	882.4608	$C_{44}H_{64}N_7O_{12}$	[1]
2	[D-Asp³]MC-VA	882.4608	$C_{44}H_{64}N_7O_{12}$	[1]
3	[D-Asp³]MC-LA	896.4764	$C_{45}H_{66}N_7O_{12}$	[2]
4	[Dha⁷]MC-LA	896.4764	$C_{45}H_{66}N_7O_{12}$	[1]
5	[DMAdda⁵]MC-LA	896.4764	$C_{45}H_{66}N_7O_{12}$	[3]
6	MC-VA	896.4764	$C_{45}H_{66}N_7O_{12}$	[1]
7	MC-LA	910.4921	$C_{46}H_{68}N_7O_{12}$	[2, 4, 5]
8	MC-LAba or MC-LAib	924.5077	$C_{47}H_{70}N_7O_{12}$	[2, 6]
9	MC-LBu	924.5077	$C_{47}H_{70}N_7O_{12}$	[7]
10	[D-Asp³, D-Glu(OMe)⁶]MC-LAba	924.5077	$C_{47}H_{70}N_7O_{12}$	[8]
11	[D-Asp³]MC-LV	924.5077	$C_{47}H_{70}N_7O_{12}$	[1]
12	[D-Asp³]MC-FA	930.4608	$C_{48}H_{64}N_7O_{12}$	[9]
13	[Asp³, Dha⁷]MC-YA	932.4400	$C_{47}H_{62}N_7O_{13}$	[10]
14	[Asp³]MC-LL	938.5234	$C_{48}H_{72}N_7O_{12}$	[2]
15	MC-LV	938.5234	$C_{48}H_{72}N_7O_{12}$	[1, 11]
16	[D-Asp³]MC-RA	939.4935	$C_{45}H_{67}N_{10}O_{12}$	[1, 9]
17	MC-FA	944.4764	$C_{49}H_{66}N_7O_{12}$	[12]
18	MC-LL	952.5390	$C_{49}H_{74}N_7O_{12}$	[2, 11]
19	MC-AR	953.5091	$C_{46}H_{69}N_{10}O_{12}$	[13]
20	MC-RA	953.5091	$C_{46}H_{69}N_{10}O_{12}$	[14, 15]
21	[D-Asp³]MC-RAba	953.5091	$C_{46}H_{69}N_{10}O_{12}$	[9]
22	MC-FAba	958.4921	$C_{50}H_{68}N_7O_{12}$	[9]
23	MC-YA	960.4713	$C_{49}H_{66}N_7O_{13}$	[15, 16]
24	MC-AHar	967.5248	$C_{47}H_{71}N_{10}O_{12}$	[17]
25	[D-Asp³, Dha⁷]MC-LR	967.5248	$C_{47}H_{71}N_{10}O_{12}$	[18, 19]
26	[D-Asp³, DMAdda⁵]MC-LR	967.5248	$C_{47}H_{71}N_{10}O_{12}$	[20]
27	[Asp³, DMAdda⁵, Dhb⁷]MC-LR	967.5248	$C_{47}H_{71}N_{10}O_{12}$	[21]
28	[Gly¹, Asp³, Dhb⁷]MC-LR	967.5248	$C_{47}H_{71}N_{10}O_{12}$	[22]
29	MC-RAba	967.5248	$C_{47}H_{71}N_{10}O_{12}$	[15]
30	[D-Asp³]MC-WA	969.4717	$C_{50}H_{65}N_8O_{12}$	[9]
31	[D-Asp³, Dha⁷]MC-EE(OMe)	970.4404	$C_{46}H_{64}N_7O_{16}$	[23]
32	MC-LM	970.4954	$C_{48}H_{72}N_7O_{12}S$	[11]
33	[D-Asp³]MC-LF	972.5077	$C_{51}H_{70}N_7O_{12}$	[5, 24]
34	MC-VF	972.5077	$C_{51}H_{70}N_7O_{12}$	[25]
35	[Asp³, Dha⁷]MC-LY	974.4870	$C_{50}H_{68}N_7O_{13}$	[5]
36	MC-YAba	974.4870	$C_{50}H_{68}N_7O_{13}$	[15]
37	[D-Asp³]MC-LR	981.5404	$C_{48}H_{73}N_{10}O_{12}$	[26–28]
38	[D-Asp³,(*E*)-Dhb⁷]MC-LR	981.5404	$C_{48}H_{73}N_{10}O_{12}$	[29]
39	[D-Asp³,(*Z*)-Dhb⁷]MC-LR	981.5404	$C_{48}H_{73}N_{10}O_{12}$	[30]
40	[Dha⁷]MC-LR	981.5404	$C_{48}H_{73}N_{10}O_{12}$	[13, 18]
41	[DMAdda⁵]MC-LR = *O*-demethyl-MC-LR	981.5404	$C_{48}H_{73}N_{10}O_{12}$	[13]
42	[Gly¹, Asp³, Dhb⁷]MC-LHar	981.5404	$C_{48}H_{73}N_{10}O_{12}$	[22]
43	MC-RApa	981.5404	$C_{48}H_{73}N_{10}O_{12}$	[15]
44	MC-VR	981.5404	$C_{48}H_{73}N_{10}O_{12}$	[31]

(Continued)

Table A3.1 *(Continued)*

Number	Microcystin	Calculated* $[M+H]^+$	Corresponding composition*	References
45	MC-AW	983.4873	$C_{51}H_{67}N_8O_{12}$	[32]
46	MC-WA	983.4873	$C_{51}H_{67}N_8O_{12}$	[12]
47	[Ser¹, D-Asp³, Dha⁷]MC-LR	983.5197	$C_{47}H_{71}N_{10}O_{13}$	[1]
48	[Dha⁷]MC-EE(OMe)	984.4561	$C_{47}H_{66}N_7O_{16}$	[23]
49	[D-Asp³, Dha⁷]MC-E(OMe)E(OMe)	984.4561	$C_{47}H_{66}N_7O_{16}$	[23]
50	Demethyl [L-Ser⁷]MC-LR	985.5353	$C_{47}H_{73}N_{10}O_{13}$	[1]
51	MC-FL	986.5234	$C_{52}H_{72}N_7O_{12}$	[9]
52	MC-LF	986.5234	$C_{52}H_{72}N_7O_{12}$	[2, 33]
53	MC-KynA (oxidized W)	987.4822	$C_{50}H_{67}N_8O_{13}$	[34]
54	[Asp³]MC-LY	988.5026	$C_{51}H_{70}N_7O_{13}$	[5]
55	[Asp³, Dhb⁷]MC-LY [Asp³, (*E*)-Dhb⁷]MC-LY	988.5026	$C_{51}H_{70}N_7O_{13}$	[21, 35]
56	[D-Asp³, ADMAdda⁵, Dha⁷]MC-LR	995.5197	$C_{48}H_{71}N_{10}O_{13}$	[1]
57	[Gly¹, Asp³, ADMAdda⁵, Dhb⁷]MC-LR	995.5197	$C_{48}H_{71}N_{10}O_{13}$	[22]
58	[D-Asp³, ADMAdda⁵]MC-VR	995.5197	$C_{48}H_{71}N_{10}O_{13}$	[1]
59	[Asp³, Dhb⁷]MC-AhaR	995.5561	$C_{49}H_{75}N_{10}O_{12}$	[21]
60	[D-Asp³]MC-HilR/HleR [D-Asp³]MC-HilR	995.5561	$C_{49}H_{75}N_{10}O_{12}$	[1, 36]
61	[D-Asp³, (*E*)-Dhb⁷]MC-HilR = (*E*)-[Dhb⁷]MC-HilR	995.5561	$C_{49}H_{75}N_{10}O_{12}$	[37]
62	[Dha⁷]MC-HilR	995.5561	$C_{49}H_{75}N_{10}O_{12}$	[38]
63	[DMAdda⁵]MC-HilR	995.5561	$C_{49}H_{75}N_{10}O_{12}$	[1]
64	[DMAdda⁵]MC-LHar	995.5561	$C_{49}H_{75}N_{10}O_{12}$	[1]
65	[D-Asp³, D-Glu(OMe)⁶]MC-LR	995.5561	$C_{49}H_{75}N_{10}O_{12}$	[39]
66	MC-LR	995.5561	$C_{49}H_{75}N_{10}O_{12}$	[16, 26, 40]
67	[6Z-Adda⁵]MC-LR	995.5561	$C_{49}H_{75}N_{10}O_{12}$	[27, 41]
68	MC-RL	995.5561	$C_{49}H_{75}N_{10}O_{12}$	[15]
69	MC-WAba	997.5030	$C_{52}H_{69}N_8O_{12}$	[34]
70	[Dha⁷]MC-E(OMe)E(OMe)	998.4717	$C_{48}H_{68}N_7O_{16}$	[23]
71	MC-OiaA (oxidised W)	999.4822	$C_{51}H_{67}N_8O_{13}$	[34]
72	[Asp³]MC-MR	999.4968	$C_{47}H_{71}N_{10}O_{12}S$	[42]
73	[seco-4/5] [Asp³]MC-LR	999.5510	$C_{48}H_{75}N_{10}O_{13}$	[42]
74	[Asp³, MeSer⁷]MC-LR	999.5510	$C_{48}H_{75}N_{10}O_{13}$	[36, 42]
75	[L-Ser⁷]MC-LR	999.5510	$C_{48}H_{75}N_{10}O_{13}$	[19]
76	MC-LHph	1000.5390	$C_{53}H_{74}N_7O_{12}$	[43]
77	MC-KynAba	1001.4979	$C_{51}H_{69}N_8O_{13}$	[34]
78	[D-Asp³, Dha⁷]MC-FR	1001.5091	$C_{50}H_{69}N_{10}O_{12}$	[1]
79	[L-Ser⁷]MC-EE(OMe)	1002.4666	$C_{47}H_{68}N_7O_{17}$	[23]
80	[D-Asp³, Ser⁷]MC-E(OMe)E(OMe)	1002.4666	$C_{47}H_{68}N_7O_{17}$	[23]
81	[Asp³]MC-HilY	1002.5182	$C_{52}H_{72}N_7O_{13}$	[5]
82	MC-LY	1002.5182	$C_{52}H_{72}N_7O_{13}$	[8, 44]
83	MC-YL	1002.5182	$C_{52}H_{72}N_7O_{13}$	[8]
84	[Asp³, MSer⁷]MC-LY	1006.5132	$C_{51}H_{72}N_7O_{14}$	[5]
85	[D-Asp³, ADMAdda⁵, Dha⁷]MC-HilR	1009.5353	$C_{49}H_{73}N_{10}O_{13}$	[1]

Table A3.1 (*Continued*)

Number	Microcystin	Calculated* $[M+H]^+$	Corresponding composition*	References
86	[Gly[1], Asp[3], ADMAdda[5], Dhb[7]] MC-LHar	1009.5353	$C_{49}H_{73}N_{10}O_{13}$	[22]
87	[D-Asp[3], ADMAdda[5]]MC-LR	1009.5353	$C_{49}H_{73}N_{10}O_{13}$	[45, 46]
88	[ADMAdda[5], Dha[7]]MC-LR	1009.5353	$C_{49}H_{73}N_{10}O_{13}$	[1]
89	[Asp[3], ADMAdda[5], Dhb[7]]MC-LR	1009.5353	$C_{49}H_{73}N_{10}O_{13}$	[47]
90	[MeAla[1]]MC-LR or [MeLeu[2]]MC-LR	1009.5717	$C_{50}H_{77}N_{10}O_{12}$	[48]
91	MC-HilR	1009.5717	$C_{50}H_{77}N_{10}O_{12}$	[28]
92	MC-LHar	1009.5717	$C_{50}H_{77}N_{10}O_{12}$	[49]
93	[D-Glu(OCH_3)[6]]MC-LR	1009.5717	$C_{50}H_{77}N_{10}O_{12}$	[39, 50]
94	[Mdhb[7]]MC-LR	1009.5717	$C_{50}H_{77}N_{10}O_{12}$	[45]
95	[Leu[1], Asp[3], DMAdda[5]]MC-LR	1009.5717	$C_{50}H_{77}N_{10}O_{12}$	[51]
96	[D-Asp[3], Dha[7]]MC-RR	1010.5418	$C_{47}H_{72}N_{13}O_{12}$	[26, 52]
97	[Asp[3], DMAdda[5],]MC-RR	1010.5418	$C_{47}H_{72}N_{13}O_{12}$	[53]
98	[Gly[1], Asp[3], Dhb[7]]MC-RR	1010.5418	$C_{47}H_{72}N_{13}O_{12}$	[22]
99	Demethyl-LW = [D-Asp[3]]MC-LW	1011.5186	$C_{53}H_{71}N_{8}O_{12}$	[24]
100	[Asp[3], (*E*)-Dhb[7]]MC-LW	1011.5186	$C_{53}H_{71}N_{8}O_{12}$	[35]
101	MC-OiaAba (oxidised W)	1013.4979	$C_{52}H_{69}N_{8}O_{13}$	[34]
102	MC-MR	1013.5125	$C_{48}H_{73}N_{10}O_{12}S$	[42]
103	[L-MeSer[7]]MC-LR	1013.5666	$C_{49}H_{77}N_{10}O_{13}$	[2, 13, 28]
104	Acyclo-LR: Adda-D-Glu(γ)-Mdha-D-Ala-L-Leu-D-MeAsp(β)-L-Arg-OH = [*seco*-4/5]MC-LR	1013.5666	$C_{49}H_{77}N_{10}O_{13}$	[28, 42, 54]
105	L-Leu-D-MeAsp(β)-L-Arg-Adda-D-Glu(γ)-Mdha-D-Ala-OH = [*seco*-1/2] MC-LR	1013.5666	$C_{49}H_{77}N_{10}O_{13}$	[28, 54, 55]
106	[Asp[3]]MC-M(O)R (oxidation artefact)	1015.4917	$C_{47}H_{71}N_{10}O_{13}S$	[42]
107	[D-Asp[3], Dha[7]]MC-HphR	1015.5248	$C_{51}H_{71}N_{10}O_{12}$	[1]
108	[D-Asp[3]]MC-FR	1015.5248	$C_{51}H_{71}N_{10}O_{12}$	[56]
109	[Dha[7]]MC-FR	1015.5248	$C_{51}H_{71}N_{10}O_{12}$	[57]
110	[DMAdda[5]]MC-FR	1015.5248	$C_{51}H_{71}N_{10}O_{12}$	[20]
111	[Asp[3]]MC-RF	1015.5248	$C_{51}H_{71}N_{10}O_{12}$	[5]
112	MC-NfkA (oxidised W)	1015.4771	$C_{51}H_{67}N_{8}O_{14}$	[34]
113	[L-Ser[7]]MC-E(OMe)E(OMe)	1016.4823	$C_{48}H_{70}N_{7}O_{17}$	[23]
114	MC-LHty	1016.5339	$C_{53}H_{74}N_{7}O_{13}$	[43]
115	[Asp[3], Dha[7]]MC-RY	1017.5040	$C_{50}H_{69}N_{10}O_{13}$	[5]
116	[Asp[3], DMAdda[5]]MC-RY	1017.5040	$C_{50}H_{69}N_{10}O_{13}$	[21]
117	MC-YM	1020.4747	$C_{51}H_{70}N_{7}O_{13}S$	[8]
118	[D-Asp[3], ADMAdda[5]]MC-LHar	1023.5510	$C_{50}H_{75}N_{10}O_{13}$	[45, 58]
119	[ADMAdda[5]]MC-LR	1023.5510	$C_{50}H_{75}N_{10}O_{13}$	[45, 46]
120	[Leu[1], Asp[3]]MC-LR	1023.5874	$C_{51}H_{79}N_{10}O_{12}$	[50, 51]
121	[Leu[1], Dha[7]]MC-LR	1023.5874	$C_{51}H_{79}N_{10}O_{12}$	[43]
122	[Val[1]]MC-LR	1023.5874	$C_{51}H_{79}N_{10}O_{12}$	[43]

(*Continued*)

Table A3.1 (*Continued*)

Number	Microcystin	Calculated* $[M+H]^+$	Corresponding composition*	References
123	[Gly[1], Asp[3], Dhb[7]]MC-RHar	1024.5574	$C_{48}H_{74}N_{13}O_{12}$	[22]
124	[D-Asp[3]]MC-RR	1024.5574	$C_{48}H_{74}N_{13}O_{12}$	[59, 60]
125	[Dha[7]]MC-RR	1024.5574	$C_{48}H_{74}N_{13}O_{12}$	[61]
126	[D-Asp[3], (*E*)-Dhb[7]]MCYST-RR	1024.5574	$C_{48}H_{74}N_{13}O_{12}$	[62]
127	MC-WL	1025.5343	$C_{54}H_{73}N_{8}O_{12}$	[34]
128	MC-LW	1025.5343	$C_{54}H_{73}N_{8}O_{12}$	[25]
129	H-L-Hil-MeAsp-L-Arg-Adda-D-Glu-Mdha-D-Ala-OH	1027.5828	$C_{50}H_{79}N_{10}O_{13}$	[28]
130	MC-NfkAba (oxidised W)	1029.4928	$C_{52}H_{69}N_{8}O_{14}$	[34]
131	MC-M(O)R (oxidation artefact)	1029.5074	$C_{48}H_{73}N_{10}O_{13}S$	[13]
132	MC-FR	1029.5404	$C_{52}H_{73}N_{10}O_{12}$	[6, 13]
133	MC-RF	1029.5404	$C_{52}H_{73}N_{10}O_{12}$	[15]
134	[Asp[3]]MC-HphR	1029.5404	$C_{52}H_{73}N_{10}O_{12}$	[36]
135	[Dha[7]]MC-HphR	1029.5404	$C_{52}H_{73}N_{10}O_{12}$	[63]
136	[Asp[3]]MC-M(O_2)R (oxidation artefact)	1031.4865	$C_{47}H_{71}N_{10}O_{14}S$	[42]
137	[D-Asp[3], Dha[7]]MC-HtyR	1031.5197	$C_{51}H_{71}N_{10}O_{13}$	[63]
138	[Asp[3], DMAdda[5]]MC-HtyR	1031.5197	$C_{51}H_{71}N_{10}O_{13}$	[36]
139	[Asp[3]]MC-RY	1031.5197	$C_{51}H_{71}N_{10}O_{13}$	[5, 64]
140	[Dha[7]]MC-RY	1031.5197	$C_{51}H_{71}N_{10}O_{13}$	[15]
141	[Asp[3], Dhb[7]]MC-RY	1031.5197	$C_{51}H_{71}N_{10}O_{13}$	[21]
142	[D-Asp[3]]MC-YR	1031.5197	$C_{51}H_{71}N_{10}O_{13}$	[60]
143	[Dha[7]]MC-YR	1031.5197	$C_{51}H_{71}N_{10}O_{13}$	[65]
144	[Asp[3], (E)-Dhb[7]]MC-YR	1031.5197	$C_{51}H_{71}N_{10}O_{13}$	[66]
145	[DMAdda[5]]MC-YR	1031.5197	$C_{51}H_{71}N_{10}O_{13}$	[3]
146	MC-LY(OMe)	1032.5288	$C_{53}H_{74}N_{7}O_{14}$	[5]
147	[Asp[3]]MC-(H_4)YR	1035.5510	$C_{51}H_{75}N_{10}O_{13}$	[3]
148	[Dha[7]]MC-(H_4)YR = [Dha[7]]MC-ThTyrR	1035.5510	$C_{51}H_{75}N_{10}O_{13}$	[38]
149	[DMAdda[5]]MC-(H_4)YR	1035.5510	$C_{51}H_{75}N_{10}O_{13}$	[20]
150	MC-YM(O)	1036.4696	$C_{51}H_{70}N_{7}O_{14}S$	[8, 16, 67]
151	[ADMAdda[5]]MC-HilR	1037.5666	$C_{51}H_{77}N_{10}O_{13}$	[1]
152	[ADMAdda[5]]MC-LHar	1037.5666	$C_{51}H_{77}N_{10}O_{13}$	[45, 46]
153	MC-AnaR	1037.6030	$C_{52}H_{81}N_{10}O_{12}$	[3]
154	[D-Leu[1]]MC-LR	1037.6030	$C_{52}H_{81}N_{10}O_{12}$	[68, 69]
155	[Asp[3]]MC-YY	1038.4819	$C_{54}H_{68}N_{7}O_{14}$	[5]
156	[Gly[1], Asp[3], ADMAdda[5], Dhb[7]]MC-RR	1038.5367	$C_{48}H_{72}N_{13}O_{13}$	[22]
157	MC-RR	1038.5731	$C_{49}H_{76}N_{13}O_{12}$	[13, 70]
158	[(6Z)-Adda[5]]MC-RR	1038.5731	$C_{49}H_{76}N_{13}O_{12}$	[27, 41]
159	[D-Asp[3], D-Glu(OCH_3)[6]]MCYST-RR	1038.5731	$C_{49}H_{76}N_{13}O_{12}$	[71]
160	[D-Ser[1], ADMAdda[5]]MC-LR	1039.5459	$C_{50}H_{75}N_{10}O_{14}$	[58]
161	[Met[1], Asp[3]]MC-LR	1041.5437	$C_{50}H_{77}N_{10}O_{12}S$	[43]
162	[ADMAdda[5], MeSer[7]]MC-LR	1041.5615	$C_{50}H_{77}N_{10}O_{14}$	[58]
163	[L-Ser[7]]MC-RR	1042.5680	$C_{48}H_{76}N_{13}O_{13}$	[19]

Table A3.1 (*Continued*)

Number	Microcystin	Calculated* $[M+H]^+$	Corresponding composition*	References
164	[D-Asp^3, MeSer^7]MC-RR	1042.5680	$C_{48}H_{76}N_{13}O_{13}$	[72]
165	[Asp^3, Thr^7]MC-RR	1042.5680	$C_{48}H_{76}N_{13}O_{13}$	[73]
166	*seco*[D-Asp^3]MC-RR	1042.5680	$C_{48}H_{76}N_{13}O_{13}$	[71]
167	MC-HphR	1043.5560	$C_{53}H_{75}N_{10}O_{12}$	[42, 74]
168	MC-M(O$_2$)R (oxidation artefact)	1045.5021	$C_{48}H_{73}N_{10}O_{14}S$	[42]
169	[D-Asp^3]MC-HtyR	1045.5353	$C_{52}H_{73}N_{10}O_{13}$	[75]
170	[Dha^7]MC-HtyR	1045.5353	$C_{52}H_{73}N_{10}O_{13}$	[63]
171	[D-Asp^3, (*E*)-Dhb^7]MC-HtyR	1045.5353	$C_{52}H_{73}N_{10}O_{13}$	[29]
172	[D-Asp^3, (*Z*)-Dhb^7]MC-HtyR	1045.5353	$C_{52}H_{73}N_{10}O_{13}$	[30]
173	MC-RY	1045.5353	$C_{52}H_{73}N_{10}O_{13}$	[15, 64]
174	MC-YR	1045.5353	$C_{52}H_{73}N_{10}O_{13}$	[13, 16]
175	L-Phe-D-MeAsp(β)-L-Arg-Adda-D-Glu(γ)-Mdha-D-Ala-OH = [*seco*-1/2]MC-FR	1047.5353	$C_{52}H_{75}N_{10}O_{13}$	[28, 54, 55]
176	MC-HphHph	1048.5390	$C_{57}H_{74}N_{7}O_{12}$	[43]
177	[D-Asp^3, L-Ser^7]MC-HtyR	1049.5302	$C_{51}H_{73}N_{10}O_{14}$	[1]
178	[Asp^3, MSer^7]MC-RY	1049.5302	$C_{51}H_{73}N_{10}O_{14}$	[5]
179	[Ser^7]MC-YR	1049.5302	$C_{51}H_{73}N_{10}O_{14}$	[48]
180	MC-(H$_4$)YR = MC-ThTyrR	1049.5666	$C_{52}H_{77}N_{10}O_{13}$	[13, 28]
181	[ADMAdda^5]MC-HilHar	1051.5822	$C_{52}H_{79}N_{10}O_{13}$	[1]
182	[Leu^1, Asp^3, ADMAdda^5]MC-LR	1051.5822	$C_{52}H_{79}N_{10}O_{13}$	[51]
183	[Leu^1]MC-HilR	1051.6187	$C_{53}H_{83}N_{10}O_{12}$	[43]
184	[Leu^1]MC-LHar	1051.6187	$C_{53}H_{83}N_{10}O_{12}$	[51]
185	[Leu^1,Glu(OMe)^6]MC-LR	1051.6187	$C_{53}H_{83}N_{10}O_{12}$	[50]
186	[D-Asp^3, (*E*)-Dhb^7]MC-HtyY	1052.4975	$C_{55}H_{70}N_{7}O_{14}$	[76]
187	MC-YY	1052.4975	$C_{55}H_{70}N_{7}O_{14}$	[66]
188	[Gly^1, Asp^3, ADMAdda^5, Dhb^7]MC-RHar	1052.5524	$C_{49}H_{74}N_{13}O_{13}$	[22]
189	[Asp^3, ADMAdda^5]MC-RR	1052.5524	$C_{49}H_{74}N_{13}O_{13}$	[51]
190	[Asp^3, ADMAdda^5, Dhb^7]MC-RR	1052.5524	$C_{49}H_{74}N_{13}O_{13}$	[47, 77]
191	[D-Glu-OC$_2$H$_3$(CH$_3$)OH^6]MC-LR = [D-Glu(OC$_3$H$_7$O)^6]MC-LR (not a natural product)	1053.5985	$C_{52}H_{81}N_{10}O_{13}$	[13, 28]
192	[D-Asp^3]MC-WR	1054.5357	$C_{53}H_{72}N_{11}O_{12}$	[3, 56]
193	[Dha^7]MC-WR	1054.5357	$C_{53}H_{72}N_{11}O_{12}$	[3]
194	[DMAdda^5]MC-WR	1054.5357	$C_{53}H_{72}N_{11}O_{12}$	[3]
195	[Met^1]MC-LR	1055.5594	$C_{51}H_{79}N_{10}O_{12}S$	[43]
196	[Leu^1, *N*MeSer^7]MC-LR	1055.6136	$C_{52}H_{83}N_{10}O_{13}$	[50]
197	[ADMAdda^5]MC-FR	1057.5353	$C_{53}H_{73}N_{10}O_{13}$	[20]
198	[Asp^3, ADMAdda^5]MC-HphR	1057.5353	$C_{53}H_{73}N_{10}O_{13}$	[51]
199	MC-HtyR	1059.5510	$C_{53}H_{75}N_{10}O_{13}$	[74]
200	[Asp^3, mGlu^6]MC-HtyR (m denotes *O*-methylation)	1059.5510	$C_{53}H_{75}N_{10}O_{13}$	[36]
201	[D-Glu(OMe)^6]MC-YR	1059.5510	$C_{53}H_{75}N_{10}O_{13}$	[66]

(*Continued*)

Table A3.1 (*Continued*)

Number	Microcystin	Calculated* $[M+H]^+$	Corresponding composition*	References
202	[Ser¹, Asp³]MC-HtyR	1061.5302	$C_{52}H_{73}N_{10}O_{14}$	[42]
203	[D-Asp³]MC-Y(OMe)R	1061.5302	$C_{52}H_{73}N_{10}O_{14}$	[3]
204	[DMAdda⁵]MC-Y(OMe)R	1061.5302	$C_{52}H_{73}N_{10}O_{14}$	[3]
205	[*seco*-4/5][Asp³]MC-HtyR	1063.5459	$C_{52}H_{75}N_{10}O_{14}$	[42]
206	[L-Ser⁷]MC-HtyR	1063.5459	$C_{52}H_{75}N_{10}O_{14}$	[63]
207	[Asp³, MSer⁷]MC-HtyR	1063.5459	$C_{52}H_{75}N_{10}O_{14}$	[36]
208	[D-Asp³, ADMAdda⁵]MC-(H_4)YR	1063.5459	$C_{52}H_{75}N_{10}O_{14}$	[20]
209	[MeSer⁷]MC-RY	1063.5459	$C_{52}H_{75}N_{10}O_{14}$	[15]
210	[Asp³, MeSer⁷]MC-YHar	1063.5459	$C_{52}H_{75}N_{10}O_{14}$	[66]
211	[MeSer⁷]MC-YR	1063.5459	$C_{52}H_{75}N_{10}O_{14}$	[3]
212	[**N**MeSer⁷]MC-YR	1063.5459	$C_{52}H_{75}N_{10}O_{14}$	[78]
213	MC-HphHty	1064.5339	$C_{57}H_{74}N_7O_{13}$	[43]
214	[Leu¹, ADMAdda⁵]MC-LR	1065.5979	$C_{53}H_{80}N_{10}O_{13}$	[51]
215	[Leu¹, Glu(OMe)⁶]MC-HilR	1065.6343	$C_{54}H_{85}N_{10}O_{12}$	[50]
216	[D-Asp³, (*E*)-Dhb⁷]MC-HtyHty	1066.5132	$C_{56}H_{72}N_7O_{14}$	[76]
217	[9-AcO-Adda⁵]MC-RR = [ADMAdda⁵] MC-RR	1066.5680	$C_{50}H_{76}N_{13}O_{13}$	[51, 79]
218	MC-hRhR = MC-HarHar	1066.6044	$C_{51}H_{80}N_{13}O_{12}$	[80]
219	[Leu¹, Asp³]MC-RR	1066.6044	$C_{51}H_{80}N_{13}O_{12}$	[43]
220	MC-WR	1068.5513	$C_{54}H_{74}N_{11}O_{12}$	[13]
221	[Leu¹, Glu(OMe)⁶, **N**MeSer⁷]MC-LR	1069.6292	$C_{53}H_{85}N_{10}O_{13}$	[50]
222	[M(O)¹]MC-LR	1071.5588	$C_{51}H_{79}N_{10}O_{13}S$	[50]
223	[Leu¹, Asp³]MC-HphR	1071.5874	$C_{55}H_{79}N_{10}O_{12}$	[51]
224	[Phe¹]MC-LR	1071.5874	$C_{55}H_{79}N_{10}O_{12}$	[43]
225	MC-KynR (oxidised W)	1072.5462	$C_{53}H_{74}N_{11}O_{13}$	[34]
226	[Asp³, ADMAdda⁵]MC-HtyR	1073.5302	$C_{53}H_{73}N_{10}O_{14}$	[81]
227	[Asp³, ADMAdda⁵, Dhb⁷]MC-HtyR	1073.5302	$C_{53}H_{73}N_{10}O_{14}$	[47]
228	[ADMAdda⁵]MC-YR	1073.5302	$C_{53}H_{73}N_{10}O_{14}$	[1]
229	MC-MhtyR	1073.5671	$C_{54}H_{77}N_{10}O_{13}$	[82]
230	[Asp³, (*E*)-Dhb⁷]MC-HtyW	1075.5135	$C_{57}H_{71}N_8O_{13}$	[35]
231	MC-RY(OMe)	1075.5459	$C_{53}H_{75}N_{10}O_{14}$	[15]
232	[Asp³]MC-Hty(OMe)R	1075.5459	$C_{53}H_{75}N_{10}O_{14}$	[42]
233	MC-Y(OMe)R	1075.5459	$C_{53}H_{75}N_{10}O_{14}$	[3]
234	[*seco*-4/5]MC-HtyR	1077.5615	$C_{53}H_{77}N_{10}O_{14}$	[42]
235	[ADMAdda⁵]MC-(H_4)YR	1077.5615	$C_{53}H_{77}N_{10}O_{14}$	[20]
236	[MSer⁷]MC-HtyR	1077.5615	$C_{53}H_{77}N_{10}O_{14}$	[42]
237	[Leu¹, ADMAdda⁵]MC-LHar	1079.6136	$C_{54}H_{83}N_{10}O_{13}$	[51]
238	[ADMAdda⁵]MC-RHar	1080.5837	$C_{51}H_{78}N_{13}O_{13}$	[51]
239	[Leu¹]MC-RR	1080.6200	$C_{52}H_{82}N_{13}O_{12}$	[43]
240	MC-OiaR (oxidised W)	1084.5462	$C_{54}H_{74}N_{11}O_{13}$	[34]
241	[Leu¹]MC-HphR	1085.6030	$C_{56}H_{81}N_{10}O_{12}$	[43]
242	[M(O)¹, Glu(OMe)⁶]MC-LR	1085.5700	$C_{52}H_{81}N_{10}O_{13}S$	[50]
243	[ADMAdda⁵]MC-(H_4)YHar	1091.5772	$C_{54}H_{79}N_{10}O_{14}$	[20]

Table A3.1 (Continued)

Number	Microcystin	Calculated* $[M+H]^+$	Corresponding composition*	References
244	MC-NfkR (oxidised W)	1100.5411	$C_{54}H_{74}N_{11}O_{14}$	[34]
245	[Leu[1]]MC-HtyrR	1101.5979	$C_{56}H_{81}N_{10}O_{13}$	[51]
246	[L-MeLan[7]]MC-LR	1116.5758	$C_{52}H_{82}N_{11}O_{14}S$	[28]

*Monoisotopic masses were calculated with the aid of Stirling and Miles [55] http://www.toxinology.no/Downloads.aspx or with BMRB molecular mass calculator http://www.bmrb.wisc.edu/metabolomics/mol_mass.php. The use of "seco" in linear peptides is described in http://www.chem.qmul.ac.uk/iupac/AminoAcid/AAad.html.

Abbreviations: Aha, aminoheptanoic acid, Aba, L-2-aminobutanoic acid, aminobutyrate, Aib, aminoisobutyric acid [6], aminoisobutyrate, Bu, 2-amino-*N*-butyric acid, L-butyrine, ADMAdda, *O*-acetyl-*O*-demethylAdda, Apa, aminopropionic acid, Dha, dehydroalanine, Dhb, dehydrobutyrine, DMAdda, *O*-demethylAdda, E(OMe), glutamic acid methylester, (H4)Y, 1,2,3, 4-tetrahydrotyrosine, Har, homoarginine, Hil, homoisoleucine, Hph, homophenylalanine, Hty, homotyrosine, Kyn, kynurenine, Me, CH_3, MeLan, *N*-methyllanthionine, M(O), methionine-*S*-oxide, MeSer or Mser, *N*-methylserine, Nfk, *N*-formylkynurenine, Oia, oxindolyalanine, Y(OMe), methoxytyrosine

Note: Two additional microcystins were reported recently:
247 [Asp[3]]MC-HarAba, calculated $[M+H]^+$, 967.5247, corresponding composition $C_{47}H_{71}N_{10}O_{12}$, [83]
248 [Glu(OCH$_3$)[6]]MC-FR], calculated $[M+H]^+$, 1043.5560, corresponding composition, $C_{53}H_{74}N_{10}O_{12}$, [83]

References

[1] Fewer, D.P., Rouhiainen, L., Jokela, J. *et al.* (2007) Recurrent adenylation domain replacement in the microcystin synthetase gene cluster. *BMC Evolutionary Biology* **7**:183. (In suppl.)

[2] Diehnelt, C.W., Dugan, N.R., Peterman, S.M. and Budde W.L. (2006) Identification of microcystin toxins from a strain of *Microcystis aeruginosa* by liquid chromatography introduction into a hybrid linear ion trap-Fourier transform ion cyclotron resonance mass spectrometer. *Analytical Chemistry* **78**, 501–512.

[3] Ballot, A., Sandvik, M., Rundberger, T. *et al.* (2014) Diversity of cyanobacteria and cyanotoxins in Hartbeespoort Dam, South Africa. *Marine and Freshwater Research* **65**, 175–189.

[4] Botes, D.P., Tuinman, A.A., Wessels, P.L. *et al.* (1984) The structure of cyanoginosin-LA, a cyclic heptapeptide toxin from the cyanobacterium *Microcystis aeruginosa. Journal of the Chemical Society Perkin Transactions* **1** 2311–2318.

[5] Miles, C.O., Sandvik, M., Nonga, H.E. *et al.* (2012) Thiol derivatization for LC-MS identification of microcystins in complex matrices. *Environmental Science & Technology* **46**, 8937–8944.

[6] Gathercole, P.S. and Thiel, P.G. (1987) Liquid chromatographic determination of the cyanoginosins, toxins produced by the cyanobacterium *Microcystis aeruginosa. Journal of Chromatography* **408**, 435–440.

[7] Mahakhant, A., Sano, T., Ratanachot, P. *et al.* (1998) Detection of microcystins from cyanobacterial water blooms in Thailand fresh water. *Phycological Research* **46**, 25–29.

[8] del Campo, F.F. and Ouahid Y. (2010) Identification of microcystins from three collection strains of *Microcystis aeruginosa. Environmental Pollution* **158**, 2906–2914.

[9] Puddick, J. (2013) Spectroscopic investigations of oligopeptides from aquatic cyanobacteria. Ph. D. thesis, University of Waikato, New Zealand.

[10] Vasas, G., Szydlowska, D., Gaspar, A. *et al.* (2006) Determination of microcystins in environmental samples using capillary electrophoresis. *Journal of Biochemical and Biophysical Methods* **66**, 87–97.

[11] Craig, M., McGready, T.L., Luu, H.A. *et al.* (1993) Identification and characterization of hydrophobic microcystins in Canadian freshwater cyanobacteria. *Toxicon* **31**, (12), 1541–1549.

[12] Puddick, J., Prinsep, M.R., Wood, S.A. *et al.* (2013) Isolation and structure determination of two new hydrophobic microcystins from *Microcystis* sp. (CAWBG11). *Phytochemistry Letters* **6**, 575–581.

[13] Namikoshi, M., Rinehart, K.L., Sakai R. *et al.* (1992) Identification of 12 hepatotoxins from a Homer Lake bloom of the cyanobacteria *Microcystis aeruginosa*, *Microcystis viridis*, and *Microcystis wesenbergii*: Nine new microcystins. *Journal of Organic Chemistry* **57**, 866–872.

[14] Lee, T., Chen, Y. and Chou, H. (1998) First report of microcystin in Taiwan. *Toxicon* **36**, (2) 247–255.

[15] Miles C.O., Sandvik, M., Nonga, H.E. *et al.* (2013) Identification of microcystins in a Lake Victoria cyanobacterial bloom using LC-MS with thiol derivatization. *Toxicon* **70**, 21–31.
[16] Botes, D.P., Wessels, P.L., Kruger, H. *et al.* (1985) Structural studies on cyanoginosin-LR,-YR,-YA, and -YM. Peptide toxins from *Microcystis aeruginosa. Journal of the Chemical Society Perkin Transactions* **1**, 2747–2748.
[17] Prakash, S., Lawton, L.A. and Edwards, C. (2009) Stability of toxigenic *Microcystis* blooms. *Harmful Algae* **8**, 377–384.
[18] Harada, K-I., Ogawa, K., Matsuura, K. *et al.* (1991) Isolation of two toxic heptapeptide microcystins from an axenic strain of *Microcystis aeruginosa*, K-139. *Toxicon* **29**, (4/5), 479–489.
[19] Namikoshi, M., Sivonen, K., Evans, W.R. *et al.* (1992) Two new L-serine variants of microcystin-LR and -RR from *Anabaena* sp. strains 202 A1 and 202 A2. *Toxicon* **30** (11), 1457–1464.
[20] Kaasalainen, U., Jokela, J., Fewer, D.P. *et al.* (2009) Microcystin production in the tripartite cyanolichen *Peltigera leucophlebia. Molecular Plant-Microbe Interactions* **22** (6), 695–702
[21] Miles, C.O., Sandvik, M., Haande S. *et al.* (2013) LC-MS analysis with thiol derivatization to differentiate [dhb^7] from [Mdha^7]microcystins: Analysis of cyanobacterial blooms, *Planktothrix* cultures and European crayfish from Lake Steinsfjorden, Norway. *Environmental Science & Technology* **47**, 4080–4087.
[22] Puddick, J., Prinsep, M.R., Wood, S.A. *et al.* (2015) Further characterization of glycine-containing microcystins from the McMurdo Dry Valleys of Antarctica. *Toxins* **7**, 493–515.
[23] Namikoshi, M., Yuan, M., Sivonen K. *et al.* (1998) Seven new microcystins possessing two L-glutamic acid units, isolated from *Anabaena* sp. strain 186. *Chemical Research in Toxicology* **11**, 143–149.
[24] Robillot, C., Vinh J., Puiseux-Dao, S. *et al.* (2000) Hepatotoxin production kinetics of the cyanobacterium *Microcystis aeruginosa* PCC 7820, as determined by HPLC-mass spectrometry and protein phosphatase bioassay. *Environmental Science & Technology* **34**, 3372–3378.
[25] Bateman, K.P., Thibault, P., Douglas, D.J. and White, R.L..(1995) Mass spectral analyses of microcystins from toxic cyanobacteria using on-line chromatographic and electrophoretic separations. *Journal of Chromatography A* **712**, 253–268.
[26] Krishnamurthy T., Szafraniec, L., Hunt, D.F. *et al.* (1989) Structural characterization of toxic cyclic peptides from blue-green algae by tandem mass spectrometry. *Proceedings of the National Academy of Sciences USA* **86**, 770–774.
[27] Harada, K-I., Matsuura, K., Suzuki, M. *et al.* (1990) Isolation and characterization of the minor components associated with microcystins LR and RR in the cyanobacterium (blue-green algae). *Toxicon* **28** (1), 55–64.
[28] Namikoshi, M., Sun, F., Choi, B.W. *et al.* (1995) Seven more microcystins from Homer Lake cells: Application of the general method for structure assignment of peptides containing α,β-dehydroamino acid unit(s). *Journal of Organic Chemistry* **60**, 3671–3679.
[29] Sano, K. and Kaya. K. (1998) Two new (E)-2-amino-2-butenoic acid (dhb)-containing microcystins isolated from *Oscillatoria agardhii. Tetrahedron* **54**, 463–470.
[30] Sano, K., Beattie, K.A., Codd, G.A. and Kaya, K. (1998) Two (Z)-dehydrobutyrine containing microcystins from a hepatotoxic bloom of *Oscillatoria agardhii* from Soulseat Loch, Scotland. *Journal of Natural Products* **61** (6), 851–853.
[31] Mazur-Marzec, H., Browarczyk-Matusiak, G., Forycka, K. *et al.* Morphological, genetic, chemical and ecophysiological characterisation of two *Microcystis aeruginosa* isolates from the Vistula Lagoon, southern Baltic. *Oceanologia*, **52** (1), 127–146.
[32] Kaya, K., Mahakhant, A. and Keovara, L. (2000) Microcystin-AW, a new microcystin variant isolated from cyanobacterial waterblooms in Thailand. Poster presented at the 9th International Conference on Harmful Algal Blooms, Hobart, Tasmania, Australia February 7-11, 2000.
[33] Azevedo, S.M.F.O., Evans, W.R., Carmichael, W.W. and Namikoshi, M. (1994) First report of microcystins from Brazilian isolate of the cyanobacterium *Microcystis aeruginosa. Journal of Applied Phycology* **6**, 261–265.
[34] Puddick, J., Prinsep, M.R.,Wood, S.A. *et al.* (2013) Structural characterization of new microcystins containing tryptophan and oxidized tryptophan residues. *Marine Drugs* **11**, 3025–3045.
[35] Niedermeyer, T.H.J., Schmieder, P. and Kurmayer, R. (2014) Isolation of microcystins from the cyanobacterium *Planktothrix rubescens* strain No80. *Natural Products and Bioprospecting* **4**, 37–45.
[36] Welker, M., Christiansen, G. and von Döhren, H., (2004) Diversity of coexisting *Planktothrix* (cyanobacteria) chemotypes deduced by mass spectral analysis of microcystins and other oligopeptides. *Archives of Microbiology* **182**, 288–298.

[37] Sano, T., Takagi, H. and Kaya, K. (2004) A dhb-microcystin from the filamentous cyanobacterium *Planktothrix rubescens*. *Phytochemistry* **65**, 2159–2162.

[38] Sano, T., Takagi, H., Nishikawa, M. and Kaya, K. (2008) NIES certified reference material for microcystins, hepatotoxic cyclic peptide toxins from cyanobacterial blooms in eutrophic water bodies. *Analytical and Bioanalytical Chemistry* **391**, 2005–2010.

[39] Sivonen, K., Skulberg, O.M., Namikoshi, M. *et al.* (1992) Two methyl ester derivatives of microcystins, cyclic heptapeptide hepatotoxins, isolated from *Anabaena flos-aquae* strain CYA 83/1. *Toxicon* **30** (11), 1465–1471.

[40] Krishnamurthy, T., Carmichael, W.W. and Sarver, E.W. (1986) Toxic peptides from freshwater cyanobacteria (blue-green algae). 1. Isolation, purification and characterization of peptides from *Microcystis aeruginosa* and *Anabaena flos-aquae*. *Toxicon* **24** (9), 865–873.

[41] Harada, K-I., Ogawa, K., Matsuura, K. *et al.* (1990) Structural determination of geometrical isomers of microcystin LR and RR from cyanobacteria by two-dimensional NMR spectroscopic techniques. *Chemical Research in Toxicology* **3**, 473–481.

[42] Miles, C.O., Melanson, J.E. and Ballot, A. (2014) Sulfide oxidations for LC-MS analysis of methionine-containing microcystins in *Dolicospermum flos-aquae* NIVA-CYA 656. *Environmental Science & Technology* **48**, 13307–13315.

[43] Shishido T.K., Kaasalainen, U., Fewer, D.P. *et al.* (2013) Convergent evolution of [D-Leucine[1]] microcystin-LR in taxonomically disparate cyanobacteria. *BMC Evolutionary Biology* **13**:86 (In suppl).

[44] Stoner, R.D., Adams, W.H., Slatkin, D.N. *et al.* (1989) The effects of single L-amino acid substitutions on the lethal potencies of the microcystins. *Toxicon* **27** (7), 825–828.

[45] Sivonen, K., Carmichael, W.W., Namikoshi, M. *et al.* (1990) Isolation and characterization of hepatotoxic microcystin homologs from the filamentous freshwater cyanobacterium *Nostoc* sp strain 152. *Applied and Environmental Microbiology* **56** (9), 2650–2657.

[46] Namikoshi, M., Rinehart, K.L., Sakai, R. *et al.*(1990) Structures of three new cyclic heptapeptide hepatotoxins produced by the cyanobacterium (blue-green algae) *Nostoc* sp. strain 152. *Journal of Organic Chemistry* **55**, 6135–6139.

[47] Beattie, K.A., Kaya, K., Sano, T. and Codd, G.A. (1998) Three dehydrobutyrine-containing microcystins from *Nostoc*. *Phytochemistry* **47** (7), 1289–1292.

[48] Mayumi, T., Kato, H., Imanishi S. *et al.* (2006) Structural characterization of microcystins by LC/MS/MS under ion trap conditions. *Journal of Antibiotics* **59** (11), 710–719.

[49] Brittain, S., Mohamed, Z.A., Wang, J. *et al.* (2000) Isolation and characterization of microcystins from a River Nile strain of *Oscillatoria tenuis* Agardh ex Gomont. *Toxicon* **38**, 1759–1771.

[50] Qi, Y., Rosso, L., Sedan, D. *et al.* (2015) Seven new microcystin variants discovered from native *Microcystis aeruginosa* strain - unambigous assignment of product ions by tandem mass spectrometry. *Rapid Communications in Mass Spectrometry* **29**, 220–224.

[51] Kaasalainen, U., Fewer, D.P., Jokela, J. *et al.* (2012) Cyanobacteria produce a high variety of hepatotoxic peptides in lichen symbiosis. *Proceedings of the National Academy of Sciences USA* **109** (15), 5886–5891.

[52] Sivonen, K., Namikoshi, M., Evans, W.R. *et al.* (1992) Isolation and characterization of a variety of microcystins from seven strains of the cyanobacterial genus *Anabaena*. *Applied and Environmental Microbiology* **58** (8), 2495–2500.

[53] Christiansen, G., Fastner, J., Erhard, M., *et al.* (2003) Microcystin biosynthesis in *Planktothrix*: Genes, evolution and manipulation. *Journal of Bacteriology* **185** (2), 564–572.

[54] Choi, B.W., Namikoshi, M., Sun, F. *et al.* (1993) Isolation of linear peptides related to the hepatotoxins nodularin and microcystins. *Tetrahedron Letters* **34** (49), 7881–7884.

[55] Stirling, D.J. and Miles, C.O. (2014) Marine algal toxin and cyanobacterial toxin mass lists. http://www.toxinology.no/Downloads.aspx

[56] Lee, T. and Chou, H. (2000) Isolation and identification of seven microcystins from a cultured M.TN-2-strain of *Microcystis aeruginosa*. *Botanical Bulletin of Academia Sinica* **41**, 197–202.

[57] Luukkainen, R., Namikoshi, M., Sivonen, K., *et al.* (1994) Isolation and identification of 12 microcystins from four strains and two bloom samples of *Microcystis* spp.: Structure of a new hepatotoxin. *Toxicon* **32** (1), 133–139.

[58] Sivonen, K., Namikoshi, M., Evans, W.R. *et al.* (1992) Three new microcystins, cyclic heptapeptide hepatotoxins, from *Nostoc* sp. strain 152. *Chemical Research in Toxicology* **5**, 464–469.

[59] Meriluoto, J.A.O., Sandström, A., Eriksson, J.E. *et al.* (1989) Structure and toxicity of a peptide hepatotoxin from the cyanobacterium *Oscillatoria agardhii*. *Toxicon* **27** (9), 1021–1034.

[60] Namikoshi, M., Sivonen, K. Evans, W.R. *et al.* (1992) Isolation and structures of microcystins from a cyanobacterial water bloom (Finland). *Toxicon* **30** (11), 1473–1479.

[61] Kiviranta, J., Namikoshi, M., Sivonen, K. *et al.* (1992) Structure determination and toxicity of a new microcystin from *Microcystis aeruginosa* strain 205. *Toxicon* **30** (9), 1093–1098.

[62] Sano, T. and Kaya, K. (1995) A 2-amino-2-butenoic acid (dhb)-containing microcystin isolated from *Oscillatoria agardhii*. *Tetrahedron Letters* **36** (47), 8603–8606.

[63] Namikoshi, M., Sivonen, K., Evans, W.R. *et al.* (1992) Structures of three new homotyrosine-containing microcystins and a new homophenylalanine variant from *Anabaena* sp. strain 66. *Chemical Research in Toxicology* **5** (5), 661–666.

[64] Okello, W., Portmann, C., Erhard, M., *et al.* (2009) Occurrence of microcystin-producing cyanobacteria in Ugandan fresh waters. *Environmental Toxicology* **25**, 367–380.

[65] Sivonen, K., Namikoshi, M., Evans, W.R. *et al.* (1992) Isolation and structures of five microcystins from a Russian *Microcystis aeruginosa* strain CALU 972. *Toxicon* **30** (11), 1481–1485.

[66] Niedermeyer, T.H.J., Daily, A., Swiatecka-Hagenbruch, M. and Moscow, J.A. (2014) Selectivity and potency of microcystin congeners against OATP1B1 and OATP1B3 expressing cancer cells. *PLOSone* **9**, e91476.

[67] Elleman, T.C., Falconer, I.,R., Jackson, A.R.B. and Runnegar, M.T. (1978) Isolation, characterization and pathology of the toxin from *Microcystis aeruginosa* (=*Anacystis cyanea*) bloom. *Australian Journal of Biological Sciences* **31**, 209–218.

[68] Matthiensen, A., Beattie, K.A., Yunes, J.S. *et al.* [D-Leu^1]microcystin-LR, from the cyanobacterium *Microcystis* RST 9501 and from a *Microcystis* bloom in the Patos Lagoon estuary, Brazil. *Phytochemistry* **55**, 383–387.

[69] Park, H., Namikoshi, M., Brittain, S.M. *et al.* (2001) [D-Leu^1]microcystin-LR, a new microcystin isolated from waterbloom in a Canadian prairie lake. *Toxicon* **39**, 855–862.

[70] Kusumi, T., Ooi, T., Watanabe, M.M. *et al.* (1987) Cyanoviridin RR, toxin from the cyanobacterium (blue-green alga) *Microcystis viridis*. *Tetrahedron Letters* **28** (40), 4695–4698.

[71] Grach-Pogrebinsky, O., Sedmak, B. and Carmeli, S. (2004) Seco[D-Asp^3]microcystin-RR and [D-Asp^3, D-Glu(OMe)^6]microcystin-RR, two new microcystins from a toxic water bloom of the cyanobacterium *Planktothrix rubescens*. *Journal of Natural Products* **67**, 337–342.

[72] Luukkainen, R., Sivonen, K., Namikoshi, M. *et al.* (1993) Isolation and identification of eight microcystins from thirteen *Oscillatoria agardhii* strains and structure of a new microcystin. *Applied and Environmental Microbiology* **59** (7), 2204–2209.

[73] Welker, M. and Erhard, M. (2007) Consistency between chemotyping of single filaments of *Planktothrix rubescens* (cyanobacteria) by MALDI-TOF and the peptide patterns of strains determined by HPLC-MS. *Journal of Mass Spectrometry* **42**, 1062–1068.

[74] Fewer, D.P., Tooming-Klunderud, A., Jokela, J., *et al.* (2008) Natural occurrece of microcystin synthetase deletion mutants capable of producing microcystins in strains of the genus *Anabaena* (Cyanobacteria). *Microbiology* **154**, 1007–1014.

[75] Harada, K.-I., Ogawa, K., Kimura, Y. *et al.* (1991) Microcystins from *Anabaena flos-aquae* NRC 25-17. *Chemical Research in Toxicology* **4**, 535–540.

[76] Christiansen, G., Yoshida, W.Y., Blom, J.F. *et al.* (2008) Isolation and structure determination of two microcystins and sequence comparison of the McyABC adenylation domains in *Planktothrix* species. *Journal of Natural Products* **71**, 1881–1886.

[77] Kleinteich, J., Wood, S.A., Puddick, J. *et al.* (2013) Potent toxins in Arctic environments - presence of saxitoxins and an unusual microcystin variant in Arctic freshwater ecosystems. *Chemico-Biological Interactions* **206**, 423–431.

[78] Okello, W., Ostermaier, V., Portmann, C. *et al.* (2010) Spatial isolation favours the divergence in microcystin net production by *Microcystis* in Ugandan freshwater lakes. *Water Research* **44**, 2803–2814.

[79] Ferranti, P., Fabbrocino, S., Nasi, A. *et al.* (2009) Liquid chromatography coupled to quadrupole time-of-flight tandem mass spectrometry for microcystin analysis in freshwaters: Method performances and characterisation of a novel variant of microcystin-RR. *Rapid Communications in Mass Spectrometry* **23**, 1328–1336.

[80] Frias, H.V, Mendes M.A., Cardozo, K.H.M *et al.* (2006) Use of electrospray tandem mass spectrometry for identification of microcystins during a cyanobacterial bloom event. *Biochemical and Biophysical Research Communications* **344**, 741–746.

[81] Laub, J. Henriksen, P., Brittain, S.M. *et al.* (2002) [ADMAdda[5]]microcystins in *Planktothrix agardhii* strain PH-123 (cyanobacteria) - Importance for monitoring of microcystins in the environment. *Environmental Toxicology* **17**, 351–357.

[82] Teta, R., Della Sala, G., Glukhov, E. *et al.* (2015) Combined LC-MS/MS and molecular networking approach reveals new cyanotoxins from the 2014 cyanobacterial bloom in Green Lake, Seattle. *Environmental Science & Technology*, **49** (24), 14301–14310.

[83] Bouhaddada, R., Nélieu, S., Nasri, H. *et al.* (2016) High diversity of microcystins in a *Microcystis* bloom from an Algerian lake. *Environmental Pollution*, in press, http://dx.doi.org/10.1016/j.envpol.2016.06.055

Table A3.2 Nodularins

Number	Nodularin	Calculated* $[M+H]^+$	Corresponding composition*	References
1	[L-Val^2]NOD = motuporin	768.4184	$C_{40}H_{58}N_5O_{10}$	[1]
2	[D-Asp^1]NOD	811.4354	$C_{40}H_{59}N_8O_{10}$	[2, 3]
3	[DMAdda^3]NOD	811.4354	$C_{40}H_{59}N_8O_{10}$	[2, 3]
4	[dhb^5]NOD	811.4354	$C_{40}H_{59}N_8O_{10}$	[3]
5	NOD-R	825.4511	$C_{41}H_{61}N_8O_{10}$	[3–6]
6	[6(Z)-Adda^3]NOD	825.4511	$C_{41}H_{61}N_8O_{10}$	[2, 3]
7	[Glu^4(OMe)]NOD	839.4667	$C_{42}H_{63}N_8O_{10}$	[3]
8	[L-Har^2]NOD	839.4667	$C_{42}H_{63}N_8O_{10}$	[6–8]
9	[MeAdda^3]NOD	839.4667	$C_{42}H_{63}N_8O_{10}$	[3]
10	Adda-D-Glu(γ)-Mdhb-D-MeAsp(β)-L-Arg-OH (linear nodularin)	843.4616	$C_{41}H_{63}N_8O_{11}$	[2, 3, 9]

*Monoisotopic masses calculated with BMRB molecular mass calculator www.bmrb.wisc.edu/metabolomics/mol_mass_ php.

References

[1] Dilip de Silva, E., Williams, D.E., Andersen, R.J. *et al.* (1992) Motuporin, a potent protein phosphatase inhibitor isolated from the Papua New Guinea sponge *Theonella swinhoei* Gray. *Tetrahedron Letters* **33** (12), 1561–1564.

[2] Namikoshi, M., Choi, B.W., Sakai, R., *et al.* (1994) New nodularins: A general method for structure assignment. *Journal of Organic Chemistry* **59**, 2349–2357.

[3] Mazur-Marzec, H., Meriluoto, J., Plinski M., and Szafranek, J. (2006) Characterization of nodularin variants in *Nodularia spumigena* from the Baltic Sea using liquid chromatography/mass spectrometry/mass spectrometry. *Rapid Communications in Mass Spectrometry* **20**, 2023–2032.

[4] Carmichael, W.W., Eschedor, J.T., Patterson, G.M.L. and Moore, R.E. (1988) Toxicity and partial structure of a hepatotoxic peptide produced by the cyanobacterium *Nodularia spumigena* Mertens emend. L575 from New Zealand. *Applied and Environmental Microbiology* **54**, 2257–2263.

[5] Rinehart, K.L., Harada, K.-I., Namikoshi, M. *et al.* (1988) Nodularin, microcystin, and the configuration of Adda. *Journal of the American Chemical Society* **110**, 8557–8558.

[6] Gehringer, M.M., Adler, L., Roberts, A.A. *et al.* (2012) Nodularin, a cyanobacterial toxin, is synthesized in planta by symbiotic *Nostoc* sp. *The ISME Journal* **6**, 1834–1847.

[7] Beattie, K.A., Kaya, K. and Codd, G.A. (2000) The cyanobacterium *Nodularia* PCC 7804, of freshwater origin, produces [L-Har[2]]nodularin. *Phytochemistry* **54**, 57–61.

[8] Saito, K., Konno, A., Ishii, H. *et al.* Nodularin-Har: a new nodularin from *Nodularia*. *Journal of Natural Products* **64**, 139–141.

[9] Choi, B.W., Namikoshi, M., Sun, F. *et al.* (1993) Isolation of linear peptides related to the hepatotoxins nodularin and microcystins. *Tetrahedron Letters* **34** (49), 7881–7884.

Index

Handbook of Cyanobacterial Monitoring and Cyanotoxin Analysis, First Edition.
Edited by Jussi Meriluoto, Lisa Spoof and Geoffrey A. Codd.

Printed and bound by CPI Group (UK) Ltd, Croydon, CR0 4YY
17/09/2026
14964303-0006